XXVI IUPAP Conference on Computational Physics (CCP 2014)

Journal of Physics: Conference Series Volume 640

Boston, Massachusetts, USA
11 - 14 August 2014

ISBN: 978-1-5108-1317-5
ISSN: 1742-6588

Printed from e-media with permission by:

Curran Associates, Inc.
57 Morehouse Lane
Red Hook, NY 12571

Some format issues inherent in the e-media version may also appear in this print version.

Printed by Curran Associates, Inc. (2016)

For permission requests, please contact the Institute of Physics
at the address below.

Institute of Physics
Dirac House, Temple Back
Bristol BS1 6BE UK

Phone: 44 1 17 929 7481
Fax: 44 1 17 920 0979

techtracking@iop.org

Additional copies of this publication are available from:

Curran Associates, Inc.
57 Morehouse Lane
Red Hook, NY 12571 USA
Phone: 845-758-0400
Fax: 845-758-2634
Email: curran@proceedings.com
Web: www.proceedings.com

Table of contents

Volume 640

XXVI IUPAP Conference on Computational Physics (CCP2014)
11–14 August 2014, Boston, USA

Accepted papers received: 28 August 2015
Published online: 28 September 2015

Papers

Statistical Physics

Materials and Nano Science

Fluid Dynamics

Quantum Many-Body Physics

Astrophysics

Computational-Physics Education

High-Performance Computing and General Computational Physics

Journal of Physics: Conference Series **640** (2015) 011001 doi:10.1088/1742-6596/640/1/011001

26th IUPAP Conference on Computational Physics

A. W. Sandvik[1], **D. K. Campbell**[1], **D. F. Coker**[2], **and Y. Tang**[1,3]

[1]Department of Physics, Boston University, Boston, Massachusetts 02215,USA
[2]Department of Chemistry, Boston University, Boston, Massachusetts 02215, USA
[3]Department of Physics, Gordon College, Wenham, Massachusetts 01984, USA

E-mail: sandvik@bu.edu

1. Editorial

The 26th IUPAP Conference on Computational Physics, CCP2014, was held in Boston, Massachusetts, during August 11-14, 2014. Almost 400 participants from 38 countries convened at the George Sherman Union at Boston University for four days of plenary and parallel sessions spanning a broad range of topics in computational physics and related areas.

The first meeting in the series that developed into the annual Conference on Computational Physics (CCP) was held in 1989, also on the campus of Boston University and chaired by our colleague Claudio Rebbi. The express purpose of that meeting was to discuss the progress, opportunities and challenges of common interest to physicists engaged in computational research. The conference having returned to the site of its inception, it is interesting to reflect on the development of the field during the intervening years. Though 25 years is a short time for mankind, computational physics has taken giant leaps during these years, not only because of the enormous increases in computer power but especially because of the development of new methods and algorithms, and the growing awareness of the opportunities the new technologies and methods can offer. Computational physics now represents a "third leg" of research alongside analytical theory and experiments in almost all subfields of physics, and because of this there is also increasing specialization within the community of computational physicists. It is therefore a challenge to organize a meeting such as CCP, which must have sufficient depth in different areas to hold the interest of experts while at the same time being broad and accessible. Still, at a time when computational research continues to gain in importance, the CCP series is critical in the way it fosters cross-fertilization among fields, with many participants specifically attending in order to get exposure to new methods in fields outside their own.

As organizers and editors of these Proceedings, we are very pleased with the high quality of the papers provided by the participants. These articles represent a good cross-section of what was presented at the meeting, and it is our hope that they will not only be useful individually for their specific scientific content but will also represent a historical snapshot of the state of computational physics that they represent collectively.

The remainder of this Preface contains lists detailing the organizational structure of CCP2014, endorsers and sponsors of the meeting, plenary and invited talks, and a presentation of the 2014 IUPAP C20 Young Scientist Prize.

We would like to take the opportunity to again thank all those who contributed to the success of CCP214, as organizers, sponsors, presenters, exhibitors, and participants.

Anders Sandvik, David Campbell, David Coker, Ying Tang

2. Conference Organization

Chair
Anders Sandvik (Boston University)

Vice Chairs
David Campbell (Boston University)
David Coker (Boston University)

Secretary
Cheryl Endicott (Boston University)

Local organizing committee
James Adler (Tufts University)
Alfredo Alexander-Katz (MIT)
Arun Bansil (Northeastern University)
Richard Brower (Boston University)
Bulbul Chakraborty (Brandeis University)
Claudio Chamon (Boston University)
Edward Farhi (MIT)
Adrian Feiguin (Northeastern University)
Harvey Gould (Clark University)
Andrzej Herczynski (Boston College)
Efthimios Kaxiras (Harvard University)
Jonathan Machta (University of Massachusetts Amherst)
Hossein Mosallaei (Northeastern University)
Merav Opher (Boston University)
Nikolai Prokofev (University of Massachusetts Amherst)
Claudio Rebbi (Boston University)
Emily Ryan (Boston University)
Gene Stanley (Boston University)
Bala Sundaram (University of Massachusetts Boston)
Mark Vogelsberger (MIT)

International advisory board
Joan Adler (Technion, Haifa, Israel)
Constantia Alexandrou (University of Cyprus)
Amanda Barnard (CSIRO, Australia)
Stefano Baroni (SISSA, Italy)
Bruce Boghosian (American University of Armenia and Tufts University, USA)
Nithaya Chetty (University of Pretoria, South Africa)
Giovanni Ciccoti (University of Rome 1, Italy)
Stefano Curtarolo (Duke University, USA)
Ronald Dickman (City University of New York, USA)
Hans Fangor (University of Southampton, UK)
James Gubernatis (Los Alamos National Lab, USA)
Guang-Yu Guo (National Taiwan University, Taipei, Taiwan)
Karen Hallberg (Centro Atomico Bariloche, Argentina)
Alex Hansen (University of Trondheim, Norway)
Masatoshi Imada (University of Tokyo, Japan)
Vicky Kalogera (Northwestern University, USA)
David Landau (University of Georgia, USA)
Hai-Qing Lin (Beijing Computational Science Research Center, China)

Joaquin Marro (University of Granada, Spain)
Roger Melko (University of Waterloo, Canada)
Andrew Millis (Columbia University, USA)
Adriana Moreo (Oak Ridge National Lab, USA)
Kai Nordlund (University of Helsinki, Finland)
Lev Shchur (Landau Institute, Moscow, Russia)
Malcolm Stocks (Oak Ridge National Lab, USA)
Hideaki Takabe (University of Osaka, Japan)
Matthias Troyer (ETH Zürich, Switzerland)
Umesh Waghmare (JNCASR, Bangalore, India)
Philipp Werner (University of Freiburg, Switzerland)
Anthony Williams (The University of Adelaide, Australia)
Daoxin Yao (Sun Yat-Sen University, Guangzhou, China)
Naoki Yoshida (University of Tokyo, Japan)

Program Committee

Chair
Anders Sandvik (Boston University)

Vice Chairs
David Campbell (Boston University)
David Coker (Boston University)

Program coordinator
Ying Tang (Boston University, Gordon College)

Track subcommittees:

Classical statistical mechanics and complex systems

Wolfhard Janke (University of Leipzig, Germany), Chair
Jonathan Machta (University of Massachusetts Amherst, USA), Vice Chair
Bulbul Chakraborty (Brandeis University, USA)
Harvey Gould (Clark University, USA)
Alex Hansen (NTU, Trondheim, Norway)
Nobuyasu Ito (University of Tokyo, Japan)
Bala Sundaram (University of Massachusetts Boston, USA)

Soft-matter, polymer, and biological physics

Joerg Rottler (UBC, Vancouver, Canada), Chair
Alfredo Alexander-Katz (MIT, USA), Vice Chair
Ralf Everaers (ENS Lyon, France)
Frauke Graeter (Heidelberg Institute for Theoretical Studies, Germany)
Mikko Karttunen (University of Waterloo, Canada)

Materials science and nano science

Talat Rahman (University of Central Florida, USA), Chair
Arun Bansil (Northeastern University, USA), Vice Chair
Hai-Ping Cheng (University of Florida, USA)
Mei-Yin Chou (Georgia Institute of Technology, USA, and Academia Sinica, Taiwan)
Matthias Scheffler (Fritz Haber Institute, Berlin, Germany)
Malcolm Stocks (Oak Ridge National Lab, USA)
Umesh Waghmare (J Nehru Centre for Advanced Scientific Research, Bangalore, India)

Fluid dynamics, magnetohydrodynamics

James Adler (Tufts University, USA), Chair
Emily Ryan (Boston University, USA), Vice Chair
Barry Koren (Eindhoven University of Technology, The Netherlands)
Alexandre Tartakovsky (University of South Florida, USA)

Quantum many-body physics

Rajiv Singh (UC Davis, USA), Chair
Adrian Feiguin (Northeastern University, USA), Vice Chair
Kedar Damle (Tata Institute for Fundamental Research, Mumbai, India)
Chisa Hotta (Kyoto Sangyo University, Japan)
Didier Poilblanc (Laboratoire de Physique Théorique, Toulouse, France)
Nikolai Prokofev (University of Massachusetts Amherst, USA)

Quantum Computing

Matthias Troyer (ETH, Zürich, Switzerland), Chair
Edward Farhi (MIT, USA), Vice Chair
Claudio Chamon (Boston University, USA)
Krysta Svore (Microsoft Research, USA)
Barbara Terhal (RWTH Aachen, Germany)
Rodney van Meter (Keio University, Japan)

Lattice field theory in particle and nuclear physics

Anna Hasenfratz (University of Colorado, Boulder, USA), Chair
Claudio Rebbi (Boston University, USA), Vice Chair
Richard Brower (Boston University, USA)
Yasumichi Aoki (Nagoya University, Japan)

Astrophysics, space-plasma physics, gravitation, cosmology

Tiziana DiMateo (Carnegie Mellon University, USA), Chair
Merav Opher (Boston University, USA), Vice Chair
Phil Armitage (JILA, University of Colorado, USA)
Stephan Rosswog (Stockholm University, Sweden)
Romain Teyssier (University of Zürich, Switzerland)
Mark Vogelsberger (MIT, USA)

Novel hardware and software paradigms

Barry Schneider (NIST, USA), Chair
Richard Brower (Boston University, USA), Vice Chair
Mark Jarrell (Louisiana State University, USA)
Jim Sexton (IBM, USA)

Computational-physics education

Wolfgang Christian (Davidson College, USA), Chair
Harvey Gould (Clark University, USA), Vice Chair
Nithaya Chetty (The University of Kwazulu-Natal, South Africa)

3. IUPAP Young Scientist Price

The 2014 IUPAP Young Scientist Prize for Computational Physics was awarded at CCP2014 by the C20 Commission to Professor Mathieu Salanne of Université Pierre et Marie Cure, Paris, France, in recognition of his development of appropriate methods to allow realistic atomistic simulation of molten salts and ionic liquids in situations of relevance to electrochemistry. For further information on the Prize and the nomination process, please see information on line at `http://phycomp.technion.ac.il/~C20/prizes.html`.

4. Endorsers and Sponsors of CCP2014

4.1. Endorsers

CCP2014 was endorsed by the following professional organizations

- The International Union of Pure and Applied Physics (IUPAP)
- The American Physical Society (APS)
- The European Physical Society (EPS)
- The Association of Asia Pacific Physical Societies (AAAPS)

4.2. Sponsors

CCP2014 was sponsored financially by the IUPAP, APS, and the following universities in the Boston area

- Boston University
- Boston College
- Clark University
- Harvard University (Institute for Applied Computational Science)
- Northeastern University
- University of Massachusetts Amherst
- University of Massachusetts Boston

as well as by these corporations

- Intel Corporation
- Cambridge University Press
- Elsevier
- Institute of Physics Publishing
- American Institute of Physics Publishing

We would like to take this opportunity to thank all the sponsors for their generous support, which was critical to the success of CCP2014.

To secure IUPAP sponsorship, the organizers have provided assurance that CCP2014 will be conducted in accordance with IUPAP principles as stated in the ICSU document *Universality of Science* (sixth edition, 1989) regarding the free circulation of scientists for international purposes. In particular, no bona fide scientist will be excluded from participation on the grounds of national origin, nationality, or political considerations unrelated to science.

Journal of Physics: Conference Series **640** (2015) 011001 doi:10.1088/1742-6596/640/1/011001

5. Plenary Presentations

A total of 15 plenary talks were presented over the four days of the meeting. They were formally organized into two sections, with presenters and titles as follows (in order of appearance):

General Computational Physics

- Sauro Succi (IAC-CNR, Italy), *Lattice Boltzmann simulations of complex flows across scales: turbulence, soft- glasses and quark-gluon plasmas*
- Yuko Okamoto (Nagoya University, Japan), *Enhanced configurational sampling methods for spin systems and biomolecular systems*
- Nandini Trivedi (Ohio State University, USA), *Topology and Correlations driving new materials, phases and phenomena*
- Helmut Katzgraber (Texas A&M University, USA), *Four decades of frustration in spin-glass physics: Advances and applications*
- A. Peter Young (University of California, Santa Cruz, USA), *Numerical Studies of the Quantum Adiabatic Algorithm*
- Ursula Rothlisberger (Ecole Polytechnique Federale de Lausanne, Switzerland), *Mixed Quantum Mechanical/Molecular Mechanical (QM/MM) Simulations of Biological Systems: From Understanding to Control*
- Brian Granger (California Polytechnic State University, USA), *Open source tools for exploratory and reproducible computational physics*
- Steven Louie (University of California, Berkeley, and Lawrence Berkeley National Lab, USA), *GW-based Methods for ab initio Studies of Electronic Excited-State Phenomena in Condensed Matter*
- Steven White (University of California, Irvine, USA), *Solving frustrated magnetic systems with the density matrix renormalization group*
- Luigi Del Debbio (University of Edinburgh, UK), *Recent progress in simulations of gauge theories on the lattice: QCD at the physical point and new strongly-interacting dynamics beyond the Standard Model*
- Romeel Davé (University of the Western Cape, South Africa), *Simulations of Galaxy Formation*

Enabling Technologies for Computational Science

- John Danskin (NVIDIA), *The Physics of Computation and GPU Architecture*
- Karl Schulz (Intel), *Enabling Technology Trends in High Performance Computing*
- Thomas Sterling (Center for Research in Extreme Scale Technologies, Indiana University, USA), *Computational Physics at Extreme Scale*
- James Sexton (IBM), *A Vision for Data Centric Systems*

6. Parallel and Poster Sessions

The parallel sessions were organized as ten different tracks with a mix of invited (30 minute) and contributed (15 minute) talks, and one session on General Computational Physics with eight contributed talks. In total, 52 invited talks and almost 100 contributed presentations were given. A poster session with almost 100 presenters was held during the first evening of the conference. The invited talks are listed below.

Statistical Physics

- Mark Newman, (University of Michigan, USA), *Large-scale structure in networks*
- Lenka Zdeborova, (CEA Saclay and CNRS, France), *Module detection in networks: phase transitions and optimal algorithms*
- Werner Krauth, (ENS Paris, France), *Rejection-free, Irreversible, and Infinitesimal Monte Carlo Algorithms and Melting in two dimensions*
- Salvatore Torquato (Princeton University, USA), *New Algorithm to Generate Jammed Sphere Packings*
- Youjin Deng (University of Science and Technology of China), *Universal amplitudes in the canonical ensemble*
- Koji Hukushima (University of Tokyo, Japan), *Equilibrium-state simulations of some (spin) glass models in finite dimensions*

Soft Matter and Biological Physics

- Mark Robbins (Johns Hopkins University, USA), *Welding and healing of polymer interfaces: Strength from entanglements*
- Marcus Müller, (Georg-August-Universität, Germany), *Studying the kinetics of copolymer self-assembly*
- Ivet Bahar, (University of Pittsburgh, USA), *Structure-Encoded Dynamics of Proteins: Learning from Network Models and Experiments*
- Normand Mousseau (Université de Montréal, Canada), *Computational challenges for the study of amyloid processes*
- Marina Guenza (University of Oregon, USA), *A coarse-graining method that preserves the free energy, structural correlations, and thermodynamic state of polymer melts from the atomistic to the mesoscale*
- Celeste Sagui (North Carolina State University, USA), *Free energy methods for biomolecular simulations*

Materials Science and Nanoscience

- Markus Eisenbach (Oak Ridge National Laboratory, USA), *Magnetic Materials at finite Temperatures: thermodynamics and combined spin and molecular dynamics derived from first principles calculations*
- Lin-lin Wang (Ames Laboratory, USA), *Computational Modeling of Transition-Metal Alloyed Nanoparticles in Working Condition*
- Hsin Lin (National University of Singapore), *Topological Crystalline Insulators: A New Phase of Quantum Matter*
- Caterina Cocchi (Humboldt-Universität zu Berlin, Germany), *From Molecules to Organic Crystals: Optical Excitations from First Principles*
- Volodymyr Turkowski (University of Central Florida, USA), *Development and application of DFT+DMFT and TDDFT+DMFT techniques for nanosystems*
- Luca Ghiringhelli, (Fritz Haber Institute, Germany), *Big Data of Materials Science - Critical Role of the Descriptor*

Fluid Dynamics

- Monika Nitsche (University of New Mexico, USA), *Vortex Shedding and Low Order Models*
- George Karniadakis (Brown University, USA), *Microscopic theory of Brownian motion: Effects of memory and confinement*
- Marc Gerritsma (TU Delft, The Netherlands), *Structure preserving discretizations for computational physics*
- Chun Liu, (Penn State University, USA), *Energetic Variational Approaches in Complex Fluids*

Quantum Many-Body Physics

- Ribhu Kaul (University of Kentucky, USA), *Deconfined quantum criticality in SU(N) magnets*
- Boris Svistunov (University of Massachusetts Amherst, USA), *Diagrammatic Monte Carlo for Fermionic and Fermionized Systems*
- Federico Becca (CNR and SISSA, Trieste, Italy), *Variational wave functions for strongly-correlated models*
- Philippe Corboz (University of Amsterdam, The Netherlands), *Recent progress in simulating strongly correlated systems with tensor network methods*
- Tao Xiang (Institute of Physics, Chinese Academy of Sciences, China), *Renormalization of quantum many-body systems by the projected entangled simplex states*
- Corinna Kollath (University of Bonn, Germany), *Spreading of correlations in strongly correlated (dissipative) quantum gases*

Quantum Computing

- David Clader (Johns Hopkins University, USA), *Preconditioned quantum linear system algorithm*
- Bryan Clark (University of Illinois at Urbana-Champaign, USA), *The cost of simulating quantum mechanics on a quantum computer*
- Sergio Boixo (Google, USA), *Experiments with the D-Wave prototype*

Lattice Field Theory

- Taku Izubuchi (RIKEN, Japan, and BNL Research Center, USA), *Lattice QCD calculations for particle physics*
- Aida El-Khadra (University of Illinois at Urbana-Champaign, USA), *Lattice QCD and Quark Flavor Physics*
- John Negele (MIT, USA), *Understanding the Structure of Nucleons using Lattice QCD*
- William Detmold (MIT, USA), *Dark Nuclei*
- Simon Catterall, (Syracuse University, USA), *Supersymmetry on a lattice*
- Frithjof Karsch (Brookhaven National Laboratory, USA), *Conserved charge fluctuations in strong interaction matter*

Novel Computing Paradigms

- Thomas Cheatham (University of Utah, USA), *Molecular dynamics simulation of nucleic acids: Convergence, reproducibility, assessment/validation, and data dissemination enabled by GPUs on XSEDE and Blue Waters*
- Ying-jer Kao (National Taiwan University, Taipei, Taiwan), *Uni10: the Universal Tensor Network Library*
- Martin Berzins (University of Utah, USA), *Multiscale and Multiphysics Computations on Present and Future Architectures*
- Erik Schnetter (Perimeter Institute, Canada), *Automated Code Generation for Solving PDEs on Modern HPC Architecture*
- Norbert Attig (Jülich Supercomputing Centre, Germany), *The Path to Exascale: A European Perspective*
- Lars Korsterke (Texas Advanced Computing Center, USA), *Heterogeneous computing. What is it and do we need it?*

Astrophysics

- Mike Boylan-Kolchin (University of Maryland, USA), *The Local Universe as a Dark Matter Laboratory*
- Claude-Andre Faucher-Giguere (Northwestern University, CIERA, USA), *The Universe on a computer: Cosmological simulations of galaxy formation*
- Christian David Ott (Caltech, USA), *Petascale Simulations of Core-Collapse Supernovae*
- Deirdre Shoemaker (Georgia Tech, USA), *Numerical Relativity and Gravitational Waves*

Computational Physics Education

- Ruth Chabay (North Carolina State University, USA), *Computation and Conceptual Understanding in Introductory Physics*
- Francisco Esquembre, (Universidad de Murcia, Spain), *Facilitating programming computational physics simulations for tablets*
- Beate Schmittmann (Iowa State University, USA), *K-12 outreach and student recruitment with computational science*
- Spencer Wheaton (University of Cape Town, South Africa), *Infusing Computational Physics throughout the Undergraduate Curriculum*
- Shobhana Narasimhan (Jawaharlal Nehru Centre for Advanced Scientific Research, India), *Teaching Density Functional Theory through Experiential Learning: Examples from the Developing World*

Peer review statement

All papers published in this volume of *Journal of Physics: Conference Series* have been peer reviewed through processes administered by the proceedings Editors. Reviews were conducted by expert referees to the professional and scientific standards expected of a proceedings journal published by IOP Publishing.

Functionality limit of classical simulated annealing

M Hasegawa

Faculty of Engineering, Information and Systems, University of Tsukuba, Tsukuba 305-8573, Japan

E-mail: hasegawa@esys.tsukuba.ac.jp

Abstract. By analyzing the system dynamics in the landscape paradigm, optimization function of classical simulated annealing is reviewed on the random traveling salesman problems. The properly functioning region of the algorithm is experimentally determined in the size-time plane and the influence of its boundary on the scalability test is examined in the standard framework of this method. From both results, an empirical choice of temperature length is plausibly explained as a minimum requirement that the algorithm maintains its scalability within its functionality limit. The study exemplifies the applicability of computational physics analysis to the optimization algorithm research.

1. Introduction

With the rapid development of computer capabilities, there is a growing interest in the understanding of physical and other phenomena in the energy landscape paradigm [1]. Searching for the ground state of a molecular system is minimizing its energy function, and to solve this problem, one has used general-purpose optimization methods such as simulated annealing (SA) [2, 3]. In recent computer experiments for equilibrium systems, more sophisticated methods like parallel tempering [4, 5] are generally employed instead of the conventional Metropolis algorithm (MA) [6]. While these improvements are needed to meet the computational requirement for systems with very rugged landscapes (e.g. glasses), they expand the range of options for stochastic optimization [7]. In such an interaction, we find a positive relationship between computational physics methodology and optimization algorithm design. Concerning the molecular systems, the understanding of the system dynamics in the landscape is expected to be useful, for example, for the selective and efficient formation of crystals [8]. A similar understanding should be informative in the effective and efficient implementation of the local search (LS) optimization algorithms [9].

This paper revisits classical SA with particular attention to its functionality, which is assessed through the observation of the search dynamics hidden in the landscape of the cost function. In a previous study [10], the author verified the physical analogy of SA in the solution of the traveling salesman problem (TSP) [11], a commonly-used benchmark in combinatorial optimization. There, we found that the effectiveness of this method comes not from equilibrium sampling at low temperatures, though described in its original scenario, but from downward interbasin dynamics occurring in the rugged landscape before equilibrium. Subsequent studies on the same problem [12–17] showed that the optimization characteristics of SA and SA-related algorithms can be understood more properly in a unified manner by focusing on the search function of the above relaxation dynamics. Needless to say, identification of actual success factors should be conducive to fair competition among existing solution methods and further to practical evaluation of related new approaches like quantum annealing [18].

Here, we consider a commonly recommended, but not necessarily well-founded strategy for "temperature length [11]" in the implementation of classical SA and attempt to obtain a clear picture of its validity as an extension of the series of studies [12–17]. This strategy specifically states that the number of steps allocated at each temperature should be at least proportional to the neighborhood size in the standard geometric cooling scheme. In the practical application of SA, it has been rightly recognized and accepted that the method has an unavoidable imperfection due to the finiteness of its implementation time. In addition to this defect, the second but preventable malfunction is detected in the solution of the random TSPs [13]. As discussed below, the present empirical strategy can in fact plausibly be explained as a minimum requirement that the algorithm maintains its scalability within its functionality limit.

This work is intended neither to improve the existing solution methods nor to propose a novel algorithm. The purpose is to show that a function-based analysis using the computational physics method helps to refine our knowledge of optimization. For this reason, the algorithm is implemented even under the condition unsuited for practical use. This extensive analysis tells us that the optimization study in the landscape paradigm provides fresh insight even into the present old-established method. The paper is organized as follows: In Sec.2, the method and results of functionality assessment are presented with detailed descriptions of the problem and the algorithm. Although the same assessment has already been made in the framework of generalized SA [14], the results for classical SA are shown for the first time for completeness in the present context. In Sec.3, the influence of the second defect (described in Sec.2) on the scalability test is demonstrated using the standardized SA algorithm, and the empirical strategy for temperature length is reviewed from the functionality point of view. The study is summarized and concluded in Sec.4.

2. Functionality assessment of the relaxation dynamics

To begin with, the test-bed problem, notation, and terminology are summarized. This paper examines the solution process of the TSP, the goal of which is to find the shortest tour that passes through each of the given cities once and returns to the starting city. We consider here a random Euclidean TSP, where the locations of N cities are sampled uniformly in the unit square region and the intercity distance is computed under the Euclidean metric. (The characteristic features described below were similarly observed in a qualitative sense for a random distance matrix TSP [11], the details of which are omitted here.) Let x be a tour (a feasible solution), $f(x)$ be the cost function defined by the tour length, and $\mathcal{N}(x)$ be the neighborhood function. Throughout the study, we use a 2-opt neighborhood [11], which is defined as a set of tours constructed by any change of two intercity paths from the tour x. We write x_n $(n = 0, 1, 2, \ldots)$ to represent the (actual) search history and x^* to represent the incumbent solution (the best-so-far solution). A locally optimal solution is a solution that does not have any improved solution in its neighborhood. A basin is defined as a set of solutions attracted to the same locally optimal solution by a simple LS, that is, a repetition of the move to the best solution in the neighborhood. Let $y(x)$ be the locally optimal solution inside the basin including the solution x and we refer to its cost $f(y(x))$ as the cost of the basin. (Obviously, $y(x)$ can be found by a simple LS starting from the solution x; this identification method is analogous to the mapping-onto-minima approach used in studies on liquid and glass [19–21].) We denote the temperature by T; however, we often use the logarithmic temperature $\Theta(= \log_{10} T)$.

In the assessment experiment, a simple geometric cooling scheme was adopted: The logarithmic temperature schedule $\Theta(t)$ is described as $\Theta_{\rm s} + (t/t_{\rm e})(\Theta_{\rm e} - \Theta_{\rm s})$, where $\Theta_{\rm s}$ is the initial logarithmic temperature, $\Theta_{\rm e}$ is the final logarithmic temperature, and $t_{\rm e}$ is the total search time. In what follows, the total number of search steps, $n_{\rm e}$, is often referred to instead of

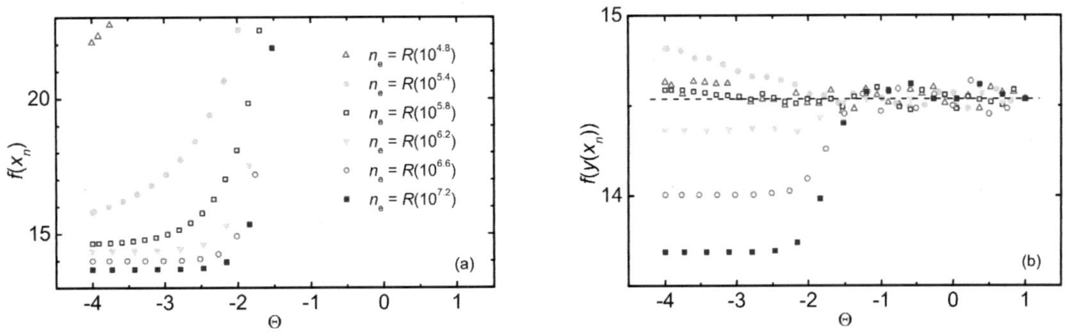

Figure 1. Actual [(a) $f(x_n)$] and hidden [(b) $f(y(x_n))$] search dynamics ($N = 316$, $I = 2^5$). The abscissa represents the logarithmic temperature Θ and $R(z)$ denotes that z is rounded to the nearest integer. Points are the results at some representative time steps and the dashed line depicted in (b) represents the performance of a simple local search starting from a random solution. Plotted are the results averaged over I search processes.

t_e. The values of Θ_s and Θ_e were always fixed at 1 and -4, respectively. These are sufficiently high and low temperatures and the cooling range includes the characteristic temperatures found previously [10]. The SA algorithm for this experiment is described as follows:

[SA$_1$1] Determine the logarithmic temperature schedule $\Theta(t)$; that is, select the value of t_e $(= n_e)$.

[SA$_1$2] Set $n := 0$ and generate an initial solution x_0 randomly. Set $\Theta := \Theta(0)$ and $x^* := x_0$.

[SA$_1$3] Implement a single step of the MA and increment n by one.

[SA$_1$4] If the termination condition, $n = n_e$, is satisfied, output x^* and stop. Otherwise, set $\Theta := \Theta(n)$ and return to [SA$_1$3].

In a single step of the MA in [SA$_1$3], the logarithmic temperature Θ is converted to the temperature T and the following [MA1] is implemented:

[MA1] Choose a trial solution $x'_n \in \mathcal{N}(x_n)$ randomly and set $\Delta := f(x'_n) - f(x_n)$. If $\Delta < 0$, accept the trial solution and set $x_{n+1} := x'_n$; furthermore, if $f(x_{n+1}) < f(x^*)$, renew the incumbent solution, $x^* := x_{n+1}$. If $\Delta \geq 0$, accept the trial solution with the probability $\exp(-\Delta/T)$ and set $x_{n+1} := x'_n$; with the complementary probability, reject the trial solution and set $x_{n+1} := x_n$.

First, we look at the relevant features of the relaxation dynamics. In Fig. 1, the actual search dynamics described by $f(x_n)$ $(n = 0, 1, 2, \ldots)$ and the hidden search dynamics described by $f(y(x_n))$ $(n = 0, 1, 2, \ldots)$ are shown for various time scales for $N = 316$ $(\approx 10^{2.5})$; the results averaged over $I(= 2^5)$ independent runs are plotted against the logarithmic temperature. Though not fully shown in the figure, the actual dynamics [Fig.1(a)] start from a higher level ($f(x_0) \approx 159$ at $\Theta = 1$) and the cost decreases monotonically with cooling. As n_e decreases, this mechanism for optimization becomes less sufficient and as found in the hidden dynamics [Fig.1(b)], the overall change in the basin's cost is reversed at some time scale. The monotonic decrease of $f(x_n)$ means that the incumbent solution is renewed throughout the optimization process. Consequently, the optimization performance evaluated by $f(y(x^*))$ deteriorates with decreasing n_e and below the critical point, SA is defeated by a simple LS, the performance of which is represented here by $f(y(x_0))$. This is the occurrence of the second malfunction touched

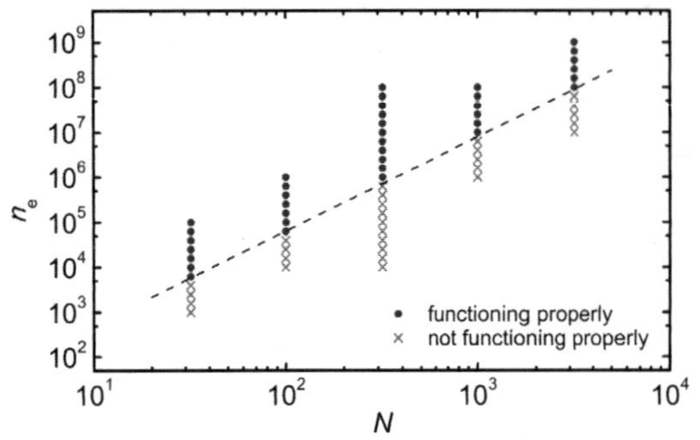

Figure 2. Results of the functionality assessment. • : functioning properly $[f(y(x_{n_e})) < f(y(x_0))]$; × : not functioning properly $[f(y(x_{n_e})) \geq f(y(x_0))]$. The assessment is based on the average result over I search processes, where the value of I is 2^7 (for $N = 32$), 2^6 (100), 2^5 (316), 2^4 (1000), and 2^3 (3162). The dashed line is obtained by a least-squares fit (in double logarithmic scale) of the critical points, each of which was estimated by linear interpolation in the logarithmic scale of n_e.

upon in the introduction. The amount of the deterioration is maximized at some lower time scale, below which the performance recovers to the level of the simple LS.

To describe the functioning condition of the algorithm precisely, we define its functionality by the overall displacement in the hidden search dynamics. That is, if the cost of the basin including the final solution, $f(y(x_{n_e}))$, is smaller than that including the initial random solution $f(y(x_0))$, we regard the relaxation dynamics as functioning properly, and otherwise, not functioning properly. Although this definition is too crude for the performance evaluation of modern optimizers, it is appropriate here because, as seen below, elucidating the minimum requirements is essential. Repeating the above experiment for different system sizes, we assess the functionality of the algorithm for various combinations of the values of the system size N and the total number of search steps, n_e. In this way, the functioning region is determined in the N-n_e plane.

This analysis was done for the other four sizes, $N = 32(\approx 10^{1.5})$, 100, 1000, and $3162(\approx 10^{3.5})$, in addition to the above $N = 316$, and all the results are shown in Fig. 2. The assessment was based on the average result over I independent runs; the value of I was selected from the range from 2^3 to 2^7 so that the amount of computation is kept within a reasonable range. Note that the number of cost evaluations instead of the Monte Carlo step is used as a unit of time. This is because the former is the usual choice in optimization studies [11] and relevant to the following discussion. We see from this figure that irrespective of the system size, there is a critical time at which the functionality definitely changes. The functioning region appears to be bounded by a straight line with a slope approximately equal to two in double logarithmic scale. Obviously, the algorithm should be implemented within the functioning region for the utilization of its intrinsic optimization function; however, this can be recognized as an unreasonable demand for the solution of larger-size instances on a given time scale. Also, it can readily be expected that the emerging limitation due to the second defect restricts the scalable implementation of the algorithm. To observe the influence of this limitation, a scalability test was actually performed on the same problem instances as in the above assessment experiment. As discussed in the next section, the test results provide a possible explanation of the empirical strategy for the temperature length.

3. Influence on the scalability test

The test was performed using the standardized algorithm with stepwise cooling. This was introduced for the baseline implementation in Ref.[11] and is described as follows:

[$SA_2 1$] Select an initial temperature T_s, a cooling ratio γ, and a temperature length L.

[$SA_2 2$] Set $n := 0$ and generate an initial solution x_0 randomly. Set $T := T_s$ and $x^* := x_0$.

[$SA_2 3$] Implement a single step of the MA, namely, [MA1] and increment n by one.

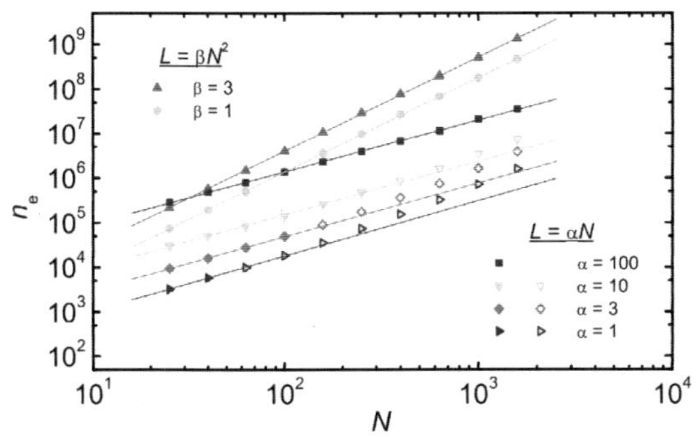

Figure 3. Total number of search steps, n_e, spent in the scalability test. Plotted are the results averaged over I search processes, where the value of I is 2^7 (for $N = 25$ and 40), 2^6 (63, 100, and 158), 2^5 (251 and 398), and 2^4 (631, 1000, and 1585). A solid symbol is used for the case that the relaxation dynamics function properly, and an open symbol is used for the case that the second malfunction occurs. Each line is obtained for each individual test by a least-squares fit (in double logarithmic scale) of the data points represented by solid symbols.

[$SA_2$4] If the stopping criterion for the MA search, $n = L$, is satisfied, go to [$SA_2$5]. Otherwise, return to [$SA_2$3].

[$SA_2$5] If the termination condition of the algorithm is satisfied, output x^* and stop. Otherwise, set $T := \gamma T$, $x_0 := x_n$, and $n := 0$; and return to [$SA_2$3].

The termination condition in [$SA_2$5] is that five consecutive reductions of temperature proceed without any renewal of the incumbent solution and without the acceptance ratio going above 2%. Here, the acceptance ratio is the ratio of the case that a trial solution is accepted as the next solution in the MA search.

For the sake of discussion, we consider two forms of L: One is proportional to the system size ($L = \alpha N$) and the other is to the square of the system size ($L = \beta N^2$). After fixing the proportionality constant, the above algorithm was implemented for various system sizes; the value of N was selected from the range from 25 ($\approx 10^{1.4}$) to 1585 ($\approx 10^{3.2}$). The value of T_s was fixed at 10, which is equal to that used in the first experiment (described in Sec.2), and the value of γ was taken to be 0.95.

The test results are summarized in Fig. 3, where the total number of search steps (specifically the accumulated number of search steps throughout the cooling process), n_e, is plotted against the system size. The results were averaged over I independent runs; the value of I was selected from the range from 2^4 to 2^7 by reference to the first experiment. The functionality was assessed in the same manner except that $f(y(x^*))$ was used instead of $f(y(x_{n_e}))$. The figure shows that in the case of $L = \beta N^2$, all the points are almost in a line and the upward trend is uniform for both $\beta = 1$ and 3. The average final temperature was always higher than 10^{-4} ($\Theta_e = -4$). This, taken together with the average values of n_e, means the cooling rate in the average run always meet the properly functioning condition of the algorithm. Each test could be conducted appropriately beyond the selected maximum size because the slope of the depicted line is almost equal to that of the estimated boundary line in Fig.2. In the case of $L = \alpha N$, by contrast, a change in the upward trend draws our attention in the cases $\alpha \leq 10$. This change appears to occur with the onset of the second defect and once it occurs, n_e increases more rapidly. In this runaway phase, it is considered that whereas the actual search dynamics do not descend to the expected cost level, the system is intensively cooled according to the temperature schedule without ever satisfying the termination condition. The intrinsic optimization function is lost in this situation and such a test makes no sense. It is highly predictable that even in the case $\alpha = 100$, a similar change occurs at a larger value of N.

As mentioned in the introduction, the length L is recommended to be at least proportional to the neighborhood size. If we follow this strategy, only the form βN^2 is acceptable in the present test because the neighborhood size is $O(N^2)$. This setting, however, has been explained in the literature merely as a need for worthwhile tour quality or for sufficient exploration in the neighborhood. As seen here, these explanations can be refined, at least for the present TSP instances, from the functionality point of view. That is, the form βN^2 is required for the algorithm not to be implemented outside its properly functioning region, which is a prerequisite for its scalable implementation maintaining the quality of the outcome.

As noted previously [13], the second malfunction occurs when the temperature is rapidly decreased over the effective intermediate range; therefore, it will occur even when one adopts some elaborated cooling schedule. It is uncertain whether the same defect occurs in the solution of other problems (including other TSP instances) because the functionality of the relaxation dynamics could be affected by the landscape topography. Despite the lack of generality, however, it is worth pointing out that the present function-based analysis in the landscape paradigm can complement the conventional procedure-based algorithm study. The present justification for the parameter setting of the algorithm relies on the determination of the functionality limit, which was recognized neither by the observation of the actual search dynamics depicted in Fig. 1(a) nor by the usual performance evaluation, for example, by $f(y(x^*))$ alone.

4. Conclusion

The functionality of the relaxation dynamics in optimization by classical SA was assessed for the solution of the random TSPs. The assessment was performed through the observation of the hidden search dynamics analyzed in the rugged landscape of the cost function. The properly functioning condition of the method was experimentally clarified and the scalability in its standard implementation was tested on the same problem instances. The results refined our knowledge of SA: The recommended choice of the temperature length is explained as a minimum requirement that the algorithm maintains its scalability under its properly functioning condition. The study draws our attention to an illuminating aspect of computational physics analysis in the optimization algorithm research.

Acknowledgments

This work was supported in part by JSPS KAKENHI (Grant Nos.20656017 and 24656071).

References

[1] Wales D J 2003 *Energy Landscapes* (Cambridge: Cambridge University Press)
[2] Kirkpatrick S *et al* 1983 *Science* **220** 671
[3] Černý V 1985 *J.Optim.Theory Appl.* **45** 41
[4] Hukushima K and Nemoto K 1996 *J.Phys.Soc.Jpn.* **65** 1604
[5] Coluzzi B and Parisi G 1998 *J.Phys.* A **31** 4349
[6] Metropolis N *et al* 1953 *J.Chem.Phys.* **21** 1087
[7] Schneider J J and Kirkpatrick S 2006 *Stochastic Optimization* (Berlin: Springer)
[8] Hoffmann K H and Schön J C 2013 *Eur.Phys.J.* B **86** 220
[9] Aarts E H L and Lenstra J K (ed) 1997 *Local Search in Combinatorial Optimization* (Chichester: Wiley)
[10] Hasegawa M 2011 *Phys.Rev.* E **83** 036708
[11] Johnson D S and McGeoch L A 1997 *Local Search in Combinatorial Optimization* ed E H L Aarts and J K Lenstra (Chichester: Wiley) p 215
[12] Hasegawa M 2011 *Comput.Phys.Commun.* **182** 229
[13] Hasegawa M 2012 *Phys.Rev.* E **85** 056704
[14] Hasegawa M 2012 *J.Phys.Soc.Jpn.* **81** 085001
[15] Hasegawa M 2013 *AIP Conf.Proc.* **1518** 733
[16] Hasegawa M and Hiramatsu K 2013 *Physica* A **392** 4491
[17] Hasegawa M and Kim C 2014 *JPS Conf.Proc.* **1** 019009
[18] Finnila A B *et al* 1994 *Chem.Phys.Lett.* **219** 343
[19] Stillinger F H and Weber T A 1982 *Phys.Rev.* A **25** 978
[20] Stillinger F H and Weber T A 1983 *Phys.Rev.* A **28** 2408
[21] Stillinger F H and Weber T A 1984 *Science* **225** 983

Finite-size scaling and latent heat at the gonihedric first-order phase transition

Wolfhard Janke[1], **Marco Mueller**[1] **and Desmond A. Johnston**[2]

[1] Institut für Theoretische Physik, Universität Leipzig, Postfach 100 920, 04009 Leipzig, Germany

[2] Department of Mathematics, School of Mathematical and Computer Sciences, Heriot-Watt University, Riccarton, Edinburgh EH14 4AS, Scotland

E-mail: `Wolfhard.Janke@itp.uni-leipzig.de`

Abstract. A well-known feature of first-order phase transitions is that fixed boundary conditions can strongly influence finite-size corrections, modifying the leading corrections for an L^3 lattice in $3d$ from order $1/L^3$ under periodic boundary conditions to $1/L$. A rather similar effect, albeit of completely different origin, occurs when the system possesses an exponential low-temperature phase degeneracy of the form 2^{3L} which causes for periodic boundary conditions a leading correction of order $1/L^2$ in $3d$. We discuss a $3d$ plaquette Hamiltonian ("gonihedric") Ising model, which displays such a degeneracy and manifests the modified scaling behaviour. We also investigate an apparent discrepancy between the fixed and periodic boundary condition latent heats for the model when extrapolating to the thermodynamic limit.

Generic features of first-order phase transitions in numerical studies are rarely discussed compared to the universal aspects of second-order phase transitions [1]. Numerical investigations always rely on observations of finite systems, but often one is really interested in the properties of macroscopic systems. The thermodynamic limit is approached by finite-size scaling, a theory that was developed in the 1980's [2, 3, 4] and refined [5, 6, 7, 8, 9]. Employing a simple heuristic two-phase model [10] rather than the full-blown Pirogov-Sinai theory [11], corrections to typical quantities due to finiteness of the lattice can be calculated already to some extent. For a *periodic* $3d$ system residing on a cube with volume $V = L^3$, the maximum of the specific heat $C_V(\beta, L) = -\beta^2 \partial e(\beta, L)/\partial\beta$ with $e(\beta, L) = \langle E\rangle(\beta, L)/L^3$ denoting the energy per site is according to the two-phase model [10] given by

$$C_V^{\max}(L) = L^3(\beta^\infty \Delta\hat{e}/2)^2 \,. \qquad (1)$$

Here, β^∞ is the inverse transition temperature of a system with infinitely many sites and $\Delta\hat{e} = \hat{e}_\mathrm{d} - \hat{e}_\mathrm{o}$ is its latent heat, the difference between the energy of the disordered and ordered phases. The position of the maximum is calculated using an expansion around β^∞ [10, 12] and found to be located at inverse temperature

$$\beta^{C_V^{\max}}(L) = \beta^\infty - \frac{\ln q}{L^3 \Delta\hat{e}} + \cdots \,, \qquad (2)$$

where a dependency on the degeneracy q (as in a q-states Potts model) enters the first correction along with the inverse latent heat. Similarly, the inverse temperature $\beta^{B^{\min}}(L)$ where one finds the minimum of Binder's energy parameter $B(\beta, L) = 1 - \langle E^4 \rangle / 3 \langle E^2 \rangle^2$ is predicted [10, 12],

$$\beta^{B^{\min}}(L) = \beta^{\infty} - \frac{\ln(q\hat{e}_{\mathrm{o}}^2/\hat{e}_{\mathrm{d}}^2)}{L^3 \Delta \hat{e}} + \dots \,, \tag{3}$$

and apparently yields a $1/L^3$ correction to scaling too.

When a macroscopic degeneracy is present, we find that the scaling laws for periodic boundaries have to be adjusted, effectively enlarging the leading corrections [13]. A $3d$ plaquette (4-spin) interaction Ising model on a cubic lattice, where the Ising spins $\sigma_i = \pm 1$ live on the vertices of the lattice,

$$\mathcal{H} = -\frac{1}{2} \sum_{[i,j,k,l]} \sigma_i \sigma_j \sigma_k \sigma_l \,, \tag{4}$$

is a suitable candidate for investigating such behaviour, since its ground-state degeneracy of $q = 2^{3L}$ on an L^3 lattice is unbroken throughout the low-temperature phase [14]. The plaquette Hamiltonian is a member of a family of gonihedric Ising models which were originally formulated as a lattice discretization of a particular string-theory action [15]. It has attracted attention in its own right because it has a strong first-order transition [16] and displays evidence of glass-like behaviour at low temperatures [15, 17], in spite of the absence of quenched disorder.

If we take $q = 2^{3L}$ in Eqs. (2) and (3), as is the case for the plaquette Hamiltonian, we find

$$\beta^{C_V^{\max}}(L) = \beta^{\infty} - \frac{\ln(2^{3L})}{L^3 \Delta \hat{e}} + \mathcal{O}\left(\frac{[\ln(2^{3L})]^2}{L^6}\right) = \beta^{\infty} - \frac{3 \ln 2}{L^2 \Delta \hat{e}} + \mathcal{O}\left(L^{-4}\right) \tag{5}$$

and

$$\beta^{B^{\min}}(L) = \beta^{\infty} - \frac{\ln(2^{3L}\hat{e}_{\mathrm{o}}^2/\hat{e}_{\mathrm{d}}^2)}{L^3 \Delta \hat{e}} + \mathcal{O}\left(\frac{[\ln(2^{3L}\hat{e}_{\mathrm{o}}^2/\hat{e}_{\mathrm{d}}^2)]^2}{L^6}\right) = \beta^{\infty} - \frac{3 \ln 2}{L^2 \Delta \hat{e}} - \frac{\ln(\hat{e}_{\mathrm{o}}^2/\hat{e}_{\mathrm{d}}^2)}{L^3 \Delta \hat{e}} + \mathcal{O}\left(L^{-4}\right) \tag{6}$$

so the leading contribution to the finite-size corrections is now $\propto 1/L^2$.

The modified scaling [13] has been extensively investigated by multicanonical simulations and found to be consistent, even with higher-order corrections for the inverse temperatures [18] and the maximum value of the specific heat [19]. The multicanonical Monte Carlo algorithm [20, 21] allows the system to quickly flip between the ordered and disordered phases, because the probability of unlikely states that hinder such fast transit is increased artificially. A weight function $W(\mathcal{H})$ that replaces the canonical Boltzmann weight $\exp(-\beta\mathcal{H})$ is improved recursively [22] so that we eventually get a flat multicanonical energy probability density. The weights are then fixed and the actual measurements are taken. The simulation converges towards the auxiliary multicanonical distribution as a fixed point and when the system is in equilibrium we can use the inverse of the weights to relate the measurements to their canonical Boltzmann weights. In this way, we achieve a double benefit from the algorithm. Firstly, the system is able to traverse highly improbable states to give a rapid oscillation between the ordered and disordered phases. Secondly, it is now possible to apply reweighting techniques over a much broader range of temperatures, since these need sufficient statistics in each bin of the histogram, which is naturally obtained from a flat histogram method.

For several lattice sizes L, we have simulated the model (4) with both fixed (all outer spins are set to $+1$) and periodic boundary conditions in Ref. [18] and measured observables such as the specific heat and Binder's energy parameter. Reweighting techniques allow the evaluation of those quantities as a function of temperature, and enable us to determine the positions of their peaks, $\beta^{C_V^{\max}}(L)$ and $\beta^{B^{\min}}(L)$ numerically with high precision. Along with the peak locations, we also calculated additional estimators $\beta^{\mathrm{eqw}}(L)$ and $\beta^{\mathrm{eqh}}(L)$ that relate to the double-peaked

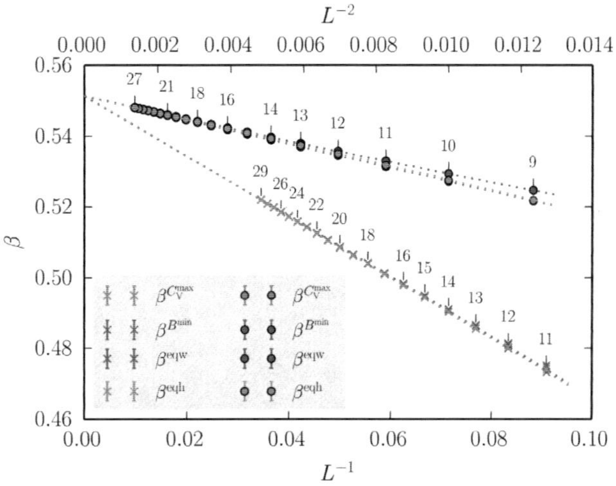

Figure 1. Finite-size scaling for periodic (upper abscissa) and fixed boundary conditions (lower abscissa) for the four inverse temperatures of the extremal values of the specific heat $C_V^{\max}$, Binder's energy parameter $B^{\min}$, or the difference of the peak weights or heights of the double-peaked energy probability density, D^{eqw} and D^{eqh} for various lattice sizes. Dotted lines are the extrapolations from the best fits, and small numbers give the linear lattice sizes for convenience.

canonical energy probability distribution $p(E,\beta)$. The energy of the minimum between the two peaks, $E_{\min}$, is determined to distinguish between the different phases and then used to minimize the difference of the peak *weights* or *heights*, respectively, namely

$$D^{\mathrm{eqw}}(\beta)=\left(\sum_{E<E_{\min}} p(E,\beta)-\sum_{E\geq E_{\min}} p(E,\beta)\right)^2 \text{ and } D^{\mathrm{eqh}}(\beta)=\left(\max_{E<E_{\min}}\{p(E,\beta)\}-\max_{E\geq E_{\min}}\{p(E,\beta)\}\right)^2 . \tag{7}$$

The data and fits in Fig. 1 for the estimates of the inverse transition temperature using Eqs. (5), (6) clearly display the non-standard scaling laws with $1/L^2$ corrections for periodic boundary conditions. The extrapolated values for the inverse temperature of the infinite system ($\beta^\infty = 0.551\,332(8)$) are visually and numerically in very good agreement with the somewhat less accurate values from the system with fixed boundaries ($\beta^\infty = 0.551\,38(5)$), where fits on the latter are conducted with a leading $1/L$ correction accounting for surface effects. The temperatures minimizing $D^{\mathrm{eqw}}(\beta)$ and $D^{\mathrm{eqh}}(\beta)$ in Eq. (7) also agree with the transmuted scaling behaviour for periodic boundaries and with $1/L$ scaling for fixed boundaries [18].

Apart from the transition temperature, another characteristic quantity that determines the dynamics at a first-order phase transition is the latent heat. The straightforward approach of measuring this in multicanonical simulations is to calculate the energies per site of the individual phases,

$$e_{\mathrm{o}}(L)=\sum_{E<E_{\min}} e\,p(E,\beta^{\mathrm{eqw}})\Big/\sum_{E<E_{\min}} p(E,\beta^{\mathrm{eqw}})\,,\quad e_{\mathrm{d}}(L)=\sum_{E\geq E_{\min}} e\,p(E,\beta^{\mathrm{eqw}})\Big/\sum_{E\geq E_{\min}} p(E,\beta^{\mathrm{eqw}})\,, \tag{8}$$

take their difference $\Delta e(L)=e_{\mathrm{d}}(L)-e_{\mathrm{o}}(L)$ and then extrapolate to the thermodynamic limit $L\to\infty$. This yields for periodic boundary conditions (bc) and lattice sizes up to $L=27$ [18]

$$\Delta\hat{e} = 0.851\,48(5) \qquad \text{periodic bc from (8)}\,, \tag{9}$$

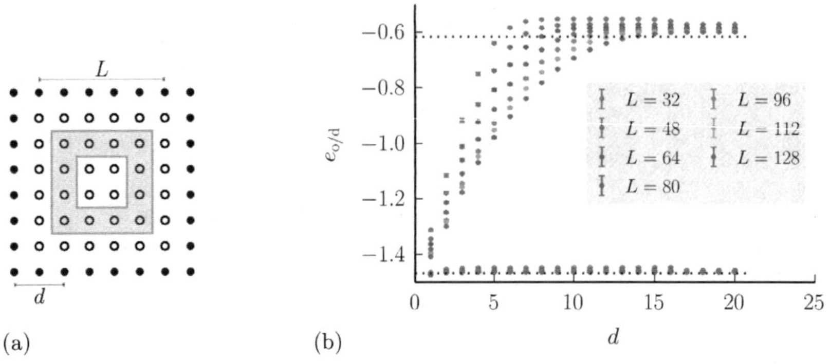

Figure 2. (a) Illustration of the setup of the simulations in a two-dimensional analogue. A cube of edge length L of free spins (represented by open circles) is surrounded by fixed spins with value $+1$ each (filled circles) and measurements are taken within a layer of width 1 at a distance d to the boundary (shaded area). (b) Energies of the disordered phase (upper data points) and ordered phase (lower data points) for different lattice sizes L. The dotted lines are the energies of the infinite system obtained from the multicanonical simulations. Both the ordered and disordered energies of the layers are overshooting the infinite lattice energies. Towards the boundary, the disordered phase becomes heavily perturbed by the fixed spins of the boundary.

and can be compared to Eqs. (5), (6), where we read off that the first correction of the non-standard finite-size scaling law only depends on the latent heat. The latent heat obtained by the fitting parameters was found to be consistent [18], although tending slightly to higher values.

For the fixed boundary conditions, measuring similarly the energy from an extrapolation of lattice sizes up to $L = 29$ [18] gives a much smaller latent heat of

$$\Delta\hat{e} = 0.694(4) \qquad \text{fixed bc from (8)}\ , \tag{10}$$

which is puzzling since this is not in accordance with the expectation that contributions from the boundary vanish in the thermodynamic limit.

To clarify this curious feature, we prepared short (canonical) Metropolis simulations for systems up to much larger linear sizes $L \in \{32, 48, 64, 80, 96, 112, 128\}$ and fixed boundaries. The systems were prepared in the disordered or ordered phases at a fixed temperature given by

$$\beta^{\mathrm{eqw}}(L) = 0.551\,19(12) - 0.8486(25)/L\ , \tag{11}$$

where we used the best fit of Ref. [18] obtained for fixed boundaries. In a first approximation, the system then has an equal probability to be in the disordered or ordered phases and, taking advantage of the huge free-energy barrier between them [18], we made sure that *no* "tunneling" between the phases occurs during the canonical simulations. We perform a number of $2^{16} = 65\,536$ sweeps and discard the first half of the time series, where thermalization effects may hamper the measurements. Effectively, we have a number of $2^{15} = 32\,768$ (correlated) measurements of the energy in each of the different phases. The statistical errors are then computed by Jacknife analysis using 16 blocks.

The influence of the fixed boundary is investigated by taking measurements in layers of unit-width at several distances d to the fixed boundary around a bigger cube of linear system size L as illustrated in Fig. 2(a). To have the energy of the layers normalized without trivial bias, the local energy at a given spin (the sum over all plaquettes touching that spin) is divided by

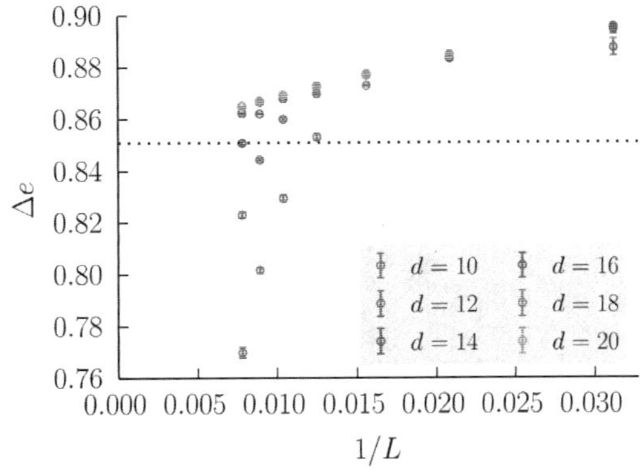

Figure 3. The latent heat over inverse lattice length L in layers at a distance d for the measurements from Fig. 2(b) at several d. The expected latent heat as $1/L \to 0$ obtained from multicanonical simulations with periodic boundary conditions is denoted by the dotted line.

the 12 possibly activated plaquettes. This is averaged over all spins in the layer and multiplied by a factor of 3 to keep the same energetic scale[1] as in Ref. [18]. The energies of the ordered and disordered phases $e_{\mathrm{o/d}}$ obtained in this way as a function of the distance d are collected in Fig. 2(b). With decreasing distance, the layers get increasingly ordered, which is reflected by both curves bending towards the ground-state energy $e = -1.5$. We find that far from the boundary the energies are overshooting the infinite volume limits. This is in accordance with the Taylor series expansion of the energies around $\beta{=}\beta^{\infty}$,

$$e_{\mathrm{o/d}}(\beta) = \hat{e}_{\mathrm{o/d}} + \left.\frac{\partial e_{\mathrm{o/d}}}{\partial \beta}\right|_{\beta=\beta^{\infty}} (\beta - \beta^{\infty}) + \mathcal{O}\left((\beta - \beta^{\infty})^2\right), \tag{12}$$

which leads to

$$e_{\mathrm{o/d}}(\beta^{\mathrm{eqw}}) = \hat{e}_{\mathrm{o/d}} - \hat{C}_{\mathrm{o/d}} \left(\frac{1}{\beta^{\infty}}\right)^2 (\beta^{\mathrm{eqw}} - \beta^{\infty}) + \mathcal{O}\left((\beta^{\mathrm{eqw}} - \beta^{\infty})^2\right). \tag{13}$$

Here, the specific heats $\hat{C}_{\mathrm{o/d}} = -(\beta^2 \partial e_{\mathrm{o/d}}/\partial\beta)|_{\beta=\beta^{\infty}}$ of the pure phases enter and the (empirical) first correction $\beta^{\mathrm{eqw}} - \beta^{\infty} < 0$ from Eq. (11) gives

$$e_{\mathrm{o/d}}(\beta^{\mathrm{eqw}}) = \hat{e}_{\mathrm{o/d}} + 0.8486(25)\hat{C}_{\mathrm{o/d}} \left(\frac{1}{\beta^{\infty}}\right)^2 /L + \dots \quad . \tag{14}$$

Therefore the first correction to the energies is a positive value and we would expect the energies of the ordered and disordered phase to be larger than those for infinite volume, as measured.

The spatial range of the perturbation of the disordered phase by the fixed boundary spins is somewhat unexpected. For the simulated lattice sizes, the disordered energy is retrieved first at a distance of at least $d = 14$. At this distance, already 52% of the total number of spins reside in the boundary of a $L = 128$ cube, which means that one must neglect more than

[1] For periodic boundaries, dividing by the number of spins L^3 instead of the number of plaquettes $3L^3$ leaves an additional factor of 3 in the energy scale. For fixed boundaries the number of plaquettes is $3L(L+1)^2$ when L is the linear lattice length of free spins and only those plaquettes that are composed of at most three *fixed* spins are added to the system's energy.

half of the interactions in the system to find an unperturbed disordered energy. This implies that the extrapolation of the latent heat in [18] is clearly obscured, where only lattices up to linear size $L = 29$ were accessible. The latent heat contained in layers at several distances from the boundary as a function of the inverse linear lattice size is shown Fig. 3, to qualitatively understand the scaling of the energies within the layers. With increasing distance d, fits would yield increasingly higher values as $L \to \infty$, at least asymptotically bigger than that of 0.694(4) in Eq. (10). To further elaborate on this, a fit on the asymptotics of the latent heat contained in the layer at $d = 20$ yields $\Delta e(L) = 0.8530(3) + 1.56(3)/L$ with $\chi^2_{\text{dof}} = 1.15$, which is much closer to 0.85148(5) from the simulations with periodic boundary conditions.

Any model with an exponentially degenerate low-temperature phase will display the modified scaling at a first-order phase transition described for the $3d$ gonihedric model here. Apart from higher-dimensional variants of the gonihedric model or its dual [18], there are numerous other fields where the scenario could be realized. Examples range from ANNNI models [23] to topological "orbital" models in the context of quantum computing [24] such as the $3d$ classical compass or t_{2g} orbital model where a highly degenerate ground state is well known and signatures of a first-order transition into the disordered phase have recently been found numerically [25].

Acknowledgements
This work was supported by the Deutsche Forschungsgemeinschaft (DFG) through the Collaborative Research Centre SFB/TRR 102 (project B04) and by the Deutsch-Französische Hochschule (DFH-UFA) under Grant No. CDFA-02-07.

[1] Janke W 2003 *First-order phase transitions* in *Computer Simulations of Surfaces and Interfaces* (NATO Science Series, II. Math., Phys. and Chem. 114) eds B Dünweg, D P Landau and A I Milchev (Dordrecht: Kluwer) pp 111–135, and references therein
[2] Imry Y 1980 *Phys. Rev.* B **21** 2042
[3] Binder K 1981 *Z. Phys.* B **43** 119
[4] Fisher M E and Berker A N 1982 *Phys. Rev.* B **26** 2507
[5] Privman V and Fisher M E 1983 *J. Stat. Phys.* **33** 385
[6] Binder K and Landau D P 1984 *Phys. Rev.* B **30** 1477
[7] Challa M S S, Landau D P and Binder K 1986 *Phys. Rev.* B **34** 1841
[8] Peczak P and Landau D P 1989 *Phys. Rev.* B **39** 11932
[9] Privman V and Rudnik J 1990 *J. Stat. Phys.* **60** 551
[10] Janke W 1993 *Phys. Rev.* B **47** 14757
[11] Borgs C and Kotecký R 1990 *J. Stat. Phys.* **61** 79; 1992 *Phys. Rev. Lett.* **68** 1734
[12] Borgs C and Janke W 1992 *Phys. Rev. Lett.* **68** 1738
[13] Mueller M, Janke W and Johnston D A 2014 *Phys. Rev. Lett.* **112** 200601
[14] Pietig R and Wegner F 1996 *Nucl. Phys.* B **466** 513; 1998 *Nucl. Phys.* B **525** 549
[15] Johnston D A, Lipowski A and Malmini R P K C 2008 *The gonihedric Ising model and glassiness* in *Rugged Free-Energy Landscapes – An Introduction* (Springer Lecture Notes in Physics vol 736) ed W Janke (Berlin/Heidelberg: Springer) pp 173–199, and references therein
[16] Espriu D, Baig M, Johnston D A and Malmini R P K C 1997 *J. Phys. A: Math. Gen.* **30** 405
[17] Lipowski A 1997 *J. Phys. A: Math. Gen.* **30** 7365
[18] Mueller M, Johnston D A and Janke W 2014 *Nucl. Phys.* B **888** 214
[19] Mueller M, Janke W and Johnston D A 2014 *Physics Procedia* **57** 68
[20] Berg B A and Neuhaus T 1991 *Phys. Lett.* B **267** 249; 1992 *Phys. Rev. Lett.* **68** 9
[21] Janke W 1992 *Int. J. Mod. Phys.* C **03** 1137; 1998 *Physica* A **254** 164
[22] Janke W 2003 *Histograms and all that* in *Computer Simulations of Surfaces and Interfaces* (NATO Science Series, II. Math., Phys. and Chem. 114) eds B Dünweg, D P Landau and A I Milchev (Dordrecht: Kluwer) pp 137–157, and references therein
[23] Selke W 1988 *Phys. Rep.* **170** 213
[24] Nussinov Z and van den Brink J 2013 Compass and Kitaev models – Theory and physical motivations *e-print* `arXiv:1303.5922`
[25] Gerlach M H and Janke W 2015 *Phys. Rev.* B **91** 045119

Journal of Physics: Conference Series **640** (2015) 012003 doi:10.1088/1742-6596/640/1/012003

Generalized ensemble method applied to study systems with strong first order transitions

E. Małolepsza[1], J. Kim[2] and T. Keyes[1]

[1] Department of Chemistry, Boston University, Boston, MA 02215-2521

[2] Broad Institute of MIT and Harvard, Cambridge, MA 02142, USA

E-mail: keyes@bu.edu

Abstract. At strong first-order phase transitions, the entropy versus energy or, at constant pressure, enthalpy, exhibits convex behavior, and the statistical temperature curve correspondingly exhibits an S-loop or back-bending. In the canonical and isothermal-isobaric ensembles, with temperature as the control variable, the probability density functions become bimodal with peaks localized outside of the S-loop region. Inside, states are unstable, and as a result simulation of equilibrium phase coexistence becomes impossible. To overcome this problem, a method was proposed by Kim, Keyes and Straub [1], where optimally designed generalized ensemble sampling was combined with replica exchange, and denoted generalized replica exchange method (gREM). This new technique uses parametrized effective sampling weights that lead to a unimodal energy distribution, transforming unstable states into stable ones. In the present study, the gREM, originally developed as a Monte Carlo algorithm, was implemented to work with molecular dynamics in an isobaric ensemble and coded into LAMMPS, a highly optimized open source molecular simulation package. The method is illustrated in a study of the very strong solid/liquid transition in water.

1. Introduction

First order phase transitions (solid↔liquid, liquid↔gas, and solid↔gas) are defined by a discontinuity in the free energy with respect to density. The phase transition starts in one phase and ends up in another, passing through coexisting region with both phases present. It involves heat absorption or release, the so called latent heat, where the temperature does not change with added heat until the phase transition is completed. Due to their ubiquitous presence, phase transitions have been extensively studied. However, computer simulation of phase transitions is a challenging task and here we focus on the source of the problem and applications of a proposed solution.

The problem in simulating first order phase transitions centers on the phase coexistence region and it is caused by unusual behaviour of entropy, S [1, 2, 3, 4, 5, 6, 7]. Entropy is a function that monotonically increases with energy or enthalpy, but in the transition region its second derivative changes sign twice, and so S changes curvature from concave to convex and back to concave, forming "convex intruder". Consequently, the statistical temperature, T_S,

$$T_S(E) = \left[\frac{dS}{dE}\right]^{-1}, \tag{1}$$

also does not monotonically grow, but instead exhibits a back-bending, called an S-loop, as shown in black in Fig. 1.

Journal of Physics: Conference Series **640** (2015) 012003 doi:10.1088/1742-6596/640/1/012003

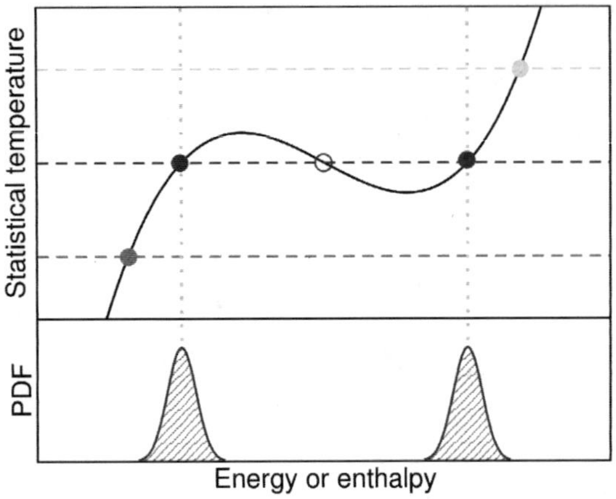

Figure 1. Above: Cartoon of statistical temperature exhibiting an S-loop in first order phase transition region is shown in black. Color dashed lines represents three temperatures chosen for isothermal simulations, with blue one as transition temperature. **Below**: canonical probability density function (PDF) for equilibrium transition temperature indicated by blue line on the plot above, shows two peaks, each corresponding to a pure phase system, with a region in between that is not sampled.

Isothermal simulations (either as NVT or NPT) equilibrate readily in regions where only one phase is present, and the system is well sampled satisfying the ergodic hypothesis. Such ensembles correspond to red and green dots on Fig. 1, where red and green lines representing chosen temperatures intersect the statistical temperature. For both cases, below and above transition temperature, the constant temperature defining the ensemble intersects the statistical temperature only once. The intersection locates the most probable energy and minimum in free energy.

The situation changes when one would like to study the transition region and sample configurations with coexisting phases. A constant temperature in the vicinity of the transition temperature crosses the statistical temperature three times (see blue line on Fig. 1). Two filled circles correspond to minima in free energy where only one phase is present in the system, and maxima in the energy probability distribution function (PDF). The result is bimodal behavior of the PDF (see lower panel of Fig. 1). The empty circle in the middle indicates a barrier in free energy profile that might lead to broken ergodicity. Depending on the barrier height, sampling in the phase coexisting region is either inefficient or, in fact, impossible, with system in simulations stuck on either side of the barrier in minima indicated by filled blue circles.

2. The gREM

While constant-temperature ensembles are the most widely used, we would like to suggest a generalized ensemble that is characterized by a temperature function depending on energy, $T_g(E)$ (or H in case of constant pressure simulations) [1]. The location of the maxima of the PDF is given by the generalized relation,

$$T_g(E) = T_S(E). \tag{2}$$

The gREM utilizes generalized ensembles with a linear $T_g(E)$,

$$T_\alpha(E) = \lambda_\alpha + \eta(E - E_0), \tag{3}$$

with negative slope, η, chosen to be large enough so that Eq. 2 has a single solution in the coexistence region, see orange lines on Fig. 2. E_0 is a reference energy. The result is an easily-simulated unimodal PDF in the transition region [1].

Replicas are selected with varying intercepts of the $T_g(E)$ function, λ_α, such that the coexistence region is spanned. Replicas exchange configurations with an acceptance probability,

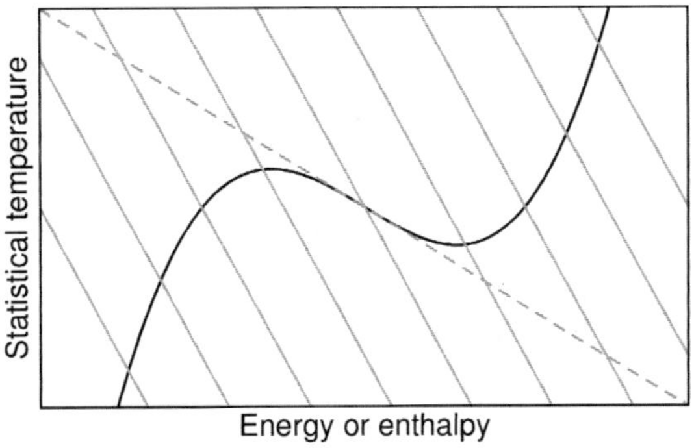

Figure 2. Cartoon of statistical temperature in black with orange lines as examples of generalized effective temperature $T_\alpha(E)$ defined in Eq. 3, each crossing the statistical temperature only once. Dashed blue line represents the tangent to the statistical temperature that defines marginal slope for effective temperature to ensure single solution of Eq. 2.

Acc, calculated to preserve detailed balance of the generalized ensemble,

$$Acc = \min[1, e^{\Delta}] \tag{4}$$

$$\Delta = \omega_{\alpha+1}(E') - \omega_{\alpha+1}(E) + \omega_\alpha(E) - \omega_\alpha(E') \tag{5}$$

where ω_α is the gREM sampling weight of the α-th replica and is calculated as follows:

$$\omega_\alpha = \int_{\mathrm{E}} \frac{1}{T_g(E')} dE'. \tag{6}$$

The original gREM method employed Monte Carlo, while here the approach was combined with the more efficient molecular dynamics method. The method was implemented by our group as a fix type module in LAMMPS [8, 9]. The required sampling is obtained from isothermal-isobaric MD by fixing the temperature at a constant "kinetic temperature", T_0, and scaling the forces by the factor, $s(H) = T_0/T_g(H)$.

The new MD gREM algorithm was applied to pure water. A cubic box containing 3337 water molecules in configuration of hexagonal ice was initially equilibrated in the isobaric-isothermal ensemble at 200 K and 1 atm, and then used as the starting configuration in each generalized replica. Periodic boundary conditions were applied. The kinetic temperature was chosen as 330 K and pressure was kept as 1 atm using Nose-Hoover style non-Hamiltonian equations of motion [10, 11, 12, 13] as implemented in LAMMPS.

Water interactions were modeled by Stillinger-Weber potential [14] with parameters adjusted by Molinero [15] and denoted the mW model. In this approach each water molecule is represented as a single bead interacting via two- and three-body potential terms. Such coarse-graining reduces computational time while retaining a good reproduction of the properties of water [15].

Data gathered in all replicas were combined using the statistical temperature weighted histogram analysis method (ST-WHAM) [16] that is efficient and iteration-free method in contrast to the usual WHAM method [17].

3. Solid↔liquid transition in water

Three of the most important properties of the gREM ensemble are that (1) it can successfully sample the phase coexisting region and overcome barriers, (2) though structures in this region are unstable in canonical simulations, the gREM ensemble stabilizes them and allows for equilibration [1] and (3) by using λ to control the intersection of $T_g(H)$ with $T_S(H)$, enthalpy, not temperature, becomes the control variable for the simulation.

Characteristic structures observed around the solid↔liquid transition in water are shown in Fig. 3. The lowest enthalpy structure corresponds to hexagonal ice occupying the entire

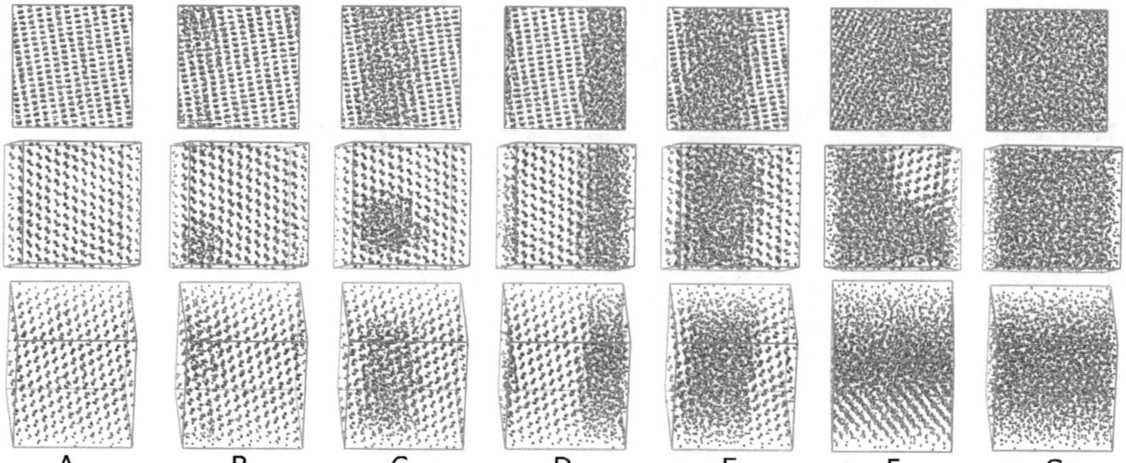

Figure 3. Snapshots of representative configurations observed during the solid↔liquid transition in water, shown from three points of view for easier identification of structural topology and arranged in columns. Blue dots correspond to water molecules, simulation box shown in gray. From left to right: (**A**) ice, (**B**) ice with liquid ball, (**C**) ice with liquid cylinder, (**D**) and (**E**) two structures with ice and liquid slabs differing in the ratio of the phases, (**F**) ice cylinder in liquid, (**G**) liquid.

simulation box (structure denoted as **A**). Next, when ice starts melting, a small liquid ball (**B**) appears followed by a liquid cylinder (**C**), both surrounded by hexagonal ice. Once there is a sufficiently large amount of liquid water in the system, topology changes to liquid/ice slabs (**D, E**), where the interface between these two phases stays flat while the ratio of liquid to ice increases with increasing enthalpy. The next type of structures appears when there is not enough ice to form a slab and we observe instead an ice cylinder in equilibrium with bulk water (**F**). Finally, ice fully melts and the system consists of only bulk water (**G**). Topologies of the observed structures are in agreement with predictions from other studies [18].

Every change in the topology is connected to a feature in the calculated statistical temperature presented on Fig. 4. Therefore the expected S-loop in $T_S(H)$ for ice↔liquid water exhibits a complex structure in the phase coexistence region. Plateaus (or regions where $T_S(H)$ flattens) correspond to structures that belong to the same topology, while every change of topology type is in fact a transition and can be identified by an almost vertical change in $T_S(H)$.

There are two special points of interest on the $T_S(H)$ curve, namely spinodals, that define the limits of stability of both phases. The hexagonal ice in the mW model in our study can be heated up to 325.8 K before melting process starts. The experimental value of this limit equals 330±10 K [19]. In an earlier study of the mW model [20], a box of hexagonal ice was heated in canonical simulations for 100 ns until it decomposed to liquid water. The estimated limit of stability was found to be between 330 and 340 K. This value is higher compared to our result and is most likely caused the problem of overcoming the melting barrier when a liquid nucleus appears in an ice block.

The other limit of stability describes the situation where ice fully melts and changes into supercooled water at 223.4 K. The homogeneous nucleation temperature of mW water is equal to 202 K under 1 atm [15], below our estimated spinodal temperature, which is nor surprising considering fact that we started the simulation from hexagonal ice, not liquid water, and as such we monitored ice melting instead of ice nucleation.

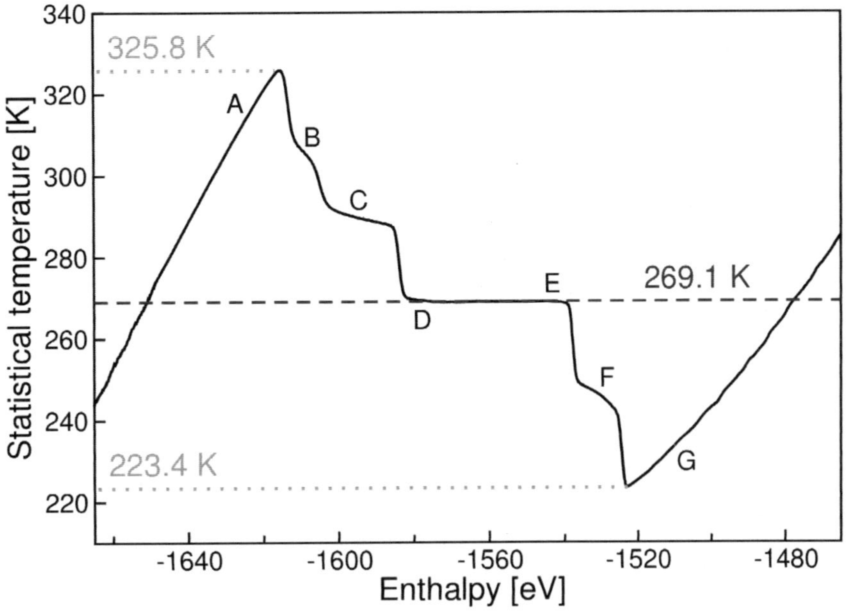

Figure 4. The statistical temperature for the hexagonal ice↔water transition, where each replica started from hexagonal ice, is shown as black line. The transition temperature from the Maxwell construction equals 269.1 K and is shown as a dashed blue line, while the limits of stability are shown in cyan and are equal to 223.4 K and 325.8 K. Letters A to G denote locations of structures shown on Fig. 3.

With the statistical temperature available one can apply the Maxwell construction to the inverse of $T_S(H)$ to obtain the transition temperature. In this procedure one is looking for a temperature that yields equal areas above and below it. The calculated equilibrium temperature of ice melting in mW model is equal to 269.1 K, compared to temperature of 274.6 K obtained by Molinero [15] using the phase coexistence method. Interestingly, the calculated melting temperature corresponds to a large plateau on the statistical temperature where structures consist of liquid and ice slabs with increasing amount of liquid towards higher enthalpies. This topology is identical to those used in the phase coexistence method where half of the simulation box is filled with one phase and the other half with the second phase. Location of this plateau can in fact be used to estimate the transition temperature.

In the present paper we showed preliminary results from the new MD version of an enhanced sampling algorithm, gREM, applied to the strong first order solid↔liquid transition in water modeled by the mW potential. We successfully calculated the statistical temperature for this transition as well as the transition temperature. We observed a series of structures with coexisting phases in several distinct topologies. Though usage of the gREM simulation method each of the observed structures was in equilibrium, in contrast to results from canonical simulations where the transition region is barely sampled.

Acknowledgments
The research was supported by the U.S. Department of Energy, Office of Basic Energy Sciences, Division of Chemical Sciences, Geosciences and Biosciences under Award DE-SC0008810. We thank Profs. Bill Klein and Harvey Gould for discussions, Dr. Marek Orzechowski for help with graphics, and Dr. Christopher Knight for help with implementation.

Journal of Physics: Conference Series **640** (2015) 012003 doi:10.1088/1742-6596/640/1/012003

[1] Kim J, Keyes T and Straub J E 2010 *J. Chem. Phys.* **132** 224107
[2] Labastie P and Whetten R L 1990 *Phys. Rev. Lett.* **65** 1567
[3] Wales D J and Berry R S 1994 *Phys. Rev. Lett.* **73** 2875
[4] Bachmann M 2014 *Thermodynamics and Statistical Mechanics of Macromolecular Systems* (Cambridge University Press, Cambridge, UK)
[5] Gross D H E 2006 *Physica A* **365** 138
[6] Junghans C, Bachmann M and Janke W 2008 *J. Chem. Phys.* **128** 085103
[7] Behringer H and Pleimling M 2006 *Phys. Rev. E* **74** 011108
[8] Plimpton S 1995 *J. Comp. Phys.* **117** 1–19
[9] `http://lammps.sandia.gov`
[10] Martyna G L, Tobias D J and Klein M L 1994 *J. Chem. Phys.* **101** 4177
[11] Parrinello M and Rahman A 1981 *J. Appl. Phys.* **52** 7182
[12] Tuckerman M E, Alejandre J, Lopez-Rendon R, Jochim A L and Martyna G J 2006 *J. Phys. A: Math. Gen.* **39** 5629
[13] Shinoda W, Shiga M and Mikami M 2004 *Phys. Rev. B* **69** 134103
[14] Stillinger F H and Weber T A 1985 *Phys. Rev. B* **31** 5262
[15] Molinero V and Moore E B 2009 *J. Phys. Chem. B* **113** 4008
[16] Kim J, Keyes T and Straub J E 2011 *J. Chem. Phys.* **135** 161103
[17] Ferrenberg A M and Swendsen R H 189 *Phys. Rev. Lett.* **63** 1195
[18] Schrader M, Virnau P and Binder K 2009 *Phys. Rev. E* **79** 061104
[19] Schmeisser M, Iglev H and Laubereau A 2007 *J. Phys. Chem. B* **111** 11271
[20] Jacobson L C, Hujo W and Molinero V 2009 *J. Phys. Chem. B* **113** 10298

Journal of Physics: Conference Series **640** (2015) 012004 doi:10.1088/1742-6596/640/1/012004

Transport with hard-core interaction in a chain of asymmetric cavities.

G. P. Suárez[1,2], **M. Hoyuelos**[1,2], **H. O. Mártin**[1,2]

[1]Instituto de Investigaciones Físicas de Mar del Plata (IFIMAR – CONICET)
[2]Departamento de Física, Facultad de Ciencias Exactas y Naturales, Universidad Nacional de Mar del Plata. Deán Funes 3350, 7600 Mar del Plata, Argentina

E-mail: gsuarez@mdp.edu.ar

Abstract. In this paper we investigate the diffusion of particles inside a chain of asymmetric cavities. We are considering particles that interact through a hard–core potential and are driven by an external force. We show that the difference in the current when the force is applied to the left and to the right strongly depends on the concentration inside the cavity. We found that, when the concentration is high enough, the hard–core interaction vanishes and inverts the asymmetric effect of the cavity. We also introduce a new equation, a modification to the Fick–Jacobs equation, to describe this system analytically. Finally, we used numerical simulations to verify the analytic results, finding a good agreement between theory and simulations.

1. Introduction

The diffusion of particles is a subject that has been studied extensively in many different contexts (see [1] and cites within). One of this contexts is the diffusion of particles inside narrow channels [2, 3, 4, 5, 6]. However, the diffusion of interacting particles inside a confined environment has received much less attention, even when there are numerous examples in nature [7, 8, 9, 10, 11, 12].

In this paper we studied the transport of Brownian particles in a channel of variable transverse width. On the one hand, we studied how the current of particles that cross an asymmetric cavity is modified by increasing the concentration, considering particles that interact through a hard-core potential. On the other hand, we introduced a modified Fick–Jacobs equation to describe this process. This equation has been solved using numerical methods for the case of asymmetric cavities. We mainly studied how the particles arrange inside the cavity, when a small force is applied. Numerical simulations with the method of Monte Carlo where carried out. The results obtained with the equation are in complete agreement with the numerical simulations.

This paper is organized as follows. In Sec. 2, we analyze the current of particles in a triangular cavity as a function of the concentration and we expose some of the characteristics of the numerical simulations. In Sec. 3, we explain the most important steps to derive the non–linear Fick–Jacobs equation. In Sec. 4, we show how the theoretical solution matches with the results obtained with the numerical simulations. Some final remarks are stated in Sec. 5.

Journal of Physics: Conference Series **640** (2015) 012004 doi:10.1088/1742-6596/640/1/012004

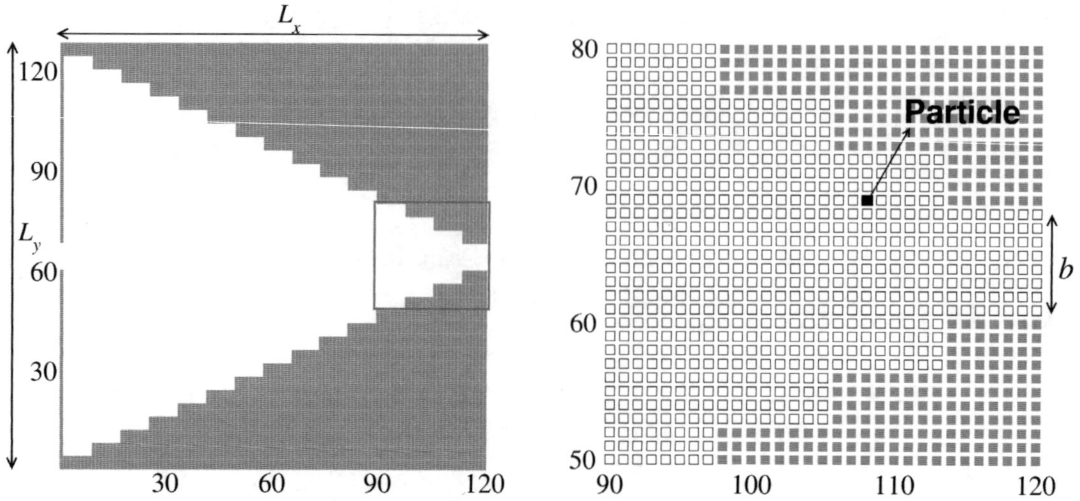

(a) Cavity used to carry out simulations. (b) Zoom of the red squared region.

Figure 1: Asymmetric cavity used in simulations. The white region represents the allowed sites where the particles can diffuse. Distances are expressed in terms of the lattice constant, a. $L_x = 120; L_y = 128; b = 8$.

2. Current of particles

In the first part of this work we analyzed the diffusion of hard–core interacting particles inside a triangular cavity. To perform the numerical simulations we depicted the triangular cavity inside a regular square lattice of $L_x \times L_y$ sites, as seen in Fig. 1. The distance between neighboring sites is a (a value $a = 1$ was considered). There is a small exit of size b on each side. Periodic boundary conditions were set in those exits to simulate a long chain of identical cavities. Each site inside the cavity can be either occupied by a particle or empty. The hard–core interaction is realized by allowing only one particle on each site of the lattice.

We define the mean concentration $c = N/S$, where S is the total number of sites inside the cavity and N is the number of particles. At the beginning of the simulation, the particles are distributed randomly inside the cavity. The size of a particle is exactly one site, and they move by jumping to one nearest neighbor site. They are allowed to jump freely inside the cavity, except when the new site is occupied with another particle. In this case the particle remains on its site. We say that one Monte Carlo step occurs when, in average, every particle has had the opportunity to jump once.

The particles are driven by a dimensionless external force, $\delta = \beta F a$ where $\beta^{-1} = k_B T$. At any given time, a particle has four possible directions to move. One of the them has a a higher jump rate, due to the effect of the external force applied. The other three have the same (lower) jump rate. Thus, if the force is applied to the right, the jump rates in different directions are

$$\begin{aligned} p_\uparrow = p_\downarrow = p_\leftarrow &= P \\ p_\rightarrow &= P(1+\delta), \end{aligned} \tag{1}$$

where $P = D/a^2$, and D is the diffusion coefficient. In the same way, if the force is applied to the left $p_\leftarrow = P(1+\delta)$, and $p_\rightarrow = P$. In the first part of this paper we will analyze the difference in the current of particles of a system when the force is applied first to the right, and then to the left. We let the system evolve to the stationary state, and then we start to measure the relevant quantities, like the distribution of the particles inside the cavity, the distance covered by each one or the mean velocity, among others.

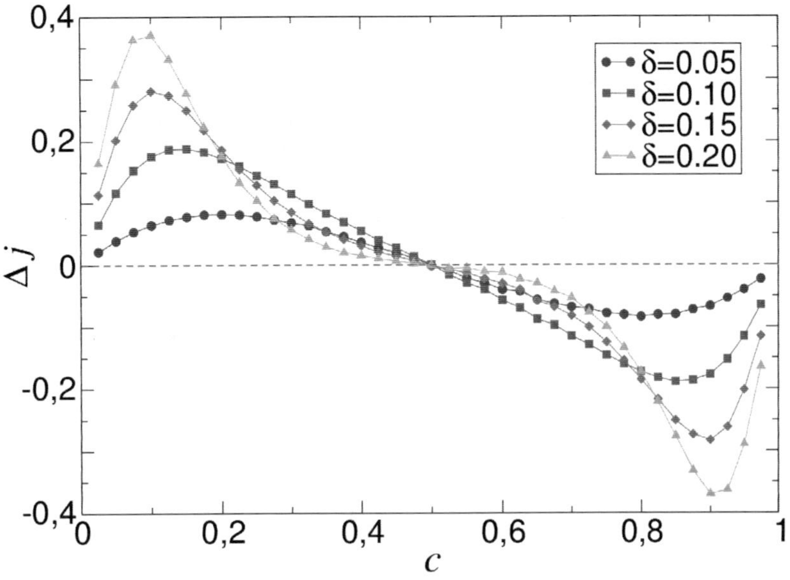

Figure 2: Difference in the current, Δj a a function of the mean concentration c in the cavity of Fig. 1. The values of the force are: $\delta = 0.05(\bullet); \delta = 0.10(\blacksquare); \delta = 0.15(\blacklozenge); \delta = 0.20(\blacktriangle)$. Average over 10^6 samples.

Here we are interested in the total current of particles that cross the cavity, j. If there were no interaction between particles, the current will be higher as the concentration or the force becomes higher. But that is not necessarily true for interacting particles. Also, because of the particular shape of the cavity, the current will not be the same when the external force is applied to the right, j_+, or to the left, j_-. Since this asymmetric effect depends on the size of the particles, it has been proposed by Reguera *et al.*[13] as a possible way to separate a mixture of different size particles.

In Fig. 2 we show the difference of currents of particles, $\Delta j = |j_+| - |j_-|$ as a function of the concentration for a wide range of forces. There are a few things that we can notice from this plot. Around $c = 0.1$ we can see there is a maximum value in the difference of the current Δj. There is a value of the mean concentration, c, for each value of the force, δ, for which the difference in the current is maximum. Increasing the concentration after this point, will only decrease the asymmetric effect of the cavity. Particularly, if the mean concentration is $c = 0.5$ the asymmetric effect vanishes, and the current of particles is the same when the force is applied to the left or to the right, i.e. $\Delta j = 0$, independently of its magnitude. Finally, if the concentration is higher than 0.5, we found that the current of particles when the force is applied to the left is always higher than the current of particles when the force is applied to the right. This is specially interesting because the asymmetry in the cavity is designed to favor the diffusion to the right. After this we can conclude that the hard–core interaction, whose influence increases as c is increased, is the cause of an inversion of the asymmetry effect.

3. Fick-Jacobs Equation

The Fick-Jacobs equation describes the diffusion of non–interacting particles inside a channel of variable width. It has been used to describe the diffusion of not–interacting particles or in the low concentrations regime [14, 15]. In this work we have extended this equation to include the case of diffusion of particles with hard–core interaction. With this we mean that two particles can not occupy the same space. The excluded volume effect of the other particles has a considerable influence on the diffusion of a tagged particle. This equation allows us to describe a rather complex system, of many interacting particles inside a two– or three–dimensional environment, with a simple uni-dimensional equation.

To derive the modified Fick–Jacobs equation, let us consider a set of hard–core interacting particles in a one–dimensional discretized system, the so called single file diffusion [16, 17, 18]. In a given configuration, the current of particles between the sites i and $i+1$ is given by,

$$\begin{aligned} J_i^o &= n_i^o P_{i,i+1} - n_{i+1}^o P_{i+1,i} \\ &= P(1-\beta\Delta U_i)\, n_i^o\,(1-n_{i+1}^o) - P\, n_{i+1}^o\,(1-n_i^o), \end{aligned} \tag{2}$$

where $n_i^o = 0$ or 1 is the occupation number of site i, $\Delta U_i = U_{i+1} - U_i$ is the variation of the external potential and $\beta^{-1} = k_B T$. P is the jumping rate, $P = D/a^2$ where D is the diffusion constant and $a = 1$ is the length of the jump. We are assuming that $\beta|\Delta U_i| \ll 1$. The terms $(1 - n_{i+1}^o)$ and $(1 - n_i^o)$ take into account the hard–core interaction.

Averaging over configurations one has $n_i = \langle n_i^o \rangle$ and $J_i = \langle J_i^o \rangle$. In the continuous limit $n_i \to n(x_i)$, $U_i \to U(x_i)$ and $J_i \to J(x_i)$, where $x_i = i$. Assuming the decorrelation $\langle n_i^o n_{i+1}^o \rangle \simeq n(x_i)n(x_{i+1})$ and small variation of $n(x)$ we obtain

$$J = -D\left[\frac{\partial n}{\partial x} + \beta\frac{dU}{dx}n(1-n)\right]. \tag{3}$$

Let us now consider a two–dimensional channel composed by a chain of cavities (as the one shown in Fig. 3b) where the external potential $U(x)$ is independent of the transverse direction, y. Assuming local equilibrium, the site concentration $m(x,y)$ does not depend on the transverse direction either. That is, $m(x,y) = n(x)$. Then, the total current is $j = JA$, where $A(x)$ is the transverse width of the channel and from Eq. (3) one obtains

$$-\frac{j}{D} = A\left[\frac{\partial n}{\partial x} - \beta F n(1-n)\right], \tag{4}$$

where $F = -dU/dx$ is the external force. Now from the continuity equation $\frac{\partial m}{\partial t} + \frac{\partial j}{\partial x} = 0$ we get

$$\frac{\partial m}{\partial t} = D\frac{\partial}{\partial x}\left[\frac{\partial m}{\partial x} - \frac{dA}{dx}\frac{m}{A} - \beta F m\left(1 - \frac{m}{A}\right)\right], \tag{5}$$

which corresponds to our extension of the Fick-Jacobs equation where the hard-core interaction is taken into account. For non–interacting particles $(1-\frac{m}{A}) \to 1$ and the well known Fick–Jacobs equation is recovered.

4. Results and Discussion

To test how accurate Eq. (4) is, we solved it for a particular case using a smoother asymmetric cavity than the one shown in Fig. 1, see Fig. 3b.

In Fig. 3a we show the stationary state of site concentration along the direction of the channel in an asymmetric cavity, for different values of the force. In the three curves the force is applied to the right. However, as a consequence of the analogy in the definition of a particle and a vacancy, they also represent the system behavior when the force is applied to the left. For every particle that move to the right there is a vacancy that moves to the left. All we have to do is to change $n \to 1-n$ (then $c \to 1-c$), $\delta \to -\delta$ and $j \to -j$. Note that in Eq. (4) this symmetry holds, see also Fig. 2. Further details on this matter may be found in [19, 20, 21].

The results show that the agreement between theory and simulations is very good almost everywhere. However, it can be noticed that, in some regions the agreement is better than others. This is due to the fact that we assumed m to be constant along the y axis. When this condition is not fulfilled, the theoretical curves slightly differs from simulations (see Fig. 3).

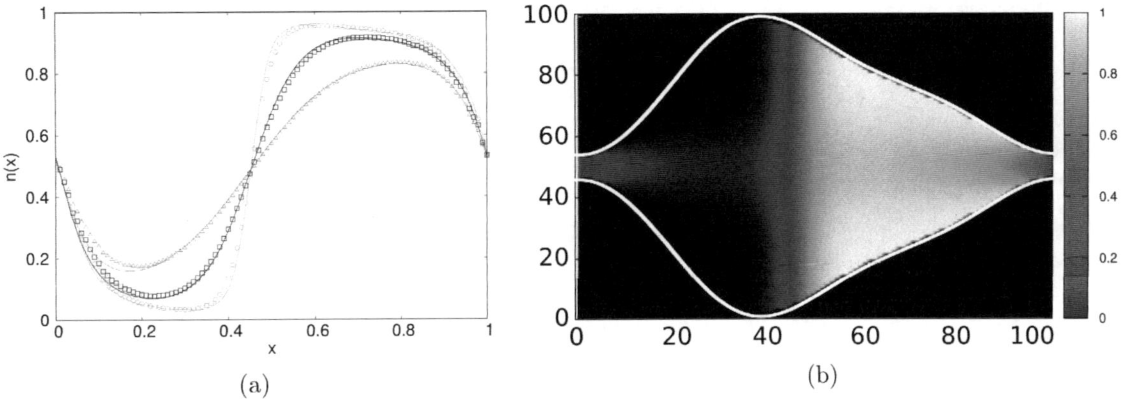

Figure 3: (a) Concentration along the x–axis, $n(x)$. Comparison between analytic solution to Eq. (4) and simulations for $c = 0.5$. In each case, the force is applied to the right and the following values were used: $\delta = 0.1$ (blue triangles), $\delta = 0.2$ (black squares) and $\delta = 0.5$ (red circles). Dots were obtained with numerical simulations (10^5 samples between $t = 10^4$ and $t = 10^7$ Monte Carlo time-steps). (b) Site concentration inside the cavity. Corresponds to the case $c = 0.5$ and $\delta = 0.5$ (red circles in (a)).

5. Conclusion

In this paper we studied how the diffusion of interacting particles is modified by increasing the concentration. We showed that the interaction between particles plays a fundamental role in this kind of systems. One reason is the formation of clogs where the available space is limited.

Apart from that, we were able to obtain a non–linear Fick-Jacobs equation to describe the transport of hard–core particles inside a cavity with variable cross section. This equation correctly predicts the steady state of the concentration of particles along the direction of the channel.

The agreement of theory and simulations confirms that the modified equation of Fick–Jacobs here introduced is a useful tool to describe the transport of particles when the interactions between them can not be neglected.

Acknowledgments

This work was partially supported by Consejo Nacional de Investigaciones Científicas y Técnicas (CONICET, Argentina, PIP 0041 2010-2012). GPS wishes to thank the organizing committee of the XXVI IUPAP Conference on Computational Physics for financial support to assist to the conference.

References

[1] ed Heitjans P and Kärger J 2005 *Diffusion in Condensed Matter: Methods, Materials, Models* (Springer)
[2] Rubí J M and Reguera D 2010 *Chem. Phys.* **375** 518
[3] Hänggi P and Marchesoni F 2009 *Rev. of Mod. Phys.* **81** 387
[4] Burada S, Schmid G, Talkner P, Hänggi P, Reguera D and Rubí J M 2008 *Bio Systems* **93** 16
[5] Ai B Q and Liu L G 2006 *Phys. Rev. E* **74** 051114
[6] Dagdug L, Berezhkovskii A M, Makhnovskii Y A, Zitserman V Y, Bezrukov S M 2012 *J. Chem. Phys.* **136** 214110
[7] Pries A R, Secomb T W and Gaehtgens P 1996 *Cardiovasc. Res.* **32**(4) p 654–667
[8] McHedlishvili G and Maeda N 2001 *Jpn. J. Physiol.* **51**(1) p 19–30
[9] Zhou H X, Rivas G and Minton A P 2008 *Annual Review of Biophysics* **37** p 375–397
[10] Kärger J and Ruthven D M 1992 *Diffusion in Zeolites and Other Microporous Solids* (New York: John Wiley)
[11] Keil F J , Krishna R and Coppens M O 2000 *Rev. Chem. Eng.* **16**(2) p 71–197

[12] Beerdsen E , Dubbeldam D and Smit B 2006 *Phys. Rev. Lett.* **96**(4) p 044501
[13] Reguera D, Luque A, Burada P S, Schmid G, Rubí J M and Hänggi P 2012 *Phys. Rev. Lett.* **108** 020604
[14] Jacobs J H 1967 *Diffusion Processes* (New York: Springer) p 68
[15] Zwanzig R 1992 *J. Phys. Chem.* **96** 3926
[16] Harris T E 1965 *J. Appl. Prob.* **2** 323
[17] Richards P M 1977 *Phys. Rev. B* **16** 1393
[18] Van Beijeren H, Kehr K W and Kutner R 1983 *Phys. Rev. B* **28** 5711
[19] Suárez G P, Hoyuelos M and Mártin H O 2013 *Phys. Rev. E* **88** 052136
[20] Dierl M, Einax M and Maass P 2013 *Phys. Rev. E* **87** 062126
[21] Torrez Herrera J J, Ranzuglia G A, Manzi S J and Pereyra V D 2013 *Phys. Rev. E* **87** 052101

Spin glass behavior of the antiferromagnetic Heisenberg model on scale free network

Tasrief Surungan[1,3], Freddy P. Zen[2,3], Anthony G. Williams[4]

[1]Department of Physics, Hasanuddin University, Makassar 90245, Indonesia
[2]Department of Physics, Bandung Institute of Technology, Bandung 40132, Indonesia
[3]Indonesian Center for Theoretical and Mathematical Physics (ICTMP), Bandung Institute of Technology, Bandung 40132, Indonesia
[4]Special Research Center for the Subatomic Structure of Matter (CSSM), The University of Adelaide, Adelaide, SA 5005, Australia

E-mail: tasrief@unhas.ac.id, fpzen@fi.itb.ac.id, anthony.williams@adelaide.edu.au

Abstract. Randomness and frustration are considered to be the key ingredients for the existence of spin glass (SG) phase. In a canonical system, these ingredients are realized by the random mixture of ferromagnetic (FM) and antiferromagnetic (AF) couplings. The study by Bartolozzi *et al.* [Phys. Rev. B**73**, 224419 (2006)] who observed the presence of SG phase on the AF Ising model on scale free network (SFN) is stimulating. It is a new type of SG system where randomness and frustration are not caused by the presence of FM and AF couplings. To further elaborate this type of system, here we study Heisenberg model on AF SFN and search for the SG phase. The canonical SG Heisenberg model is not observed in d-dimensional regular lattices for ($d \leq 3$). We can make an analogy for the connectivity density (m) of SFN with the dimensionality of the regular lattice. It should be plausible to find the critical value of m for the existence of SG behaviour, analogous to the lower critical dimension (d_l) for the canonical SG systems. Here we study system with $m = 2, 3, 4$ and 5. We used Replica Exchange algorithm of Monte Carlo Method and calculated the SG order parameter. We observed SG phase for each value of m and estimated its corersponding critical temperature.

1. Introduction

Spin glass (SG) is one of the most complex systems in condensed matter physics and has been intensively studied in the last four decades[1, 2, 3, 4, 5, 6]. It is a randomly frustrated magnetic system with frozen disordered spin orientation at low temperatures. This unusual configuration is regarded as a temporal ordered phase [7], different from the spatially ordered phase found in regular magnets. The complexity of the system is due to the presence of frustration and randomness, which are the key ingredients for the existence of SG phase. Frustration is a state where spins can not find fixed orientations to fully satisfy all the interactions with their neighboring spins. This can be caused either by the conflicting interaction between FM and AF couplings, or between among AF coupling due to the topological factors. Frustration alone can not lead a system to an SG phase, firmly exemplified by the fully frustrated AF planar spin systems which have spatially ordered phase at low temperatures[8, 9].

Most SGs studied are canonical system where both FM and AF couplings exist. The examples of these are Sherrington-Kirkpatrick model[2], Edward-Anderson model[3] and p-spin interaction model[4]. Bartolozzi *et al.* first reported SG behavior of the Ising model with AF interaction on scale free network (SFN)[10]. This is a new type of SG system without random distribution of FM and AF couplings. The nodes of the network do not have homogeneous number of links and frustration is fully due to the topological factor. The work has brought a new insight into the study of SGs, suggesting that the irregular connectivity can also be one of the ingredients of SGs, different from the previous notion insisting the presence of random mixture of FM and AF interactions.

An SFN consists of abundance of triangular units on which spins are frustrated if the couplings are AF. While the frustration is caused by the topological factor, the randomness is due to irregular connectivity. There are some vertices having very large number of connections, acting as the hubs as in the internet connection. In fact, the structure of the internet follows a scale free behavior. The AF Ising model on random networks without a scale free behavior was also reported to exhibit SG phase [11]. We believe that random connectivity can generally be an alternative for the usual randomness, together with frustration, as the ingredients of SGs.

Here we study the Heisenberg SG model on SFN. The prevalent controversy on existence of SG phase for the 3D canonical Heisenberg model [5] over the last three decades is one of the main motivations. Most SGs studied, such as CuMn, AgMn or CdMnTe, are Heisenberg-like systems. Early numerical study by Coluzzi observed SG phase transition on 4D Heisenberg model. Kawamura and Nishikawa pointed out the absence of Heisenberg SG phase on D-dimensional space, for D $\leq$ 3 [5]. These works suggested that the lower critical dimension d_l might be a fractional number, $3 < d_l < 4$.

For a regular lattice, spatial dimension is related to the coordination number, i.e., the number of neighbors of each spin. We can associate the coordination number with connectivity density m, which is the average number of links, of SFNs. Due to the presence of short-cut between spins, the notion of spatial dimension is not an appropriate term for SFNs. Nonetheless, an SFN can still be associated with a high dimensional regular lattice. Therefore, we assume that there may also be a critical value of m, analogous to d_l of regular lattice, for the random and SFNs. The paper is organized as follows: Section 2 describes the models and method. The results are discussed in Section 3. Section 4 is devouted for the summary and concluding remarks.

2. Model and Method of Simulation

The Heisenberg model on an SFN can be written with the following Hamiltonian,

$$H = -J \sum_{\langle ij \rangle} \vec{s}_i \cdot \vec{s}_j \tag{1}$$

where for an AF system, the coupling constant is set to be negative ($J < 0$); and $\vec{s}_i$ are the Heisenberg spins residing on the nodes of the network. The summation is performed over all directly connected neighbors. In an SFN, the number of neighbors of each spin is not homogeneous. It is distinguished from a random network due to its scale free behavior, $P(k) = k^{-\gamma}$, where k is the number of links of each node and γ the decay exponent of its link distribution[12]. Networks with large γ have very famous nodes (or hubs), i.e., those having direct links to most other nodes. This type of network found many realizations in real world, from World Wide Webs, power grids, neural and cellular networks, till routers of the internet and citation network of scientists.

In addition, an SFN is also characterized by a clustering coefficient, C, defined as the average of local clustering, C_i,

$$C_i = \frac{2y_i}{z_i(z_i - 1)} \tag{2}$$

where z_i is the total number of nodes linked to the i-th site and y_i is the total number of links connecting those nodes. Both C_i and C lie in the interval [0,1]. For a fully connected network,

$C = C_i = 1$, all nodes connect to each other. The parameter C is related to the number or the density of triangles. Because spins on the vertices of triangular units are frustrated, C is also related to the frustration degree of the network. More comprehensive review on the ubiquitous of SFNs can be found for example in [13].

Thermal averages of the physical quantities of interest are calculated using the Replica Exchange algorithm of MC method[16]. This algorithm is implemented to overcome the slow dynamics due to the presence of local minima in the energy landscape of the system. Slow dynamics is a phenomenon commonly found in dealing with SGs, where a random walker can easily get trapped in one particular local minimum. It is an extended Metropolis algorithm where a system is duplicated into K replicas. All replicas are simulated in parallel and each is in equilibrium with a heat bath of an inverse temperature. Given K inverse temperatures, $\beta_1, \beta_2, \cdots, \beta_K$, the probability distribution of finding the whole system in a state $\{X\} = \{X_1, X_2, \ldots, X_K\}$ is given by,

$$P(\{X, \beta\}) = \prod_{m=1}^{K} \tilde{P}(X_m, \beta_m), \tag{3}$$

with

$$\tilde{P}(X_m, \beta_m) = Z(\beta_m)^{-1} \exp(-\beta_m H(X_m)), \tag{4}$$

and $Z(\beta_m)$ is the partition function of the m-th replica. We can define an exchange matrix between replicas as $W(X_m, \beta_m | X_n, \beta_n)$, which is the probability of switching configuration X_m at the temperature β_m with configuration X_n at β_n.

In order to keep the entire system at equilibrium, by using the detailed balance condition

$$\begin{aligned} &P(\ldots, \{X_m, \beta_m\}, \ldots, \{X_n, \beta_n\}, \ldots) \cdot W(X_m, \beta_m | X_n, \beta_n) \\ &= P(\ldots, \{X_n, \beta_m\}, \ldots, \{X_m, \beta_n\}, \ldots) \cdot W(X_n, \beta_m | X_m, \beta_n), \end{aligned} \tag{5}$$

along with Eq. (4), we have

$$\frac{W(X_m, \beta_m | X_n, \beta_n)}{W(X_n, \beta_m | X_m, \beta_n)} = \exp(-\Delta), \tag{6}$$

where $\Delta = (H(X_m) - H(X_n))(\beta_n - \beta_m)$. With the above constraint we can choose the matrix coefficients according to the standard Metropolis method, therefore

$$W(X_m, \beta_m | X_n, \beta_n) = \begin{cases} 1 & \text{if} \quad \Delta \leq 0, \\ \exp(-\Delta) & \text{if} \quad \text{Otherwise.} \end{cases} \tag{7}$$

As the acceptance ratio decays exponentially with $(\beta_n - \beta_m)$, the exchange is performed only to the replicas next to each other, i.e., $W(X_m, \beta_m | X_{m+1}, \beta_{m+1})$. The replica exchange method is extremely efficient for simulating systems such as SGs. This method has been widely implemented in many complex systems, including the AF-SFN Ising SG[10]. In the next section, we discuss the search for SG behaviour of Heisenberg antiferrmagnetic system on SFN.

3. Results and Discussion

3.1. Equilibrium Behavior

We have simulated AF Heisenberg model on SFN with various connectivity densities, $m = 2, 3, 4$ and 5. For each density, we consider several system sizes, $L = 1024, 2048$, and 4096, which are the number of spins. Since this is a random system, we took many realizations of the networks for each system size, then averaged the results over the number of realizations. Each realization corresponds to one particular connectivity distribution. We have to take reasonable

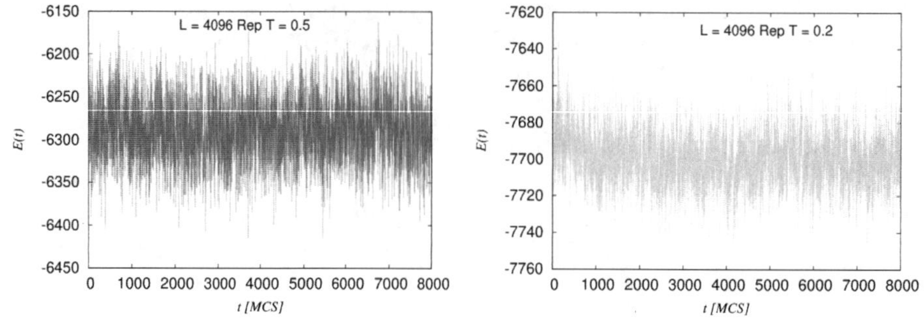

Figure 1. Energy time series of system with $L = 4096$ and $m = 5$ at $T = 0.2$ and 0.5.

number of realizations N_r for the better statistics of the results. Previous study on Ising system took $N_r = 1000$ realizations[10]. Here, due to the less fluctuation of the results from different realizations, we took moderate number of realizations, i.e., $N_r = 500$.

To check for the equilibrium behavior of the systems we evaluated the energy time series. In Monte Carlo (MC) simulation, time is assigned as a series of MC steps (MCSs). One MCS is defined as visiting each spin once, either randomly or consecutively, and performing a prescribed spin update, i.e., Metropolis update. We performed M MCSs for each temperature and took N samples out of M MCSs. The time series plot of energy for two different temperatures for system size $L = 4096$ is shown in Fig. 1. The phenomenon of slow dynamics at lower temperature ($T = 0.2$) is shown in Fig. 1 where the average value for the first half of the plot is larger than that of the rest half. To overcome this problem, the equilibration process was increased up to 8000 MCSs.

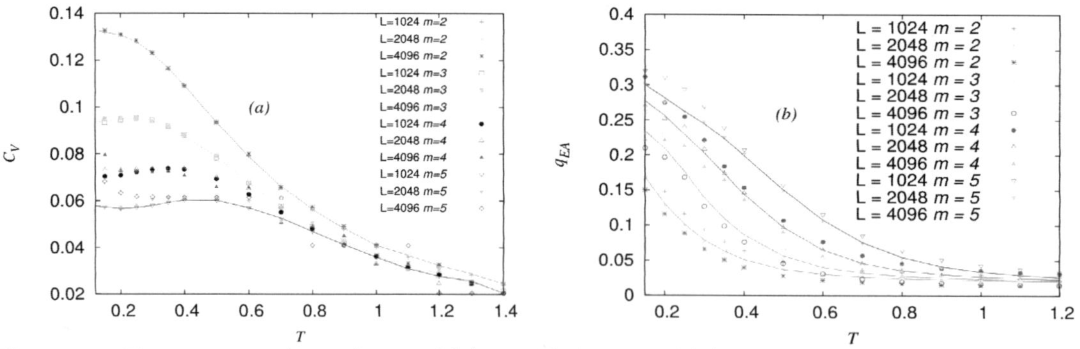

Figure 2. Temperature dependence of (a) specific heat and (b) SG order parameter for different sizes with m =2,3,4 and 5. The solid lines are guides to the eye for the clarity of peaks.

We also calculated the specific heat which is defined as follows

$$C_v = \frac{N}{kT^2}\left(\langle E^2\rangle - \langle E\rangle^2\right) \tag{8}$$

where L, k and $\langle E\rangle$ are respectively the number of nodes, Boltzmann constant and the ensemble average of energy. The temperature dependence of the specific heat for various system sizes is shown in Fig. 2(a), where the statistical errors are comparable to the size of the symbols. Although the specific heat plot exhibits no singularity, there is a maximum value at intermediate temperatures. As indicated, the peaks of the specific heat shift to higher temperatures as the increase in m. The presence of peak may signify the existence of phase transition. A clear sign

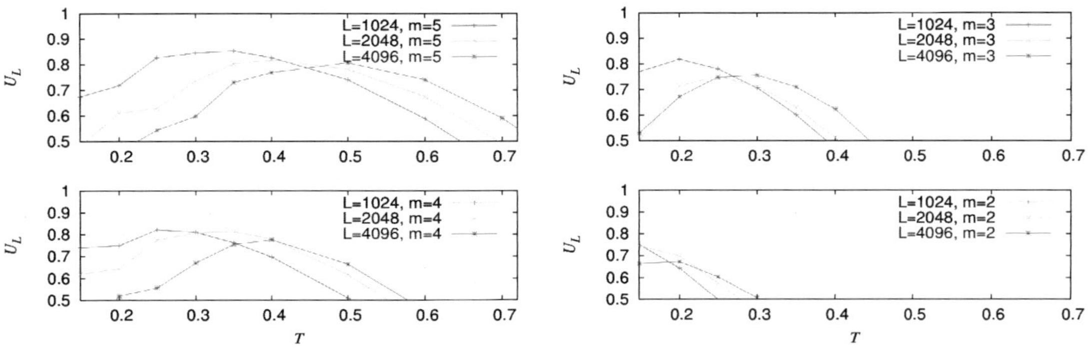

Figure 3. Binder parameter of q_{EA} for various values of m. Solid lines are guide to the eyes.

of the SG phase can be observed from the plot of overlapping order parameter which will be presented in the next subsection.

3.2. Spin Glass Order Parameter
To search for the SG phase transition, we calculate SG order parameter defined as follows

$$q_{EA} = \langle \left| \sum_i \vec{s}_i^{\alpha} \otimes \vec{s}_i^{\beta} \right| \rangle_{av} \tag{9}$$

where $\vec{s}_i$ is the spin on node i-th while α and β denote two sets of replicas. The origin of this quantity is from the scalar product of the vector spins. A scalar product of two vectors will be maximum (zero) if they are parallel (perpendicular). This idea is implemented to capture the frozeness of the spin configuration. If a system is frozen, the spin configurations of two replicas with the same inverse temperature from different sets will be more or less similar, regardless of globally rotational invariant. As explained in Sec. 2, system is replicated into K replicas, each belongs to inverse temperature. For the sake of q_{EA} calculation, the whole replicas are duplicated, therefore we have two sets of replicas. Each set contains K replicas which are exchanged during the simulation. Only replicas from the same set were exchanged during the simulation.

The overlapping parameter will give finite value if system in SG phase. This applies for any SG model, including the Ising and the Heisenberg model. For Ising model, q_{EA} is simply the multiplication of the overlapped spins. In contrast for the Heisenberg model where spins are allowed to rotate in any direction, we take the tensor product instead of the scalar product, resulting q_{EA} with nine components. The plot of temperature dependence of q_{EA} for various different system sizes is shown in Fig. 2(b). As indicated, this parameter increases as temperature decreases, which is the evidence for the existence of SG phase at lower temperatures.

To clarify that this a true SG phase, we calculate the cumulant ratio (Binder parameter) [18] of q_{EA} defined as follows

$$U_L = \frac{1}{2}\left[11 - \frac{9\langle q^4\rangle}{\langle q^2\rangle^2}\right] \tag{10}$$

The plot of U_L for different system sizes is shown in Fig. 3. It is clear that there is a single crossing point, emphasizing the existence of SG phase transition. However, due to the anomaly of the crossing pattern, which is different from the standard ones, such as in the previous study on Ising model[10], we did not perform scaling plot of U_L for the estimate of critical temperature and exponent. Instead, we roughly estimate the critical temperature for each connectivity density as listed in Table 1, where numbers in bracket are the uncertainty for the last digits. There

Table 1. The estimate of critical temperature T_c for each value of m.

Connectivity density (m)	T_c
5	0.44(3)
4	0.35(5)
3	0.24(6)
2	0.19(6)

is a systematic increase of critical temperature as the connectivity density increases. This is consistent with the Mean Field theory argument, where an ordered phase for a system with large connectivity tend to be more stable.

4. Summary and Conclusion

In summary, we have studied AF Heisenberg model on scale free network, using Replica Exchange MC method. We simulated several different connectivity densities ($m = 2, 3, 4$ and 5) and calculated such physical quantities as ensemble average of energy, the specific heat, the overlapping parameter and its cumulant ratio (Binder parameter). A sign for finite SG phase transition was observed for all values of m. There is a systematic increase of T_c as m increases. This is related to the fact that systems with large connectivity tend to be more robust against thermal fluctuation. The existence of SG phase even in the system with $m = 2$ is generally in a good agreement with the case for canonical systems where Heisenberg SG phase was clearly observed in system with large dimension, e.g. the 4D system[14]. It is therefore interesting to search for the lower critical value of m by studying systems with fractional value of m. We will consider such system in our future study.

Acknowledgments

The authors wish to thank K. Hukushima, M. Troyer and Y. Okabe for valuable discussions. The computation of this work was performed using parallel computing facility in the Department of Physics Hasanuddin University and the HPC facility of the Indonesian Institute of Science. The work is supported by the HIKOM research grant 2014 of the Indonesian Ministry of Education and Culture.

References

[1] V. Cannella and J. A. Mydosh, Phys. Rev. B, **6**,4220 (1972)
[2] D. Sherrington and S. Kirkpatrick, Phys. Rev. Lett. **35**, 1792 (1975).
[3] S. F. Edwards and P. W. Anderson, J. Phys. F 5, 965, (1975).
[4] T. R. Kirkpatrick and D. Thirumalai, Phys. Rev. Lett. **58**, 2091, (1987).
[5] H. Kawamura and S. Nishikawa, Phys. Rev. B **85**, 134439 (2012).
[6] M. Wittmann and A. P. Young, Phys. Rev. E **85**, 041104, (2012)
[7] H. Nishimori, *Statistical Physics of Spin Glasses and Information Processing*, Oxford Univ. Press, (2001).
[8] J. Villain, J. Phys. C **10**, 1717 and 4793 (1977).
[9] T. Surungan, Y. Okabe, and Y. Tomita, J. Phys. A **37**, 4219, (2004).
[10] M. Bartolozzi, T. Surungan, D.B. Leinweber and A.G. Williams, Phys. Rev. B**73**, 224419 (2006).
[11] C. P. Herrero, Phys. Rev. E **77**, 041102 (2008).
[12] A.-L. Barabasi and R. Albert, Science 286, 509 (1999).
[13] M. Newman, A-L. Barabasi and D. J. Watts, *The Structure and Dynamics of Networks*, Princeton Uni. Press, (2006).
[14] B. Coluzzi, J. Phys. A: Math. Gen. 28 , 747 (1995).
[15] F. Matsubaral, T. Shirakura, S. Endoh and S. Takahashi, J. Phys. A: Math. Gen. 36 10881, (2003).
[16] K. Hukushima and K. Nemoto, J. Phys. Soc. Jpn. **65**, 1863, (1996).
[17] N. Kawashima and A. P. Young, Phys. Rev. B **53**, R484, (1996).
[18] K. Binder, Z. Phys. B**43**, 119 (1981); Phys. Rev. Lett. **47**, 693 (1981).

Exploring Replica-Exchange Wang-Landau sampling in higher-dimensional parameter space

Alexandra Valentim[1,2], Julio C. S. Rocha[1,3], Shan-Ho Tsai[1,4], Ying Wai Li[5], Markus Eisenbach[5], Carlos E. Fiore[2,6], David P. Landau[1]

1 Center for Simulational Physics, University of Georgia, Athens, GA 30602, USA
2 Departamento de Física, Universidade Federal do Paraná, Curitiba-PR 81531-980, Brazil
3 Departamento de Física, Universidade Federal de Minas Gerais, Belo Horizonte-MG 31270-901, Brazil
4 Enterprise Information Technology Services, University of Georgia, Athens, GA 30602, USA
5 National Center for Computational Sciences, Oak Ridge National Laboratory, Oak Ridge, TN 37831, USA
6 Instituto de Física, Universidade de São Paulo, São Paulo-SP 05315-970, Brazil

E-mail: alexandra@fisica.ufpr.br

Abstract. We considered a higher-dimensional extension for the replica-exchange Wang-Landau algorithm to perform a random walk in the energy and magnetization space of the two-dimensional Ising model. This hybrid scheme combines the advantages of Wang-Landau and Replica-Exchange algorithms, and the one-dimensional version of this approach has been shown to be very efficient and to scale well, up to several thousands of computing cores. This approach allows us to split the parameter space of the system to be simulated into several pieces and still perform a random walk over the entire parameter range, ensuring the ergodicity of the simulation. Previous work, in which a similar scheme of parallel simulation was implemented without using replica exchange and with a different way to combine the result from the pieces, led to discontinuities in the final density of states over the entire range of parameters. From our simulations, it appears that the replica-exchange Wang-Landau algorithm is able to overcome this difficulty, allowing exploration of higher parameter phase space by keeping track of the joint density of states.

1. Introduction

The Wang-Landau (WL) Monte Carlo method [1] is a sampling technique that can be used to obtain the density of states (DOS) of a system. It has the property of generating a flat histogram in some random walk space, where the parameters for the random walk and the flatness criterion can be chosen according to the system of interest [2, 3]. Due to these characteristics and to the simplicity of the algorithm, the Wang-Landau approach has been studied and applied successfully to different kinds of systems (see, for examples, [4, 5, 6, 7]), including those with rough free energy landscapes and strong first-order phase transitions, which cannot be easily studied with traditional methods.

Recently an efficient parallel scheme of the Wang-Landau method was implemented [8, 9, 10, 11], whereby independent random walks are carried out in different overlapping windows of energy and a replica-exchange scheme [12, 13] is used to exchange system configurations of neighboring energy windows. This parallel replica exchange Wang-Landau (REWL) method was shown to scale extremely well up to several thousands of computing cores. Therefore, it can be used to study more complex systems and much larger system sizes than it is possible with the serial Wang-Landau method.

While the density of states as a function of a single parameter, such as the total energy, suffices to provide insight into many systems, there are a number of physical properties and systems whose investigation relies on a joint density of states dependent on additional parameters [14, 15, 16]. Therefore, it would be of great interest to extend the one-dimensional REWL method [8, 9, 10] to a higher-dimensional parameter space, and implement a similar method that combines enhanced sampling techniques [12, 13, 17] and the advantages of parallel computing to obtain a joint density of states.

In this exploratory study we consider the Ising model, for which the density of states as a function of total energy is known exactly [18]. The phase space of this model, as a function of energy and magnetization, has cusps and empty regions with no realizable configurations. Simulations in this higher-dimensional parameter space are challenging because, depending on the sampling method used, the lower and higher energy edges of the phase space are hard to sample and thus have poor statistics. Moreover, the empty region between cusps can be hard to overcome in a simulation.

2. Two-dimensional Replica-Exchange Wang-Landau sampling

The REWL method [8] is arguably the most successful parallelization attempt to the WL method so far, and it consists of only three simple modifications to the serial version of the algorithm. Taking the Ising model as an example, a serial WL sampling of the total energy (E) and magnetization (M) phase space employs a two-dimensional random walk in the entire E and M ranges. In the process a random walker keeps track of the joint density of states $g(E, M)$ and the histogram of visited states $H(E, M)$. Let $A'(E', M')$ and $A''(E'', M'')$ denote, respectively, the current and trial spin configurations of the system. The WL probability that the trial move is accepted is given by

$$P(A' \to A'') = \min\left[1, \frac{g(E', M')}{g(E'', M'')}\right]. \tag{1}$$

If $A''(E'', M'')$ is accepted, the density of states $g(E'', M'')$ is adjusted by the modification factor $f > 1$, that is, $g(E'', M'') \to g(E'', M'') * f$; and the histogram is updated with $H(E'', M'') \to H(E'', M'') + 1$. If the trial state is not accepted, we update $g(E', M') \to g(E', M') * f$ and $H(E', M') \to H(E', M') + 1$. One Monte Carlo step is defined as $L \times L$ trial state considerations.

The random walker proceeds with choosing a new trial spin configuration and checking whether it is accepted, until $H(E, M)$ is considered flat, in accordance with a chosen criterion. At this point the current iteration is ended and the histogram is reset to zero. The next iteration begins with a finer f, reduced by some predefined rule (e.g. $f \to \sqrt{f}$). More iterations are performed until the modification factor reaches a chosen final value f_{final}, such as $f_{\text{final}} = \exp(10^{-8})$. The density of states is never reset and, at the end of the simulation, $g(E, M)$ is expected to converge to the true value with an accuracy proportional to $\sqrt{f_{\text{final}}}$.

A replica-exchange Wang-Landau sampling starts by splitting the phase space into smaller overlapping subspaces (windows), as schematically shown in Fig. 1. For simplicity, only the energy range is split and the subspaces were artificially displaced vertically (in Fig.1(b)) for clarity.

Inside each subspace one or more independent random walkers perform a standard WL sampling, with independent $g(E, M)$ and $H(E, M)$. After a predetermined number of Monte Carlo steps, a configurational exchange between two walkers in adjacent windows is proposed. The walker pairs are chosen at random. If either of them is not in the overlapping region, the exchange of configurations will not be attempted. However, if the walker pairs are in the same

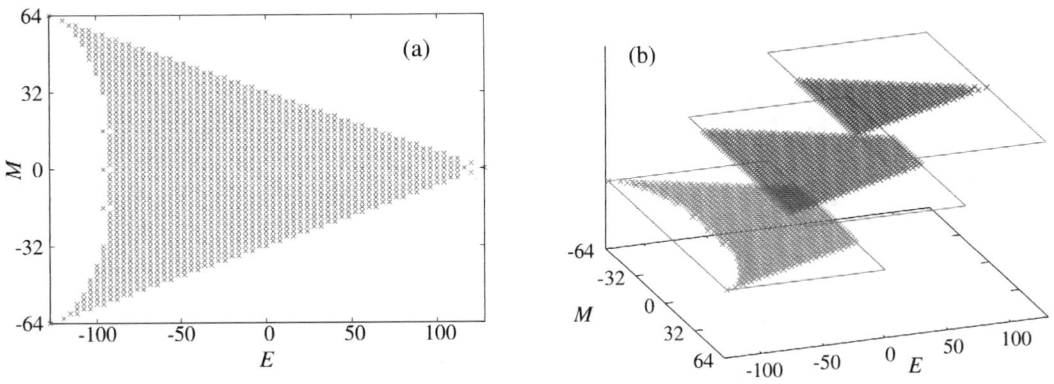

Figure 1. (color online) (a) Complete phase space for the Ising model on a 8×8 square lattice. (b) An example of splitting of the parameter space into overlapped, equal-sized energy windows. Energy windows are shifted upward progressively for clarity.

overlapping region, the configuration swap between them (walkers i and j) will be accepted with the probability:

$$P_{\text{acc}} = \min\left[1, \frac{g_i(E(X), M(X))}{g_i(E(Y), M(Y))} \frac{g_j(E(Y), M(Y))}{g_j(E(X), M(X))}\right], \tag{2}$$

where X and Y denote, respectively, the spin configurations of walkers i and j before the exchange. Note that P_{acc} satisfies detailed balance. All walkers (in all regions) move to the next iteration at the same time and the parallel simulation ends when the modification factor for every walker reaches f_{final}.

At this stage of the simulation, each walker is left with a joint DOS fragment (Fig.2(a)) and one needs to combine all the fragments to recover $g(E, M)$ over the entire phase space. To do so, we first sum over the magnetization and obtained the density of states as a function of energy:

$$g(E) = \sum_M g(E, M), \tag{3}$$

as shown in Fig.2(b). After that we calculate the derivative of $\ln[g(E)]$ as a function of E in the overlapping region. In this procedure, one finds a point E_{mat} where the difference between the inverse of the microcanonical temperature $\beta(E) = \mathrm{d}\ln[g(E)]/\mathrm{d}E$ is a minimum. This E_{mat} point is used to match and adjust the two adjacent surfaces. That is, the whole DOS surface $g(E, M)$ has to be rescaled at $g(E_{\text{mat}}, M)$ and the joint DOS is obtained by:

$$g(E, M) = \begin{cases} g_i(E, M), & \text{if} \quad E < E_{\text{mat}}, \\ g_j(E, M), & \text{if} \quad E \geq E_{\text{mat}}. \end{cases} \tag{4}$$

where i and j are, respectively, walkers in the lower and higher energy range and, $g_j(E, M)$ is the shifted value of $g(E, M)$ for the jth walker.

In principle, other ways of matching the 2D-DOS could be applied [10, 17] and other ways to split the phase space might be desired to increase the performance of the simulation (see, e.g., Ref. [8]). For example, one might be interested in splitting the magnetization instead of the energy range, or splitting both energy and magnetization ranges. In this case both dimensions of the overlapping phase space will need to be analyzed, one at a time, to find the E_{mat} and

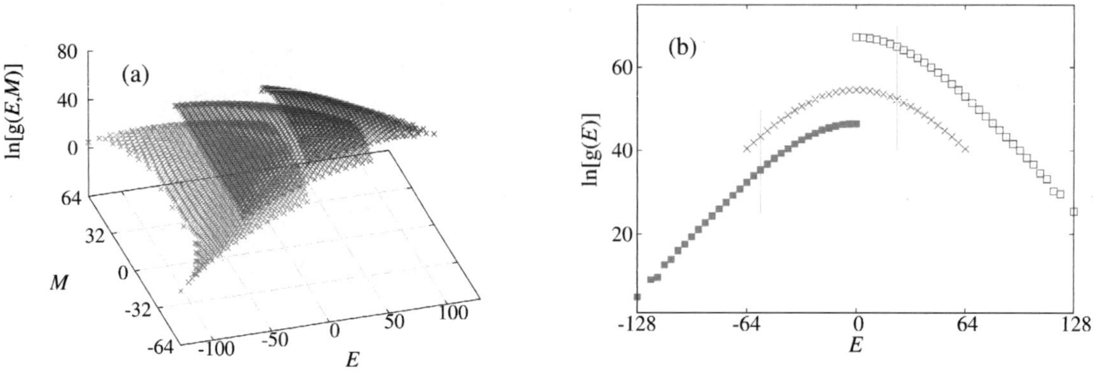

Figure 2. (color online) (a) Final $g(E, M)$ pieces of the two-dimensional Ising model on a 8×8 square lattice. DOS pieces are shifted upward progressively for visualization purpose. (b) Individual pieces of $g(E)$ obtained by a summation over the magnetization range (Eq. (3)). Vertical lines indicate the matching points (E_{mat}) for joining the DOS pieces together. E_{mat} is determined as the point with minimal difference in the inverse microcanonical temperatures of the two DOS pieces. See text for more details.

M_{mat} matching points. Moreover, since multiple walkers can be used to sample a window simultaneously, it is possible to employ jackknifing or bootstrapping techniques to compute the statistical errors from a unique production run.

3. Preliminary results for the two-dimensional Ising model

From our simulations for the two-dimensional Ising model we examine (*i*) the generalization of the REWL approach to higher dimensions and (*ii*) the accuracy of the algorithm. We carried out serial 2D-WL and 2D-REWL simulations by performing random walks in total energy E and magnetization M space until the modification factor reaches $f_{\mathrm{final}} = \exp(10^{-8})$. We adopted the 80% flatness criterion, i. e., all entries in the histogram $H(E, M)$ are no less than the chosen percentage of the average height of the histogram.

For the 2D-REWL simulations we divided the parameter space into equal-sized windows with restricted values of E and unrestricted M. We used a constant 50% overlap between neighboring windows and multiple random walkers to keep track of the joint DOS within each window. Fig.3(a) shows the resulting DOS that is obtained in a few hours by 9 parallel walkers (the energy is split into 3 windows with three walkers per window). An equivalent serial WL sampling over the entire two-dimensional parameter space with a single random walker takes several days to complete.

To compare the DOS obtained from WL and REWL sampling (denoted by $g_{\mathrm{WL}}(E, M)$ and $g_{\mathrm{REWL}}(E, M)$, respectively), we calculated the relative difference:

$$\Delta_{\mathrm{diff}}(E, M) = \frac{|g_{\mathrm{WL}}(E, M) - g_{\mathrm{REWL}}(E, M)|}{\bar{g}(E, M)}, \qquad (5)$$

where $\bar{g}(E, M) = (g_{\mathrm{WL}}(E, M) + g_{\mathrm{REWL}}(E, M))/2$. The relative difference is plotted in Fig. 3(b). Over the entire phase space, there is no apparent sign of systematic errors due to the matching of windows. The order of magnitude of the relative difference is comparable to the relative errors (as shown in Fig. 4(a)); they both share the same behavior of having larger values in the lower and upper energy edges. This is conceivable because of the rareness of configurations as well as the presence of cusps and empty regions in the phase space.

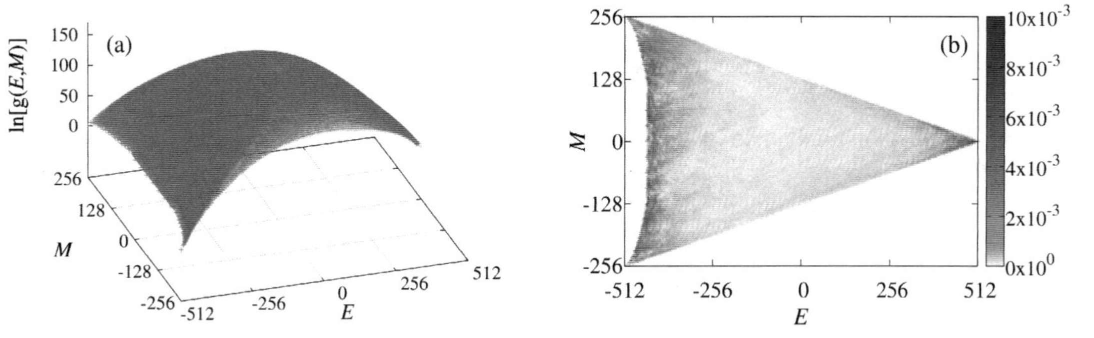

Figure 3. (color online) Logarithm of the complete joint DOS for the Ising model on a 16×16 square lattice obtained by the two-dimensional REWL approach. (b) Relative difference between the joint DOS obtained from WL and REWL sampling.

To verify the accuracy of the algorithm, we compute $g(E)$ with Eq. (3) so that we can compare with the exact results. Fig.4(a) shows the logarithmic DOS obtained by using 3 walkers per window in a 2D-REWL simulation along with the exact values, the relative errors are shown in the insert. We can see that the results obtained from our simulations agree extremely well with the analytic solutions. Fig.4(b) shows the relative errors for the simulations employing 2 walkers per window and 20 walkers per window respectively. We observe that increasing the number of walkers, which update the DOS independently, decreases the sampling errors correspondingly.

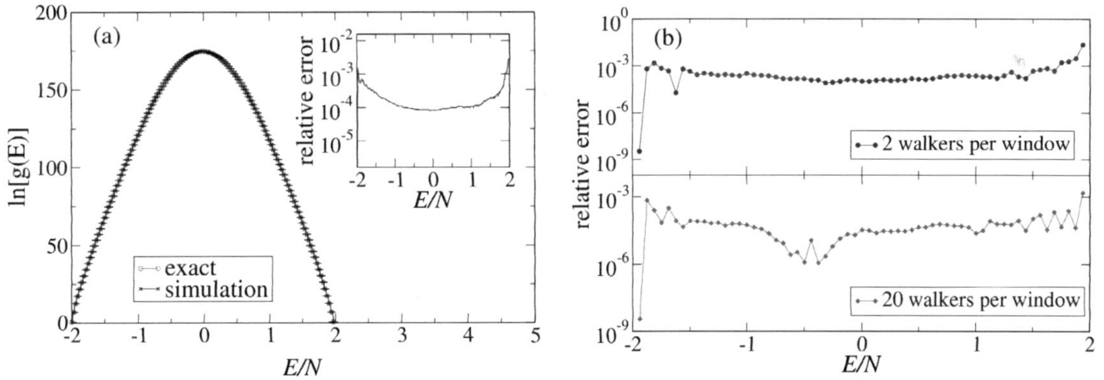

Figure 4. (color online) (a) Comparison of the density of states from 2D-REWL sampling and the exact results for a 16×16 square lattice (relative errors are shown in the insert). (b) Relative errors for a 8×8 lattice square using 2 walkers per window (upper panel) and 20 walkers per window (lower panel) respectively.

4. Summary

We implemented a two-dimensional replica-exchange Wang-Landau sampling method and applied it to determine the joint density of states $g(E, M)$ for the Ising model on a square lattice. The joint DOS over the entire two-dimensional parameter space, obtained by combining the joint DOS in the different sampled windows, agrees very well with the results obtained with a serial WL sampling over the entire two-dimensional parameter space. No artifacts in the final joint DOS were observed due to the parallelization and DOS joining procedures. We also showed that errors can be reduced by employing multiple random walkers in each window. Similar to the original REWL method, the higher-dimensional implementation of the REWL method is very efficient and it scales well with the number of CPUs. Hence, it can be used to study much larger systems than what is feasible using a serial WL method.

Acknowledgments

This research was sponsored in part by the National Science Foundation under Grant OCI-0904685, by the Office of Science, Office of Basic Energy Sciences, Materials Sciences and Engineering Division (M.E.) and by the Office of Advanced Scientific Computing Research (Y.W.L.) of U.S. Department of Energy, under Contract DE-AC05-00OR22725. The simulations were carried out on resources provided by the Georgia Advanced Computing Resource Center at the University of Georgia. A. Valentim was sponsored by CAPES Foundation, Ministry of Education of Brazil, Brasilia - DF 70.040-020, Brazil.

References

[1] Wang F and Landau D P 2001 *Phys. Rev. Lett.* **86** 2050
[2] Wang F and Landau D P 2001 *Phys. Rev. E.* **64** 056101
[3] Landau D P, Tsai S -H and Exler M 2004 *Am. J. Phys.* **72** 1294
[4] Rathore N and de Pablo J J 2002 *J. Chem. Phys.* **116** 7225
[5] Yamaguchi C and Kawashima N 2002 *Phys. Rev. E* **65** 056710
[6] Taylor M P, Paul W and Binder K 2009 *J. Chem. Phys.* **131** 114907
[7] Wüst T and Landau D P 2009 *Phys. Rev. Lett.* **102** 178101
[8] Vogel T, Li Y W, Wüst T and Landau D P 2013 *Phys. Rev. Lett.* **110** 210603
[9] Vogel T, Li Y W, Wüst T and Landau D P 2014 *Phys. Rev. E* **90** 023302
[10] Li Y W, Vogel T, Wüst T and Landau D P 2014 *J. Phys.:Conf. Ser.* **510** 012012
[11] Vogel T, Li Y W, Wüst T and Landau D P 2014 *J. Phys.:Conf. Ser.* **487** 012001
[12] Marinari E and Parisi G 1992 *Europhys. Lett.* **19** 451
[13] Hukushima K and Nemoto K 1996 *J. Phys. Soc. Jpn.* **65** 1604
[14] Zhou C, Schulthess T C, Torbrügge S and Landau D P 2006 *Phys. Rev. Lett.* **96** 120201
[15] Tsai S -H, Wang F and Landau D P 2009 *Int. J. Mod. Phys. C* **20** 1357
[16] Silva C J, Caparica A A and Plascak J A 2006 *Phys. Rev. E* **73** 036702
[17] Sugita Y and Okamoto Y 2000 *Chem. Phys. Lett.* **329** 261
[18] Beale P D 1996 *Phys. Rev. Lett.* **76** 78

Epidemic spreading and immunization strategy in multiplex networks

Lucila G. Alvarez Zuzek[1], Camila Buono[1] and Lidia A. Braunstein[12]

[1] Instituto de Investigaciones Físicas de Mar del Plata UNMdP-CONICET and Departamento de Física FCEyN, Universidad Nacional de Mar del Plata, Funes 3350 (7600) Mar del Plata, Argentina.

[2] Center for Polymer Studies, Boston University, Boston, Massachusetts 02215, USA.

E-mail: lgalvere@mdp.edu.ar

Abstract. A more connected world has brought major consequences such as facilitate the spread of diseases all over the world to quickly become epidemics, reason why researchers are concentrated in modeling the propagation of epidemics and outbreaks in multilayer networks. In this networks all nodes interact in different layers with different type of links. However, in many scenarios such as in the society, a multiplex network framework is not completely suitable since not all individuals participate in all layers. In this paper, we use a *partially overlapped multiplex* network where only a fraction of the individuals are shared by the layers. We develop a mitigation strategy for stopping a disease propagation, considering the Susceptible-Infected-Recover model, in a system consisted by two layers. We consider a random immunization in one of the layers and study the effect of the overlapping fraction in both, the propagation of the disease and the immunization strategy. Using branching theory, we study this scenario theoretically and via simulations and find a lower epidemic threshold than in the case without strategy.

1. Introduction

In the last years the complex networks analysis has been focused in no further considering networks as isolated entities, but characterizing how networks interact with other networks and how this interaction affects processes that occurs on top of them. A system composed of interconnected networks is called a *Network of Networks* (NoN) [1, 2, 3, 4]. In NoN there are connectivity links within each individual network, and external links that connect each network to other networks in the system. Very recently physicists have begun to consider a particular class of NoN in which the nodes have multiple types of links across different *layers* called *multiplex or multilayer* networks [5, 6, 7, 8, 9, 10, 11].

Recently, the study of the effect of multiplexity of networks in propagation processes such as epidemics has been the focus of many recent researches [12, 13, 14, 15]. In Ref [16] the research concentrated in the propagation of a disease in partially overlapped multilayer networks, owing to the fact that individuals are not necessarily present in all the layers of a society and this has an impact in the epidemic propagation. For the epidemic model they used the susceptible-infected-recovered (SIR) model [17, 18, 19] that describes the propagation of non recurrent diseases in which infected individuals either die or, after recovery, become immune to future infections. In the SIR model each individual of the population can be in one of three different states:

Susceptible, Infected, or Recovered. Infected individuals transmit the disease to its susceptible neighbors with a probability β and recover after a fixed period of time t_r. The spreading process stops when there is only susceptible and/or recovered nodes. The dynamic of the epidemic is controlled by the transmissibility $T = 1 - (1-\beta)^{t_r}$, which is a measure of the disease virulence, i.e., the effective probability that the disease will be transmitted by an infected individual across any given link. At the final state of this process, the fraction of recovered individuals R is the order parameter of a second order phase transition with a control parameter T. For $T < T_c$, where T_c is the epidemic threshold, there is an epidemic-free phase with only small outbreaks. However, for $T \geq T_c$, an epidemic phase develops. In isolated networks the epidemic threshold is given by $T_c = 1/(\kappa - 1)$, where κ is the branching factor that is a measure of the heterogeneity of the network. The branching factor is defined as $\kappa = \langle k^2 \rangle / \langle k \rangle$, where $\langle k^2 \rangle$ and $\langle k \rangle$ are the second and first moment of the degree distribution, respectively. Since the SIR model presents a local tree structure we can employ the branching theory approach within a generating function formalism [20, 21] that holds in the thermodynamic limit. In [16] the SIR model was studied, with β and t_r constant, in a system composed of two overlapping layers in which only a fraction q of individuals can act in both layers. In their model, the two layers represent contact networks in which only the overlapping nodes enable the propagation between layers, and thus the transmissibility T is the same in both layers. They found that decreasing the overlap decreases the risk of an epidemic compared to the case of full overlap ($q = 1$). They also found that the critical threshold increases as q decreases, and that in the limit of small overlapping fraction, the epidemic threshold is dominated by the most heterogeneous layer, this effect could have important implications in the implementation of mitigation strategies.

Motivated by this, in this work we study a disease spreading process in overlapped multiplex networks and an immunization strategy for the epidemic spreading. For the strategy, we use a random immunization of individuals in one layer of the network. Those immunized overlapped individuals will remain immunized in all layers of the network.

2. Epidemic propagation process

In our model we use an overlapping multiplex network formed by two layers, A and B, of the same size N, where an overlapping fraction q of *shared* individuals is active in both layers. Figure 1(a) shows schematically the partially overlapped network. The dashed lines that represent the fraction q of shared individuals should not to be interpreted as interacting or interdependent links but as the shared nodes and their counterpart in the other layer.

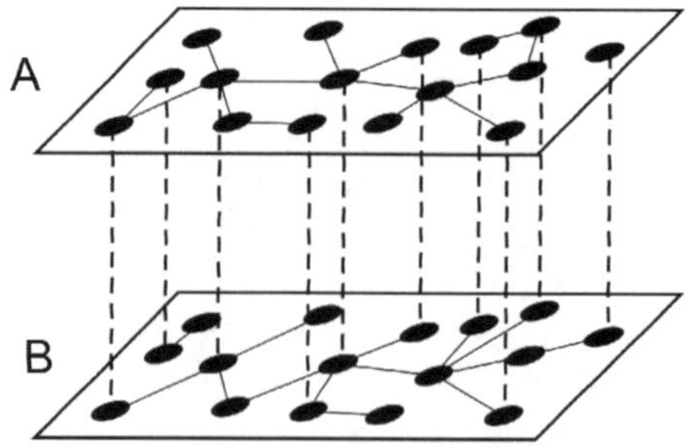

Figure 1. Partially overlapped multiplex network with layer size $N = 16$ and fraction of shared nodes $q = 0.625$. The total size of the network is $(2-q)N = 22$ individuals. The dashed lines are used as a guide to show the fraction q of shared nodes. Before the spreading dynamics, all individuals are in the susceptible stage represented by black circles.

For the simulation, we construct each layer using the Molloy Reed algorithm [22], and we choose randomly a fraction q of nodes in each of the layers that represent the same nodes. In our model we assume that the transmissibility is the same in both layers because there is only one disease and all individuals in the system spread with the same probability. We begin by infecting a randomly chosen individual in layer A. The spreading process then follows the SIR dynamics in both layers, the overlapped nodes in both layers have the same state because they represent the same individuals. After all infected nodes infect their susceptible neighbors with probability β in both layers, the time is increased in one, and the states of the nodes are updated simultaneously. Note that because there are shared nodes the branches of infection can cross between the two layers. Thus the probability that, following a random link, a node belonging to the infected branch will be reached in each layer can be written as,

$$f_A = (1-q)\,[1 - G_1^A(1-Tf_A)] + q\,[1 - G_1^A(1-Tf_A)\,G_0^B(1-Tf_B)]\,, \tag{1}$$
$$f_B = (1-q)\,[1 - G_1^B(1-Tf_B)] + q\,[1 - G_1^B(1-Tf_B)\,G_0^A(1-Tf_A)]\,, \tag{2}$$

where $G_0^i(x) = \sum_{k=k_{\min}}^{k_{\max}} P_i(k)x^k$ is the generating function of the degree distribution and $G_1^i = \sum_{k=k_{\min}}^{k_{\max}} P_i(k)\,kx^{k-1}$ is the generating function of the excess degree distribution in layer i, with $i = A, B$ [21].

Equation (1) has two terms, since the probability f_A to expand an infected branch following a random chosen link in layer A, can be written as the probability to reach one of the $(1-q)$ non-overlapped individuals and that the branch of infection expands through the $k-1$ remaining connections of the individual in layer A, combined with the probability of reaching one of the q overlapped individuals and that the branch of infection expands through the $k-1$ remaining connections of the individual in layer A and through the k connections of the individual in layer B. An analogous interpretation holds for the equation (2).

The solution of the system of equations (1) and (2) for all T above and at criticality is given by the intersection of the curves f_A and f_B. At criticality, this intersection can be derived by solving the determinant equation $|J - I| = 0$, where I is the identity and J is the Jacobian matrix of the system of equations (1) and (2). The only possibility to have a non-epidemic regime is that none of the branches of infection spread, *i.e.* $f_A = f_B = 0$, therefore below and at criticality $f_A = f_B = 0$. The evaluation of the Jacobian matrix $J_{ij} = (\partial f_i/\partial f_j)|_{f_A=f_B=0}$ allow us to obtain a quadratic equation for T_c with only one stable solution [23] given by,

$$T_c = \frac{[(\kappa_A-1)+(\kappa_B-1)] - \sqrt{[(\kappa_A-1)-(\kappa_B-1)]^2 + 4q^2\langle k_A\rangle\langle k_B\rangle}}{2(\kappa_A-1)(\kappa_B-1) - 2q^2\langle k_A\rangle\langle k_B\rangle}, \tag{3}$$

where $\kappa = 1 + 1/T_c$ is the total branching factor of the system and κ_A, κ_B are the isolated branching factors of layer A and B respectively. For $q \to 0$ we recover the isolated network result $T_c = 1/(\kappa_A - 1)$, which is compatible with our model in which the infection starts in layer A and the disease never reaches layer B. In contrast, when $q \to 1$, we find that $T_c = 1/\sqrt{[(\kappa_A - \kappa_B)]^2 + 4\langle k_A\rangle\langle k_B\rangle}$. Note that $T_c(q \to 1) < T_c(q \to 0)$. In general, T_c decreases with q. This is the case because an increase in the overlapping between layers increases the total branching factor, and therefore the total system becomes more heterogeneous in degree, *i.e.*, the total branching factor is equal to or bigger than the branching factor of the isolated layers.

3. Immunization strategy

We study a random immunization strategy on the partially overlapped multiplex network. We start by immunizing a random fraction m of individuals in layer A, before the epidemic spreading take place. An immunized individual will be immune to the disease in all layers, and therefore can not be infected or infect during all the propagation process. Note that, due to the presence of the overlapped individuals, in layer B there will be a random fraction mq of immunized individuals.

After immunizing, we spread a disease in the network, starting by infecting a random susceptible non-immunized individual in layer A (patient zero). Thus, the probability that reaching a node by following a randomly chosen link, it belongs to a branch of infection is given by the system of equations (1) and (2), using a node diluted degree distribution in each layer [24] due to the immunization strategy. Thus with the diluted degree distribution we have that the branching factor of the diluted layers are,

$$\widetilde{\kappa_A} = (1-m)\,\kappa_A \quad (4)$$
$$\widetilde{\kappa_B} = (1-qm)\,\kappa_B\,, \quad (5)$$

where κ_A and κ_B are the branching factor of the original layers respectively. Note that the branching factor is reduced due to the immunization strategy increasing the epidemic threshold and thus hindering the diseases propagation.

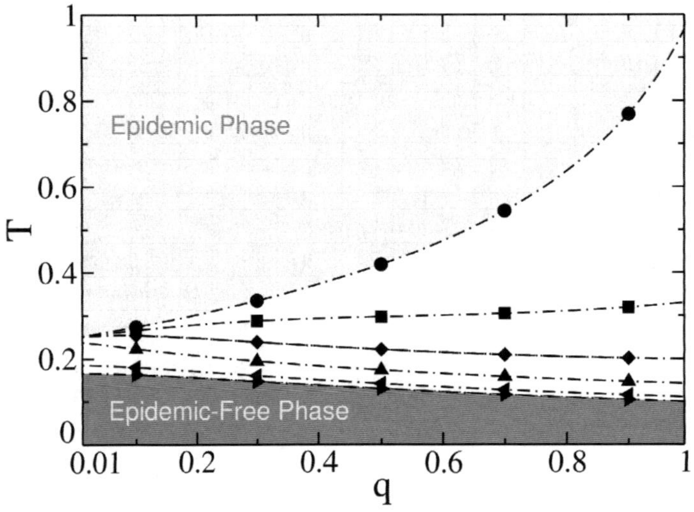

Figure 2. Phase diagram in the plane $T-q$ for the SIR model in the multiplex network, when the random immunization strategy is applied, for different values of the immunized fraction m. Both, layer A and B, have Erdős Rényi degree distributions with mean values of connectivity $\langle k_A\rangle = 6$ and $\langle k_B\rangle = 4$ for layer A and B respectively. Symbols corresponds to the value of T_c for different values of m obtained by numerical simulation with layer size $N = 10^5$, while the lines denote the theoretical results obtained numerically from Eqs. (1) and (2) using $\widetilde{\kappa_A}$ and $\widetilde{\kappa_B}$ given by Eqs. (4) and (5). From top to bottom $m = 0.9; 0.7; 0.5; 0.3; 0.1; 0$. Above the lines the system is in the epidemic phase for each value of m, and below it is in the epidemic-free phase where the disease can not propagate. All simulations were done over 10^5 network realizations.

In Figure 2 we show the phase diagram in the plane $T-q$ for different values of the immunization fraction m. We consider that both layers have Erdős Rényi degree distributions with mean values of connectivity $\langle k_A\rangle = 6$ and $\langle k_B\rangle = 4$ for layer A and B respectively, and we use $k_{min} = 1$ and $k_{max} = 40$ as the minimum and maximum connectivity respectively in each layer. The lines represent T_c for many values of m obtained theoretically from Eqs. (1) and (2) while symbols denote the numerical simulation results. Above T_c there is an epidemic phase and below T_c only outbreaks exists (non-epidemic phase). Fig. 2 shows that T_c has different behaviors with q depending on the value of m. From Figure 2 we can see a good agreement between the theoretical predictions and the numerical simulation results.

Note that when $q = 0$ (not shown) the critical threshold corresponds to an isolated layer in which the disease starts, *i.e.* layer A and where the critical threshold is given by $T_c = 1/(\kappa_A - 1)$. For $q \to 0$ ($q = 0.01$) the epidemic threshold converges to the threshold of the layer with the bigger branching factor, since in this limit the process is dominated by the most heterogeneous layer [16]. We can observe from Fig. 2 that as the immunization fraction increases, the epidemic-free phase widens. When $m < 0.7$ (see Fig. 2) T_c decreases with q owing to the fact that as the overlapping between the layers increases the total branching factor of the network increases. However, for $m \geq 0.7$ T_c increases as q increases. This last effect can be understood taking into account that for $m > 0.7$ layer A is very diluted, thus the disease spreads mostly through layer B, as q increases the immunization fraction mq of layer B increases, hindering the propagation through that layer. It is expected that for more heterogeneous networks this strategy has less impact in the spreading process, due to the fact that the more heterogeneous the network is, the more harder it is to dilute with this strategy.

4. Discussion

In this work we study, theoretically and via simulations, an epidemic spreading and a random immunization strategy in a partially overlapped multiplex network composed by two layer with an overlapping fraction q. We immunize a fraction m of individuals in one layer of the network and study how this process affects the propagation of the disease through all layers. We found that for $q \to 0$ the critical threshold of the epidemic is dominated by the threshold of the most heterogeneous layer for all $m > 0$. We found that there is a regime in which T_c decreases with q due to the fact that the total branching factor of the system increases. This behavior stands for $m < 0.7$, however for bigger values of m, T_c increases as q increases, hindering the disease propagation. This last effect can be understood taking into account that when $m > 0.7$, layer A is diluted, and as q increases the immunization fraction mq of layer B increases, and the effect of the immunized individuals in that layer is stronger.

We can observe from Fig. 2 that as the immunization fraction increases, the epidemic-free phase widens. When $m < 0.7$ we can see that T_c decreases with q owing to the fact that as the overlapping between the layers increases the total branching factor of the network increases. However, for $m \geq 0.7$, T_c increases as q increases, hindering the disease propagation. This last effect can be understood taking into account that as q increases the immunization fraction mq of layer B increases and for $m > 0.7$ the effect of the immunized individuals in that layer is stronger. Our study suggests that vaccinating or isolating only in one layer with the higher propagation capacity, can reduce drastically the total branching factor of the network. As a consequence, the epidemic threshold of the system increases significantly, reducing the risk of a disease epidemic in the system.

5. Acknowledgments

The authors wish to thank to UNMdP and FONCyT (Pict 0429/2013) for financial support. LGAZ wishes to thank CCP 2014 Sponsors for financial support to assist to the conference. LGAZ also wishes to thank Professor H. E. Stanley to host her in the Center for Polymer Studies. The authors wish to thank Lucas D. Valdez for his useful comments and discussions.

References

[1] Gao J, Buldyrev S V, Havlin S and Stanley H E 2011 *Phys. Rev. Lett.* **107** 195701

[2] Gao J, Buldyrev S V, Stanley H E and Havlin S 2012 *Nature Physics* **8**
[3] Dong G, Gao J, Du R, Tian L, Stanley H E and Havlin S 2013 *Phys. Rev. E* **87** 052804
[4] Valdez L D, Macri P A, Stanley H E and Braunstein L A 2013 *Phys. Rev. E* **88** 050803(R)
[5] Lee K M, Kim J Y, Cho W K, Goh K I and Kim I M 2012 *New J. Phys.* **14** 033027
[6] Brummitt C D, Lee K M and Goh K I 2012 *Phys. Rev. E* **85** 045102(R)
[7] Gómez S, Díaz-Guilera A, Gómez-Gardeñes J, Pérez-Vicente C J, Moreno Y and Arenas A 2013 *Phys. Rev. Lett.* **110** 028701
[8] Kim J Y and Goh K I 2013 *Phys. Rev. Lett.* **111** 058702
[9] Cozzo E, Arenas A and Moreno Y 2012 *Phys. Rev. E* **86** 036115
[10] Gómez-Gardeñes J, Reinares I, Arenas A and Floria L M 2012 *Nature Scientific Reports* **10.1038** srep00620
[11] Kivelä M, Arenas A, Barthelemy M, Gleeson J P, Moreno Y and Porter M A 2013 Multilayer Networks http://arxiv.org/abs/1309.7233
[12] Dickison M, Havlin S and Stanley H E 2012 *Phys. Rev. E* **85** 066109
[13] Marceau V, Noël P, Hébert-Dufresne L, Allard A and Dubé L J 2011 *Phys. Rev. E* **84**(2) 026105
[14] Yagan O, Qian D, Zhang J and Cochran D 2013 *IEEE JSAC Special Issue on Network Science* **31** 1038
[15] Cozzo E, Baños R A, Meloni S and Moreno Y 2013 *Phys. Rev. E* **88** 050801(R)
[16] Buono C, Zuzek L G A, Macri P A and Braunstein L A 2014 *PLoS ONE* **9** e9220
[17] Bailey N T J 1975 *The Mathematical Theory of Infectious Diseases* (Griffin, London)
[18] Colizza V, Barrat A, Barthlemy M and Vespignani A 2006 *Proc. Natl. Acad. Sci. USA* **103** 2015
[19] Colizza V and Vespignani A 2007 *Phys.Rev.Lett.* **99** 148701
[20] Callaway D, Newman M E J, Strogatz S H and Watts D J 2000 *Phys. Rev. Lett.* **85** 5468
[21] Newman M E J, Strogatz S H and Watts D J 2001 *Phys. Rev. E* **64** 026118
[22] Molloy M and Reed B 1995 *Random structures and algorithms* **6** 161
[23] Alligood K T, Sauer T D and Yorke J A 1997 *CHAOS: An Introduction to Dynamical Systems* (Springer)
[24] Cohen R and Havlin S 2010 *Complex Networks: Structure, Robustness and Function* (Cambridge University Press)

Journal of Physics: Conference Series **640** (2015) 012008 doi:10.1088/1742-6596/640/1/012008

Aggregation process of Aβ_{1-40} with non-Aβ amyloid component of α-synuclein

Cindie Eugene, Normand Mousseau

Département de Physique and Groupe de recherche sur les protéines membranaires (GEPROM), Université de Montréal, C.P. 6128, succursale Centre-ville, Montréal (Québec), Canada

E-mail: normand.mousseau@umontreal.ca

Abstract. Many neurodegenerative diseases, such as Alzheimer's and Parkinson's diseases, are characterized by the presence of amyloid fibers. Recently, attention has turned from the fibers to the early stages of oligomerization where toxicity could be highest. Here, we focus on the interactions between non-Aβ amyloid component of α-synuclein (NAC) and Aβ_{1-40}, two proteins found in amyloid fibrils associated with Alzheimer's disease. We combine the coarse-grained OPEP potential with a Hamiltonian and temperature replica exchange molecular dynamics simulation (HT-REMD) to identify mechanisms leading to the formation of secondary structures promoting fibrillation. We observe that the propensity to form beta-sheet remains the same for Aβ_{1-40} whereas is decreases significantly for NAC. In particular, the 25-35 region of Aβ_{1-40} is a significant area of secondary structure stabilization with NAC. The ionic interactions between salt-bridge D23 and K28 in Aβ_{1-40} and K20 and E23 in NAC of the heterogeneous dimer are consistent with the salt-bridges found in NAC and Aβ_{1-40} homogenous dimers and allow us to see that these interactions don't necessarily dominate the interchain stabilizations. Our numerical simulation also show the formation of interaction between the early oligomer of NAC and Aβ_{1-40}.

1. Introduction

Amyloid proteins are known for their implications in many neurodegenerative diseases [1, 2]. While research focus was initially on large amyloid fibrils, first associated with the various diseases, attention has turned, over the last decade, to small metastable aggregates believed to be the main toxic agent [3, 4]. In spite of considerable experimental efforts, these structures remain difficult to characterize due to their metastability and a very rich energy landscape leading computer simulations to play a major role to fill some of the knowledge gap regarding these systems (for a recent review of the problem see, for example, Ref. [5]).

Alzheimer's disease is one of the prototypical amyloid diseases and, has such, it has received considerable attention. Experimentally and numerically, for example, the aggregation of full length β-amyloid (Aβ), associated with this diseases, and related fragments has been at the center of efforts for understanding the onset of amyloid aggregation. The interest for this protein is due, in part, to the richness of *in vivo* Aβ aggregation, which is associated with a range of cleaved lengths ranging from 39 to 43 residues, even though two lengths dominate [6]: Aβ_{1-40} and A β_{1-42}. Onset of aggregation studies have therefore been mostly done with the former sequence, the dominant component of amyloid fibrils, which shows a slower and easier to control aggregation mechanism.

In addition to Aβ, however, it has been known for 20 years that a second protein is present in amyloid fibrils associated with Alzheimer's disease: the non-amyloid component (NAC), a 35-residue fragment corresponding to α-synclein's strongly hydrophobic central region, a protein that is also a major component of Lewy bodies and neurites associated with Parkinson's disease but is also found in the brain of 60 % of Alzheimers disease patients. [7, 8, 9].

Relatively little attention has been given to the NAC and its interaction with Aβ, both experimentally and numerically, over the last years. [10, 11] This is due, in part, to the already difficult task of characterizing the role of Aβ alone and because the full α-synuclein protein is, in itself, an amyloid protein associated with a major degenerative disease.

Nevertheless, there are a number of reasons to look in more details into NAC aggregation mechanisms and its relation to Aβ's aggregation. First, toxic oligomers share some common structural properties that have yet to be identified and can only be resolved by comparing the oligomerization process of as many sequences as possible, since most turn out to be very difficult to characterize experimentally and computationally; a recent study of the NAC monomer, dimer and trimer confirms the computational interest of this sequence for the study of the onset of aggregation [12]. Second, most studies of heterogeneous aggregation have focused on various length or mutated Aβ-sequences. Looking at the interactions between NAC and Aβ offers a more complex but still manageable system to study this question. Finally, recent experimental evidence show evidence that α-synuclein increases Aβ's toxicity, although the mechanism is still unclear. [13, 10]

In this article, we focus on the first steps in the formation of heterogenous oligomers of Aβ_{1-40} and NAC using replica-exchange molecular dynamics (REMD) coupled with the coarse-grained OPEP potential.[14, 15, 16] This setup was already used successfully to characterize NAC's onset of homogeneous oligomerization [12]. More precisely, we look at the evolution taking place in the three-dimensional structures of Aβ_{1-40} monomer and dimer with NAC monomer to speculate on the role of stabilizer accelerating the onset of beta-sheet organized structure earlier in the polymerization mechanism. We find that these two sequences do associate to form interconnected β-rich structures, with an interaction that supports Paik and al. [17] experimental results regarding the role of Aβ's 25-35 region in coupling with the NAC peptide, and lead to enhanced β-sheet formations that is also observed with the NAC + Aβ dimer system.

2. Methods

We simulate the structural and thermodynamical properties of small amyloid aggregates composed of a α-synuclein non-Aβ amyloid component (NAC) monomer coupled with Aβ_{1-40} monomer or dimer. The amino acid sequence of the 37-residue NAC and Aβ_{1-40} are given by EQVTNVGGAVVTGVTAVAQKTVEGAGSIAAATGFV and DAEFRHDSGYEVHHQ-KLVFFAEDVGSNKGAIIGLMVGGVV respectively. Intra and interpeptide interactions are described with the coarse-grained Optimized Potential for Efficient peptide structure Prediction (OPEP) version 3.2. This potential focuses on the heavy backbone atoms, reducing the amino acid representation to six beads: Cα, N, H, C, and O and one bead for the side chain except for the proline amino acid which is represented by all heavy atoms [18].

The propensities of each residue to adopt α and β conformations are taken into account through four-body hydrogen bonding terms. These generic terms are essential to stabilize the appropriate secondary structure with having successfully been applied to a number of systems including many amyloid-forming proteins such as polyglutamine [19], amylin [20, 14] and various segments of Aβ [21, 22, 23, 24, 25, 26, 27].

Following our previous work on NAC oligomers [12], simulations are performed using the Hamiltonian-temperature replica exchange molecular dynamics, [19, 14, 26] HT-REMD, a hybrid of temperature replica exchange (T-REMD) and Hamiltonian replica exchange (H-REMD) is combined with OPEP. For the system studied here, we launch a total of 40 replicas: 35

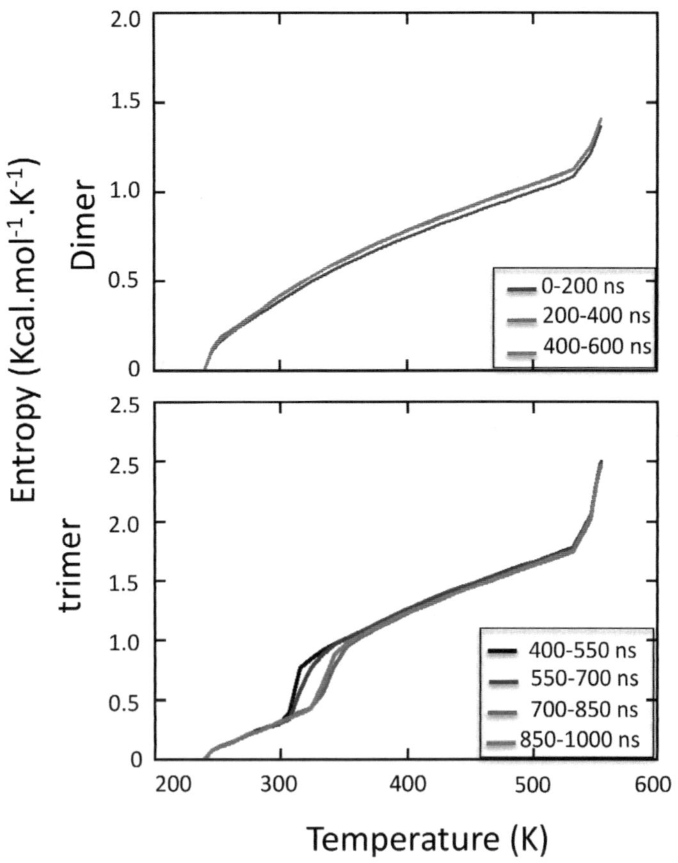

Figure 1. Entropy as a function of temperature taken over three subsequent 200-ns time intervals for the NAC and $A\beta_{1-40}$ system and the NAC plus 2 $A\beta_{1-40}$ systems. Equilibrium is reached after the first 200 ns for the two-chain system and after 700 ns for the three-chain one. Statistical error (not shown for legibility) is about ± 0.01 kcal/(mol*K)

replicas are launched at temperatures ranging from 240 to 555 K and distributed according to a logarithmic scale while the remaining 5 replicas are launched, at the highest temperature, with reduced non-bonded attractive forces corresponding to 0.8, 0.7, 0.6, 0.4 and 0.2 of the initial interaction.

Newton's equations of motion are integrated with a 1.5 fs timesteps with the hydrogen high-frequency vibrations stabilized by the Rattle algorithm timesteps [28]. The temperature is controlled with the Berendsen thermostat using an external coupling constant of 100 fs [29]. Replica exchanges are attempted every 7.5 ps. The dimeric system composed of the monomer NAC of α-synuclein and $A\beta_{1-40}$ is simulated for 700 ns per replica (total 28 μs) in a 40 Å-radius sphere with reflecting boundary conditions. The trimeric system, one NAC and two $A\beta_{1-40}$ chains, is run for 1000 ns per replica, for a total of 40 μs.

The replica-exchanged adapted weighted histogram method, PTWHAM [30], is used for the reconstruction of the thermodynamical properties of the oligomer. STRIDE, a protein secondary structure assignment code based on the combined use of hydrogen bond energy and statistically derived backbone tortional angle informations [31], is selected to predict the secondary and tertiary structures. For clustering we use the Cα root-mean square deviations (rmsd) and follow Daura's procedure [32]: we first identify the largest cluster with a rmsd of 2.5 Å; all configurations belonging to this cluster and then removed and the procedure is repeated to identify the second and following clusters.

The entropy, obtained from the free energy computed with PTWHAM, is used to determine when the system's convergence as it is a direct measure of the number of visited configurations. To estimate convergence, we compute the entropy over a non-overlapping sequence of time intervals and consider that equilibrium is reached when the entropy, computed over the whole temperature range, becomes constant. As seen in Fig. 1, the entropy converges after 200 ns/replica for the NAC+Aβ dimer. For the NAC+ 2 Aβ, we observe a jump in in the entropy near room temperature, as discussed below, this is caused by the dissociation into a monomeric NAC with a dimeric Aβ in most of replicas. Some results are presented here, however, but they must be confirmed by further simulation. We present results at 350 K, to avoid the problematic region found between 300 and 340 K, and will focus mostly on the NAC+Aβ. Statistical error bars are evaluated using the bootstrapping algorithm. [33]

3. Results

Table 1. Percentage of secondary structure taken at 300 K for the dimer NAC+Aβ and 350K for the trimer NAC+ 2Aβ . Results for the dimer and the trimer are averaged over their interval of convergence (see Section 2). Statistical error is estimated at ±1% but see the methodology for a discussion regarding the trimer.

Sec. struct. (%)	Dimer		Trimer	
	per species	average	per species	average
α-helix	11 (NAC) - <1(Aβ_{1-40})	6	27 (NAC)- <1(Aβ_{1-40})	14
β-strand	8 (NAC) - 9 (Aβ_{1-40})	9	<1 (NAC)- 29 (Aβ_{1-40})	14
Turn	58 (NAC) - 53 (Aβ_{1-40})	56	46 (NAC) - 27 (Aβ_{1-40})	37
Random coil	23 (NAC) - 38 (Aβ_{1-40})	31	26(NAC) - 43 (Aβ_{1-40})	35

We first present the detailed results for the two systems under study: a heterogeneous dimeric compound formed of a NAC and a Aβ_{1-40} monomers and a heterogeneous trimeric compound formed of one NAC and two Aβ_{1-40} monomers. In the two cases, we first place fully stretched peptides at random in the box and first relax the configuration at zero Kelvin. Configurations are then thermalized at the various temperatures used in the replica-exchange molecular dynamics and the run is started.

3.1. The dimer of Aβ_{1-40} with non-Aβ amyloid component of α-synuclein

We simulated the dimer for 700 ns per replica and collected the last 500 ns for analysis. In this interval, we find relatively little overall secondary structure : α-helix propensity is about 6 % while β-stand propensity is 50 % higher, at 9% (Table 1). As shown in Fig. 2, α-helices are concentrated in the NAC chain (with a 11 % propensity), with two peaks at residues Q2-T12 and G13-A29 showing a ~22% and ~30% propensity, respectively. We also observe a small helical tendency at residues 56-63 (A21-G28 in Aβ_{1-40}) with a propensity of ~6%, for an overall α-helical propensity of only 1%.

β-strands are slightly more common overall with no dominating region: the highest per-residue propensity barely reaches 25 %. In NAC, two regions dominate at residues Q2-T12 and

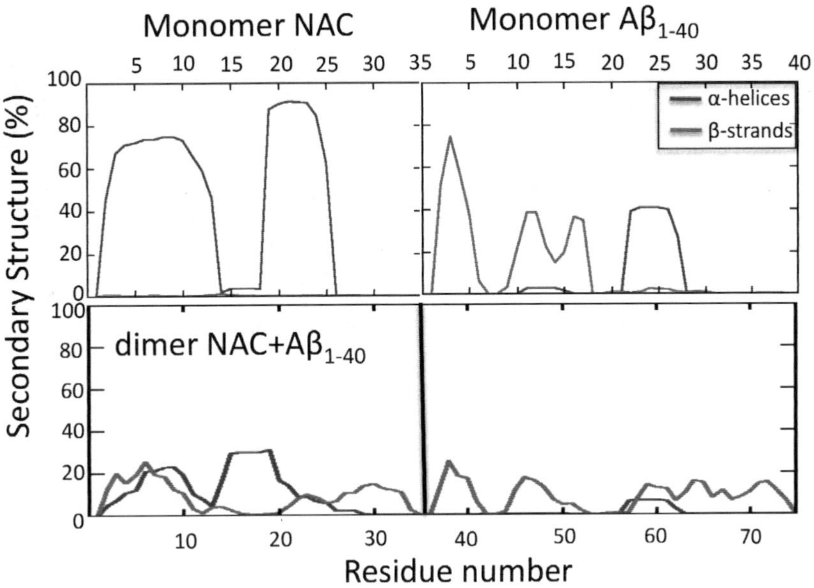

Figure 2. Top: α-helix and β-strand propensity as a function of residue number at 300K for isolated NAC and Aβ monomers (taken, respectively, from Refs. [12] and [26]). Bottom: α-helix and β-strand propensity for the NAC/Aβ dimer calculated over the 200-700 ns time interval.

K20-F34 with probabilities of about 25 and 14 %, respectively for a total β-strand propensity of 9 %, slightly less than for α-helices.

Aβ shows, for its part, three regions with non-zero probability of forming β-sheets: residues A2-D7 (37-42 in Fig. 2), S8-H18 (43-53) and D23-V40 (58-75) of Aβ_{40}, with a maximum propensity per residue of ~25%, ~17% and 15%, respectively. From Table 1, we note that the total β-strand propensity of Aβ_{40} is at 9%, small but much higher than the α-helical propensity.

Maps showing total, intrachain and interchain contacts are presented in Fig. 3. Here again, we observe a wide distribution of contacts, both interchain and interchain. While most of these contacts occur with a probability below 50 %, a number of long-range interchain interactions are present in a majority of structures. We note, in particular, V10-I66 (I31 in Aβ's sequence) and V17-V53 (V18) and V35-F54 (F19) (respectively with ~65% for the first and second and ~80% for the third interaction).

The relatively small probability of forming well-defined contacts and secondary structure elements is explained by the diversity of tertiary structures that are visited by dimer. Figure 4 shows the six dominant clusters that represent, together, only about 30 % of all configurations, as defined with a 2.5 Å-RMSD cut-off. These six clusters show a very diverse set of secondary and tertiary elements that we describe below in some details.

D1, the first cluster, with an occurrence of 8.7 %, shows no β-strand or α-helical secondary structure. However, it displays many interchain contacts responsible for maintaining the tertiary structure in a mostly antiparallel arrangement (see Fig. 5). D2, that represents 4.7 % of all visited structures, is more structured. It forms a short anti-parallel interchain β-sheet, composed of residues N5-V6 of NAC and A2-E3 of Aβ_{1-40}. We also see an α-helix in the region 14-20. Looking at contacts (Fig. 5), we observe that the few interchains contacts are dominated, beside the β-sheet, by residues V6 of NAC and F4 of Aβ (F39 in the figures).

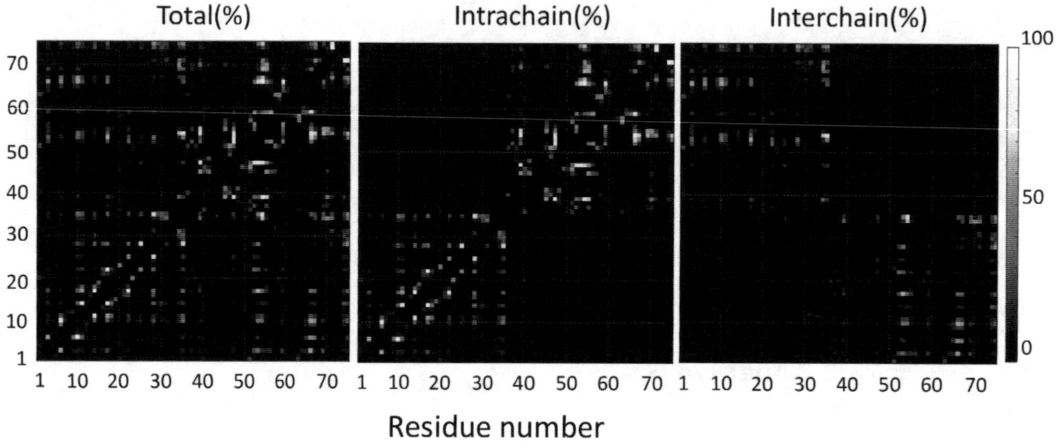

Figure 3. Map of the contact propensity of total, intrachain and interchain contacts the NAC/Aβ_{1-40} dimer. The NAC runs from residues 1 to 35 and Aβ from residues 36 to 75 over the 200-700 ns time interval. The total contact propensity is calculated from the intrachain and interchain contacts.

D3, with a formation probability of 4.6 %, shows a 3-stranded intrachain antiparallel β-sheets at residues E3-R5/E11-H14/V24-D26 in Aβ_{1-40}, with few interchain contacts. D4, representing 3,9 % of all visited configurations, also forms a short antiparallel interchain β-sheet, like D2, but between NAC's Q2-V3 and Aβ's D23-V24. It also forms an α-helix in NAC between residues V6 and A9. Looking at contacts (Fig. 5), we note that residues I31 and I32 in Aβ_{1-40} have a high propensity to form interchain contacts with some regions of NAC: the region V3-V6 and V10-V14 and F34-V35. Residues I31 and I32 are important in the interchain stabilization of the structure.

D5, that occurs 3.5 % of the time, is the only structure with both an intrachain and an interchain β-sheet. The antiparallel interchain β-sheet is found between residues T4-G8 of NAC and K28-I32 of Aβ. A short intrachain antiparallel sheet is also observed at residues A2-E3/Q16-K17 of Aβ. D6, finally, with an occurrence probability of 3.5 %, shows, like for D3 and D4, relatively few interchain contacts, with the exception of an antiparallel region between residues V17-V22 of NAC and L17-E22 of Aβ. D6 is also rich in secondary structure with an intrachain antiparallel β-sheet formed by NAC's residues G7-V10 and A29-T32, as well as a short α-helix at residues E11-H13 of Aβ.

In terms of salt-bridges, while the intramolecular bridge at K20-E23 is present on NAC for all dominant clusters, we also observe a contact between E11-K16, K16-E22, E22-K28 and D23-K28 of Aβ_{1-40}) with a percentage close to 80 in a majority of the six clusters, suggesting that these few charged residues, particularly K16 and E22 (51 and 57 in the total sequence) play an important role for stabilize intrachain structures. There are no interchain salt-bridges.

3.2. The trimer : two Aβ_{1-40} and a non-Aβ amyloid component of α-synuclein peptides

We also simulated a trimer composed of two Aβ_{1-40} and one NAC peptides for 1000 ns per replica and collected the last 300 ns for analysis. While the entropy is relatively stable over this time range, we find that most replicas sampled a decomposed system with an isolated NAC monomer and a Aβ_{1-40} dimer. Properties for the full configurational ensemble are given in Table 1, where we note the same amount of secondary structure: α-helix and β-stand propensity, at 14%.

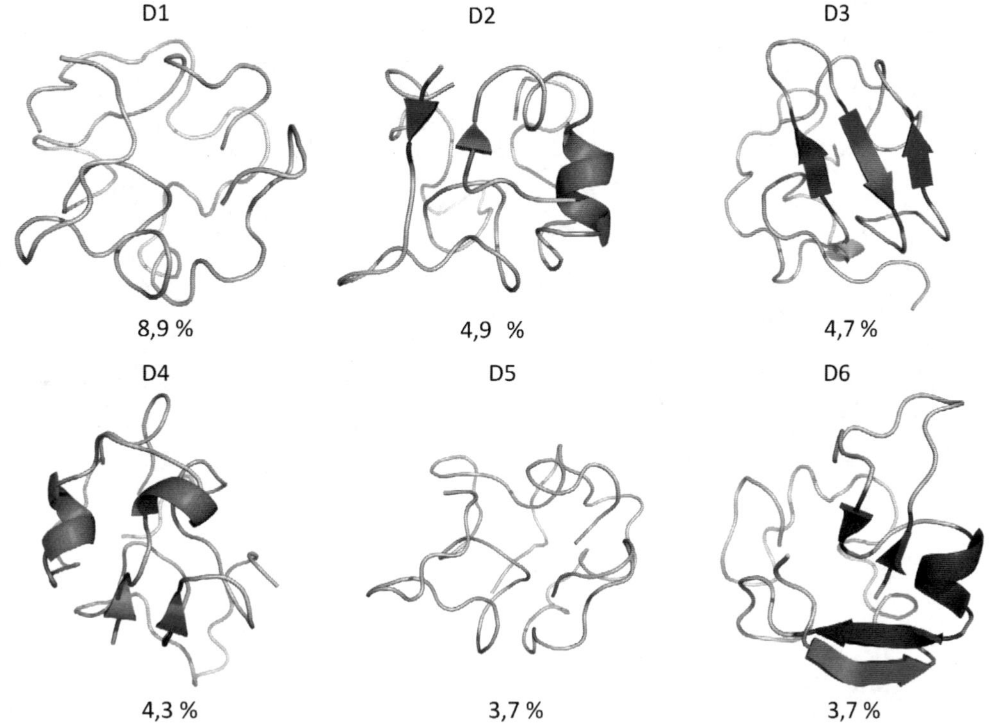

Figure 4. Dominant clusters' centers for the dimer NAC/Aβ_{1-40} averaged over the 200-700 ns time interval. The N-terminal and the C-terminal are shown in green and orange, respectively. The NAC chain is shown in red, Aβ_{1-40} in blue.

α-helices are concentrated in the NAC chain and β-stand in Aβ_{1-40}.

A few replicas produce a bound trimer however. The dominant cluster for these runs, that represent only 7 % of all configurations, and defined with a 3 Å-RMSD cut-off is presented in Fig.6. This cluster show a interchain contact β-stand, between residue A31-G33 (NAC) /V36-G38(Aβ_{1-40}) and A30-G33(NAC)/V36-V39(Aβ_{1-40}).

Clearly, while these results show that interactions between the two species are possible among larger aggregates, further simulations are necessary to evaluate whether or not the dislocated structure is due to the initial setting and the fact that OPEP's relatively short cutoff prevents peptides, once far from each other, to attract, or is characteristic of an important part of the configurational space of this setup.

4. Discussion

As discussed above, the NAC represents about a 10 % of the proteins found in amyloid fibrils associated with Alzheimer's disease. [7] Its role in aggregation and toxicity remains to be understood and it is not clear whether it is simply a contaminant or whether it impacts the fibril growth. This impact could be multiple. It could help stabilize the Aβ monomers or

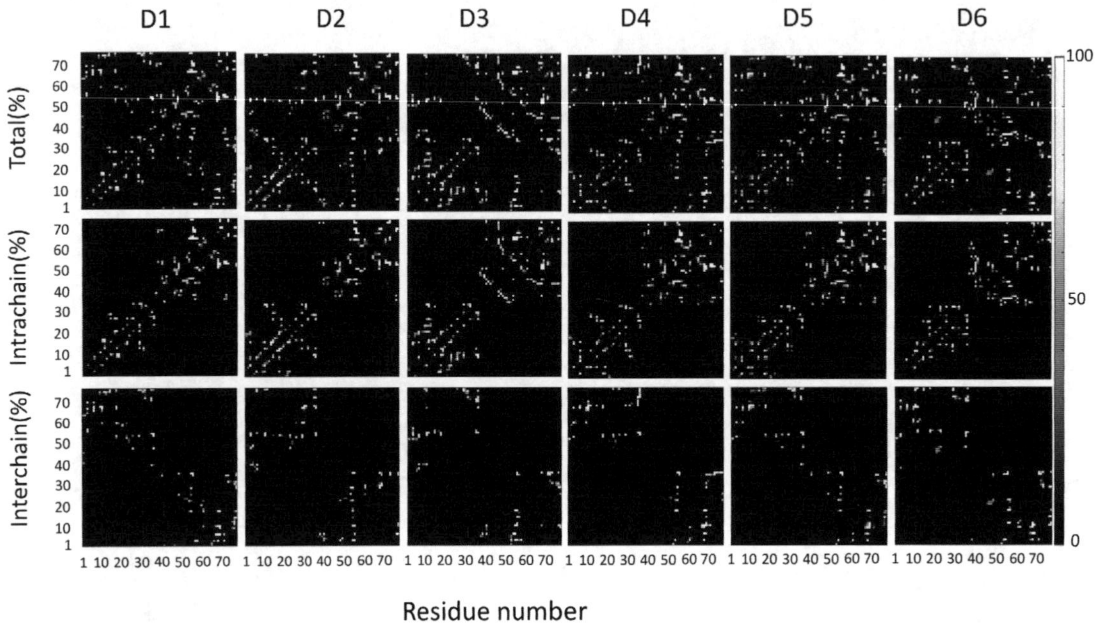

Figure 5. Map of the total, intrachain and interchain contact propensity for the six dominant clusters in the mixed dimer. The NAC runs from residues 1 to 35 and Aβ from residues 36 to 75. The total contact propensity is calculated by summing the intrachain and interchain contacts.

oligomers as they grow into fibril or play a more active role into the oligomeric structures.

There exists very few experimental papers that attempt to characterize the interaction between the NAC and Aβ. Indeed, amyloid oligomers are particularly difficult to control and, even for simply monospecies oligomers, very little precise structural information exists on the structure of the dimer or small assemblies.

In *in vitro* growth studies, Han *et al.* explored the nucleation dependent polymerization mechanism between NAC and Aβ_{1-40}. [34] They show, first, that the NAC peptide does aggregate into amyloid fibrils. Han *et al.*, moreover, demonstrate that NAC fibrils can seed Aβ fibrils' formation and, *vice-versa*, that Aβ fibrils can seed NAC amyloid aggregation. While this heterogeneous seeding is possible, it remains less efficient than homogeneous seeding as determined by the lag time before fibril growth takes place, suggesting that the structural match between these two sequences is less amyloid prone that with a pure system. [34] While this work supports a positive interaction between NAC and Aβ leading to amyloid formation, it is not clear whether or not the two chains can interact at the monomeric level or only through preformed fibril structures. Indeed, Han *et al.* find no evidence of an increase in the formation of insoluble aggregates in mixed NAC and Aβ solutions compared with pure ones. In some cases, the rate is even slowed down, although this results says nothing about the formation of soluble oligomers and their relative toxicity. The first part of the question is partially answered by Paik *et al.* in an experiment on the self-oligomerization of the full-length α-synuclein with Aβ_{25-35} [17]. Paik and collaborators confirm that the 25-35 region of the Aβ is sufficient for promoting α-synuclein aggregation. They also show that, although Aβ_{25-35} could interact with other regions of α-synuclein, the NAC region seems to be a preferred binding site. Repeating their experiments with a reverse 35-25 sequence, Paik *et al.* showed that the orientation was essential for promoting oligomerization [17].

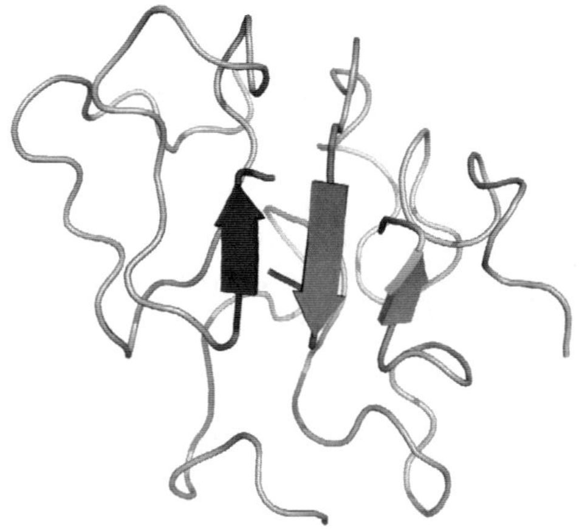

Figure 6. The dominant cluster's center for the aggregated trimer NAC+ 2 Aβ_{1-40} over the 700-1000 ns time interval. The N-terminal and the C-terminal are shown in green and orange, respectively. The NAC chain is shown in red, the Aβ_{1-40} in dark and light blue.

Our simulations, that show binding of both peptides, are compatible with these observations. For example, Fig. 3 shows that contacts between Aβ and NAC are dominated by residues V17(NAC) with V53(Aβ's V18) and V35(NAC) with F54(Aβ's F19), showing that, indeed, Aβ's N-terminal is relatively inactive here. Moreover, interchain β-sheets involve residues in this region : D3(24-26), D4(23-24), D5(28-32) and D6(26-32). These results confirm the interaction and the co-stabilization of secondary structure in or around region 25-35 of Aβ and NAC.

Comparison between our results and the formation of pure NAC and Aβ dimers allows us to link our simulations to Han *et al.* results. Côté *et al.* studied the assembly of Aβ_{1-40}, Aβ_{1-40}(D23N) and Aβ_{1-42} dimers using the same OPEP and HT-REMD methodology, allowing a direct comparison with results presented here. [16] They show that while β-strand propensity increase comparing with the monomer, particularly at the fibril-loop region(22-28) and C-terminal(29-40), the overall change in secondary structure remains small for Aβ_{1-40}: α-helical propensity goes from 6 to 1.3 % while β-strand propensity moves form 12 to 13 %, similar to what is observed here. For the NAC peptide, however, the difference is more important: while the β-sheet content reaches 24 % for the NAC dimer, it is only 9 % here.

Moreover, if the propensity is lower, β-sheets observed in the heterogeneous dimer are similar to those observed in the homogenous dimers, with a relatively high β-sheet propensity in the N-terminal for Aβ and at the N-terminal and C-terminal for NAC, even though some of these β-sheets are formed by the two chains. We also observe an important presence for the salt-bridge between D23 and K28 of Aβ_{1-40}, and K20-E23 is present on NAC, in agreement with Côté *et al.* results. [26, 16] and our study of NAC dimer.

5. Conclusion

In this study, we simulated the dimer and trimer of $A\beta_{1-40}$ with non-$A\beta$ amyloid component of α-synuclein. We focused on this system in order to understand the possible role of NAC in the amyloid fibrils associated with Alzheimer's disease, for which $A\beta_{1-40}$ is the major component. Our simulations confirm experimental observations by Paik *et al.*[17] that the 25-35 region is a significant area of secondary structure stabilization for the dimer with NAC. Comparing with NAC and $A\beta$ previous study using the same simulation and analysis procedure, we find the beta-sheet propensity remains the same for $A\beta_{1-40}$ and decrease for NAC. The same ionic interaction between D23-K28 in $A\beta_{1-40}$ and K20-E23 in NAC in the homogenous oligomer are found in heterogenous and confirm ionic interaction is not involve in the interchain stabilization.

Clearly, more similations and experiments are needed to establish the full interaction between NAC and $A\beta$. Our results, however, confirm that NAC could be used as one of the numerical standards for studying protein aggregation both in a pure and a mixed form.

6. Acknowledgments

This work is supported in part by the Natural Science and Engineering Research Council of Canada, the Canada Research Chair Foundation and the Fonds de recherche du Québec - Santé. This work was made possible through generous allocation of computer time by Calcul Québec/Compute Canada.

References

[1] Ross C and Poirier M 2004 *Nature medicine* **10 Suppl** S10–7
[2] Ross C and Poirier M 2005 *Nature reviews. Molecular cell biology* **6** 891–8
[3] Bodles A, Guthrie D, Greer B and Irvine G 2001 *Journal of neurochemistry* **78** 384–395
[4] Taylor J, Hardy J and Fischbeck K 2002 *Science (New York, N.Y.)* **296** 1991–1995
[5] Derreumaux P (ed) 2013 *Alzheimer's disease: Insights into Low Molecular Weight and Cytotoxic Aggregates from In Vitro and Computer Experiments* (Imperial College Press)
[6] Haass C and Selkoe D J 2007 *Nat. Rev. Mol. Cell Biol.* **8** 101–112
[7] Uéda K, Fukushima H, Masliah E, Xia Y, Iwai A, Yoshimoto M, Otero D A, Kondo J, Ihara Y and Saitoh T 1993 *Proc Natl Acad Sci U S A* **90** 11282–6
[8] Duda J, Lee V and Trojanowski J 2000 *Journal of neuroscience research* **127** 121–127
[9] Mikolaenko I, Pletnikova O, Kawas C H, O'Brien R, Resnick S M, Crain B and Troncoso J C 2005 *J Neuropathol Exp Neurol* **64** 156–62
[10] Kazmierczak A, Strosznajder J B and Adamczyk A 2008 *Neurochem Int* **53** 263–9
[11] Kazmierczak A, Czapski G A, Adamczyk A, Gajkowska B and Strosznajder J B 2011 *Neurochem Int* **58** 206–14
[12] Eugène C, Laghaei R and Mousseau N 2014 *J. Chem. Phys.* **141** 135103
[13] Mandal P K, Pettegrew J W, Masliah E, Hamilton R L and Mandal R 2006 *Neurochem Res* **31** 1153–62
[14] Laghaei R, Mousseau N and Wei G 2011 *J. Phys. Chem. B* **115** 3146–3154
[15] Liang G, Zhao J, Yu X and Zheng J 2013 *Biochemistry* **52** 1089–100
[16] Côté S, Laghaei R, Derreumaux P and Mousseau N 2012 *J. Phys. Chem. B* **116** 4043–4055
[17] Paik S, Lee J H, Kim D H, Chang C S and Kim Y S 1998 *FEBS 19685* **421** 73–76
[18] Maupetit J, Tuffery P and Derreumaux P 2007 *Proteins : Struct., Funct., Bioinf.* **69** 394–408
[19] Laghaei R, Mousseau N and Wei G 2010 *J. Phys. Chem. B* **114** 7071–7077
[20] Laghaei R and Mousseau N 2010 *J. Chem. Phys.* **132** 165102
[21] Chen W, Mousseau N and Derreumaux P 2006 *J. Chem. Phys.* **125** 84911
[22] Wei G, Mousseau N and Derreumaux P 2007 *Prion* **1** 3–8
[23] Dong X, Chen W, Mousseau N and Derreumaux P 2008 *J. Chem. Phys.* **128** 125108
[24] Chebaro Y, Mousseau N and Derreumaux P 2009 *J. Phys. Chem. B* **113** 7668–7675
[25] Melquiond A, Dong X, Mousseau N and Derreumaux P 2008 *Curr. Alzheimer Res.* **5** 244–250
[26] Côté S, Derreumaux P and Mousseau N 2011 *J. Chem. Theory Comput.* **7** 2584–2592
[27] Lu Y, Wei G and Derreumaux P 2011 *J. Phys. Chem. B* **115** 1282–1288
[28] Andersen H 1983 *J. Comput. Phys.* **52** 24–34
[29] Berendsen H, Postma J, van Gunsteren W, DiNola A and Haak J 1984 *J. Chem. Phys.* **81** 3684–3690
[30] Chodera J, Swope W, Pitera J, Seok C and Dill K 2007 *J. Chem. Theory Comput.* **3** 26–41

[31] Frishman D and Argos P 1995 *Proteins: Struct., Funct., Genet.* **23** 566–579
[32] Daura X, Suter R and van Gunsteren WF 1999 *J. Chem. Phys.* **110** 3049–3055
[33] Newman M E J and Barkema G T 1999 *Monte Carlo Methods in Statistical Physics.* (Clarendon Press (Oxford))
[34] Han H, Weinreb P H and Lansbury P T 1995 *Chemistry & biology* **2** 163–169

Journal of Physics: Conference Series **640** (2015) 012009 doi:10.1088/1742-6596/640/1/012009

Structural and thermodynamic consistency in coarse-grained models of macromolecules

M.G. Guenza

Department of Chemistry and Biochemistry, Institute of Theoretical Science, University of Oregon, Eugene, Oregon 97403

E-mail: mguenza@uoregon.edu

Abstract. Coarse-grained models of complex liquids are becoming increasingly important in simulations of complex fluids because they substantially reduce computational time while bridging information between simulations performed with atomistic resolution and continuum models. However, to be useful coarse-graining approaches need to provide physically consistent representations of the system under study, independent of the degree of coarse-graining resolution, i.e. they need to ensure consistency in both the thermodynamic and structural properties.

Developing coarse-graining approaches that ensure consistency of thermodynamic and structural properties across variable coarse-graining resolution is a challenge. The difficulty is in properly mapping many-body interactions into effective pair interactions, which are input to the fast mesoscale molecular dynamic simulations of the coarse-grained representations.

In this paper we review a Coarse-Grained approach, based on Integral Equation theory. This model conserves thermodynamics while accurately reproducing the structure of polymer liquids across variable levels of resolution. As it is solved analytically, it provides a formalism to reconstruct *a posteriori* the correct dynamics from the accelerated dynamics of the coarse-grained simulation. This approach has the advantage of being largely analytically solved and it is useful to highlight some important and general points that need to be considered when developing coarse-graining models.

1. Introduction

In recent years the field of coarse-graining (CG) has been in rapid expansion. The interest in coarse-graining methods is motivated by the need of extending the range of time and length scales that can be covered by simulations, for example molecular dynamics. Coarse-graining methods help bridge the region in resolution between atomistic simulations and continuum models by partitioning the intermediate region into subparts where the system is described at variable levels of resolution. [1, 2, 3, 4, 5, 6, 7]

During a coarse-graining procedure a number of atoms are grouped together defining a new "fictitious" coarse-grained unit that is centered in their center of mass. In doing this simple process, the dimensionality of the configurational space is reduced, smoothing the probability distribution. The new statistical unit represents now a number of microstates that have been averaged out during coarse-graining. This modifies the entropy of the system as coarser descriptions represent a larger number of "ghost" microstates, which are not present in the final coarse-grained representation. The "shape" and size of the molecule are also modified as a consequence of coarse-graining and so is its friction coefficient.

With increasing the level of coarse-graining an increasing number of fast processes are eliminated in the reduced description in favor of random fluctuations for the dynamics of the coarse-grained units. In this sense the process of coarse-graining is analogous to the transformation by projection operator techniques, e.g. Mori-Zwanzig, of the Liouville equation into a Langevin equation. The potential that in the Liouville equation is the Hamiltonian of the system becomes a potential of mean-force, i.e. a free energy of the reduced variables. And while the dynamics in the Liouville equation is deterministic it becomes dissipative in the Langevin equation, where the fluctuations of the suppressed particles enter as random forces and their correlation defines the memory function.[8]

With changes in the number of microstates that are sampled and in the "shape" of the molecule, i.e. changing the extent of surface available to the random collisions of the "solvent" and so its friction coefficient, the dynamics of the CG units accelerates. The coarser the CG description adopted the faster the simulation, and the higher the computational gain.[9]

While a coarser description is computationally more useful, all the problems and uncontrolled consequences of coarse-graining emerge more dramatically the higher the level of coarse-graining. These can affect the mesoscale CG simulations leading to unwanted incorrect predictions of structure, thermodynamics, and dynamics. Clearly it is quite important to develop coarse-grained descriptions that are statistically well controlled, or alternatively to develop methods to evaluate the extent of errors that emerge from the use of poorly controlled coarse-graining method.

In this paper we review a coarse-grained method for macromolecular liquids that is solved analytically, and in this way provides a mean to understand the consequences of coarse-graining in relation to the consistency of structural and thermodynamic quantities, and to the dynamics. The Integral Equation Coarse-Grained (IECG) model is solved for a variable level of coarse-graining, representing the macromolecule in the liquid as a collection of soft blobs whose number can be arbitrarily chosen.[10, 11] The theory ensures an analytical solution of the effective coarse-grained potential, and structural and thermodynamic consistency at variable CG resolution.[12] It provides also information about the general properties of coarse-graining and a method for rescaling the dynamics obtained from mesoscale simulations.[9, 13, 14]

2. Numerical versus analytical coarse-grained methods

Most coarse-grained methods are solved numerically. Numerically solved CG models assume a trial function as the starting potential, and rely on data from high resolution atomistic simulations or from experiments to optimize self-consistently the parameters of the potential until the test function is reproduced.[1, 2, 3, 4, 5, 15]

As mentioned befoere, because through the coarse-graining procedure the interaction potential becomes a free energy of the system, it depends on all the parameters that define the system including thermodynamic as well as molecular parameters.[16] In this way a numerically optimized potential is specific of the system for which it has been optimized, i.e. the molecular structure of the system and thermodynamic parameters, and of the physical quantities that have been used to optimize the model. Furthermore, because the free energy potential depends on so many parameters, a "brutal" optimization of all the quantites of interest is not a viable strategy.

In principle, a numerically optimized potential does not apply to any other system in different thermodynamic conditions or for different molecular parameters than the ones for which the potential has been optimized. Even if the calculations are performed at the correct state conditions numerical potentials optimized to reproduce one quantity, e.g. the correct pair distribution function, do not necessarily reproduce any other physical quantity, such as the pressure or energy. Furthermore, an effective pair potential optimized at one set of conditions (molecular structure and thermodynamic) will not generally be transferable to another set of conditions. These are commonly referred to as the "representability problem" and the "transferability problem" respectively. Finally, numerically optimized potentials are precise to the level of the physical quantity used to optimize them, and if the data are truncated or contain some experimental errors, the CG potential will carry this "hidden error" in any following application.

The hope for numerically optimized potential is that they are applicable in a region of the phase diagram close to the one for which they have been optimized, and for systems that are similar from a molecular point of view. We see that this is an acceptable way to proceed for CG models that group together a small number of atoms. These Numerical High Resolution CG models (NHRCG), once parameterized to reproduce the total correlation function, predict increasingly incorrect thermodynamic quantities with decreasing coarse-graining resolution.[7, 10, 11] This is the reason behind the success of fine-graining models like the united atom model. If the model is very coarse-grained the potential is specific, and it is convenient to have an analytical form of the potential because all the parameters entering the potential are explicit.

The ultimate purpose of developing coarse-graining methods is to have a mesoscale potential that can be directly applied in molecular dynamics simulations without the need of performing initial atomistic simulations to parameterize them. In this way, analytical coarse-graining methods when they can be derived appear to be a quite promising strategy to model complex systems on a low resolution scale.

3. An analytical and transferable coarse-graining method for polymer liquids

Starting from integral equation theory, we developed a coarse-graining method called the Integral Equation Coarse-Graining (IECG) approach, which allows for variable coarse-graining resolution in the molecular representation while conserving thermodynamics and structure. The IECG potential is fully transferable and the method is predictive in the sense that it is not necessary to perform atomistic simulations to parameterize the potential. The IECG formalism has been solved analytically for systems where the distribution of CG sites is Markovian and the distribution of monomers in the coarse-grained unit is Gaussian. These requirements are fulfilled in polymer liquids when the CG length scale is chosen to be larger than a number of persistence lengths. This is the length scale where the local molecular structure of the polymer in the form of localized conformational barrier and rigidity is averaged out and the central limit theorem applies to the intrablob site distribution.

The IECG method is not particularly tied to any of the approximations just described. It can be applied to a large number of systems, for example where the structure of the molecule is characterized by non-Gaussian statistics, by adopting a numerical solutions of the equation. It is important to point out that the numerical solution of the coarse-graining equation is not related at all to the numerical optimization of a parametric potential, which is typical of numerically optimized CG models. In the IECG formalism, even when the equations are solved numerically, the parameter dependence is clearly defined.

The analytical IECG model has been developed so far for homopolymer and block-copolymer liquids, and for polymer mixtures.[6, 7, 10, 17, 18, 19] The analytical solution pertains to isotropic liquids of polymers, and when simulating system undergoing phase separation it does not describe the strong segregation regime where the liquid is strongly anisotropic and macromolecules are stretched. For sake of simplicity we summarize in this paper the IECG approach for a homogeneous liquid of homopolymer chains because for this system the mathematical formalism is the simplest and a number of essential aspects of the procedure of coarse-graining already clearly emerges.

The approach has several steps. Starting from the Ornstein-Zernike integral equation we derived the pair distribution function of the coarse-grained model and the effective potential acting between units. Using the potential we performed simulations of the coarse-grained systems, and then compared thermodynamic quantities and structural quantities of interest from these coarse-grained simulations with united atom simulation data. The agreement of structural and thermodynamic properties between CG and atomistic descriptions is quantitative, while the analytical expressions depend on one non-trivial parameter, the direct correlation function at large distances, $c(k \to 0) = c_0$. This function is evaluated either from experiments or from theory. Other parameters are the themodynamic properties of temperature, T, density, ρ, as well as the structural parameters of the number of monomers in a chain, N, and the effective segment length, σ. The molecular specificity enters the theory through the effective segment length, which depends on the stiffness of the polymer, and the direct correlation function at large separation distances, which is also specific of the macromolecule that is simulated.

The formalism is then used to calculate the analytical correction factors that are needed to rescale the dynamics measured from the fast molecular dynamics simulations of the coarse-grained system and obtain the correct dynamical quantities for the atomistic scale. For example, the reconstructed diffusion coefficients show quantitative agreement with united atom simulations and with experiments performed using a number of different techniques. The one parameter needed to be adjusted in the dynamic reconstruction procedure is the effective hard-sphere diameter in the Lennard-Jones mapping of the molecule from atomistic simulation into a chain of hard-spheres.[9, 13, 14]

The model applies to any type of polymer, because the length scale of coarse-graining has only to be assumed to be larger than its local persistence length. By selecting a length scale larger than the persistence length, which is specific of the polymer considered, the coarse-grained units are statistically uncorrelated and follow a random walk in space. The chain of blobs can then be modeled as freely jointed. This model has unique characteristics because being analytical is fully transferable: it applies to different points in the phase diagram, and represents well any type of homopolymer liquid, independent of the molecular structure of the monomer. In this model atomic-scale simulations are used only as a test and not to optimize the numerical form of the CG potential.

4. The Integral Equation Coarse-Graining Approach for a liquid of homopolymers

In developping a coarse-grained model it is common to start from an atomistic simulation and to build the new coarse-grained description in a reduced representation by matching physical quantities in the atomistic and reduced descriptions. These numerically optimized models, however, suffer from several problems, as discussed earlier on in this paper, including the lack of transferability. To overcome this problem, we developped a coarse-graining model that approach the problem from a different perspective.

It is well known that for each system a pair distribution function is uniquely defined[20] and that when this pair distribution function is known it is possible to calculate from this function all the thermodynamic quantities of interest using the statistical mechanics of liquids.[21] A method to calculate the pair distribution function of a molecular liquid is to use integral equation theory. Specifically we start from the Ornstein-Zernike integral equation theory for the coarse-grained description and for the monomer-level description. The latter is the Polymer Reference Interaction Site Model (PRISM) approach.[22] The coarse-grained sites are represented as fictitious sites and the monomer/atomistic sites are represented as real sites. This roughly corresponds to assuming strongly repulsive or Lennard-Jones interactions between atomistic sites and bound soft potentials of much smaller amplitude than the atomistic one between fictitious and real sites, and between fictitious sites. In a low resolution coarse-grained model where the number of atomic sites represented by the coarse-grained unit is high, each CG unit represents the center-of-mass of a polymer subchain. Because two center-of-mass have some probability of superimposing their interaction potential is bound and soft.[10]

The solution of the pair distribution function from the Integral Equation theory does not require performing atomistic simulations. Instead the atomistic pair distribution function is defined starting from the molecular and the thermodynamic parameters of the system. Once the pair distribution function is derived, then the effective pair potential between the coarse-grained units is calculated using the appropriate closure, which for soft bound potentials is the HyperNetted Chain (HNC) approximation.[16] The closure approximation has the function of mapping three-body and higher-order many-body contributions, which are important in dense liquids, into the effective pair interaction potential.

The derived pair potential is an input to mesoscale simulations of the coarse-grained system, which once they are performed provide all the molecular and thermodynamic quantities of interest. Those are, for example, the pair distribution function of the coarse-grained description, the pressure, the excess free energy, as well as the internal energy and the entropy.[12] The structural and thermodynamic quantities show quantitative agreement with the ones measured in atomistic simulations for the system under study, but atomistic simulations are not used to optimize the potential. Only one non-trivial parameter, which is related to the compressibility of the liquid, needs to be adjusted as discussed later on in this paper.

The homopolymer liquid is composed of n molecules in a volume V, with each chain including a total number of monomer N. The density of chains is $\rho_{ch} = n/V$ and is related to the liquid monomer density $\rho = \rho_{ch}N$. Every chain in the liquid is partitioned in a variable number of coarse-grained units or blobs, n_b, each containing a number $N_b = N/n_b$ of monomers, with the blob density $\rho_b = \rho/N_b$. Given N monomers with a chain density ρ_{ch}, and an effective segment length $\sigma = \sqrt{6/N}R_g$ with R_g being the polymer radius of gyration and $R_{gb} = R_g/\sqrt{n_b}$ the blob radius of gyration.

We derived specific expressions for the potential for the single soft sphere representation, where each chain is represented as one sphere, as well as for the three blob and the five blob representations. For the three-blob and five-blob representations the blobs are not all equivalent, as the blobs at the end of the chains are different than the one(s) in the internal part of the chain. For chains with a number of blobs larger than five end effects become negligible and it is possible to use a blob-averaged description.[11].

The intramolecular distributions in the blob-averaged limit are normalized as $\Omega(k) = \omega(k)/N$, for the blob-blob (bb), blob-monomer (bm), and monomer-monomer (mm) distributions. The normalized blob-monomer and the blob-blob distributions are given respectively as

$$\hat{\Omega}^{bm}(k) = \frac{1}{n_b}\left[\frac{\sqrt{\pi}}{kR_{gb}}Erf\left(\frac{kR_{gb}}{2}\right)e^{-\frac{k^2R_{gb}^2}{12}} - 2\left(\frac{e^{-n_bk^2R_{gb}^2} - n_be^{-k^2R_{gb}^2} + n_b - 1}{k^2R_{gb}^2n_b(e^{-k^2R_{gb}^2} - 1)}\right)e^{-k^2R_{gb}^2/3}\right] , \tag{1}$$

and

$$\hat{\Omega}^{bb}(k) = \frac{1}{n_b} + 2\left[\frac{e^{-n_bk^2R_{gb}^2} - n_be^{-k^2R_{gb}^2} + (n_b - 1)}{n_b^2(e^{-k^2R_{gb}^2} - 1)^2}\right]e^{-2k^2R_{gb}^2/3} . \tag{2}$$

The monomer distribution $\Omega^{mm}(k)$ is normalized as

$$\Omega^{mm}(k) = \hat{\omega}^{mm}(k)/N = \frac{2}{n_b^2k^4R_{gb}^4}(k^2R_{gb}^2n_b - 1 + e^{-n_bk^2R_{gb}^2}) . \tag{3}$$

Given the Ornstein-Zernike relation for the coarse-grained blob representation

$$\hat{h}^{bb}(k) = n_b\hat{\Omega}^{bb}(k)\hat{c}^{bb}(k)\left[n_b\hat{\Omega}^{bb}(k) + \rho_b\hat{h}^{bb}(k)\right] , \tag{4}$$

the bob-blob direct correlation function is given as

$$\hat{c}^{bb}(k) = \frac{\hat{h}^{bb}(k)}{n_b\hat{\Omega}^{bb}(k)\left[n_b\hat{\Omega}^{bb}(k) + \rho_b\hat{h}^{bb}(k)\right]} . \tag{5}$$

The effective pair potential between blob CG units is calculated from the Fourier transform of Eq. 4 and Eq. 5 and the Hyper Netted Chain closure equation

$$\frac{v^{bb}(r)}{k_B T} = -\ln\left[h^{bb}(r) + 1\right] + h^{bb}(r) - c^{bb}(r) \ . \tag{6}$$

The solution of Eq.6 can be obtained analytically when the intramolecular distributions are represented by simple formal equations, or numerically in the other cases.

When $|h^{bb}(r)| << 1$, which always holds at large separations ($r >> 1$ in units of R_{gb}) and at any separation for representations with large N_b and high densities, the potential further simplifies to

$$V^{bb}(r) \approx -k_b T c^{bb}(r) \ . \tag{7}$$

This formula is referred to as the mean spherical approximation (MSA) in the literature, and applies to low compressible polymer liquids.[23] If this formalism is improperly used to treat low density liquids, where the mean spherical approximation does not hold, this approximation would lead to unphysical behavior. For low density liquids the full closure approximation has to be used.

We now focus on the effective direct correlation function, $c^{bb}(r)$, and the related MSA potential. In the limit of large separations in real space, where $r >> 1$ (in R_{gb} units), the inverse transform integral is sufficiently dominated by $\hat{c}^{cc}(k)$ (in the small wave vector limit) that the large wave vector contribution can be entirely neglected. Furthermore, since the expansion for small wave vectors is bounded at large k, the error incurred in using the small k form with the integral bounds extended to infinity is small. This approximation leads, for $r >> 1$ (in R_{gb} units), to

$$\begin{aligned} c^{bb}(r) \ &\approx \ \frac{-N_b\Gamma_b}{2\pi^2\rho_m R_{gb}^3 r}\int_0^\infty \left(k\sin(kr)\left[\frac{45}{45+\Gamma_b k^4} + \frac{5k^2}{28}\frac{13\Gamma_b k^4-3780}{(\Gamma_b k^4+45)^2}\right]\right)dk \\ &= \ [-\left(\frac{45\sqrt{2}N_b\Gamma_b^{1/4}}{8\pi\sqrt{3}\sqrt[4]{5}\rho_m R_{gb}^3}\right)\frac{\sin(Q'r)}{Q'r}e^{-Q'r} + \left(\frac{\sqrt{5}N_b}{672\pi\rho_m\Gamma_b^{1/4}R_{gb}^3}\right)[(13Q^3(Q'r-4))cos(Q'r) \\ &\quad + \left(\frac{945+13Q^4}{\Gamma_b^{1/4}}\right)r\sin(Q'r) + \frac{945r}{\Gamma_b^{1/4}}cos(Q'r)]\frac{e^{-Q'r}}{Q'r}] \ , \end{aligned} \tag{8}$$

where $Q' = 5^{1/4}\sqrt{3/2}\Gamma_b^{-1/4}$ and $Q \equiv Q'\Gamma_b^{1/4}$. This is the main result of our theory as it provides an anlytical, albeit approximated, expression for the effective pair potential between CG units in a liquid of homopolymers. Some relevant points emerge from the analysis of this equation.

The key quantity of interest here is the universal parameter $\Gamma_b = N_b\rho|c_0|$, which is defined once we decide the level of coarse-graining, N_b, as well as the molecular and thermodynamic parameters of our system. This quantity also depends on the direct correlation function at $k = 0$, c_0, which is in principle not known. This function is however related to the potential between atomistic units and to the isothermal compressibility of the liquid. It can be determined numerically or from experiments as we discuss later in this paper.

The range of the potential, in units of the radius of gyration of the blob, scales as $\Gamma_b^{1/4} \propto N_b^{1/4}$. This scaling behavior describes how the interaction between effective units propagates through the liquid. By increasing the density of the liquid or decreasing the resolution in the coarse-grained description, i.e. increasing the number of monomeric units in the blob, the potential becomes increasingly longer ranged. The characteristic scaling with the number of monomers in the blob emerges from the many-body interaction pathway following a random walk ($N_b^{1/2}$) in the space defined by the lengthscale of the blob-blob interpenetration, which also scales with the degree of polymerization as $N_b^{1/2}$.

Interestingly, the range of the potential increases with the number of monomers comprised in the coarse-grained unit, i.e. blob or soft sphere, while the potential at contact decreases.

However the interblob potential does not vanish even when the length of the polymer chain becomes infinity, indicating that intermolecular interactions between polymers are important even for infinitely long chains. This result disagrees with the conventional assumption in polymer physics that intramolecular interactions are dominant over intermolecular contributions, and a polymer melt can simply be described by mean-field approaches of the single chain.[24]

The potential becomes longer-ranged with increasing the lengthscale of coarse-graining. For highly coarse-grained models the gain in computational time is maximized. However the presence of long-ranged forces in the molecular dynamic simulations makes the use of a large box necessary, given that the simulation box is usually chosen to be at least twice the range of the effective potential. Then the simulation is slowed down by the need of simulating a larger number of CG units. In practice the two effects tend to compensate and their balance will dictate the choice of the level of coarse-graining. Finally, the inconvenience of having long-range interactions and large boxes can be alleviated by the use of simulations in reciprocal space, as it is conveniently done in the case of electrostatic interactions using the Ewald summation.[16]

The potential has a long-range, slowly decaying repulsive component and a second attractive part that is smaller in comparison with the repulsive part. This attractive contribution is important when one evaluates the thermodynamic properties of the system and cannot be discarded. Higher order terms, which are present in the equation of the potential, tend to give increasingly more negligible contributions. The potential in our simulations is often truncated after the first attractive well, or more rarely the second repulsive contribution, depending on error minimization.

It is interesting to note that the attractive contribution is present in the effective potential even when the intermolecular atomistic potential, from which the coarse-grained potential is derived, is purely repulsive. This indicates that the attractive contribution to the intermolecular potential is, at least partially, a consequence of coarse-graining and of the propagation of the interactions through the liquid. Being the resultant of the projection of many-body interactions onto the pair of coarse-grained units, the attractive component of the potential contains contributions that are of entropic origin.

When already at the atomistic-scale the interaction potential includes an attractive part, i.e. for example it is a Lennard-Jones potential, the attractive part of the atomistic potential also provides a contribution to the total attractive component of the CG potential. In that case the attractive CG component is enhanced with respect to purely entropic one arising from the hard-sphere monomer-monomer interaction.

5. Thermodynamics: equation of state, compressibility, and Helmholtz free energy

Given the analytical form of the potential for the multiblob coarse-grained description we formally derived the thermodynamic quantities of interest, starting form the equation of state.[12] The normalized pressure is derived using the virial expression of a fluid under the assumption of pairwise additivity as

$$\frac{P}{\rho_{ch} k_B T} = 1 - \frac{2\pi\rho_b}{3k_B T}\int_0^\infty \frac{dV^{bb}(r)}{dr} g^{bb}(r) r^3 dr \approx 1 - \frac{N\rho c_0}{2} , \tag{9}$$

with ρ_b the number blob density.[25] The equation of state Eq.9 is independent of the level of coarse-graining adopted in the multiblob description, as it is expected given that the pressure is a macroscopic thermodynamic property of the liquid. All the non-ideal contributions that arise from system-specific interactions are contained in the non-trivial parameter c_0, which is specific of the system under consideration, depends on the density of the system, and is independent of the resolution of the coarse-grained model.

The isothermal compressibility is derived from the equations for the blob and monomer total correlation functions. For the different CG models, which have variable level of resolution, the isothermal compressibility is independent of the level of coarse-graining and is given by

$$\rho k_B T \kappa_T = \frac{N}{1 - \rho N c_0} \, . \tag{10}$$

The excess Helmholtz free energy per monomer, calculated relatively to the energy of the system in its gas phase, is simply given by

$$\frac{F - F_0}{n k_B T} = \frac{\rho_b}{2 k_B T} \int_0^\infty V^{bb}(r) g^{bb}(r) 4\pi r^2 dr = -\frac{N c_0 \rho}{2} \, , \tag{11}$$

with ρ_b the number density of blobs. Both the isothermal compressibility and the excess free energy are depending on the density and number of monomer in the polymer, as well as on the non-universal parameter c_0, however they are independent of the level of coarse-graining, as it is reasonable to expect for macroscopic liquid properties.

It is important to note that these equations are derived from the approximated analytical solution of the effective potential, which is accurate in the mean-field limit of a nearly incompressible liquid. The potential so derived should not be used to describe systems that are highly compressible, for example system at low density. In that case the equation would correctly predict an instability of the liquid, but the predictions will be qualitative.

Because in the range of thermodynamic parameters, temperature and density, of a regular polymeric liquid the theory predicts thermodynamic properties that are in quantitative agreement with the atomistic simulations, this result is indicative that many-body interactions, which are typical of liquids, are well accounted for in the two-body effective potential. While intergal equation theories are expressed as expansions in density of the many-body interactions, the closure approximation projects onto effective pair interactions the many-body terms. In this case the HNC closure equation, which is appropriate for bound soft potentials, is shown to be a valid approximation.

Mostly because of the need of a closure equation an integral equation theory inspired coarse-grained model like the one we developed is by its nature approximated. However, the corrections to the closure approximation are contained in the effective parameter of the direct correlation function at $k \to 0$. Once the c_0 parameter is defined for the system under study, we see that the thermodynamic quantities are found to be fully consistent across different levels of coarse-graining, and with the atomistic description. The equation of state was tested against united atom simulations of polyethylene liquids at variable chain length and different state points. The consistency between atomistic and coarse-grained descriptions both at the formal level and when compared with simulation data support the approximations adopted in the analytical solution of the potential.

From the equation of state of the pressure, all the thermodynamic quantities can be calculated.[16] However to do so it is necessary to know the dependence of the non-trivial parameter c_0 on the thermodynamic parameters for the given system that is simulated. To do so, we derived an equation for c_0 as a function of the liquid packing fraction, as described in the following section.

6. Determination of the monomer direct correlation function $c(k \to 0) = c_0$.

6.1. Soft-sphere Equation of State

As mentioned earlier the only non-trivial parameter in our theory is the direct correlation function c_0, which is related through the Ornstein-Zernike equation to the compressibility of the liquid and to the equation of state. The direct correlation function is a parameter specific of the system that is simulated and has a non-trivial dependence on the liquid density, while it is independent of the level of coarse-graining.

Because in the IECG fromalism the equation of states and the compressiblity and the excess free energy do not depend on the number of coarse-grained units in which the molecule is partitioned, it is convenient to fromally derived an equation of state for the polymer liquid in the most coarse-grained, and easiest to model, representation, i.e. the soft sphere model. This is the simplest, most reduced, representation where the whole macromolecule is described as a point particle interacting through an effective long-ranged potential. Effectively a liquid of soft spheres is a simple liquid, and expressions for the thermodynamic properties of this liquid can be easily derived following and further implementing the existing literature of hard-sphere fluids.

We derived for the soft-sphere liquid an equation of state in of the form of a Carnahan-Starling expression in analogy with the equation for the hard-sphere liquid. The equation of state is expressed as a function of an effective soft-sphere packing fraction

$$\eta_{eff} = \frac{\pi}{6}\rho d^3 \ , \tag{12}$$

where d the effective soft sphere diameter, as

$$\frac{P}{\rho k_B T} = \frac{4(\eta_{eff} + c_1\eta_{eff}^2 + c_2\eta_{eff}^3)}{(1-\eta_{eff})^3} \ . \tag{13}$$

The equation of state includes three numerical parameters (d, c_1, and c_2), which reflect the specificity of the macromolecular structure, e.g. the chain connectivity, and the fact that the real potential is not of the simple hard-sphere form.

As an example, the numerical values of these parameters c_1, c_2, and d have been calculated for liquids of polyethylene chains. Specifically we displayed data of normalized pressure as a function of an effective packing fraction. We used data for polyethylene from simulations performed at $T = 400\ K$ and increasing degree of polymerization (N = 44, 66, 78, 100, and 200) and variable density, as well as simulations for polyethylene samples at $T = 509\ K$, density $\rho = 0.03153\ sites/A^3$, and degree of polymerization N = 36, 44, 66, 78, 100, 192, 224, and 270. All the data fall onto an universal curve, which is well represented by the equation of state for soft spheres, Eq.13 when the parameters are defined as following: the optimized effective sphere diameter is $d = 2.5$ Å, while the other two parameters are $c_1 = -11.9$ and $c_2 = 31.11$. It is interesting to see that the soft sphere effective diameter is actually of the order of the Lennard-Jones sigma parameter, indicating that polymer chains can highly interpenetrate and the centers-of-mass of a pair of polymers can superimpose in space.

Following the same procedure the direct correlation function at $k \to 0$ is also defined as a function of the same parameters that enter the virial equation of state

$$c_0 = -\frac{4\pi d^3}{3}\frac{1 + c_1\eta_{eff} + c_2\eta_{eff}^2}{(1-\eta_{eff})^3} \ . \tag{14}$$

Once the parameters of the equation of state are defined, the non-trivial parameter c_0 is also defined. Eq.14 shows how the density dependence of c_0, through the effective packing fraction, is not simply linear. When thermodynamic properties are calculated as a function of a variable liquid density, the dependence of c_0 on the thermodynamic and structural parameters ensures the prediction of the correct thermodynamic quantities of interest.

It is worth noticing that a number of simulations had to be perfomed to calibrate the curve of the equation of state as a function of the effective packing fraction, and to define in this way the parameters in the equation of state. However once this procedure is completed for a given macromolecular liquid in a range of temperatures and densities, the parameters are defined and can be used to calculate c_0 for any other liquid of the macromolecule of interest and in the range of temperatures and densities where the optimized parameters apply. In this limits the theory, once it is calibrated, is quantitative and fully predictive.

The direct correlation function, c_0, appears to be a function of the density and is specific of the monomeric structure of the polymer, but it does not depend on the degree of coarse-graining selected, i.e. it is not a function of the number of blobs or the number of monomers in a blob. In this way the formalism ensures full structural and thermodynamic consistency across the different coarse-graining resolution.

6.2. Other methods to evaluate c_0

While the definition of the equation of state for soft spheres is a way to calculate c_0, this is by no means the only possible way to do so. For example we have shown how to calculate c_0 from the numerical solution of the PRISM equation starting from a realistic representation of the polymer chain. The solution of the PRISM equation provides results that are consistent with the equation of state method described above.[12]

A third method to evaluate the direct correlation function at $k = 0$ is to directly use the experimental isothermal compressibility of the liquid under study. Using this method the theory directly connects the coarse-graining method with the experimental data, without the need of performing atomistic simulations.

The isothermal compressibility, κ_T, which is also preserved during coarse-graining, is related to the static structure factor $\hat{S}(k = 0)$ as

$$\rho k_B T \kappa_T = \hat{S}(k = 0) = [N + \rho \hat{h}^{mm}(0)] \ , \tag{15}$$

with $\hat{S}(k = 0)$ the $k \to 0$ limit of $\hat{S}(k) = [\hat{\omega}^{mm}(K) + \rho \hat{h}^{mm}(K)]$. The isothermal compressibility in the blob description is formally identical to the isothermal compressibility in the monomer description. The value of c_0 is then determined, for example in the monomer description, as

$$c_0 = \frac{\hat{h}^{mm}(0)}{\rho N \hat{h}^{mm}(0) + N^2} \ , \tag{16}$$

with $\hat{h}^{mm}(0)$ related to the isothermal compressibility through Eq. 15.

7. Potential energy and entropy

In the multiblob description the *potential energy* has both inter- and intra-molecular contributions. The intermolecular part is calculated as an extension of the formula for the soft sphere

$$\frac{E^{bb}}{nk_BT} = \frac{2\pi \rho_b}{k_BT} \int_0^\infty v^{bb}(r) g^{bb}(r) r^2 dr \approx -\frac{\rho N c_0}{2} \ , \tag{17}$$

and gives a contribution that even in the multiblob description is a constant. The potential energy, however, contains in this case also contributions from the intramolecular structure.

Since the bond energy is a harmonic potential with a Gaussian probability distribution, the average bond energy is simply the equipartition result,

$$\left\langle \frac{E_{bond}}{nk_BT} \right\rangle = \frac{3(n_b - 1)}{2} \ . \tag{18}$$

For the angular contribution to the energy we add an additional $E_{angle} \approx (n_b - 2)/2k_BT$ contribution per chain. In this way the potential energy depends on the level of coarse-graining. Tests of the equation for the potential energy as measured from the united atom simulations show that the theory and the simulations are quantitatively in agreement.[12]

The basic procedure of any coarse-graining formalism is the averaging of the microscopic states that are then represented by effective units, with the consequence that the *entropy* of the system in a given coarse grained representation is different with respect to the atomistic description. The extent of the change in entropy depends on the level of detail maintained in the coarse-grained representation, which determines the number of atomistic configurations that can be mapped into a single coarse-grained configuration. It can be quite large when the level of coarse-graining is extreme and the underlying chain is flexible. This is commonly called as the "mapping entropy," and is simply the difference in entropy of the atomistic model when viewed from the atomistic configurations and the coarse-grained configurations.[26]

If the chain is assumed to have a statistical distribution of monomers in space that follows a Gaussian form, the entropy associated with increasing the number of blobs in the coarse-graining procedure is given as

$$\frac{S_{bb}}{nk_B} \approx \frac{3}{2}n_b + \frac{3}{2}(n_b - 1) - \frac{3}{2}(n_b - 1)\ln\left(\frac{3n_b}{8\pi R_g^2}\right) + \ln\left(\frac{Ve}{\Lambda^3 n}\right). \tag{19}$$

The first two terms in Eq. 19 arise from the kinetic energy and bond potential energy, while the final two terms are the ideal translational and vibrational free energy.

Importantly, there is no contribution in Eq. 19, or in the intramolecular contribution to the potential energy, from the potential or c_0. In both cases the increasing entropy and potential energy with the number of blobs n_b is due solely to the increasing configurational degrees of freedom and not the interaction potential itself.

Another type of entropy of interest is the relative entropy.[5] This function is based on the "information" that is lost during corse-graining, which has to be minimized to optimize the coarse-grained model. Our coarse-grained formalism, based on liquid state theory, is devised to reproduce the correct distribution function, so that the relative entropy between the coarse-grained sites and monomer sites is minimized, and the potential is optimized, without need for any variational approach. This is equivalently to say that the relative entropy, whch is based on the information function that discriminates between coarse-grained configurations sampled in the two levels of representation, is zero.

8. Concluding remarks

We have presented a brief overview of the thermodynamic properties in the Integral Equation Coarse-Graining theory (IECG) for macromolecular systems. The Hyper-Netted Chain closure equation has been adopted to account for the projection of many-body interactions, typical of liquids, into effective pair interactions between coarse-grained sites. Pair distribution functions and effective potentials for the coarse-grained units have been calculated for models that have variable levels of coarse-graining resolution. For isotropic macromolecular liquids the IECG approach can be solved analytically and provides the analytical coarse-grained pair potential input to mesoscale simulations.

From the potential structural and thermodynamic properties of the coarse-grained description can be calculated as a function of the resolution of the model, or level of coarse-graining. Structural and thermodynamc properties, such as pressure and excess free energy, are consistently described across variable levels of coarse-graining. Internal energy and entropy are instead depending on the degree of resolution of the coarse-grained model.

All the quantities are functions of one non-trivial parameter, i.e. the total correlation function at $k \to 0$, c_0, which can be directly determined from the experimental isothermal compressibility of the liquid. This parameter is system specific, depends on the thermodynamic conditions, but does not depend on the resolution of the selected coarse-grained description.

The IECG theory differs from most alternative coarse-graining approches because it does not require performing high-resolution simulations to numerically parametrize the coarse-grained model. Atomistic simulations are used only *a posteriori* as a test of the consistency of the coarse-grained description. In this way the IECG approach is fully predictive of the structural and thermodynamic properties of the system simulated on the length scale of the coarse-graining description and larger.

9. Acknowledgements

This material is based upon work partially supported by the National Science Foundation under Grant No. CHE-1362500. This work used the Extreme Science and Engineering Discovery Environment (XSEDE), which is supported by National Science Foundation grant number ACI-1053575.

10. References

[1] P. Ganguly, and N. F. A. van der Vegt J. Chem. Theory and Computation **9**, 5247(2013).
[2] W. Shinoda, R. Devane, M. L. Klein Mol. Sim. **33**, 27 (2007)
[3] J. F. Dama, A. V. Sinitskiy, M. McCullagh, J. Weare, B. Roux, A. R. Dinner, and G. A. Voth J. Chem. Theory Comput. **9**, 2466 (2013).
[4] P. Ganguly, and N. F. A. van der Vegt J. Chem. Theory Comput. **9**, 5247 (2013).
[5] M. S. Shell, J. Chem. Phys. **129**, 144108 (2008).
[6] G. Yatsenko, E. J. Sambriski, M. A. Nemirovskaya, and M. Guenza, Phys. Rev. Lett **93** 257803 (2004).
[7] A. J. Clark; J. McCarty; I. Y. Lyubimov; M. G. Guenza, Phys. Rev. Lett., **109** 168301 (2012).
[8] R. Zwanzig *Nonequilibrium Statistical Mechanics* (Oxford University Press, New York, 2001).
[9] I. Y. Lyubimov, M. G. Guenza J. Chem. Phys. **138**, 12A546 (2013).
[10] A. J. Clark, and M. G. Guenza J. Chem. Phys. **132**, 044902 (2010).
[11] A. J. Clark, J. McCarty, and M. G. Guenza, J. Chem. Phys. **139** 124906 (2013).
[12] J. McCarty, A. Clark, J. Copperman, and M. G. Guenza J. Chem. Phys. **140**, 204913 (2014).
[13] I. Y. Lyubimov, J. McCarty, A. Clark, and M. G. Guenza, J. Phys. Chem. **133**, 094904 (2010).
[14] I. Y. Lyubimov, and M. G. Guenza, Phys. Rev. E, **84**, 031801 (2011).
[15] C. Dalgicdir, O. Sensoy, C. Peter, M. Sayar, J. Chem. Phys. **139**, 234115 (2013).
[16] J. -P. Hansen and I. R. McDonald,*Theory of Simple Liquids*, Academic Press, New York, N.Y. 2nd Ed., 1990.
[17] E. J. Sambriski and M. G. Guenza, Phys. Rev. E **76** 051801 (2007).
[18] G. Yatsenko, E. J. Sambriski, and M. G. Guenza, J. Chem. Phys. **122** 054907 (2005).
[19] E. J. Sambriski, G. Yatsenko, M. A. Nemirovskaya, M. G. Guenza, J. Chem. Phys., **125**, 234902 (2006).
[20] R. L. Henderson Phys. Lett. A **49**, 197 (1974).
[21] D. A. McQuarrie *Statistical Mechanics*; University Science: Sausalito, CA, 2000.
[22] K. S. Schweizer and J. G. Curro, Chemical Physics **149**, 105 (1990).
[23] C. N. Likos, A. Lang, M. Watzlawek and H. Lowen, Phys Rev. E **63**, 031206 (2001).
[24] M. Doi and S. F. Edwards *The Theory of Polymer Dynamics* (Oxford University Press, New York, 1986).
[25] J. McCarty; A. J. Clark; I. Y. Lyubimov; M. G. Guenza, Macromol, **45** 8482 (2012).
[26] J. F. Rudzinski and W. G. Noid, J. Chem. Phys. 135 214101 (2011).

Journal of Physics: Conference Series **640** (2015) 012010 doi:10.1088/1742-6596/640/1/012010

Kinetics of directed self-assembly of block copolymers on chemically patterned substrates

Marcus Müller, Weihua Li, Juan Carlos Orozco Rey, and Ulrich Welling

Institute for Theoretical Physics, Georg-August University, D37073 Göttingen, Germany

E-mail: mmueller@theorie.physik.uni-goettingen.de

Abstract. Chemically patterned surfaces have been successfully employed to direct the kinetics of self-assembly of block copolymers into dense, periodic morphologies ("chemoepitaxy"). Significant efforts have been directed towards understanding the kinetics of structure formation and, particularly, the formation and annihilation of defects. In the present manuscript we use computer simulations of a soft, coarse-grained polymer model to study the kinetics of structure formation of lamellar-forming block copolymer thin films on a chemical pattern of lines and spaces. The case where the copolymer material replicates the surface pattern and the more subtle scenario of sparse guiding patterns are considered. Our simulation results highlight (1) the importance of the early stages of pattern-directed self-assembly that template the subsequent morphology and (2) the dependence of the free-energy landscape on the incompatibility between the two blocks of the copolymer.

1. Introduction

Block copolymers are comprised of two or multiple blocks of chemically distinct repeat units joined by covalent bonds into a flexible macromolecule. By virtue to the small translational entropy, a minuscule repulsion between unlike blocks gives rise to phase separation. The connectivity of the blocks along the chain molecule prevents the formation of macroscopically extended domains. Instead the different blocks form spatially modulated structures. The symmetry and connectivity of these microphase-separated structures can be controlled by the macromolecular architecture, i.e., the volume fraction f of the blocks, their spatial extension, and their incompatibility, χN [1]. The characteristic spatial dimension of the microphase-separated structure is dictated by the end-to-end distance R_{e0} of the molecules, and it is of the order of 1-100 nanometers [2]. Therefore, these polymeric materials have attracted abiding interest for templating nanostructures in the context of filtration, catalysis, or microelectronic device manufacturing [3, 4].

The latter application exploits a combination of extreme UV lithography and block copolymer self-assembly ("block copolymer lithography" [3, 5, 6, 7]). Extreme UV lithography is employed to fabricate a chemical guiding pattern. In the following we restrict ourselves to a pattern of lines and spaces (L/S). This chemical guiding pattern directs the self-assembly ("directed self-assembly/DSA") of symmetric, lamella-forming AB diblock copolymer in a thin film on top of the chemical L/S guiding pattern. If the lines and spaces of the L/S guiding pattern coincide with the period L_0 of the lamellar structure of the block copolymers in the bulk and if both blocks have similar affinity to the free surface of the film, the copolymer will replicate the L/S guiding pattern and form standing lamellae orientated perpendicular to the film surfaces [8, 9, 10]. The stripe structure of the block copolymer mitigates the line-edge roughness [11, 12] or small imperfections of the guiding pattern. Alternatively the periodicity of the L/S guiding pattern can be a small multiple of the bulk lamellar structure of the copolymer ("density multiplication" [13, 14, 15]). In this case the copolymer structure will "fill-in" the missing lines and multiply the density of the L/S guiding pattern fabricating a pattern with periodicity of a few tens of nanometers.

Defects in block copolymer lithography refer to localized structures of the copolymer material that deviate from the ideal L/S pattern [16]. The types of defects that are observed in experiments encompass *inter alia* broken lines or spaces ("micro-bridges"), dislocations and disclinations. The integration of block copolymer lithography into microelectronic device manufacturing poses extreme demands of the areal density of defects – industry standards require that the DSA should result in less than 0.01 defect per square centimeter [17]!

Thermodynamically, such a strict requirement can be fulfilled because the excess free energy $\Delta F_{\rm d}$ of a defect exceeds the thermal energy scale $k_{\rm B}T$ by about two orders of magnitude [18]. This large defect free energy can be rationalized as follows: Since a macromolecule in a polymer melt adopts a fractal, random-walk-like conformation, ${R_{\rm e0}}^2 = b^2(N-1)$ where b and N denote the statistical segment length and number of segments, respectively, one molecule interacts with many neighbors. The number of interaction partners, $\sqrt{\bar{\mathcal{N}}} \equiv \rho {R_{\rm e0}}^3/N \sim \sqrt{N}$ (ρ being the number density of segments), typically adopts large values of the order 10^2, and the free-energy scale is set by $k_{\rm B}T\sqrt{\bar{\mathcal{N}}}$. Locally the distortion of the internal AB interfaces resembles a grain boundary and the typical grain-boundary free energy scales like $\gamma_{\rm gb} \sim 0.1 k_{\rm B}T\sqrt{\bar{\mathcal{N}}}/{R_{\rm e0}}^2$ [19, 20] and the typical defect area is of the order of a few lamellar spacings L_0 times the film thickness $D \sim L_0$. The concomitant defect density scales like $\sigma_{\rm d} \sim \exp(-\Delta F_{\rm d}/k_{\rm B}T)/L_0^2 \sim 0.001/{\rm cm}^2$ for $\Delta F_{\rm d} = 30 k_{\rm B}T$.

Nevertheless, the fact that a significantly higher defect density has been observed in DSA experiments demonstrates that defects cannot be conceived as equilibrium fluctuations around a perfectly ordered structure [18, 21]. Instead, defects are rather metastable localized structures, into which the system got trapped in the course of structure formation. Therefore it is important to understand and control the kinetics of self-assembly and defect-annihilation mechanisms in the DSA of block copolymer materials. This is a computational challenge because (1) structure formation evolves on the time scale of minutes or hours that are inaccessible to atomistic modeling and (2) the free-energy landscape, which provides a basic ingredient for the understanding of the adiabatic structure formation, is complex.

In this manuscript we use a soft, coarse-grained polymer model [22, 23] that is able to describe the kinetics of structure formation on the pertinent time and length scales. Then we discuss the kinetics of structure formation for pattern replication, i.e., when there is a one-to-one correspondence between the chemical guiding pattern and copolymer structure, and for two-fold density multiplication. We highlight the role of the guiding pattern at the initial stage of self-assembly and the dependence of the free-energy landscape on incompatibility. The manuscript closes with a brief summary and discussion.

2. Model and simulation technique

The large time and length scales – minutes and micrometer – require the use of highly coarse-grained models, in which a single interaction center (segment) represents several tens of monomeric repeat units of a chemically realistic description. On the length scale of such a highly coarse-grained segment, the macromolecule already adopts its random-walk like statistics and the distances are Gaussian distributed. Therefore we use a discretized Edwards Hamiltonian to model the connectivity along the polymer backbone [22, 24, 23]

$$\frac{\mathcal{H}_\mathrm{b}}{k_\mathrm{B}T} = \sum_{i=1}^{N} \frac{3(N-1)}{2{R_\mathrm{e0}}^2} \sum_{s=1}^{N-1} \left[\mathbf{r}_i(s+1) - \mathbf{r}_i(s)\right]^2 \tag{1}$$

where the first sum, $i = 1, \cdots, n$ runs over all diblock copolymers that are comprised of $N = 32$ segments. $\mathbf{r}_i(s)$ denotes the spatial position of the sth segment of polymer i. The molecular density is $\sqrt{\bar{\mathcal{N}}} = 128$ in all following simulations.

Since each coarse-grained segment corresponds to a multitude of monomeric repeat units, their center-of-mass positions may overlap without violating the excluded volume condition for the atomistic constituents. Therefore we describe the non-bonded interactions between coarse-grained segments by soft potentials. The non-bonded interactions represent two relevant interactions: (1) the low compressibility of a dense polymer liquid that stems from the excluded volume of the constituent atoms and (2) the repulsion between A and B segments that gives rise to microphase separation.

$$\frac{\mathcal{H}_\mathrm{nb}}{k_\mathrm{B}T\sqrt{\bar{\mathcal{N}}}} = \int \frac{\mathrm{d}^3\mathbf{r}}{{R_\mathrm{e0}}^3} \left\{ \frac{\kappa N}{2} \left[\hat{\phi}_A(\mathbf{r}|\{\mathbf{r}\}) + \hat{\phi}_B(\mathbf{r}|\{\mathbf{r}\}) - 1\right]^2 - \frac{\chi N}{4} \left[\hat{\phi}_A(\mathbf{r}|\{\mathbf{r}\}) - \hat{\phi}_B(\mathbf{r}|\{\mathbf{r}\})\right]^2 \right\} \tag{2}$$

where $\kappa N = 50$ is the dimensionless inverse isothermal compressibility of the dense liquid, and χN denotes the product of Flory-Huggins parameter and number of segments per polymer that parameterizes the incompatibility of the distinct blocks. $\hat{\phi}_A(\mathbf{r}|\{\mathbf{r}\})$ denotes the local volume fraction of A segments that depends on the microscopic configuration $\{\mathbf{r}\}$.

$$\hat{\phi}_A(\mathbf{r}|\{\mathbf{r}\}) = \frac{1}{\rho} \sum_{i=1}^{n} \sum_{s=1}^{fN} \delta(\mathbf{r} - \mathbf{r}_i(s)) \tag{3}$$

where $f = 1/2$ is the fraction of A segments. A similar expression holds for the dimensionless B density. In order to numerically evaluate the density, Eq. (3), and the integral in Eq. (2) we use a collocation grid with spacing $\Delta L \approx R_\mathrm{e0}/6$ and use a linear assignment of the particles to the mesh. Measuring the anisotropy of the bonded stress parallel and perpendicular to the lamellae we identified the equilibrium lamellar spacing, L_0, in the bulk yielding $L_0 = 1.68R_\mathrm{e0}$ and $1.58R_\mathrm{e0}$ for $\chi N = 30$ and 20, respectively.

We consider a thin film with geometry $D \times L_y \times L_z$. Periodic boundary positions are applied in the two lateral directions y and z. The confining substrate at $x = 0$ and the air/vacuum surface at $x = D$ are represented by hard, impenetrable walls. This is a good approximation because the surface tension of the polymer with air is about an order of magnitude larger than the tension of the internal AB interfaces and, consequently, there is hardly any deformation of the free surface of the polymer film for standing lamellae. Moreover, the limitation of our highly coarse-grained model in representing the narrow polymer-substrate and polymer/air interfaces does not warrant a more detailed modeling of the film surfaces [25].

We represent the chemical guiding pattern by a short-range potential [26]

$$\frac{\mathcal{H}_\mathrm{s}}{k_\mathrm{B}T\sqrt{\bar{\mathcal{N}}}} = -\frac{\Lambda N R_\mathrm{e0}}{\epsilon} \int \frac{\mathrm{d}^3\mathbf{r}}{{R_\mathrm{e0}}^3}\, e^{-\frac{x^2}{2\epsilon^2}} f_\mathrm{s}(y,z) \left[\hat{\phi}_A(\mathbf{r}|\{\mathbf{r}\}) - \hat{\phi}_B(\mathbf{r}|\{\mathbf{r}\})\right] \tag{4}$$

where f_s adopts the value +1 on A-attracting lines of the guiding pattern, 0 on neutral patches of the substrate, and negative values on B-attractive areas. $\epsilon = 0.15R_\mathrm{e0}$ characterizes the perpendicular range of the surface potential and ΛN denotes its selectivity. Since the two blocks of the copolymer are structurally symmetric, they suffer the same entropy loss at the substrate, and we can approximate the surface free energy difference, $\Delta\gamma$, of the two components by the surface energy difference [27]

$$\frac{\Delta\gamma {R_{\rm e0}}^2}{k_{\rm B}T\sqrt{\bar{\mathcal{N}}}} \approx \sqrt{2\pi}\Lambda N \tag{5}$$

The top surface to air/vacuum is non-preferential in all simulations.

We use Smart Monte-Carlo (SMC) updates [28] in conjunction with the Single-Chain-in-Mean-Field algorithm [29]. The former scheme biases the proposed, local, random displacements of coarse-grained segments according to the strong bonded forces. In conjunction with the soft, non-bonded interactions it results in Rouse-like dynamics [30, 31]. From the single-chain diffusion coefficient $D_{\rm diff}$ we obtain a characteristic time $\tau = {R_{\rm e0}}^2/D_{\rm diff} = 8990$ SMC moves per segment. The Single-Chain-in-Mean-Field algorithm is the Monte-Carlo analog of multiple-time step algorithm in molecular dynamics simulations [32] and temporarily replaces the weak, non-bonded forces by external, fluctuating fields that are recomputed from the instantaneous microscopic densities, Eq. (3), after each SMC sweep over all segments [22, 29, 33]. This quasi-instantaneous approximation of the pairwise interactions is accurate and quantitatively describes composition fluctuations. Since the molecular conformations are independently updated, our algorithm is well suited for parallel computations. Taking generic values for molten polymers, $D_{\rm diff} \approx 10{\rm nm}^2/{\rm s}$ and $R_{\rm e0} \approx 20$nm, we obtain an order-of-magnitude estimate $\tau \approx 40$s. Typically simulations last for $10^5 - 10^6$ SMC corresponding to minutes or hours, respectively. The lateral spatial extensions of our system are $L_y \sim L_z \approx 10L_0 \approx 0.3\mu$m, and our systems are comprised of more than 10^6 coarse-grained segments.

The model and simulation technique provides a computationally efficient and realistic description of single-chain dynamics and collective kinetics of structure formation of dense multi-component polymer melts of low and intermediate molecular weight. The particle dynamics duly accounts for the coupling between molecular conformations, single-chain dynamics and the kinetics of the collective density that is difficult to accomplish in field-theoretic models. Entanglement effects due to the non-crossability of the chain molecules, strong disparities in the segmental mobilities, or hydrodynamic flow, however, are not captured by our simulation technique. The former effects are only thought to be important for large molecular weights or systems with very disparate glass transition temperatures, or at the late stages of structure formation, respectively.

3. Results

In DSA experiments a mixture of block copolymer and low-molecular-weight co-solvent is spin-coated on the patterned substrate and subsequently annealed at elevated temperature. As the solvent evaporates the polymer concentration increases and structure formation begins. Unfortunately, we lack detailed knowledge about the initial state of structure formation within the first $\tau \approx 40$s. Therefore we crudely represent the experimental procedure by instantaneously quenching a disordered structure, $\chi N = 0$, to the final value of the incompatibility $\chi N = 30$

The structure formation after this quench can be divided into qualitatively distinct stages:

(i) *initial surface-directed spinodal self-assembly:* The initial kinetics of collective density fluctuations can be qualitatively described by the dynamic Random-Phase Approximation (RPA) [34]. Since the composition is conserved it follows diffusive Cahn-Hilliard dynamics [35, 36]. In Fourier space the time evolution of the local A-density takes the form

$$\frac{\partial\phi_A(\mathbf{q},t)}{\partial t} = -q^2\Lambda(q)\mu(q) \tag{6}$$

where $\Lambda(q) \sim {R_{\rm e0}}^2 g_0(q)/\tau$ denotes the wave-vector-dependent Onsager coefficient and the Debye function, $g_0(q)$, characterizes the Gaussian chain structure [37, 38]. The chemical

potential μ is comprised of two contributions. The first, bulk contribution is proportional to $S(q)\phi_A(\mathbf{q},t)$, where $S(q)$ is the RPA structure factor at $\chi N = 30$. Since the homogeneous structure is absolutely unstable, $S(q)$ is negative in an interval of wave vectors, resulting in an exponential growth of the concomitant Fourier modes of composition fluctuations ("spinodal self-assembly"). The initial amplitude of composition fluctuations, however, is rather small because the large molecular density, $\sqrt{\bar{\mathcal{N}}}$, strongly suppresses composition fluctuations in the initial state. The second contribution to μ arises from the chemical guiding pattern, and it is proportional to the Fourier transform, $h(\mathbf{q})$, of the surface potential $\exp(-\frac{x^2}{2\epsilon^2})f_s(y,z)$. It gives rise to a linear growth of composition fluctuations with time [39]. In this way the chemical guiding pattern directs the structure formation in the initial stage. [1]
After a time of order τ, one observes the formation of well-segregated domains, inside of which the local densities of A and B segments adopt their saturation values 0 or 1. At the end of the first stage, the overall structure of the copolymer melt is roughly aligned and registered with the guiding pattern but there may be numerous defects.

(ii) *Local defect annihilation and grain formation:* The second stage of structure formation consists in the removal of defects. Defects interact through their distortion of the internal AB interfaces away from the perfectly ordered lamellar structure. These long-range strain fields result in the movement of defects. For instance, dislocations with opposite Burgers vectors attract each other and collide [40, 41, 42]. Upon collision they either annihilate each other or they form a tightly bound, metastable dislocation pair. These metastable defects have a protracted lifetime because the free-energy barrier, ΔF_b, which has to be overcome in order to remove the metastable defect, scales like $\sqrt{\bar{\mathcal{N}}}k_BT$. After the local removal of defects, grains of perfectly ordered regions form. This local defect annihilation completes the DSA process because the chemical guiding pattern registers and aligns the grains with the guiding pattern.

(iii) *Grain coarsening:* In the absence of a guiding pattern, grains are characterized by different orientations and meet at grain boundaries. In that case, there exists a third regime, where the grain structure coarsens by the motion of grain boundaries giving rise to universal power-law behaviour of grain growth [43, 44].

Since we are chiefly interested in the kinetics of DSA we focus on the two initial stages of structure formation in the following.

3.1. Pattern replication

First we study the structure formation of a symmetric diblock copolymer on a L/S pattern whose periodicity exactly matches that of the block copolymer in the bulk at $\chi N = 30$. We use the film thickness $D = \frac{3}{4}L_0 = 1.26R_{e0}$ to maximally frustrate lying lamellae (sandwich structures).

Top-down snapshots and averaged two-dimensional xz density profiles of the copolymer structure at various times are depicted in Figs. 1 and 2. In accord with previous experiments and simulations [9] and the rationale of the previous section, we observe that the guiding pattern rapidly attracts the corresponding blocks and directs the structure formation into a perfectly ordered and registered structure at the chemically patterned substrate within $t = 1000$ SMC

[1] Alternative one could assume that the initial state of structure formation consisted in a solvent-swollen compatible system, $\chi N = 0$, subjected to the interaction with the chemical guiding pattern. In this scenario, the chemical guiding pattern would impose density fluctuations of the form $\phi_A(\mathbf{q}, t = 0) \sim h(\mathbf{q})/S_0(q)$ where S_0 is the structure factor of the disordered system at $\chi N = 0$. These initial density fluctuations would be amplified by the spontaneous structure formation and result in a similar structure evolution, $\phi_A(\mathbf{q},t) - \phi_A(\mathbf{q}, t = 0) \sim q^2 g_0(q)h(\mathbf{q})t$ during the initial stage (neglecting the wave-vector dependence of the structure factor around its extremum). For selected examples we have verified that this alternative starting configuration yields very similar results.

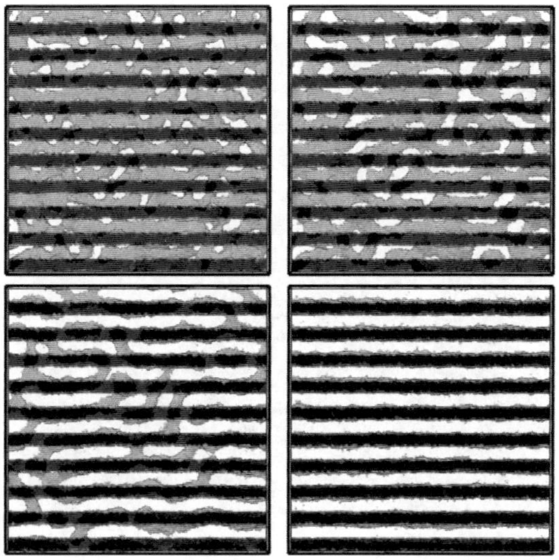

Figure 1. Top-down images of the ordering on a symmetric L/S guiding pattern at $\chi N = 30$. The lateral dimensions are $L_y = L_z = 10L_0$ and the guiding pattern consists of alternating A and B-attractive lines of widths $W/L_0 = 1/2$ and strength $\Lambda N = 2$. The different panels present snapshots at times $t = 1000, 3000, 8000$, and 14000 SMC.

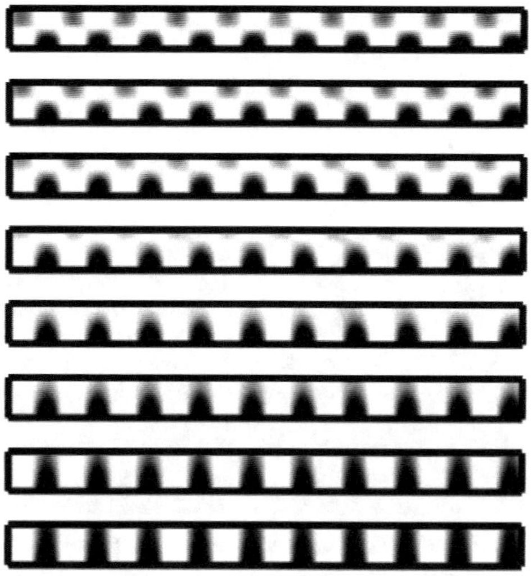

Figure 2. Two-dimensional density profiles for the same system as in Fig. 1 obtained by averaging along the L/S pattern and over 10 independent simulations. The panels correspond to times $t = 1000, 2000, 3000, 4000, 5000, 6000, 8000,$ and 10000 SMC from top to bottom.

$\approx 0.1\tau$. The enrichment of A segments on top of the A-attracting guiding lines gives rise to a concomitant depletion of A segments further away from the substrate. In the averaged side views this checkerboard structure is clearly visible. It consists of a perfectly ordered and registered lamellar grain at the substrate and a top grain that is shifted by half a lamellar period and contains some defects. The perfectly ordered bottom grain and the misaligned top grain are separated by a horizontal grain boundary. From the side views we observe that this horizontal grain boundary gradually moves upwards, i.e., the perfectly ordered bottom grain growths in thickness whereas the misaligned top grain becomes thinner. This process can be conceived as the wetting of the perfectly ordered and registered bottom grain on the chemically guiding pattern. Note that the strength of the chemical guiding pattern is large. The surface free-energy difference $\Delta\gamma$ in Eq. 5 is much larger than the tension of the internal AB interface, $\gamma_{AB}R_{e0}{}^2/k_BT\sqrt{\bar{\mathcal{N}}} \approx \sqrt{\chi N/6}$ and therefore even the liquid of segments will wet the corresponding lines of the chemical pattern. The grain boundary tension typically adopts much smaller values [19, 20]. Eventually the misaligned top grain becomes so thin that the shifted stripes break up into spots [9], and defect-free standing lamellae are established at $t = 14000$ SMC $\approx 1.1\tau$.

This initial spontaneous self-assembly is directed by the guiding pattern. During this ideal pattern replication no long-lived metastable defects are formed, and the entire process illustrates the first stage of DSA – surface-directed spinodal self-assembly.

3.2. Density multiplication

Instead of merely replicating (and improving) the guiding pattern, sparse guiding patterns have been successfully used to produce structures with a smaller characteristic dimension than the guiding pattern [13]. In this second example we study the same system as before but the guiding pattern comprises only 5 A-attracting lines with width $W/L_0 = 1/2$ and a period of $2L_0$, i.e., the line density of the guiding pattern is half than that of the copolymer structure in the bulk. The strength of the A-attracting lines is $\Lambda_s N = 2$ and the substrate area between the lines ("background") does not exhibit any preference for the segment species, $\Lambda_b N = 0$.

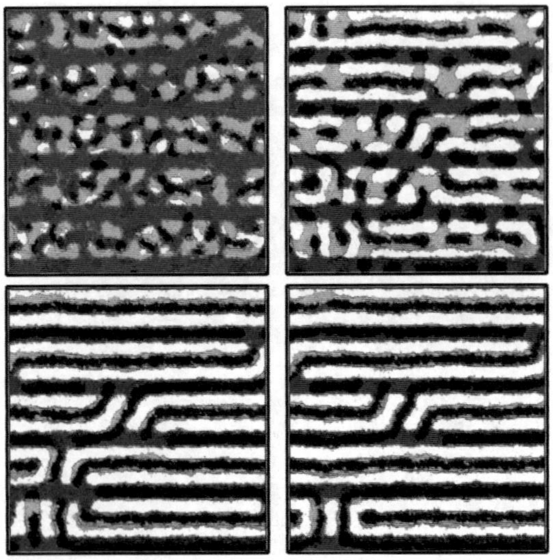

Figure 3. Top-down images of the ordering on a L/S guiding pattern with 2×density multiplication at $\chi N = 30$. The guiding pattern consists of A-attractive lines with $W/L_0 = 1/2$ and $\Lambda N = 2$ on a neutral background. The different panels present snapshots at times $t = 1000, 10000, 100000$, and $2 \cdot 10^6$ SMC.

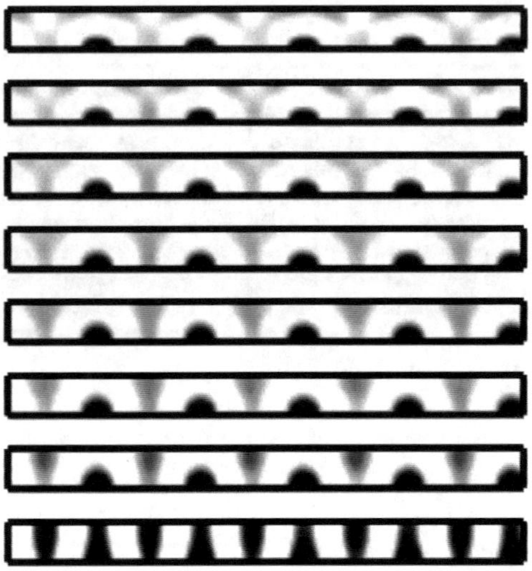

Figure 4. Two-dimensional density profiles for the same system as in Fig. 3 obtained by averaging along the L/S pattern and over 10 independent simulations. The panels correspond to times $t = 1000, 2000, 3000, 4000, 5000, 6000, 10000$, and 100000 SMC from top to bottom.

The kinetics of structure formation is presented in Figs. 3 and 4. The guiding pattern directs the initial spinodal self-assembly. Within 1000 SMC a thin A-domain is established on top of the A-attracting guiding line and the diffusion of A segments towards the lines gives rise to a cylindrical depletion zone that is clearly visible in Fig. 4. Within a molecular relaxation time, τ, a thin, defect-free and registered A-domain is formed on each A-attracting guiding line, and a preference for additional, standing A-lamellae is established *on average* in the middle between the guiding lines.

In the subsequent course of structure formation the thickness of the defect-free and registered A-domains on the A-attracting guiding line increases and standing A-lamellae are established. On the neutral background, however, the lamellae are riddled with defects forming multiple bridges to the neighboring A-lamellae on top of the guiding lines. Upon annealing, the defect density decreases but some defects are persistent. At $t = 300000$ SMC $\approx 33\tau$ none of the 10 independent systems formed a defect-free lamellar structure. We annealed two systems for up to $2 \cdot 10^6$ SMC $\approx 222\tau$ (see also Fig. 8) but tight dislocation pairs and small dislocations remained (cf. last panel of Fig. 3). This observation provides a clear evidence that these defect structures are metastable and that the thermally activated defect-removal process entails a large free-energy barrier, ΔF_b. This is in good agreement with recent self-consistent field calculations studying the minimum free-energy path of defect removal [21].

The equilibrium properties of copolymers on sparse guiding patterns have been carefully investigated by Detcheverry *et al* [14] and Ramirez-Hernandez and co-workers [15]. Efficiently equilibrating the structures by a sophisticated variety of Monte-Carlo moves – single-bead displacements, reptation, whole-chain translation, and swapping of the two blocks – they observed that a weakly B-attractive background increases the stability of defect-free standing lamellae in comparison to alternate structures, and this finding has been corroborated by experiments [14, 15].

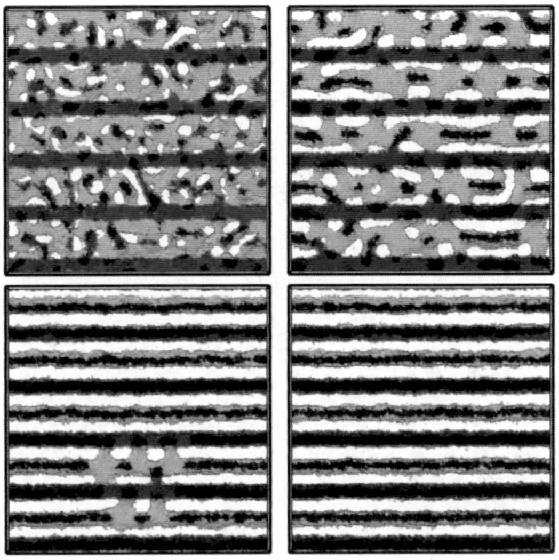

Figure 5. Top-down images of the ordering on a L/S guiding pattern with 2×density multiplication at $\chi N = 30$. The guiding pattern consists of A-attractive lines with $W/L_0 = 1/2$ and $\Lambda N = 2$ on a weakly B-attractive background, $\Lambda_s N = -0.2$. The different panels present snapshots at times $t = 1000, 10000, 100000$, and 200000 SMC.

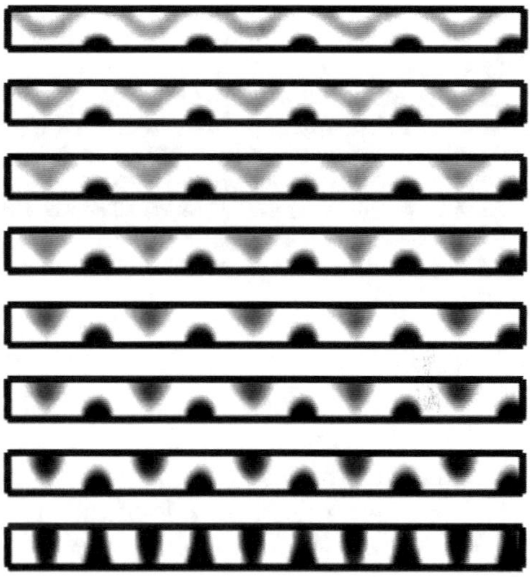

Figure 6. Two-dimensional density profiles for the same system as in Fig. 5 obtained by averaging along the L/S pattern and over 10 independent simulations. The panels correspond to times $t = 1000, 2000, 3000, 4000, 5000, 6000, 10000$, and 100000 SMC from top to bottom.

In Figs. 5 and 6 we present the kinetics of structure formation on a weakly B-attractive background, $\Lambda_b N = -0.2$. Qualitatively the structure formation proceeds similar to the situation with a neutral background. However, the averaged side views in Fig. 6 reveal that at the end of the surface-directed spinodal self-assembly, $t = 10000$ SMC $\approx 1.1\tau$, the copolymer structure is significantly more symmetric. There is an A-domain on top of the strongly A-attractive guiding lines and a similar A-domain at the top of film between the guiding lines (instead of a weak preference for a standing A-lamella for $\Lambda_s N = 0$). This more symmetric averaged structure is less prone to form persistent defects. Typically, defects consists in localized connections between neighboring lines ("micro-bridges", cf. penultimate panel of Fig. 5) that rather rapidly anneal. 7 out of 10 independent simulations achieved defect-free assembly at $t = 200000$ SMC $\approx 22\tau$. The remaining 3 systems featured (i) a micro-bridge, (ii) a partially formed dislocation dipole (see Fig. 7), and (iii) a tight dislocation pair (similar to Fig. 8). The system (i) with the micro-bridge converted into a defect-free structure at time $t = 220000$ SMC, whereas the system (iii) with the tight dislocation pair still retained the defect at $t = 575000$ SMC.

The defect annihilation pathway of the system (ii) with a partial dislocation dipole is

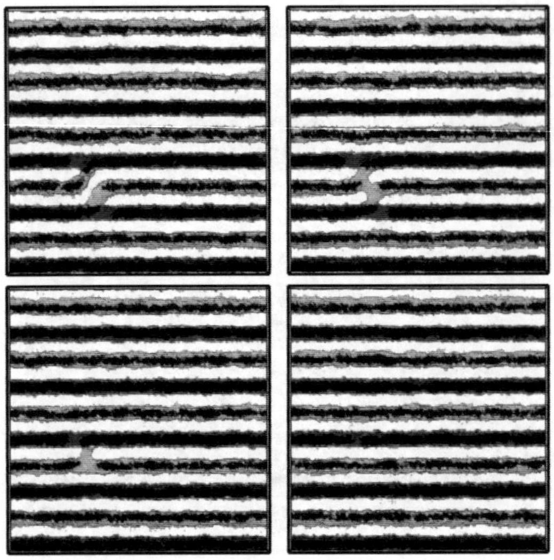

Figure 7. Top-down images of the annihilation of a partially formed dislocation dipole on a L/S guiding pattern with 2×density multiplication (same system as in Fig. 5). The top-down images illustrate the sequence of structures along the pathway of defect annihilation. The panels correspond to times $t = 200000, 388000, 398000$, and 404000 SMC.

illustrated in Fig. 7. At $t = 200000$ SMC there is a dislocation dipole but the lower one of the two A-connections between the A-lamella on the A-attractive line of the guiding pattern and the A-stripe on the weakly B-attractive background is broken at the bottom substrate. Although the dislocation dipole is only partially formed it is quite persistent. At time $t = 388000$ SMC, however, the two ends of the A-stripe on the weakly B-attractive background join at the neutral top surface, and the second, upper A-connection that bridges across the weakly B-attractive background is broken at the bottom substrate. Subsequently, at $t = 398000$ SMC, the lower A-connection vanishes completely and, finally, at $t = 404000$ SMC the defect-free lamellar structure is established. This sequential pathway of defect removal, which we observe in our dynamic particle simulations, is very good agreement with our results on defect annihilation mechanisms using the string method in conjunction with self-consistent field theory [21].

3.3. Process-directed self-assembly

Our self-consistent field theory study of defect annihilation mechanisms in thin films of lamella-forming diblock copolymers [21] demonstrated that there exists an interval of incompatibilities, $\chi_{\mathrm{ODT}}N \approx 10.5 < \chi N < \chi N_*$, between the order-disorder transition (ODT) and a threshold χN_* where two conditions are met: (1) The excess free energy of a defect is large compared to the thermal energy scale, $\Delta F_{\mathrm{d}} \gg k_{\mathrm{B}}T$, and consequently defects are not spontaneously generated in a defect-free structure by thermal fluctuations and (2) the free-energy barrier ΔF_{b} for defect annihilation is comparable to $k_{\mathrm{B}}T$ and thus defects are not even metastable but spontaneously annihilate. This interval of incompatibility in the vicinity of the order-disorder transition is very favorable for defect-free directed self-assembly. In the absence of a guiding pattern, the minimum free-energy path calculations predict $\chi N_* \approx 18$, and a chemical guiding pattern extends this favorable parameter window to large values of incompatibility.

To test this prediction of mean-field theory by our particle simulations we quench the two structures that formed persistent defects on guiding patterns with 2×density multiplication on a neutral background (see Fig. 3) at $t = 1.3 \cdot 10^6$ SMC $\approx 145\tau$ from $\chi N = 30$ to 20. Although there is a significant mismatch of -6% between the half period of the sparse guiding pattern and the bulk lamellar spacing at $\chi N = 20$, defects readily annihilated within 30000 SMC $\approx 3.3\tau$. Reestablishing the original incompatibility $\chi N = 30$ at $t = 1.33 \cdot 10^6$ SMC we observe that the segregation between A and B-domains increases and the line-edge roughness of the AB interfaces decreases but no defects are formed.

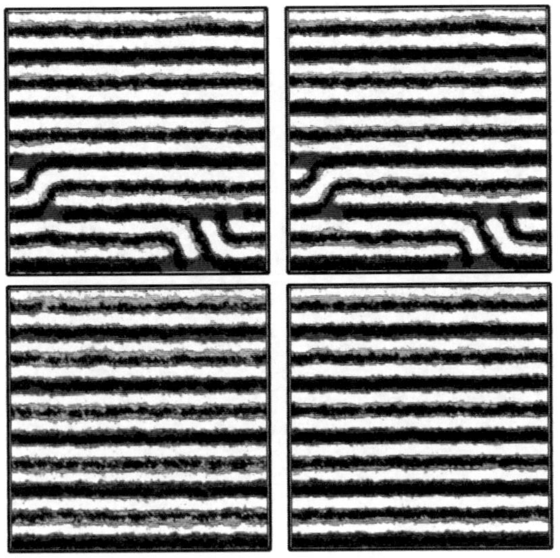

Figure 8. Top-down images of the ordering on a L/S guiding pattern with 2×density multiplication. The guiding pattern consists of A-attractive lines with $W/L_0 = 1/2$ and $\Lambda N = 2$ on a neutral background, $\Lambda_s N = 0$. The system was quenched from a disordered state to $\chi N = 30$ far below the ODT, and subsequently annealed for $1.3 \cdot 10^6$ SMC. This long annealing did not result in defect free-ordering. Then the system was quenched closer to the ODT, $\chi N = 30 \to 20$. Annealing for only $3 \cdot 10^4$ SMC in this DSA-favorable parameter window removed all defects. After increasing the incompatibility back to its original value $\chi N = 30$ t $t = 1.33{\cdot}10^6$ SMC the structure remains defect-free. The different panels present snapshots at $t = 5 \cdot 10^5, 1.3 \cdot 10^6, 1.33 \cdot 10^6$, and $1.5 \cdot 10^6$ SMC.

4. Summary and discussion

Using a soft, coarse-grained model of block copolymer melts in conjunction with efficient, parallel simulation techniques we have studied the kinetics of structure formation of lamella-forming copolymer films on chemical guiding patterns. Both the replication of the guiding pattern and two-fold density multiplication have been investigated in the limit of perfect commensurability between the bulk lamellar structure and the L/S guiding pattern. Defect-free pattern replication is driven by the pattern-directed spinodal self-assembly. It is robust and only requires a time, τ, that is comparable to the longest time of conformational relaxation. Density multiplication, in turn, requires significantly longer annealing times at $\chi N = 30$, and it sensitively depends on the DSA process parameters. In our simulation study we have illustrated the dependence on incompatibility and strength of the surface interactions. In agreement with experiments and previous simulations that focused on the equilibrium properties [14, 15], we observe that sparse guiding patterns of A-attractive lines with a background that is weakly attractive to the other species, B, are preferable to neutral backgrounds. Our simulations indicate that such a weak background preference results in a more symmetric structure at the end of the initial pattern-directed spinodal self-assembly (i.e. after a time τ), and thereby reduces the number of metastable, persistent defects. Our simulations also highlight the benefits of performing the DSA process in the vicinity of the order-disorder transition and thereby corroborate our recent self-consistent field study on defect annihilation mechanisms [21].

The tailoring of the time dependence of thermodynamic process variables ("process-directed self-assembly" [45]) – here, the incompatibility χN – opens additional opportunities for directing the structure formation in block copolymer systems and has also been found to be a promising strategy for fabricating bicontinuous network structures [46]. This strategy relies on the observation that the free-energy landscape changes with the external thermodynamic process conditions and that pathways of collective structure changes sensitively depend on the structure of the complex free-energy landscape.

In an experiment the incompatibility can be changed by temperature if the Flory-Huggins parameter χ exhibits a significant temperature dependence or by adding a co-solvent. Solvent tempering is a common strategy for defect removal. In the simplest case, the solvent uniformly dilutes the repulsive interactions between unlike segment species [47, 48]. This results in a reduction of the effective incompatibility and, additionally, speeds up the segmental dynamics ("plasticizing effect") and may also alter the equilibrium period of the microphase-separated structure. In an experiment it may be difficult to disentangle these different effect, and our simulation study highlights the specific role of the free-energy landscape.

Acknowledgments

It is a great pleasure to thank Juan J. de Pablo, Paul F. Nealey, and Mark P. Stoykovich for stimulating discussions and fruitful collaborations. The authors wish to acknowledge financial support from the European Union FP7 under grant agreement No.619793 CoLiSA.MMP (Computational Lithography for Directed Self-Assembly: Materials, Models, and Processes) and the Deutsche Forschungsgemeinschaft under grant Mu1674/11. The simulations have been enabled by the generous access to the computing facilities at the GWDG Göttingen, the HLRN Hannover/Berlin and the Jülich Supercomputing center.

References

[1] Leibler L 1980 *Macromolecules* **13** 1602–1617
[2] Fredrickson G H and Bates F S 1996 *Ann. Rev. Mater. Sci.* **26** 501–550
[3] Stoykovich M P and Nealey P F 2006 *materials today* **9** 20–29
[4] Bita I, Yang J K W, Jung Y S, Ross C A, Thomas E L and Berggren K K 2008 *Science* **321** 939–943
[5] Hawker C J and Russell T P 2005 *MRS Bulletin* **30** 952–966
[6] Bang J, Jeong U, Ryu D Y, Russell T P and Hawker C J 2009 *Advanced Materials* **21** 4769–4792
[7] Bates C M, Maher M J, Janes D W, Ellison C J and Willson C G 2014 *Macromolecules* **47** 2–12
[8] Kim S O, Solak H H, Stoykovich M P, Ferrier N J, de Pablo J J and Nealey P F 2003 *Nature* **424** 411–414
[9] Edwards E W, Stoykovich M P, Müller M, Solak H H, de Pablo J J and Nealey P F 2005 *J. Polym. Sci. B: Polymer Physics* **43** 3444–3459
[10] Edwards E W, Müller M, Stoykovich M P, Solak H H, de Pablo J J and Nealey P F 2007 *Macromolecules* **40** 90–96
[11] Daoulas K C, Müller M, Stoykovich M P, Kang H, de Pablo J J and Nealey P F 2008 *Langmuir* **24** 1284–1295
[12] Stoykovich M P, Daoulas K C, Müller M, Kang H M, de Pablo J J and Nealey P F 2010 *Macromolecules* **43** 2334–2342
[13] Ruiz R, Kang H M, Detcheverry F A, Dobisz E, Kercher D S, Albrecht T R, de Pablo J J and Nealey P F 2008 *Science* **321** 936–939
[14] Detcheverry F A, Liu G L, Nealey P F and de Pablo J J 2010 *Macromolecules* **43** 3446–3454
[15] Liu C C, Ramirez-Hernandez A, Han E, Craig G S W, Tada Y, Yoshida H, Kang H, Ji S, Gopalan P, de Pablo J J and Nealey P F 2013 *Macromolecules* **46** 1415–1424
[16] Li W H, Müller M 2015 *Annu. Rev. Chem. Biomol. Eng* in press
[17] Jeong S-J, Kim J Y, Kim B H, Moon H-S and Kim S O 2013 *Materials Today* **16** 468–476
[18] Nagpal U, Müller M, Nealey P F and de Pablo J J 2012 *ACS Macro Letters* **1** 418–422
[19] Duque D, Katsov K and Schick M 2002 *J. Chem. Phys.* **117** 10315–10320
[20] Müller M, Daoulas K C and Norizoe Y 2009 *Phys. Chem. Chem. Phys.* **11** 2087
[21] Li W H, Nealey P F, de Pablo J J and Müller M 2014 *Phys. Rev. Lett.* **113** 168301
[22] Daoulas K C, Müller M, de Pablo J J, Nealey P F and Smith G D 2006 *Soft Matter* **2** 573–583
[23] Müller M 2011 *J. Stat. Phys.* **145** 967–1016
[24] Detcheverry F A, Pike D Q, Nealey P F, Müller M and de Pablo J J 2009 *Phys. Rev. Lett.* **102** 197801
[25] Müller M, Steinmüller B, Daoulas K C, Ramirez-Hernandez A and de Pablo J J 2011 *Phys. Chem. Chem. Phys.* **13** 10491–10502
[26] Geisinger T, Müller M and Binder K 1999 *J. Chem. Phys.* **111** 5241–5250
[27] Müller M 2012 *Phys. Rev. Lett.* **109** 087801
[28] Rossky P J, Doll J D and Friedman H L 1978 *J. Chem. Phys.* **69** 4628–4633
[29] Daoulas K C and Müller M 2006 *J. Chem. Phys.* **125** 184904
[30] Müller M and Daoulas K C 2008 *J. Chem. Phys.* **129** 499–503
[31] Ramirez-Hernandez A, Müller M and de Pablo J J 2013 *Soft Matter* **9** 2030–2036
[32] Tuckerman M, Berne B and Martyna G 1992 *J. Chem. Phys.* **97** 1990
[33] Müller M and Daoulas K C 2011 *Phys. Rev. Lett.* **107** 227801
[34] deGennes P G 1979 *Scaling concepts in polymer physics* (Cornell University Press, Ithaca)
[35] Hohenberg P C and Halperin B I 1977 *Rev. Mod. Phys.* **49** 435–479

[36] Cross M C and Hohenberg P C 1993 *Rev. Mod. Phys.* **65** 851–1112
[37] Binder K 1983 *J. Chem. Phys.* **79** 6387–6409
[38] Reister E, Müller M and Binder K 2001 *Phys. Rev. E* **64** 041804
[39] Reister E and Müller M 2003 *J. Chem. Phys.* **118** 8476–8488
[40] Peach M and Koehler J S 1950 *Phys. Rev.* **80** 436–439
[41] Pershan P S 1974 *J. Appl. Phys.* **45** 1590–1604
[42] Hahm J and Sibener S J 2001 *J. Chem. Phys.* **114** 4730–4740
[43] Harrison C, Adamson D H, Cheng Z, Sebastian J M, Sethuraman S, Huse D A, Register R A and Chaikin P M 2000 *Science* **290** 1558–1560
[44] Harrison C, Cheng Z, Sethuraman S, Huse D A and Chaikin P M 2002 *Phys. Rev. E* **66** 011706
[45] Müller M and de Pablo J J 2013 *Annu. Rev. Mater. Sci.* **43** 1–34
[46] Müller M and Sun D W 2013 *Phys. Rev. Lett.* **111** 267801
[47] Naughton J R and Matsen M W 2002 *Macromolecules* **35** 5688–5696
[48] Rudov A A, Patyukova E S, Neratova I V, Khalatur P G, Posselt D, Papadakis C M and Potemkin I I 2013 *Macromolecules* **46** 5786–5795

Apparent Viscosity of Active Nematics in Poiseuille Flow

Zhenlu Cui[1] Jianbing Su[2]and Xiaoming Zeng[3]
[1] The Department of Mathematics and Computer Science, Fayetteville State University, Fayetteville, NC 28301, USA
[2]School of Mathematics and Statistics, Jiangsu Normal University, Xuzhou, 221116, China
[3]School of Mathematical Sciences, Xiamen University, Xiamen, 361005, China

E-mail: zcui@uncfsu.edu

Abstract. A Leslie-Erickson continuum hydrodynamic for flowing active nematics has been used to characterize active particle systems such as bacterial suspensions. The behavior of such a system under a plane pressure-driven Poiseuille flow is analyzed. When plate anchoring is tangential and normal, we find the apparent viscosity formula indicating a significant difference between tangential anchoring and normal anchoring conditions for both active rodlike and discoid nematics.

1. Introduction

Active nematics are non-equilibrium complex fluids composed of active or self-propelled particles showing a nematic phase and have attracted much attention in both theoretical modeling [1-3, 5-8] and experiments [9, 11] in the past decade. In these systems, active particles exert forces on the surrounding fluid, resulting in local extensile or contractile stresses proportional to the amount of orientational tensor: $t^{a} = \delta \mathbf{nn}$ where is the orientation director in the nematic phase which is generally described by a unit vector, and δ is the amplitude of the dipolar forces exerted by the particle to swim. The sign of δ determines whether the dipolar flow field generated by the active particles is extensile ($\delta < 0$) or contractile ($\delta > 0$). In the swimmer literature, the former situation describes "pushers", i.e., most bacteria including E. Coli and Bacillus Subtilis, while the latter corresponds to "pullers" including Chlamydomonas.

Although theoretical modeling efforts [1-3, 5-8] have been made in understanding active nematics, almost all these studies are only focused on sheared systems. Active nematics in Poiseuille flow are

Corresponding author, Email:zcui@uncfsu.edu

scarcely studied [10, 12]. Here, we will address rheology of active nematics in Poiseuille flow and focus on the apparent viscosity.

2. The governing equations for flowing active nematics

We adopt the Erickson-Leslie-Parodi model [1, 2], in which the nematic order parameter is a fixed magnitude unit vector field $\vec{\mathbf{n}}$ which evolves according to

$$\frac{\partial \vec{\mathbf{n}}}{\partial t} + \vec{\mathbf{v}} \cdot \nabla \vec{\mathbf{n}} = \lambda \mathbf{D} \cdot \vec{\mathbf{n}} - \mathbf{\Omega} \cdot \vec{\mathbf{n}} + \Gamma \vec{\mathbf{h}} \tag{1}$$

Where $\vec{\mathbf{v}}$ is the velocity field of the solvent; λ is the flow alignment parameter, Γ is a rotational viscosity; $\vec{\mathbf{h}} = K \Delta \vec{\mathbf{n}}$ is the molecular field and K is the Frank elasticity constant (single constant approximation), $\mathbf{\Omega} = \frac{1}{2}(\nabla \vec{\mathbf{v}} - \nabla \vec{\mathbf{v}}^{\mathrm{T}})$ and $\mathbf{D} = \frac{1}{2}(\nabla \vec{\mathbf{v}} + \nabla \vec{\mathbf{v}}^{\mathrm{T}})$ are the rate-of vorticity and the rate-of -strain tensors, respectively.

The first two terms on the right-hand side of Eq. (1) describe alignment (or tumbling) of the director field by local shear flow. The third term accounts for the tendency of the ordered nematic to resist distortions, and arises ultimately from excluded volume interactions between individual particles. The flow field $\vec{\mathbf{v}}$ obeys the Navier-Stokes equation

$$\rho \left(\frac{\partial \vec{\mathbf{v}}}{\partial t} + \vec{\mathbf{v}} \cdot \nabla \vec{\mathbf{v}} \right) = -\frac{\partial p}{\partial x} \mathbf{I} + \nabla \cdot (\boldsymbol{\tau} + 2\eta \mathbf{D}) \tag{2}$$

with the continuity equation $\nabla \cdot \vec{\mathbf{v}} = 0$ to guarantee incompressibility, the stress tensor given by the passive and active contributions, $\boldsymbol{\tau} = t^p + t^a$

$$t^p = -\frac{\lambda}{2}[\vec{\mathbf{n}}\vec{\mathbf{h}} + (\vec{\mathbf{n}}\vec{\mathbf{h}})^T] + \frac{1}{2}[\vec{\mathbf{n}}\vec{\mathbf{h}} - (\vec{\mathbf{n}}\vec{\mathbf{h}})^T \tag{3}$$

where ρ is the fluid density and η is the viscosity and λ is the flow-alignment parameter. The magnitude of λ controls how the director field responds to a shear flow. $|\lambda| > 1$ corresponds to flow aligning regime in which the director tends to align to the flow direction at the Leslie alignment angle $\theta_L = \frac{1}{2}\cos^{-1}\frac{1}{\lambda}$ while $|\lambda| < 1$ corresponds to flow tumbling regime in which the director continuously rotates under shear. The value of λ is mainly determined by the shape of the active particles. $\lambda > 0$ corresponds to a rod-shaped particle, $\lambda < 0$ for a disc-shaped particle and $\lambda = 0$ for a spherical particle. The active contribution can be expressed as [2, 5],

$$t^a = \delta \vec{\mathbf{n}}\vec{\mathbf{n}} \tag{4}$$

The sign of δ determines whether the particles are extensile ($\delta < 0$) or contractile ($\delta > 0$).

3. Plane Poiseuille flow

We consider a plane Poiseuille flow between two parallel plates located at $y = \pm H$ with an imposed small pressure gradient $\frac{\partial p}{\partial x}$, where $\frac{\partial p}{\partial x}$ is constant and negative. We assume strong particle anchoring and the velocity boundary condition no-slip at the plates given by

$$\mathbf{n}_0 = (\cos\varphi_0, \sin\varphi_0, 0),$$
$$v(\pm H) = 0 \tag{5}$$

where $\mathbf{n}_0$ is the initial director across the plates in absence of Poiseuille flow. Figure 1 depicts the cross section of the shear flow on the (*x*, *y*) plane. We consider the in-plane orientation of an active nematic confined to the plane (*x*, *y*) and strong anchoring boundary conditions. The director $\vec{\mathbf{n}}$ confined to the plane (*x*, *y*) and parameterized by the director angle, $\vec{\mathbf{n}} = (cos\varphi, sin\varphi, 0)$. Variations in the direction of flow (*x*) and primary vorticity direction (*z*) are suppressed.

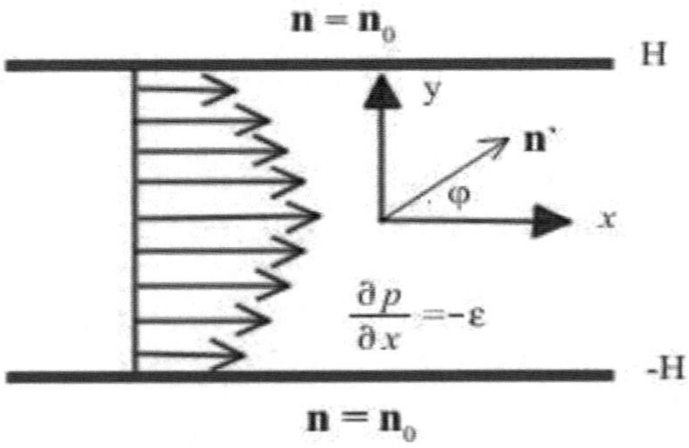

Figure 1. The geometry of the plane Poiseuille flow. The boundary anchoring for the director is $\mathbf{n}_0 = (\cos\varphi_0, \sin\varphi_0, 0)$ and the velocity boundary condition is no-slip $v(\pm H) = 0$.

We consider a uniform plate anchoring boundary condition, either parallel to the flow direction, called tangential anchoring, i.e. $\varphi_0 = 0$, or perpendicular to the plates, called normal or homeotropic anchoring, i.e. $\varphi_0 = \pi/2$. We linearize the governing system of equations with respect to $-\frac{\partial p}{\partial x}$ with the boundary conditions given by (5). The linearized system is

$$\frac{\partial \varphi}{\partial t} = A\frac{\partial^2 \varphi}{\partial y^2} + B\frac{\partial v_x}{\partial y},$$
$$\rho\frac{\partial v_x}{\partial t} = 1 + \frac{\partial \tau_{xy}}{\partial y}, \tag{6}$$
$$\tau_{xy} = C\frac{\partial^2 \varphi}{\partial y^2} + D\frac{\partial v_x}{\partial y} + \delta\varphi$$

Where $A=\Gamma K$, $C=-KB$, $D=\eta$, and $B = \begin{cases} \frac{\lambda-1}{2}, & if\ \varphi_0 = 0 \\ -\frac{\lambda+1}{2}, & if\ \varphi_0 = \frac{\pi}{2} \end{cases}$

3. Steady states and apparent viscosity

In order to establish the apparent viscosity formula, we need the steady state solutions of the system (6) in the confined geometry which are given by

$$v_x = \frac{AHr\coth ry}{B\delta}\left(1-\frac{\cosh ry}{\cosh rH}\right), \qquad \varphi = \frac{H}{\delta}\left(\frac{\sinh ry}{\sinh rH}-\frac{y}{H}\right) \tag{7}$$

where $r = \begin{cases} \sqrt{\frac{(\lambda-1)\delta}{2(AD-BC)}}, & if\ \varphi_0 = 0 \\ \sqrt{\frac{-(\lambda+1)\delta}{2(AD-BC)}}, & if\ \varphi_0 = \frac{\pi}{2} \end{cases}$. The apparent viscosity defined by $\eta_{app} = \dfrac{2H^3}{3F}$ [4], where $F = \int_0^H v_x(y)dy = \dfrac{AH}{B\delta}(rH\coth rH - 1)$ is the flow rate per unit length. The resulting apparent viscosity is given by

$$\eta_{app} = \frac{2H^2B\delta}{3A(rH\coth rH - 1)} \tag{8}$$

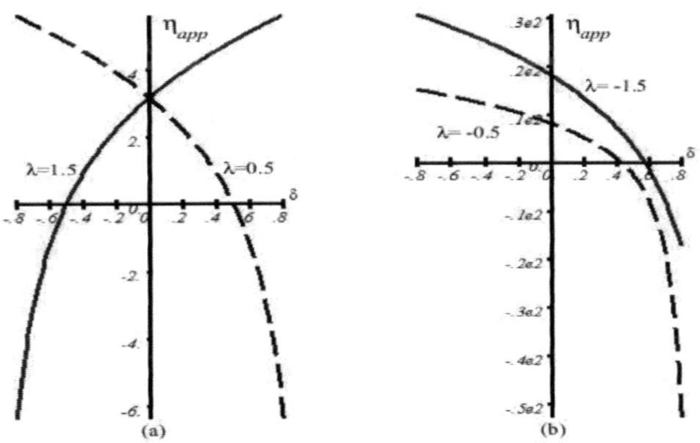

Figure 2. The apparent viscosity versus the activity for rodlike and discoid swimmers in flow-aligning (solid line) and tumbling regimes (dashed line) at tangential anchoring conditions. The values of parameters are: $\Gamma = 0.2$, $K = 0.04$, $\eta = 1.27$ and $H = 1$.

Figure 2 shows how the apparent viscosity depends on the activity for rodlike and discoid swimmers at tangential anchoring boundary conditions. We summarize the noticeable features below.

- The apparent viscosity of a flow-aligning rodlike puller system or a tumbling rodlike pusher system is thickened by the activity while the apparent viscosity of a flow-aligning rodlike pusher or a tumbling rodlike puller system is thinned by the activity.

- The apparent viscosity of a discoid system is always thinned by the activity.
- The apparent viscosity could be negative for all rodlike systems but only for discoid fuller system in a high activity regime.

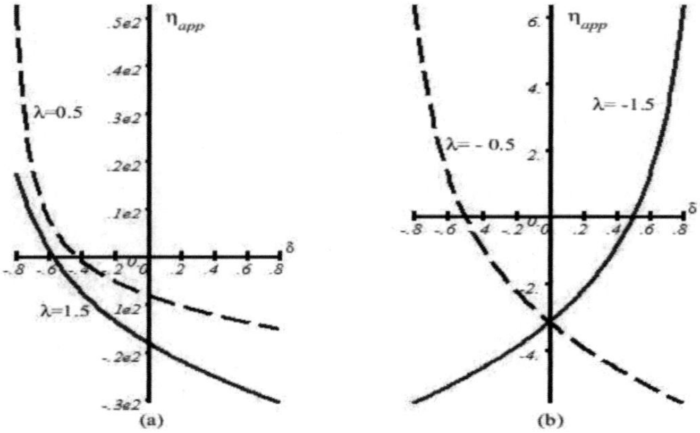

Figure 3. The apparent viscosity versus the activity for rodlike and discoid swimmers in flow-aligning (solid line) and tumbling regimes (dashed line) at normal anchoring conditions. The values of parameters are: $\Gamma = 0.2$, $K = 0.04$, $\eta = 1.27$ and $H = 1$.

Figure 3 shows how the apparent viscosity depends on the activity for rodlike and discoid swimmers at normal boundary conditions. We summarize the noticeable features below.

- The apparent viscosity of a rodlike system is always thinned by the activity.
- The apparent viscosity of a flow-aligning discoid system is thickened by the activity while the apparent viscosity of a tumbling discoid system is thinned by the activity.
- The apparent viscosity could be negative for all rodlike and discoid system in a high activity regime.

We notice that the apparent viscosity of rodlike systems is enhanced by one order of magnitude for normal relative to parallel anchoring; however the apparent viscosity of discoid systems is diminished one order of magnitude for normal relative to parallel anchoring.

Common swimming bacteria, such as Bacillus Subtilis, E. Coli, and many others, rodlike swimmers (length about ~ $5\mu m$, diameter of the order of ~$1\mu m$) are flow-aligning rodlike pushers. The apparent viscosity of these systems is thinned by the activity for tangential anchoring. In contrast, Chlamydomonas Reinhardtii are tumbling pullers. The apparent viscosity of such systems is thinned for a tangential anchoring.

4. Conclusions

We have used the Leslie-Ericken –Parodi continuum theory to model flowing active nematics. We establish the apparent viscosity formula active nematics subject to a plane Poiseuille flow. The results are consistent with the previous results on rodlike swimmers and discoid swimmers, experimentally and theoretically. We look forward to tests of other predictions in experiments on active nematics in future.

Acknowledgment

Zhenlu Cui's work is partially grant supported by the National Science Foundation through grant HRD-1436120. Jianbian Su's work is supported in part by the National Science Foundation of China through rant No. 11171285. Xiaoming Zeng's work is supported in part by the National Science Foundation of China through grant No. 61170324.

References

[1] Cui Z Zeng X and Su J 2014 *Steady States of Sheared Active Nematics* Advances in Applied Mathematics and Mechanics 6 p 75-86

[2] Cui Zhenlu and Zeng X 2012 *Rheology of Sheared Bacterial Suspensions* IMA 155 p 217

[3] Cui Zhenlu 2011 *Weakly Sheared Active Suspensions: Hydrodynamics, Stability and Rheology* Phys. Rev. E 83 p 031911

[4] Cui Z Calderer M C and Wang Q 2006 *Mesoscale structure in flows of weakly sheared cholesteric liquid crystalline polymers* Discrete and Continuous Dynamical Systems-Series B 6 p 291

[5] Fielding SM Marenduzzo D Cates ME 2011 *Nonlinear dynamics and rheology of active fluids: Simulations in two dimensions* Phys. Rev. E 83 p 041910

[6] Giomi L et al 2010 *Sheared active fluids: Thickening, thinning, and vanishing viscosity* Phys. Rev. E 81 p 051908

[8] Hatwalne Y et al 2004 *Rheology of active-particle suspensions* Phys. Rev. Lett. 92 p 118101

[9] Rafaï S Jibuti L and Peyla P 2010 *Effective Viscosity of Microswimmer Suspensions* Phys. Rev. Lett. 104 p 098102

[10] Ravnik M and Yeomans J M 2013 *Confined Active Nematic Flow in Cylindrical Capillaries* Phys. Rev. Lett. 110 p 026001

[11] Sokolov A et al 2009 *Reduction of Viscosity in Suspension of Swimming Bacteria* Phys. Rev. Lett. 103 p 148101

[12] Zöttl A and Stark H 2012 *Nonlinear dynamics of a microswimmer in Poiseuille flow* Phys. Rev. Lett. 108 p 218104

Journal of Physics: Conference Series **640** (2015) 012012 doi:10.1088/1742-6596/640/1/012012

Monte Carlo parameterization in the VirtualLeaf framework

A Dzhurakhalov[1,2], D De Vos[1,2], G Beemster[2] and J Broeckhove[1]

[1]Department of Mathematics and Computer Science, University of Antwerp, Belgium
[2]Department of Biology, University of Antwerp, Belgium

E-mail: abdiravuf.dzhurakhalov@uantwerpen.be

Abstract. The recently developed simulation framework VirtualLeaf uses Metropolis Monte Carlo dynamics for studying plant tissue morphogenesis. Minimizing the energy of the tissue is done by an energy evaluation-only method. We developed a more robust criterion for the energy minimization method for the multivariable and complex systems where the use of a gradient norm is impossible. The proposed criterion is based on checking energy changes in a sliding window of successive energy steps against a threshold value. The advantages of the sliding window criterion are discussed and results obtained by this method are presented. The impact of the choice threshold value on the energy minimization has been studied.

1. Introduction

1.1 Modeling the cell behavior in VirtualLeaf

As indicated in [1, 2], the process of the equilibration of the water relations in plant tissue is significantly faster than turgor regulation and growth processes. That is why the system can be considered in a state of equilibrium at any given time [3, 1, 4]. The process of the equilibration is the balance between intracellular turgor pressure (e.g. due to the water uptake) and counteracting of the cell wall to this pressure. For modelling this state one can directly use the stress-strain relationship using the tensile force within cell walls as done in the CellModeller framework [5] or the more general and widely used energy formalism [1, 3, 6] based implicitly on the force as applied in the VirtualLeaf framework [3].

In VirtualLeaf a tissue is represented by a mesh of cells where a cell is a polygon surrounded by walls (Fig. 1). The wall consists of edges linked by nodes. In VirtualLeaf the cell and tissue behaviors are represented within a generalized energy, i.e. Hamiltonian [3]:

$$E = \lambda_A \sum_{cells} (a_i - A_i)^2 + \lambda_E \sum_{edges} (l_j - L_j)^2, \qquad (1)$$

where the first term represents the turgor in the cell and the second term represents the resistance of the edges to the turgor, a_i - actual cell area and l_j - actual edge length, A_i - cell area in the absence of counteracting cell wall forces (target area) and L_j - edge lengths of cell wall in the absence of turgor pressure, λ_A - cell compression or expansion resistance parameter and λ_E is a spring constant of the edge.

1.2. Energy minimization in VirtualLeaf

The stable state of the system, i.e. the balance between turgor pressure and cell wall resistance at a certain time is defined by the minimization of the generalized potential. In VirtualLeaf, minimizing the energy of the system is done using the Metropolis Monte Carlo method [7, 8] as shown by the flowchart in Fig.1. One node from the set of nodes is randomly selected and is displaced by distance D in a random direction. It causes geometric changes both in actual area of cells and in actual length of edges that leads to an energy change of the system. The displacement is always accepted if it leads to an energy drop, i.e. $\Delta E_i < 0$. Otherwise, it is rejected. Displacements resulting in an energy increase are accepted with the Boltzmann probability $P(\Delta E_i) = e^{-\Delta E_i/T}$, where T is a parameter of random fluctuation as a noise and rejected otherwise. The conditional acceptance of some positive energy displacements helps avoid getting stuck in local minima.

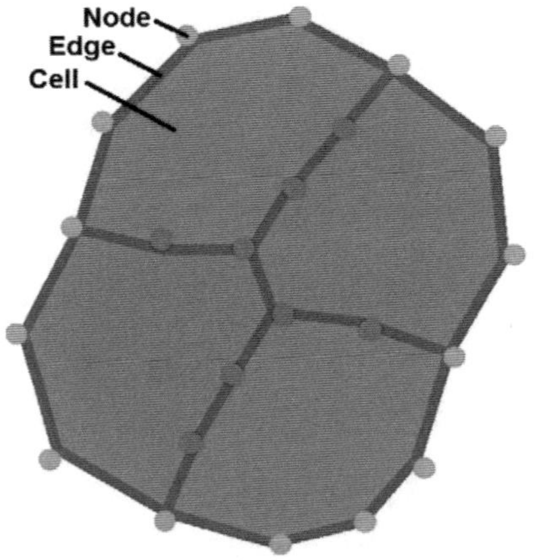

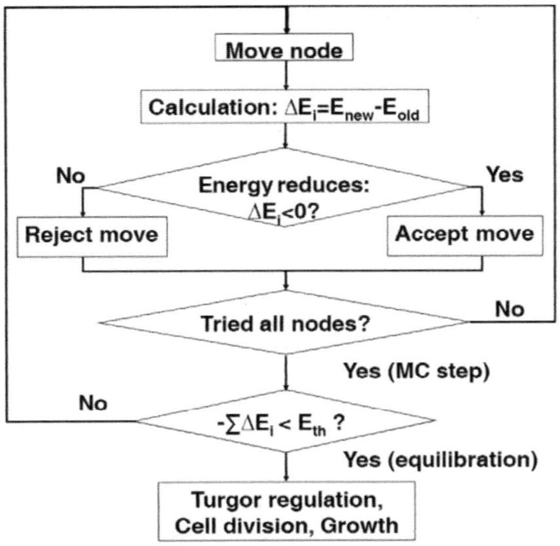

Figure 1. Representation of the tissue (left) and MC energy minimization flowchart (right) in VirtualLeaf.

In this way, iteratively all nodes are attempted to be displaced and it is called one Monte Carlo (MC) step if the number of attempts is equal to the number of nodes in the whole system. The MC step is repeated until the desirable equilibration state of the system, i.e. a sufficient balance between the turgor pressures and cell walls resistances, is reached. At the end of each MC steps the energy change is evaluated by

$$-\sum_{nodes} \Delta E_i < E_{th}, \qquad (2)$$

where E_{th} represents the tolerance of the $|\sum \Delta E_i|$ convergence. If the system satisfies this criterion, then the slower biological processes such as the turgor pressure increase, cell division and cell growth are performed. Otherwise, again the displacements of all nodes are attempted to reach the desirable state of the equilibration.

However, this criterion based on an energy evaluation only method cannot result in the desirable equilibration of the system. That is why our main aim is to elaborate a more robust method for energy minimization in VirtualLeaf.

2. Results and discussion

2.1. Analysing the criteria for energy minimization

One of the robust methods typically used for the energy minimization of a system is the derivative method based on the first derivative of the energy with respect to the geometry of the system when it is available. However, the energy derivatives are not available for the Hamiltonian in Eq. (1) due to its multi-variability, as it consists of the contributions to the energy from various terms, the large number of cells and cell walls. That is why in VirtualLeaf a non-derivative method, i.e. an energy evaluation only method or, more precisely, a "function-only method" [9] is used as a convergence criterion for minimization.

Let's analyse the possible consequences of the VirtualLeaf criterion in an actual simulation. In addition to the VirtualLeaf criterion, we also use the criterion of a predefined maximum number of MC steps that can be configured by the VirtualLeaf user. We have studied the minimization convergence of small (4-cell) and big (287-cell) tissues with 11000 MC steps for the case of the Auxin Growth model of VirtualLeaf. In Fig. 2 we present the total energy of these tissues versus the number of MC steps during one time step. One can see the decrease of the total energy of the tissue during energy minimization, i.e. when the number of MC steps increases. We should note that when the minimization is close to convergence the energy changes from MC step to MC step should be very small, i.e. the curve in Fig. 2 comes to a plateau. We observed quite fast convergence (~100 MC steps) in the case of the small tissue and slower convergence for the big tissue.

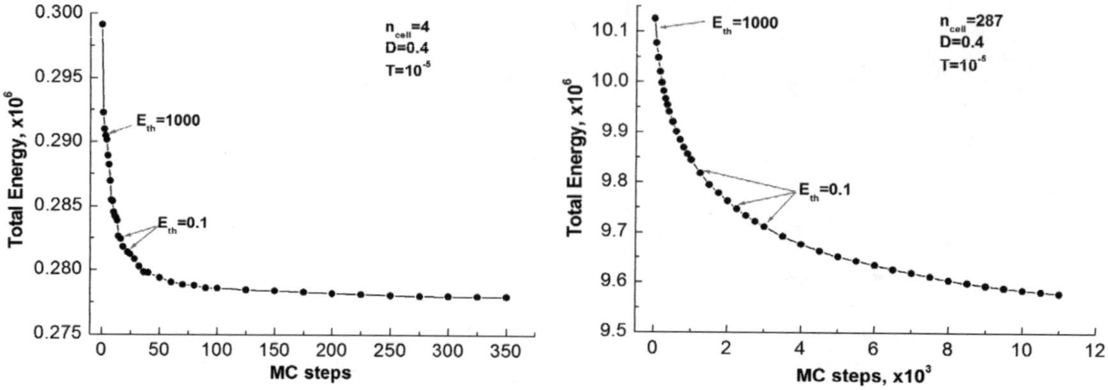

Figure 2. Total energy of the 4-cell (left) and 287-cell (right) tissues versus the MC steps. Arrows correspond to different random seeds.

We also performed MC simulation using the energy evaluation only criterion. When we use E_{th}=1000, it resulted in very early ending of the MC calculation in both cases. With this criterion the system supposedly has reached convergence, though in fact it was far from a converged minimum. As this value of E_{th} was quite large, we have also used E_{th} = 0.1 which is sufficiently small. In this case the system was closer to the converged minimization. However, the convergence obtained with this tolerance became more dependent on the random seeds, which causes some doubt about the robustness of the criterion used in VirtualLeaf.

To find the reason of such sensitivity we studied the energy minimization and convergence of the 287-cell tissue with 11000 MC steps for three random seeds. We present the results of this study in Fig. 3, left. There is almost no difference in these dependencies within the entire interval of MC steps performed which is, indeed, inherent to the stochastic processes. Then we analysed the evolution of the energy difference during energy minimization cycle which we present in Fig. 3, right. We would observe a gradual decreasing of the curve, if we use a gradient of energy.

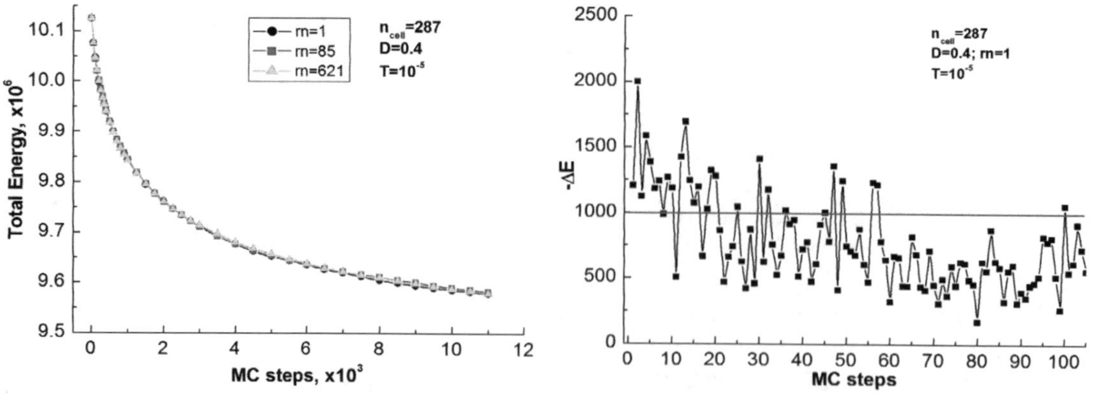

Figure 3. Total energy at various random number (*rn*) seeds (left) and the energy difference (right) versus the number of MC steps for the 287-cell tissue.

However, as the $\sum \Delta E_i$ is used directly instead of any gradient, it increases and decreases with a large deviation even if its mean value decreases with increasing MC steps. As a result, the convergence criterion can be met at any MC steps where $|\Delta E| < E_{th}$. In VirtualLeaf the convergence criterion is considered satisfied at very early MC steps, for instance at *MC step* = 8 (which is sensitive to the random seed), though until *MC step* = 40 the mean value of $|\Delta E|$ is large than E_{th} =1000.

2.2. Modified criteria for energy minimization

As the energy difference is changed around its mean value, we assume that a more correct criterion for convergence in this case is more systematically satisfying the condition of $|\Delta E| < E_{th}$, for instance its satisfaction at a number of consecutive MC steps. For this aim we developed a *sliding window* method. In this method we use the average value of $|\Delta E|$ in *m* consecutive MC steps instead of its single value:

$$\frac{1}{m}\sum^{m}\left(-\sum_{nodes}\Delta E_i\right) < E_{th}. \quad (3)$$

This energy window with width *m* slides step-by-step together with the increasing MC steps as shown in Fig. 4 (left) until the condition in Eq.(3) is satisfied, e.g. until *MC step* = 40 in this case. We have demonstrated that this is a more robust criterion for minimization convergence as the average has a much less erratic behaviour and is not prone to having "false positives" in exhibiting convergence in contrast to (2).

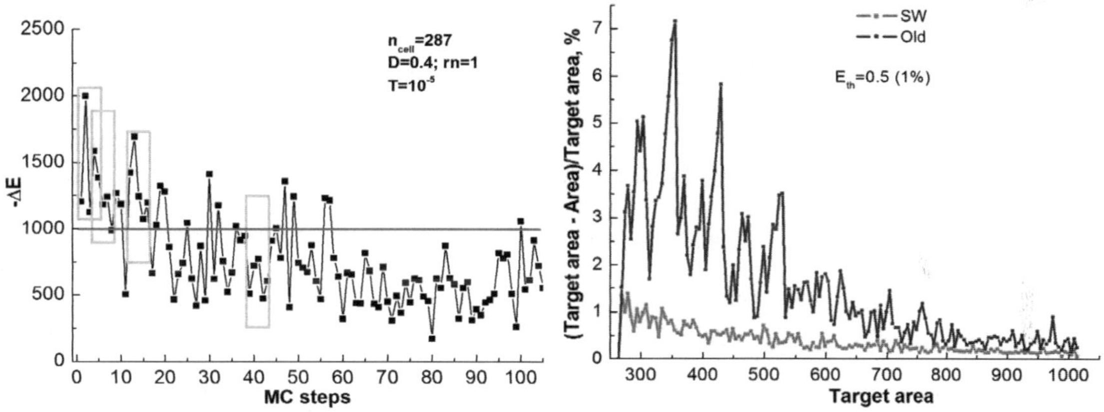

Figure 4. Sliding window (green rectangle) in energy difference versus the MC steps for the 287-cell tissue (left) and comparison of the sliding window (SW) and old criteria for minimization with respect to convergence for the growth of the 2-cell tissue in the Demo1 model of VirtualLeaf (right).

To demonstrate the effect of the modified criterion we plotted in Fig. 4 (left) the difference between the target and actual area versus the target area during tissue growth without any cell division. We took only the first term in Eq. (1) and allowed expansion of the tissue by the systematic increase of the target area at each time step. The width of the sliding window was *m* = 5 MC steps. As threshold value we used E_{th} = 0.5 which corresponds to 1 % of the initial total energy.

In the case of the sliding window we observed a very nice agreement between the predicted minimum and the tolerance value E_{th} at the initial stage of growth. Indeed, one can see the difference between the target and actual area reaches 1 % quickly at the beginning of simulation. Then it decreases as the relation of E_{th} to the total energy decreases due to the increase of the total energy during the tissue growth. However, when we used the old criterion, i.e. (2), there was no agreement in results with the used values of E_{th}: the predicted values were significantly far from the tolerance, with very big deviations in the entire interval of the growth studied. These results show the disadvantage of the criterion (2) resulting in unpredictable and wrong (unstable) states of the system. At the same time, they demonstrate the robustness of the sliding window criterion in the case of this multivariable system.

2.3. Exploration of MC parameters

As shown in Sec. 2.1, the energy convergence is sensitive to the change of threshold energy E_{th}. We also studied the sensitivity of energy convergence to other parameters such as the length of node displacement *D* and the random fluctuation parameter *T*. For the right choice of the values of these parameters we have to answer the next questions: which values are more suitable for the MC energy minimization and what is the criterion for the choice of suitable values?

One can propose the analysis of the system state during time evolution is suitable for this aim. Nevertheless, due to the continuous dynamic changes of the system studied here with processes such as cell division and growth and various terms in Hamiltonian such integrated studies cannot provide adequate answers.

We believe that the best method is the study of the system state directly during the energy minimization cycle. In particular, the total energy evolution of the system used in Sec. 2.1 (see Fig. 2) is a robust way to evaluate the suitability of the values of the parameters. We therefore used this method in our study.

We studied the influence of the length of the node displacement *D* on the MC energy minimization. We present the results in Fig. 5 (left). Note that until now we presented results with *D* =0.4. In Fig. 5 (left) one can see that the increase of *D* (e.g. *D* =0.7) results in the regression (worsening) of the convergence to the equilibrated state: the total energy of the system decreases more slowly with a large slope in the entire range of the MC steps calculated. However, the decrease of *D* (e.g. *D* =0.1) drastically changes the behaviour of this dependence. Indeed, in this case the total energy of the tissue decreases very fast at the beginning MC steps, then it is almost not changed having a plateau of curve in the large range of MC steps which indicates convergence to an energetically stable state of the tissue approximately after 4×10^3 MC steps. From this plateau one can conclude that the energy of the stable state of this tissue at this time is approximately 9.45×10^6.

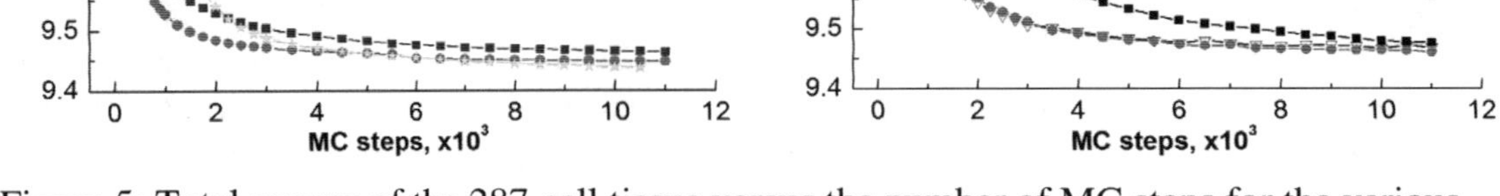

Figure 5: Total energy of the 287-cell tissue versus the number of MC steps for the various values of *D* (left) and *T* (right).

Changing *D* from *D* =0.1 to *D* =0.05 does not give a significant effect, just slightly decreases the number of MC steps for reaching convergence. However, a further decrease of *D* (in particular, *D* =0.005) results again in the regression of the reaching of the equilibrated state. The reason is that too small node displacements cause too small decrease of the total energy at per MC step. Hence it requires more MC steps to decrease the energy until to its equilibrium value.

In Fig. 5 (right) we show the influence of T on the energy minimization. One can see that for small values (T $\leq$10) the increase of T does not give a desirable effect: reaching an equilibrated state requires very long MC calculation. However, further increasing T results in the steepest decreasing energy in early MC steps and thus faster reaching an equilibrated state. We conclude that the use of T =35 is enough in the studied case.

By evaluation of the influence of both D and T on the state of the system we can summarize that the values of E_{th} =0.5, D =0.05 and T =35 are optimal for relatively fast convergence to the equilibrated state in the studied case. Note that though the values of MC parameters are well suited for fast and good convergence, they should be explored in each specific cases, e.g. for different cell sizes, for the cell expansion regime (without division).

3. Conclusions

We analysed the criteria for energy minimization in the Monte Carlo phase in VirtualLeaf simulations. We developed a robust criterion for MC energy minimization in VirtualLeaf - the sliding window criterion. Without this criterion the state of the system found during time evolution and at the final stage can be unpredictable and too far from equilibrium. We have found the sliding window criterion is well suitable for a multi-variable and complex system where the derivative methods are not available such as in this case.

We found that suitable values for the MC parameters can be determined by studying the evolution of the total energy of the system during MC energy minimization. This is a direct, robust and very fast method in comparison with integrated analysis methods. Finding the universal values of parameters suitable for various cases of studies is not easy task or even impossible. However, we developed the robust method for the evaluation of the MC parameter values in order to be sure finding suitable values for MC minimization.

References

[1] Corson F, Hamant O, Bohn S, Traas J, Boudaoud A and Couder Y 2009 *P. Natl. Acad. Sci. USA* **106** 8453.
[2] Peters W S and Tomos A D 2000 *Plant Physiol.* **123** 605.
[3] Merks R M H, Guravage M, Inzé D and Beemster G T S 2011 *Plant Physiol.* **155** 656.
[4] Rudge T and Haseloff J 2005 *Lect. Notes Comput. Sci.* **3630** 78
[5] Dupuy L, Mackenzie J, Rudge T and Haseloff J 2008 *Ann. Bot.-London* **101** 1255.
[6] Grieneisen V A, Xu J, Marée A F M, Hogeweg P and Scheres B 2007 *Nature* **449** 1008.
[7] Frenkel D and Smit B 2002 *Understanding Molecular Simulation* (Academic Press).
[8] Kalos M H and Whitlock P A 2008 *Monte Carlo Methods* (Wiley-VCH).
[9] H. Bernhard Schlegel H B 2007 *Optimization of equilibrium geometries and transition structures AB INITIO Methods in Quantum Chemistry* (Adv. Chem. Phys. vol 67) ed K P Lawley (Hoboken, NJ, USA: Wiley) chapter 4 pp 249–286.

Major component analysis of dynamic networks of physiologic organ interactions

Kang K. L. Liu[1,2], **Ronny P. Bartsch**[3], **Qianli D. Y. Ma**[1,4] **and Plamen Ch. Ivanov**[1,5]

[1] Department of Physics, Boston University, 590 Commonwealth Ave, Boston, MA 02215, USA
[2] Department of Neurology, Beth Israel Deaconess Medical Center, Harvard Medical School, Boston, MA 02115, USA
[3] Department of Physics, Bar-Ilan University, Ramat-Gan, 52900, Israel
[4] College of Geographical and Biological Information, Nanjing University of Posts and Telecommunications, Nanjing, China
[5] Division of Sleep Medicine, Brigham and Women's Hospital, Harvard Medical School, Boston, MA 02115, USA

E-mail: kangliu@buphy.bu.edu, plamen@buphy.bu.edu

Abstract. The human organism is a complex network of interconnected organ systems, where the behavior of one system affects the dynamics of other systems. Identifying and quantifying dynamical networks of diverse physiologic systems under varied conditions is a challenge due to the complexity in the output dynamics of the individual systems and the transient and non-linear characteristics of their coupling. We introduce a novel computational method based on the concept of time delay stability and major component analysis to investigate how organ systems interact as a network to coordinate their functions. We analyze a large database of continuously recorded multi-channel physiologic signals from healthy young subjects during night-time sleep. We identify a network of dynamic interactions between key physiologic systems in the human organism. Further, we find that each physiologic state is characterized by a distinct network structure with different relative contribution from individual organ systems to the global network dynamics. Specifically, we observe a gradual decrease in the strength of coupling of heart and respiration to the rest of the network with transition from wake to deep sleep, and in contrast, an increased relative contribution to network dynamics from chin and leg muscle tone and eye movement, demonstrating a robust association between network topology and physiologic function.

1. Introduction

A defining feature of the human organism is the presence of complex signaling and interaction processes between organ systems and sub-systems. These interactions occur at multiple levels of integration and across spatio-temporal scales to optimize and coordinate physiologic organ functions. Coordinated network interactions of organ systems are essential to maintain health and generate distinct physiologic states, e.g., rest and exercise; wake and sleep; light and deep sleep; dreams; consciousness and unconsciousness. Disrupting these communications can lead to dysfunction of individual systems or to a collapse of the entire organism.

A fundamental question is how organ systems dynamically interact and collectively behave as a network to produce health or disease. A system-wide integrative approach is needed to address this fundamental question. We have recently initiated a new area of research, Network Physiology[1, 2], with the aim to uncover the nature and basic characteristics of organ-to-organ interactions, and to associate specific networks of organ interactions with physiologic function during distinct states and conditions.

Identifying and quantifying organ interactions is a major challenge due to several levels of complexity inherent to the dynamics of organ systems: (i) each organ is a multi-component system with its own regulatory mechanisms, leading to complex emerging behavior characterized by intermittent, scale-invariant and nonlinear output signals [3, 4]; (ii) organ systems have different types of output dynamics — oscillatory, stochastic or mixed [5, 6]; (iii) systems operate on different time scales from msec to hours; (iv) the output dynamics and coupling links between systems vary in time and with different physiologic states [7, 8, 9, 10]; (v) as we have recently discovered certain pairs of organ systems can communicate through several forms of coupling that switch on/off and can simultaneously coexist [11, 12, 13]; (vi) due to non-linear feedback interactions, global network dynamics of the entire organism can not be simply expressed as a sum of the behaviors of individual systems, and can be strongly influenced by minor changes in the relative strength of systems' interactions, even when the network topology between these systems remains unchanged.

Currently, there are no analytic instrumentarium and theoretical framework suitable to probe diverse organ systems with different output dynamics, and to quantify their interactions from continuous streams of noisy and transient signals. The current framework of complex network theory is also not directly applicable to address challenges in this new field, since it mainly addresses (i) network growth or disintegration by adding/removing nodes and links, or (ii) the flow of information on networks with fixed structure. The fundamental question we address in the field of Network Physiology relates to a different aspect, namely, to characterize how global network behavior and function emerge from complex output dynamics and coupling interactions between diverse systems (nodes), and how global behavior changes in response to variations in the strength of interactions (links), even when the topology of the network remains unchanged.

We propose a novel approach based on the concept of time delay stability, to identify and quantify pair-wise interactions between diverse systems (network nodes) with different types of output dynamics, combined with a major component analysis designed to dissect complex networked interactions between multiple organ systems.

2. Methods

Time Delay Stability (TDS): The structural and neuronal networks that control physiological systems lead to a high degree of complexity, which is further compounded by various coupling and feedback interactions that continuously vary in time, and the nature of which is not understood. These systems operate on different time scales, from msec to hours, and exhibit multiple coexisting forms of coupling. To quantify these interactions and characterize how they change in time under different physiological conditions, we study the time delay with which modulations in the output dynamics of a given system are consistently followed by corresponding modulations in the signal output of another system (time delay stability, TDS). Periods with constant time delay indicate a stable physiological interaction, and stronger coupling between systems results in longer periods of TDS.

Physiological output signals are first re-sampled at 1 Hz and normalized to zero mean and unit standard deviation within overlapping time windows of $\Delta t = 60$s with a moving step of 30 seconds (effectively de-trending the signals). Synchronous bursts in the normalized signals lead to pronounced cross-correlation calculated in windows of $\Delta t = 60$s in steps of $30s$. The time delay τ_0 is determined by the position of the maximum of the cross-correlation function in each moving window Δt. We identify two systems as linked if their corresponding signals exhibit a time delay that does not change by more than ±1 sec for several consecutive time windows Δt.

We track the values of τ_0 along the series $\{\tau_0\}_{\Delta t}$ — when for at least four out of five consecutive time windows Δt (corresponding to a period of 5×30 sec) the time delay remains in the interval $[\tau_0 - 1, \tau_0 + 1]$ these segments are labeled as stable. The procedure is repeated for a sliding time window with a step size one along the entire series $\{\tau_0\}_{\Delta t}$. Longer periods of TDS between the output signals of two systems reflect more stable interaction/coupling between these systems. Thus, the strength of coupling is determined by the percentage of time when TDS is observed: higher percentage of TDS corresponds to stronger coupling. The % TDS is calculated as the fraction of segments with stable time delay out of the entire time series $\{\tau_0\}_{\Delta t}$.

Major Component Analysis (MCA): To dissect the complex network of physiologic organ interactions and to identify (i) which subset of links play major role and (ii) what is the relative contribution of individual organs (nodes) to the entire network dynamics, we employ matrix theory and the concept of Major Component Analysis. Utilizing the matrix of network interactions, where each matrix element represents the pair-wise coupling between organ systems as measured by the % TDS, we perform matrix diagonalization to extract the spectrum of eigenvalues and associated eigenvectors, where each eigenvector element represents the relative contribution of each organ to the corresponding eigenmode of the network. The global network dynamics is thus a linear superposition of the different eigenmodes.

First, we obtain the TDS matrix of pair-wise interactions, where matrix element $A_{i,j}$ represents the strength of dynamic coupling between organ i and organ j, measured by the % TDS derived from the output signals of these two systems.

Matrix A is symmetric and can be diagonalized to determine its spectrum of real eigenvalues.

$$D \equiv UAU^{\intercal} = \begin{pmatrix} V^1 \\ V^2 \\ \vdots \\ V^n \end{pmatrix} \begin{pmatrix} A_{1,1} & A_{1,2} & \cdots & A_{1,n} \\ A_{2,1} & A_{2,2} & \cdots & A_{2,n} \\ \vdots & \vdots & \ddots & \vdots \\ A_{n,1} & A_{n,2} & \cdots & A_{n,n} \end{pmatrix} \begin{pmatrix} V^{1\intercal} & V^{2\intercal} & \cdots & V^{n\intercal} \end{pmatrix} = \begin{pmatrix} \lambda_1 & 0 & \cdots & 0 \\ 0 & \lambda_2 & \cdots & 0 \\ \vdots & \vdots & \ddots & \vdots \\ 0 & 0 & \cdots & \lambda_n \end{pmatrix}, \tag{1}$$

where U is a unitary matrix; λ_i is the i-th eigenvalue of A (with $\lambda_1 \leq \lambda_2 \leq \cdots \leq \lambda_n$), and $V^i \equiv (r^i_1, r^i_2, \cdots, r^i_n)$ is the eigenvector corresponding to λ_i.

The original TDS matrix A can be presented as a spectrum decomposition $A = U^{\intercal}DU = \sum_{i=1}^{n} \lambda_i V^{i\intercal} V^i$, where the major component M of matrix A that corresponds to the largest eigenvalue λ_n is:

$$M \equiv V^{n\intercal} V^n = \begin{pmatrix} r^n_1 r^n_1 & r^n_1 r^n_2 & \cdots & r^n_1 r^n_n \\ r^n_2 r^n_1 & r^n_2 r^n_2 & \cdots & r^n_2 r^n_n \\ \vdots & \vdots & \ddots & \vdots \\ r^n_n r^n_1 & r^n_n r^n_2 & \cdots & r^n_n r^n_n \end{pmatrix}. \tag{2}$$

Because λ_n is the largest eigenvalue of A, the matrix M represents the first-order approximation for the links configuration of the original TDS network, where network links strength is quantified by the off-diagonal matrix elements $r^n_i r^n_j$ for $i \neq j$, and the relative weight of each node i in the network dynamics is represented by the diagonal elements $r^n_i r^n_i$.

3. Data

We analyse continuously recorded multichannel physiological data obtained from 36 healthy young subjects (18 female and 18 male; 20 - 40 years old, average age 29) during night-time sleep [14] (average record duration is 7.8 h). This allows us to track the dynamics and evolution of the network of physiological interactions during different sleep stages and sleep-stage transitions (Fig. 1). We focus on physiological dynamics during sleep as sleep stages are well-defined physiological states, and external influences due to physical activity or sensory inputs are reduced

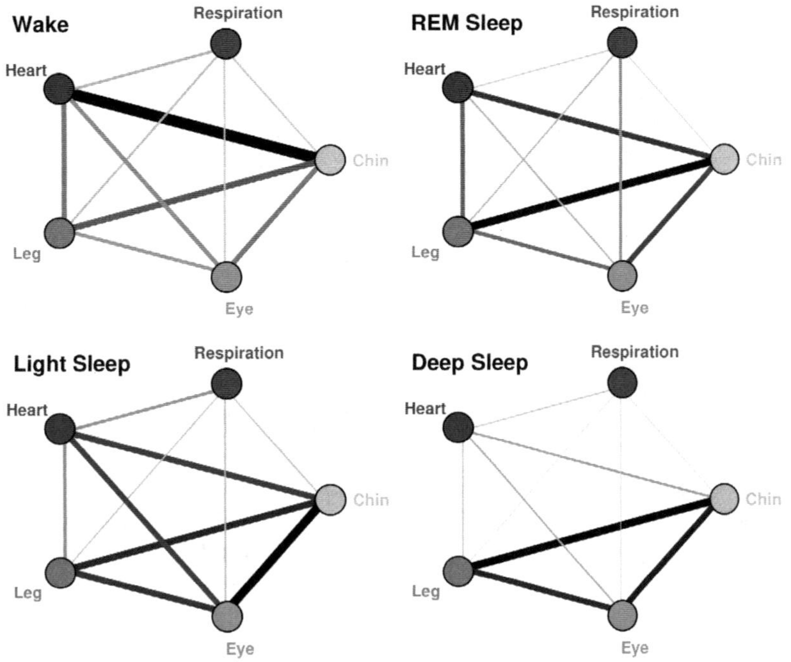

Figure 1. Network of interactions between key organ systems and transitions across different physiologic states. Interactions between organ systems are represented by weighted undirected graphs, where links reflect the strength of dynamic coupling as measured by the stability in the time delay between bursts of activities in the systems output signals (% TDS, see Methods). Darker and thicker links correspond to stronger coupling and higher time delay stability. Results are obtained from 8-hour recordings during sleep by weighted averaging over all segments of different sleep stages pooled from 36 healthy subjects. A pronounced re-organization in the configuration of network links strength is observed with transitions from one sleep stage to another, demonstrating a clear association between network structure and physiologic function.

during sleep. Sleep stages are scored in 30s epochs by sleep lab technicians based on standard criteria. We study the electrocardiogram (ECG), respiration, the electrooculogram (EOG), and the electromyogram (EMG) of the chin and leg. In order to compare these very different signals with each other and to study interrelations between them, we extract the following time series from the raw signals: the variance of the EOG and EMG signals in moving windows of 2s with a 1s overlap; heartbeat RR intervals and interbreath intervals are both re-sampled to 1 Hz (1s bins) after which values are inverted to obtain instantaneous heart and respiratory rate. Thus, all time series have the same time resolution of 1s before the TDS method of analysis is applied.

4. Results and Summary

We find that different physiologic states are characterized by significantly different average strength of network links as measured by the proportion of data segment with time delay stability (% TDS, see Methods): stronger links during Wake ($\approx$ 8.4% TDS) and Light Sleep ($\approx$ 8.4% TDS), and much weaker links in REM ($\approx$ 6.8% TDS) and Deep Sleep ($\approx$ 5.5% TDS). This sleep-stage stratification pattern is observed across all individual subjects, indicating a universal behavior related to the underlying mechanism of neural regulation. The observed similarity in the average links strength between Wake and Light Sleep is intriguing, since traditionally physiologic dynamics are considered similar during Light Sleep and Deep Sleep (Non-REM phase), and very different from Wake and REM.

We find that different physiologic states are characterized by different largest eigenvalues λ_n,

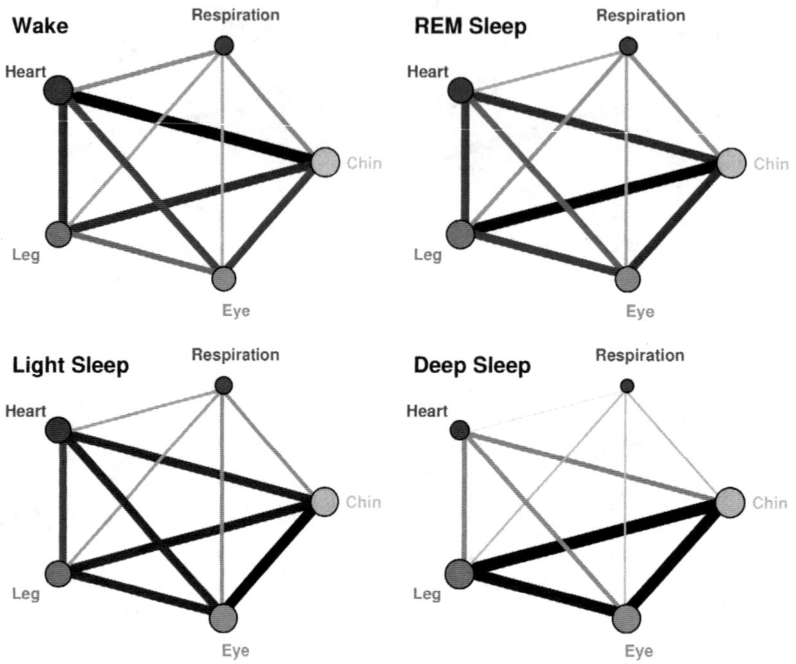

Figure 2. Weighted undirected graphs represent the major component of the networks of organ systems interactions for different physiologic states. The relative contribution of each organ system to the global network dynamics is presented by the size of each network node — the radius of each network node i is proportional to the component r_i^n of the eigenvector V^n corresponding to the largest eigenvalue λ_n of the original TDS network for each physiologic state (Fig.1). Links strength (indicated by different line thickness) is proportional to the off-diagonal elements $M_{i,j} = r_i^n r_j^n$ $(i \neq j)$ of the major component of the original TDS networks in Fig.1 (see Methods). Graphs show a gradual decrease of the role of heart and respiration in the physiologic network dynamics with transition from wake to REM, to light and deep sleep, while in contrast, the relative contribution of leg and chin muscle tone and eye movement increases.

Table 1. Largest eigenvalues λ_n and corresponding eigenvectors V^n of the TDS matrix A representing the network of organ interactions shown in Fig. 1 for different physiologic states (Wake, REM, Light Sleep (LS) and Deep Sleep (DS)). Eigenvector components $r_i (i = 1, \cdots, 5)$ correspond to different organ systems: chin, respiration, heart, leg and eye, respectively.

	V^n_{Wake} $(\lambda_n = 0.37)$	V^n_{REM} $(\lambda_n = 0.30)$	V^n_{LS} $(\lambda_n = 0.36)$	V^n_{DS} $(\lambda_n = 0.27)$
r_1: Chin	0.5714	0.5711	0.5421	0.5805
r_2: Respiration	0.2251	0.2124	0.1959	0.1137
r_3: Heart	0.5494	0.4171	0.4307	0.2347
r_4: Leg	0.4251	0.5284	0.4462	0.5601
r_5: Eye	0.3744	0.4189	0.5321	0.5304

which form a pronounced stratification pattern across physiologic states (higher values for Wake and LS, and lower values for REM and DS, Table. 1), indicating transitions in the connectivity strength of the TDS network with physiologic states consistent with earlier observations [1]. Further, our results of the major component analysis of the TDS networks shown in Fig. 1 demonstrate that different physiologic systems have very different relative contribution to global network dynamics, as quantified by the components r_i of the eigenvector V^n corresponding to λ_n of the TDS matrix for each state (Table. 1), as well as by the size of network nodes (Fig. 2).

Our analyses show that the relative weight of the heart in network dynamics is higher than respiration. Further, we find that for both heart and respiration, the relative contribution to the global dynamics gradually decreases from Wake to REM, to Light and Deep Sleep, while in contrast, contributions from chin, leg and eye movement increase and are highest during Deep Sleep (Fig. 2, and Table. 1). Notably, the reduced role and link strength of the heart and respiratory system during Light Sleep and Deep Sleep compared to REM and Wake is consistent with earlier findings of reduced sympathetic input and associated loss of long-range auto-correlation in cardiac and respiratory dynamics during Light and Deep Sleep [6, 8, 15, 16].

In summary, we demonstrate that a network approach to physiological interactions is necessary to understand how modulations in the regulatory mechanism of individual organ systems translate into reorganization of physiological interactions across the human organism. We find that the bursting activity in the output dynamics of key organ systems exhibits a robust pattern characterized by a stable time delay that persist in time during a given physiologic state. This pattern is consistently observed in all subjects for the same physiologic state, indicating a universal association between network topology and physiologic function. Specifically, we find that each physiologic state is characterized by a distinct configuration and distribution of links strength, and that transitions from one physiologic state to another are associated with rapid reorganization (within 30s to 1min) in network links strength and configuration. The proposed novel approach allows to dissect dynamic networks and to the quantify relative contributions from different nodes to global network dynamics. This is essential to address a fundamental question in the new field of Network Physiology [1, 2], namely, how physiologic organ systems interact as a network to synchronize and optimize their function to produce health and disease.

5. Acknowledgement

We acknowledge support from NIH (Grant 1R01- HL098437), ONR (Grant 000141010078), BSF (Grant 2012219), EC- FP7 Marie Curie Fellowship (IIF 628159), NSF China (61271082, 61201029), NSF Jiangsu Province (BK2011759).

References

[1] Bashan A, Bartsch R P, Kantelhardt J W, Havlin S and Ivanov P Ch 2012 *Nat. Commun.* **3** 702
[2] Ivanov P Ch and Bartsch R P 2014 *Network Physiology: Mapping Interactions Between Networks of Physiologic Networks* (Springer International Publishing) chap 10, pp 203–222
[3] Ivanov P Ch, Rosenblum M G, Peng C K, Mietus J, Havlin S, Stanley H E and Goldberger A L 1996 *Nature* **383** 323–327
[4] Ivanov P Ch, Amaral L A N, Goldberger A L, Havlin S, Rosenblum M G, Struzik Z R and Stanley H E 1999 *Nature* **399** 461–465
[5] Schmitt D T and Ivanov P Ch 2007 *Am. J. Physiol.-Regul. Integr. Comp. Physiol.* **293** R1923–R1937
[6] Kantelhardt J W, Havlin S and Ivanov P Ch 2003 *Europhys Lett* **62** 147–153
[7] Ivanov P Ch, Bunde A, Amaral L A N, Havlin S, Fritsch-Yelle J, Baevsky R M, Stanley H E and Goldberger A L 1999 *Europhys. Lett.* **48** 594–600
[8] Kantelhardt J W, Ashkenazy Y, Ivanov P Ch, Bunde A, Havlin S, Penzel T, Peter J H and Stanley H E 2002 *Phys. Rev. E* **65** 051908
[9] Hu K, Ivanov P Ch, Hilton M F, Chen Z, Ayers R T, Stanley H E and Shea S A 2004 *Proc. Natl. Acad. Sci. USA* **101** 18223–18227
[10] Ivanov P Ch, Hu K, Hilton M F, Shea S A and Stanley H E 2007 *Proc. Natl. Acad. Sci. USA* **104** 20702–20707
[11] Bartsch R P, Schumann A Y, Kantelhardt J W, Penzel T and Ivanov P Ch 2012 *Proc. Natl. Acad. Sci. USA* **109** 10181–10186 URL `http://dx.doi.org/10.1073/pnas.1204568109`
[12] Bartsch R P and Ivanov P Ch 2014 *Communications in Computer and Information Science* **438** 270–287
[13] Bartsch R P, Liu K K L, Ma Q D Y and Ivanov P Ch 2014 *Computing in Cardiology* **41** in press
[14] Klösch G, Kemp B, Penzel T, Schlögl A, Rappelsberger P, Trenker E, Gruber G, Zeitlhofer J, Saletu B, Herrmann W M, Himanen S L, Kunz D, Barbanoj M J, Röschke J, Värri A and Dorffner G 2001 *IEEE Eng. Med. Biol. Mag.* **20** 51–57
[15] Schmitt D T, Stein P K and Ivanov P Ch 2009 *IEEE Trans. Biomed. Eng.* **56** 1564–1573
[16] Schumann A Y, Bartsch R P, Penzel T, Ivanov P Ch and Kantelhardt J W 2010 *Sleep* **33** 943–955

Calculating transition and reaction rates with nonequilibrium work measurements

Mahmoud Moradi[1], **Celeste Sagui**[2,*], **Christopher Roland**[2,*]

1. Beckman Institute, University of Illinois, Urbana, IL 61801
2. Center for High Performance Simulations (CHiPS) and Department of Physics, North Carolina State University, Raleigh, NC 27695

E-mail: `cmroland@ncsu.edu, sagui@ncsu.edu`

Abstract. By combining elements of the nonequilibrium work theorem with Transition Path Theory (TPT), we have developed a formalism for investigating transition pathways and probabilities that may effectively be implemented by means of Steered Molecular Dynamics (SMD) simulations. The workings of this formalism are illustrated by means a simple example based on a diproline peptide.

Introduction

Computational investigations of transitions and reactions are hampered by the fact that they typically constitute so-called rare events. Rare events of this type occur on timescales beyond the feasible simulation time of current molecular dynamics (MD) simulations, and hence, special methods are needed to probe them. This situation is particularly acute in biomolecular systems, which are characterized by complex and rugged free energy surfaces with different competing mechanisms. The microscopics of transitions of such systems are not described in terms of a single mechanism, but rather as a reaction ensemble of mechanisms whose importance one typically wants to quantify. Here, we summarize a formalism [1] that we have recently developed for the study of such transitions and reaction rates [2]. The formalism is based on combining the results of the nonequilibrium work theorem [3] with the main elements of Transition Path Theory (TPT) [4], and numerically may easily be realized in terms of nonequilibrium Steered Molecular Dynamics (SMD) simulations [5]. The workings of this formalism are illustrated via the cis-trans isomerization in a proline dipeptide [1,2].

In studying a transition, typical information sought includes knowledge of the stable and metastable states of the system, the transition state (TS), the reaction rate (ω), and microscopics of mechanisms driving the transition. To answer these questions, an understanding of the underlying free energy surface (FES) is important, which is usually expressed as a function of a suitable set of collective variables or reaction coordinates. These variables are chosen in such a manner as to reflect the physics of the transition in as transparent of a manner as possible. Currently, there are many different methods available for calculating a FES, but over the past decade nonequilibrium methods such as metadynamics and the Adaptively Biased Molecular Dynamics (ABMD) method [6] have become popular. Having determined a suitable FES surface, it becomes popular to identify the different minima and asssociate them with different states of the system. The question then becomes: how does the system move between two different states. In principle, all paths over the FES are possible, although some paths are of course more likely than others and different methods such as the nudged elastic band method, the least free energy path method and others have been developed to identify the most probable paths. Here, transition path theory (TPT) provides for an alternate viewpoint. Rather than focusing on the most likely path through the TS, it takes into account an ensemble average of so-called reactive paths (see Fig.1), and uses this to identify transition mechanisms and calculate reaction rates. This kind of approach implies that some kind of path sampling must be performed in order to calculate the quantities of interest. We find that this may be efficiently achieved by means of nonequilibrium SMD simulations, and show that the transition or reaction rate calculated by means of such driven simulations (represented by $\omega^{dr}_{A\to B}$ for a forced system moving from minima A to B) is related to the correponding equilibrium reaction rate ($\omega^{eq}_{A\to B}$) by

$$\omega^{dr}_{A\to B} = \omega^{eq}_{A\to B}\eta_{AB} \tag{1}$$

with η_{AB} representing a corrective biasing factor that depends on the details of how the SMD simulations are carried out. We will show that η_{AB} is conveniently calculated by an ensemble average of the dissipative work performed as the system moves between minima A and B. We also note that this equation strongly resembles Evans' relation for steady state systems, which may be viewed as a nonequilibrium counterpart of a detailed balance relation [7].

1. Theory

Since we are primarily interested in applying the formalism to biomolecular systems, we consider a FES which is defined in terms of some suitable reaction coordinates with (meta)stable states A and B. Suppose that for such a system, the transition between these two points takes place via n distinct mechanisms. Associated with each mechanism is a distinct transition tube and a well-defined sets of non-overlapping transition pathways, as illustrated in Fig.1. While we are ultimately interested in calculating the equilibrium reaction rates associated with each mechanism, we find it somewhat easier to focus on the relative transition rates, which are easier to compute. This result gives us a quantitative measure of the importance of each mechanism. We approach the problem by using the TPT formalism – specifically as outlined in the finite temperature string (FTS) method – and then calculate the needed quantities by means of nonequilibrium work measurements.

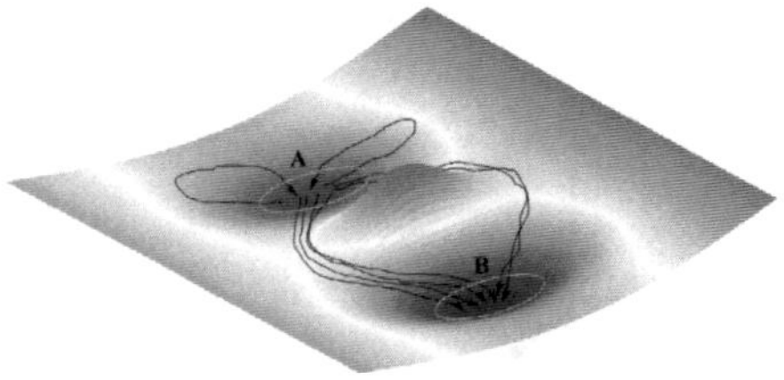

Figure 1. Schematic illustrating possible system trajectories for system moving between minima A to minima B over a two-dimensional free energy landscapes. Reactive trajectories are those that start at A and actually reach B, while unreactive trajectories are those that start from A and end up at A after some time. An equilibrium ensemble of trajectories generated by unbiased simulations, most trajectories will be unreactive. In a driven simulation under SMD conditions, the trajectories will be reactive.

1.1. TPT and FTS theory

The FTS theory [8] focuses on an ensemble of pathways in a suitable configuration space. Consider following one such pathway, where the system starts out at minimum A and reaches minimum B. As the system moves along such a pathway, one can calculate the probability that system will end up at B for any point along the given trajectory. This in turn may be used to define a committor function via an ensemble average of the different trajectories as they cross a given isocommittor surface (An isocommittor surface is defined as the hypersurface formed in

Journal of Physics: Conference Series **640** (2015) 012014 doi:10.1088/1742-6596/640/1/012014

configuration space on which each trajectory has the same probabilty of reaching completion). Clearly, such a committor function may be formed for each of the different mechanisms associated with the transition with the trajectories contained within each of the different transition tubes. FTS theory shows that the committor function $f_i(\alpha)$ associated with the ith mechanism and parameterized by variable α (which is normalized such that $\alpha = 0$ when the system is at minimum A and $\alpha = 1$ when is system is at B), is defined in terms of the free energy F_i via:

$$f_i(\alpha) = \kappa_i \int_0^{\alpha} exp(\beta F_i(\alpha'))d\alpha' \tag{2}$$

in which κ_i is proportional to the associated transition rate

$$\kappa_i = 1/\int_0^1 exp(\beta F_i(\alpha)d\alpha \tag{3}$$

with $\beta = 1/k_B T$ with Boltzmann constant k_B and temperature T. This defines the equilibrium probability associated with the ith mechanism as – *i.e.*, that the transition takes place with the ith channel as

$$\pi_i^{eq} = \kappa_i/\Sigma_j^n \kappa_j. \tag{4}$$

Under FTS, the TS is readily found by identifying a given α^* such that $f_i(\alpha^*) = 1/2$. As may be expected, the TS is typically associated with the largest value of $F_i(\alpha)$.

1.2. Nonequilibrium work relations

The required FTS quantities may readily be calculated by means of SMD simulations as follows. The second law of thermodynamics asserts that the average work (W) performed when a system moves between two states cannot be less than the free energy difference ΔF between the two states – *i.e.*, $\langle W \rangle \geq \Delta F$, with the equality holding when the process is reversible. However, the Jarzynski equality [3] is an exact equality that holds even when the transformation is irreversible:

$$exp(-\beta\Delta F) = \langle exp(-\beta W)\rangle, \tag{5}$$

where the angular brackets denote the average over all possible realizations of the process connecting the two states. This, in principle, ΔF can be calculated from finite-time nonequilibrium simulations by measuring the work. This may conveniently be accomplished by means of SMD simulations. In SMD simulations, a system is driven out of equilibrium by means of a time-varying force, which typically is achieved by adding a harmonic biasing potential that pulls on the collective variable ξ. The system is thus forced to trace out a trajectory in phase space between minima A and B over some time interval T'. If properly chosen, the Feynman-Kac theorem shows that the distribution of states associated with the collective variable ξ in equilibrium may be represented by the driven ensemble in which each trajectory carries a time-dependent statistical weight $exp(-\beta w_d^t)$:

$$\langle \delta(\mathbf{x} - \mathbf{x^t})\rangle_\xi^{eq} = \langle \delta(\mathbf{x} - \mathbf{x^t})exp(-\beta w_d^t)\rangle^{dr}, \tag{6}$$

where $w_d^t = w^t - (F(\xi) - F(\xi_A))$ represents the total dissipative work generated during the course of the trajectory and $F(\xi)$ is the free energy. Clearly, for an ensemble of trajectories describing the evolution of a system driven out of equilibrium, the equilibrium distribution may be reconstructed by giving each realization the time-dependent weight $exp(-\beta w_d^t)$. Since this is true for all phase space, its also true for the parts of phase space associated with the ith mechanism. One can then readily show:

$$exp(-\beta F_i(\xi)) = \langle \delta(\xi - \xi^t)exp(-\beta\Delta w^t)\rangle_i^{dr} \pi_i^{dr} \tag{7}$$

with $\Delta w^t = w^t - U(\xi^t, t)$ and w^t representing the work and U the harmonic biasing potential, and π_i^{dr} the probability that the transition takes place via the ith mechanism or transition tube.

1.3. Applicaton of nonequilibrium work relations to TPT

The last relation shows that one can estimate $F_i(\xi)$ from measurements of the nonequilibrium work. Hence, we are now in a position to calculate all the TPT quantities in terms of these relations. Thus, the committor function associated with a given mechanism becomes:

$$f_i(\xi) = \frac{\int_0^\xi (\langle \delta(\xi' - \xi^t) exp(-\beta\Delta w^t)\rangle_i)^{-1} d\xi'}{\int_0^1 (\langle \delta(\xi' - \xi^t) exp(-\beta\Delta w^t)\rangle_i)^{-1} d\xi'} \tag{8}$$

(assuming that one can relate ξ to α, as is usually the case). From this, one readily obtains:

$$\pi_i^{eq} = \frac{\pi_i^{dr}(\int_0^1 (\langle \delta(\xi - \xi^t) exp(-\beta\Delta w^t)\rangle_i)^{-1} d\xi)^{-1}}{\Sigma_{j=1}^n \pi_j^{dr}(\int_0^1 (\langle \delta(\xi - \xi^t) exp(-\beta\Delta w^t)\rangle_j)^{-1} d\xi)^{-1}}. \tag{9}$$

However, rather than calculate the true probabilities, it is much easier to work with relative rates so that:

$$\frac{\pi_i^{dr}}{\pi_j^{dr}} = \frac{\pi_i^{eq}}{\pi_j^{eq}}\frac{\eta_i}{\eta_j} \tag{10}$$

with η_i representing a biasing factor

$$\eta_i = \int_0^1 \langle \delta(\xi - \xi^t) exp(-\beta\Delta w^t)\rangle_i^{-1} d\xi. \tag{11}$$

The above equations constitute our main result.

It is clear that our result may be cast in the form $\omega_{A\to B}^{dr} = \omega_{A\to B}^{eq}\eta_{AB}$. In this form, our result closely resembles that of Evans' relation [7] for steady state transition rates. Specifically, Evans' relation shows that the transition rates for a nonequilibrium ensemble in a steady state, with mean energy $\langle E\rangle$ and mean flux $\langle J\rangle$, the biasing factor is given by

$$\eta_{AB} = lim_{\tau\to\infty}\frac{p_\tau^{eq}(\langle J\rangle | A \to B, \langle E\rangle)}{p_\tau^{eq}(\langle J\rangle, A, \langle E\rangle)}, \tag{12}$$

where $p_\tau(X|Y)$ represents the normalized probability that system property X is observed in time interval τ when the system is subject to condition Y. Thus, the enhancement factor associated with Evans' formula represents the ratio of the probability that a system starting out at A reaches B subject to conditions of mean flux and energy over time interval τ to the total probability of the system starting at A with similar conditions.

To elucidate the connection between our result and Evans' formula, consider Bayes' theorem $p(X|Z)p(Y|X,Z) = p(Y|Z)p(X|Y,Z)$, which expresses the joint probability of two outcomes X and Y given condition Z in two different ways. This theorem may be applied to an ensemble of trajectories that take the system from A to B via different transition mechanisms. We associate with Z the number of reactive trajectories in equilibrium $(A \to B))$, while with X we pair the number of driven trajectories $(A \xrightarrow{dr} B)$. We assume that any member of the $A \xrightarrow{dr} B$ set is also contained in $A \to B$, so that any driven trajectory is also a member of the set of equilibrium trajectories although the latter is assumed to be much larger than the former. Finally, we let i represent the trajectories that hit the TS associated with the ith mechanism. Then $p(A \xrightarrow{dr} B|A \to B)p(i|A \xrightarrow{dr} B, A \to B) = p(i|A \to B)p(A \xrightarrow{dr} B|i, A \to B)$. Noting that $\pi_i^{eq} = p(i|A \to B)$ and $\pi_i^d r(i|A \xrightarrow{dr} B, A \to B)$, we obtain $\pi_i^{dr} p(A \xrightarrow{dr} B|A \to B) = \pi_i^{eq} p(A \xrightarrow{dr} B|A \xrightarrow{i} B)$ which shows that $\eta_i \propto p(A \xrightarrow{dr} B|A \xrightarrow{i} B)$. This conditional probability therefore represents the fraction of driven trajectories in the reactive ensemble associated with the ith mechanism. Reworking this expression in terms of the flux of reactive trajectories makes the connection between our expression and Evans' result explicit.

Finally, we note that the formalism outlined here is based on some of the simplest free energy estimators from the nonequilibrium work theorem. More complicated formulas – based on a combination of both forward and backward SMD simulations – are possible and even desirable because of enhanced convergence [9]. However, because the resulting expressions are long and cumbersome, we do not reproduce those formulas here, but simply refer the reader to the original publications [1,2].

2. Applications

The outlined formalism has been applied to a number of typical applications, including the so-called circular random walk, a diproline peptide, a triproline peptide and dialanine. For brevity, we summarize here our results for the diproline application only.

Polyproline is known to form two types of idealized helical structures: an extended left-handed polyproline helix (PPII) and a more compact right-handed helix (PPI). The former helix is based on an all trans (T) conformation of the prolyl angle ($\omega = 180^o$), while the latter is associated with the all-cis (C) form ($\omega = 0^o$). Recently, we have examined the structural characteristics and free energies of short polyproline peptides both *in vacuo* and in implicit solvent environments with the Adaptively Biased Molecular Dynamics (ABMD) method [6]. The capped diproline peptide $Ace - (Pro)_2 - Nme$ has two prolyl bonds, and can therefore exist in four different conformations: CC, TC, CT, TT. The $PPII \rightarrow PPI$ transition can therefore take place via two different pathways: $CC \rightarrow CT \rightarrow TT$ or $CC \rightarrow TC \rightarrow TT$. Generally speaking, it is difficult to study this transition with regular MD simulations because of the high barriers associated with the C to T transition. However, the transition is readily probes with SMD simulations. Clearly, the $CC \rightarrow TT$ involves two possible and competing transition pathways, with intermediates TC or CT.

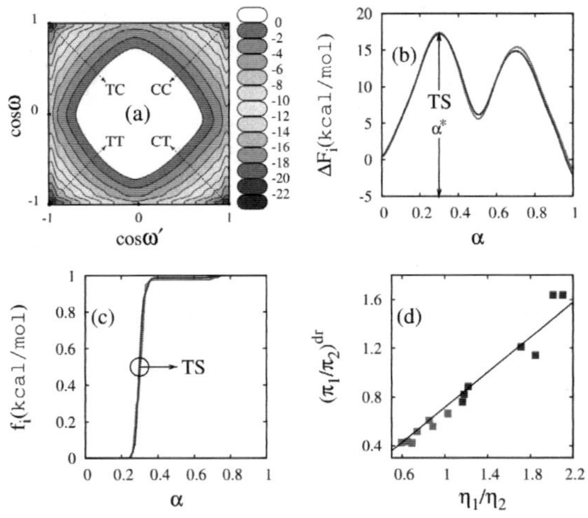

Figure 2. (a) Free energy landscape of diproline peptide in terms of $cos(\omega)$ and $cos(\omega')$; (b) one-dimensional free energy landscape in transition tubes; (c) the committor function along the transtion tube as red (blue) as obtained from 1000 forward SMD runs over a 10 ns runtime; and (d) π_1^{dr}/π_2^{dr} plotted against η_1/η_2.

To investigate this system, we have carried out extensive SMD simulations in an implicit solvent environment, with simulation details found in Ref.[10]. The SMD simulations were carried out in the $\Omega = cos(\omega) + cos(\omega')$ space, where the two ω's represent the Ace-Pro and Pro-Pro prolyl bonds respectively. For this particular collective variable, $\Omega = 2$ (-2) represents the CC (TT) state, while $\Omega = 0$ is associated with both TC or CT. Steering the system with Ω results in the system choosing the $i = 1, 2$ path in a quasi-random manner.

Figure 2a shows the ABMD free energy landscape, while Figs.2b,c show the ensemble average of the free energy and the committor function, respectively. For the case shown, we measured a π_1^{dr} of 53% with an estimated π_1^{eq} of 39%. Keeping the spring constant the same for all the different simulations (K=100kcal/mol), we performed a number of SMD simulations in both forward and backward directions. The calculated π_1^{dr}'s varied between 30-60%, but the estimated π_1^{eq} obtained from our formalism was $40.0 \pm 2\%$. This small difference in the reaction probability is natural given the asymmetry and shortness of the proline peptide. It suggests that the cis/trans isomerization starts at the acetylated as opposed to the amidated end of the peptide. Figure 2d plots the ratio π_1^{dr}/π_2^{dr} versus η_1/η_2. As expected, the resulting variation is linear, in agreement with the main result of our work [1,2].

3. Summary
This paper briefly summarizes recent work on combining the TPT with results for the nonequilibrium work theorem in order to estimate equilibrium transition rates using SMD simulations.

4. Acknowledgements
We thank NSF (grants MCB-1021883 and SI2/SEE-1148144) for funding, and the NCSU HPC for extensive computer support.

5. References
[1] Moradi M, Sagui C, and Roland C 2011 *Chem. Phys. Lett.* **518** 109.
[2] Moradi M, Sagui C, and Roland C 2014 *J. Chem. Phys.* **140** 034114; *ibid* 034115.
[3] Jarzynski C 1997 *Phys. Rev. Lett.* **78** 2690.
[4] E W, Ren W, and Vanden-Eijnden E. 2005 *Chem. Phys. Lett.* **413** 242.
[5] See, for example: Izrailev B, Stepaniants S, Isralewitz D, Kosztin D, Lu H, Molnar F, Wriggers W and Schulten K (1998) *Steered Molecular Dynamics. Computational Molecular Dynamics: Challenges, Methods, Ideas* (Springer Verlag, Berlin Germany).
[6] Babin V, Roland C, and Sagui C 2008 *J. Chem. Phys.* **128** 134101.
[7] Evans RML 2004 *Phys. Rev. Lett.* **92** 150601.
[8] E W, Ren W, and Vanden-Eijenden E 2010 *Annu. Rev. Chem.* **61** 391.
[9] Minh D and Abid A 2008 *Phys. Rev. Lett.* **100** 180602.
[10] Moradi M, Babin V, Roland C, Darden T, and Sagui C 2009 *Proc. Natl. Acad. Sci. U.S.A.* **106** 20746.

Effect of surface attractive strength on structural transitions of a confined HP lattice protein

Busara Pattanasiri[1], Ying Wai Li[2], Thomas Wüst[3], and David P. Landau[1]

[1]Center for Simulational Physics, Department of Physics and Astronomy, The University of Georgia, Athens, GA 30602, USA

[2]National Center for Computational Sciences, Oak Ridge National Laboratory, Oak Ridge, TN 37831, USA

[3]Scientific IT Services, ETH Zürich IT Services, 8092 Zürich, Switzerland

E-mail: busara@physast.uga.edu

Abstract.
We investigate the influence of surface attractive strength on structural transitions of a hydrophobic-polar (HP) lattice protein confined in a slit formed by two parallel, attractive walls. We apply Wang-Landau sampling together with efficient Monte Carlo updates to estimate the density of states of the system. The conformational transitions, namely, the debridging process and hydrophobic core formation, can be identified by analyzing the specific heat together with several structural observables, such as the numbers of surface contacts, the number of hydrophobic pairs, and radii of gyration in different directions. As temperature decreases, we find that the occurrence of the debridging process is conditional depending on the surface attractive strength. This, in turn, affects the nature of the hydrophobic core formation that takes place at a lower temperature. We illustrate these observations with the aid of a HP protein chain with 48 monomers.

1. Introduction

Confinement is one of the major effects that impacts protein folding in a cellular environment. In a living cell, there is a highly crowded and complex setting containing multiple biological macromolecules, such as lipids, carbohydrates, and proteins [1]. For example, proteins and nucleic acids inside the cytoplasm occupy about $30 - 40\%$ of the total cell volume [2, 3, 4]. This complexity can hinder the correct folding of a protein to its native structure, which leads to various pathologies. Incorrect protein folding is shown to be associated with a number neurodegenerative disorders, including Alzheimer's, Huntington's, and Mad Cow diseases [5, 6].

The behavior of confined proteins that deviates from bulk properties has drawn much attention in a wide variety of research fields, ranging from the understanding of biological processes [7, 8, 9], improving material properties [10, 11], enhancing biological technologies [12, 13], and designing nanodevices and nanomaterials [14, 15]. Chemical properties of the confining surfaces govern the interactions with proteins, which can be exploited to enhance its selectivity. An understanding of the effects of confinement along with the confining surface properties on conformational or structural changes of proteins is thus essential.

In this work, we study the effect of surface attractive strength on structural transitions of

Journal of Physics: Conference Series **640** (2015) 012015 doi:10.1088/1742-6596/640/1/012015

a confined protein with the use of a simplified coarse-grained model - the hydrophobic-polar (HP) lattice protein model [16, 17]. In spite of its simplicity, the ground state search of a HP model has been shown to be NP-complete for both 2- and 3-dimensional lattices [18, 19]. Traditional Metropolis sampling [20] is extremely inefficient in approaching the problem, due to the fact that the random walker is easily trapped in local free energy minima. Obtaining correct low temperature thermodynamics is then impractical within a reasonable time frame. To circumvent the problem, we employed a modern algorithm called Wang-Landau (WL) sampling [21, 22, 23], which allows us to deal with rugged free energy landscapes efficiently. Additionally, two inventive Monte Carlo moves, pull moves [24] and bond-rebridging moves [25], have been utilized in combination with WL sampling. This has been proven to be robust in ground state searches and accurate estimation of the density of states for lattice protein models [26, 27].

2. Model and simulation methods

2.1. Hydrophobic-polar (HP) protein model with surface confinement

The HP model [16, 17] is a well-known coarse-grained lattice model for the study of protein folding problems. This model is motivated by the fact that the hydrophobic effect, i.e., the tendency of hydrophobic (non-polar) amino acids to assemble together in an aqueous environment, is a dominant driving force for tertiary structure formation. Amino acids in a protein are classified as hydrophobic (H) or polar (P) according to the interactions of their side chains with water. The protein chain is modeled as a self-avoiding walk on a 2-dimensional square or a 3-dimensional cubic lattice, where an amino acid is located at each lattice site along the walk without overlaps. The hydrophobic effect is mimicked by assigning an attraction, ε_{HH}, between two non-bonded, nearest-neighbor H monomers. In order to study a confined protein in our system, two attractive walls are constructed in parallel to the $xy-$plane and distance h_w apart. For simplicity, we only considered surfaces that attract all monomers with the same strength, ε_S, and this interaction is identical for both the top and bottom surfaces. The total energy E of a confined protein is thus defined as

$$E = -\varepsilon_{HH} n_{HH} - \varepsilon_S (n_{SH} + n_{SP}), \qquad (1)$$

where n_{HH}, n_{SH} and n_{SP} are the numbers of H-H contacts and surface contacts with H and P monomers, respectively.

2.2. Thermodynamics and Wang-Landau sampling

In general, the transitions between conformational phases can be investigated via the temperature dependence of specific heat C_V/N, where N is the system size. The heat capacity C_V can be calculated by the energy fluctuation:

$$C_V = \frac{\langle E^2 \rangle - \langle E \rangle^2}{kT^2}, \qquad \text{where} \qquad \langle E \rangle = \frac{1}{Z} \sum_E E g(E) e^{-E/kT}. \qquad (2)$$

Here, k is the Boltzmann constant, T is the temperature and Z is the partition function, which can be obtained by summing the density of states, $g(E)$, over all energies weighted by the Boltzmann factor:

$$Z = \sum_E g(E) e^{-E/kT}. \qquad (3)$$

In our simulations, we employed Wang-Landau (WL) sampling [21, 22, 23] to determine the $g(E)$ of the system. The simulation starts with an initial estimate of $g(E) = 1$ for all possible energies E. In each Monte Carlo update, a random configuration has been generated 20% of the time from pull moves [24] and 80% of the time from bond-rebridging moves [25]. The probability of accepting such an update is given by

$$p(E_i \to E_j) = \min\left(\frac{g(E_i)}{g(E_j)}, 1\right), \tag{4}$$

where E_i is the initial energy and E_j is the energy of the trial configuration. Every time a new configuration is accepted, the density of states is adjusted by $g(E_j) \to g(E_j) \times f$; otherwise $g(E_i) \to g(E_i) \times f$. f is a modification factor which is initially set to $f_0 = e^1 \approx 2.71828$. With each energy update, the corresponding entry in the histogram $H(E)$ is incremented by $H(E) \to H(E) + 1$. This procedure repeats until the histogram $H(E)$ is sufficiently flat, i.e., all the entries of $H(E)$ are greater than 80% of the average number of total visits. The histogram is then reset to $H(E) = 0$ and a new iteration begins with a reduced modification factor, $f \to \sqrt{f}$. The simulation continues until f reaches $f_{\text{final}} = \exp(10^{-8})$. Finally we obtain a good estimate for $g(E)$ up to an undetermined constant.

2.3. Structural observables and production run

Observing C_V alone is no longer sufficient as it merely gives overall energy fluctuations that could be induced by a combination of coexisting structural transitions. In some cases, this may result in enhanced or suppressed energy fluctuations. Supplementary structural quantities are then necessary for a reliable identification of conformational transitions. The canonical averages of the number of nearest-neighbor H-H contacts, $\langle n_{HH}\rangle$, and the numbers of monomer-surface contacts, $\langle n_{SH}\rangle$ and $\langle n_{SP}\rangle$, are useful quantities for monitoring energy contributing interactions. The average radii of gyration in the x, y, and z directions, $\langle R_g^x\rangle$, $\langle R_g^y\rangle$ and $\langle R_g^z\rangle$, are informative properties about the HP chain conformations. R_g^x measures the degree of the spreading of monomers in the x direction and is defined by

$$R_g^x = \sqrt{\frac{1}{N}\sum_{i=1}^{N}(r_i^x - r_{CM}^x)^2}, \qquad \text{with} \qquad r_{CM}^x = \frac{1}{N}\sum_{i=1}^{N} r_i^x. \tag{5}$$

r_i^x and r_{CM}^x are the x coordinates of the ith monomer and the center of mass, respectively. R_g^y and R_g^z are defined likewise.

To obtain these structural observables, we perform a multicanoncial simulation [28] as a production run using the $g(E)$ that we obtained from the previous simulation as the sampling weights. We estimate a 2-dimensional joint density of states $g(E, Q)$ in energy E and structural quantity Q. In this process, the acceptance probability is the same as in Eq. (4), except that $g(E)$ is held fixed during this stage. Every time a configuration is selected, the structural quantities Qs are calculated while the corresponding 2-dimensional histograms $H(E, Q)$s are updated. At the end of the simulation, $g(E, Q)$s are obtained by reweighting $H(E, Q)$s with $g(E)$:

$$g(E, Q) = g(E)H(E, Q). \tag{6}$$

The average of the structural quantity Q can then be calculated by

$$\langle Q\rangle = \frac{1}{Z}\sum_{E,Q} Q g(E, Q) e^{-E/kT}. \tag{7}$$

3. Results and discussion

We performed simulations on a 48mer ($PHPHP_4HPHPHP_2HPH_6P_2H_3PHP_2\ HPH_2P_2HPH_3P_4H$), which is one of the ten sequences proposed by Yue *et al.* for algorithm efficiency testing [29]. In

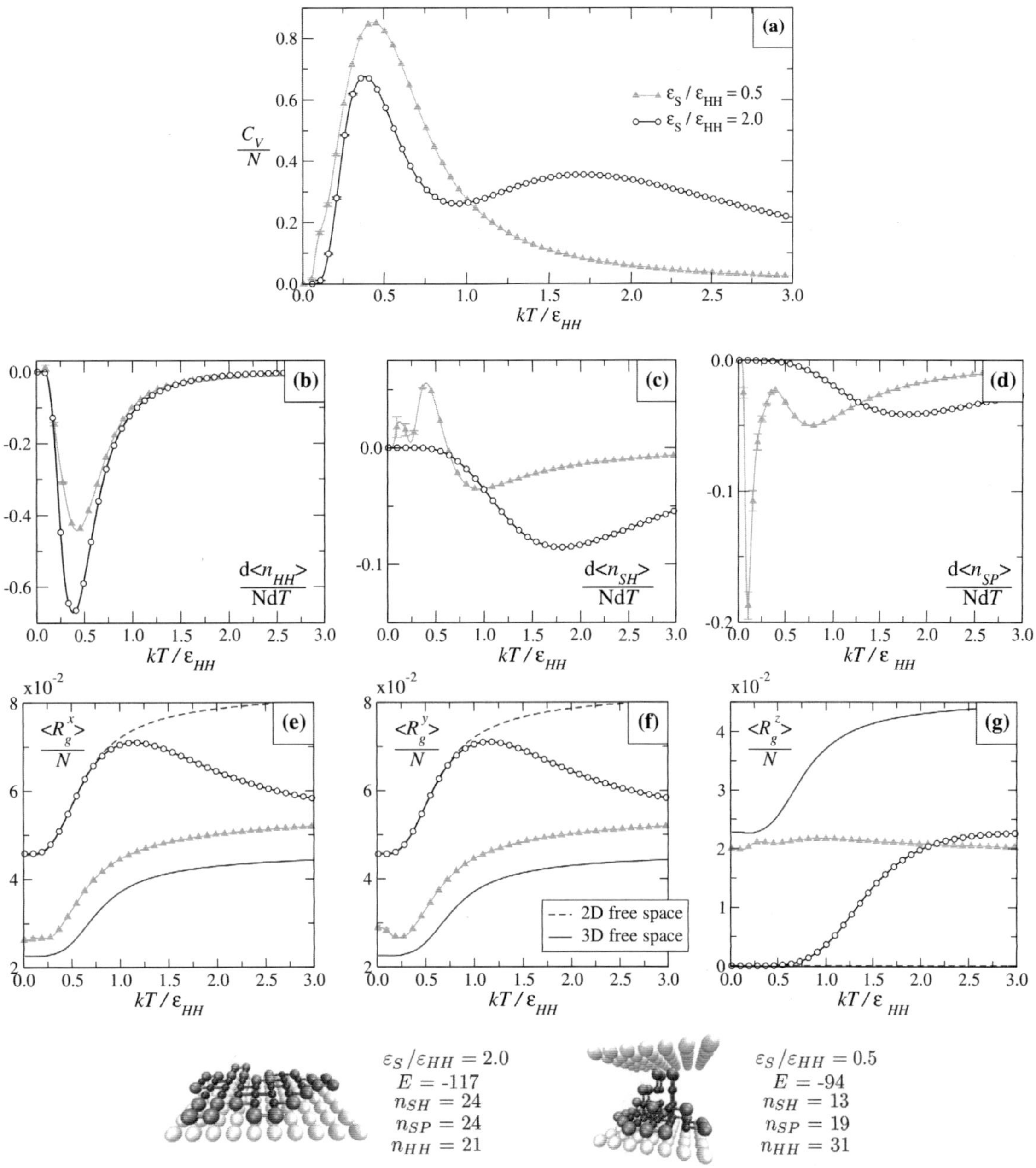

Figure 1. (color online) Thermodynamic and structural quantities of the 48mer confined between two surfaces with a separation of $h_w = 5$ and different surface attractive strength, ε_S. In all figures, error bars smaller than the data points are not shown. (a) Specific heat; (b)-(d) thermal derivatives of $\langle n_{HH}\rangle$, $\langle n_{SH}\rangle$, and $\langle n_{SP}\rangle$; (e)-(g) average radii of gyration in x, y, and z directions. The 2D and 3D surface-free cases (i.e., without confinement) are also plotted for comparisons. Typical ground state structures are displayed at the bottom.

this sequence, there are equal numbers of H- and P-monomers. The chain is placed between two attractive surfaces with a small surface separation $h_w = 5$. The effect of surface attraction was studied by varying the ratio between the surface attractive strength and the attraction between the H-monomers, $\varepsilon_S/\varepsilon_{HH}$. Two cases were considered in this study: $\varepsilon_S/\varepsilon_{HH} = 0.5$ and 2.0. 15 independent runs were performed to obtain statistical errors in both cases.

Depending on the surface attractive strength, structural transitions occur at different temperatures. Fig. 1 shows the specific heat and the thermodynamic properties we defined in Sec. 2.2 and 2.3. Since the surface separation is much smaller than the HP chain length, i.e. $h_w \ll N$, parts of the chain form contacts with both surfaces and the chain forms a "bridge" between the two surfaces at high temperature for all cases. As temperature lowers, the HP chain tends to maximize the surface contacts. This process is signaled by negative peaks in the thermal derivatives of $\langle n_{SH} \rangle$ and $\langle n_{SP} \rangle$ due to the increase in the numbers of surface contacts. This occurs at $kT/\varepsilon_{HH} \approx 1.75$ for $\varepsilon_S/\varepsilon_{HH} = 2.0$, and at $kT/\varepsilon_{HH} \approx 0.8$ for $\varepsilon_S/\varepsilon_{HH} = 0.5$.

When the surface attraction is relatively strong, i.e. $\varepsilon_S/\varepsilon_{HH} = 2.0$, the chain in fact detaches from one surface to adsorb onto the other until most of the monomers sit on the same surface. Since most of the monomers need to spread out on the same surface to form a film-like structure as a result of this debridging process, this is accompanied by an increase in $\langle R_g^x \rangle$ and $\langle R_g^y \rangle$, but a decrease in $\langle R_g^z \rangle$ around the transition temperature. In terms of the conformational changes and the signals observed in relevant structural properties, the debridging process resembles an adsorption transition as if there was only one adsorbing surface. In this case, the same analysis techniques as for the HP adsorption model [30] could be applied for the understanding of structural transitions.

Note that this debridging process is completely missing for the weakly attractive surface case $\varepsilon_S/\varepsilon_{HH} = 0.5$. The slight increase in surface contacts (as signaled by small peaks in $d\langle n_{SH} \rangle/dT$ and $d\langle n_{SP} \rangle/dT$ at $kT/\varepsilon_{HH} \approx 0.8$), the relatively constant $\langle R_g^z \rangle$ but the decreasing $\langle R_g^x \rangle$ and $\langle R_g^y \rangle$ as temperature lowers give the hint of a slight unwinding of the chain in preparation for a more energetically favorable hydrophobic core (H-core) formation that follows.

Further decrease in temperature induces H-core formation, where the extended chain collapses to form a compact structure with hydrophobic monomers forming a central core and polar monomers forming the outer shell. Regardless of the surface attraction strength, this occurs at $kT/\varepsilon_{HH} \approx 0.4$ as shown by the peaks in $d\langle n_{HH} \rangle/dT$. However, we stress that these H-core formations are not identical. While the one in the $\varepsilon_S/\varepsilon_{HH} = 0.5$ case is a 3-dimensional process in which the globular structure retains contacts to both surfaces, the one in the $\varepsilon_S/\varepsilon_{HH} = 2.0$ case is a 2-dimensional process where it occurs on a single adsorbing surface.

4. Conclusion

We studied the influence of surface attractions on the structural transitions of a HP lattice protein confined between two attractive surfaces, combining Wang-Landau sampling with efficient Monte Carlo moves. We focused on the case where the surface separation is small compared to the chain length, and depending on the surface attraction strength, the debridging process may or may not be present. Decreasing the surface attraction shifts this process to a lower temperature (as proven by our simulations but results were not shown in this work); until it vanishes when the surface attraction becomes too weak. At low temperature, there is a hydrophobic core formation. With a strongly attractive surface where debridging has taken place at high temperature, H-core formation is a 2-dimensional process that results in a film-like ground state. For a weakly attractive surface, the H-core formation occurs while the HP chain is in contact with both the top and bottom surfaces, leading to a 3-dimensional, globular ground state.

Structural transitions are also be affected by other factors or system parameters as discussed in our previous studies [31, 32, 33]. The debridging process can disappear when the separation between the two attractive surfaces is relatively large or sufficiently small compared to the chain length. Moreover, both the surface types and the proportion of the H and P monomers in the chain can influence the transitions. It implies that the size of confining space, as well as the chemical properties of the confining surfaces, are the dominant factors of structural transitions.

Acknowledgments

B. Pattanasiri was supported by The Royal Thai Government Scholarship under the Development and Promotion of Science and Technology Talent Project (DPST), and the grant from The Institute for the Promotion of Teaching Science and Technology (IPST). Y.W.L. was sponsored by the Oak Ridge Leadership Computing Facility, Office of Advanced Scientific Computing Research, U.S. Department of Energy under Contract DE-AC05-00OR22725. Computing time was provided by the Georgia Advanced Computing Resource Center at the University of Georgia.

References

[1] Goodsell D S 1991 *Trends Biochem. Sci.* **16** 203–6
[2] Zimmerman S and Trach S 1991 *J. Mol. Biol.* **222** 599–620
[3] Zimmerman S and Minton A 1993 *Annu. Rev. Biophys. Biomol. Struct.* **22** 27–65
[4] Ellis R 2001 *Trends Biochem. Sci.* **26** 597–604
[5] Taubes G 1996 *Science* **271** 1493–95
[6] Sewell R 2007 *Protein Misfolding in Neurodegenerative Diseases: Mechanisms and Therapeutic Strategies* Enzyme Inhibitors Series (Taylor & Francis)
[7] Minton A P 2000 *Curr. Opin. Struct. Biol.* **10** 34–9
[8] Ellis R J 2001 *Curr. Opin. Struct. Biol.* **11** 114–9
[9] Lucent D, Vishal V and Pande V S 2007 *Proc. Natl. Acad. Sci. U.S.A.* **104** 10430–4
[10] Ruckenstein E and Byungip Chung D 1988 *J. Colloid Interface Sci.* **123** 170–85
[11] Martins D C, Chu V and Conde J P 2013 *Biomicrofluidics* **7** 034111
[12] Sang J, Du H, Wang W, Chu M, Wang Y, Li H, Zhang H A, Wu W and Li Z 2013 *Biomicrofluidics* **7** 024112
[13] Grieshaber D, MacKenzie R, Vörös J and Reimhult E 2008 *Sensors* **8** 1400–58
[14] Kim S J and Han J 2008 *Anal. Chem.* **80** 3507–11
[15] Duan C, Wang W and Xie Q 2013 *Biomicrofluidics* **7** 026501
[16] Dill K A 1985 *Biochemistry* **24** 1501–09
[17] Dill K 1990 *Science* **250** 297–8
[18] Berger B and Leighton T 1998 *J. Comp. Biol.* **5** 27–40
[19] Crescenzi P, Goldman D, Papadimitriou C, Piccolboni A and Yannakakis M 1998 *J. Comp. Biol.* **5** 423–65
[20] Metropolis N, Rosenbluth A W, Rosenbluth M N, Teller A H and Teller E 1953 *J. Chem. Phys* **21** 1087
[21] Wang F and Landau D 2001 *Phys. Rev. Lett.* **86**(10) 2050–2053
[22] Landau D P and Wang F 2002 *Comput. Phys. Commun.* **147** 674–7
[23] Landau D P, Tsai S H and Exler M 2004 *Am J Phys* **72** 1294–1302
[24] Lesh N, Mitzenmacher M and Whitesides S 2003 A complete and effective move set for simplified protein folding *Proc. 7th Annual Inter. Conf. on Research in Computational Molecular Biology (RECOMB)* (Berlin: ACM Press) pp 188–95
[25] Deutsch J M 1997 *J. Chem. Phys* **106** 8849–54
[26] Wüst T and Landau D 2008 *Comput. Phys. Commun.* **179** 124–7
[27] Wüst T and Landau D P 2009 *Phys. Rev. Lett.* **102** 178101
[28] Berg B A and Neuhaus T 1992 *Phys. Rev. Lett.* **68** 9–12
[29] Yue K, Fiebig K M, Thomas P D, Chan H S, Shakhnovich E I and Dill K A 1995 *Proc. Natl. Acad. Sci. U.S.A.* **92** 325–9
[30] Li Y W, Wüst T and Landau D P 2013 *Phys. Rev. E* **87** 012706
[31] Pattanasiri B, Li Y W, Landau D P and Wüst T 2012 *Int. J. Mod. Phys. C* **23** 1240008
[32] Pattanasiri B, Li Y W, Landau D P, Wüst T and Triampo W 2012 *J. Phys.: Conf. Series* **402** 012048
[33] Pattanasiri B, Li Y W, Landau D P, Wüst T and Triampo W 2013 *J. Phys.: Conf. Series* **454** 012071

Multiscale Modelling of UniMolecular FRET Probes Using Monte Carlo Simulations

Shourjya Sanyal [1,2], Donal MacKernan[1], and David F. Coker[1,3]

[1] CECAM-IRL and School of Physics and Complex Adaptive Systems Laboratory, University College Dublin, Dublin 4, Ireland

[2] Systems Biology Ireland, University College Dublin, Ireland

[3] Department of Chemistry, Boston University, 590 Commonwealth Avenue, Boston, MA 02215, USA

E-mail: shourjya.sanyal@ucdconnect.ie

Abstract. Föster Resonance Energy Transfer (FRET) based biosensors are widely used in experimental biology to understand spatiotemporal dynamics of protein pairs both in-vivo and in-vitro. In our study we have developed an idealised model consisting of two macroparticles, representing proteins attached to fluorescent chromophores, linked with an idealized flexible linker consisting of N point-like beads joined by harmonic springs. Monte Carlo simulations with these models are used to investigate the influence of flexible linkers on unimolecular FRET based bio-sensors. The results provide qualitative insight into the efficacy of designing flexible peptide linkers.

1. Introduction

Genetically encoded FRET bio-sensors are extensively used in molecular biology both in-vivo and in-vitro to observe via optical microscopy, binding between key bio-molecules due to biochemical signals, and thereby deduce, identify and understand biochemical reactions and pathways[1, 2, 3]. Despite the vast variety of forms that such probes take, the basic operating principle is essentially the same. Given a suitable pair of chromophores, when light of an appropriate frequency range is absorbed by the donor, and if the distance r between it and the other chromophore (the acceptor) is sufficiently close (1-10nm)[4], non-radiative energy transfer can take place from the donor to the acceptor[5]. Föster showed that this distance $\vec{r}$ dependent non-radiative energy transfer efficiency E varies as r^{-6}. The corresponding transfer rate is measurable in a variety of ways, including through donor-to-acceptor emission intensity ratios $\mathcal{I}$, the optical emission intensity of the acceptor, or alteration of donor fluorophore fluorescence lifetime.

The basic structure of the FRET biosensor developed and used in experiment by Komatsu et al[6], that we wish to model, comprises of a ligand domain attached to the donor chromophore, and a sensor domain attached to the acceptor chromophore connected by a flexible linker (see fig. 1.A.). The sensor domain changes its conformation upon perception of a bio-chemical signal. This effectively switches on an attractive interaction between the sensor and ligand domains, increasing the likelihood of their binding and is known as the "ON" state. In absence of the bio-chemical signal there is a lesser likelihood of the two chromophores coming closer and is known as the "OFF" state. FRET based microscopy is feasible provided the FRET sensitised donor-to-acceptor intensity ratio $\mathcal{I}$ in the "ON" state is measurable, and is distinguishably greater than the "OFF" state basal/background value $\mathcal{I}_0$. The performance of a linker is optimized in terms of maximising the signal to noise ratio of those two quantities, defined hereafter as $\Delta\mathcal{I}/\mathcal{I}_0 \equiv \frac{\mathcal{I}-\mathcal{I}_0}{\mathcal{I}_0}$. To increase the measurable $\mathcal{I}$ and $\mathcal{I}_0$, the chromophores are joined by suitable peptide linkers of length $\mathcal{L}$, giving origin to the so called unimolecular FRET bio-sensor systems. Decreasing linker length leads to an increase in both $\mathcal{I}$ and $\mathcal{I}_0$, but a decrease in $\Delta\mathcal{I}/\mathcal{I}_0$. The goal of our study is to see if the simplified models can correctly predict values of $\mathcal{I}$, $\mathcal{I}_0$and $\Delta\mathcal{I}/\mathcal{I}_0$.

A great deal of work has succeeded in designing FRET unimolecular probes such as KAR, focused on improving their sensitivity and dynamic range (i.e. $\Delta\mathcal{I}/\mathcal{I}_0$). Allen and Zhang[7, 8], motivated by the desire to track cAMP activity, for example at the plasma membrane, cytoplasm, nucleus, and mitochondria in living cells, systematically optimized probes through circular permutation(cp). Lissandron et al[9] improved cAMP unimolecular probes combining molecular simulation with experiments to optimise linkers through mutation of selective residues, and thereby almost doubled FRET efficiency and substantially improved dynamic range. Pertz et al[10] focusing on the Rho family of GTPases (G proteins) which regulate the actin and adhesion dynamics that control cell migration, developed a fluorescent biosensor to visualize the spatiotemporal dynamics of RhoA activity during cell migration, and subsequently engineered a library of probes[3] by varying several geometrical parameters, such as fluorophore distance (using linkers of different length), dipole orientation (using cp mutants in both the acceptor and donor fluorophores), and sensing module domain topology. This approach accommodated structural constraints imposed by different signaling proteins to produce sensors with increased dynamic range.

The approach of the present paper builds on the probe designs of Komatsu et al[2]. By developing long, flexible linkers ranging from 116 to 244 amino acids in length, they demonstrated that they could optimize the corresponding gain (i.e. the $\Delta\mathcal{I}/\mathcal{I}_0$). Evers et al[11] also investigated similar linker systems made of flexible glycine and serine containing peptide linkers, with enhanced cyan and yellow fluorescent protein terminals (ECFP-linker-EYFP) in which the linker length was systematically varied by incorporating between 1 and 9 GGSGGS repeats. As expected, both steady-state and time-resolved fluorescence measurements showed a decrease in energy transfer with increasing linker length. They also explained how these effects could be quantitatively understood by simple models that describe the flexible linker as a worm-like chain.

The paper is organised as follows. The models, FRET observable and simulation method are detailed in section 2. The simulation results are given in section 3 and compared where appropriate with experimental findings. The conclusions of this work are presented in section 4.

2. Models and Methods

To explore idealized design motifs for chromophore - linker - chromophore systems, we have built simplified models of unimolecular FRET probes, represented by two macroparticles joined by a mechanistic flexible linker (modeling the sort of system investigated by Komatsua et al). One spherical macroparticle represents the donor fluorophore bound to ligand domain, another represents the acceptor fluorophore bound to the signal domain. We use the Monte Carlo simulation approach introduced by Metropolis et al[12, 13] to explore the statistical properties of each model.

The spherical macromolecules interact through a pair potential of the form $V(\vec{r}_1, \vec{r}_2) = V_s(\vec{r}_1, \vec{r}_2) + V_\ell(\vec{r}_1, \vec{r}_2)$, where the first term denotes binding between the macroparticles due to a biochemical signal, and the second term an interaction specific to each linker type. When the signal domain is in "ON" state, V_s is represented by an attractive square well interaction of depth ϵ for $\sigma < r < \sigma + \Delta$, where $r = |\vec{r}_2 - \vec{r}_1|$. When the signal domain is in the "OFF" or basal state V_s is set uniformly to zero.

The protein diameter defined as σ will hereafter be used as a measure of length and the equivalent of $\sim 24\mathring{A}$ representing the diameter of the Fluorescent Proteins. The Föster Radius R_0 is assumed to be 2.5 times larger than σ [14], and the macroparticles in the "ON" state interact within and a narrow attractive well Δ of width 0.5σ. In our models the distance dependence of the FRET efficiency is approximated by the expression $E(r) = \frac{1}{1+(r/R_0)^6}$ [5]. The binding in the "ON" state is measured in terms of ϵ. Energy is measured in terms of k_BT and in this units $\epsilon\sigma^2/k_BT = 1$ can be equivalently written as $\epsilon = 1k_BT$.

2.1. Bead and Spring linker

In the case of the bead and spring linker model the macroparticles are joined by a discrete chain made of N point-like beads at position $r_i, i = 1, ..., N$. These point particles are connected to the macroparticles and among themselves by harmonic springs, and interact with a simple isotropic pair potential [15].

$$V_\ell = V_{chain} = \frac{1}{2}k_s[|r_1 - R_1| - d_1]^2 + \sum_{i=1}^{N-1} \frac{1}{2}k_s[|r_{i+1} - r_i| - d_0]^2 + \frac{1}{2}k_s[|R_2 - r_n| - d_1]^2 \quad (1)$$

The equilibrium distance d_0 between successive beads along the chain is assumed to be $3.8\mathring{A}$ corresponding to the distance between two C_α atoms in a peptide. The equilibrium distance d_1 between the macroparticle and the bead is defined as sum of radius of the macroparticle and the bead i.e. $13.9\mathring{A}$. We set the spring constant $k_s = 80$ in the units of k_BT. This value of k_s was chosen to reproduce experimentally observed values of FRET intensity as a function of number of residues.

2.2. Bead and Spring linker with Bending Forces

These models consider an additional bending contribution attributed to triplets of consecutive beads in a chain, $V_\ell = V_{chain} + V_{bend}$ where the reference bend angle $\phi_0 = 116°$ often used to model angular bonds between two C_α atoms in a peptide [16].

$$V_{bend} = \frac{1}{2}k_{bend} \sum_{i=2}^{N-1} [cos\phi - cos\phi_0]^2 \quad (2)$$

where

$$cos\phi = \frac{(r_{i+1} + r_i) \cdot (r_{i-1} + r_i)}{|r_{i+1} + r_i||r_{i-1} + r_i|} \quad (3)$$

2.3. Effective potential linker

These macroparticles interact with a simple isotropic pair potential,

$$V_\ell(|\vec{r}_2 - \vec{r}_1|) = \begin{cases} 0 & \text{if } r > \sigma \text{ or } r < L \\ \infty & \text{otherwise} \end{cases} \quad (4)$$

Geometrically this can be visualized as two non-overlapping particles free to move inside a sphere of diameter L.

3. Results

3.1. Flexible Linker systems

Simulations have been carried out for systems of two macroparticles linked by a harmonic chain of $10 \leq N \leq 250$ discrete beads, each of them qualitatively representing an amino acid unit in the linker peptide [17]. All systems have been equilibrated for 10^7 Monte Carlo moves, and statistics have been collected over an equal number of steps. The binding energy ϵ representing attractive potential between the macroparticle has been varied over the range $0 \leq \epsilon/k_BT \leq 6$. A remarkable correspondence between the experimentally observed behavior and the bead and spring linker is presented in the Fig.1.B. Where the 0ϵ represents the "OFF" or basal state, and the binding energy in the case of "ON" state is predicted to be 2.5ϵ. This illustrates the possibility to predict the complex protein - linker - protein system using such simple models.

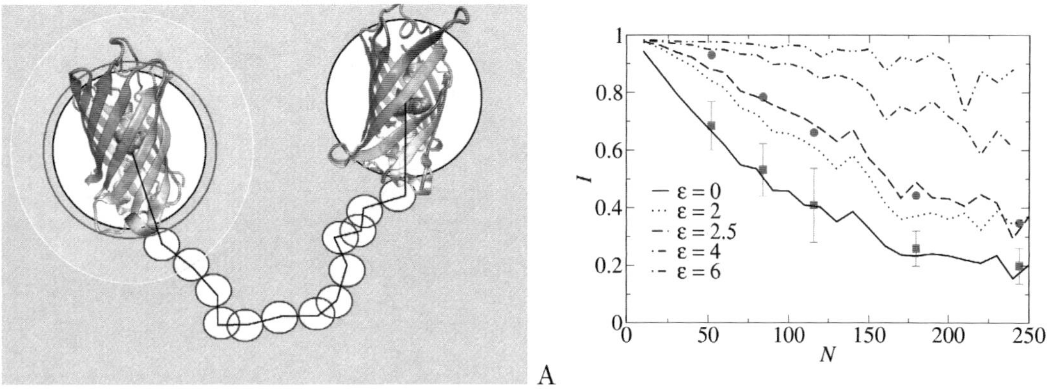

A B

Figure 1. (A) Schematic representation of the unimolecular FRET probe with flexible linker used in experiment reported by Komatsu et al. (B) FRET sensitised donor-to-acceptor intensity ratio of the flexible linker model as a function of number of linker residues. Results presented are for the model "OFF" state $\mathcal{I}_0$ (full line, black), and in the "ON" state $\mathcal{I}$ for four different values of the macroparticle interaction strength ϵ. Experimental data from Komatsu et al[6], in the "OFF" or basal state (filled red squares) and in the "ON" state (filled red circles) are superimposed on the theoretical predictions.

In order to compare the bead and spring linker with the effective potential linker, similarities in configuration space is illustrated by plotting the end to end distribution of the macroparticles in the basal or "OFF state for the four linker lengths as shown in Fig.2.A. Our simulation shows that the linker length N in the case of the bead and spring linker, can be mapped to the sphere diameter L (in the case of effective linker potential) using a simple relation $N \sim 40L - 70$

3.2. Effects of Binding Energy ϵ

The thermodynamics state of the system is characterised primarily by the mean square distance between the macroparticles, and by the equilibrium fluctuation of this same quantity. The major feature revealed by this figure is the transition from a Gaussian chain at low ϵ, to a closed ring at high value of ϵ as shown in Fig.2.B. In the former case, the average separation scales like $N^{\frac{1}{2}}$, in agreement with well-known universal properties of Gaussian chains, in the second case the average distance is independent of N. As shown in Fig.1.B. the FRET sensitised donor-to-acceptor intensity is low in the open chain regime, and high when the system is in the ring configuration. As a result, the intensity monotonically increases with increasing ϵ, and decreases with increasing N, at a rate that increases with decreasing ϵ.

3.3. Effects of Bending and Torsion Forces on Chain Dynamics

The addition of a weak bending term k_{bend} does not change significantly the end-to-end distribution and thus the FRET results, as shown in Fig.3.A. Simulations have been carried

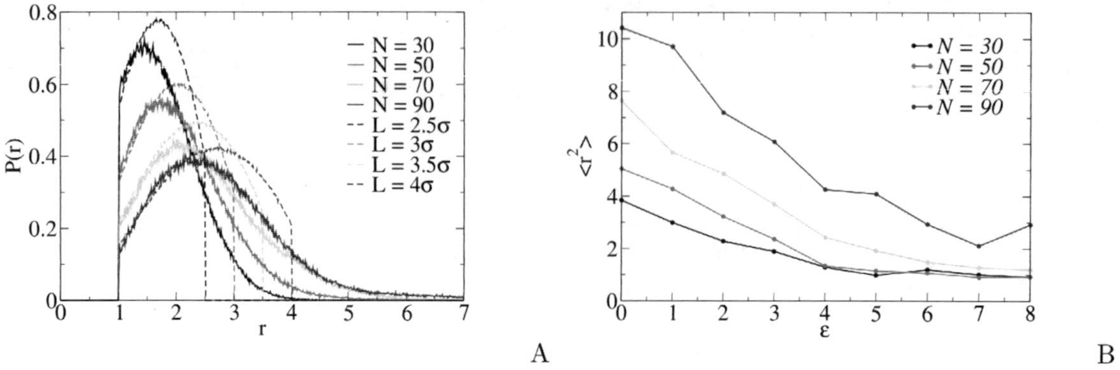

Figure 2. (A) Distribution of end-to-end distances in the "OFF" case for the bead and spring (full line) and the effective potential linker (dashed line). (B) Mean square displacement as a function of binding energy ϵ for the bead and spring linker.

out with $k_{bend}/k_BT = 40$, corresponding to a relatively weak perturbation of the harmonic chain. A few simulations with higher k_{bend} values have confirmed the insensitivity of the results on the choice of k_{bend}. The insensitivity of the end-point distribution to bending is consistent with general results on the structural and thermodynamic properties of Gaussian chains [18].

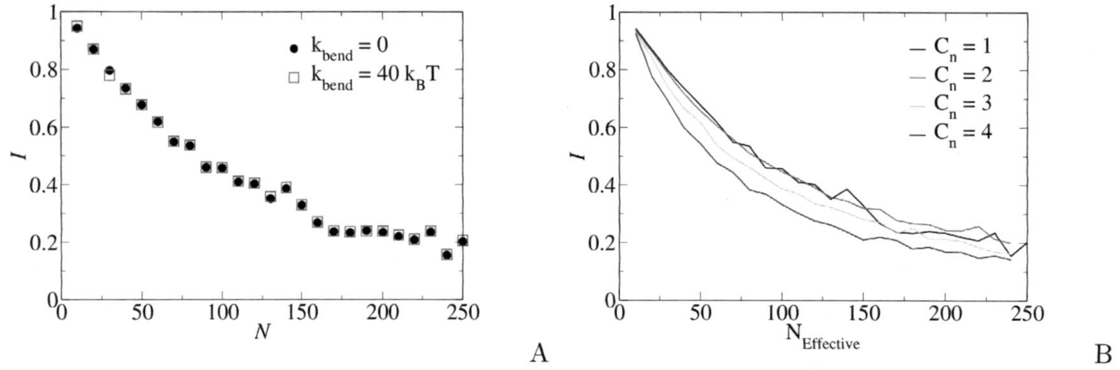

Figure 3. (A) FRET sensitised donor-to-acceptor intensity ratio in the "OFF" case as a function of linker length for a bead and spring linker with bending energy term ($k_{bend}/k_BT = 0$, filled black dots) and ($k_{bend}/k_BT = 40$, empty red squares). (B) FRET sensitised donor-to-acceptor intensity ratio in the "OFF" case as a function of effective linker length for a for a bead and spring linker with varying C_n from 1 to 4.

3.4. Relation between the Characteristic Ratio C_n and Spring Constant k_s

Two different approaches are widely used to model flexible polymer systems. On one hand there is the worm like harmonic chain with uniform flexibility determined by a spring constant k_s, and the segmentally flexible chain [19, 20] where the segment size is determined by the characteristic ratio C_n. In Fig.3.B. we have used the same effective linker length $N_{Effective}$ of 240 residues but varied the segment size from $C_n = 1$ at $N = 240$ to $C_n = 4$ at $N = 60$. As increasing the value of spring constant k_s gives higher FRET sensitised donor-to-acceptor intensity, the results show that by increasing the characteristic ratio C_n one can increase the effective spring constant k_s.

4. Conclusions

The idealised nature of the models used in this study implies that our observations and conclusions might have fairly general validity. The simulation results show that the bead and spring model linkers are able to reproduce the experimental properties of unimolecular FRET probes. The agreement between experiment and simulation suggests that this approach could be used to predict experimental values of binding energy between the ligand and signal domain. Since these effects are due to configuration phase properties, the results of the bead and spring model can be reproduced by the simple effective potential linker model.

Acknowledgments

Acknowledgments Support from SimSci (Funded under the programme for Research in Third-level Institutions and co-funded Under the European Regional Development Fund) (S.S), Systems Biology Ireland, Dublin (S.S), and SFI Grant No.08-IN.1-I1869 (D.M.K) is acknowledged. The authors thank Pietro Ballone for his insightful contributions to this work .

References

[1] Miyawaki A 2003 *Dev. Cell.* **4** 295-305
[2] Miyawaki A 2011 *Annu. Rev. Biochem.* **80** 357-73
[3] Fritz RD, Letzelter M, Reimann A, Martin K, Fusco L, Ritsma L, Ponsioen B, Fluri E, Schulte-Merker S, van Rheenen J and Pertz O 2013 *Sci. Signal.* **6** rs12
[4] Berney C and Danuser G 2003 *Biophys. J.* **84** 3992-4010
[5] Lakowicz J R 2010 *Principles of Fluorescence Spectroscopy* vol 3 (New York : Springer) pp 443-72
[6] Komatsu N, Aoki K, Yamada M, Yukinaga H, Fujita Y, Kamioka Y and Matsuda M 2011 *Mol. Biol. Cell.* **22** 4647-56
[7] Allen MD and Zhang J 2006 *Biochem. Biophys. Res. Commun.* **348** 716-21
[8] Zhang J and Allen MD 2007 *Mol. BioSyst. 3* 759-65
[9] Lissandron V, Terrin A, Collini M, D'alfonso L, Chirico G, Pantano S and Zaccolo M 2005 *J. Mol. Biol.* **354** 546-55
[10] Pertz O, Hodgson L, Klemke RL and Hahn KM 2006 *Nature* **440** 1069-72
[11] Evers TH, van Dongen EM, Faesen AC, Meijer EW and Merkx M 2006 *Biochemistry* **45** 13183-92
[12] Metropolis N, Rosenbluth A W, Rosenbluth M N, Teller A H and Teller E 1953 *J. Chem. Phys.* **21** 1087-92
[13] Frenkel D and Smit B 1996 *Understanding Molecular Simulation: From Algorithms to Applications* vol 1 (Orlando, USA: Academic Press) chapter 2 pp 23-58
[14] Hink M A, Visser N V, Borst J W, van Hoek A and Visser A J W G 2003 *J. Fluoresc.* **13** 185-187
[15] Strobl G 2007 *The Physics of Polymers* (Berlin Heidelberg : Springer-Verlag)
[16] Bereau T and Deserno M 2009 *J. Chem. Phys.* **130**
[17] Caracciolo S, Parisi G and Pelissetto A 1994 *J.Statist. Phys.* **77** 519-543
[18] de Gennes P G 2007 *Journal of Polymer Science: Polymer Symposia* **61** 313-15
[19] Yguerabide J, Epstein H F and Stryer L 1970 *J. Mol. Biol.* **51** 57388
[20] Garca de la Torre J, Pérez Sánchez HE, Ortega A, Hernández JG, Fernandes MX, Díaz FG and López Martnez MC 2003 *Eur. Biophys. J.* **32** 477-86

Protein folding of the H0P model: A parallel Wang–Landau study

G Shi[1], T Wüst[2], Y W Li[3] and D P Landau[1]

[1]Center for Simulational Physics, The University of Georgia, Athens, GA 30602, USA
[2]Scientific IT Services, ETH Zürich IT Services, 8092 Zürich, Switzerland
[3]National Center for Computational Sciences, Oak Ridge National Laboratory, Oak Ridge, TN 37831, USA

E-mail: sgjerry@physast.uga.edu

Abstract. We propose a simple modification to the hydrophobic-polar (HP) protein model, by introducing a new type of monomer, "0", with intermediate hydrophobicity of some amino acids between H and P. With the replica-exchange Wang–Landau sampling method, we investigate some widely studied HP sequences as well as their H0P counterparts and observe that the H0P sequences exhibit dramatically reduced ground state degeneracy and more significant transition signals at low temperature for some thermodynamic properties, such as the specific heat.

1. Introduction

For decades, scientists have endeavored to understand the secret code of protein folding, which is believed to be essential in determining the shape of the native state and thus the biological function of a protein. Beside fundamental research, the motivation for studying the process of protein folding also comes from the large number of diseases in humans resulting from protein misfolding, e.g., Alzheimer's, Parkinson's, etc. [1]. Despite the astronomical number of possible configurations, a protein folds into its functional, three-dimensional structure (native state) amazingly fast; nevertheless, the mechanism behind folding is still unclear [2].

Generic, coarse-grained protein models have been developed to help understand the properties of protein folding [3, 4]. Among these, the hydrophobic-polar (HP) lattice protein model [5, 6] is arguably the simplest, and thus has been called the "Ising model" for protein folding [7]. The HP model attracts researchers from various disciplines because of its simplicity yet challenging computational complexity. It has been proven that finding the lowest energy structure of a given sequence is an NP-complete problem [8]; however, many folding algorithms and Monte Carlo methods have been developed for approaching this problem. Some representative examples include, but are not limited to, chain-growth methods [9, 10], e.g. the pruned-enriched Rosenbluth method (PERM) and its variants [11, 12, 13, 14, 15], sequential importance sampling [16], fragment regrowth Monte Carlo [17], multidomain sampler [18], genetic algorithm [19, 20], evolutionary Monte Carlo [21], ant colony models [22] and Wang–Landau sampling [23, 24].

The HP model has been used in several problems of biological interest, such as protein surface adsorption [25, 26, 27, 28], protein folding in membranes [29] or confined environments [30, 31], and the effects of sequence mutations [32, 33]. However, simplification into only H and P monomers yields large ground state degeneracy which stands in contrast to the generally unique native state of natural proteins. More advanced models such as the HPNX model [34, 35], which considers not only the hydrophobicity but also the charges of amino acids, has been shown to be effective in reducing degeneracy. From a different perspective, we propose a simple modification to the original HP model, by introducing a new type of "neutral" monomer, "0", i.e., neither hydrophobic nor polar, thus rendering the model more realistic without significantly increasing the difficulties of sampling.

2. Model and methods

2.1. H0P lattice protein model

The HP lattice model classifies amino acids into just two types, hydrophobic (H) and polar (P), and the structure of a protein is simulated by a chain of connected beads, each of which represents a single monomer, on a simple cubic lattice. The only interaction considered in the HP model is an effective monomer-monomer coupling ϵ_{HH} between non-bonded, nearest-neighbor H monomers. However, most HP sequences suffer from the drawback of quite large ground-state degeneracy.

Since, in reality, different amino acids possess different levels of hydrophobicity [36], simplification into only two types might be insufficient to characterize the features of the hydrophobic interaction, which is believed to be the driving force of protein folding and tertiary structure formation. Therefore, we introduce a new level of "neutral" monomers, "0", which neither favors nor dislikes water as much as H or P monomers do. In the H0P model, two different non-bonded interactions are considered: ϵ_{HH} and ϵ_{H0}, where ϵ_{H0} denotes the interaction strength between the H and the "0" monomers, and $\epsilon_{\mathrm{HH}} > \epsilon_{\mathrm{H0}}$. The Hamiltonian is then given by

$$\mathcal{H} = -\epsilon_{\mathrm{HH}} n_{\mathrm{HH}} - \epsilon_{\mathrm{H0}} n_{\mathrm{H0}}, \qquad (1)$$

where n_{HH} is the number of non-bonded HH contacts (n_{H0} has corresponding meaning). We used $\epsilon_{\mathrm{HH}} = 2$ and $\epsilon_{\mathrm{H0}} = 1$ for this work, but other interactions involving "0" monomers could be easily added.

2.2. Replica-exchange Wang–Landau sampling

We have used a recently developed generic, parallel Wang–Landau sampling scheme, namely Replica-exchange Wang–Landau sampling (REWL) [37, 38]. In this parallel scheme, the energy range of the system is split into multiple, overlapping windows, each of which is simulated by independent processes (random walkers, running serial Wang–Landau (WL) sampling [39, 40]). The replica exchanges between overlapping windows are attempted during the simulation at a fixed time interval, such that each replica can travel through the entire energy space of interest.

WL sampling is a serial Monte Carlo simulation method that has proven to be powerful for studying problems with complex free energy landscapes. The ultimate goal of WL sampling is to estimate the density of states $g(E)$, using a random walker in an energy range $[E_{\min}, E_{\max}]$. During the simulation, a MC trial move that changes the system from configuration A (with energy E_A) to configuration B (with energy E_B) will be accepted with probability

$$P(A \to B) = \min\left\{1, \frac{g'(E_A)}{g'(E_B)}\right\}, \qquad (2)$$

where $g'(E)$ is an instantaneous estimator for the density of states. Upon acceptance, this estimator is updated via $g'(E_B) \to f \times g'(E_B)$, where f is a modification factor, and the histogram of visited energies is increased by $H(E_B) \to H(E_B) + 1$. If the trial move was rejected, $g'(E_A)$ and $H(E_A)$ are updated instead in the same way. The modification factor will

Journal of Physics: Conference Series **640** (2015) 012017 doi:10.1088/1742-6596/640/1/012017

be reduced if the histogram is sufficiently "flat" and at the same time, all the histogram entries are reset to zero. The above procedures will be performed iteratively, until the modification factor is less than some predefined threshold value f_{final}. In this work, the initial modification factor is set to $f_{\text{init}} = e^1$ and decreased via $f \to \sqrt{f}$, and $f_{\text{final}} = \exp(10^{-8})$. Initially, $g'(E)$ is set to 1 and we use the "80%" flatness criterion for the histogram, i.e., every entry in the histogram is no less than the 80% of the mean histogram height. Multiple, independent runs are performed to test for systematic errors and to determine statistical errors. As Monte Carlo trial moves, we applied pull and bond-rebridging moves which have proven very efficient together with WL sampling [23, 24].

In order to study the ground-state properties of lattice proteins, we proposed a ground-state sampling method [33] to estimate the ground-state degeneracy and collect the ground-state structures. The basic idea behind this method is to perform a multicanonical sampling using $1/\tilde{g}(E)$ as the sampling weight, where $\tilde{g}(E)$ is the final estimator obtained from the preceding WL or REWL run. During the process, each encountered ground-state structure, i.e. the structure obtains the lowest energy found during the WL or REWL run, will be converted into a unique sequence of directions (SoD) containing five elements: Forward(F), Left(L), Right(R), Up(U) and Down(D), and will be stored in a tree data structure for efficient searching. When a ground-state structure is generated, its SoD will first be compared with existing ones. Any newly discovered structure will be added to the database. The simulation terminates when the ground-state degeneracy has "converged" (i.e., no further ground-state is discovered for a period of time). Since a SoD uniquely represents a ground-state structure with symmetries of a simple cubic lattice taken into account, this method provides an effective means of identifying and storing different ground-state structures and determine the degeneracy.

3. Results and discussion

3.1. Randomized Sequences

Our first step is to test the H0P model on one of the widely studied 48mers ("Harvard sequences"), 48.1 [41], which has been shown to have enormous ground-state degeneracy in previous studies [33, 42]. This sequence was originally designed for testing of folding algorithms and, thus, there is not a defined way to map it into a H0P sequence. Therefore, we randomly generated four H0P sequences based on 48.1 by converting each monomer (either H or P) to the new type 0 with a probability of one third, as shown in Table 1. After performing REWL sampling as well as the ground-state degeneracy estimation procedure, we list the ground-state degeneracies for all four sequences in Table 1. The ground-state degeneracies are quite limited for all four H0P sequences, and are orders of magnitude less than that for the equivalent HP 48mer.

Table 1. Estimation of ground-state degeneracy of four randomly generated H0P sequences with 48 monomers. Sequences with converged degeneracy do not have statistical errors.

ID	sequence	ground-state degeneracy
48.1	$HPH_2P_2H_4PH_3P_2H_2P_2HPH_3PHPH_2P_2H_2P_3HP_8H_2$	$1.03(4) \times 10^7$ [33]
48.1-1	$0_3HP0_2H_20PH_30P(0HP)_2H0(0P)_2H(H00)_2P(P0)_20P_30_2H_2$	269
48.1-2	$HP0HP_2(H0)_2PH_20P_20HP(PH)_20_5H_2P0_2HP(0P0)_20P_4H0$	1561
48.1-3	$HPH0P0H_4P0H_2P0H0_2P0(0H)_3P0_3P0H0P0_2(PP0)_20_3H$	34
48.1-4	$HPH_2P(0H)_20_2H_20_4HP_2HP0H_20HPH0_2P0H0P_2HP_60PH0$	3695

3.2. A Real Protein

Table 2. Amino acid sequence of Crambin, as well as its HP and H0P mappings.

ID	sequence
Crambin	TTCCPSIVARSNFNVCRLPGTPEALCATYTGCIIIPGATCPGDYAN
HP3D46 [43]	PPHHPPHHHPPPHPHHPHPPPPPHHHHPHPPHHHHPPHPHPPPHHP
H0P3D46	00HH00HH0P0PHPHHPH0000P0HH00000HHHH0000H00P00P

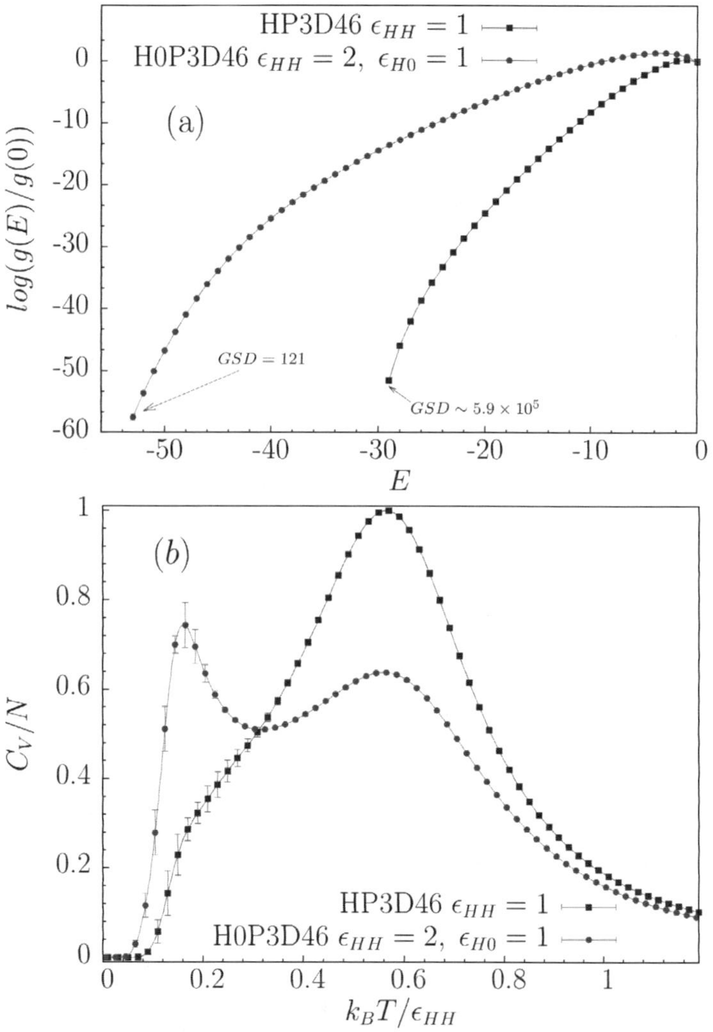

Figure 1. Physical observables from WL/REWL simulations of HP3D46 and H0P3D46 lattice models for Crambin: (a) the densities of states with their ground-state degeneracies (GSD) and (b) the specific heat of HP3D46 and H0P3D46 sequences. In both figures, error bars smaller than data points are not shown.

As another example, we choose a real protein, Crambin, which has 46 residues and has been converted into a HP sequence by Lattman et al. [43]. Based on the hydrophobic index table in [36], we defined Ile, Val, Leu, Phe and Cys as "H", Met, Ala, Gly, Thr, Trp, Ser, Tyr and Pro as "0", and the rest as "P". The resulting H0P sequence (H0P3D46) is shown in Table 2, along with the residue sequence of Crambin and its (simpler) HP mapping (HP3D46).

The densities of states, as the direct results of our simulations, are shown in Fig. 1 (a), in which we also label the ground-state degeneracies. Along with a lower ground-state energy, the H0P model also has a significantly lower ground-state degeneracy than the HP model. For comparing the ground-state structures, we calculate the contact maps for ground states of HP3D46 and H0P3D46, by assigning +1 to energy-contributing contacts (HH contacts for HP model; HH and H0 contacts for H0P model) and −1 to all the other (zero-energy) contacts. These contact maps are averaged to obtain the contact map density and is shown in Fig. 2. From this figure, we could tell that the H0P model compared to the HP model, has many fewer zero-energy contacts, which also explains the lower ground-state degeneracy. The specific heat (C_V/N) in Fig. 1 (b), is quite different for HP3D46 and H0P3D46. While H0P3D46 shows very clear signals of a two-step folding process as observed in all-atom simulations [44], the coarseness of the lattice model does not allow formation of the clear, connected two-helix structure found by experiments (see [45] and the references therein).

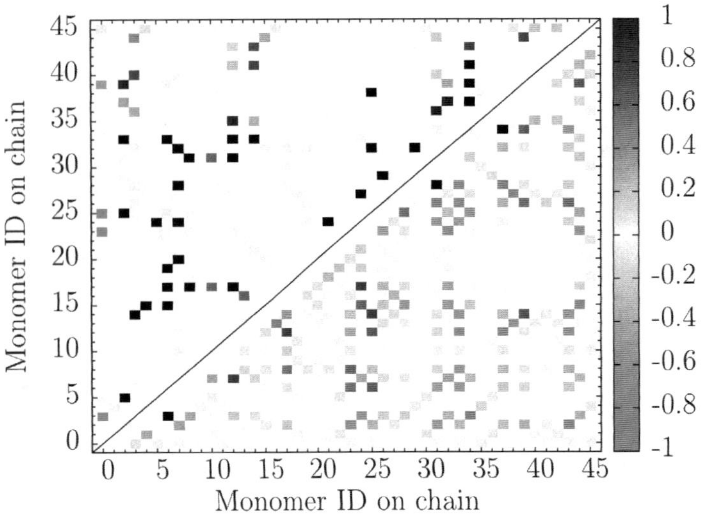

Figure 2. Contact map density of ground states for sequence H0P3D46 (upper) and HP3D46 (lower). See text for details.

4. Conclusion

In conclusion, we improved the original HP model by considering more complex hydrophobicity levels and thus introduced a new type of monomer "0". With replica-exchange Wang–Landau sampling, we investigated some widely studied HP sequences and their H0P counterparts. As expected, H0P model shows more pronounced transition signals (e.g. C_V/N) in the low temperature region, as a result of its lower ground-state degeneracy.

Acknowledgments

This work was supported by NSF under Grant No. OCI-0904685. Y.W.L. was sponsored by the Oak Ridge Leadership Computing Facility, the Office of Advanced Scientific Computing Research; U.S. Department of Energy. Part of the work was carried out at the Oak Ridge National Laboratory, which is managed by UT-Battelle, LLC under Contract No. De-AC05-00OR22725. Computing resources were partially provided by the Georgia Advanced Computing Resource Center and the Texas Advanced Computing Center under XSEDE Grant No. PHY130014.

References

[1] Chaudhuri T K and Paul S 2006 *The FEBS J.* **273** 1331–49
[2] Dill K A and MacCallum J L 2012 *Science* **338** 1042–6
[3] Dill K A 1999 *Protein Sci.* **8** 1166–80
[4] Kolinski A and Skolnick J 2004 *Polymer* **45** 511–524
[5] Dill K A 1985 *Biochemistry* **24** 1501–9
[6] Lau K F and Dill K A 1989 *Macromolecules* **22** 3986–3997
[7] Li Y W, Wüst T and Landau D P 2011 *Comput. Phys. Commun.* **182** 1896–1899
[8] Irbäck A and Troein C 2002 *J. Biol. Phys.* **28** 1–15
[9] O'Toole E M and Panagiotopoulos A Z 1992 *J. Chem. Phys.* **97** 8644
[10] Beutler T C and Dill K A 1996 *Protein Sci.* **5** 2037–2043
[11] Grassberger P 1997 *Phys. Rev. E* **56** 3682–3693
[12] Frauenkron H, Bastolla U, Gerstner E, Grassberger P and Nadler W 1998 *Phys. Rev. Lett.* **80** 3149–3152
[13] Bachmann M and Janke W 2003 *Phys. Rev. Lett.* **91** 208105
[14] Hsu H P, Mehra V, Nadler W and Grassberger P 2003 *J. Chem. Phys.* **118** 444
[15] Prellberg T and Krawczyk J 2004 *Phys. Rev. Lett.* **92** 120602
[16] Zhang J L and Liu J S 2002 *J. Chem. Phys.* **117** 3492
[17] Zhang J, Kou S C and Liu J S 2007 *J. Chem. Phys.* **126** 225101
[18] Tang W and Zhou Q 2012 *Phys. Rev. E* **86** 031909
[19] Unger R and Moult J 1993 *J. Mol. Biol.* **231** 75
[20] König R and Dandekar T 1999 *BioSystems* **50** 17–25
[21] Liang F and Wong W H 2001 *J. Chem. Phys.* **115** 3374
[22] Shmygelska A and Hoos H H 2005 *BMC Bioinf.* **6** 30
[23] Wüst T and Landau D P 2009 *Phys. Rev. Lett.* **102**(17) 178101
[24] Wüst T and Landau D P 2012 *J. Chem. Phys.* **137** 064903
[25] Castells V, Yang S and Van Tassel P 2002 *Phys. Rev. E* **65** 031912
[26] Swetnam A and Allen M P 2012 *Phys. Rev. E* **85** 062901
[27] Radhakrishna M, Sharma S and Kumar S K 2012 *J. Chem. Phys.* **136** 114114
[28] Li Y W, Wüst T and Landau D P 2013 *Phys. Rev. E* **87** 012706
[29] Bonaccini R and Seno F 1999 *Phys. Rev. E* **60** 7290–8
[30] Ping G, Yuan J M, Vallieres M, Dong H, Sun Z, Wei Y, Li F Y and Lin S H 2003 *J. Chem. Phys.* **118** 8042
[31] Pattanasiri B, Li Y W, Landau D P, Wüst T and Triampo W 2012 *J. Phys.: Conf. Ser.* **402** 012048
[32] Holzgräfe C, Irbäck A and Troein C 2011 *J. Chem. Phys.* **135** 195101
[33] Shi G, Vogel T, Wüst T, Li Y W and Landau D P 2014 *Phys. Rev. E* **90** 033307
[34] Bornberg-Bauer E 1997 *Proceedings of the First Annual International Conference on Computational Molecular Biology* RECOMB '97 (New York, NY, USA: ACM) pp 47–55 ISBN 0-89791-882-7
[35] Hoque T, Chetty M and Sattar A 2009 *J. Comput. Biol* **16** 85–103
[36] Kyte J and Doolittle R F 1982 *J. Mol. Biol.* **157** 105 – 132
[37] Vogel T, Li Y W, Wüst T and Landau D P 2013 *Phys. Rev. Lett.* **110** 210603
[38] Vogel T, Li Y W, Wüst T and Landau D P 2014 *Phys. Rev. E* **90** 023302
[39] Wang F and Landau D P 2001 *Phys. Rev. Lett.* **86** 2050–2053
[40] Wang F and Landau D P 2001 *Phys. Rev. E* **64** 1–16
[41] Yue K, Fiebig K M, Thomas P D, Chan H S, Shakhnovich E I and Dill K A 1995 *PNAS* **92** 325–329
[42] Bachmann M and Janke W 2004 *J. Chem. Phys.* **120** 6779–91
[43] Lattman E E, Fiebig K M and Dill K A 1994 *Biochemistry* **33** 6158–6166
[44] Shimada J, Kussell E L and Shakhnovich E I 2001 *J. of Mol. Biol.* **308** 79 – 95
[45] Schmidt A, Teeter M, Weckert E and Lamzin V S 2011 *Acta Crystallographica Section F* **67** 424–428

Journal of Physics: Conference Series **640** (2015) 012018 doi:10.1088/1742-6596/640/1/012018

Lattice Boltzmann Model for Electronic Structure Simulations

M. Mendoza[1], **H.J. Herrmann**[1,3], **S. Succi**[2]

[1]Computational Physics for Engineering Materials, Institute for Building Materials, ETH Zürich, Schafmattstrasse 6, HIF, CH-8093 Zürich (Switzerland).
[2]Istituto per le Applicazioni del Calcolo C.N.R., Via dei Taurini, 19 00185, Rome (Italy), and Institute for Advanced Computational Science, Harvard University, Oxford Street, 29, MA 0213879, Cambridge, (USA).
[3]Departamento de Física, Universidade Federal do Ceará, Campus do Pici, 60455-760 Fortaleza, Ceará, (Brazil).

E-mail: `mmendoza@ethz.ch`, `hans@ifb.baug.ethz.ch`, `succi@iac.cnr.it`

1. Abstract
Recently, a new connection between density functional theory and kinetic theory has been proposed. In particular, it was shown that the Kohn-Sham (KS) equations can be reformulated as a macroscopic limit of the steady-state solution of a suitable single-particle kinetic equation. By using a discrete version of this new formalism, the exchange and correlation energies of simple atoms and the geometrical configuration of the methane molecule were calculated accurately. Here, we discuss the main ideas behind the lattice kinetic approach to electronic structure computations, offer some considerations for prospective extensions, and also show additional numerical results, namely the geometrical configuration of the water molecule.

2. Introduction
The calculation of physical properties of interacting many-body quantum systems is one of the major challenges in chemistry and condensed matter. In principle, this requires the solution of the Schrödinger or Dirac equations for $3N$ spatial variables and N spin electronic variables, where N is the number of particles in the system. Usually, for atoms, the number of electrons is in the range $N \sim 1 - 100$, for small molecules often more than 100, and for solids, one has around $N \sim 10^{23}$. Therefore, developing approximate models to describe these systems becomes crucial. In many cases, the most important goal is to calculate several measurable quantities, such as the bonding energy, polarisability, conductivity, etc., rather than the wave function itself. A formalism which can treat the systems mentioned with way less computational effort, while still being formally *exact*, is the density functional method [1], developed by Hohenberg and Kohn [2] and Kohn and Sham [3]. The Kohn-Sham approach to density functional theory allows an exact description of the interacting many-particle systems in terms of an effective non-interacting particle system. The effective potential can be shown to be completely determined by the total electron density of the interacting system, which is the reason why it is called a density functional. In particular, the ground state energy of the system is a density functional, whose exact expression is not known due to the complicated nature of the quantum many-body problem. However over the years, thanks to an intense and thoughtful work, many increasingly more accurate approximations have continued to appear [4, 5].

On an apparently very different front, fluid dynamics, the Lattice Boltzmann (LB) method has been used with considerable success to simulate a variety of physical systems [6, 7, 8]. In the recent years, however, it has become increasingly apparent that the LB methodology is not confined to fluid dynamics and can indeed be extended to a variety of complex dynamical systems described by linear and non-linear partial differential equations, such as electromagnetism, quantum mechanics, Bose-Einstein condensation and many other [9, 10, 11]. In a very recent paper, Mendoza *et al.* [12] showed for the first time that the LB extends to the case of electronic structure simulations; more precisely, these authors developed and validated a lattice kinetic formulation of the Kohn-Sham equations of electrons density functional theory. In this paper, we review the main ideas and motivations behind the above formulation and discuss main challenges ahead.

3. Electron density functional theory: Kohn-Sham equations

The quantum mechanical behaviour of a molecule consisting of N_I ions and N_e electrons is described by the Schrödinger equation for the N-body wave function $\Psi(r_1 \dots r_{N_e}; R_1 \dots R_{N_I}; t)$, that is

$$i\hbar\partial_t\Psi = H\Psi \quad , \tag{1}$$

where H is the N-body Hamiltonian of the system, i.e. kinetic energy plus electron-electron and electron-ion Coulomb interactions. This equation is computationally unviable for all but the smallest molecules owing to the exponential barrier of complexity with the number of electrons [13]. Therefore, approximated models have been developed to study such systems.

3.1. Kohn-Sham formulation

In the early 60's Kohn and Hohenberg proved that the ground-state energy of a quantum many-body system is a unique functional of the one-body electron density $\rho(\vec{r})$ [2]. Starting from this basic result, shortly later Kohn and Sham, proceeded to derive a set of effective one-body time-independent Schrödinger equations (SE) of the form [3]:

$$H_{KS}\phi_j = E_j\phi_j \quad , \tag{2}$$

where $\phi_j(\vec{r})$ is the wave function of the j-th orbital with energy E_j and the KS Hamiltonian reads as follows:

$$H_{KS} = -\frac{\hbar^2}{2m}\nabla_j^2 + \int \frac{\rho(\vec{r}')}{|\vec{r} - \vec{r}'|}d^3r' + \sum_I Z_I \int \frac{\rho(\vec{r}')}{|\vec{R}_I - \vec{r}'|}d^3r' + V_{xc}[\rho] \quad , \tag{3}$$

where $\vec{R}_I$ and Z_I are the coordinates and charges of the nuclei, respectively. In the above, V_{xc} *defines* the exchange-correlation energy, the term collecting all of the unknown effects of many-body interactions and Pauli exclusion principle. Note that, by effect of the Kohn-Hohenberg theorem, this term is guaranteed to depend only on the total electron density

$$\rho(\vec{r}) = \sum_j |\phi_j(\vec{r})|^2 \quad , \tag{4}$$

which represents an enormous simplification of the quantum many-body problem. If such dependence could be sorted out, the Kohn-Sham equations (KSE) would be *exact*.

From the mathematical point of view, the KSE's represent a set of single-particle time-independent Schrödinger equations, coupled only through the total density $\rho(\vec{r})$. Furthermore, the orbitals must obey the orthogonality condition

$$\int \phi_j(\vec{r})\phi_k(\vec{r})d\vec{r} = \delta_{jk} \quad , \tag{5}$$

which is a set of $N_e(N_e+1)/2$ global constraints. To be noted that the above orbitals are a mere computational device to perform variational minimisation of the many-body ground-state energy, with no specific physical meaning.

4. The kinetic approach to Kohn-Sham theory

The main observation is that, upon performing a Wick rotation, $t \to t' = -it$, the time-dependent Schrödinger equation is basically an one-particle diffusion-reaction equation, where the reaction term contains all the details of the potential energy, i.e.,

$$\frac{\partial \psi}{\partial t'} = \frac{\hbar}{2m}\nabla^2\psi - \frac{V}{\hbar}\psi \quad , \tag{6}$$

where ψ is the wave function, and V is the potential energy. Since reaction-diffusion equations are known to emerge as a macroscopic limit of an underlying Boltzmann equation, it is natural to treat the wave function as a "fluid" and propose a (lattice) kinetic representation for the wave function, such that

$$\sum_p f_p(\vec{r};t') = \psi(\vec{r};t') \quad , \tag{7}$$

In the above, $f_p(\vec{r})$, with $p = 0,...,b$, is the probability of finding a "particle" with velocity $\vec{v}_p$ at position $\vec{r}$ and time t, where b is the total number of discrete velocity vectors. For the moment, "particle" means simply a computational quasi-particle, with no physical implications for the existence of an underlying kinetic theory behind the Schrödinger representation. Note that assuming an arbitrary initial condition for the wave function ψ, one can expand it using a basis of orthogonal orbitals,

$$\psi(\vec{r},t') = \sum_k a_k\phi_k(\vec{r})\exp(-E_k t'/\hbar) \quad , \tag{8}$$

such that after some time, it will be $\psi(\vec{r},t') \simeq a_0\phi_0(\vec{r})\exp(-E_0t'/\hbar)$, which upon normalisation leads to the ground state solution of the time-independent Schrödinger equation,

$$E_0\phi_0 = -\frac{\hbar^2}{2m}\nabla^2\phi_0 + V\phi_0 \quad . \tag{9}$$

The distribution function f_p evolves according to the lattice Boltzmann equation (time step made unity for simplicity),

$$f_p(\vec{r}+\vec{v}_p;t+1) = f_p(\vec{r};t) - \omega(f_p - f_p^e) + V_p \quad , \tag{10}$$

where ω is a typical relaxation frequency which controls the quantum diffusivity $D = \hbar/m$, f_p^e is a suitable local equilibrium ensuring mass conservation, and V_p is a suitable source term representing the particle loss/gain associated with the potential energy scattering. In lattice units $\Delta x = \Delta t = 1$, one has:

$$D = c_s^2(1/\omega - 1/2) \quad , \tag{11}$$

where c_s is the lattice sound speed. Thus, by changing ω, one has a handle on the effective mass of the electrons. The explicit expression for the equilibrium distribution is as follows:

$$f_p^e = \xi_p\psi \quad , \tag{12}$$

where $\xi_p = (1 + (D - c_s^2)q_p)$, with w_p being suitable lattice weights normalised to unity, and $q_p = (v_p^2 - 3c_s^2)/2c_s^4$ the second order lattice Hermite polynomial. The scattering source V_p reads as follows:

$$V_p = \chi_p\frac{V}{\hbar}\psi \quad , \tag{13}$$

where $\chi_p = w_p[1 - (v_p^2 - 3c_s^2)/2c_s^2)]$ ensures that the first and second order moments are zero, namely $\sum_p f_p \vec{v}_p = 0$ and $\sum_p f_p \vec{v}_p \vec{v}_p = 0$, so that momentum and energy of the wave function are unaffected.

In order to solve the KS system, we need a distribution function for each jth-orbital, i.e. $f_p \to f_{jp}$ and consequently $\psi \to \psi_j$, and the potential energy V contains the electron-electron and ion-electron interactions, plus the exchange-correlation potential. Thus, the corresponding electronic Lattice Boltzmann (LB) equation reads as follows:

$$f_{jp}(\vec{r} + \vec{v}_p; t+1) = f_{jp}(\vec{r}; t) - \omega(f_{jp} - f_{jp}^e) + V_{jp} + W_{jp} \quad , \tag{14}$$

where, W_{jp} is the source term in charge of securing the orthogonality constraints (5). The explicit expressions for the equilibrium distributions and scattering potentials are as follows:

$$f_{jp}^e = \xi_p \psi_j, \quad \text{and} \quad V_{jp} = \chi_p \frac{V}{\hbar} \psi_j \quad , \tag{15}$$

and the orthogonalisation potential takes the form

$$W_{jp} = -\xi_p \omega \sum_{k<j} \Lambda_{jk} \psi_k \quad , \tag{16}$$

where $\Lambda_{jk} = \langle \psi_j | \psi_k \rangle / \langle \psi_k | \psi_k \rangle$ is the cosine of the angle between orbitals j and k in Hilbert space. Clearly, once two orbitals are mutually orthogonal, $\Lambda_{jk} = 0$, so that they no longer contribute to the potential W_{jp}.

In order to understand the role of the orthogonalisation potential we will introduce the following derivation. Let us first assume that all orbitals are initialised with the same wave function, $\psi_j(\vec{r}) = \psi(\vec{r})$, and we do not consider any orthogonalisation potential in equation (14), i.e. $W_{jp} = 0$. According to our previous explanation, the wave functions can be written as expansions in the orthogonal basis $\phi_k(\vec{r})$,

$$\psi_j(\vec{r}, t') = \sum_p f_{jp} = \sum_k a_{kj} \phi_k(\vec{r}) \exp(-E_k t'/\hbar) \quad . \tag{17}$$

where $a_{kj} = \langle \phi_k | \psi_j \rangle$ are projection coefficients. After some time, the only leading term in the sum will be the ground state orbital, ϕ_0, such that $\psi_j(\vec{r}, t') \simeq a_{0j} \phi_0(\vec{r}) \exp(-E_0 t'/\hbar)$. At this stage, all orbitals in our lattice kinetic approach will reach the same ground state. Let us now subtract from the wave function ψ_1, the wave function ψ_0, such that

$$\psi_1(\vec{r}, t') = \sum_k a_{k1} \phi_k(\vec{r}) \exp(-E_k t'/\hbar) - \Lambda_{10} \psi_0(\vec{r}, t') \quad . \tag{18}$$

This equation can be rewritten explicitly as,

$$\psi_1(\vec{r}, t') = \sum_k \left(\langle \phi_k | \psi_1 \rangle - \frac{\langle \psi_1 | \psi_0 \rangle}{\langle \psi_0 | \psi_0 \rangle} \langle \phi_k | \psi_0 \rangle \right) \phi_k(\vec{r}) \exp(-E_k t'/\hbar) \quad . \tag{19}$$

The evolution equation for the orbital ψ_0 implies that it will converge eventually to the ground state ϕ_0. Therefore, the first term in the sum, for $k = 0$, vanishes and the remaining contributions to the sum are

$$\psi_1(\vec{r}, t') = \sum_{k>0} a_{k1} \phi_k(\vec{r}) \exp(-E_k t'/\hbar) \quad . \tag{20}$$

Thus, the orbital ψ_1 will eventually converge to the first excited state ϕ_1. Repeating this procedure for each wave function, one can write

$$\psi_j(\vec{r},t') = \sum_k \left(\langle\phi_k|\psi_j\rangle - \sum_{l<j} \frac{\langle\psi_j|\psi_l\rangle}{\langle\psi_l|\psi_l\rangle} \langle\phi_k|\psi_l\rangle \right) \phi_k(\vec{r}) \exp(-E_k t'/\hbar) \quad , \tag{21}$$

or

$$\psi_j(\vec{r},t') = \sum_k a_{kj}\phi_k(\vec{r}) \exp(-E_k t'/\hbar) - \sum_{l<j} \Lambda_{jl}\psi_l(\vec{r},t') \quad , \tag{22}$$

and show that each wave function will eventually converge, upon normalisation, to the respective KS orbital, i.e. $\psi_j \to \phi_j$. In order to include this procedure into the evolution of the wave functions, we need to force the jth wave function to converge to

$$\psi_j^*(\vec{r},t') = \psi_j(\vec{r},t') - \sum_{l<j} \Lambda_{jl}\psi_l(\vec{r},t') = \sum_p f_{jp} - \sum_{l<j} \Lambda_{jl}\psi_l(\vec{r},t') \quad . \tag{23}$$

This can be done by replacing in the equilibrium distribution ψ_j by ψ_j^*, obtaining the following lattice kinetic equation,

$$f_{jp}(\vec{r}+\vec{v}_p;t+1) = f_{jp}(\vec{r};t) - \omega(f_{jp} - \xi_p\psi_j^*) + V_{jp} \quad . \tag{24}$$

By reorganising terms, this equation leads to

$$f_{jp}(\vec{r}+\vec{v}_p;t+1) = f_{jp}(\vec{r};t) - \omega(f_{jp} - \xi_p\psi_j) + V_{jp} - \xi_p\omega \sum_{l<j} \Lambda_{jl}\psi_l(\vec{r},t') \quad , \tag{25}$$

which is precisely our previous definition of the lattice kinetic equation with the orthogonalisation potential. Note that after some time, we have $\psi_j \to \psi_j^* \to \phi_j$. This process is a kind of time-dependent Gram-Schmidt procedure. Indeed, one can also evolve first the wave function ψ_0, and once the ground state is found, evolve ψ_1, and so on. However, this will lead to longer computational times to achieve the total electronic ground state. The scaling properties and convergency speed of the orthogonalisation procedure described here will be a subject of future research.

4.1. Concurrent Dynamics

The above electronic LB only applies to ground-state calculations, where the electronic distribution is systematically adjusted to the Born-Oppenheimer surface defined by the actual positions of the ions, $R_I(t)$, these latter being obtained by a standard Molecular Dynamics integration. The total energy is thus evolving in (imaginary) time until the total ground state value is attained. This is the standard Born-Oppenheimer (BO) scenario.

The KS-LB can be extended to the concurrent-dynamic (CD) scenario, whereby the electrons and the ions are moved simultaneously, according to the celebrated Car-Parrinello picture [14]. In this scenario, energy owes to be conserved, and consequently the irreversible relaxation operator in Eq. (14) needs to be augmented with a potential, U, such that density and momentum are conserved, while energy is supplied to (removed from) the system so as to keep it constant.

The corresponding LB equation takes then the form:

$$f_{jp}(\vec{r}+\vec{v}_p h;t+h) = f_{jp}(\vec{r}-\vec{v}_p h;t-h) + 2h(U_{jp} + iV_{jp} + iW_{jp}) \quad , \tag{26}$$

where h is the time-step. The specific expression of U_{jp} is as follows:

$$U_{jp} = \kappa w_p q_p E_j \phi_j \quad , \tag{27}$$

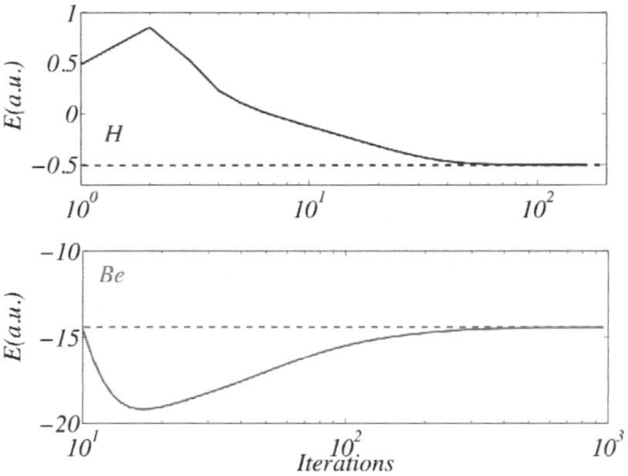

Figure 1. Convergence of the total energy for the atoms H and Be as a function of the number of iterations performed by our model. The dashed-line corresponds to the final value, $E = 0.500$ a.u. (1 a.u. $\simeq 27.2$ eV) and $E = 14.45$ a.u., for the hydrogen and beryllium atoms, respectively. For Beryllium, we found an error of 1% with respect to the expected value in Refs. [15, 16], while for hydrogen the error is of the order of 0.1%. We have taken $\tau_K = \delta t$.

Atom	$\mathcal{V}_x$	Exp. $\mathcal{V}_x$	$\mathcal{V}_c$	Exp. $\mathcal{V}_c$	Time	Iterations
H	−0.310	−0.310	0	0	9 sec	158
He	−1.025	−1.025	−0.044	−0.044	1 min	906
Be	−2.658	−2.658	−0.095	−0.095	3 min	965
B	−3.728	−3.728	−0.128	−0.128	29 min	2704
C	−5.032	−5.032	−0.161	−0.161	33 min	2305

Table 1. Exchange-Correlation energies for H, He, Be, B, and C in atomic units (a.u.). Computational time and the number of iterations performed by the model to reach the ground state are also shown. The expected values of exchange, $\mathcal{V}_x$, and correlation, $\mathcal{V}_c$, energies are taken from Refs. [4, 5]. The simulations have run on an Intel Core i7 at 2.3 GHz. We have taken $\tau_K = \delta t$.

κ being an adjustable parameter in charge of keeping the energy at E_j. To be noted that the algorithm is now a two-step time-centered scheme, in compliance with the requirement of energy conservation. This is the analogue, in kinetic language, of Car-Parrinello's inertial term in the electron equation, i.e. a second order time derivative to comply with the overarching electron-ion Lagrangian. The fact that the present LB scheme operates successfully also in the CD regime hints at the existence of an overarching kinetic Lagrangian. At the time of this writing, however, such Lagrangian remains to be found.

5. Results

In order to show the validity of our approach, we present the calculations of the exchange and correlation energies for different atoms, H, He, Be, B, and C. For H, Be, B, and C, one can find more details about the calculations and parameter values in Ref. [12]. For those cases, we have used lattice sizes of 32^3, 40^3, 56^3, and 58^3, respectively. For He, which has not been reported before using our model, we have taken a system size of 36^3, Bohr radius $a_0 = 6.5$, $\hbar/m = 1$, and $\tau = 1$ (all values are in numerical units). We have also used fourth order in the Hermite expansion for the equilibrium distribution and source term (for details about the

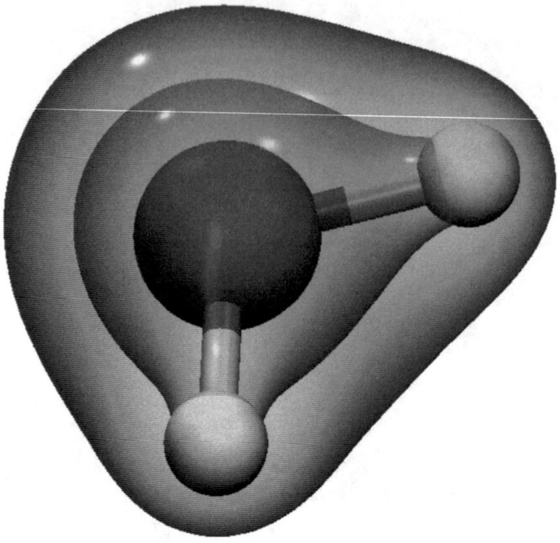

Figure 2. Water molecule, H_2O. The blue and red isosurfaces denote low and high electronic density, respectively. Using our model, we have obtained for the angles between bonds, 104.4 degrees, and a O-H bond distance of 0.95 Å.

Hermite expansion, see Ref. [12]). The measured values for the exchange and correlation energies are given in Table 1, together with the computational time and number of iterations. Note that the reported number of iterations might exceed the one of existing iterative methods. However, as we shall show shortly, each of our iterations takes typically much less computational time, thus making our method competitive. In Fig. 1, we see that the convergence of the total energy, for H and Be, to the ground state is very fast in the beginning and slows down when it approaches the exact solution. The differences between the obtained and expected exchange and correlation energies are of the order of 1%. The simulations are stopped when the energy of the orbitals presents changes of less than $10^{-5}\%$ between two subsequent steps.

As an additional application, we build the water molecule from scratch, by using the *bare* Coulomb potential. For this simulation, we use a lattice size of 74^3, $\hbar/m = 1$, $a_0 = 10.5$ cells, and $\mathcal{N} = 5$, and place the oxygen atom at the center of the lattice. The hydrogen atoms are located randomly in space, and we let the system evolve to the ground configuration. After 4 hours (10447 iterations), we achieved the configuration shown in Fig. 2. The angle, 104.4 degrees, and the bond distance of 0.95 Å are in excellent agreement with expected and experimental values [17, 18].

Finally, we compare BO and CD versions of our kinetic scheme against each other. Here, we assume that the BO dynamics provides the correct results, and will be used to validate the CD. To this purpose, we excite the H_2 molecule and let it vibrate in its first mode. We have used a lattice size of 24^3. Further details about the values of the model parameters can be found in Ref. [12]. For the BO dynamics we use the equation of motion for the ions and, at each time step, we calculate the ground state of the electronic density using Eq. (14). On the other hand, for CD we evolve simultaneously the ions and the electronic orbitals, using the discrete version of Eq. (26). Since in CD there is no relaxation process to the ground state, we must use a smaller time step such that electrons can accommodate according to the new position of the ions at each time step. The time step of the BO molecular dynamics is set to $\Delta t_{BO} = 0.09$ fs, while for CD, $\Delta t_{CD} = 0.0014$ fs. Although the time step in CD is smaller by nearly two orders

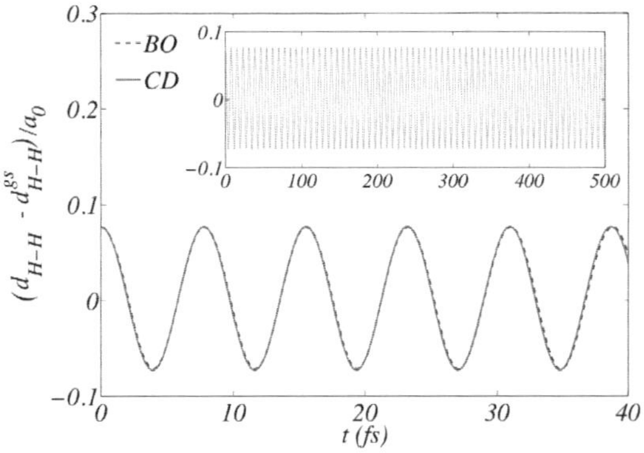

Figure 3. Vibrational mode of the diatomic molecule H_2. Here BO and CD denote Born-Oppenheimer and Concurrent Dynamics, respectively. The inset shows that up to 500 fs, the amplitude of the oscillations does not show any appreciable decay in time. Here, a_0 is the Bohr radius, d_{H-H} the distance between H atoms and d^{gs}_{H-H} the ground state distance.

of magnitude, it runs 17 times faster (taking 17 sec to simulate one fs) than BO dynamics. In fig. 3, we report the first vibrational mode of the hydrogen molecule with BO and CD after 40 fs, from which one can appreciate that the trajectories of CD are pretty close to the respective BO ones. In the inset of this figure, we can also see that after 500 fs, the amplitude of the oscillations remains constant, without any appreciable loss of energy. The oscillation frequency of the vibrational mode obtained by the simulation is 4206 cm^{-1}, which departs by 1% from the experimental value of 4161 cm^{-1} [19].

Interestingly, the LB simulations do resolve the full Coulomb interactions, without making any use of soft pseudo-potentials [12]. This is possibly due to the high-order of spatial accuracy achieved by the source terms. The amount of computational memory that our model uses can be roughly estimated as $M_T = 8(36 + 2N_O)V$ in bytes, where N_O is the number of orbitals and V the lattice size. For instance, for the water molecule the total memory used is about 150 MB, and by using a standard computer (or a single node of a cluster) with 8 GB of RAM, one could study about 8 of these molecules. This small system size implies that parallel computing is a crucial step in order to study larger system sizes, and the above relation for the memory requirements provides a reasonable estimate of the maximum system size that one can simulate at a given level of computing power.

Our simulations have been performed using open boundary conditions. However, the use of periodic boundary conditions is also possible at the price of introducing corrections due to Ewald summation for the electric potential. Indeed, using periodic boundary conditions can lead to more stable simulations, since the distribution functions do not need to be extrapolated at the boundaries, and therefore, their dynamics or relaxation to ground state follows the correct equations holding in the bulk.

6. Prospective developments

The kinetic representation discussed in this paper lends itself to a number of prospective developments. First, one should note that such representation is not restricted to the time-independent Kohn-Shan formulation. Indeed, in the recent past Runge and Gross have extended Kohn-Hohenberg's theorem to the case of time-dependent potentials [20]. In this case, the KS formalism still applies, although in real-time form, i.e:

$$i\hbar\partial_t\phi_j(\vec{r};t) = H_{KS}(t)\phi_j(\vec{r};t) \quad , \tag{28}$$

where the KS Hamiltonian now contains a time-dependent potential, $H_{KS} = K + V(\vec{r};t)$.

The above equations can be integrated using consolidated quantum LB schemes [10], exporting the very same upgrades described in this paper. Another direction, possibly even closer in spirit to kinetic theory, is the so-called *density-current* formulation of TDDFT, in which the hamiltonian depends not only on the electron density, but also on the electron current:

$$\vec{J}(\vec{r};t) = \sum_j \vec{v}_j |\phi_j(\vec{r};t)|^2 \tag{29}$$

where $\vec{v}_j \equiv i\frac{\hbar}{m}\nabla_j$ is the velocity operator.

The density and current functionals evolve according to the continuity equation:

$$\partial_t \rho + \nabla \cdot \vec{J} = 0 \tag{30}$$

which provides $\rho(\vec{r};t)$ once the current functional $\vec{J} = \vec{J}(\rho(\vec{r};t))$ is prescribed. Of course, the latter is not known exactly, but sensible approximations have been developed in the specialised literature, e.g the Vignale-Kohn functional [21], which could be implemented in the corresponding electronic LB exactly along the lines described in this paper.

Finally, one may also envisage a more radical strategy, whereby the energy functional would be expressed in terms of the Boltzmann distribution f itself rather than by its lower-order kinetic moments, such as density, current, and momentum flux. The prospective Boltzmann functional would symbolically read as

$$E = \int \mathcal{E}[f] d\vec{r} d\vec{v} \tag{31}$$

where the double integration extends over the six-dimensional phase space.

What would one gain out of such functional? Possibly, the non-perturbative incorporation of inhomogeneity effects at *all orders*. Indeed, the Boltzmann distribution splits naturally into an equilibrium and non-equilibrium component, $f = f^e + f^{ne}$, the former depending parametrically only on invariants, density and momentum, while the latter, by definition, collects the effects of space-time inhomogeneity at all orders. Of course, such functional would make the difference only at the level where further moments on top of current-density, are required.

7. Conclusions

Summarising, we have presented the main ideas behind the kinetic formulation of electronic density functional theory, along with the ensuing implementation within the Lattice Boltzmann formalism. We have described in detail the effects of the orthogonalisation potential and shown how it contributes to find the system of orthogonal KS orbitals. In order to show the validity of the lattice Boltzmann approach for KS, we have shown numerical data for the computation of the exchange and correlation energies for simple atoms, finding excellent agreement with the previous literature. The results of numerical simulations for the first vibrational mode of the hydrogen molecule are also presented. The prospects, including future extensions to time-dependent DFT, look exciting.

Acknowledgements

Illuminating discussions with K. Bravaya, D. Coker and E. Kaxiras are kindly acknowledged. Financial support from the European Research Council (ERC) Advanced Grant 319968-FlowCCS is kindly acknowledged.

Journal of Physics: Conference Series **640** (2015) 012018 doi:10.1088/1742-6596/640/1/012018

References

[1] Gross E and Dreizler R 1995 *Density Functional Theory* NATO ASI Series: Physics (Springer)
[2] Hohenberg P and Kohn W 1964 *Phys. Rev.* **136**(3B) B864–B871
[3] Kohn W and Sham L J 1965 *Phys. Rev.* **140**(4A) A1133–A1138
[4] Becke A D 1988 *Phys. Rev. A* **38**(6) 3098–3100
[5] Lee C, Yang W and Parr R G 1988 *Phys. Rev. B* **37**(2) 785–789
[6] Succi S 2008 *The European Physical Journal B-Condensed Matter and Complex Systems* **64** 471–479
[7] Mendoza M, Herrmann H and Succi S 2013 *Scientific Reports* **3**
[8] Mendoza M, Karlin I, Succi S and Herrmann H 2013 *Physical Review D* **87** 065027
[9] Mendoza M and Muñoz J D 2010 *Phys. Rev. E* **82**(5) 056708
[10] Succi S and Benzi R 1993 *Physica D: Nonlinear Phenomena* **69** 327–332
[11] Palpacelli S and Succi S 2008 *Physical Review E* **77** 066708
[12] Mendoza M, Succi S and Herrmann H J 2014 *Phys. Rev. Lett.* **113**(9) 096402
[13] Kohn W 1999 *Reviews of Modern Physics* **71** 1253–1266
[14] Car R and Parrinello M 1985 *Phys. Rev. Lett.* **55**(22) 2471–2474
[15] Mori-Sánchez P, Wu Q and Yang W 2005 *The Journal of chemical physics* **123** 062204
[16] Hirata S, Ivanov S, Grabowski I and Bartlett R J 2002 *The Journal of chemical physics* **116** 6468–6481
[17] Sprik M, Hutter J and Parrinello M 1996 *The Journal of chemical physics* **105** 1142–1152
[18] Benedict W, Gailar N and Plyler E K 1956 *The Journal of Chemical Physics* **24** 1139–1165
[19] Stoicheff B P 1957 *Canadian Journal of Physics* **35** 730–741
[20] Runge E and Gross E K 1984 *Physical Review Letters* **52** 997
[21] Vignale G and Kohn W 1996 *Phys. Rev. Lett.* **77**(10) 2037–2040

XXVI IUPAP Conference on Computational Physics (CCP2014) IOP Publishing
Journal of Physics: Conference Series **640** (2015) 012019 doi:10.1088/1742-6596/640/1/012019

Magnetic Materials at finite Temperatures: thermodynamics and combined spin and molecular dynamics derived from first principles calculations[1]

Markus Eisenbach[1], Dilina Perera[2], David P. Landau[2], Don M. Nicholson[3], Jungqi Yin[4], Gregory Brown[5]

[1] National Center for Computational Sciences, Oak Ridge National Laboratory, Oak Ridge, TN 37831, USA

[2] Center for Simulational Physics, University of Georgia, Athens, GA 30602, USA

[3] Department of Physics, University of North Carolina, Asheville, NC 28804, USA

[4] National Institute for Computational Sciences, University of Tennessee, Oak Ridge, TN 37831, USA

[5] Department of Physics, Florida State University, Tallahassee, FL 32306, USA

E-mail: eisenbachm@ornl.gov

Abstract. We present a unified approach to describe the combined behavior of the atomic and magnetic degrees of freedom in magnetic materials. Using Monte Carlo simulations directly combined with first principles the Curie temperature can be obtained ab initio in good agreement with experimental values. The large scale constrained first principles calculations have been used to construct effective potentials for both the atomic and magnetic degrees of freedom that allow the unified study of influence of phonon-magnon coupling on the thermodynamics and dynamics of magnetic systems. The MC calculations predict the specific heat of iron in near perfect agreement with experimental results from 300K to above Tc and allow the identification of the importance of the magnon-phonon interaction at the phase-transition. Further Molecular Dynamics and Spin Dynamics calculations elucidate the dynamics of this coupling and open the potential for quantitative and predictive descriptions of dynamic structure factors in magnetic materials using first principles derived simulations.

1. Introduction

Understanding magnetic materials has been a long standing problem in condensed matter physics. While Hund's rule explains the magnetic moments of isolated atoms, it is the collective behavior of the electrons that is required to understand the interactions that lead to the formation of moments in bulk materials and their finite temperature behavior. In this paper we will concentrate on materials that can be mapped onto Heisenberg like models. Firstly we will describe our approach to calculating the energy of an arbitrary magnetic configuration from first principles using constrained density functional theory, then we will introduce the Mont-Carlo

[1] This manuscript has been authored by UT-Battelle, LLC, under Contract No. DE-AC0500OR22725 with the U.S. Department of Energy. The United States Government retains and the publisher, by accepting the article for publication, acknowledges that the United States Government retains a non-exclusive, paid-up, irrevocable, world-wide license to publish or reproduce the published form of this manuscript, or allow others to do so, for the United States Government purposes.

sampling methods we employ to calculate the partition functions for these materials and for calculating effective interactions that will be used in model simulations that allow us to explore the coupling of the magnetic and atomic degrees of freedom that are presently still beyond the capabilities of purely first principles approaches. In the last two sections we will use the parametrizations of the magnetic interactions to investigate the effects of coupling between the atomic and magnetic degrees of freedom on the thermodynamics and on the thermalization dynamics of the magnetization in iron systems.

2. Constrained First Principles Calculations

For the energy evaluation, we employ the first principles framework of density functional theory (DFT) in in the local spin density approximation (LSDA). To solve the Kohn-Sham equations arising in this context, we use a real space implementation of the multiple scattering formalism. The details of this method for calculating the Green function and the total ground state energy $E[n(\vec{r}), \vec{m}(\vec{r})]$ are described elsewhere [1, 2]. Most importantly for the application in the hybrid Wang-Landau LSMS method, our Locally Self-consistent Multiple Scattering (LSMS) method allows the possibility of non-collinear magnetism [3].

The orientation $\hat{e}^i$ of the magnetic moment for each site is determined by

$$\hat{e}^i = \int_{\Omega^i} d\vec{r}\,\vec{m}^i(\vec{r}) / |\int_{\Omega^i} d\vec{r}\,\vec{m}^i(\vec{r})|. \tag{1}$$

Since an arbitrary arrangement is not a DFT ground state we will have to deal with a constrained general state as presented by Stocks *et al* [4, 5]. In the constrained local moment (CLM) model the LSDA equations are solved subject to a constraint

$$\int_{\Omega_i} \vec{m}_i(\vec{r}) \times \hat{e}_i d\vec{r} = 0 \tag{2}$$

that ensures that the local magnetizations lie along the directions prescribed by $\{\hat{e}_i\}$. The result is that, in order to maintain the specific orientational configuration, a local transverse constraining field must be applied at each site. The constraining field is obtained from the condition

$$\frac{\delta E^{con}[\{\hat{e}_i\}, \{\vec{B}_i^{con}\}]}{\delta \hat{e}_i} = 0 \tag{3}$$

applied to all sites and where E the generalized energy functional in the presence of the constraining field. Thus this method enables the calculation of the energies of arbitrary orientational states that will form the input of the following finite temperature calculations.

3. Wang Landau Sampling and Model Parameter Sampling

In the preceding section we described how we can calculate the constrained ground state energy $E(\{\hat{e}_i\})$ of an arbitrary magnetic ordering. In the approximation we employ in this paper, this represents the energy of a point in the system's phase space that is described by this set magnetization directions $\{\hat{e}_i\}$. As we are interested in the thermodynamics of this system our goal is to calculate the partition function of this system of N sites:

$$Z(T) = \int_{S_2^N} e^{-E(\{\hat{e}_i\})/(k_B T)} d^{2N}\{\hat{e}_i\} \tag{4}$$

Even if atomic motions are not taken into account this represents a $2N$ dimensional integral, that in general can not be evaluated efficiently with standard numerical integration techniques. Also the energy calculation inside the integrand is orders of magnitude more computationally

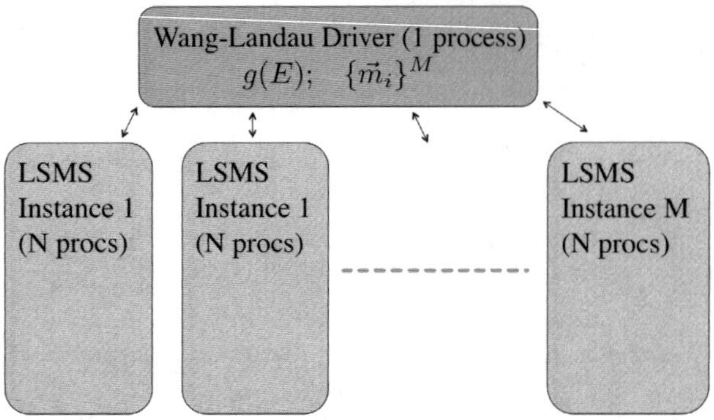

Figure 1. Parallelization strategy of the combined Wang-Landau/LSMS algorithm. The Wang-Landau process generates random spin configurations for M walkers and updates a single density of states $g(E)$. The energies for these N atom systems are calculated by independent LSMS processes. This results in two levels of communication, between the Wang-Landau driver and the LSMS instances, and the internal communication inside the individual LSMS instances spanning N processes each.

expensive than a simple Ising or Heisenberg model. In the next two subsections we will first describe the direct combination of the first principles energy calculation with a Monte Carlo sampling algorithm and secondly we will describe the fitting of model parameters in a way that will allow the construction of effective models that incorporate complex spin-spin interactions that depend on the atomic environment.

3.1. First Principles WL-LSMS

The Wang-Landau method [6, 7] has proven itself as a very efficient method to calculate the density of states $g(E)$ by flat energy histogram sampling of a system's phase space. This allows the partition function to be written as a one dimensional integral:

$$Z(T) = \int g(E)e^{-E/(k_BT)}dE \tag{5}$$

We have combined the Wang-Landau algorithm with with the first principles LSMS method described above in a very efficient highly scalable parallel code (WL-LSMS) [8]. The WL-LSMS code uses a hybrid parallelization scheme. At the top level, the code parallelizes over concurrent random walkers, where we use a master-slave scheme, with a master that accumulates the density of states of the system, and the slaves that execute the random walks, each running its own instance of the LSMS method. The second parallelization level is the LSMS portion of the code, where domain decomposition is used with one atom per processing core. In typical production runs, the WL method would use a few hundred concurrent walkers, and the LSMS portion would be parallelized over up to a few thousand processing cores. The method hence will scale to hundreds of thousands or millions of processing cores. The schematics of the parallelization structure are shown in fig. 1.

3.2. Fitting Model Parameters for Monte Carlo and Spin Dynamics simulations

The procedure used to extract exchange parameters from first principle calculations is proposed in Ref. [10]. Considering a $4 \times 4 \times 4$ BCC lattice with thermal displacements generated using the Finnis-Sinclair potential [11] near the transition temperature, for each atom pair at site i and j within the first five shells (58 pairs for each atom), we calculate the pair energy E^{ij} using the locally self-consistent multiple scattering (LSMS) first principles method [9]. A Heisenberg

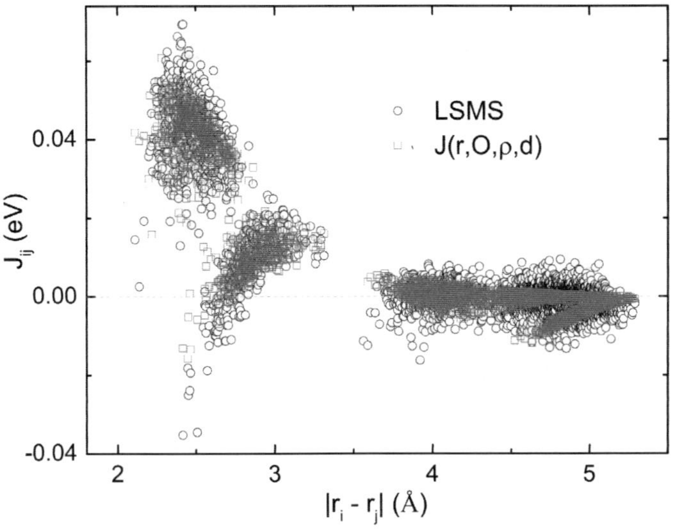

Figure 2. Pair exchange parameters J_{ij} calculated for all pairs (up to fifth nearest neighbor) in a 128 atom cell using the method described in the text (black circle) for a representative configuration (positions and spins) of BCC iron near T_c. The values for each pair is also shown as approximated by the function described in the text, $J(r, O, \rho, d)$ (red squares).

type of Hamiltonian is used to approximate the LSMS energies we obtained,

$$\begin{aligned} \mathcal{H} &= -\sum_{i<j} J_{ij}\vec{e}_i \cdot \vec{e}_j \\ &= -\sum_{k;k\neq i} J_{ik}\vec{e}_i \cdot \vec{e}_k - \sum_{k;k\neq j} J_{jk}\vec{e}_j \cdot \vec{e}_k \\ &\quad - \sum_{k<l;k,l\neq i,j} J_{kl}\vec{e}_k \cdot \vec{e}_l - J_{ij}\vec{e}_i \cdot \vec{e}_j \end{aligned} \tag{6}$$

where $\vec{e}_i$ is a unit vector for the atomic spin at site i. Then The exchange interaction parameter J_{ij} can be evaluated by considering 4 pair configurations for $(\vec{e}_i, \vec{e}_j)$: $(\hat{z}, \hat{z})$, $(-\hat{z}, \hat{z})$, $(\hat{z}, -\hat{z})$, $(-\hat{z}, -\hat{z})$, for a total of $14848 = 4 * 128 * 58/2$ spin configurations, with corresponding energies E^{ij}, $E^{\bar{i}j}$, $E^{i\bar{j}}$, $E^{\bar{i}\bar{j}}$ as follows

$$J_{ij} = \frac{1}{4}\left(E^{ij} + E^{\bar{i}\bar{j}} - E^{\bar{i}j} - E^{i\bar{j}}\right). \tag{7}$$

Values of J_{ij} pairs are shown in Fig. 2. The large fluctuations indicate the instantaneous local magnetic moments become very inhomogeneous in the presence of displacements, and near the T_c the displacements are large, e.g., fluctuations in first nearest neighbor separations exceed the separation of second nearest neighbors, resulting in exchange parameters that depend on the local environment in a complicated manner that cannot be captured by a simple separation dependent interaction.

Note that one of the assumptions of our formula is that the Hamiltonian can be written in terms of pair-spin interactions. To eliminate the effect of the higher order terms, it is important that the energy calculation is performed in a background paramagnetic spin configuration (generated by randomly assigning spin directions) so that the cavity fields for the atoms considered are close to zero. As a tradeoff, the resulting J values could contribute to the underestimation of the magnetization in the ferromagnetic states [10], where effective many site interactions are becoming more important. We also investigated the direct least-square fitting of a representative set of LSMS energies to a Heisenberg model [12]; the data generally agree with the above procedure but with much larger variances.

4. Curie Temperature of Fe and Fe_3C

Utilizing the methods and algorithm described above for WL-LSMS we calculated the Curie temperatures of bulk bcc iron and cementite (Fe_3C) [9]. For the underlying LSMS calculations the atoms are placed on lattices with lattice parameters corresponding to the experimental room temperature values ($5.42a_0$ for bcc iron), and the local interaction zone has a radius of $11.5a_0$. The self-consistently converged potentials for the ferromagnetic ground states were used for all the individual frozen-potential energy calculations in the combined Wang-Landau/LSMS algorithm. The calculation was performed by randomly choosing a site in the supercell and randomly picking a new moment direction. We chose as convergence criterion for the Wang-Landau density of states the convergence of Curie temperature. Using the density of state thus obtained we can calculate the specific heat. The peak in the specific heat allows us to identify the Curie temperature to be 980K for iron, in good agreement with the experimental value of 1050K. The Curie temperature obtained for (Fe_3C is 425K which again is in good agreement with the experimental value of 480K. Note that compared with both experimental and model results the specific heat does not show the expected gradual rise below T_c but exhibits a sudden divergence whereas the high temperature behavior has the qualitatively correct behavior. This peculiarity is an artifact of the convergence criterion we have chosen for the Wang-Landau DOS. The method will first discover the high DOS areas and gradually explore the lower lying regions. To limit the expense of the calculation, it was terminated when no further changes in T_c were observed. This will yield good values for the DOS at energies that correspond to the transition temperature and above. Conversely, for lower energies the DOS will be far from convergence, resulting in the wrong qualitative behavior of thermodynamic properties calculated below T_c. A full convergence of the DOS for lower energies would have required an order of magnitude more samples, which would have made this calculation impractical on the computer resources currently available.

5. Monte Carlo combining magnetic excitations with atomic motion

In Ref. [10], we presented a parametrization of the exchange interaction for each shell as a function of the atomic separation, local volume, and local displacement. In general, the local atomic environment can also be described in terms of quantities already used in classical force fields, e.g., the density function employed in embedded-atom potentials is related to the coordination of the atom. To make the model applicable to molecular dynamics simulations, after comparing several popular iron potentials [11, 16, 17, 19], we re-parameterize the exchange parameter using the magnetic potential proposed by Dudarev and Derlet [19], which is given by

$$E^{\mathrm{eam}} = \sum_i F[\rho_i] + \frac{1}{2}\sum_{ij,i\neq j} V(r_{ij}) \tag{8}$$

where the magnetic embedded function follows

$$F[\rho_i] = -A\sqrt{\rho_i} - \frac{B}{ln2}(1-\sqrt{\rho_i})ln(2-\rho_i)\Theta(1-\rho_i) \tag{9}$$

the embedded density is $\rho_i = \sum_{j,i\neq j} f(r_{ij})$, and the pairwise functions for the density and pair interaction are defined as

$$\begin{aligned} f(r) &= \sum_{n=1}^{N^f} f_n(r-r_n^f)^3\Theta(r-r_n^f) \\ V(r) &= \sum_{n=1}^{N^V} V_n(r-r_n^V)^3\Theta(r-r_n^V) \end{aligned} \tag{10}$$

Table 1. Parameters for the Dudarev-Derlet [19] potential.

A	3.527586256672234	r_n^V	V_n
B	1.642855167616477	4.1	$1.7533861115604772\times10^{-3}$
r_n^f	f_n	3.8	-0.9321219572059338
3.0	0.5055681753757052	3.5	1.696463955030590
2.86	-0.4255552831136833	3.2	0.6638478725109788
2.73	-0.5629408109339820	2.9	-1.914559267568704
2.6	0.4318530885665762	2.6	3.193687184255540

Θ is the step function, and parameters for $f(r)$ and $V(r)$ are listed in Table 1. One of the important reasons we choose the Dudarev-Derlet potential is that its pair density function is relatively short-range, and hence can better represent the local geometric properties.

Considering that pairs of atoms from different shells share different numbers of neighbors and $f(r)$ measures the "neighborness" for each pair, we define the "overlapness" of atom i and j as

$$O_{ij} = \sum_{k,k\neq i,j} f(r_{ik})f(r_{jk}) \tag{11}$$

Where the index k runs through the remaining atoms. The quantity O can statistically distinguish atom pairs from different shells, hence is a good parameter to represent the shell.

Similarly, the density ρ_i is a measure of the coordination of the atom i that depends on the positions of neighboring atoms. Therefore, it strongly correlates with the local volume which is defined as the volume of the Voronoi polyhedra formed by the surrounding atoms. In fact, the two quantities are inversely proportional. There is a small spread in the distribution, but overall, ρ is a good substitute for the local volume. Since the relative displacement from the perfect BCC structure results a force acting on the atom, we can replace the local displacement variable with a force related term defined as

$$d_i = \sum_{k,k\neq i} |\vec{F}_{ik} \cdot \vec{r}_{ik}| \tag{12}$$

With only the separation distance as the parameter, the fitting is statistically significant up to the cubic term [10]. We choose a cutoff radius to bound the r term. After exhausting combinations of above explanatory variables and keeping terms that are significant in statistical tests, our model $J(r, O, \rho, d)$ for the exchange interaction between atom i and j follows,

$$\begin{aligned} J_{ij}^{\mathrm{eff}} &= (3.0035724\times10^{-3}\rho - 7.6680166\times10^{-7}d^2 \\ &+ 4.1178080\times10^{-4})(5.5-r_{ij})^3\Theta(5.5-r_{ij}) \\ &+ (8.1213771\times10^{1}r_{ij} - 1.9304247\times10^{1}r_{ij}^2 \\ &+ 1.4927816r_{ij}^3 + 9.8187830\times10^{-4}d^2 \\ &- 1.1340527\times10^{2} - 3.5371412\times10^{2}O_{ij} \\ &+ 3.9653967\times10^{4}O_{ij}^2)O_{ij} \end{aligned} \tag{13}$$

where $\rho = \rho_i + \rho_j$ and $d = d_i + d_j$. In Fig. 2, we plot J values from LSMS calculations and fitted values from our model for comparisons. The goodness-of-fit is about 0.92. Since the cutoff of $f(r)$ is at 3Å, $J^{\mathrm{eff}}(r, O, \rho, d)$ is bound by the O at 6Å. Our Hamiltonian is in the form,

$$\mathcal{H} = -\sum_{i<j} J_{ij}^{\mathrm{eff}}(\vec{e}_i\cdot\vec{e}_j - \langle\vec{e}_i\cdot\vec{e}_j\rangle) + E^{\mathrm{eam}} \tag{14}$$

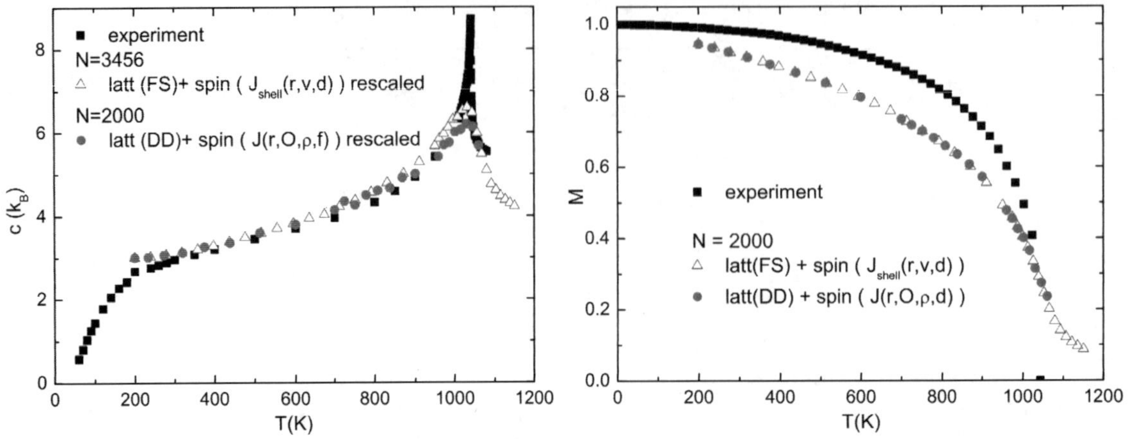

Figure 3. Comparison of specific heat and magnetization for experiment results[13, 14, 15], $J_{\text{shell}}(r, v, d)$ [10] and $J(r, O, \rho, d)$ models. Finnis-Sinclair (FS) potential is used with $J_{\text{shell}}(r, v, d)$ model for a $12 \times 12 \times 12$ supercell, and Dudarev-Derlet (DD) potential with $J(r, O, \rho, d)$ model for a $10 \times 10 \times 10$ supercell. The specific heat data from model calculations are rescaled according to Eq. 15.

We subtract the spin-spin correlation to correct the double counting of the ferromagnetic energy and obtain the same average energy as in E^{eam} potential. In practice, the correlation term is estimated self-consistently, and usually is converged during the process of equilibrating spins in Monte Carlo simulations. Since the parametrization of J^{eff} is based on quantities already employed in embedded-atom potentials, the model is a natural augmentation of the existing iron potentials, and is applicable to molecular dynamics simulations.

The specific heat for a $10\times10\times10$ supercell with Monte Carlo generated thermal displacements is calculated using the $J(r, O, \rho, d)$ model. The $\frac{3}{2}k_B$ contribution from the kinetic degree of freedom is added. Therefore, at zero temperature, the specific heat goes to $4k_B$ with $1k_B$ coming from the classical spin contribution. For small spin numbers ($S < \frac{7}{2}$), there is a discrepancy between classical and quantum calculations even at finite temperatures [18]. To obtain the corresponding quantum specific heat, a scaling factor $f = c_{\text{quantum}}/c_{\text{classical}}$ for the spin contribution was given by Körmann *et al.* [18] as follows,

$$f(t, s) = \left(\frac{2t_s/t}{e^{t_s/t} - e^{-t_s/t}}\right)^2 \qquad (15)$$
$$1/t_s = 0.54S + 0.54$$

where $t = T/T_c$ and $S = 1.1$ for iron. For comparisons, in Fig. 3, we plot the scaled specific heat and magnetization for our model, as well as the experimental data [13, 14, 15] and results from Ref. [10] in which a similar $J_{\text{shell}}(r, v, d)$ model was used. The magnetization curves of the two models overlap with each other, and the small discrepancy near specific heat peaks is simply due to the finite size effect in the simulation near the transition. The generalized shell-independent model agrees with $J_{\text{shell}}(r, v, d)$ model within error bars, and can reproduce iron's specific heat from near T_c down to about $400K$. It also confirms our previous estimate of the spin-lattice contribution to the specific heat for BCC iron, which is significant near the T_c. Since the same procedure as in Ref. [10] is employed in obtaining LSMS energies, the magnetization of $J(r, O, \rho, d)$ model is underestimated comparing with experimental data away from T_c.

6. Combining Molecular Dynamics and Spin Dynamics calculations

As shown in the preceding section, Monte Carlo is an ideal approach for effectively exploring the phase space and revealing the thermodynamic behavior of the system. However, due to the inherent stochastic nature of the sampling, MC methods are incapable of describing the time-dependent properties governed by the deterministic equations of motion (EOM). To gain qualitative insight into the dynamic behavior of the system, one needs to numerically solve the EOM and obtain the phase space trajectories in "real" time.

To adapt our coupled spin-lattice approach into a dynamical simulational model, we incorporate kinetic energy into Eq. 5 and obtain the extended Hamiltonian [20, 21]

$$\mathcal{H} = \sum_{i=1}^{N} \frac{m{v_i}^2}{2} + E^{\mathrm{eam}} - \sum_{i<j} J(r_{ij})\,(\mathbf{e}_i \cdot \mathbf{e}_j - 1) \quad , \tag{16}$$

where m is the mass of an atom and $\{\mathbf{v}_i\}$ are the velocities. For simplicity, we have replaced the spin-spin correlation $\langle \mathbf{e}_i \cdot \mathbf{e}_j \rangle_T$ by 1. In addition, J_{ij}^{eff} has been replaced by a simple pairwise representation, $J(r_{ij})$. [20]

The EOM governing the dynamics of the phase variables $\{\mathbf{r}_i\}$, $\{\mathbf{v}_i\}$ and $\{\mathbf{e}_i\}$ are given by

$$\frac{d\mathbf{r}_i}{dt} = \mathbf{v}_i \ , \qquad \frac{d\mathbf{v}_i}{dt} = \frac{\mathbf{f}_i}{m} \qquad \text{and} \qquad \frac{d\mathbf{e}_i}{dt} = \frac{1}{\hbar S_i}\mathbf{H}_i^{eff} \times \mathbf{e}_i \ , \tag{17}$$

where $\mathbf{f}_i = -\nabla_{\mathbf{r}_i}\mathcal{H}$ is the interatomic force and $\mathbf{H}_i^{eff} = \nabla_{\mathbf{S}_i}\mathcal{H}$ is the effective field, respectively. Although the spin lengths $\{S_i\}$ are not considered as phase variables, they are dynamically varied according to the local atomic environment via a regression relation [22] between the magnetic moment and the embedded density ρ_i of the Dudarev-Derlet potential.

The goal of the combined molecular and spin dynamics (MD-SD) approach is to numerically integrate the EOM starting from a given initial configuration. To achieve the highest possible numerical accuracy and stability, we use an advanced integration algorithm based on the second order Suzuki-Trotter (ST) decomposition for non-commuting operators [23, 24, 25].

6.1. Phonon-magnon interactions

MD-SD approach provides us with a unified framework for investigating the collective excitations at finite temperatures and explore the implications of the spin-lattice coupling. To characterize phonon and magnon modes in our model, we calculate dynamic structure factors $S(\mathbf{q}, \omega)$ [25] for momentum transfer $\mathbf{q}$ and frequency transfer ω, for a $10 \times 10 \times 10$ simulation cell. EOM were integrated up to a total time of 1 ns using a timestep of 1 fs, with periodic boundary conditions imposed. Initial states for the time integration were drawn from the canonical ensemble using MC simulations.

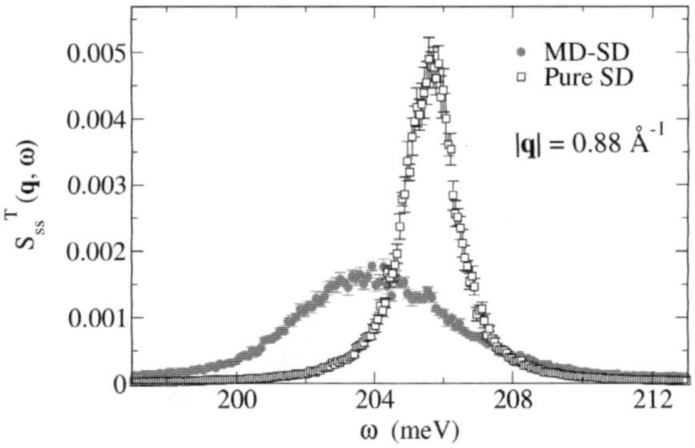

Figure 4. Transverse component of the spin-spin dynamic structure factor $S_{ss}^{T}(\mathbf{q}, \omega)$ at $T = 293$ K for $\mathbf{q} = (0.88\,\text{Å}^{-1}, 0, 0)$, for a $10 \times 10 \times 10$ simulation cell. Full circles represent data obtained from combined molecular and spin dynamics (MD-SD) while open squares show pure spin dynamics (SD) results.

Fig. 4 shows the transverse component of the spin-spin dynamic structure factor at the room temperature $T = 293\,\text{K}$, for $\mathbf{q} = (0.88\,\text{Å}^{-1}, 0, 0)$. For comparison, we have also shown the results obtained from pure spin dynamics (SD) simulations with atoms fixed at perfect BCC lattice positions. With the inclusion of lattice vibrations, we observe that the peak shift to lower frequencies. Moreover, half-width of the peak increases while the amplitude decreases, which are clear indications of the damping of the spin waves.

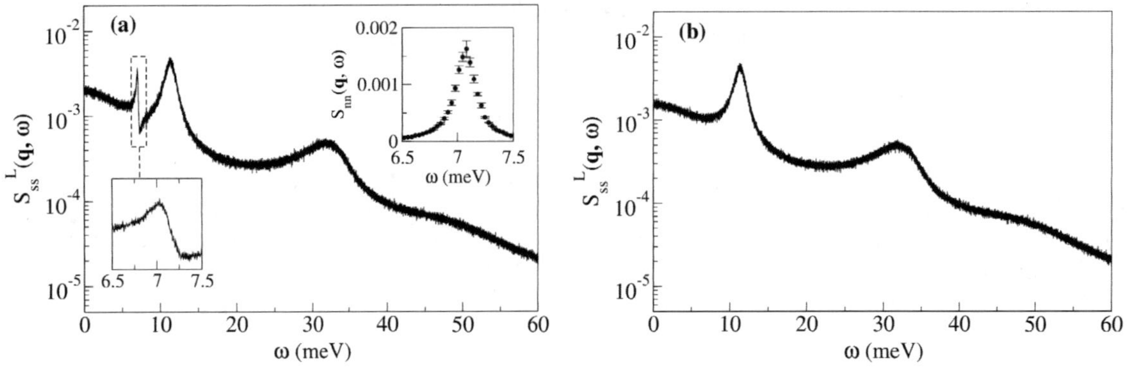

Figure 5. Longitudinal component of spin-spin dynamic structure factor $S^L_{ss}(\mathbf{q}, \omega)$ at $T = 800\,\text{K}$ for $\mathbf{q} = (0.22\,\text{Å}^{-1}, 0, 0)$ for a $10 \times 10 \times 10$ simulation cell. (a) Combined molecular and spin dynamics simulations (b) Pure spin dynamics simulations on a rigid lattice. Inset of (a) shows the density-density dynamic structure factor for the same $\mathbf{q}$.

Fig. 5 shows the longitudinal spin-spin dynamic structure factor at $T = 800\,\text{K}$ for $\mathbf{q} = (0.22\,\text{Å}^{-1}, 0, 0)$; *with* [subset (a)] and *without* [subset (b)] the influence of the phonons. Surprisingly, the additional excitation peak we observe in the subset (a) at $\omega \approx 7\,\text{meV}$ coincides with the peak position of the density-density dynamic structure factor (shown in the inset) for the same $\mathbf{q}$. This, undoubtedly proves the existence of coupled phonon-magnon modes.

6.2. Modeling spin-orbit interactions

Hamiltonian described by Eq. 16 suffers from a major deficiency in regard to accurately establishing the coupling between the lattice and the spin subsystems. Although the coordinate-dependent exchange interaction enables the exchange of energy between the two subsystems, it does not facilitate the transfer of angular momentum in a similar manner [20].

In conventional molecular dynamics, system temperature can be externally controlled via a thermostat coupled to the lattice degrees of freedom. In MD-SD formalism, controlling the lattice temperature, in principle, should allow the spin subsystem to thermalize and subsequently brought into equilibrium with the lattice subsystem. However, the conservation of spin angular momentum strictly prohibits the thermal excitation of the spin orientations, imposing an artificial entropic barrier between the two subsystems.

As the first step towards solving this issue, we identify that the "missing ingredient" in our model is a mechanism that incorporates spin-orbit coupling. Since the notion of orbital angular momentum ($\mathbf{L}$) does not exist in the MD-SD approach, one cannot introduce spin-orbit interactions directly into the Hamiltonian as an atomistic level abstraction of the form, $\mathcal{H}_{\text{SO}} \sim \mathbf{L} \cdot \mathbf{S}$. Therefore we model the underlying effects of spin-orbit coupling in terms of its emergent phenomenon, magnetocrystalline anisotropy.

In the bulk phase of 3d transition metals, crystal-field interaction largely surpasses spin-orbit coupling, leading to quenched orbital magnetic moments and negligibly small anisotropy energies

[27, 28]. However, in low symmetry environments such as surfaces and thin films, spin-orbit coupling is greatly enhanced [29, 30] due to the changes in the effective potential experienced by the electrons. Following the same argument, we claim that the lattice symmetry breaking that occurs due to phonons at finite temperatures may also strengthen spin-orbit coupling by a significant amount. Consequently, this will induce magnetic anisotropies locally on an atomic scale. Magnitudes and easy axes of these anisotropies are completely determined by the local atomic environment and change dynamically as the system evolves in time.

Expanding the anisotropy energy of the ith spin $\mathbf{S}_i$ in successive powers of its components we obtain

$$E_i^{\text{anis}}(S_{ix}, S_{iy}, S_{iz}) = \varepsilon + \sum_{\alpha} \kappa_\alpha S_{i\alpha} + \sum_{\alpha}\sum_{\beta} \lambda_{\alpha\beta} S_{i\alpha} S_{i\beta} + \sum_{\alpha}\sum_{\beta}\sum_{\gamma} \xi_{\alpha\beta\gamma} S_{i\alpha} S_{i\beta} S_{i\gamma} + \cdots, \tag{18}$$

where $\alpha, \beta, \gamma = x, y, z$. Coefficients of this expansion solely depend on symmetry of the local atomic environment surrounding the ith atom. Note that we have completely neglected the cubic anisotropy in bcc iron. Hence all anisotropy coefficients become identically zero as the atoms occupy perfect crystal lattice positions.

Ignoring the terms of third order and above, we write the anisotropic contribution to the Hamiltonian in the following concise form

$$\mathcal{H}_{\text{anis}} = -C_1 \sum_{i=1}^{N} \mathbf{K}_i \cdot \mathbf{S}_i - C_2 \sum_{i=1}^{N} \mathbf{S}_i^{\mathsf{T}} \cdot \mathbf{\Lambda}_i \cdot \mathbf{S}_i, \tag{19}$$

where C_1 and C_2 are constants. To establish the connection to the local environment, we write vector $\mathbf{K}_i$ and tensor $\mathbf{\Lambda}_i$ in terms of the derivatives of a scalar function $\varrho_i(\{\mathbf{r}_k\})$ as

$$\mathbf{K}_i = \nabla_{\mathbf{r}_i} \varrho_i \ , \qquad \mathbf{\Lambda}_i = \begin{pmatrix} \frac{\partial^2 \varrho_i}{\partial {x_i}^2} & \frac{\partial^2 \varrho_i}{\partial x_i \partial y_i} & \frac{\partial^2 \varrho_i}{\partial x_i \partial z_i} \\ \frac{\partial^2 \varrho_i}{\partial y_i \partial x_i} & \frac{\partial^2 \varrho_i}{\partial {y_i}^2} & \frac{\partial^2 \varrho_i}{\partial y_i \partial z_i} \\ \frac{\partial^2 \varrho_i}{\partial z_i \partial x_i} & \frac{\partial^2 \varrho_i}{\partial z_i \partial y_i} & \frac{\partial^2 \varrho_i}{\partial {z_i}^2} \end{pmatrix}. \tag{20}$$

Function $\varrho_i(\{\mathbf{r}_k\})$ is chosen such that its derivatives quantitatively reflect the local symmetry surrounding the ith atom. For convenience, we construct ϱ_i in a manner similar to which the embedded density ρ_i is defined in the embedded-atom formalism. That is, $\varrho_i = \sum_{j(j\neq i)} \phi(r_{ij})$, where $\phi(r_{ij})$ is a pairwise function with a short range cutoff distance. This particular functional form guarantees that under perfect crystalline symmetry, $\nabla_{\mathbf{r}_i} \varrho_i$ and the off-diagonal elements of $\mathbf{\Lambda}_i$ vanish. Although the diagonal elements of $\mathbf{\Lambda}_i$ are non-zero, they become identical, which only leads to a constant shift in energy. $\phi(r_{ij})$ was chosen to be a rapidly decaying function with an arbitrary form of $\phi(r_{ij}) = (1 - r_{ij}/r_c)^4 \exp(1 - r_{ij}/r_c)$, where the cutoff distance $r_c = 3.5$ Å lies between the second and the third nearest neighbor distances. Fourth order polynomial contribution ensures that all interatomic forces that arise due to the coordinate-dependence of $\mathcal{H}_{\text{anis}}$ smoothly go to zero at the cutoff distance.

Eq. 16 combined with the anisotropy terms in Eq. 19 constitutes a complete MD-SD model that fully couples the translational and spin degrees of freedom. To show that our refined model eradicates the shortcomings of the original approach, we perform NVT simulations to investigate the thermal relaxation of a $10 \times 10 \times 10$ lattice. Anisotropy coefficients C_1 and C_2 were arbitrarily set to 2.0 and 1.0, respectively. Initially, all atoms were arranged on a perfect BCC lattice with zero velocities while all spins were aligned in the z direction. Lattice temperature was explicitly controlled by replacing the dynamical equation for the velocities with the stochastic Langevin equation [31]

$$m\frac{d\mathbf{v}_i}{dt} = -\eta \mathbf{v}_i + \mathbf{f}_i + \mathbf{R}_i(t), \tag{21}$$

where η is the damping constant and $\mathbf{R}_i(t)$ is the random force that satisfies the condition $\langle R_{i\alpha}(t)R_{j\beta}(t')\rangle = 2\eta k_B T\delta(t-t')\delta_{ij}\delta_{\alpha\beta}$, with T being the temperature and $\alpha, \beta = x, y, z$. Time evolution of the spins is governed by the conventional spin dynamics equation. That is, NO external torques were applied to the spins in order to control the spin temperature. Fig. 6 (upper panel) shows the time evolution of the instantaneous lattice temperature (T_{L}) and spin temperature (T_{S}) with target temperatures set to 300 K [subset (a)] and 800 K [subset (b)]. Subsets (c) and (d) show the net magnetization per atom for the same respective simulations. T_{L} was measured using the familiar average kinetic energy expression whereas for T_{S} we have used the formula developed by Nurdin et al. [32]

$$T_{\mathrm{S}} = \frac{\sum_i |\mathbf{S}_i \times \nabla_{\mathbf{S}_i}\mathcal{H}|^2}{k_B \sum_i [\mathbf{S}_i \times \nabla_{\mathbf{S}_i}] \cdot [\mathbf{S}_i \times \nabla_{\mathbf{S}_i}]\mathcal{H}}. \tag{22}$$

As expected, we observe a gradual decrease in the magnetization accompanied by an increase in the spin temperature, as the spin subsystem loses angular momentum via the anisotropy terms in the Hamiltonian. Eventually, these quantities converge to their corresponding equilibrium values as the spin and lattice subsystems mutually reach the equilibrium with the thermostat. Moreover, we observe that the relaxation rate of the spin subsystem increases noticeably with the target temperature. As the thermostat temperature increases, thermal fluctuations of the atomic positions also increase. This would further intensity the locally induced anisotropies, hence leading to an increase in the relaxation rate.

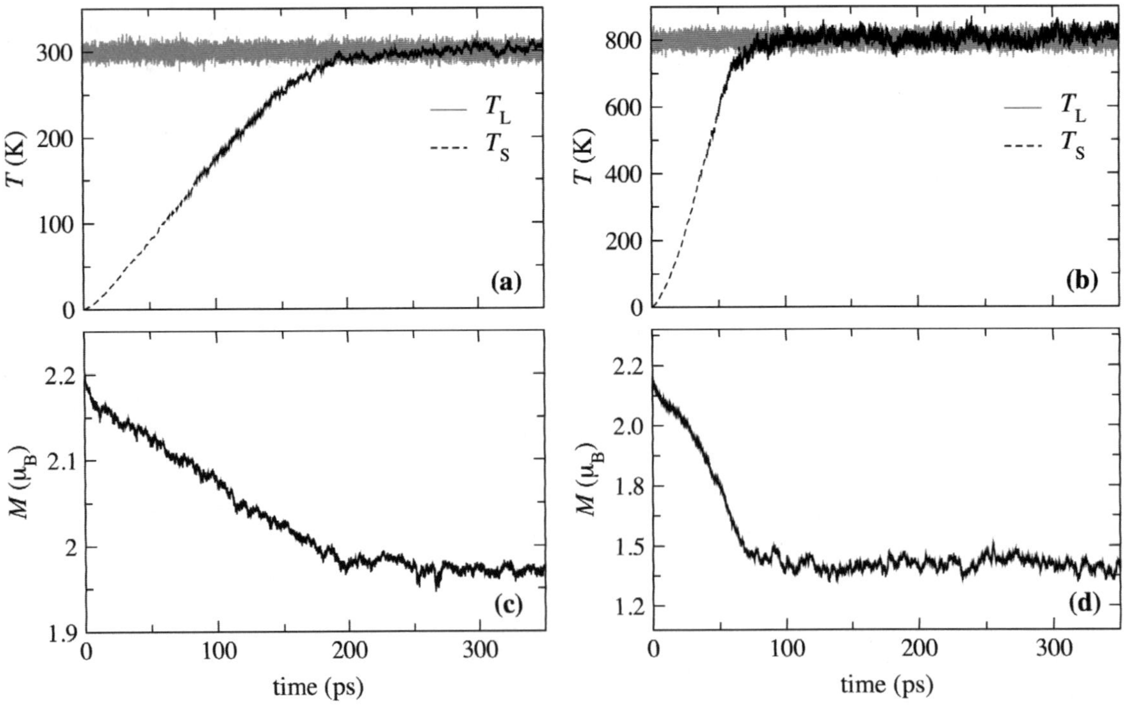

Figure 6. Thermalization of a $10 \times 10 \times 10$ simulation cell via lattice thermostats set to target temperatures 300 K [(a) and (c)] and 800 K [(b) and (d)]. Upper panel [(a) and (b)] shows the time evolution of the lattice temperature (T_{L}) and spin temperature (T_{S}) while the lower panel [(c) and (d)] shows the net magnetization per atom.

7. Conclusion

We have developed the capability to calculate thermodynamic properties of magnetic systems purely from first principles. While this direct first principles approach is still too computationally expensive to be applied as a routine method, we have demonstrated a method to use these first principles calculations to derive classical models that encapsulate the essential physics. These models allow us to explore material specific consequences of magnon-phonon coupling both on the equilibrium thermodynamics as well as on the dynamics of magnetic materials.

Acknowledgments

This Work is sponsored by the U.S Department of Energy, Office of Science, Basic Energy Sciences, Materials Sciences and Engineering Division (Ab initio and Monte-Carlo) and by the *Center for Defect Physics*, an Energy Frontier Research Center (Spin Dynamics-Molecular Dynamics). This research used resources of the Oak Ridge Leadership Computing Facility at ORNL, which is supported by the Office of Science of the U.S. Department of Energy under Contract No. DE-AC05-00OR22725.

References

[1] Wang Y, Stocks G M, Shelton W A, Nicholson D M, Szotek Z and Temmerman W M 1995 *Phys. Rev. Lett.* **75** 2867
[2] Eisenbach M, Györffy B L, Stocks G M and Ujfalussy B 2002 *Phys. Rev. B* **65** 144424
[3] Stocks G M, Wang Y, Nicholson D M C, Shelton W A, Szotek Z, Temmerman W M, Harmon B N and Antropov V P 1996 *Mater. Res. Soc. Symp. Proc.* **408** 157
[4] Stocks G M, Ujfalussy B, Wang X, Nicholson D M C, Shelton W A, Wang Y, Canning A and Györffy B L 1998 *Philos. Mag. B 78* 665
[5] Ujfalussy B, Wang X, Nicholson D M C, Shelton W A, Stocks G M, Wang Y and Györffy B L 1999 *J. Appl. Phys.* **85** 4824
[6] Wang F and Landau D P 2001 *Phys. Rev. Lett.* **86** 2050
[7] Wang F and Landau D P 2001 *Phys. Rev. E* **64** 056101
[8] Eisenbach M, Zhou C-G, Nicholson D M, Brown G, Larkin J, Schulthess T C 2009 *SC09: Proceedings of the Conference on High Performance Computing, Networking, Storage and Analysis* ACM
[9] Eisenbach M, Nicholson D M, Rusanu A and Brown G 2011 *J. Appl. Phys.* **109** 07E138
[10] Yin J, Eisenbach M, Nicholson D M and Rusanu A 2012 *Phys. Rev. B* **86** 214423
[11] Finnis M W and Sinclair J E 1984 *Phil. Mag. A* **50** 45
[12] Yin J, Eisenbach M, Nicholson D M and Rusanu A 2013 *J. Appl. Phys* **113** 17E112
[13] Shacklette L W 1974 *Phys. Rev. B* **9** 3789
[14] White G K and. Minges M L 1997 *Int. J. Thermophys.* **18** 1269
[15] Crangle J and Goodman G M 1971 *Proc. R. Soc. Lond. A* **321** 477
[16] Finnis M W and Sinclair J E 1986 *Phil. Mag. A* **53** 161
[17] Mendelev M I, Han S, Srolovitz D J, Ackland G J, Sun D Y and Asta M 2003 *Phil. Mag. A* **83** 3977
[18] Körmann F, Dick A, Hickel T and Neugebauer J 2010 *Phys. Rev. B* **81** 134425
[19] Dudarev S L and Derlet P M 2005 *J. Phys.:Condens. Matter* **17** 7097
[20] Ma P W, Woo C H and Dudarev S L 2008 *Phys. Rev. B* **78** 024434
[21] Perera D, Landau D P, Nicholson D M, Malcolm Stocks G, Eisenbach M, Yin J and Brown G 2014 *Journal of Applied Physics* **115** 17D124
[22] Derlet P M and Dudarev S L 2007 *Prog. Mater. Sci.* **52** 299
[23] Krech M, Bunker A and Landau D P 1998 *Comput. Phys. Commun.* **111** 1
[24] Omelyan I P, Mryglod I M and Folk R 2001 *Phys. Rev. Lett.* **86** 898
[25] Perera D, Landau D P, Nicholson D M, Malcolm Stocks G, Eisenbach M, Yin J and Brown G 2014 *J. Phys.: Conf. Ser.* **487** 012007
[26] Aharoni A 2000 *Introduction to the theory of ferromagnetism* (New York: Oxford University Press)
[27] Daalderop G H O, Kelly P J and Schuurmans M F H 1990 *Phys. Rev. B* **41** 11919
[28] Skomski R 2008 *Simple Models of Magnetism* Oxford Graduate Texts (New York: Oxford University press)
[29] Wang D S, Wu R and Freeman A J 1993 *Phys. Rev. B* **47** 14932
[30] Lessard A, Moos T H and Hübner W 1997 *Phys. Rev. B* **56** 2594
[31] Kubo R 1966 *Rep. Prog. Phys.* **29** 255
[32] Nurdin W B and Schotte K D 2000 *Phys. Rev. E* **61** 3579

Journal of Physics: Conference Series **640** (2015) 012020 doi:10.1088/1742-6596/640/1/012020

The Adaptively Biased Molecular Dynamics method revisited: New capabilities and an application

Mahmoud Moradi[1], Volodymyr Babin[2], Christopher Roland[2,*], and Celeste Sagui[2,*]

1. Beckman Institute, University of Illinois, Urbana, IL 61801
2. Center for High Performance Simulations (CHiPS) and Department of Physics, North Carolina State University, Raleigh, NC 27695

E-mail: cmroland@ncsu.edu, sagui@ncsu.edu

Abstract. The free energy is perhaps one of the most important quantity required for describing biomolecular systems at equilibrium. Unfortunately, accurate and reliable free energies are notoriously difficult to calculate. To address this issue, we previously developed the Adaptively Biased Molecular Dynamics (ABMD) method for accurate calculation of rugged free energy surfaces (FES). Here, we briefly review the workings of the ABMD method with an emphasis on recent software additions, along with a short summary of a selected ABMD application based on the B-to-Z DNA transition. The ABMD method, along with current extensions, is currently implemented in the AMBER (ver.10-14) software package.

Introduction

The central role of atomistic free energy calculations for basic chemical and biological research is now well established. The free energy is one of the central quantities that determines the behavior of a system at or near equilibrium. For instance, the free energy determines the delicate three dimensional structure of biomolecules, their conformations, their binding, and the reactions they catalyze. Unfortunately, the FES of most biomolecular systems is multi-dimensional, often quite rough and complicated, so that very long time-scales are typically needed in order to explore a FES with any reasonable accuracy. This is generally precluded in a regular molecular dynamics (MD) simulation, even under present computer capabilities. Given the central importance of this problem, considerable effort has been invested in developing and refining new free energy methods, especially those tailored for complex FES when entropic contributions cannot be neglected.

Over the past decade, methods targeting the computation of free energy using non-equilibrium simulations have become popular. These methods all estimate the free energy as a function of a set of "collective variables" from an evolving ensemble of realizations [1,2], and use that estimate to bias the system dynamics and thereby flatten the effective free energy surface. Perhaps the first method to implement this idea was the Local Elevation Method [3]. More recent approaches include the adaptive force bias method [4], the Wang-Landau approach [5], the metadynamics method [6], and the Adaptively Biased Molecular Dynamics (ABMD) method [7] which was developed by our group. Collectively, these methods may be considered to be umbrella sampling methods. In ABMD it is the biasing potential that eventually reproduces the negative of the unknown FES in the long time limit.

This short paper briefly reviews the current state of the ABMD method. The focus will be on a review of the method, as well as on the recent additions based on the idea of coupled multiple walkers [8] and the multi-dimensional string method [9]. An application of the ABMD method to the B-Z DNA transition [10] will also be considered. Currently, the ABMD method is part of the AMBER software package [11], with the implementation of the newer methodological additions pending.

1. ABMD theory

To investigate the equilibrium properties of a complex multi-atomic system, it typically is customary to identify a suitable set of collective variables $\sigma(r_1, ..., r_N)$ which maps the atomic positions $r_1, ..., r_N$ onto a manifold, and then to study its equilibrium probability density:

$$p(\xi) = \langle \delta[\xi - \sigma(r_1, ..., r_N)] \rangle,$$

with the angular brackets indicating an ensemble average. This probability density $p(\xi)$ provides useful information as to the relative stability of the states corresponding to different values of ξ. In practice, the Landau free energy

$$f(\xi) = -k_B T log[p(\xi)],$$

(with k_B representing the Boltzmann constant and T the temperature) is typically preferred. Typically, one thinks of $p(\xi)$ or $f(\xi)$ as providing for a reduced dimensionality description of the system in terms of the chosen collective variables only: the other degrees of freedom have all been effectively integrated out. It is therefore important that the collective variables investigated are chosen in as judicious of a manner as possible, thereby bringing out the underlying physics in the most transparent manner possible. Often, the collective variable is associated with the slowest degrees of freedom, although this is not a formal requirement.

The idea behind the ABMD method [7] is the following. Consider an unknown FES of a system as a function of a suitable set of collective variables. Typically, free energy barriers for biomolecular systems are quite large, and so its difficult for the system to escape the different minima and explore other regions of phase space. To facilitate this, the system dynamics is determined by an *effective* free energy, which is given by the sum of the unknown free energy of the system and a time-dependent biasing potential. Initially, the effective free energy coincides with the unknown free energy of the system. However, with time a biasing potential is built up, which penalizes the locations that the system has visited. The net effect of this is to force the system out of any given minimum and to explore other regions. Ultimately, for very long times, the effective free energy is flattened and the biasing potential gives the negative of the unknown free energy.

The ABMD method is formulated in terms of the following equations:

$$m_a \frac{d^2 r_a}{dt^2} = F_a - \frac{\partial}{\partial r_a} U\left[t|\sigma\left(r_1, \ldots, r_N\right)\right],$$

$$\frac{\partial U(t|\xi)}{\partial t} = \frac{k_B T}{\tau_F} G\left[\xi - \sigma\left(r_1, \ldots, r_N\right)\right],$$

where the first one contains the Newton's equations that govern ordinary MD (temperature and pressure regulation terms not shown), augmented with an additional force coming from the time-dependent biasing potential $U(t|\xi)$ (with $U(t = 0|\xi) = 0$), whose time evolution is given by the second equation. We refer to τ_F as flooding timescale and to $G(\xi)$ as the kernel (in analogy to the kernel density estimator widely used in statistics. The kernel should be positive definite ($G(\xi) > 0$) and symmetric ($G(-\xi) = G(\xi)$). It may be viewed as a smoothed Dirac delta function. For large enough τ_F and small enough kernel width, the biasing potential $U(t|\xi)$ converges towards $-f(\xi)$ as $t \to \infty$.

Our numerical implementation of the ABMD method [7] involves the use of cubic B-splines (or products of thereof) to discretize $U(t|\xi)$. We use the biweight kernel for $G(\xi)$, and an Euler-like discretization scheme for the time evolution of the biasing potential, a very simple scheme that satisfies the goal of flattening $U(t|\xi) + f(\xi)$ in the long time limit. Note that the numerical cost of evaluating the time-dependent potential is constant over time, and so ABMD scales trivially as $O(t)$. Storage requirements are also relatively mild, especially since sparse arrays may be used because only the nonzero elements of U_m need to be stored explicitly. In addition, ABMD is characterized by only *two* control parameters: the flooding timescale τ_F and the kernel width $4\Delta\xi$.

We have also implemented two important extensions to ABMD. The first is identical in spirit to the multiple walkers metadynamics [1,12]. It amounts to carrying out several different MD simulations biased by the same $U(t|\xi)$, which evolves via:

$$\frac{\partial U(t|\xi)}{\partial t} = \frac{k_B T}{\tau_F} \sum_{\alpha} G[\xi - \sigma(r_1^{\alpha}, \ldots, r_N^{\alpha})] \,,$$

where α labels different MD trajectories. A second extension is to gather several different MD trajectories, each bearing its own biasing potential and, if desired, its own distinct collective variable, into a generalized ensemble for "replica exchange" with modified "exchange" rules [13,14,15]. Both extensions are advantageous and lead to a more uniform flattening of $U(t|\xi) + f(\xi)$. This enhanced convergence to $f(\xi)$ is due to the improved sampling of the "evolving" canonical distribution.

We have implemented the ABMD method in the AMBER simulation package (ver.10-14) [11], with support for both replica exchange and multiple-walkers. In pure "parallel tempering" replica exchange (same collective variable for all replicas), N_r replicas are simulated at different temperatures T_n, $n = 1, \ldots, N_r$. Each replica has its own biasing potential $U^n(t|\xi)$, $n = 1, \ldots, N_r$, that evolves according to its own dynamical equation. Exchanges between neighboring replicas are attempted at a prescribed rate, with an exchange probability given by [13]:

$$w(m|n) = \begin{cases} 1, & \Delta \le 0, \\ \exp(-\Delta), & \Delta > 0, \end{cases}$$

$$\Delta = \left(\frac{1}{k_B T_n} - \frac{1}{k_B T_m}\right)\left(E_p^m - E_p^n\right) + \frac{1}{k_B T_m}\Big[U^m(\xi^n) - U^m(\xi^m)\Big] - \frac{1}{k_B T_n}\Big[U^n(\xi^n) - U^n(\xi^m)\Big],$$

where E_p denotes the atomic potential energy. The biasing potentials are temperature-bound and converge in the $t \to \infty$ limit to the free energies at their respective temperatures.

In a general replica exchange scheme, we allow different replicas to have different collective variables and/or temperatures. We also allow for either an evolving or a static biasing potential (the latter obviously includes the case of $U^n(t|\xi) = 0$) on a per-replica basis (when all biasing potentials are static, the method reduces to a peculiar variant of the "Hamiltonian Replica Exchange" described in Ref.[13]). Exchanges between random pairs of replicas are then attempted at a prescribed rate. To illustrate the point, consider the *in vacuo* folding of the hydrophobic peptide $Ace - (Gly)_2 - Pro - (Gly)_3 - Nme$, which has a stable β-hairpin in the 1999 version of the Cornell *et al* [16] force field. First, accurate ABMD free energy calculations at eight different temperatures as a function of the radius of gyration R_g as the collective variable were carried out (see the inset in Fig.1). The global minimum corresponding to the smaller value of R_g is associated with a coil structure, while the metastable minimum associated with the larger value corresponds to a β-hairpin structure. The relative depth of these wells changes, as expected, in the presence of explicit solvent environments [7,17]. However, there is more information that can be gleaned from these simulations. For instance, if one wants additional information with regards to the number of hydrogen bonds formed along the peptide backbone N_{OH}, it is possible to carry out an ABMD simulation with the modified replica exchange in such a way that one reuses the previously computed free energies for R_g in order to obtain the *two-dimensional* free energy landscape in the (N_{OH}, R_g) space. Here, the initial simulations play a double role in serving as a "sampling enhancement device" for different replicas, thereby greatly speeding up the calculation. The resulting *two-dimensional* free energy landscape shown in Fig.1 now conveys additional information not contained in the *one-dimensional* plots. In particular, the "coil minimum" associated with the *one-dimensional* plot is degenerate, and actually corresponds to two different minima for which the globular structures have a different number of hydrogen bonds. This example illustrates not only the type of information that may be obtained with ABMD, but also shows, that as one learns more about a given biomolecular system, one can "dynamically" change the questions asked. Previous calculations may often be readily reused, in order to speed up the new simulations.

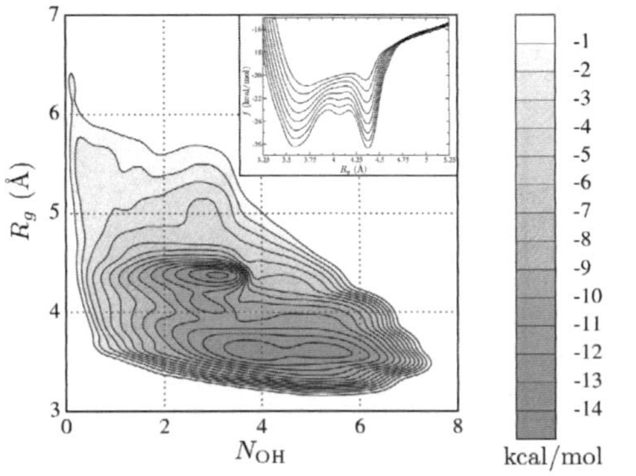

Figure 1. Two-dimensional free energy landscape for *in vacuo* Ace-(Gly)$_2$-Pro-(Gly)$_3$-Nme peptide at T=300 K as a function of collective variables R_g and N_{OH}. Inset shows the corresponding one-dimensional free energy as a function of only R_g for eight different temperatures ranging from T=300 (bottom) to T=600 K (top). See Ref.[7].

Finally, we note two important points. First, if one uses the ABMD method alone, one needs follow-up equilibrium umbrella sampling runs to improve the accuracy of the results, according to the recipe that we provided in the context of the metadynamics method [18]. Second, if in addition to the free energy map, one desires equilibrium, unbiased properties of the system, it is possible to set up a generalized REMD scheme. In this scheme, the choice of the Hamiltonians determines the performance of the method. If the exchange rate between the replicas is sufficiently high to guarantee a random walk between the replicas, then one needs only to consider the details of the so-called "hot" replica whose purpose is to facilitate barrier crossing (or, more formally, to decrease the ergodic time scale). One possibility for this is to run the hot replica at high temperature. Another possibility [19] is to construct the hot replica by adding a biasing potential to the original Hamiltonian that acts on some collective variable that describes one of the slow modes of the system that needs "acceleration". ABMD is therefore first used to generate potentials for these "hot" replicas. A combination of such Hamiltonian and Temperature based replica exchange molecular dynamics (HT-REMD) [18,19,20] provides for a practical way to reduce the computational costs associated with sampling, since it facilitates the sampling in the "hottest" replica by both means, and therefore also allows for a better "tuning" of the entire setup. These procedures ultimately can yield high accuracy free energy curves and equilibrium properties, and have already been applied to a variety of different biomolecular systems including small peptides [7,18,19], sugar puckering [19], polyproline systems [19,20,21], guest-host systems [22,23], polyglutamine systems [24], and the B-to Z DNA transition [10].

In addition, we have implemented Steered Molecular Dynamics (SMD) simulations to examine transition pathways and mechanisms, as well as to estimate free energy differences [20]. The SMD method is based on the following ideas. Consider a thermodynamic process that changes a system by means of a control parameter ξ, which runs from ξ_0 to ξ_t over a time interval t. The second law of thermodynamics states that the work W done on the system cannot be smaller than the free energy difference: $\Delta F = f(\xi_t) - f(\xi_0) \leq W$. Thus, a non-equilibrium process can only provide us with an upper bound for the free energy difference. However, the remarkable Jarzynski [25] identity (JE) is an exact equality that holds even when the transformation is irreversible:

$$\exp(-\Delta F/k_B T) = \langle \exp(-W/k_B T) \rangle$$

(the angular brackets denote the average over all possible realizations of the process connecting the states ξ_0 and ξ_t). Thus, in principle ΔF can be estimated using a number of finite-time *non-equilibrium* simulations. Unfortunately, because the average of the exponential function is dominated by rare realizations, convergence is often poor unless a very large number of runs is used. It is, however, possible to improve the free energy estimates by using modifications based on the maximum likelihood estimator (MLE) and Bennet's acceptance ratio (BAR) method, which are considered to be equivalent [26,27,28]. These methods make use of both forward (F) and reverse (R) non-equilibrium pulling processes, in order to move the system between an initial and final state. Typically, the system is moved along a specific pathway in its configurational space, which is readily accomplished through the addition of a harmonic restraining potential to the Hamiltonian in the stiff-spring limit [20]. This approximation ensures that the calculated potential of mean force is well represented by the free energy of the system with the harmonic restraint. In our system, we carried out a number of non-equilibrium "pushing" and "pulling" simulations by changing the value of the collective variable ξ at a constant velocity over a prescribed pathway. The numerical values of the work for the forward (W^F) and reverse (W^R) pullings are then collected and ΔF is estimated from the transcendental equation [26,27,28]:

$$\sum_{i=1}^{n_{\mathrm{F}}} \frac{1}{1 + n_{\mathrm{F}}/n_{\mathrm{R}} \exp\left(W_i^{\mathrm{F}} - \Delta F\right)} - \sum_{i=1}^{n_{\mathrm{R}}} \frac{1}{1 + n_{\mathrm{R}}/n_{\mathrm{F}} \exp\left(W_i^{\mathrm{R}} + \Delta F\right)} = 0,$$

with $n_{\mathrm{F,R}}$ denoting the numbers of forward and reverse simulations, respectively. As may be expected, we find that ΔF calculated with this two-sided method displays an improved convergence over estimates obtained with the straightforward one-sided application of the Jarzynski equality [20]. To date, we have extensively tested SMD on proline and DNA-based systems. We have been able to use SMD to steer the system between different states over the ABMD FES, thereby being able to probe different transition mechanisms, their associated barriers, and to estimate free energy differences. It comes as no surprise that the modified formula based on the BAR method shows the fastest convergence and gives the best results [20].

2. New Methodological Developments

Recently, we have augmented our implementation of the ABMD method in a number of significant ways, including the addition of a multiple-walker selection mechanism [8], the well-tempered ABMD [29], driven ABMD [30], and a version of the string method [9]. These methods, which are currently being built into the latest release of AMBER, are briefly described in this section.

2.1. ABMD with Selection Algorithm

The multiple-walker ABMD method is based on a set of *non-interacting* walkers or replicas, all of which persist for the same duration. However, this can lead to problems because not all walkers are equally effective in sampling the configuration space. Often, these end up being clustered together, and kept near local metastable regions because of hidden barriers resulting from degrees of freedom orthogonal to the reaction coordinate. Thus, for maximum efficiency, one would like to "encourage" walkers that are wandering in the under-explored regions, and penalize walkers in the oversampled regions, and achieve this on-the-fly. This requires a dynamic monitoring of the walkers, and their updating according to suitable, preselected fitness functions. A typical algorithm of this kind starts [8] by assigning the same weights to every walker, and lets them run for a preselected period of time. The "fitness" of each walker is then tested, which is a *de facto* measure of overlap of the explored phase space of each walker. Walkers who are wandering in under-sampled regions are then given a stronger weight and replicated a number of times, while walkers found to be exploring oversampled regions are correspondingly killed. This kind of procedure is then repeated periodically, thereby accelerating the convergence to a uniform distribution of walkers in the reaction coordinate space. This "selection" algorithm is thus a bootstrapping-based resampling of conformations in so far that the walkers are bootstrapped periodically to flatten the free energy landscape more efficiently. We note that a variation of selection algorithm similar to the one implemented here has previously been implemented in the NAMD [8] software package.

We have implemented an ABMD multiple-walker selection algorithm as follows. At fixed simulation time intervals τ, a weight w_i is assigned to walker i proportional to $\exp(\int_{(n-1)\tau}^{n\tau} S(\zeta_t^i)\,dt)$ in which ζ_t^i is the instantaneous value of reaction coordinate vector. Furthermore, we use $S(\zeta) = c\frac{\nabla^2\rho(\zeta)}{\rho(\zeta)}$, with $\rho(\zeta)$ representing the density of ζ, c is a constant, and index n denotes the time interval. This update is performed either for the duration of the simulation, or until some stopping criterion is satisfied. The stopping criterion is based on the entropy of weights $H = \sum_i w_i log(w_i)$. The selection algorithm will stop if $E_w = H - \log(\frac{1}{N})$ falls below $\epsilon \log(\frac{1}{N})$ in which N is the total number of walkers and $0 \le \epsilon \le 1$ is a constant. Note that $\log(\frac{1}{N})$ represents the entropy of the uniform weights.

Comparing to the original method [8], our implementation also incorporates the following technical features: (i) We use the biasing potential instead of a histogram to approximate the density; (ii) We use a generalized Laplacian instead of a simple second order derivative for multidimensional cases for added stability; and (iii) Since ABMD uses B-splines, we build up analytic functions for the biasing potential and therefore can calculate the needed derivatives analytically for additional stability and speedup. Tests indicate that our current implementation of the described selection mechanism for multiple walkers leads to a speed-up in the convergence by a factor ranging from 2 to 5 over a corresponding system with noninteracting multiple walkers.

2.2. The well-tempered ABMD

In its original implementation, ABMD uses a history-dependent biasing potential that is updated at a fixed rate:

$$U_a(\zeta, t) = U^0(\zeta) + \int_0^t dt'\,\omega\,K(\zeta - \zeta^{t'}), \tag{1}$$

in which K is a kernel function, U^0 is an arbitrary function (*e.g.*, this typically represents the initial guess for the biasing potential and is typically assumed to be flat; ω represents the contant, unbiased rate). It has been shown [45] that once converged, $\langle U_a(\zeta, t \to \infty)\rangle_a \approx U_a^s(\zeta) + u(t)$ (here $u(t)$ is an additive constant at each t) in which $U_a^s(\zeta)$ (*i.e.*, converged biasing potential) equals $-F(\zeta)$.

It turns out that using the same rate of updating the biasing potential throughout the simulation often leads to poor convergence to the free energy. Instead, rather than converging smoothly to the desired result, the biasing potential ends up fluctuating around $-F(\zeta)$ (with an amplitude which depends on ω). A solution to this problem has been proposed [29] via the use of a "well-tempered" ω:

$$U_a(\zeta, t) = U^0(\zeta) + \int_0^t dt'\,\omega(\zeta^{t'}, t')\,K(\zeta - \zeta^{t'}), \tag{2}$$

in which $\omega(\zeta,t)$ is a time-dependent, non-uniform rate $\omega_0 \mathrm{e}^{-\beta' U_a(\zeta,t)}$ ($1/\beta' = k_B T'$ and T' is a pseudo-temperature) that reduces to a constant ω_0 in the $\beta' \to 0$ limit (*i.e.*, conventional ABMD). It has been shown [29] that $\langle U_a(\zeta, t \to \infty)\rangle_a \approx U_a^s(\zeta) + u(t)$, and so $U_a^s(\zeta)$ and $F(\zeta)$ are related via $U_a^s(\zeta) = -(1+\frac{\beta'}{\beta})^{-1}F(\zeta)$ or $F(\zeta) = -(1+\frac{T}{T'})U_a^s(\zeta)$. This way of updating leads to smoother convergence of the biasing potential to the desired free energy.

2.3. Driven ABMD

Both ABMD and SMD schemes are both powerful nonequilibrium free energy methods. However, each method has its own practical limitations associated with it. For instance, SMD is often associated with a very slow convergence when used to calculate free energies. However, it is very useful for the qualitative exploration of transition paths. ABMD, on the other hand, is very useful in mapping out free energy landscapes, even though this may take a considerable computational effort depending on the complexity of the FES. It turns out, however, that ABMD and SMD may be combined and integrated into a new driven adaptively biased method [30]. Such a driven ABMD (D-ABMD) takes advantage of both methods for faster convergence. D-ABMD has an advantage over conventional (or well-tempered) ABMD in that it ensures the exploration of the transition pathway (from one end to the other) in the early stages of the simulation and gradually improves the estimate of the free energies almost uniformly along the reaction coordinate. D-ABMD has also an advantage over the conventional SMD in that the effective free energy surface gradually becomes smooth and flat such that the system can move along the reaction coordinate with progressively less amount of work. The D-ABMD method implemented is a straighforward generalization of the so-called driven metadynamics method [30].

To combine the two schemes, we have developed a driven adaptive-bias scheme that adds an adaptive ($U_a(\zeta,t)$) and a driving ($U_d(\zeta,t)$) potential to the Hamiltonian. We use an iterative approach in which an independent simulation is performed from time $t=0$ to $t=T$ in the nth iteration ($n = 1, 2, \ldots$), biased by the potential $U_d(\zeta,t)+U_a^n(\zeta,t)$ in which $U_d(\zeta,t) = \frac{k}{2}(\zeta-\eta(t))^2$ for all n ($\eta(t)$ is the moving center of the SMD harmonic potential in ζ space), and:

$$U_a^n(\zeta,t) = U^{n-1}(\zeta) + \int_0^t dt'\, \omega(\zeta^{t'},t')\, K(\zeta-\zeta^{t'})\, \mathrm{e}^{-\beta w^{t'}}, \tag{3}$$

in which w^t represents either the accumulated work or the transferred work. The accumulated and transferred works are defined as $w_{ac}^t = \int_0^t dt' \frac{\partial}{\partial t'} U_d(\zeta^{t'},t')$ and $w_{tr}^t = w_{ac}^t - U_d(\zeta^t,t)$ (note that this is actually the transferred work shifted by $-U_d(\zeta^0,0)$). Theoretically the $\mathrm{e}^{-\beta w_{tr}^{t'}}$ factor or "constant weight" is more accurate but for practical reasons the $\mathrm{e}^{-\beta w_{ac}^{t'}}$ factor or "pulling weight" is preferred. For a discussion on these weight functions used for the free energy estimation from conventional pulling experiments, please see Ref. [31]. Particularly, in our algorithm, the constant weight $\mathrm{e}^{-\beta w_{tr}^{t'}} = \mathrm{e}^{-\beta w_{ac}^{t'}} \mathrm{e}^{\beta U_d(\zeta^t,t)}$ may become unstable for large biasing potentials. To avoid this instability a cutoff for w^t is used. This method also provides for faster convergence and greater stability to the desired free energy.

2.4. The string method

The swarms-of-trajectories string method (STSM) [9] is a path-finding algorithm that refines a putative transition pathway (or a string of conformations) iteratively until the pathway is converged. The string is defined by a number of nodes or images parameterized in a high-dimensional space of collective variables, whose position is updated in each iteration. The drift applied to each image is estimated by averaging over the drifts of a swarm of short unbiased trajectories all starting at the current position of the image. The parallel variation of the method – which is considerably more efficient than the original variation – requires hundreds to thousands of replicas of the system to run simultaneously [32].

For a string of N nodes, each consisting of M copies, $N \times M$ replicas will be required. Each unbiased drifting iteration is followed by a biased equilibration stage to move the replicas to their updated position. Unlike the ABMD simulations, STSM simulations can easily employ many collective variables since the sampling always stays along a 1D curve. Therefore one may use all collective variables that are somewhat relevant to the transition of interest, particularly those associated with slow dynamics. The method therefore enables the mapping out of Least Free Energy Pathways (LFEP) in a multi dimensional phase space. The number of images used is important for the accuracy of the pathway and the number of copies used is important in determining the convergence behavior.

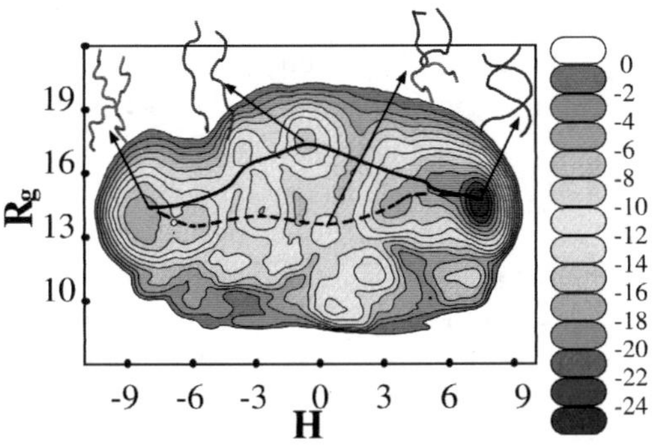

Figure 2. Free energy landscape of $(CG)_6$ DNA as a function of the radius of gyration R_g and handedness H. Ribbon diagrams associated with the four minima shown in blue (B-DNA), red (Z-DNA) and grey (transition structures). Two different Least Free Energy Paths (LFEPs) are identified with a solid line (stretch-collapse mechanism) and with a dashed line (zipper-like mechanism. See Ref.[10] for details.

Our parallel implementation of STSM method is based on iterative unbiased and restrained MD simulations followed with reparametrization of the restraint centers defined in the multidimensional collective variable space of ζ. At iteration s, M copies of image i restrained around old center η_i^{s-1} equilibrate for τ_E steps ($t = 0, \dots, \tau_E$) such that $\zeta_{i,l}^t$ for each copy l of image i is expected to be close to η_i^{s-1} at $t = \tau_E$ assuming the restraint is stiff enough. The restraint is then released to allow drifting for τ_R steps. The drifted center η_i^s for image i is first determined by averaging over all drifted copies $\zeta_{i,l}^t$ at $t = \tau_E + \tau_R$. The resulting string of images is smoothed using the protocol proposed in Ref.[33]. The smoothing parameter $0 \leq \epsilon \leq 1$ determines the smoothness of the curve and is recommended to be on the order of $1/(N-1)$. The last step is the reparametrization which again follows the same protocol proposed [33] to generate a set of N sequentially equidistant centers along the same string formed by the smoothed centers.

3. An Application: The B-Z DNA transition

As already noted, the ABMD method has been extensively applied to a number of biomolecular systems. Here, we briefly discuss a single application, namely that of the B-Z DNA transition [10]. In 1979 the publication of the crystal structure of a $d(CGCGCG)_2$ duplex revealed a left-handed double-helix DNA with two antiparallel chains joined by Watson-Crick (WC) base pairs [33]. This conformation was called Z-DNA because the sugar-phosphate backbone had a characteristic zig-zag pattern. Left-handed Z-DNA is characterized by a dinucleotide repeat, and occurs in sequences that alternate a purine-pyrimidine repeat, mainly CG (or GC). The anti-syn alternation of base pairs that underlies the zig-zag pattern is due to the rotation of the guanine residue around its glycosidic bond, resulting in its syn conformation. The paired cysteine remains in its anti conformation, but it rotates 180° jointly with its attached sugar. As a result, the base pairs in Z-DNA, as compared to B-DNA where all the bases are in the anti conformation, are stacked on opposite sides of the line connecting the two phosphate groups across the helix [34,35], and there is a different stacking pattern of the GpC and CpG sequences involving a series of two base steps along each of the two strands [36].

Since its discovery, Z-DNA has evolved from an in vitro curiosity to a challenging entity with important biological roles. The first observation of a complete inversion of the ultraviolet circular dichroism spectrum of poly(CG) DNA in the presence of a high concentration of salt took place in 1972 [36]. After the revelation of the Z-DNA structure in 1979, further evidence of the B-Z DNA transition was obtained [37]. Yet, the experimental evidence for the *in vivo* existence of Z-DNA has been slow to gather. It is now known that Z-DNA is highly immunogenic, and there are antibodies against it that have been further used to map regions favorable to Z-DNA conformations [38]. There is a high density of base sequences favoring Z-DNA, *i.e.,* bases with suitably alternating purines-pyrimidines, near transcription start sites. The formation of Z-DNA is also induced by Z-DNA binding proteins near the promoter region which boosts the transcription of the downstream genes. All in all, it is now thought that Z-DNA formation is closely related to gene expression, regulation, and recombination [39].

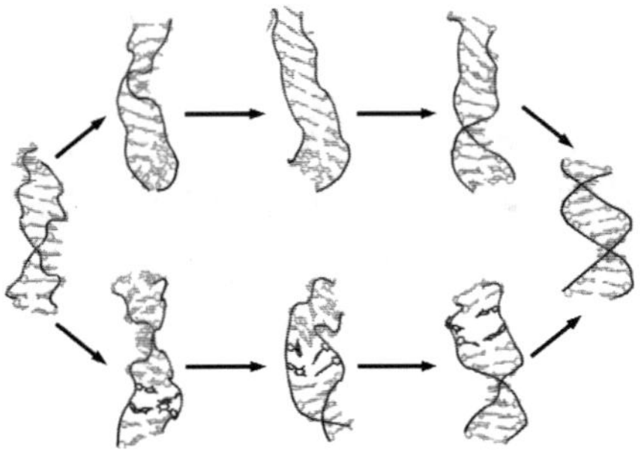

Figure 3. Alternative B-Z DNA transition pathways. Snapshots of a Z- to B-transition for the dodecamer $(CG)_6$ obtained from SMD simulations through the LFEP trajectories. Top: stretch collapse mechanism; Bottom: base extrusion mechanism. Colors in the ribbon representation are: red (Z-DNA); blue (B-DNA); grey (intermediate structures). In addition, the C and G nucleotides are colored in green and cream, respectively. See Ref.[10] for details.

Despite all the investigations about the biological role of left-handed Z-DNA, the microscopics behind the B-Z DNA transition have remained controversial, with several different mechanisms proposed in the literature [40]. Generally speaking, the models for the B-Z transition are classified into mechanisms that involve either base-pair opening or base-pair rotation without WC base-pair breaking. Each of these mechanisms may or may not have an intermediate structure. Among the models without intermediate structure, the most popular base-pair opening mechanism is the Wang model [33]. It proposes that base-pair opening occurs before base-pair plane and phosphate backbone angle rotation within the core of the helix. The Harvey model [41], on the other hand, proposes that the B-Z transition happens through the successive flipping of base-pair planes, without any disruption of the WC pairs. This process is supposed to be facilitated by breathing modes. Experimental evidence, however, seems to favor models with intermediate structures [42,43,44]. The Saenger-Heinemann model postulates that the transition takes place through two A-DNA-like intermediates (the second one with a dinucleotide unit), without breaking of the base pairs [45,46]. More recently, the extrusion of bases was observed in the crystal structure of a B-Z junction [47]. These extrusions and their propagation followed by the reformation of the base pairs in a new order also represents an alternative mechanism for the transition [47] (similarly to the Wang's model, except that the bases rotate outside the double-helix core).

We have recently reexamined the problem of the B-Z DNA transition [10] using the ABMD and SMD methods. This has the advantage of allowing for a unified description of the transition, incorporating a variety of mechanisms. A sample two-dimensional free energy for the B-Z transition is shown in Fig.2, as a function of the radius of gyration (R_g) and handedness (H). The most prominent minima are associated with B- and Z-DNA structures, respectively. Connecting these minima are two LFEPs associated with a stretch-collapse mechanism and a zipper mechanism (see Fig.3). The stretch collapse mechanism involves relatively little disruption of the base pairs. Instead, the DNA stretches and unwinds, forming two almost parallel strands that preserve their WC base pairs, and assuming an S-DNA type conformation. The conformation then rewinds to form a double helix of the opposite handedness. We note that such a mechanism has also been previously identified in other simulation studies [10]. Alternatively, the so-called zipper-like mechanism involves almost no change in R_g. However, there are large changes in handedness within the structure, and the intermediates involve at least one B-Z junction with extruded base pairs. It is interesting that very similar free energy barriers are associated with both of these transition mechanisms [10], indicating that the transition is best described in terms of a reaction path ensemble.

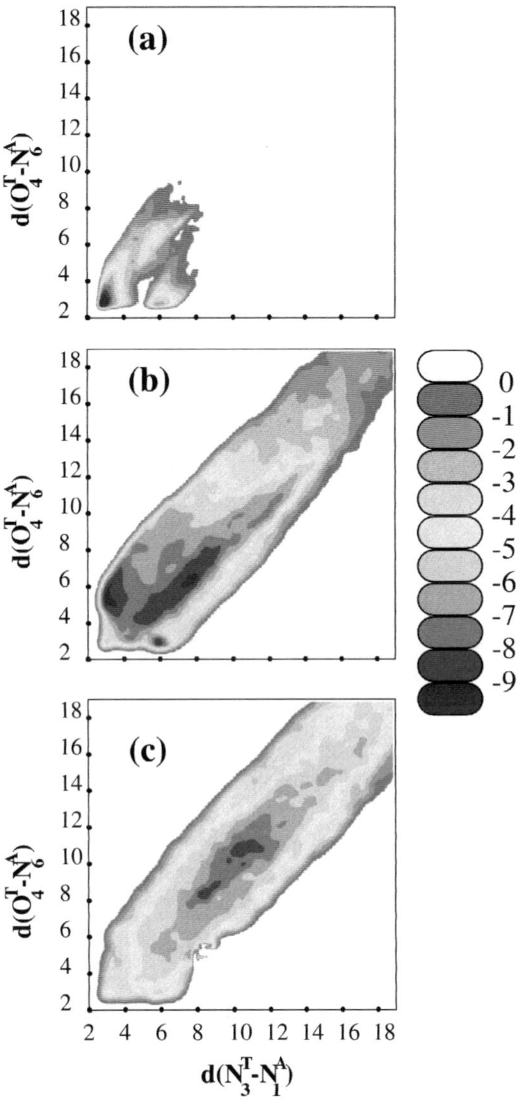

Figure 4. Free energy landscapes of an AT base in different conformation for $(CG)_3(AT)(CG)_3$ double helix DNA in terms of the collective variable $d(O_4^T - N_g^A)$ (*i.e.*, the distance between the O_4^T and N_6^A of the thymine and adenine bases) and $d(N_3^T - N_1^A)$ (*i.e.*, the N_3^T and N_1^A of the thymine and adenine bases) with the rest of the structure having a conformation corresponding to (a) B-NDA; (b) Z-Z junction; and (c) a B-Z junction. For (a) the free energy minima correspond to structures with the correct WC pairing, while in (b) the minima are associated with reverse and single WC pairing. See Ref.[10] for more details.

The crystal structure of a B-Z junction [47] has revealed the full extrusion from the helix of the two junctional bases. The formation of a B-Z junction requires little structural disruption, because it preserves the integrity of both the B- and Z-DNA helices as well as the base stacking between the two helices. While a B-Z junction is formed at the interface between B- and Z-DNA, a Z-Z junction is also commonly formed in sequences where the dinucleotide repeat is interrupted by single base-pair insertions, or deletions, that bring neighboring helices out of phase. The three-dimensional structure of a Z-Z junction has been recently described experimentally [48]. The Z-Z junction studied by de Rosa *et al* is stabilized by Zα, the Z-DNA binding domain of the RNA editing enzyme ADAR1, and consists of a single base pair AT in the middle of a CG sequence, with the resulting duplex sequence $5' - (CG)_3 - A - (CG)_3 - 3'$. The AT pair leads to a partial disruption of the helical stacking. While Z-DNA is resistant to intercalating agents, these can insert themselves into the junction region. In contrast to a B-Z junction, the basis of this structure are not fully extruded, and the stacking between the two left-handed helices is not continuous.

As part of our study of the B-Z transition, we have also investigated both the free energies of pure B-Z junctions as well as junctions with an A-T insertion. Sample free energy results for the latter are shown in Fig.4. Here the top panel (a) shows the results for a B-B junction. The free energy landscape is characterized by three minima, with the global minima corresponding – as expected – to a junction with the correct WC pairing. In (b), the free energy of a Z-Z junction and shows that the minima have shifted considerably. In this case the Z-Z junction may be conformations involving correct WC pairing, single WC pairing, and reverse WC pairing. The global minimum involves the latter, and characteristic structure is in good correspondence with the experimental results [48]. Finally, (c) shows results for a B-Z junction with minima corresponding to non-WC pairings. For these

Journal of Physics: Conference Series **640** (2015) 012020 doi:10.1088/1742-6596/640/1/012020

and further details on the B-Z DNA transition, we refer the reader to Ref.[10].

4. Summary
In summary, we have briefly reviewed the workings of the ABMD method, along with recent extensions that are currently being worked into the AMBER simulation package. To illustrate the workings of the method, a single application involving the B-Z DNA transition is briefly summarized.

5. Acknowledgements
We thank NSF (grants MCB-1021883 and SI2/SEE-1148144) for funding, and the NCSU HPC for extensive computer support.

6. References
[1] Lelievre T, Rousset M, and Stoltz G 2007 *J. Chem. Phys.* **126**, 13411.
[2] Bussi G, Laio A, and Parrinello M 2006 *Phys. Rev. Lett.* 96, 090601.
[3] Huber T, Torda A and van Gunsteren W 1994 *J. Comput. Aided Des.* **8**, 695.
[4] Darve E, and Pohorille A 2001 *J. Chem. Phys.* **115** 9169.
[5] Wang F, and Landau D 2001 *Phys. Rev. Lett.* **115** 2050.
[6] Iannuzzi M, Laio A and Parrinello M 2003 *Proc. Natl. Acad. USA* **99** 12562.
[7] Babin V, Roland C, and Sagui C 2008 *J. Chem. Phys.* **128** 134101.
[8] Minoukadeh K, Chipot C, and Lelievre T 2010 *J. Chem. Theor. and Comput.* **6** 1008.
[9] Pan A, Sezer D, and Roux B 2008 *J. Phys. Chem.* **112** 3432.
[10] Moradi M, Babin V, Roland C, and Sagui C 2013 *Nucl. Acids Res.* **41** 33.
[11] AMBER 10, 2008 (http:ambermd.org), and higher versions.
[12] Raiteri P, Laio A, Gervasio F, Micheletti C, and Parrinello M 2006 *J. Phys. Chem.* **110** 3533.
[13] Sugita Y, Kitao A and Okamoto Y 2000 *J. Phys. Chem.* **113** 6042.
[14] Bussi G, Gervasio F, Laio A, and Parrinello M 2006 *J. Am. Chem. Soc. 128* 13435.
[15] Piana S and Laio A 2007 *J. Phys. Chem. B 111* 4553.
[16] Cornell W, *et al* 1995 *J. Am. Chem. Soc.* **117** 5179.
[17] Babin V, Karpusenka V, Moradi M, Roland C, and Sagui C 2009 *Int. J. Quant. Chem.* **109** 3666.
[18] Babin V, Roland C, Darden T, and Sagui C 2006 *J. Chem. Phys.* **125** 2049096.
[19] Babin V and Sagui C 2010 *J. Chem. Phys.* **132** 104108.
[20] Moradi M, Babin V, Roland C, and Sagui C 2010 *J. Chem. Phys.* **133** 125104.
[21] Moradi M, Babin V, Roland C, Darden T, and Sagui C 2009 *Proc. Natl. Acad. Sci. USA* **106** 20746.
[22] Moradi M, Babin V, Sagui C, and Roland C 2011 *Biophys. J.* **100** 1083.
[23] Moradi M, Babin V, Sagui C, and Roland C 2011 *J. Phys. Chem. B* **115** 8645.
[24] Moradi M, Babi V, Roland C, and Sagui C 2012 *PLoS Comput. Biol.* **8** e1002501.
[25] Jarzynski C 1997 *Phys. Rev. Lett.* **78** 2690.
[26] Crooks G 2000 *Phys. Rev. E* **61** 2361.
[27] Shirts M, Bair E, Hooker G and Pande VS 2003 *Phys. Rev. Lett.* **91** 140601.
[28] Kosztin I, Barz B, and Janosi L 2006 *J. Chem. Phys.* **124** 064106.
[29] Barducci A, Bussi G, and Parrinello M 2008 *Phys. Rev. Lett.* **100** 020603.
[30] Moradi M and Tajkhorshid E. 2013 *J. Phys. Chem. Lett.* **4** 1882.
[31] Oberhofer H and Dellago C, 2009 *J. Comp. Chem.* **30**1726.
[32] Jiang W, Phillips W, Huang L, Fajer M, Meng Y, Gumbart J, Luo Y, Schulten K, and Roux B 2014 *Comp. Phys. Comm.* **185** 908.
[33] Maragliano L, Fischer A, Vanden-Eijnden E and Ciccotti G 2006 *J. Chem. Phys.* 024106.
[34] Wang A, Quigley G, Kolpak F, Crawford J, van Boom J, van der Marel G, and Rich A, 1979 *Nature* **282** 680.
[35] Rich A, Nordheim A, Wang A 1984 *Annu. Rev. Biochem.* **53** 791.
[36] Nordheim A, and Rich A, 1983 *Proc Natl. Acad. Sci. USA* **80** 1821.
[37] Pohl F and Jovin T 1971 *J. Mol. Biol.* **72** 375.
[38] Thamann T, Lord R, Wang A, and Rich A, 1981 *Nucl. Acids Res.* **9** 5443.
[39] Lipps H, 1983 *Cell* **32** 435.
[40] Oh D, Kim Y, and Rich A, 2002 *Proc. Natl. Acad. Sci. USA* **99** 16666.

[41] Fuentes M, Cepeda V, Alonso C, and Perez J, 2006 *Chem. Rev.* **106** 2045.
[42] Harvey S, 1983 *Nucl. Acids Res.* **11** 4867.
[43] Goto S, 1984 *Biopolymers* **23** 2211.
[44] Ansevin A and Wang A, 1990 *Nucl. Acids Res.* **18** 6119.
[45] Lim W, and Feng Y, 2005 *Biophys. J.* **88** 1593.
[46] Saenger W, and Hienemann U 1989 *FEBS Lett.* **257** 223.
[47] Ha S, Lowenhaupt K, Rich A, Kim Y, and Kim K 2005 *Nature* **437** 1183.
[48] de Rosa M, de Sanctis D, Rosario A, Archer M, Rich A, Athanasiasdis A, and Carrondo M, 2010 *Proc. Natl. Acad. Sci. USA* **107** 9088.

Journal of Physics: Conference Series **640** (2015) 012021 doi:10.1088/1742-6596/640/1/012021

Self-Action of Second Harmonic Generation and Longitudinal Temperature Gradient in Nonlinear-Optical Crystals

A I Baranov[1,2], A V Konyashkin[1,2], and [3]O A Ryabushkin[1,2,3]
[1]NTO "IRE-Polus", Vvedensky Sq. 1, Fryazino, 141190, Russian Federation
[2]Moscow Institute of Physics and Technology, Institutskiy per. 9, Dolgoprudnyy, 141700, Russian Federation
[3]Kotelnikov Institute of Radio-engineering and Electronics of RAS, Vvedensky Sq. 1, Fryazino, 141190, Russian Federation

E-mail: abaranov@ntoire-polus.ru

Abstract. Model of second harmonic generation with thermal self-action was developed. Second harmonic generation temperature phase matching curves were measured and calculated for periodically polled lithium niobate crystal. Both experimental and calculated data show asymmetrical shift of temperature tuning curves with pump power.

1. Introduction

Despite rapid development of laser physics and quantum optics in last decades, problem of generation spectrum broadening of high-power coherent radiation sources is still vitally important in nowadays. One of the most efficient ways to deal with this issue is application of nonlinear-optical crystals for conversion of pump laser radiation into multiple harmonics [1]. Requirement of coherence conservation specifies fulfillment of phase matching condition for pump and generated photons along nonlinear-optical crystal length [2]. Wave detuning from phase matching is linear function of refractive indices of interacting harmonics. Crystal temperature change in course of laser frequency conversion is the main factor responsible for refractive indices variations. As follows, accurate and noncontact methods of nonlinear-optical crystal temperature control are needed in case of high-power (tens or hundreds Watts) pump radiation nonlinear conversion [3].

In present paper we introduce novel method for noncontact temperature measurement of nonlinear-optical crystals during its interaction with laser radiation. Here radiofrequency (RF) impedance spectroscopy plays the key role. As every nonlinear-optical crystal possess piezoelectric properties then its response to the applied ac electric field strongly varies at frequencies that correspond to internal vibration modes of the sample. It is well known that in first approximation piezoelectric resonance frequencies linearly depend on temperature. Earlier we introduced application of impedance spectroscopy technique for precise temperature measurement of crystal heated by laser radiation [4]. Experimental determination of equivalent temperature of nonlinear-optical crystal in course of laser radiation frequency conversion was also demonstrated [5]. Moreover piezoelectric laser calorimetry was successfully applied for determination of optical absorption coefficients of crystals in wide spectral range [6, 7].

We developed mathematical model that proves validity of equivalent temperature concept, which suggests direct temperature measurement of nonuniformly heated crystal using its temperature calibrated piezoelectric resonances [8]. Temperature calibration of resonance frequencies is performed in uniform equilibrium heating conditions of the crystal. However, in initial theoretical approach, used for determination of resonance frequencies temperature dependence, just two-dimensional temperature distribution in crystal cross-section was considered. Temperature gradient along radiation propagation direction (third dimension) was assumed to be negligibly small due to very low optical absorption. In some experimental cases such simplification can be rough enough and definitely it cannot be relied on when considering nonlinear-optical processes of laser radiation frequency conversion. This is evident in case optical absorption coefficient at generated frequency is much higher than that at the pump frequency. For correct interpretation of obtained experimental data the 3D heat conduction problem with distributed heat source, conditioned by energy absorption of two light waves, ought to be solved. It should be taken into account that these waves have different values of optical absorption coefficient and besides its powers change along crystal length.

2. Theoretical model

For the interpretation of our second harmonic generation (SHG) experiments three physical processes should be considered. These are second harmonic generation, non-uniform crystal heating due to optical absorption, and piezoelectric (acoustic) vibrations of the crystal. However, as crystal vibrates in different frequency domain compared to frequencies of interacting light waves, only first two processes significantly affect each other. In addition, variation of crystal optical properties caused by its acoustic vibrations is negligibly small. So that it will be sufficient to consider optical second harmonic generation together with accompanying thermal effects and independently characteristic piezoelectric vibrations of the crystal.

2.1. Coupled equations for second harmonic generation and crystal laser heating.

Three-dimensional temperature distribution of the nonlinear-optical crystal in process of second-harmonic generation can be obtained by solving equations that describe nonlinear conversion process. We shall consider a dielectric medium with a non-zero component of the quadratic optical susceptibility. When electromagnetic waves with electric field strength E propagate in such dielectric a nonlinear polarization arises:

$$P_{NL} = \varepsilon_0 \chi^{(2)} EE \,. \qquad (1)$$

Here ε_0 – dielectric permittivity of vacuum, and $\chi^{(2)}$ - quadratic susceptibility tensor. If we consider interaction of both first and second harmonic waves, then total electric field inside crystal can be represented as:

$$E(x_3,t) = \frac{1}{2}\mathrm{e}_\mathrm{p}^1\left[A_1(x_3,t)\exp\left(j(\omega t - kx_3)\right) + c.c.\right] + \frac{1}{2}\mathrm{e}_\mathrm{p}^2\left[A_2(x_3,t)\exp\left(j(2\omega t - Kx_3)\right) + c.c.\right]. \qquad (2)$$

Here, A_1 and A_2 are complex envelopes of the first and the second harmonic respectively, k and K are corresponding wave vectors $k = n_\omega \omega / c$, $K = n_{2\omega} 2\omega / c$, e_p^1 and e_p^2 - unit vectors specifying directions of polarizations of the interacting waves. By substituting equation (1) into (2) we can obtain components of the nonlinear polarization oscillating at the fundamental and doubled frequencies. Equation describing the propagation of electromagnetic waves in a nonlinear dielectric can be obtained using Maxwell equations:

$$\nabla^2 E - \frac{1}{\varepsilon_0 c^2}\frac{\partial^2 D}{\partial t^2} = \frac{1}{\varepsilon_0 c^2}\frac{\partial^2 P_{NL}}{\partial t^2} \,. \qquad (3)$$

Here $D = \varepsilon_0(1+\chi^{(1)})E$ is electric displacement vector, c is speed of light. Performing differentiation with respect to space and time, in assumption that medium is both weakly nonlinear and low absorbing, and also neglecting the second derivatives of complex envelopes we can obtain following equation that describe interaction between the first and second harmonics [2]:

$$\begin{cases} \frac{dA_1}{dx_3} + \alpha_1 A_1 = -j\sigma_1 A_1^* A_2 \exp(-j\Delta k x_3), \\ \frac{dA_2}{dx_3} + \alpha_2 A_2 = -j\sigma_2 A_1^2 \exp(j\Delta k x_3). \end{cases} \tag{4}$$

Here value $\Delta k = K - k = 2\omega n_{2\omega}/c - \omega n_\omega/c$ is so-called wave detuning. Following designations were used:

$$\begin{aligned} \alpha_1 &= k e_p^1 \frac{\operatorname{Im}(\varepsilon(\omega)) : e_p^1}{2n_\omega^2}, & \alpha_2 &= K e_p^2 \frac{\operatorname{Im}(\varepsilon(2\omega)) : e_p^2}{2n_{2\omega}^2}, \\ \sigma_1 &= \frac{1}{2} k e_p^1 \frac{\chi^{(2)}(\omega) : e_p^1 e_p^2}{n_\omega^2}, & \sigma_2 &= \frac{1}{4} K e_p^2 \frac{\chi^{(2)}(2\omega) : e_p^1 e_p^1}{n_{2\omega}^2}. \end{aligned} \tag{5}$$

From equation (4) we can conclude that process of laser radiation nonlinear conversion occurs most efficiently when Δk value equals to zero. And in those regions where Δk is nonzero nonlinear conversion efficiency decreases due to nonzero phase difference between interacting waves. As it is mentioned above, the temperature of the crystal can strongly affect wave detuning, since the refractive index of these waves depend on crystal temperature. Thus for adequate modelling it is important to consider not only the "average" crystal temperature but also its distribution in 3D space.

In order to calculate three-dimensional temperature distribution inside nonlinear-optical crystal in process of SHG it is necessary to solve three-dimensional stationary heat conduction problem with a heat source represented by absorption of the first and second harmonic energy.

$$\begin{cases} \kappa_{cr} \nabla^2 \theta_{cr} + \alpha_1 I_\omega(x_1, x_2, x_3) + \alpha_2 I_{2\omega}(x_1, x_2, x_3) = 0, \\ -\kappa_{cr} \left. \frac{\partial \theta_{cr}}{\partial n} \right|_{\partial\Gamma} = h^{\mathrm{T}} (\theta_{cr} - \theta_{air}) \big|_{\partial\Gamma}. \end{cases} \tag{6}$$

Here κ_{cr} is thermal conductivity of the crystal, $I(x_1, x_2, x_3)$ is radiation intensity distribution; θ_{cr} and θ_{air} are temperatures of the crystal and air respectively; n is the normal vector to the crystal interface; h^{T} is heat transfer coefficient; $|\partial\Gamma$ denotes values at the interface.

To solve separately system of equations (6), one can use the Rayleigh-Ritz variation principle. It allows us to search extremum of the specific functional instead of solving differential equations directly [9, 10]. According to variation principle the solution is sought in the form of a linear combination of basis functions, which in our case can be represented by following polynomials:

$$\psi_{m,n,l} = \left(\left(\frac{x_1}{L}\right)^{2m} + \frac{2m}{\xi L} - 1 \right) * \left(\left(\frac{x_2}{L}\right)^{2n} + \frac{2n}{\xi L} - 1 \right) * \frac{\left(\left(\frac{x_3}{L_3}\right)^l + \frac{x_3}{L_3} \frac{l - \xi L_3}{\xi L_3 - 1} \right), \text{for } l \text{ odd}}{\left(\left(\frac{x_3}{L_3}\right)^l + \frac{l}{\xi L} - 1 \right), \text{for } l \text{ even}}. \tag{7}$$

Here $\xi = h^T / \kappa_{cr}$. Degree of each basis function $2m+2n+l$ does not exceed the natural number N that was set equal to 12. Every polynomial in this set satisfies boundary conditions at every facet of the crystal. In more details derivation of equations (4) and approach for solving the three-dimensional heat conduction problem (6) was introduced in our previous paper [11]. Intensities of first and second harmonic, as well as the length of the wave detuning are unknowns here. It was demonstrated that for second harmonic generation the longitudinal temperature gradient is much more substantial then transvers one. It means that the last one can be neglected. General physical considerations suggest that there is certain equilibrium state of the system when overheating of the crystal is compensated by second harmonic power decrease, conditioned by increase of phase detuning. Thus, systems (4) and (6) can be solved using iterative procedure where the calculated solutions of one system are taken as input parameters for another one. It was ascertained that several dozens of iterations are sufficient to obtain consistent solution.

2.2. Calculation of eigenfrequencies of piezoelectric crystals
Internal acoustic vibration spectrum of any elastic body can be calculated using Hamilton's principle of least action. Vibration pattern and eigenfrequencies of crystal modes are obtained by minimization of Lagrangian L of the system [12, 13]. When we deal with piezoelectric material other terms should be included into Lagrangian functional. These are electric and piezoelectric parts of potential energy [14, 15]

$$L=\iiint\left(\frac{1}{2}\rho\dot{u}_i^2-\frac{1}{8}c_{ijkl}\left(u_{i,j}+u_{j,i}\right)\left(u_{k,l}+u_{l,k}\right)-\frac{1}{2}e_{ijk}\varphi_{,i}\left(u_{j,k}+u_{k,j}\right)+\frac{1}{2}\varepsilon_{ij}\varphi_{,i}\varphi_{,j}\right)d\Omega. \quad (8)$$

Here $F_{,j}=\partial F/\partial x_j$ denotes derivative; ρ is density of the crystal; $u_i(x_1,x_2,x_3)$ are the components of mechanical displacement of the sample points; c_{ijkl}, d_{ijk}, ε_{ij} are elastic, piezoelectric and dielectric tensors respectively; φ is electric potential; Ω is the crystal volume. Lagrangian minimization can be performed using variation principle and the Rayleigh-Ritz approximation method. However, in case crystal is in the form of rectangular parallelepiped ($L_x\times L_y\times L_z$) it is more convenient to use other kind of expansion functions ψ_m - normalized Legendre polynomials $P_m(x,y,z)$ [14,15]:

$$\psi_m(x,y,z)=\left(\frac{(2m_x+1)(2m_y+1)(2m_z+1)}{L_xL_yL_z}\right)^{1/2}P_{m_x}\left(\frac{2x}{L_x}\right)P_{m_y}\left(\frac{2y}{L_y}\right)P_{m_z}\left(\frac{2z}{L_z}\right). \quad (9)$$

Applying separation of variables, every basis function can be represented as a product of Legendre polynomials each depending on the individual coordinate. Thus, we can expand unknown functions, which are three shift functions in different directions and electric potential function. By substituting these expansions into Lagrangian, variation integral problem is reduced to eigenvalue problem of specific matrix. Eigenvectors give values of corresponding expansion coefficients and the eigenvalues represent piezoelectric resonance frequencies. Figure 1 shows calculation results of typical resonance mode vibration amplitude distribution.

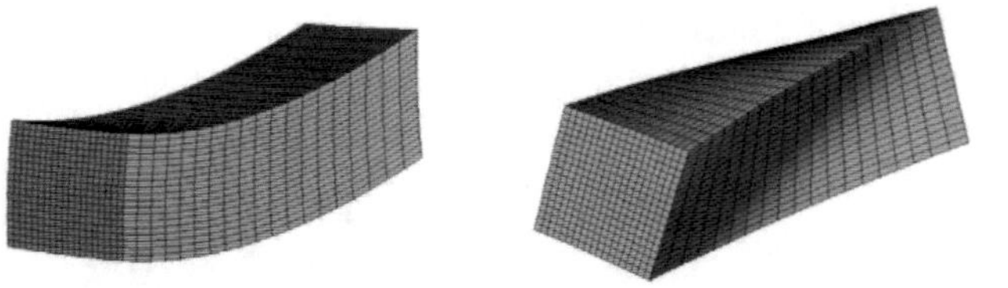

Figure 1. Calculated amplitude distribution of two piezoelectric resonances of the crystal

3. Second harmonic generation in PPLN crystal
Block scheme of experimental setup for crystal piezoelectric resonance measurement in course of second harmonic generation is shown in Fig.2. Periodically polled lithium niobate crystal (PPLN) in the form of rectangular parallelepiped with dimensions $4.9\times2.6\times1$ mm^3 was used in experiments. Polling axis was directed along 1 mm direction.

Crystal is placed inside capacitor that is connected in series with load resistor R. Crystal response to the applied ac voltage from RF generator is analyzed in wide radiofrequency range by measuring voltage on R via lock-in amplifier. Both capacitor plates are made of sets of interconnected metallic pins allocated at the radial arc. Such configuration helps to minimize uncontrollable heating of capacitor plates caused by absorption of scattered radiation. Piezoelectric thermoresonators, made of quartz crystals with small cross-section, are placed at certain distance above and under the PPLN crystal in order to control air temperature.

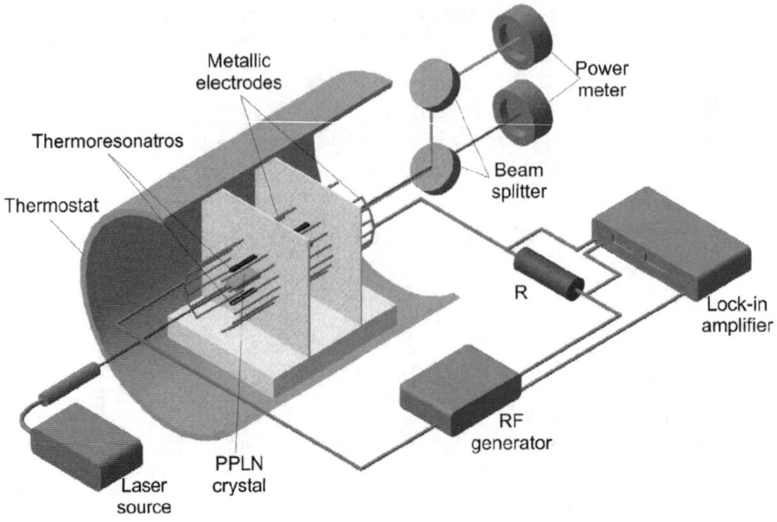

Figure 2. Block scheme of experimental setup

Temperature calibration of piezoelectric resonance frequencies $Rf_i(T)$ was performed when crystal was uniformly heated. Thus piezoelectric resonance thermal coefficients K_i^{prt} were determined. As follows crystal equivalent temperature change due to heating by laser radiation of P power can be obtained by measuring resonance frequency shift (see Ref. [4] for details):

$$\Delta\Theta_{\mathrm{eq}} = \Delta Rf_i(P)/K_i^{\mathrm{prt}} . \tag{10}$$

Equivalent temperature can be regarded as the true crystal temperature because in this case crystal itself acts as thermal probe.

SHG experiments were carried out using CW single-mode polarized ytterbium fiber laser (λ=1064 nm, $\Delta\lambda$=0.1 nm) with up to 25 W output power as a pump source. Phase matching equivalent temperature curves of PPLN were measured at fixed pump levels when wave detuning value between first and second harmonic was changed via thermostat temperature adjusting (see Fig. 3).

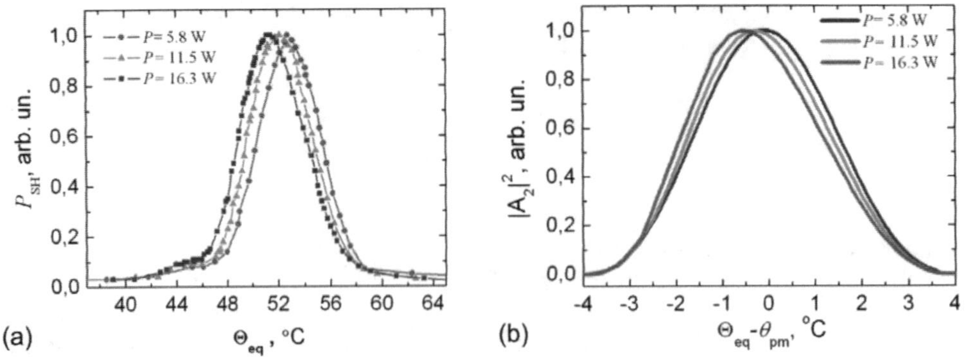

Figure 3 Measured (a) and calculated (b) equivalent temperature tuning curves of PPLN crystal at different pump levels.

Phase matching temperature that corresponds to maximum second harmonic output power P_{SH} shifts to lower temperatures with power $d\theta_{\mathrm{pm}}/dP$= -0.11 °C/W. Experimental results reveal that symmetry distortion of temperature tuning curves occurs at higher pump levels. Calculated curves exhibit the same behavior. The main reason for this is presence of the longitudinal temperature gradient inside crystal. However, characteristic width $\Delta\theta_{\mathrm{pm}}$=6 °C of the curve remains almost unchanged.

4. Conclusions

In present paper we have introduced theoretical approach that supposes consideration of SHG problem in terms of thermal self-action. Predicted by this model shift and distortion of temperature tuning curves with pump power were verified experimentally. Further development of mathematical model of nonlinear-optical crystal equivalent temperature in course of SHG should include consideration of piezoelectric (acoustical) vibrations of 3D solid in condition of nonuniform transverse and longitudinal temperature distribution.

Reference

[1] Gapontsev V P, Tyrtyshnyy V A, Vershinin O I, Davydov B L, and Oulianov D A 2013 Third harmonic frequency generation by Type-I critically phase-matched LiB3O5 crystal by means of optically active quartz crystal *Optics Express* **21** 3715

[2] Dmitriev V G and Tarasov L V 2004 *Applied nonlinear optics* (Moscow Russia: Fizmatlit)

[3] Kumar S Ch, Samanta G K, and Ebrahim-Zadeh M 2009 High-power, single-frequency, continuous-wave second-harmonic-generation of ytterbium fiber laser in PPKTP and MgO:sPPLT *Optics Express* **17** 13711

[4] Ryabushkin O A, Myasnikov D V, Konyashkin A V amd Tyrtyshnyy V A 2011 Equivalent temperature of nonlinear-optical crystals interacting with laser radiation *J. of Europ. Opt. Soc. – Rapid Publications* **6** 11032

[5] Ryabushkin O A, Konyashkin A V, Myasnikov D V, Tyrtyshnyy V A and Baranov A I 2013 Equivalent temperature of nonuniformly heated nonlinear-optical crystals in course of laser radiation frequency conversion *Proc. SPIE 8847 Photonic Fiber and Crystal Devices: Advances in Materials and Innovations in Device Applications VII* (San Diego California USA, 25 August 2013) 884711

[6] Ryabushkin O A, Myasnikov D V, Konyashkin A V and Vershinin O I 2013 Kinetics of equivalent temperature of nonlinear-optical crystals *CLEO®/Europe - IQEC 2013: Conference Digest* (Munich Germany, 12-16 May 2013) CE-P.18 TUE

[7] Ryabushkin O A, Konyashkin A V, Myasnikov D V, Tyrtyshnyy V A and Vershinin O I 2013 Piezoelectric resonance calorimetry of nonlinear-optical crystals under laser irradiation *Proc. SPIE 8847 Photonic Fiber and Crystal Devices: Advances in Materials and Innovations in Device Applications VII* (San Diego California USA, 25 August 2013) 88470Q

[8] Ryabushkin O A and Myasnikov D V 2012 Experimental determination and the theoretical model of an equivalent temperature of nonlinear-optical crystals interacting with high-power laser radiation *Quantum Electronics* **42** 539

[9] Sadiku M N O 2001 *Numerical techniques in electromagnetics* (2nd edition, Boca Raton Florida USA: CRS Press LLC)

[10] Tyn M-U Lokenath D 2007 *Linear Partial Differential Equations for Scientists and Engineers* (4th edition, Boston USA: Birkhauser)

[11] Ryabushkin O A, Myasnikov D V, Baranov A I 2014 Crystal equivalent temperature model in process of nonlinear conversion of laser radiation *25th IUPAP Conference on Computational Physics (CCP2013) J. Phys.: Conference Series* **510** 012031

[12] Tiersten H F 1968 Natural boundary and initial conditions from a modification of Hamilton's principle *J. Math. Phys.* **9** 1445

[13] Holland R 1968 Resonant properties of piezoelectric ceramic rectangular parallelepipeds *J. Acoust. Soc. Am.* **43** 988

[14] Ohno I 1990 Rectangular parallelepiped resonance method for piezoelectric crystals and elastic constants of alpha-quartz *Phys. Chem. Minerals* **17** 371

[15] Myasnikov D V, Konyashkin A V and Ryabushkin O A 2010 Identification of eigenmodes of volume piezoelectric resonators in resonant ultrasound spectroscopy *Tech. Phys. Lett.* **36** 632

Journal of Physics: Conference Series **640** (2015) 012022 doi:10.1088/1742-6596/640/1/012022

First-principles calculation of the structural and elastic properties of ternary metal nitrides $Ta_xMo_{1-x}N$ and $Ta_xW_{1-x}N$

Kh. Bouamama[1]*, P. Djemia[2] and M. Benhamida[1]
[1]LOC, Département de Physique, Université Ferhat Abbas, 19000 Sétif, Algeria.
[2]LSPM-CNRS, Sorbonne Paris Cité, Université Paris 13, 99 Avenue J.B. Clément, 93430 Villetaneuse, France.

E-mail: khaled_bouamama@univ-setif.dz

Abstract: First-principles pseudo-potentials calculations of the mixing enthalpy, of the lattice constants a_0 and of the single-crystal elastic constants c_{ij} for ternary metal nitrides $Ta_xMe_{1-x}N$ (Me=Mo or W) alloys considering the cubic B1-rocksalt structure is carried out. For disordered ternary alloys, we employ the virtual crystal approximation VCA in which the alloy pseudo-potentials are constructed within a first-principles VCA scheme. The supercell method SC is also used for ordered structures in order to evaluate clustering effects. We find that the mixing enthalpy still remains negative for $Ta_xMe_{1-x}N$ alloys in the whole composition range which implies these cubic $Ta_xMo_{1-x}N$ and $Ta_xW_{1-x}N$ ordered solid solutions are stable. We investigate the effect of Mo and W alloying on the trend of the mechanical properties of TaN. The effective shear elastic constant c_{44}, the Cauchy pressure (c_{12}-c_{44}), and the shear to bulk modulus G/B ratio are used to discuss, respectively, the mechanical stability of the ternary structure and the brittle/ductile behavior in reference to TaN, MeN alloys. We determine the onset transition from the unstable structure to the stable one B1-rocksalt from the elastic stability criteria when alloying MeN with Ta. In a second stage, in the frame of anisotropic elasticity, we estimate by one homogenization method the averaged constants $<C_{ij}>$ of the polycrystalline $Ta_xMe_{1-x}N$ alloys considering the special case of an isotropic medium with no crystallographic texture.

1. Introduction

Materials that combine both hardness and ductility are said tough materials, with a high resistance to failure [1]. In bulk materials, toughness is easily extracted from uniaxial compression or tensile tests by measuring the area under the curve of the stress-strain curve, whereas for coatings it is not well established. The design and development of such materials have always been a challenge, because an increase in hardness is usually accompanied by an increase in brittleness rather than ductility. One of the major goals of materials scientists and hard coating industries has been the development of hard materials (bulk, coatings or multilayers) with high temperature stability [2-6]. While high hardness is indispensable, most applications also demand ductility in order to avoid brittle failure due to cracking in coatings exposed to high stresses. However, hard refractory ceramics such as transition metal TM based nitrides are in general brittle. Enhancement of toughness can be obtained by the inhibition of both crack initiation and propagation. In case of polycrystalline materials, grain boundaries are the structurally weaker regions where the cracks will spread along. Since the crack size is generally proportional to the grain size, the propensity for crack propagation can also be attenuated by decreasing the size of grains or flaws to the nanoscale [7]. Such nanostructuring can be obtained by developing hard and tough ceramic coatings via the synthesis of nanoscale composites [7-9] achieved via many routes [7, 10-15]. Fundamental understanding of the origins of toughness is not widely studied, while progress in this area can only be achieved through atomic- and electronic-level understanding of the origins of brittleness *vs.* ductility. Such studies are, however, relatively few [16-19]. *Ab initio* theoretical studies, using density functional theory DFT calculations, are well adapted for such design by providing information concerning the relationship between electronic structure and mechanical properties. This work aims to benefit from existing DFT studies of toughness enhancement to predict tough TM nitrides that could be later deposited as coatings. In this paper, we investigate the stability, the structural and the elastic properties of $Ta_xMe_{1-x}N$ (Me = Mo and W) alloys by DFT calculations on both disordered ordered alloys. The evolutions of the lattice parameters a_0, of the density of electronic states DOS and of the single-crystal elastic constants c_{ij} as a function of x, are determined. The effective elastic constants and moduli of isotropic polycrystalline materials are computed using a Voigt-Reuss-Hill [20] model. Predictive ductility behavior of the polycrystalline alloys is discussed in terms of Pettifor [21] and Pugh criteria [22] in relation to their elastic properties.

2. The computational method

The calculations are performed with the density-functional perturbation theory DFPT in the plane-wave pseudo-potential method as implemented in the ABINIT code [23]. This is done with the self-consistent response function computations for electric field perturbations and the ionic displacements [24-27]. For the exchange-correlation potential, we used the generalized gradient approximation (GGA) of Perdew, Burke and Ernzerhof (PBE) [28, 29]. The atoms are represented by norm-conserving optimized [30] designed nonlocal [31] pseudo-potentials, generated with the OPIUM code [32], where the semi-core p states were included because of their importance for good transferability. The Brillouin zone integrations were replaced by discrete summations over a special set of k-points using the standard technique of Monkhorst and Pack [33], where the k-points mesh used is (16 x 16 x 16) and the plane-wave energy cutoff to expand the wave functions is set to be 80 Hartree (1 Ha = 27.2116 eV) to get a total energy accuracy better than 10^{-5} Hartree. The Fermi-Dirac smearing approach [34, 35] is used with a cold smearing of 0.005 Hartree. For the treatment of the disordered ternary alloy, we have used the virtual crystal approximation VCA implemented in the ABINIT program, in which the alloy pseudo-potentials are constructed within a first-principles VCA scheme. Elemental ionic pseudo-potentials of MeN (Me = Mo or W) and TaN are combined to construct the virtual pseudo-potential of the $Ta_xMe_{1-x}N$, *i.e.*:

$$V_{VCA} = xV_{TaN} + (1-x)V_{MeN} \qquad (1)$$

Simple cubic 8-atoms supercells, containing 1 (x=0.25), 2 (x=0.5) and 3 (x=0.75) substitutional Ta atoms on the Me sites were used to describe the $Ta_xMe_{1-x}N$ ordered compounds.

3. Results and discussion

3.1. Phase stability, structural and electronic properties analysis

The mixing energy enthalpy equivalent to the Gibbs free energy at T=0 K, is defined as:

$$\Delta E_{Ta_xMe_{1-x}N}(x) = E_{Ta_xMe_{1-x}N} - (1-x)E_{MeN} - xE_{TaN} \qquad (2)$$

with $E_{Ta_xMe_{1-x}N}$ the total energy per atom of the $Ta_xMe_{1-x}N$ alloy, E_{MeN} and E_{TaN} the total energy per atom of MeN and TaN alloys, respectively. The calculated mixing energies ΔE of $Ta_xMe_{1-x}N$ alloys with rocksalt structure using SC are shown is Fig. 1. They are negative, indicating that MeN and TaN may continuously form stable solid solutions in comparison to the phase-separated MeN+TaN. The mixing energy of $Ta_xMo_{1-x}N$ is much lower than that of $Ta_xW_{1-x}N$, indicative of a better thermal stability of $Ta_xMo_{1-x}N$ over $Ta_xW_{1-x}N$.

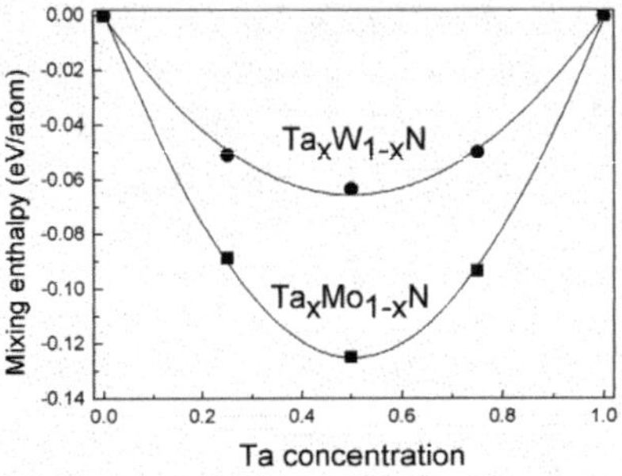

Figure 1. SC computed mixing enthalpy of $Ta_xMe_{1-x}N$ alloys as a function of Ta concentration, x.

DOS of $Ta_xMe_{1-x}N$ is presented in Fig. 2. The main features of DOSs of $Ta_xMo_{1-x}N$ and $Ta_xW_{1-x}N$ are very similar to each other. The energy region from −10 to −4 eV is characterized by the strong hybridization of the Me-d electrons of the metals with the N-p electrons, which is at the origin of the covalent-like bonding in these materials resulting to high strength and high hardness. The energy region from−4 to 7 eV shows energy states filled mainly by the Ta-d and Me-d electrons and a small contribution of the N-p electrons. These strong metal-metal bonds should contribute to a better ductility as the metallic states become more populated with the introduction of Mo or W. Such materials offer the combination of strong covalent bonds Me-N with metallic bonds Me-Ta conferring both strength and ductility, and then toughness.

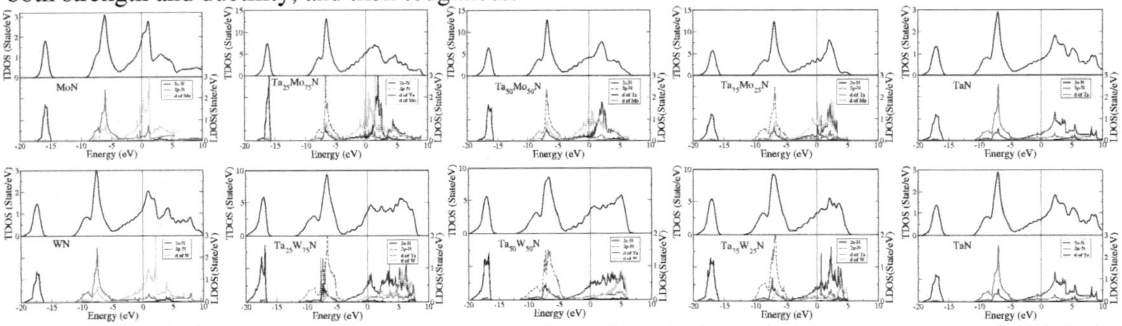

Figure 2. Total electronic density of states (TDOS) and local electronic density of states (LDOS) for the $Ta_xMe_{1-x}N$ alloys calculated with ABINIT-SC. Zero energy stands for the Fermi level and is denoted by vertical black lines.

The lattice parameters a_0 of $Ta_xMe_{1-x}N$ are shown in Fig. 3, for both VCA and SC methods, as a function of TaN composition, x. For $Ta_xMo_{1-x}N$, a_0 decreases from 0.4384 nm for MoN to 0.4316 nm for TaN with increasing x. However, for $Ta_xW_{1-x}N$, a_0 remains constant. The relative difference between VCA and SC calculations does not exceed 0.3% with a negative bowing from the Vegard's rule for VCA. We find a good agreement between our calculated lattice parameters for binary alloys WN, MoN and TaN and the calculated lattice parameters, WN (0.435 nm) [36], MoN (0.433 nm) [37] and the experimental δ-TaN (0.4336 nm) [38, 39].

3.2. Elastic properties

The single crystal elastic constants c_{ij} of $Ta_xMe_{1-x}N$ are shown in Fig. 4 for both VCA and SC methods with the bulk modulus $B = (c_{11}+2c_{12})/3$. For both alloys, the c_{11} and c_{44} elastic constants increase as a function of x, while the c_{12} decreases. B remains constant for $Ta_xW_{1-x}N$ alloys (375±3 GPa) and for $Ta_xMo_{1-x}N$ alloys B slightly increases from MoN (306 GPa) to TaN (373 GPa). The obtained values for the elastic constants and B for TaN are in good agreements with the calculated ones [40].

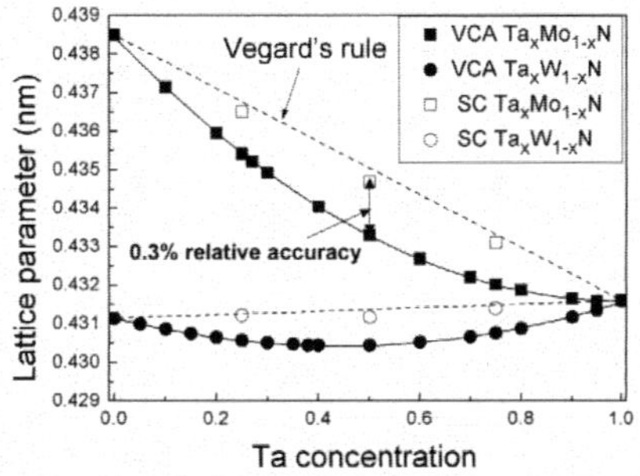

Figure 3. VCA calculated (plain symbols) and SC calculated (empty symbols) lattice parameters of $Ta_xMe_{1-x}N$ as a function of x. Dotted lines are the Vegard's rule from the binary alloys TaN and MeN.

The Born elastic stability criteria (B>0, c_{11}-c_{12}>0, c_{44}>0) are satisfied for $Ta_xMo_{1-x}N$ alloys with x>0.27 and for $Ta_xW_{1-x}N$ alloys with x>0.38 (see fig. 4). Adding Ta to MoN or WN enables to stabilize the B1-rocksalt structure above these critical amount of Ta $(x_{Ta})_{cr,MeN}$.
Using the calculated single-crystal stiffnesses c_{ij}, we performed a Voigt-Reuss-Hill average leading to the effective constants $\langle C_{11}\rangle$, $\langle C_{12}\rangle$ and $\langle C_{44}\rangle$ of polycrystalline $Ta_xMo_{1-x}N$ and $Ta_xW_{1-x}N$ considering a random orientation of crystallites [20]. Afterwards, the Young modulus E and the Poisson ratio σ of the polycrystals are calculated from the Eq. (3-4) for isotropic elastic symmetry:

$$E = C_{11} - 2\,C_{12}^2/(C_{11} + C_{12}) \quad (3)$$
$$\sigma = C_{12}/(C_{11} + C_{12}) \quad (4)$$

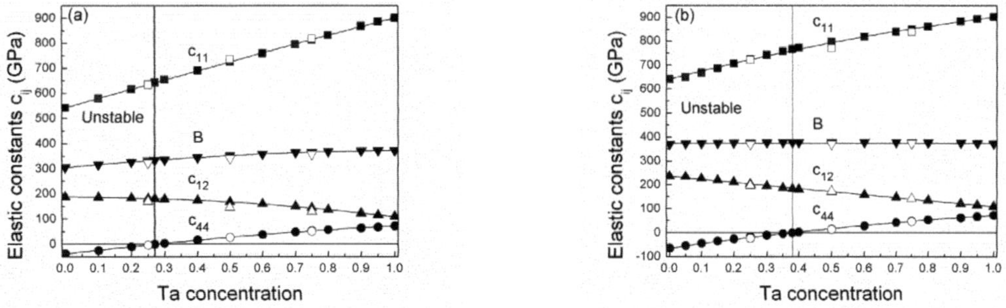

Figure 4. VCA calculated (plain symbols) and SC calculated (empty symbols) elastic constants c_{ij} and the bulk modulus B for (a) $Ta_xMo_{1-x}N$ and (b) $Ta_xW_{1-x}N$ as a function of Ta concentration, x. The vertical line indicates the transition from mechanically unstable → stable.

The shear modulus G is simply $\langle C_{44}\rangle$. The $\langle C_{ij}\rangle$ are plotted in Fig. 5 for both VCA and SC methods and the elastic moduli in Fig. 6. For both alloys, effective elastic constants remain close from each other. The $\langle C_{11}\rangle$ and $\langle C_{44}\rangle$ elastic constants increase as a function of x, while the $\langle C_{12}\rangle$ decreases.
In Fig. 6, E increases from $Ta_{0.3}Mo_{0.7}N$ (138 GPa) and from $Ta_{0.4}W_{0.6}N$ (50 GPa) to TaN (353 GPa). However, σ decreases from $Ta_{0.3}Mo_{0.7}N$ (0.432) and from $Ta_{0.4}W_{0.6}N$ (0.477) to TaN (0.333).
Pettifor [21] proposed that Cauchy pressure, defined as $p = c_{12}-c_{44}$, can be used as an indicator to classify a material as brittle or ductile. The changes in p are connected with the nature of the atomic bonding. If the bonding is more metallic, $p > 0$, while for more brittle materials with directional bonding, $p < 0$.

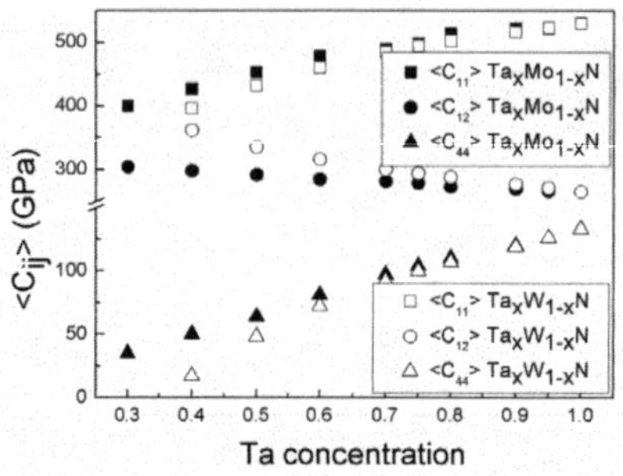

Figure 5. VCA calculated effective elastic constants $<C_{11}>$, $<C_{12}>$, $<C_{44}>$ of $Ta_xMo_{1-x}N$ (plain symbols) and $Ta_xW_{1-x}N$ (empty symbols) alloys as a function of Ta concentration, x.

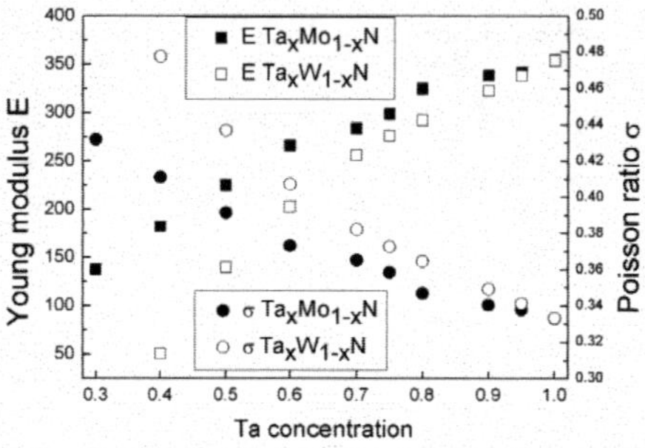

Figure 6. Effective elastic moduli (E, σ) considering non crystallographic texture for cubic polycrystalline $Ta_xMo_{1-x}N$ (plain symbols) and $Ta_xW_{1-x}N$ (empty symbols) alloys as a function of x.

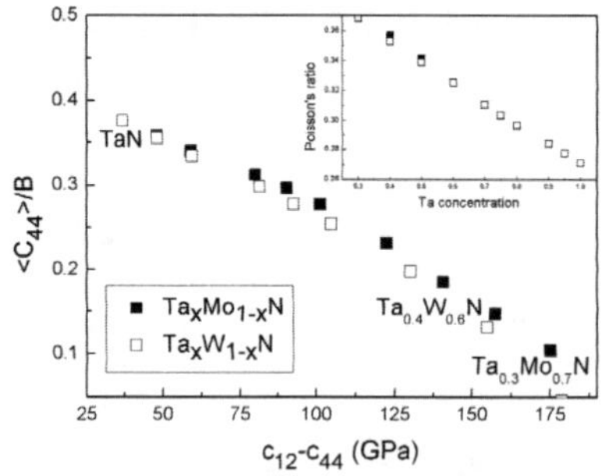

Figure 7. Map of brittleness and ductility of $Ta_xMe_{1-x}N$ disordered polycrystalline alloys.

The Fig. 7 shows that polycrystalline $Ta_xMe_{1-x}N$ have a ductile behavior, considering both the Pettifor criterion ($p > 0$) for ductile materials and the Pugh criterion [22] (G/B <0.5). One can also notice that $\sigma > 0.25$. p increases while G/B reduces with addition of Mo or W increasing ductility.

4. Summary and conclusion

The lattice constants, the mixing enthalpy, the bulk modulus, the single-crystalline elastic constants of the ternary nitrides alloys $Ta_xMe_{1-x}N$ (Me= Mo or W) are calculated as a function of x, using a first-principles pseudo-potential approach within the VCA for disordered alloys and the SC method for ordered alloys using GGA exchange-correlation potential. A negative bowing is observed for the lattice constants in comparison to the Vegard's rule for VCA calculations. SC mixing enthalpies are negative and indicate that both alloys are energetically stable in reference to their binary counterparts. We obtain close results for the elastic constants between VCA and SC calculations indicating a little influence of the ordering/clustering of metal atoms. c_{44} is positive above a critical Ta concentration $(x)_{cr,MoN}$~0.27 < $(x)_{cr,WN}$~0.38 above which the B1-rocksalt structure is mechanically stable. From Pettifor (c_{12}-c_{44}>0) and Pugh (G/B<0.5) criteria the ductility of the alloys are found to increase as more Mo or W is added. Close effective elastic properties of both alloys are observed above x~0.7.

5. References

[1] Dieter GE and Bacon DJ: *Mechanical Metallurgy* 1988 MCGRAW-HILL Higher Education.
[2] Hugosson HW, Jansson U, Johansson B and Eriksson O 2001 *Science* **293** 2434.
[3] Joelsson T, Hultman L, Hugosson HW and Molina-Aldareguia JM 2005 *Appl. Phys. Lett.* **86** 131922.
[4] Helmersson U, Todorova S, Barnett S, Sundgren JE, Markert LC and Greene J 1987 *J. Appl. Phys.* **62** 481.
[5] Veprek S, Veprek-Heijman MGJ, Karvankova P and Prochazka J 2005 *Thin Solid Films* **476** 1.
[6] Shin CS, Gall D, Hellgren N, Patscheider J, Petrov I and Greene JE 2003 *J. Appl. Phys.* **93** 6025.
[7] Zhang S, Sun D, Fu Y and Du H 2005 *Surf. Coat. Tech.* **198** 2.
[8] Zhang S, Wang HL, Ong SE, Sun D and Bui XL 2007 *Plasma Processes and Polymers* **4** 219.
[9] Abadias G, Ivashchenko VI, Belliard L and Djemia P 2012 *Acta Materiala* **60** 5601.
[10] Djemia P, Benhamida M, Bouamama Kh, Belliard L, Faurie D and Abadias G 2013 *Surf. Coat. Tech.* **215** 199.
[11] Anton D and Shah D 1990 In *MRS Proceedings*: Cambridge Univ Press: **194** 45.
[12] Abe O and Ohwa Y 2004 *Solid State Ionics* **172** 553.
[13] Zhan GD, Kuntz JD, Wan JL and Mukherjee AK 2003 *Nature Materials* **2** 38.
[14] Basu B, Venkateswaran T and Kim DY 2006 *Journal of the American Ceramic Society* **89** 2405.
[15] Zhang S, Sun D, Fu Y and Du H 2003 *Surf. Coat. Tech.* **167** 113.
[16] Hogmark S, Jacobson S and Larsson M 2000 *Wear* **246** 20.
[17] Kamran S, Chen KY and Chen L 2009 *Phys. Rev. B.* **79** 024106.
[18] Dasgupta T, Waghmare UV and Umarji AM 2007 *Phys. Rev. B.* **76** 174110.
[19] Abadias G, Kanoun MB, Goumri-Said S, Koutsokeras L, Dub SN and Djemia Ph 2014 *Phys. Rev. B.* **90** 144107.
[20] Hill R 1952 *Proc. Phys. Soc.* A **65** 349.
[21] Pettifor DG 1992 *Mater. Sci. Technol.* **8** 345.
[22] Pugh SF 1954 *Philos. Mag.* **45** 823.
[23] Gonze X *et al.* 2009 *Computer Physics Communications* **180** 2582.
[24] Giannozzi P, de Gironcoli S, Pavone P and Baroni S 1991 *Phys. Rev. B.* **43** 7231.
[25] Gonze X 1997 *Phys. Rev. B.* **55** 10337.
[26] Gonze X and Lee C 1997 *Phys. Rev. B.* **55** 10355.
[27] Ghosez Ph, Michenaud JP and Gonze X 1998 *Phys. Rev. B.* **58** 6224.
[28] Perdew JP, Burke K and Ernzerhof M 1996 *Phys. Rev. Lett.* **77** 3865.
[29] Perdew JP and A. Zunger 1981 *Phys. Rev. B.* **23** 5048.
[30] Rappe AM, Rabe KM, Kaxiras E and Joannopoulos JD 1990 *Phys. Rev. B.* **41** 1227.
[31] Ramer NJ and Rappe AM 1999 *Phys. Rev. B.* **59** (1999) 12471.
[32] http://opium.sourceforge.net
[33] Monkhorst HJ and Pack JD 1976 *Phys. Rev. B.* **13** 5189.
[34] Fu CL and Ho KM 1983 *Phys. Rev. B.* **28** 5480.
[35] Methfessel M and Paxton AT 1989 *Phys. Rev. B.* **40** 3616.
[36] Isaev EI *et al.* 2007 *J. Appl. Phys.* **101** 123519.
[37] Chen W and Jiang JZ 2010 *Journal of alloys and compounds* **499** 243.
[38] Mashimo T, Tashiro S, Nishida M, Miyahara K and Eto E 1997 *Physica* B **239** 13.
[39] Abadias G, Belliard L and Djemia P 2014 *Surf. Coat. Tech.* **257** 129.
[40] Zhao E and Wu Z 2008 *Journal of Solid State Chemistry* **181** 2814.

Journal of Physics: Conference Series **640** (2015) 012023 doi:10.1088/1742-6596/640/1/012023

Computer simulation of the interaction of ringlike carbon clusters with nanographene

A A Dzhurakhalov[1], V G Stelmakh[2] and I D Yadgarov[2]

[1]Department of Mathematics and Computer Science, University of Antwerp, Belgium
[2]Institute of Ion-Plasma and Laser Technologies, Tashkent, Uzbekistan

E-mail: abdiravuf.dzhurakhalov@uantwerpen.be

Abstract. Various cases of interaction of ringlike carbon clusters C_7, C_{12} and C_{13} with a rectangular nanographene consisting of 272 atoms were studied and presented. It was found that this interaction results in the structural change in these clusters and in the local part of nanographene. The cohesive energies of these clusters in the isolated (free) state and their binding energies with nanographene have been calculated. The results show that despite this interaction the atoms of cluster are hold together as a single cluster thanks to the significantly higher cohesive energy of cluster the its binding energy with nanographene.

1. Introduction

Graphene term was originally used to refer to an abstract model, namely, a single layer of graphite atoms which do not interact with other objects. Even at those theoretical studies, it was clear that graphene, whether in reality, would have the number of unique properties, such as high electrical and thermal conductivity. In 2004 A. Geim with his coworkers [1] received a single sheet of graphene experimentally in the laboratory, thus the term graphene has become point not only to an abstract model, but also on the nature of the real object. As the real object of nature graphene interacts with the environment and in particular with atoms, molecules and their ensembles as clusters.

There are a large number of works dealing with the interaction of graphene with atoms and molecules. Using different methods, this interaction was studied in different aspects. Nevertheless, the very first question is that you need to decide how mathematically infinite graphene according to its ideal model can be constructed to result in an object suitable for computer simulation. For this aim two approaches are usually used: periodic boundary conditions are imposed on the final size of the graphene and thus it is defined as an infinite, or piece of graphene is taken. We use the later approach in our studies as in practice the graphene with finite size called as nanoribbons are commonly used.

With regard to research methods, either the classical molecular dynamics method with the corresponding interatomic potentials, e.g. Brenner potential [2], Tersoff potential [3], Universal force field [4] or ab-inito methods (Metod of Hartree-Fock-Roothaan, Moller–Plesset perturbation theory, theory of the combined clusters and so on) are mostly used.

The choice of methods ab-inito is inevitable if you need to determine the parameters that cannot be defined using classical molecular dynamics, for example, spin, magnetic moment, etc. However, if you focus on the configuration and the cohesive energy of the atoms, the Monte Carlo and the classical molecular dynamics methods are more suitable than the ab-initio methods as later ones require significantly long time for calculation and some specific parameters. We note that the ultimate goal of studies of the interaction of atoms and molecules with graphene can be different. In particular, the graphene can be bent or rolled in a tube as a result of this interaction (for example, by deposition of particles along a single line) [5]. In other case, the local change of graphene, including the question of transition of graphene in graphane was studied when a hydrogen atom interact with it [6]. As for the atoms, which interact with graphene, the H, F, O and C atoms, metal atoms and the molecules of organic substances are more often chosen.

Our work focuses on the change in the shape of C_7, C_{12} and C_{13} clusters by interacting them with nanographene and on binding energy of clusters with nanographene.

2. Modeling and simulation method

The method of energy minimization is common technique to calculate the equilibrium configuration of molecules and clusters. The basic idea of this method is that a stable state of a molecular system corresponds to a local minimum of the potential energy [7]. The state with the lowest energy corresponds to the ground state, while other local minima are metastable states. In this paper we use the so-called Brenner second-generation reactive empirical bond order (REBO) potential [2], which is specifically parameterized for the carbon and hydrogen-carbon systems. In [6, 8] it was shown how well REBO potential describes some characteristics of carbon clusters, graphene and graphane.

As ringlike carbon clusters are the most stable among the other small carbon clusters of different configurations with the same number of atoms [8], we use ringlike conformation of C_7, C_{12} and C_{13} clusters. Their stable structures were found using the energy minimization method. The radii of these clusters and the cohesive energy E_{coh} of their atoms are shown in Table 1.

Table 1. The characteristics of the free C_7, C_{12} and C_{13} clusters.

Cluster	Radius (Å)	E_{coh} (eV/atom)
C_7	1.58	5.51
C_{12}	2.59	5.90
C_{13}	2.80	5.93

The stable state of graphene which hereinafter called nanographene with a rectangular shape of 32.7x19.9Å consisting of 272 atoms was found by the method of energy minimization. We note that the internal atoms of nanographene have cohesive energy E_{coh}=7.39 eV/atom, whereas its boundary atoms, the total number of which is 48, have cohesive energy E_{coh}=4.9÷5.2 eV/atom. The free 272-atomic rectangular nanographene and ringlike C_7, C_{12} and C_{13} clusters are presented in Fig. 1.

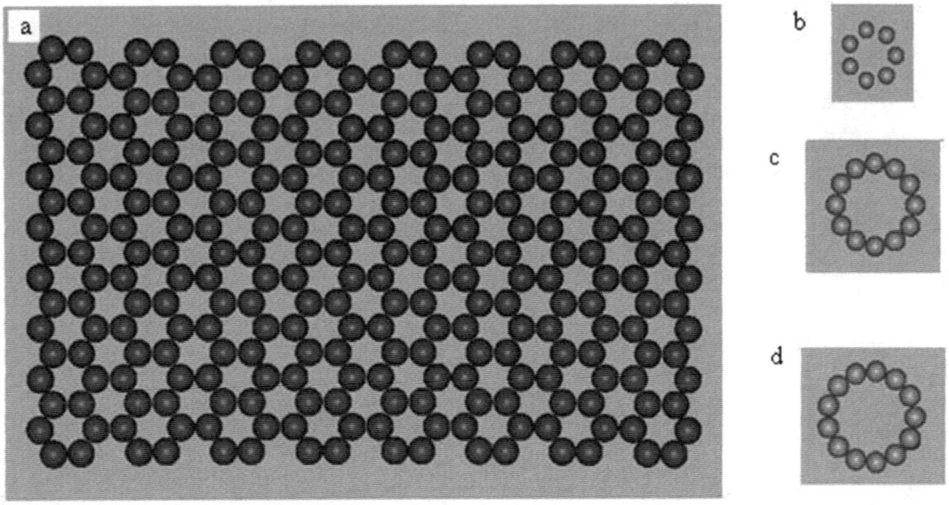

Figure 1. (Color online) Stable configurations of 272 atomic nanographene (a) and ringlike carbon clusters: C_7 (b), C_{12} (c) and C_{13} (d).

After creating of nanographene and C_7, C_{12} and C_{13} clusters their possible interactions with each other were studied. Initial position of C_7, C_{12} and C_{13} cluster was at an angle to the nanographene α = 0°, 15°, 30°, 45°, 60°, 75° and 90°, where α is the angle between the normal of nanographene and the cluster plane. The initial height of ringlike clusters, which is defined as the smallest distance from the atom of cluster to the plane of nanographene, was h_7 = 0.127Å, h_{12} = 0.149Å and h_{13} = 0.101Å. Here h_7, h_{12} and h_{13} are heights of ringlike C_7, C_{12} and C_{13} clusters, respectively. To avoid or minimize the edge effects the center of nanographene was chosen as the place of interaction with the ringlike clusters.

3. Results and discussion

The cluster interaction with nanographene takes place by a way of forming bonds between some atoms of the cluster and nanographene. In Fig. 2 we present the results of our computer simulation for the final positions of cluster-nanographene system when the ringlike clusters interact with nanographene forming bonds with their one or two atoms. These bonds define the cohesion energy of the cluster with nanographene. If one atom of the cluster forms a bond with nanographene as shown in Fig. 2 (a), then the binding energy of this atom with graphene very well estimates the cohesion energy of the whole cluster with nanographene. When several atoms of the cluster have bonds with graphene atoms, we define the maximal value of the binding energy of the cluster atom with nanographene and the sum of all the binding energies of the cluster atoms with the nanographene atoms, which makes it possible to estimate the total cohesion energy of the whole cluster with nanographene.

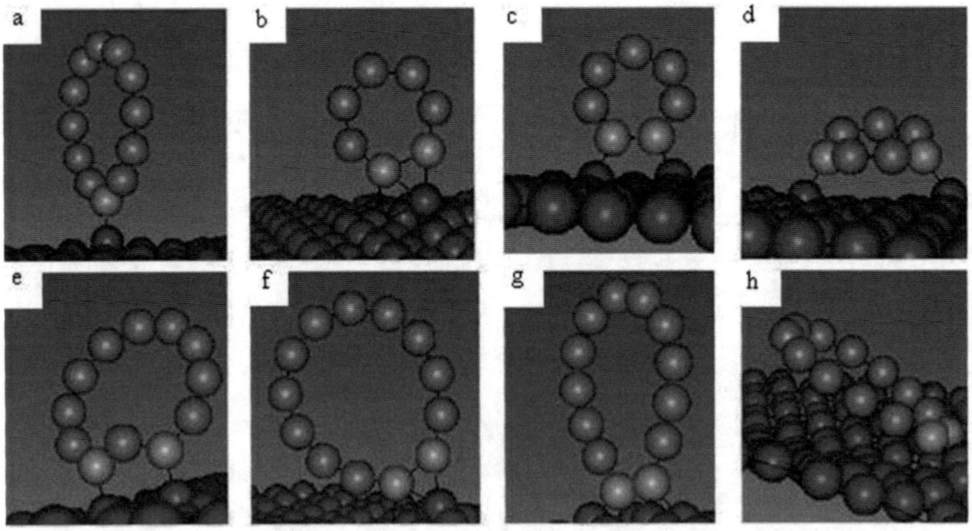

Figure 2. (Color online) Interaction of C_7, C_{12} and C_{13} clusters through their one or two atoms. Red spheres are nanographene atoms, blue spheres are cluster atoms and green ones are cluster atoms that interact directly with nanographene. a - C_{12} and α=30°, b - C_7 and α=0°, c - C_7 and α=60°, d - C_7 and α=75°, e - C_{12} and α=0° or α=75°, f - C_{13} and α=0°, g - C_{13} and α=30°, h - C_{13} and α=45°, where α indicates an initial position of cluster.

The interaction of carbon clusters with nanographene causes the changes both in the electronic structure and in geometric shape of graphene and cluster. The carbon cluster changes the hybridization from sp^2 to sp^3 for the graphene atoms bonded with cluster. This change of hybridization causes a local structural change in the graphene sheet: the graphene atoms bonded with cluster are shifted above the plane of graphene (see Fig. 2). Some atoms of cluster also can be displaced out of the cluster plane due to its interaction with nanographene (see, e.g. Fig. 2 d, e).

In Fig. 3 the final stable states of cluster-nanographene system, when the ringlike clusters interact with nanographene forming bonds through their more than two atoms, are presented. As seen, the interaction results in significant buckling the shape of clusters in these cases. The several possible cases of such interaction are observed. In particular, the ringlike cluster can interact with nanographene through its some part as shown in Fig. 3 (a, b, d-f), keeping its other part quite far from the nanographene plane. Another typical case of interaction is that when the cluster interacts through its atoms which are not neighbor each other. In this case, though cluster interacts through its few atoms, all its atoms are not far from graphene (see, Fig. 3, g-i).

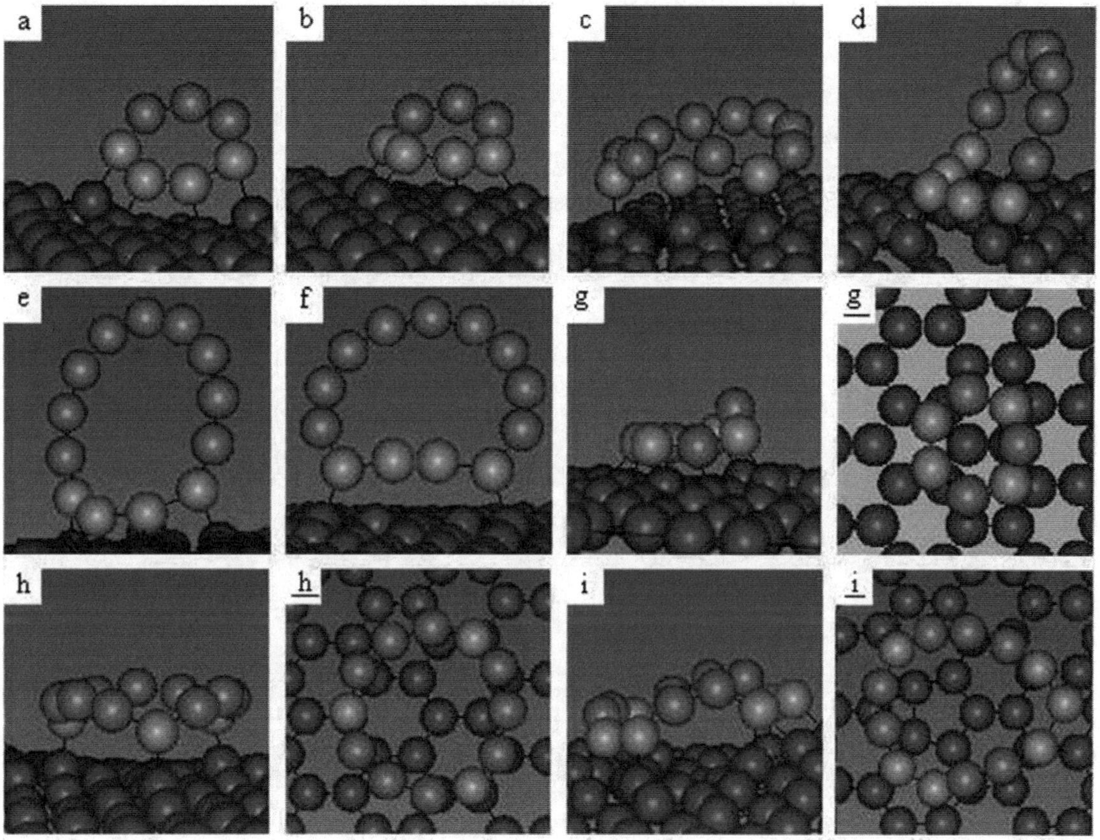

Figure 3. (Color online) Interaction of C_7, C_{12} and C_{13} clusters through its more than two atoms. Red spheres are nanographene atoms, blue spheres are cluster atoms and green ones are cluster atoms that interact directly with nanographene. a – C_7 and α=15°, b - C_7 and α=30°, c - C_{12} and α=15°, d - C_{12} and α=45°, e - C_{13} and α=15°, f - C_{13} and α=60°, g - C_7 and α=90°, h - C_{12} and α=90°, i - C_{13} and α=90°. Underlined g, h and i correspond to the top view of g, h and i images.

These results show that the different possible cases of the interaction increase with increasing the number of atoms in cluster. Note that in some cases the cluster can cause the vacancy of atoms in nanographene breaking the bonds of a few atoms of nanographene. If an initial position of cluster results in breaking the several bonds and formation of new structure near to the nanographene which are forbidden, this case can cause the failure of interaction. Therefore, there are some initial angles α, in which the computer simulation does not determine the final configuration of the atoms, i.e. these positions cause a failure of interaction. These unsuccessful angles for C_7, C_{12} and C_{13} clusters in the studied case are α=45°, 60° and 75°, respectively. However, note that these values can be depended on the place of interaction, azimuthal angle of cluster position, etc.

Journal of Physics: Conference Series **640** (2015) 012023 doi:10.1088/1742-6596/640/1/012023

In Table 2 we summarize the cohesion data of the C_7, C_{12} and C_{13} clusters interacting with nanographene versus the initial angle α of cluster position. Analyzing the results on the images of interaction and the cohesion data of clusters, we found that the maximal binding energy corresponds to the final upstanding position of clusters (i.e. when the planes of cluster and nanographene become perpendicular). As seen from Table 2, the maximal binding energy is 1.38, 1.59 and 2.93 eV for the C_7, C_{12} and C_{13} clusters interacting with nanographene, respectively. These values are considerably smaller than the cohesive energy E_{coh} of clusters presented in Table 1. That is why we can conclude that the atoms of clusters are hold together as a single cluster despite their interaction with nanographene. The cluster can be desorbed as a single cluster easily by e.g. heating the nanographene.

In the studied cases the highest value of the sum of binding energy is 4.66 eV for the C_7 cluster, 3.82 eV for the C_{12} and 5.89 eV for the C_{13} cluster with 4, 5 and 6 bonded atoms, respectively (Table 2). These configurations form the strongest bond between the ringlike cluster and nanographene.

It should be noted that in the most cases the cluster or its main part hold a ringlike structure during its interaction with nanographene. However, in the case of its lying position, the cluster is more buckled by its atoms bonded with nanographene (see, e.g. Fig. 2 d and Fig. 3 c, h, i). The interaction in the case of large clusters can result also in the change of the ringlike shape shown as in Fig. 3, i: the top view of the C_{13} cluster became more triangular shape.

Table 2. The cohesion data of the C_7, C_{12} and C_{13} clusters interacting with nanographene.

Cluster	Initial angle α (degrees)	The maximal binding energy (eV)	The sum of binding energy (eV)	The number of atoms interacting with nanographene
C_7	0	0.78	1.41	2
C_7	15	0.93	3.21	4
C_7	30	1.37	4.66	4
C_7	60	1.37	2.52	2
C_7	75	1.38	2.12	2
C_7	90	1.35	4.22	4
C_{12}	0	1.51	2.32	2
C_{12}	15	1.49	3.14	3
C_{12}	30	1.59	1.59	1
C_{12}	45	0.84	3.82	5
C_{12}	75	1.51	2.34	2
C_{12}	90	1.49	3.35	3
C_{13}	0	0.57	1.10	2
C_{13}	15	1.27	5.45	4
C_{13}	30	2.93	4.29	2
C_{13}	45	1.76	3.16	2
C_{13}	60	1.11	2.65	4
C_{13}	90	2.02	5.89	6

These results show that the interaction of carbon clusters with nanographene results in the deformation of both clusters and local part of graphene. As this interaction changes the hybridization from sp^2 to sp^3 for the graphene atoms bonded with cluster it can cause the local change of graphene properties, in particular its transition from the semimetal to the semiconductor.

3. Conclusion

Using an energy minimization method with a modified Brenner potential the stable structures of ringlike C_7, C_{12} and C_{13} clusters interacting with nanographene have been studied. Several possible configurations of these clusters were found. The cohesive energies of these clusters in the isolated (free) state and their binding energies with nanographene have been calculated. We found that these ringlike clusters can interact by one, two or more of its atoms with nanographene. It was shown that this interaction causes the structural change in both cluster and local part of graphene. As this interaction changes the hybridization from sp^2 to sp^3 for the graphene atoms bonded with cluster, the experimental investigation of this transition might be useful for the manipulation of the local changes of graphene properties.

References

[1] Novoselov K S, Geim A K, Morozov S V, Jiang D, Zhang Y, Dubonos S V, Grigorieva I V and Firsov A A 2004 *Science* **306** 666.
[2] Brenner D W, Shenderova O A, Harrison J A, Stuart S J, Ni B and Sinnott S B 2002 *J. Phys.: Condens. Matter* **14**, 783.
[3] Tersoff J 1988 *Phys. Rev.* B **61** 2879.
[4] Rappe A K, Casewit C J, Colwel K S, Goddard W A and Skiff W M 1992 *J. Am. Chem Soc.* **114** 10024.
[5] Yu D and Liu F 2007 *NanoLetters* **7** 3046.
[6] Dzhurakhalov A A and Peeters F M 2011 *Carbon* **49** 3258.
[7] Payne M C, Teter M P, Allan D C, Arias T A and Joannopoulos J D 1992 *Rev. Mod. Phys.* **64** 1045.
[8] Kosimov D P, Dzhurakhalov A A and Peeters F M 2008 *Phys. Rev. B* **78** 235433.

Journal of Physics: Conference Series **640** (2015) 012024 doi:10.1088/1742-6596/640/1/012024

Progress in characterizing submonolayer island growth: Capture-zone distributions, growth exponents, & hot precursors

Theodore L. Einstein[1,2], Alberto Pimpinelli[1,3], Diego Luis González[1,4], Josue R. Morales-Cifuentes[1,2]

[1]Dept. Physics & [2]CMTC, University of Maryland, College Park, MD 20742-4111 USA
[3]Rice Quantum Institute & MSNE Department, Rice University, Houston, TX 77005 USA
[4]Departamento de Física, Universidad del Valle, A.A. 25360, Cali, Colombia

E-mail: einstein@umd.edu; ap19@rice.edu; diego.luis.gonzalez@correounivalle.edu.co

Abstract. In studies of epitaxial growth, analysis of the distribution of the areas of capture zones (i.e. proximity polygons or Voronoi tessellations with respect to island centers) is often the best way to extract the critical nucleus size i. For non-random nucleation the normalized areas s of these Voronoi cells are well described by the generalized Wigner distribution (GWD) $P_\beta(s) = as^\beta \exp(-bs^2)$, particularly in the central region $0.5 < s < 2$ where data are least noisy. Extensive Monte Carlo simulations reveal inadequacies of our earlier mean field analysis, suggesting $\beta = i + 2$ for diffusion-limited aggregation (DLA). Since simulations generate orders of magnitude more data than experiments, they permit close examination of the tails of the distribution, which differ from the simple GWD form. One refinement is based on a fragmentation model. We also compare island-size distributions. We compare analysis by island-size distribution and by scaling of island density with flux. Modifications appear for attach-limited aggregation (ALA). We focus on the experimental system para-hexaphenyl on amorphous mica, comparing the results of the three analysis techniques and reconciling their results via a novel model of hot precursors based on rate equations, pointing out the existence of intermediate scaling regimes between DLA and ALA.

1. Introduction

During the last decade several fine reviews [1, 2, 3, 4] have been published about the early stages of thin-film growth. In this aggregation regime a major goal is to assess the size of the smallest stable cluster (denoted $i+1$, where i is the size of the critical nucleus, the largest unstable cluster [5, 6, 7, 8]). Three approaches have been taken in this pursuit: 1) rate equation theory for the scaling, with incident flux F, of island density $N \propto F^{\alpha_i}$; 2) analysis of the island-size distribution (ISD); and most recently 3) analysis of the capture-zone distribution (CZD). For ICCGE17 we recently prepared a minireview of these investigations [9], emphasizing the CZD approach. Here we include some highlights of that paper but with emphasis on a single experimental system, *para*-hexaphenyl (6P) on sputter-amorphized mica, showing results relevant to all three approaches and describing how including the participation of hot precursors reconciles the scaling data with the CZD and ISD results [10].

2. Capture-Zone Distributions and the Generalized Wigner Distribution

We begin with the most recent approach, the distribution of the areas of capture zones [3, 11, 12, 13], i.e. Voronoi (proximity) cells constructed from the islands. Cf. Fig. 1a,b. This

CZD turns out to provide information about i. For the Poisson-Voronoi (PV) case of random nucleation centers, the CZD is expected to follow a gamma distribution $P_{\Gamma}^{(\mathfrak{a})}(s)$ [14, 15],

$$P_{\Gamma}^{(\mathfrak{a})}(s) = \frac{\mathfrak{a}^{\mathfrak{a}}}{\Gamma(\mathfrak{a})} s^{\mathfrak{a}-1} e^{-\mathfrak{a}\, s}, \tag{1}$$

where s is the CZ area divided by its average value (so that $\langle s \rangle = 1$).

More generally we have argued [13], drawing from experiences analyzing the terrace-width distributions of vicinal surfaces [17], that the CZD is better described by the single-parameter generalized Wigner distribution (GWD):

$$P_\beta(s) = a_\beta s^\beta e^{-b_\beta s^2}, \quad a_\beta = 2\Gamma\left(\frac{\beta+2}{2}\right)^{\beta+1} \Big/ \Gamma\left(\frac{\beta+1}{2}\right)^{\beta+2}, \quad b_\beta = \left[\Gamma\left(\frac{\beta+2}{2}\right) \Big/ \Gamma\left(\frac{\beta+1}{2}\right)\right]^2, \tag{2}$$

where a_β and b_β are dictated by normalization and unit mean, respectively, of $P_\beta(s)$. Our derivation of $P_\beta(s)$ from a mean-field Fokker-Planck equation implied that $\beta = i + 1$ in two dimensions (2D) [13].

Extensive simulations exposed inadequacies of this mean-field prediction [18], ultimately because nucleation occurs preferentially near CZ boundaries rather than uniformly [19, 20], so that in 2D $\beta = i + 2$ (for *diffusion-limited aggregation* (DLA)) is better. Shi et al. [21] performed kinetic Monte Carlo calculations of irreversible growth ($i = 1$) of point islands with dimensions $1 - 4$, for both square and triangular lattices in 2D and for two different point-island models. They used values of $R = 10^5 - 10^{10}$, where $R \equiv D/F$, D the diffusion rate and F the deposition flux, in order to extrapolate to asymptotic behavior. Their fits to CZDs are better with $\beta = 3$ than with $\beta = 2$. While the coverage dependence (between 0.1 and 0.4 ML) is negligible, there is some dependence on R and on which of the two models of point islands is used. They also find better scaling (with R) of the peak height of the CZD using $\beta = 3$, and comparable results for the two models. Li et al. [19] considered both $i = 1$ and $i = 0$ at 0.1 ML, finding that $\beta = 3$ accounts better for their $i = 1$ CZD curve than $\beta = 2$. The GWD describes the CZD over the values of s for which there is significant data in experiments ($0.5 < s < 2$) but has deficiencies in the tails at high and low s [19, 22]. For large s, $P(s)$ may decay exponentially or like a generalized gamma distribution $s^{\beta_\nu} \exp(-\text{const.} \cdot s^\nu)$ (with non-integer ν) rather than in Gaussian fashion. Thus, their best fit gives $\nu = 1.5$ and $\beta_\nu \approx 4$ (and $\nu = 1.3$ and $\beta_\nu \approx 3$ for $i = 0$). However, the relation between β and i should be determined by the central part of the CZD rather than power-law behavior for $s \ll 1$, so that β_ν and β are expected to differ; i.e., β

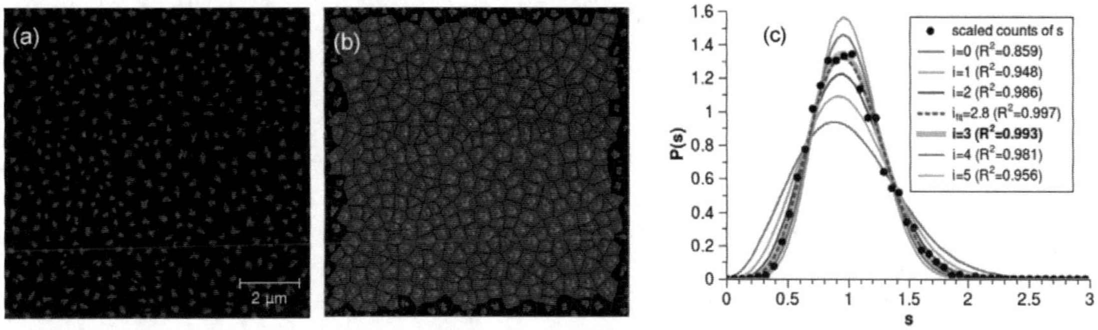

Figure 1. (a) Islands of standing *para*-hexaphenyl (6P) on sputtered mica (001) at 0.19 ML, $T = 300$K, $F = 0.04$ ML/min. (b) Construction of capture zones of (a) by Voronoi tesselation. (c) Resulting CZD from (b) compared with $P_\beta(s)$ for $\beta = 2$–7 ($i = \beta - 2$ for DLA). From [16].

is essentially an effective exponent. (Furthermore, β_ν increases to "compensate" decreasing ν. The non-Gaussian decay stems from the probability $\sim s^\nu$ that a new CZ overlaps an existing CZ of scaled area s [3, 23]; ν is typically extracted from numerical data.)

Most recently Oliveira and Aarão Reis [22] reported extensive simulations in 2D for point and extended (fractal and square) islands, with $i = 1$ and 2, for $R = 10^6 - 10^{10}$. Rather than using s as their independent (scaling) variable, they choose $u = (x - \langle x \rangle)/\sigma_x \to (s-1)[(\beta+1)/(2b_\beta) - 1]^{-1/2}$, where x is the number of lattice sites within a CZ, and σ_x is the standard deviation; the final expression for u is its form for the GWD. This scaling procedure improves data collapse (for different values of R) by reducing the corrections to scaling for small x, where the continuum model underlying the derivation of Eq. (2) becomes inaccurate. However, this scaling obscures the variation of the curves with changing values of i (and so β) in the central region [9] (but brings out differences in the tails in their log-linear plots). For point islands with $i = 1$, $\beta = 2$ and 3 both give "good fits" in the central (peak) region, with $\beta = 3$ also adequate for small s and neither doing well for large s; for $i = 2$, $\beta = 3$ fits well the center and large-s tail. For fractal and for square islands, $\beta = i + 1$ gives a good fit for four values of R, while $\beta = i + 2$, though not mentioned [22]a, is also satisfactory [22]d. Typically there is Gaussian decay for large s.

While the numerical studies leave some open questions, the preponderance of evidence points to $\beta \approx i + 2$, which we can retrieve by refining our derivation [24]: Noting that within a circular zone of radius r_0, the adatom density $n(r) \propto r_0^2 - r^2$, so that the integral over CZ area of $[n(r)]^{i+1} \propto r_0^{2i+4} \Rightarrow P(s) \propto s^{i+2}$. Furthermore, experimental data well described by $P_\beta(s)$ can also be fit with $P_\Gamma^{(\alpha \approx 2\beta+1)}(s)$ [25], progressively better for larger exponents [9].

In a refined analysis based on a widely-applicable *fragmentation model* (FM) [12], we characterized systems in terms of two physically-rooted exponents, γ and δ [20], by taking (a) the probability to nucleate, in a cell of size s, $\propto s^\gamma P(s)$, ultimately, implying $P(s) \sim \exp(-\text{const}.s^\gamma)$ for $s \gg 1$; and (b) the probability that a new center lies at a position $\mathbf{r}$, relative to the center of a preexisting cell, $\propto |\mathbf{r}|^\delta$, assuming circular isotropy for simplicity. The simplest case is the PV problem, for which $\gamma = 1$ and $\delta = 0$, with the second probability $\mathbf{r}$-independent but $\propto s^{-1}$ [20].

A harder but still tractable example is a point-island model, with irreversible attachment, in which capture zones are approximated as circular [20]. The isotropic solution of the appropriate steady-state diffusion equation gives an adatom density $n(\rho)$ that increases from 0 at the interior island edge to a smooth maximum at the outer edge. Thence we deduce that $\gamma = 3$ (more generally $\gamma = i+2$), which, with $\delta = 0$, accounts adequately for numerical data for $P(s)$, but less well for island density and the radial nearest-neighbor island probability distribution. This deficiency can be remedied by taking $\delta = 1$ near the center and $\delta = 0$ near the edge [20]. In the FM the above-noted result $\nu = 1.5$ [19] implies $\gamma = 1.5$, with $\delta \approx 1.2$ near the center and $\delta = 0$ near the edge [20]. Furthermore, the above mean-field-based argument for $\gamma = 3$ overestimates γ, which extensive simulations show to be close to 2 rather than 3 at low coverages and large R.

For $s \ll 1$, $P(s)$ is expected to behave like a power law in s, but the precise relation between this exponent and i has not yet been determined, although for $i = 1$ good agreement with simulations has been found for the (actual rather than effective) exponent $\beta = 4$ [19, 20, 22]. In this regime $P(s)$ depends on the concentration of centers and ultimately, in the FM model, on δ [20]. The skewness of the GWD also agrees well with numerical data in simulations [22]. However, earlier efforts to fit experimental data by just extracting the first few moments of the distribution were unsatisfactory [26]. Alternatively, one can allow two different values of γ: 2 for small s and 1 for large s. While such a "two-regime model" is not essential in 2D, it is in 1D. For the detailed analyses and controversies in the 1D case, see [12, 19, 20, 21, 22, 27, 28].

3. Island Size Distributions and Scaling of Island Number vs. Flux

Two alternative approaches have long been used to gauge the critical nucleus size from experiment. One is to measure the island size distribution (ISD) [3, 29, 30] and then to fit it with the Amar-Family scaling formula (at least for i = 1,2,3 and in 2D) [31],

$$f_i(u) = C_i u^i \mathrm{e}^{-ia_i u^{1/a_i}}, \quad \frac{\Gamma[(i+2)a_i]}{\Gamma[(i+1)a_i]} = (ia_i)^{a_i}; \quad f(u) \underset{R\to\infty}{=} f(0)\exp\left[\int_0^u dy \frac{2z-1-C'_{\mathrm{tot}}(y)}{C_{\mathrm{tot}}(y)-zy}\right], \tag{3}$$

with u now the island size divided by its mean and C_i a normalization constant; Amar et al. deduced $f_i(u)$ empirically from the expectations that (in the limit of large R) f_i goes like u^i for small u, cuts off exponentially for large u, and peaks at $u = 1$ (unlike the GWD, which peaks at $u < 1$, especially noticeable for small β). With scaling assumptions for capture numbers and neglect of deposition dependence, Bartelt and Evans [3, 23, 32] obtained the $R \to \infty$ limit, where C_{tot} is a linear combination of scaled capture numbers and capture zones and z is the slope of a log-log plot of mean island size vs. F (e.g., $z = 2/3$ for point islands). In principle, then, the ISD can be obtained from a rate-equation approach if one has information about the capture numbers [1, 32]; they can be measured from simulations since calculating them is intractable.

While the ISD does not generally mimic the CZD [3, 33], kinetic Monte Carlo (KMC) calculations by Fanfoni et al. on a simple model of quantum dot growth reveal a similarity between ISD and CZD at lower temperature T, when evolution of islands is dominated by atom motion along the periphery rather than attachment/detachment [34]. (However, they favor the $P_\Gamma^{(\alpha)}(s)$ over $P_\beta(s)$ in their fits.) See [22] for recent results on the high-end tails.

The second approach, based on rate equation theory, involves the scaling of the density N of [stable] islands (particularly the maximum density) with F: it is long known [5, 6] that at constant, relatively low T, there is the scaling relation [6, 35]

$$N \sim F^{\alpha_i} \exp\left(E_i/[(i+2)k_B T]\right), \qquad \alpha_i^{\mathrm{DLA}} = i/(i+2)\,, \quad \alpha_i^{\mathrm{ALA}} = 2i/(i+3) \tag{4}$$

where E_i is the cohesion energy of a cluster of size i. The two relations of α_i to i are DLA and attachment-limited aggregation (ALA) regimes [36] in 2D, respectively. There are many other regimes with signature relations for α_i [35, 37, 10]. Also, the values in 3D differ, e.g. being $2i/(2i+5)$ for DLA (with compact islands and with no desorption). In short, the value of i deduced from α_i depends strongly on the dominant mode of mass transport. In many cases one can characterize the T dependence by writing $N \sim (F/D)^{\alpha_i}$, where D has an activated, Arrhenius form; then the "activation energy" is enhanced by a term proportional to the diffusion barrier. Thus, N is expected to decrease rapidly with increasing T [8]. While a great deal of information can be gleaned from this characteristic energy, or its effective generalization $d\ln N/d(1/k_BT)$, space precludes a detailed discussion here; readers should consult [6, 7, 8, 10].

4. Experimental Applications: *Para*-hexaphenyl on Sputter-Amorphized Mica

In [9] we provided capsule descriptions of the many experimental applications of CZD analysis; here we simply give a list: pentacene on SiO_2 [38], pentacene with controlled pentacenequinone impurities [39], polar-conjugated molecule Alq_3 on passivated Si(100) [40], self-assembled Ge/Si(001) nanoislands [41], InAs quantum dots on GaAs(001) [34, 42], metallic Ga droplets on GaAs(001) [43], C_{60} on UTO [25], 6P on Ir(111), 6P on SiO_2 [44], and 6P films on amorphous mica [16, 45, 46, 47, 48]. Here we concentrate on the last in this list, to which all three approaches have been applied. Note that interest in rod-like 6P is spurred in part by its use in blue-light-emitting diodes [49], nano-optic devices [50], etc. We show that all three methods give values of i that are reasonably consistent with one other, notwithstanding some reported values.

The sample image in Fig. 1a is of islands of 6P for 0.19 ML, T = 300K, and F = 0.04 ML/min on freshly cleaved muscovite mica (001), sputter-modified to ensure the 6P are all standing, thereby precluding a competing wetting layer [16]. The Voronoi tesselation of such images, as in Fig. 1b, is used to tally the CZD. Fig. 1c then shows that the CZD is well (and best) fit with $\beta = 5$, leading to $i = 3$ (within the assumption of DLA). Fig. 2a depicts the ISD for these islands again at $T = 300$K but with half the flux. Based on a least-squares fit to Eq. (3), the result is again $i = 3\pm1$. Subsequently this group [46] found that this ISD is for large and rather dendritic islands; they report $i = 2$, but given the modest amount of data, this is viewed as consistent with the earlier value. (There are also smaller, compact islands with a smaller critical nucleus size, $i = 1$.) Third, Winkler's group's analysis of $N(F)$ data at 300K with Eq. (4) with $\alpha_i = i/(i+2)$ yielding $i = 2.5\pm0.5$ [16]. For such small i they conclude that the 6P molecules must be lying on the surface, rather than standing up. Assuming DLA and ALA Different behavior is found at low T, attributable to different kinetics [16]. Subsequent work on sputtered mica [45] showed that the 6P molecules stand up, suggesting a higher i. While no CZD analysis is given, values of α_i in both the DLA and the ALA regimes give $i = 7\pm2$ [51].

For pentacene on sputter-amorphized mica at 300 K, there is similar dogleg behavior in the plot of $\ln N$ vs. $\ln F$, with $\alpha = 0.8 \pm 0.1$ and $\alpha = 1.3 \pm 0.1$[48]. Using Eq. (4) they deduce $i = 5 \pm 1$. For the CZD they found $\beta = 5.0 \pm 0.5$ and $\beta = 4.0 \pm 0.5$ for $F = 0.08$ and 1.37 ML/min., respectively, at 300K. They deduce $\alpha\beta \approx i$ globally, with the noteworthy corollary $\beta_{\rm ALA} = (i+3)/2$.

5. Hot Precursor Model

Very recently three of us presented a model, formulated in terms of rate equations, that treats the modification of island nucleation by "hot precursors," i.e., deposited molecules which propagate ballistically for some distance before thermalizing or joining an island [10]. A key dimensionless parameter is $z \equiv \tau_h/\tau_{h\to N}$, the ratio of the thermalization time to the hot-precursor survival time before capture by an island. For $z \ll 1$ and $z \gg 1$, i.e. fast and slow thermalization, we recover, respectively, DLA and a novel hot monomer aggregation (HMA) regime with α the same as for ALA. However, the crossover between these regimes is not a smooth monotonic crossover. Two other dimensionless terms come into play, the ratio $\mathcal{R}_n$ of hot to thermalized monomers and a ballistic scale $\mathcal{R}_B$ related to the distance hot monomers travel to the spacing between islands). (A third ratio, expressing whether ballistic or thermalized channels dominate decay, also plays

Figure 2. ISD for 6P islands vs. normalized area s, after deposition of three different coverages (Θ = 0.22, 0.32, and 0.60ML) at 300K with $F = 0.02$ ML/min, measured with *ex situ* AFM at room temperature. The island density is scaled by $\langle s\rangle^2/\Theta$. The least-squares best fit using Eq. (3) yields $i = 3\pm1$. From Fig. 2 of [16] A similar plot with less data in Fig. 10a of [46] for 0.147 ML at 400K reports $i = 2$.

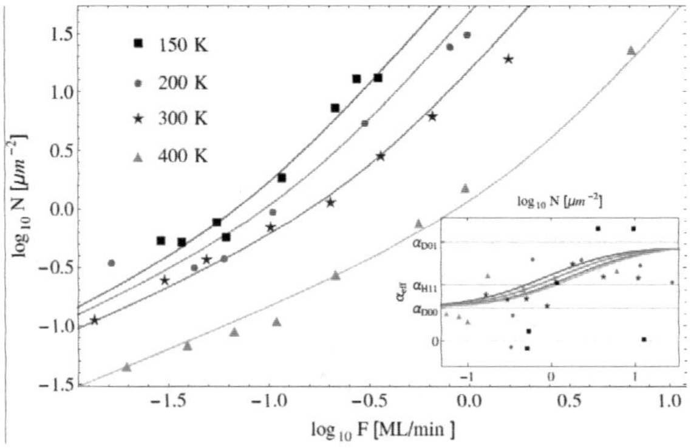

Figure 3. Island density vs. deposition rate in the aggregation regime for 6P on sputter-modified mica (001) [45] at T = 150K (black, square dots), 200K (red, round dots), 300K (blue, star dots), and 400K (green, triangular dots), with $i = 4$ and several other dimensionless best-fit parameters given in [10]. Inset: $\alpha_{\rm eff}$ (curves) and values of $d\ln N/d\ln F$ from the experimental data (dots). Adapted from Fig. 3 of [10]

a role if we consider the effective activation energy.) In plots of the effective $\alpha \equiv d\ln N/d\ln F$ vs. z, one finds the expected flat regions for small (DLA) and large (HMA/ALA) z but the expected non-monotonic crossover with extended plateaus corresponding to extreme values of $\mathcal{R}_n$ and $\mathcal{R}_B$. (See Fig. 1 of [10].) Likewise, there are such plateaus for the effective activation energy. As a test of this model, we used the solution of the rate-equation formulation to fit simultaneously data for four temperatures, as shown in Fig. 3, with the result $i = 4 \pm 1$ rather than $i = 7 \pm 2$. While the data at small z does correspond to DLA, that at high z is not ALA but rather the case of small z and $\mathcal{R}_n$ but large $\mathcal{R}_B$, for which $\alpha = i/2$ rather than $2i/(i+3)$.

6. Conclusions

We have discussed the three methods to investigate the critical nucleus size in submonolayer growth. For the case of 6P on amorphous mica, we found that scaling of island density gives results consistent with CZD analysis, once hot precursors are properly taken into account, showing the importance of that mechanism. We have shown that the GWD provides an excellent accounting of CZDs in the region where the data in experiments is most reliable. While the fits with $P_\Gamma^{(\mathfrak{a})}(s)$ (or even a Gaussian, for large β) may also be adequate, only the GWD offers a fundamental connection $\beta \approx i + 2$. Further improvements in the theory, notably the fragmentation model, allow more detailed examination of tails and of other statistical functions. We have also applied CZD analysis to data from homoepitaxial growth on vicinal Cu(001), finding that codeposited impurities offer the most likely explanation of the data [52]. The approach applies to a much broader range of problems than just crystal growth: we have characterized the distribution of Metro stations in central Paris [20] and of the areas of secondary administrative units (counties in the USA, *arrondissements* in France, etc.) [20, 53].

Acknowledgments

Work at UMD currently supported by NSF Grant No. CHE 13-05892. We are grateful for collaborations on CZD analysis with theorists R. Sathiyanarayanan and A. BH. Hammouda, with the UMD experimental surface physics group, especially E.D. Williams, W.G. Cullen, B.R. Conrad, and M.A. Groce, and on scaling in growth with A. Winkler and his group at Graz.

References

[1] Ratsch C and Venables J A 2003 *J. Vac. Sci. Tech.* A **21** S96
[2] Michely T and Krug J 2004 *Islands, Mounds and Atoms: Patterns and Processes in Crystal Growth Far from Equilibrium* (Berlin: Springer)
[3] Evans J W, Thiel P A and Bartelt M C 2006 *Surf. Sci. Rep.* **61** 1
[4] Einax M, Dieterich W and Maass P 2013 *Rev. Mod. Phys.* **85** 921
[5] Stowell M 1970 *Phil. Mag.* **21** 125
[6] Venables J A, Spiller G D and Hanbücken M 1984 *Rep. Prog. Phys.* **47** 399, and references therein
[7] Brune H 1998 *Surf. Sci. Rep.* **31** 121
[8] Pimpinelli A and Villain J 1998 *Physics of Crystal Growth* (Cambridge: Cambridge University Press)
[9] Einstein T L, Pimpinelli A and D. L. González D L 2014 *J. Cryst. Growth* **401** 627
[10] Morales-Cifuentes J R, Einstein T L and Pimpinelli A 2014 *Phys. Rev. Lett.* **113** 246101; arXiv 1411.6575
[11] Mulheran P A and Blackman J A 1995 *Philos. Mag. Lett.* **72** 55; 1996 *Phys. Rev.* B **53** 10261
[12] Blackman J A and P. A. Mulheran P A 1996 *Phys. Rev.* B **54** 11681
[13] Pimpinelli A and Einstein T L 2007 *Phys. Rev. Lett.* **99** 226102
[14] Kiang T 1966 *Z. Astrophysik* **64** 433
[15] Ferenc J S and Z. Néda Z 2007 *Physica* A **385** 518
[16] Potocar T, Lorbek G, Nabok D et al. 2011 *Phys. Rev.* B **83** 075423
[17] Einstein T L and Pierre-Louis O 1999 *Surface Sci.* **424** L299; Einstein T L 2007 *Appl. Phys.* A **87** 375
[18] Mean field works better in 1D, but that case requires—and permits—a more complicated analysis [12, 27, 28]
[19] Li M, Han Y and Evans J W 2010 *Phys. Rev. Lett.* **104** 149601
[20] González D L and Einstein T L 2011 *Phys. Rev.* E **84** 051135
[21] Shi F, Shim Y and Amar J G 2009 *Phys. Rev.* E **79** 011602

[22] Oliveira T J and Aarão Reis F D A, a) 2011 *Phys. Rev.* B **83** 201405; b) 2012 **86** 115402; c) 2013 **87** 235430; d) Aarão Reis F D A, private communication, 2014
[23] Bartelt M C and Evans J W 1996 *Phys. Rev.* B **54** 17359R; 2001 **63** 235408; 2002 **66** 235410
[24] Pimpinelli A and Einstein T L 2010 *Phys. Rev. Lett.* **104** 49602
[25] Groce M et al. 2012 *Surface Sci.* **606** 53; Michelle A. Groce, Ph.D. dissertation, University of Maryland, 2013; http://search.proquest.com/docview/1464775310
[26] Giesen M and Einstein T L 2000 *Surface Sci.* **449** 191
[27] González D L, Pimpinelli A and Einstein T L 2011 *Phys. Rev.* E **84** 011601.
[28] Grinfeld M et al. 2012 *J. Phys. A: Math. Theor.* **45** 015002; ONeill K P et al. 2012 *Phys. Rev.* E **85** 021601; Mulheran P A et al. 2012 *Phys. Rev.* E **86** 051606
[29] Amar J G, Family F and Lam P M 1994 *Phys. Rev.* B **50** 8781
[30] Bartelt M C and Evans J W 1992 *Phys. Rev.* B **46** 12675; Ratsch C et al. 1994 *Phys. Rev. Lett.* **72** 3194
[31] Amar J G and F. Family F 1995 *Phys. Rev. Lett.* **74** 2066
[32] Körner M, Einax M and Maass P 2010 *Phys. Rev.* B **82** 201401(R); 2012 *Phys. Rev.* B **86** 085403
[33] P. A. Mulheran P A and Robbie D A 2000 *Europhys. Lett.* **49** 617
[34] Fanfoni M, Arciprete F, Tirabassi C et al. 2012 *Phys. Rev.* E **86** 061605
[35] Pimpinelli A, Jensen P, Larralde H and Peyla P 1998, in *Morphological Organization in Epitaxial Growth and Removal*, ed Z Zhang and M G Lagally (Singapore: World Scientific) p 121
[36] Kandel D 1997 *Phys. Rev. Lett.* **78** 499
[37] Pimpinelli A, Larralde H and Jensen P 1997 *Phys. Rev.* B **55** 2556
[38] Pratontop S, Brinkmann M, Nüesch F and Zuppiroli L 2004 *Phys. Rev. B* **69** 165201
[39] Conrad B R, Gomar-Nadal E, Cullen W G et al. 2008 *Phys. Rev.* B **77** 205338
[40] Brinkmann M, Graff S and Biscarini F 2002 *Phys. Rev.* B **66** 165430
[41] Miyamoto S, Moutanabbir O, Haller E E and Itoh K M 2009 *Phys. Rev.* B **79** 165415
[42] a) Fanfoni M et al. 2007 *Phys. Rev.* B **75** 245312; b) Placidi E et al. 2007 *J. Phys. Cond. Mat.* **19** 225006
[43] Nothern D M and Millunchick J M 2007 *J. Vac. Sci. Tech.* B **30** 060603
[44] Lorbek S, Hlawacek G and Teichert C 2011 *Eur. Phys. J. Appl. Phys.* **55** 23902
[45] Tumbek L and Winkler A 2012 *Surface Sci.* **606** L55
[46] Tumbek L, Gleichweit C, Zojer K and Winkler A 2012 *Phys. Rev.* B **86** 085402
[47] Winkler A and Tumbek L 2013 *J. Chem. Phys. Lett.* **4** 4080
[48] Pimpinelli A, Tumbek L and Winkler A 2014 *J. Phys. Chem. Lett.* **5** 995
[49] Yanagi H and Okomato S 1997 *Appl. Phys. Lett.* **71** 2563
[50] Balzer F and Rubahn H G 2005 *Adv. Funct. Mater.* **15** 17
[51] Hlawacek G, Khokar F S, van Gastel R, Teichert C and Poelsema B 2011 *IBM J. Res. & Dev.* **55** 15
[52] Hamouda A BM et al. 2011 *Phys. Rev.* B **83** 035423; Sathiyanarayanan R et al. 2011 *Phys. Rev.* B **83** 035424
[53] Rajesh Sathiyanarayanan, Ph.D. thesis, Maryland, 2009; http://pqdtopen.proquest.com/doc/304923517.html?FMT=ABS; Sathiyanarayanan R et al., preprint

Ni cluster on a silicon surface by the Density Functional Theory and X-ray absorption spectroscopy investigations

V L Mazalova[1], O V Farberovich[1,2,3], A V Soldatov[1]

[1] Research center for nanoscale structure of matter, Southern Federal University,Zorge 5, 344090, Rostov-on-Don,Russia

[2] School of Physics and Astronomy, Beverly and Raymond Sackler Faculty of Exact Sciences,Tel Aviv University, Tel Aviv 69978, Israel

[3] Voronezh State University, Voronezh 394000, Russia

E-mail: mazalova@sfedu.ru

Abstract. In this article we offer the model of the quantum solid-state N-spin qubit based on the studying of the spin structure simulation of the 3d-metal Ni clusters on the silicon surface. The comparison of the energy pattern with the anisotropic exchange models conventionally used for the analysis of this system and, with the results of the X-ray absorption spectroscopy (XAS) study, shows that our complex investigations provide a good description of the pattern of the spin levels and the spin structures of the nanomagnetic Ni_7 qubit. The results are discussed in the view of the general problem of the solid-state spin qubits and the spin structure of the Ni cluster.

1. Introduction

The promise of quantum computers to solve classically non-computable problems [1] has generated the great excitement and much research activity in different areas of physics, mathematics and engineering. Various physical systems have been proposed for implementation of quantum bits (qubits) in quantum information processing devices: trapped ions, atoms in QED cavities, magnetic molecules, etc. Among many candidates, spin-based solid-state systems, such as quantum dots [2] or spin centers in host crystals (phosphorus donors in silicon [3], NV centers in diamonds [4]), constitute attractive candidates for qubits: these systems are well scalable, can be fabricated and operated by the methods of modern microelectronics, and advanced spin-resonance techniques are well-suited for the efficient quantum state manipulation. Spins are alternative complementary to charges as degrees of freedom to encode information. Recent examples, like for instance the discovery and application of Giant magnetoresistance in Spintronics, have demonstrated the efficient use of a spins for information technologies [5]. Moreover, spins are intrinsically quantum entities and they have therefore been widely investigated in the field of the quantum-information processing. Here we suggest for construction of the spin qubit the Ni small clusters on a silicon surface. Ni is the unique among the transition-metal adatoms, because its half-filled valence configuration ($3d^8 4s^2$) yields strong interatomic bonding leading to magnetic frustration. We apply our method to the Ni clusters deposited on a Si(111) surface. We apply X-Ray Absorption Near-Edge Structure (XANES) spectroscopy for studying the adsorption geometry and getting information concerning the electronic properties of the deposited Ni_7 clusters [6, 7]. In order to extract the necessary information from the experimental XANES spectra one needs to perform theoretical analysis. In the present research the theoretical XANES spectra was simulated using FEFF9.0 program code [8, 9]. FEFF9.0 code is based on the real space full-multiple scattering theory. The code uses the cluster approach for XANES spectra calculations and, therefore (in contrast to program based on the band structure calculations), can be applied for study of compounds without long order in atoms distributions, like clusters. The electronic structure of investigated clusters was analyzed on the basis of the density functional theory (DFT) implemented in the ADF2013 code [10, 11].

2. Theoretical approach

One of the major challenges in quantum computing is to identify a system that can be scaled up to the number of the qubits needed to execute the nontrivial quantum algorithms. Peter Shors algorithm [12] for finding the prime factors of the numbers used in the public-encryption systems (numbers that typically consist of more than a hundred digits) would likely require a quantum computer with the several thousand qubits. Depending on the error correction scheme appropriate to the particular computer, the required number could be much larger. The solid-state spin quantum computers may be more likely candidates. In this work we offer the model of the quantum computer based on studying of the spin structure of the 3d-metal Ni clusters on a silicon surface.

2.1. Calculation of the spin structure of the N-spin system

In our model the total electronic structure can be written as a sum of a non-spin-polarized charge DOS $n_{DOS}(\epsilon)$, and the spin DOS, $S_{DOS}(\epsilon)$,

$$n(\epsilon) = n_{DOS}(\epsilon) + S_{DOS}(\epsilon) \tag{1}$$

The first part $n_{DOS}(\epsilon)$ is connected with a one-electron structure and can be received by any DFT method. The second part $S_{DOS}(\epsilon)$ is purely a spin structure and is defined only proceeding from the spin model of a cluster. For division in an experimental spectra of the deposits from a purely spin states and from the excitations connected with the one-electron transitions, is reasonable within uniform approach to receive electronic structure of a cluster and then, using the one-electron data for calculation of the exchange integrals, to calculate a spin structure by the ITO method within the generalized spin Hamiltonian $\widehat{H}_{spin}$. We have developed the idea of spins as a degree of freedom, with which models are built [13]. The spin magnetic moment due to the exchange interaction is

$$\mathbf{M}_s = -2\langle\hat{\mathbf{S}}\rangle\mu_B/\hbar, \tag{2}$$

where

$$\langle\hat{S}\rangle = \sqrt{\langle\hat{S}_x\rangle^2 + \langle\hat{S}_y\rangle^2 + \langle\hat{S}_z\rangle^2}$$

is the spin structure. With the spin-Hamiltonian $\widehat{H}_{spin}$ result (??) we can obtain the quantum mechanical expectation values:

$$\langle\hat{S}_x\rangle = \langle SM \mid \hat{S}_x \mid SM\rangle, \tag{3}$$

$$\langle\hat{S}_y\rangle = \langle SM \mid \hat{S}_y \mid SM\rangle, \tag{4}$$

$$\langle\hat{S}_z\rangle = \langle SM \mid \hat{S}_z \mid SM\rangle. \tag{5}$$

With the algebra of the spin operators we can obtain the expectation values for $\widehat{H}_{spin}$:

$$\langle\hat{S}_x\rangle = \frac{1}{2}\sum_{\mu=1}^{N} c_\mu^2 A_\mu, \tag{6}$$

$$\langle \hat{S}_y \rangle = -\frac{i}{2} \sum_{\mu=1}^{N} c_\mu^2 B_\mu, \tag{7}$$

$$\langle \hat{S}_z \rangle = \sum_{\mu=1}^{N} c_\mu^2 M^{(\mu)} \tag{8}$$

where

$$A_\mu = \sqrt{S^{(\mu)}(S^{(\mu)}+1) - M^{(\mu)}(M^{(\mu)}+1)} + \\ + \sqrt{S^{(\mu)}(S^{(\mu)}+1) - M^{(\mu)}(M^{(\mu)}-1)}$$

and

$$B_\mu = \sqrt{S^{(\mu)}(S^{(\mu)}+1) - M^{(\mu)}(M^{(\mu)}+1)} - \\ - \sqrt{S^{(\mu)}(S^{(\mu)}+1) - M^{(\mu)}(M^{(\mu)}-1)}$$

The spin DOS $S(\epsilon)$ can be done by inserting the following functions:

$$S_x(\epsilon) = \sum_{\mu=1}^{N} c_\mu^2 A_\mu \delta(\epsilon - \epsilon_\mu), \tag{9}$$

$$S_y(\epsilon) = \sum_{\mu=1}^{N} c_\mu^2 B_\mu \delta(\epsilon - \epsilon_\mu), \tag{10}$$

$$S_z(\epsilon) = \sum_{\mu=1}^{N} c_\mu^2 M^{(\mu)} \delta(\epsilon - \epsilon_\mu). \tag{11}$$

The spin DOS is

$$S_{DOS}(\epsilon) = \sqrt{S_x^2(\epsilon) + S_y^2(\epsilon) + S_z^2(\epsilon)} \tag{12}$$

2.2. Calculation of the density of states $n_{DOS}(\epsilon)$

The total density of states $n_{DOS}(\epsilon)$ of Ni_7 clusters was calculated by the all-electron density functional theory (DFT) approach implemented in the ADF2013 code. The main point of the density functional theory is that the potential acting on each electron of all the other electrons in the molecule or crystal, depends only on the electron density of the ground state and its gradient. Thus, we can apply the one-electron formulation of a system of N interacting electrons by introducing the appropriate local exchange-correlation potential $V_{XC}(r)$, in addition to any external potentials $V_{ext}(r)$ and the Coulomb potential of the electron cloud $V_C(r)$ and it has the form:

$$(-\frac{1}{2}\nabla^2 + V_{ext}(\vec{r}) + V_c(\vec{r}) + V_{xc}(\vec{r}))\phi_i(\vec{r}) = \varepsilon_i \phi_i(\vec{r}) \tag{13}$$

The one-electron molecular orbitals ϕ_i with the appropriate orbital energies ε_i define precise electronic charge density and, in principle, give access to all the properties, because they are all expressed in terms of density functional, especially energy. In addition, they allow us to represent the system as a set of independent electron orbitals. In our calculations the electronic configuration of the Ni_7 cluster was described by the polarized triple zeta (TZP) basis set of the Slater-type orbitals. The total DOS was obtained using the BLYP exchange functional which is equivalent to Becke (exchange)[15] and Lee-Yang-Parr (LYP correlation)[16]. The calculations were done for the high-spin state of Ni_7 cluster.

3. Results and discussion

Here we present the results of the application of our theoretical and experimental approach to study of the N-qubits from seven nickel atoms interacting with a Si (111)-surface in the terms of numerically solvable DFT-models for the exchange integrals J_{ij} and we exploit the ITO technique [17] for calculation of the spin structure of the N-qubits system. In practice, this model was applied to the nanosystem (Ni_7-cluster on Si (111)-surface) and provides an understanding of the chemical bonding of the nickel nanoparticles with the silicon substrate. In our model the total electronic structure can be written as a sum of a non-spin-polarized,charge DOS $n_{DOS}(\epsilon)$, and a spin DOS, $S_{DOS}(\epsilon)$. The Fig.1 shows the projected total $n_{DOS}(\epsilon)$ of the Ni_7 cluster calculated without taking into account the relativistic effects. As one can see from the spectrum, there are the free electronic states higher the LUMO level, which describe well the calculated spin DOS $S(\epsilon)$ of Ni_7 clusters (see the Fig.2). The comparison of the projected spin DOS $S(\epsilon)$ of Ni_7 cluster on a silicon surface and the one-electron $n_{DOS}(\epsilon)$ of Ni_7 cluster is shown on the Fig.2. In this Figure the Ni 3d-hole one-electron cluster states which coincide with low-lying (0 - 500 mV) excited spin states have been allocated.

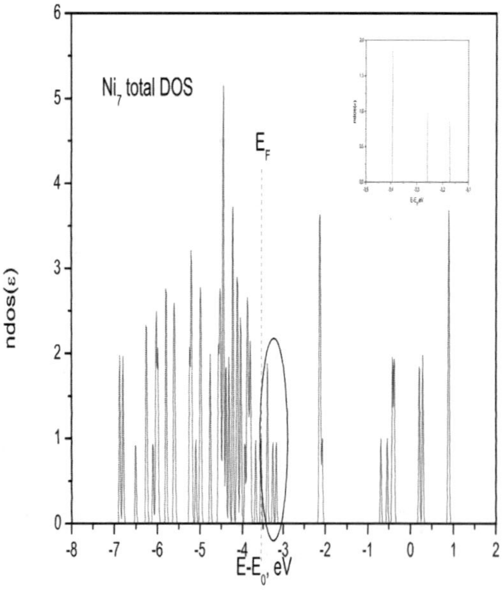

Figure 1. The projected total one-electron $n_{DOS}(\epsilon)$ of Ni_7 cluster on a silicon surface. Inset: expanded view of the first LUMOs.

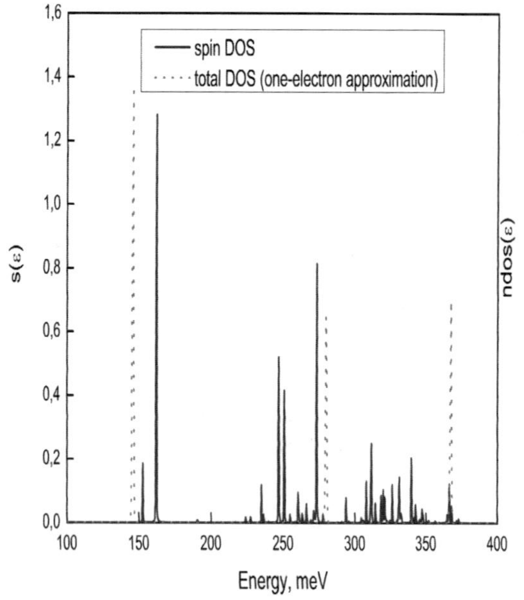

Figure 2. The comparison of the projected spin DOS $S(\epsilon)$ of Ni_7 cluster on a silicon surface and one-electron $n_{DOS}(\epsilon)$ of Ni_7 cluster.

The properties of the 3d-electrons are best probed in an X-ray absorption experiment by the excitation of $2p$ core electrons to unfilled $3d$ states. In principle, X-ray absorption spectra contain contributions from both $p \to d$ and $p \to s$ transitions, but in practice the $p \to d$ channel dominates by a factor >20. The line intensities, denoted I_{L_2} and I_{L_3}, respectively, are directly proportional to the number of d holes. The use of circularly polarized X-rays opens the door for spin studies (XMCD spectroscopy). In this methodology

$$\mathbf{M}_s = -2\langle \hat{\mathbf{S}} \rangle \mu_B/\hbar = (N_\uparrow - N_\downarrow)\mu_B, \tag{14}$$

where $N_\uparrow$ and $N_\downarrow$ are the numbers of d holes with the spin ↑ and the spin ↓.

On the Fig.3 we show the comparison between the theoretical spectrum, calculated in the one-electron approximation, with the experimental one, taken from [18]. The experimental spectrum was measured for the similar system of Ni clusters grown on the carbon nanotube layer. We have choose the experimental data as the most suitable for our system Ni clusters on the silicon surface because silicon and carbon atoms have the similar configuration of the electronic shell - [Ar]ns^2np^3, where n=2 and 3 for carbon and silicon, respectively. As one can see, there are three features on the experimental spectrum called A_1, A_2 and A_3, which represent the spin structure of the Ni clusters. This structure cant be repeated without taking into account of the multiplet effects. This fact is obvious from the theoretical spectrum calculated by the one-electron FEFF9 code. If compare the spin DOS $S(\epsilon)$ of Ni_7 cluster on a silicon surface with the post-edge region of the experimental Ni L_3-XANES spectrum for Ni catalyst nanoparticles (see Fig.4) one can notice that the energy position of the features of spin DOS $S(\epsilon)$ coincides with the post-edge features of the absorption spectrum of Ni clusters.

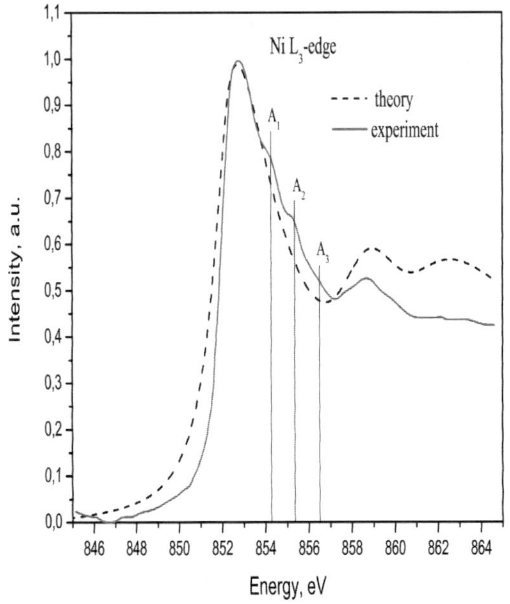

Figure 3. The theoretical one-electron and experimental NiL_3-edge XANES for Ni_7 cluster on a silicon surface.

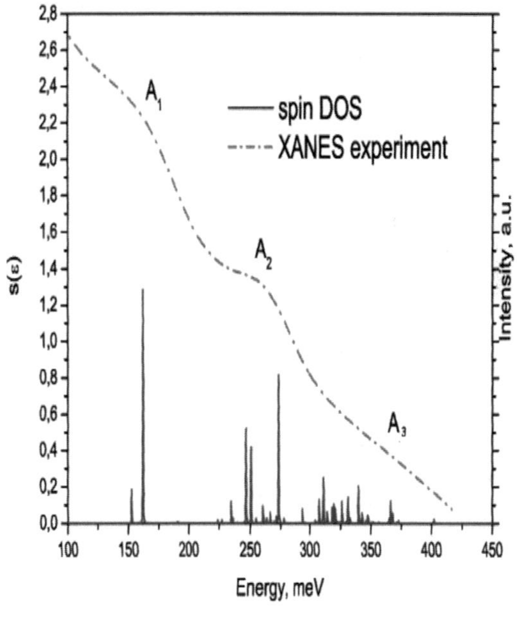

Figure 4. (Color online) The comparison of the calculated spin DOS $S(\epsilon)$ of the Ni_7 cluster on a silicon surface with the experimental Ni L_3-XANES spectra of carbon nanotube with Ni catalyst nanoparticle taken from [7]. The experimental spectrum was shifted on the value of the Ni $2p_{3/2}$ binding energy.

Under technological inputs, the cluster magnetism is now moving more and more towards surface science with the implications for the use of new theoretical and experimental techniques and with the development of new synthetic approaches. The cluster nanomagnets on a nonmagnet substrate have specific features that make them paradigmatic cases to the test models and with which we may build the novel quantum architectures. Carrying out a theoretical and experimental investigation of the quantum model of the solid-state spin qubit Ni on a nonmagnetic silicon surface for the quantum register–the quantum computer theory is developed. Within the present article, we used the numerical simulations and X-ray spectral methodology to study of a spin structure of the N-spin qubit systems. As a result, we have provided the detailed theoretical and experimental description for many important aspects of the spin structure of the systems relevant for quantum information processing and quantum computers. No doubt, the challenges we face in building a real silicon-based N-qubit with N≥2000 are significant, but our initial results offer hope that large-scale quantum computing may one day be realized.

Acknowledgments
The part of this research (for V.M.) has been supported by Russian Ministry of Science, Grant No. 213.01-11/2014-6.

4. References

[1] Nielsen M A and Chuang I L 2000 *Quantum computation and quantum information* (Cambridge University Press, Cambridge, New York)
[2] Loss D and DiVincenzo D P 1998 *Phys. Rev.* A **57** 120
[3] Kane B E 1998 *Nature* **393** 133
[4] Wrachtrup J and Jelezko F 2006 *J. Phys.: Cond. Matter* **18** 807
[5] Troiania F and Affronte M 2011 *Chem. Soc. Rev.* **40** 3119
[6] Mazalova V L, Guda A A, Smolentsev N Yu, Polozhentsev O E, Alperovich I G, Soldatov A V,Lau S P and Ji X H 2012 *In:Piezoelectrics and Related Materials: Investigations and Applications* ed I A Parinov (New York: Nova Science Pub Inc.) pp 51-77
[7] Mazalova V, Soldatov A, Johnston R, Yakovlev A, Moller T and Adam S 2009 *J. Phys. Chem.* C **113** 9086
[8] Rehr J J, Kas J J, Vila F D, Prange M P and Jorissen K 2010 *J. Phys. Chem.* **12** 5503
[9] Rehr J J, Kas J J, Prange M P, Sorini A P, Takimoto Y and Vila F D 2009 *C. R. Phys.* **10** 548
[10] te Velde G, Bickelhaupt F M, van Gisbergen S J A, Fonseca Guerra C, Baerends E J, Snijders J G and Ziegler T 2001 *J. Comp. Chem.* **22** 931
[11] Fonseca Guerra C, Snijders J G, te Velde G and Baerends E J 1998 *Theor. Chem. Accoun.* **99** 391
[12] Shor P 1994 *Proc. of the 35th Ann. Symp. on Foundations of Comduter Science* pp 116-123
[13] Loss D and DiVincenzo D P 1998 *Phys. Rev.* A **57** 120
[14] Amico L, Fazio R, Osterloch A and Vedral V 2008 *Rev. Mod. Phys.* **80** 517
[15] Becke A. 1988 *Phys. Rev.* A **38** 3098
[16] Lee C, Yang W and Parr R G 1988 *Phys. Rev.* B **37** 785
[17] Borras-Almenar J J, Clemente-Juan J M, Coronado E and Tsukerblat B S 1999 *Inorg. Chem.* **38** 6081
[18] Gao J, Zhong J, Bai L, Liu J, Zhao G and Run X 2013 *Sci. Rep.* **4** 3606

Journal of Physics: Conference Series **640** (2015) 012026 doi:10.1088/1742-6596/640/1/012026

Evaluation of the Efficiency of Clathrate Hydrates in Storing Energy Gases

N I Papadimitriou[1,3], I N Tsimpanogiannis[1,3], I G Economou[2] and A K Stubos[1]

[1] Environmental Research Laboratory, NCSR "Demokritos", 15310 Aghia Paraskevi Attikis, Greece
[2] Chemical Engineering Program, Texas A&M University at Qatar, Education City, PO Box 23847, Doha, Qatar

E-mail: nikpap@ipta.demokritos.gr (N I Papadimitriou); tsimpano@usc.edu (I N Tsimpanogiannis)

Abstract. We use Grand Canonical Monte Carlo simulations to examine the efficiency of hydrates in storing hydrogen. In this way, the hydrate lattice is treated as a solid substrate where gas adsorption takes place. Special focus is given to the study of multiple occupancy, namely the presence of more than one gas molecules in the same hydrate cavity. This phenomenon occurs in the case of hydrogen hydrates and may drastically increase their storage capacity. A parametric analysis is also carried out to quantify the correlation between the change in the lattice constant and the storage capacity since small changes in the size of the crystal unit cell may induce significant changes in the storage capacity especially in cases were multiple occupancy occurs.

1. Introduction

Clathrate hydrates are crystalline materials made of water molecules that are stable at low temperatures or high pressures. Their crystal lattice consists of cavities where gas molecules can be entrapped. The guest molecules are essential for the stability of the hydrate structures [1, 2]. As such, they have attracted extensive interest as potential materials for the storage of energy and environmental-related gases (e.g., hydrogen [3–6], methane, carbon dioxide). Storage and transportation of energy gases are of primary interest to the current study. Additional important applications of hydrates include the flow assurance in the oil and gas industry [7], gas mixture separation applications [8, 9], and the use of methane stored in geologic formation in the form of hydrates as a possible future energy source [10].

The most common hydrate structures are sI, sII, and sH. In the current study we use Monte Carlo simulations to examine the efficiency of the hydrate structure sII in storing hydrogen. Special focus is given to the study of multiple occupancy, i.e., the presence of more than one gas molecules in the same hydrate cavity. The existence of multiple occupancy was reported initially based on the experiments of Kuhs et al. [11]. Subsequently, a number of experimental [12, 13] and computational studies [14–17] confirmed the initial experimental observations. Such behaviour occurs also in the case of hydrogen hydrates and may drastically increase their storage capacity.

[3] To whom correspondence should be addressed.

A parametric analysis is carried out to quantify the correlation between the change in the lattice constant and the storage capacity since small changes in the size of the crystal unit cell may induce large changes in the storage capacity, especially in cases where multiple occupancy occurs.

Monte Carlo studies have been used extensively in order to estimate the storage capacity of different hydrate structures and guest gases [14–22]. Such studies used different water models (e.g., SPC/E [23], TIP4P [24]). Of interest to the current study is the case of TIP4P/Ice water model [25]. The particular model has been identified recently [26] as the most appropriate to predict accurately the three-phase equilibrium conditions for methane hydrate using molecular dynamics. The study identified that in order to obtain accurate predictions two requirements are needed. First, the solubility of the methane gas has to be accurately described and second, the water model should predict an accurate melting point for ice. These two requirements are met by TIP4P/Ice and therefore should be used in methane hydrate studies. It is the purpose of the current study to examine the effect of the particular water model on the storage capacity of hydrogen hydrates with respect to the size of the hydrate unit cell (expressed through the lattice constant).

2. Methodolgy

Monte Carlo simulations are performed in the Grand Canonical ensemble (GCMC) with the MCCCS Towhee code [27]. In this way, the hydrate lattice is treated as a solid substrate where gas adsorption takes place. Hydrate structure sII is examined over a wide range of temperatures (200 – 300 K) and pressures (1 – 500 MPa).

Both the lattice constant, and the space group symmetry of the examined hydrate structure have been determined from XRD measurements [28]. Oxygen atoms are placed exactly at the positions defined by the space group symmetry. For the selection of proton configuration the "ice-rules" (Bernal–Fowler [29]) have been obeyed and the proton configuration with the minimum dipole moment was adopted. The values of the lattice constants are kept fixed for the entire pressure and temperature range that was considered. In this work 3-D periodic boundary conditions are applied. Water molecules are kept at fixed positions throughout each simulation. Figure 1 shows a schematic of the sII hydrate structure.

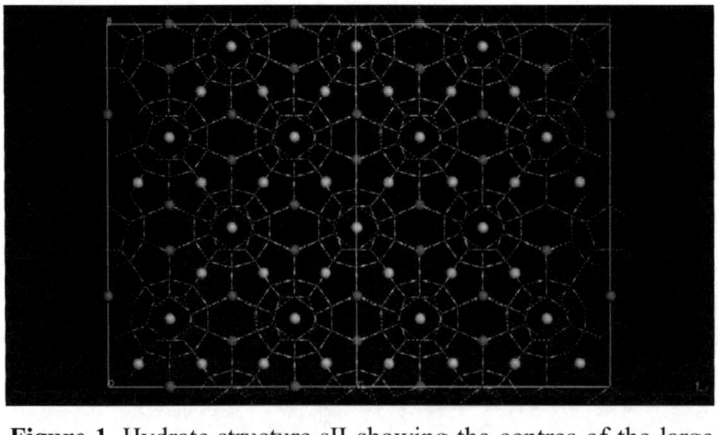

Figure 1. Hydrate structure sII showing the centres of the large and small cavities.

Water molecules in the current study are simulated by the TIP4P/Ice model [25]. The TIP4P/Ice is a rigid 4-site model in which a LJ sphere is fixed on the oxygen site. The electrostatic contributions are implemented by positive partial charges located on each hydrogen atom and a negative partial charge fixed on an "M-site", located on the bisector of the H–O–H angle at 0.1577 Å from the oxygen atom. The O–H bond length is 0.9572 Å and the H–O–H angle is 104.52°. All parameters of the TIP4P/Ice model are shown in Table 1.

H_2 molecules in the current study are simulated by the Silvera–Goldman [30] model. In this approach the length of the H–H bond in the H_2 molecule is taken as 0.7414 Å. H_2–H_2 interactions are described with an LJ potential based on the Silvera–Goldman isotropic pair potential for gas H_2. The simplest way to derive the LJ parameters from the Silvera–Goldman potential is to keep the same well depth ε and the zero-potential distance σ [31]. This set of parameters, shown in Table 1, is among the most common for simulations of hydrogen hydrates [14, 15].

Molecule (model used)	atom	σ (Å)	ε (kJ/mol)	charge (e)	d_{OH} (Å)
H_2O (TIP4P/Ice)	O	3.1668	0.8822	0.0000	d_{OH} = 0.9572 Å
	H	0.0000	0.0000	+0.5897	d_{OM} = 0.1577 Å
	Point M	0.0000	0.0000	-1.1794	θ_{HOH} = 104,52°
H_2 (Silvera-Goldman)	H	0.000	0.0000	+0.4932	d_{HH} = 0.7414 Å
	Center of mass	3.038	0.2852	-0.9864	

Interaction parameters between different types of atoms are calculated using the Lorentz–Berthelot combining rules [32] and there is a cutoff distance of 15 Å for any type of interaction. The Ewald summation [32] has been used for the long-range electrostatic interactions.

For the current study we used the GCMC approach developed by Metropolis et al. [33]. An extensive series of simulations were performed using the MCCCS Towhee open-source software package [27]. For each temperature and pressure an MC simulation is performed in the Grand-Canonical ensemble, i.e. under constant volume, temperature and gas fugacity (which is an alternative for chemical potential). The hydrogen fugacity at the specific conditions is calculated with a component-specific equation of state which gives very accurate calculations [34]. Each run consists of an equilibration stage of $2 \cdot 10^6$ MC steps that is followed by a production stage of $5 \cdot 10^6$ MC steps. Four MC moves are applied with equal probabilities: insertion 25%, destruction 25%, rotation 25%, and translation 25%. One out every 50 steps of the production stage (10^5 steps in total) are recorded and taken into account for the statistics. To determine the occupancy, for each type of cavity separately, each hydrogen molecule is assigned to the closest center of cavity. In this way, the average occupancy of every cavity during the simulation is calculated.

3. Results and Discussion

In order to examine the effect of the hydrate lattice constant on the calculation of the storage capacity of sII hydrogen hydrates we vary the lattice size by 0.6 %, and 1.5 % from the reference value which has been reported [3] to be equal to 17.047 Å. We consider only the case of increasing the lattice size, since if we decrease the lattice size will result in reducing the storage capacity. Please recall that we are motivated for this study by our interest to use hydrates as gas-storing materials. Therefore, cases that result *a priori* in reduced storage capacity are of no practical interest and are not considered in the current study.

Figure 2 shows the hydrogen storage capacity (wt. %) as a function of pressure for three different temperatures, 200 K, 250 K, and 300 K, and the three different lattice constants considered. In general, the following general observations can be made:

- Lower temperatures result in higher storage capacities.
- Increasing the pressure results in increasing the storage capacity.
- Increasing the lattice constant results in increasing the storage capacity. This result is primarily due to the increase of the storage capacity of the large cavities. The increase of the storage capacity of the small cavities is very minimal.

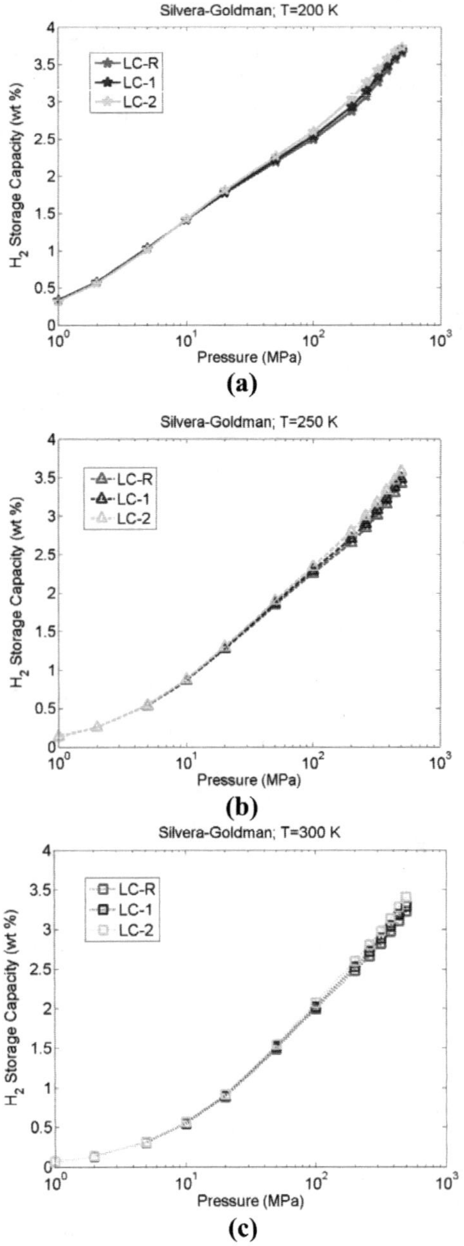

Figure 2. Hydrogen storage capacity (*wt. %*) as a function of pressure for three different temperatures: **(a)** 200 K, **(b)** 250 K, and **(c)** 300 K. Notation: LC denotes Lattice Constant; R denotes the reference case, LC–1 denotes the case with deviation + 0.6 % in the size of lattice and LC–2 deviation + 1.5 %.

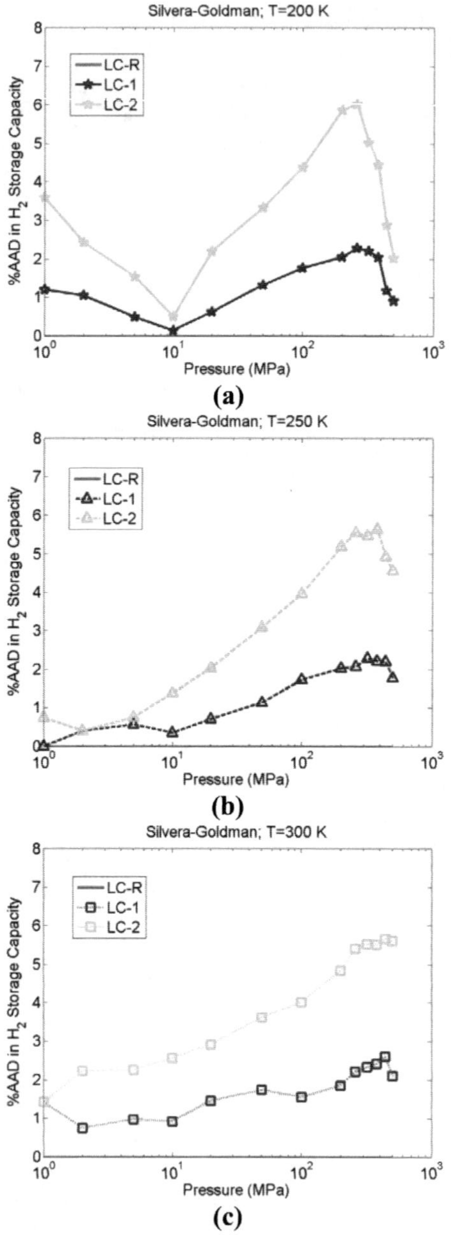

Figure 3. Percentage Average Absolute Deviation (*% AAD*) of the hydrogen storage capacity as a function of pressure for three different temperatures: **(a)** 200 K, **(b)** 250 K, and **(c)** 300 K. Notation is the same as in figure 2.

Journal of Physics: Conference Series **640** (2015) 012026 doi:10.1088/1742-6596/640/1/012026

In close agreement with the literature and regardless of which water model is used, the small cavity of hydrates cannot store more than one hydrogen molecule for most cases. From the current simulations we observe that only a very limited amount of double occupancy occurs at the higher pressures, when we use increased lattice constants. However, the large cavities of the sII structure can store up to 4 molecules. Higher occupancies have been reported [15] for the large cavities of structure sH hydrates.

In order to quantify the effect of using larger lattice constants we plot in Figure 3 the percentage Average Absolute Deviation *(% AAD)* of the hydrogen storage capacity from the reference case, as a function of pressure for the three different temperatures. We observe that when we increase the lattice constant by 0.6 % the storage capacity increases by 1–2 %, with respect to the reference case. Lower values correspond to lower pressures. On the other hand, when we increase the lattice constant by 1.5 % the storage capacity increases by 1–6 %, with respect to the reference case.

4. Conclusions

We have used Grand Canonical Monte Carlo simulations to examine the efficiency of hydrate structure sII in storing hydrogen. Special focus was given to the study of multiple occupancy, namely the presence of more than one gas molecules in the same hydrate cavity. In particular, a parametric analysis was carried out to quantify the correlation between the change in the lattice constant and the storage capacity. It was found that the major contribution to the increase of the gas-storage capacity results from the large cavities, while the contribution of the small cavities is very limited.

[1] Sloan E D and Koh C A 2008 *Clathrate Hydrates of Natural Gases*, 3rd edition ed., (Taylor & Francis, CRC Press)

[2] Sloan E D 2003 Fundamental principles and applications of natural gas hydrates *Nature* **426** 353–359

[3] Mao W L and Mao H-K 2004 Hydrogen storage in molecular compounds *Proc. Natl. Acad. Sci. USA* **101** 708–710

[4] Schuth F 2005 Hydrogen and hydrates *Nature* **434** 712–713

[5] Struzhkin V V, Militzer B, Mao W L, Mao H K and Hemley R J 2007 Hydrogen storage in molecular clathrates *Chem. Rev.* **107** 4133–4151

[6] Mao W L, Mao H K, Goncharov A F, Struzhkin V V, Guo Q, Hu J, Shu J, Hemley R J, Somayazulu M and Zhao Y 2002 *Science* **297** 2247–2249

[7] Kelland M A 2006 History of the development of low dosage hydrate inhibitors *Energy & Fuels* **20** 825–847

[8] Dabrowski N, Windmeir C and Oellrich L R 2009 Purification of natural gases with high CO_2 content using gas hydrates *Energy & Fuels* **23** 5603–5610

[9] Zhong D-L, Daraboina N and Englezos P 2013 Recovery of CH_4 from coal mine model gas mixture (CH_4/N_2) by hydrate crystallization in the presence of cyclopentane *Fuel*, **106** 425 430

[10] Boswell R and Collett T S 2014 Current perspectives on gas hydrate resources. *Energy Environmental Sci.* **4** 1206–1215

[11] Kuhs W F, Chazallon B, Radaelli P G and Pauer F 1997 Cage occupancy and compressibility of deuterated N_2–clathrate hydrate by neutron diffraction *J. Incl. Phenom. Molec. Recognit. Chem.* **29** 65–77

[12] Chazallon B and Kuhs W F 2002 In situ structural properties of N_2–O_2 and air-clathrates by neutron diffraction *J. Chem. Phys.* **117** 308–320

[13] Sasaki S, Hori S, Kume T and Smimizu H 2003 *J. Chem. Phys.* **118** 7892–7897

[14] Papadimitriou N I, Tsimpanogiannis I N, Papaioannou A Th and Stubos A K 2008 Evaluation of the hydrogen-storage capacity of pure H_2 and binary H_2–THF hydrates with Monte Carlo *J. Phys. Chem. C* **112** 10294–10302
[15] Papadimitriou N I, Tsimpanogiannis I N, Peters C J, Papaioannou A Th and Stubos A K 2008 Hydrogen storage in sH hydrates: A Monte Carlo study *J. Phys. Chem. B* **112** 14206–14211
[16] Katsumasa K, Koga K and Tanaka H 2007 On the thermodynamic stability of hydrogen clathrate hydrate *J. Chem. Phys.* **127** 044509
[17] Tanaka H, Nakatsuka T and Koga K 2004 On the thermodynamic stability of clathrate hydrates IV. Double occupancy of cages *J. Chem. Phys.* **121** 5488–5493
[18] Papadimitriou N I, Tsimpanogiannis I N, Economou I G and Stubos A K 2014 Influence of Combining Rules on the Cavity Occupancy of Clathrate Hydrates by Monte Carlo Simulations *Mol. Phys.* **112** 2258–2274
[19] Wierzchowski S J and Monson P A 2006 Calculating the phase behavior of gas-hydrate forming systems from molecular models *Ind. Eng. Chem. Res.* **45** 424–431
[20] Sizov V V and Piotrovskaya E M 2007 Computer simulation of methane hydrate cage occupancy *J. Phys. Chem. B* **111** 2886–2890
[21] Jensen L, Thomsen K, von Solms K, Wierzchowski S J, Walsh M R, Koh C A, Sloan E D, Wu D T and Sum A K 2010 *J. Phys. Chem. B* **114** 5775–5782
[22] Lasich M, Mohammadi A H, Bolton K, Vrabec J and Ramjugernath D 2014 *Fluid Phase Equilib.* **369** 47–54
[23] Berendsen H J C, Grigera J R and Straatsma T P 1987 The missing term in effective pair potentials *J. Phys. Chem.* **91** 6269–6271
[24] Jorgensen W L, Chandrasekhar J, Madura J D, Impey R W, Klein M 1983 Comparison of Simple Potential Functions for Simulating Liquid Water *J. Chem. Phys.* **79** 926–935
[25] Abascal J L F, Sanz E, Fernandez R G and Vega C 2005 A Potential Model for the Study of Ices and Amorphous Water: TIP4P/Ice *J. Chem. Phys.* **122** 234511
[26] Conde M M and Vega C 2010 Determining the three-phase coexistence line in methane hydrates using computer simulations *J. Chem. Phys.* **133** 064507
[27] Martin M. http://towhee.sourceforge.net/
[28] Yousuf M, Qadri S B, Knies D L, Grabowski K S, Coffin R B and Pohlman J W 2004 *Appl. Phys. A* **78** 925–939.
[29] Bernal J D and Fowler R H 1933 A theory of water and ionic solution, with particular reference to hydrogen and hydroxyl ions *J. Chem. Phys.* **1** 515–548
[30] Silvera I F and Goldman V V 1978 The isotropic intermolecular potential for H_2 and D_2 in the solid and gas phase *J. Chem. Phys.* **69** 4209–4213
[31] Alavi S, Ripmeester J A and Klug D D 2005 *J. Chem. Phys.* **123** 024507
[32] Allen M P and Tildesley D J 1987 *Computer Simulation of Liquids* (Oxford University Press: New York, USA)
[33] Metropolis N, Rosenbluth A W, Rosenbluth M N, Teller A H and Teller E 1953 *J. Chem. Phys.* **21** 1087–1092
[34] Leachman J W, Jacobsen R T, Penoncello S G, Lemmon E W 2009 *J. Phys. Chem. Ref. Data* **38** 721–748

Acknowledgments
This publication was made possible thanks to an NPRP award [NPRP 6–1547–2–632] from the Qatar National Research Fund (a member of The Qatar Foundation). N. I. Papadimitriou wishes to acknowledge the Greek State Scholarships Foundation for the financial support under the IKY-Siemens program. All computations were performed at the EREL (NCSR "Demokritos") computational infrastructure. The statements made herein are solely the responsibility of the authors.

Multiscale simulation of thermal disruption in resistance switching process in amorphous carbon

A M Popov[1] , **G N Shumkin**[2], **N G Nikishin**[1]

[1] Moscow State University, Faculty of Computational Mathematics and Cybernetics, Moscow, Russia

[2] IBM Systems and Technology Group, IBM Science and Technology Center, Moscow, Russia

E-mail: popov@cs.msu.su

Abstract. The switching of material atomic structure and electric conductivity is used in novel technologies of making memory on the base of phase change. The possibility of making memory on the base of amorphous carbon is shown in experiment [1]. Present work is directed to simulation of experimentally observed effects. Ab initio quantum calculations were used for simulation of atomic structure changes in amorphous carbon [2]. These simulations showed that the resistance change is connected with thermally induced effects. The temperature was supposed to be the function of time. In present paper we propose a new multiscale, self-consistent model which combines three levels of simulation scales and takes into account the space and time dependencies of the temperature. On the first level of quantum molecular dynamic we provide the calculations of phase change in atomic structure with space and time dependence of the temperature. Nose–Hover thermostats are used for MD simulations to reproduce space dependency of the temperature. It is shown that atomic structure is localized near graphitic layers in conducting dot. Structure parameter is used then on the next levels of the modeling. Modified Ehrenfest Molecular Dynamics is used on the second level. Switching evolution of electronic subsystem is obtained. In macroscopic scale level the heat conductivity equation for continuous media is used for calculation space-time dependence of the temperature. Joule heat source depends on structure parameter and electric conductivity profiles obtained on previous levels of modeling. Iterative procedure is self-consistently repeated combining three levels of simulation. Space localization of Joule heat source leads to the thermal disruption. Obtained results allow us to explain S-form of the Volt-Ampere characteristic observed in experiment. Simulations were performed on IBM Blue Gene/P supercomputer at Moscow State University.

1. Introduction

The switching of material atomic structure and electric conductivity are used in novel technologies of making memory on the base of phase change. The possibility of making memory on the base of amorphous carbon is shown in experiment [1]. Present work is directed to simulation of experimentally observed effects. Ab initio quantum calculations were used for simulation of atomic structure changes in amorphous carbon [2]. These simulations showed that the resistance change is connected with thermally induced effects. Each MD simulation was done with constant in space temperature. Also periodic boundary conditions were supposed. Temperature was maintained using Nose–Hoover thermostat. In this paper we propose a new multiscale, self-consistent model which combines three levels of simulation scales. The space and time dependencies of temperature are calculated at the macroscopic level. Transport coefficients are obtained in the molecular dynamics simulations. We show that atomic structure is localized near graphitic layers inside high temperature conducting dot with low temperature boundaries. Histogram technique for analyzing ions density is used for formulating the structure parameter and free energy profile. Structure parameter is used then in the next levels of the modeling. The heat conductivity equation for continuous media is used to calculate space-time dependence of the temperature. Joule heat source that depends on structure parameter and electric conductivity is proportional to the temperature. The space profiles of transport coefficients are calculated on previous levels of the modeling. Three dimensional heat conductivity equation is solved on the finite difference grid, using implicit finite difference scheme. Iterative procedure is self-consistently repeated combining three levels of simulation. Distributed temperature is maintained in MD calculations using Nose–Hoover thermostats for different groups of atoms. We show that the increasing of the space localization of the Joule heat source leads to the thermal disruption if perpendicular heat conductivity is small. Obtained results allow us to explain S-form of the Volt-Ampere characteristic observed in experiment. Simulations were performed on IBM Blue Gene/P supercomputer at Moscow State University.

2. Ab initio quantum simulation of atomic structure changes in amorphous carbon

We use CPMD ab initio code [3] which is based on density functional theory (DFT) and Car–Parrinello (CP) method. Within DFT and the Born–Oppenheimer approximation the movement of massive nuclei is described with a classical approximation. Forces acting on nuclei due to the interactions with the electrons are calculated solving the Kohn–Sham equations for ground state. Norm-conserving pseudopotentials are used, and Kohn–Sham orbitals are expanded in plane waves basis set up to a kinetic energy cutoff of 70 Ry. A fictitious electronic mass of 500 a.u., and a time step of 0.125 fs are used. We consider initial configuration of sp^2 sites within the sp^3 matrix. The density of system is $n = 2.34\frac{g}{cm^3}$ in the cell.

In [2] MD calculations are conducted with different temperatures. We found two different final types of atomic configurations in different MD runs. First type configurations exists at temperatures below $T = 2000$ K and has complicated amorphous structure which includes mixture of sp^2 atoms and sp^3 atoms. In all these cases the number of sp^2 atoms is increasing and the number of sp^3 is decreasing in comparison with initial state. In the work [2] MD calculations are conducted with different temperatures, but the temperature was supposed to be uniformly distributed in space. It was shown that with the temperature of order $T = 3000$ K effect of graphitic structure forming is obtained. Atomic configuration evolution at $T = 3000$ K is shown in fig. 1 obtained (a) at the beginning of heating process and (b) at the end of process. Note that all volume of unit cell is graphitic up to the boundaries.

In this paper, we suggest to define parameters, which characterize the appearance of structure from amorphous atomic state. We analized density of ions space distribution. Let us look at density of atoms evolution in the MD process and define x-direction as perpendicular direction to graphitic surfaces, and (y, z) directions are parallel to graphitic surfaces. Atomic density distributions are approximated by using histogram technique. Atomic densities profiles for atomic configuration at $T = 3000$ K obtained at the end of heating process are shown in fig. 2. Density distribution is shown in perpendicular direction. Distributions are approximated by Gaussians with the height H and semi-width $S(t)$. So, we introduce space structure parameter $S(t)$, which is characterized by semi-width of ions density distribution. The structure is localized during the evolution in the neighborhood of graphitic layers and $S(t)$ become smaller. Time evolution of structure parameter $S(t)$ is shown in fig. 3 during the time period of MD process. Time evolution in this period we call structural phase change or phase change of the second kind.

Journal of Physics: Conference Series **640** (2015) 012027 doi:10.1088/1742-6596/640/1/012027

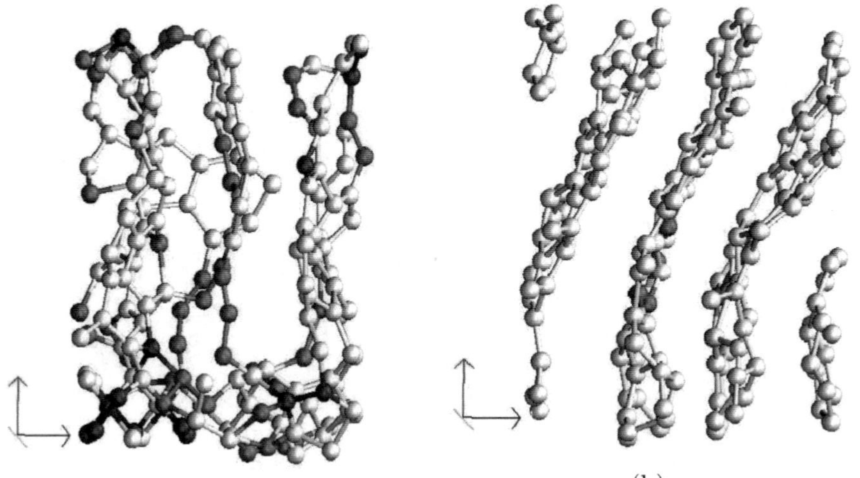

(a) (b)

Figure 1. Atomic configuration evolution at T = 3000 K, obtained (a) at the beginning of heating process and (b) at the end of the process.

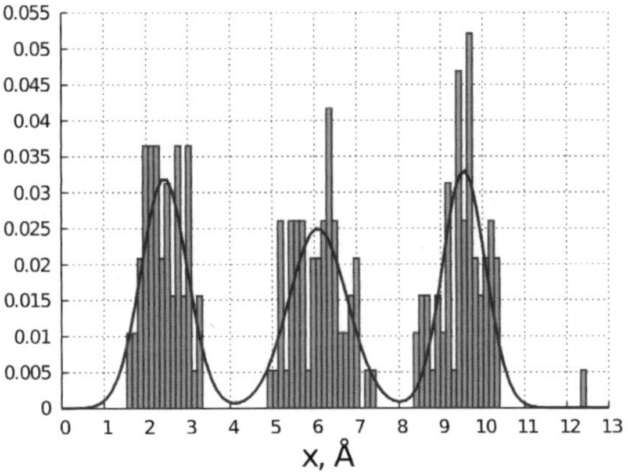

Figure 2. Atomic densities profiles at T = 3000 K at the end of process represented as histograms. Histograms are approximated by Gaussians.

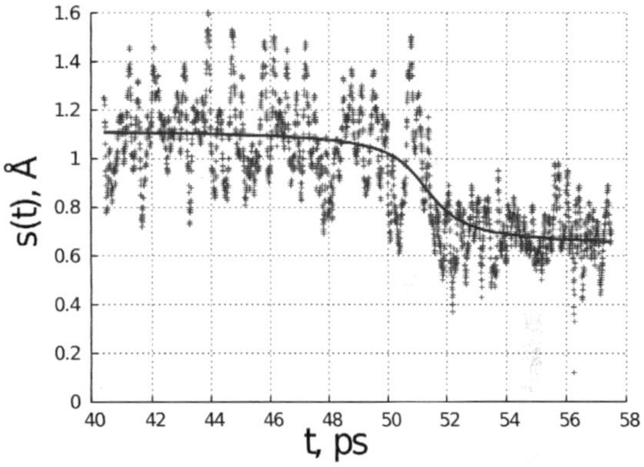

Figure 3. Time evolution of structure parameter $S(t)$ during phase change transition.

3. Electronic density evolution during atomic structure changes. Ehrenfest model

The thermal induced changes in atomic structure of a-C in the presence of an external electric field is analyzed through the usage of a multiscale model of conductivity switching.

As we can calculate electronic Kohn–Sham orbitals and we can build a reduced model on the base of Ehrenfest molecular dynamics and time dependent DFT [3]. The goal is to obtain the electronic density perturbation due to the external electric potential. We showed the appearance of total current and estimate the electric conductivity.

We expand the electronic wave functions using the full set of Kohn–Sham orbitals to get the solution of time-dependent Schroedinger equation.

$$\Psi(\mathbf{r},\mathbf{R},t) = \sum_m c_m(t)\psi_m(\mathbf{r},\mathbf{R}), \qquad \hat{H}\psi_m(\mathbf{r},\mathbf{R}) = E_m\psi_m(\mathbf{r},\mathbf{R}) \tag{1}$$

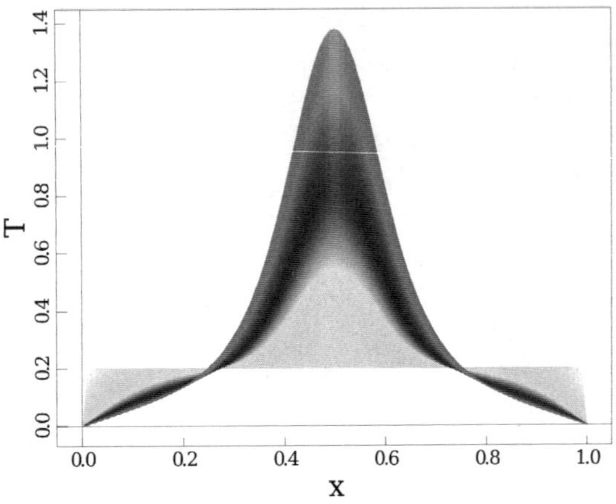

Figure 4. Temperature profiles in x-direction during time.

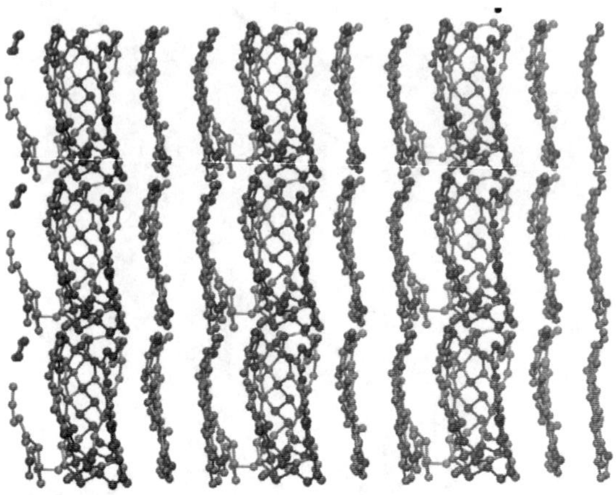

Figure 5. Structure of atoms corresponding to the localized spatial dependence, calculated in macroscopic scale.

where $\mathbf{r}$ and $\mathbf{R}$ are electronic and ionic positions. According to Ehrenfest molecular dynamics, the time dependence of the coefficients $c_m(t)$ represents the evolution of the system in time.

The main suggested approach for the computation using reduced Ehrenfest model is based on separation of scales. Long time scale CPMD calculations are used for formation of quasi-equilibrium structure of ionic positions at the given temperature. For each final equilibrium state we calculate HOMO and LUMO. Then, using excited orbitals and Ehrenfest electronic equations at the fast time scale, we determine switching to the next graphitic structure. For these positions of ions (structure parameter) we determine HOMO orbitals and excited orbitals. Process of switching is determined by overlapping integrals for HOMO and other excited orbitals.

Switching of HOMO and LUMO orbitals at fast time scale occurs around ground state — state of phase transition. This electronic switching can be define as the transition path through excited states to new graphitic atomic configuration. We introduce fictitious particle (structure parameter $\mathbf{S}(t)$) that corresponds to ions positions at the given temperature. Approximated motion equations for ions can be formulated as

$$\mu \frac{d^2\mathbf{S}}{dt^2} = -\frac{\partial U_F}{\partial \mathbf{S}} - \sum_m |c_m(t)|^2 \frac{\partial E_m}{\partial \mathbf{S}} - \sum_m \sum_{k \neq m} c_m(t) c_k(t) (E_m - E_k) \mathbf{d}_{mk}(\mathbf{S}) \tag{2}$$

The equations for electrons motion will include ions motion through structural parameter $\mathbf{S}(t)$:

$$i \frac{dc_m(t)}{dt} = c_m(t) E_m(\mathbf{S}) - i \sum_k c_k(t) \left(\frac{d\mathbf{S}}{dt} \cdot \mathbf{d}_{mk}(\mathbf{S}) \right) + \sum_k V_{mk}^{ext} c_k(t), \tag{3}$$

Connection between electrons and ions equations exists owing to overlapping integrals of orbitals d_{mk}. Oscillators connections are determined by overlapping integral V_{mk} corresponding to an external potential.

$$\mathbf{d}_{m \neq k}(\mathbf{R}(t)) = \int \psi_m \frac{d}{d\mathbf{S}} \psi_k d\mathbf{S}, V_{mk}^{ext}(s) = \langle \psi_m | V^{ext} | \psi_k \rangle . \tag{4}$$

Existence of the term corresponding to ions movement $S(t)$ makes the full system nonlinear.

We calculate the density of electric current due to switching process of atomic orbitals which are forced by external electric field. This was important for estimation of electric conductivity switching. Obtained data give us a possibility to calculate spatial model of joule source of heating, which we use on the next, macroscopic level of simulation.

4. Macroscopic transport equation

We determine spatial and time dependencies of the temperature using heat conductivity equation with the source as the Joule heat generated by **j**.

$$\rho C \frac{\partial T}{\partial t} = \nabla_\perp (K_\perp \nabla_\perp T) + \nabla_{||}(K_{||} \nabla_{||} T) + q_0 \cdot (\mathbf{j} \cdot \mathbf{E}) \quad (5)$$

where current density is determined by

$$\mathbf{j} = \sigma(\mathbf{E}, T, S(t))\mathbf{E} \quad (6)$$

Transport coefficients $\rho, C, K_\perp, K_{||}$ correspond to graphite-like structure. Temperature profiles in x-direction (perpendicular to current filament) during time are shown in fig. 4. Equation includes electric conductivity which is proportional to the temperature $T(x, y, z, t)$ and structure factor $S(t)$. Changes in the structure are taken from atomistic configuration and depend on temperature.

Heat conductivity equation is solved on the finite difference grid by implicit scheme. Founded temperature profiles are used again in MD calculations. Then, molecular dynamics of the system was recalculated with new temperature localized profile. New structure of ions is shown in fig. 5. Note that figure presents the periodic extension of unit cell. It is seen that for new distribution of temperature the boundaries are corresponded to the diamond-like sides but the central part of conducting dot are graphitized. Critical diameter of conducting dot and threshold voltage for excitation of thermal instability can be estimated analytically. These critical parameters are found close to experimental data.

5. Conclusions

Multiscale model of phase change on the base of amorphous carbon is suggested. Taking into account the spatial change dependence of temperature is fundamentally important for explanation of experiment. It is shown that localization of spatial dependence of electric conductivity leads to thermal instability, which supports the structure.

References

[1] Sebastian A, Pauza A, Rossel C, Shelby R M, Rodríguez A F, Pozidis H and Eleftheriou E 2011 *New J. Phys.* **13** 013020 URL `http://dx.doi.org/10.1088/1367-2630/13/1/013020`

[2] Shumkin G, Zipoli F, Popov A and Curioni A 2012 *Procedia Computer Science* **9** 641–650 URL `http://dx.doi.org/10.1016/j.procs.2012.04.069`

[3] Marx D and Hutter J 2009 *Ab Initio Molecular Dynamics* (Cambridge University Press) URL `http://dx.doi.org/10.1017/cbo9780511609633`

Journal of Physics: Conference Series **640** (2015) 012028 doi:10.1088/1742-6596/640/1/012028

Electronic Properties of Adsorption of Trimesic Acid Monomer on Graphene

Farzaneh Shayeganfar

Engineering Physics Department and Regroupement québécois sur les matériaux de pointe (RQMP), Polytechnique Montréal, Montréal, Québec H3C 3A7, Canada

E-mail: shayegan-far.farzaneh@polymtl.ca

Abstract. A large scale approach based on scanning tunneling microscopy STM simulation images and *ab initio* computation is used to report the adsorbed trimesic acids (TMA) monomer and monolayer behaviors on graphene. Our results unravel the interplay between the adsorption energy and band gap opening for different adsorption sites of TMA/graphene. It has been revealed that regarding to stability of different adsorption site and interaction with π electron containing surfaces (graphene), the high adsorption energy can induce maximum band gap in the system and open up possible uses of graphene in electronic device applications. A vital role on carboxyl functional groups is related to duality behaviors of combined acceptor and donor character with regard to hydrogen bonds which provides stable intermolecular self-assembly[1] and monolayers (for detailed explanation and analysis see reference [1]).

1. Introduction

A holistic understanding of the molecular self-assembly as one of the most ubiquitous driving forces in materials design in natural world,[2,3,4] engineering complex porous organic nanoarchitectures[5,6,7,8,9] is the focus of recent research interest and paramount importance. Nature molecular self-assembly such as spider silk,[10] DNA[11] and cellular membranes[12] demonstrate the potential for encoding information[2] and for producing biological function as well as permanent nanoporosity[13] can be exploited after subsequent functionaization for developing novel nanostructured materials dedicated to catalysis, gas storage, selective ion exchange, high density data storage[4] using only a small number of relatively simple molecular building blocks. Molecular self-assembly as key point which offers unique directions for the fabrication of two-dimensional organic nanoarchitectures.[5]

Moreover, carboxylic functional can significantly alter $\pi - \pi$ interactions between graphene surface and the adsorbed molecules. In the present study, we have focused on this functional group for the adsorption of trimesic acid on the high symmetry sites of graphene with using first-principals calculations based on the density functional theory (DFT). To understand the adsorption better, we compare the properties of TMA/graphene with the pristine graphene. The stability, electronic structure and adsorption profile of this molecule with different configurations were analyzed in detail. We found that the TMA/graphene can induce a gap on graphene. With the local analysis of STM simulation images and local density of states (charge accumulation) and charge transfer mechanism, we can clarify the origin of band gap opening in the system.

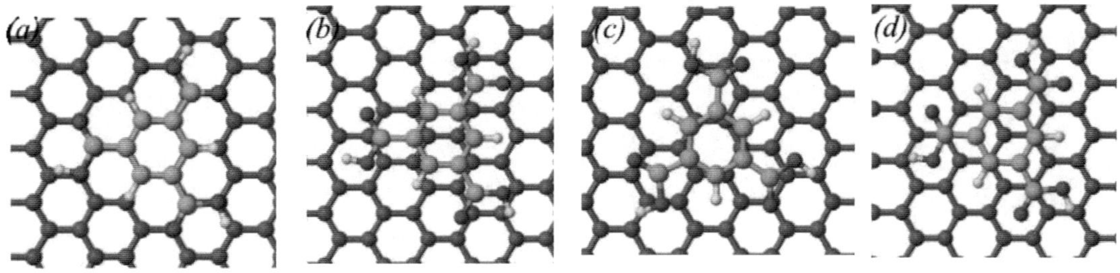

Figure 1. *High symmetry configurations, (a) top, (b) bridge, (c) cross and (d) center. For clarity, we chose pink for the carbon atoms, cyan for hydorogen atoms and red for oxygen atoms of TMA.*

2. Results and discussions

2.0.1. Stability and adsorption configuration Carboxylic groups (COOH) for intermolecular hydrogen bonding are very interesting functional units that has been used extensively to engineer crystals in 2D and 3D.[14] Rochefort and his co-workers studied effect of substituents and their interactions by comparing the adsorption of benzene, benzoic acid (1=BZA), isophthalic (2=IPA) and trimesic acid (3=TMA) on graphene.[14] However, in order to gain a more fundamental understanding and reliable predictions of the adsorption properties of the aromatic molecules like as trimesic acid (TMA) which can form more complex configuration on the graphene plane, in this feature article, we will begin and describe our theoretical analysis by comparing the adsorption of TMA on various high symmetry sites on graphene. Several positions of the TMA molecule relative to graphene sheet were considered near four symmetry sites on graphene, such as *top, bridge, cross* and *center* as can be seen in figure 1.

The first site is labeled as *top* which follows the atomic species sitting on top of the carbon atom in graphene. The other three sites are labeled as *bridge*, which means one $C-C$ bond of graphene is under the center of the TMA molecule and *cross*, which means all molecular atoms are located at top of middle of $C-C$ bond and the last one, *center* means that the one carbon atom is placed at the center of benzene ring in TMA molecule.

Our systematic study begins with the structural optimization results for adsorption of single TMA molecule on described high symmetry sites of graphene. A summary of the acquired results such as adsorption energy, equilibrium central core distance, band gap energy and net charge for different TMA adsorption sites is presented in Table 1. As shown in the Table, these different sites have significantly effect on adsorption process and electronic properties.

Further analysis of Table 1 can be performed to identify the following common trend features: (1) *top* and *cross* are the most stable adsorption sites for TMA molecule and *center* and *bridge* sites are the less stable configuration respectively, (2) the equilibrium vertical distance of aromatic ring on *top* and *cross* site is larger than *center* and *bridge*, (3) the minimum band gap is found at the *bridge* site which is 17 meV, whereas the maximum band gap is found at top site which is 48 meV, (4) *bridge* configuration shows maximal net charge transferring from graphene to TMA which amount is found -0.18e, so that charge transfer ordering for other site is arranged as center, *cross* and *top* configuration, also with increasing the adsorption energy, the charge transfer is decreased as well as band gap opening is increased.

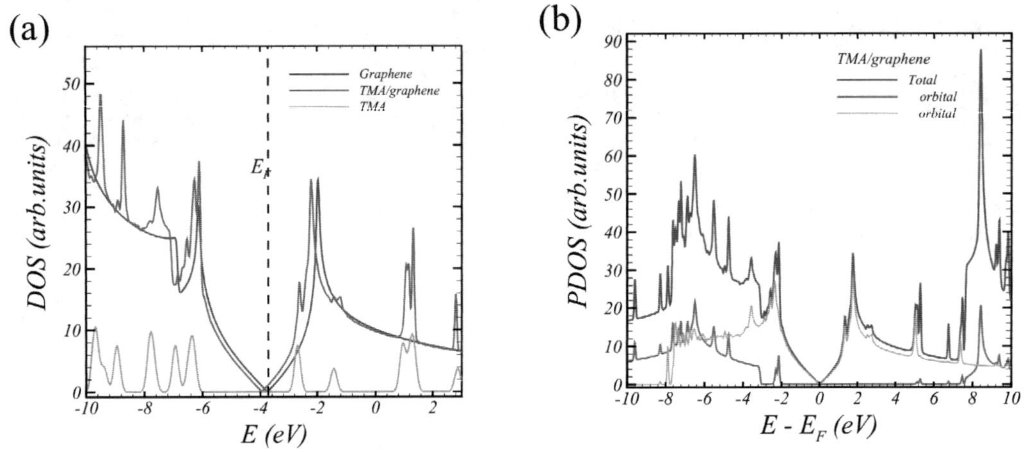

Figure 2. *(a): Density of state (DOS) of isolated graphene sheet, TMA adsorption on graphene (bridge site) and isolated trimesic acid (TMA). The Fermi-energy shifts around of* 0.1 *eV for bridge site, (b): projected density of state of total atoms for TMA/graphene.*

TABLE 1. Adsorption energy,band gaps energy, equilibrium distance and net charge transfer of isolated graphene with all sites of TMA/graphene, $\Delta Q = Q_{TMA/Ge} - Q_{Ge}$.

	Top	**Bridge**	**Cross**	**Center**
E_{ads} (eV)	-2.22	-1.25	-2.13	-1.83
ΔE_g (meV)	48	17	47	19
d (Å)	3.08	2.86	3.06	2.95
net charge ($\|e\|$)	-0.11	-0.18	-0.14	-0.17

Via carefully analyzing the band structure of all configurations of TMA adsorption on graphene which tabulated in Table 1, we infer that the TMA molecule can open a gap according to the adsorption site. The two more stable configuration i.e. *top* and *cross* adsorption site of TMA induces the larger band gap. We ascribe these behaviors to the effect of charge transfer and induced electric field between surface and adsorbed molecule, which will be discussed next. Other than that, analyzing of Mulliken charges for deep understanding of charge transfer mechanism, with defining of $\Delta Q = Q_{TMA/Ge} - Q_{Ge}$ as net charge, has been acquired and presented in Table 1. The net charge transfer from graphene to TMA changes for different configuration, the more stable adsorption sites show less charge transfer than the weaker adsorption sites. TMA molecule on *top* and *cross* receive $0.11e$ and $0.14e$, whereas charge transfer for *center* and *bridge* is $0.17e$ and $0.18e$, respectively. The measured distance of carbon atoms of graphene and oxygen atoms of $C = O$ of TMA (as mentioned previously) clearly illustrate the competing role of this distance on maximizing the charge transfer. For instance, minimum distance for *bridge* results maximum charge transfer.

2.0.2. Electronic properties Here we give results for the electronic structure of TMA/graphene. Figure 2 shows the total electronic density of states (DOS) for pristine graphene as a reference, isolated TMA and bridge adsorption site of TMA/graphene studied here. As clearly seen, the density of states (DOS) of TMA/graphene isn't be significantly changed, (DOS(TMA/graphene)≈DOS($graphene$) + $DOS(TMA)$) only the states of graphene are somehow perturbed due to adsorption of TMA, i.e. the adsorption of TMA on graphene

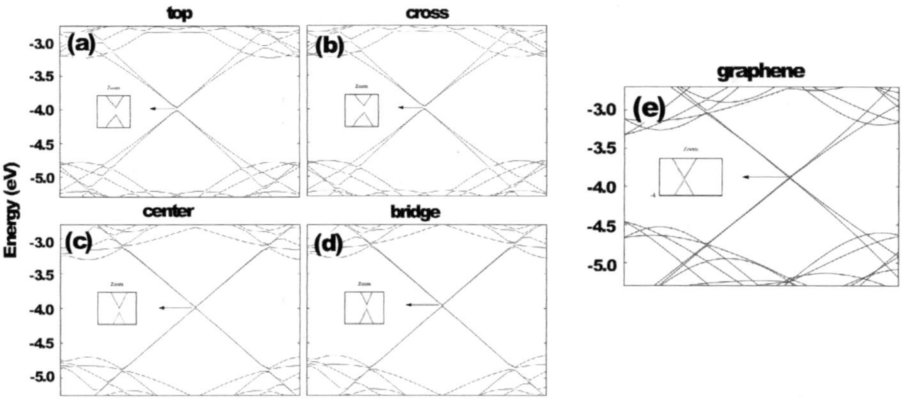

Figure 3. *Band structure of pristine graphene and TMA/graphene in isolated graphene and four different adsorption configurations. The insert provide more detailed band structures near the fermi level. The linear band property of π orbital is maintained.*

is relatively weak but due to the existence of three $COOH$ carboxylic acids level, (electron-withdrawing character of the $C = O$ in carboxyl functional groups) the electrons were moved from the valence bond of graphene to TMA molecule thereby this adsorption slightly shifts $0.1eV$ Fermi-energy into the valence band. It's worth mentioning that the shifting in Fermi energy level indicates p doping, confirms charge transferring between the molecule and graphene.

To investigate the $\pi - \pi$ interaction between the graphene and TMA, figure 2 (right panel) presents the variation of the projected density of states of π and σ orbitals of total atoms for TMA/graphene. The presented results in right panel of figure 2 suggest that π orbitals contribution dominated the DOS in the vicinity of Fermi energy.

In order to study the effect of molecule adsorption on the electronic properties of graphene, the band structure of four adsorption configurations of TMA on graphene as well as pure graphene are plotted and analyzed in figure 3. It's been clearly seen that the band structure of the TMA/graphene system is very similar to pure graphene. Thus the interaction between TMA and graphene is weak, near the Dirac point of two p_z bands, linear dispersion still exist. With providing zoom in the region near the Dirac point as the inset in figure 3, we discover the *top* and *cross* configuration induced the band gap larger than *center* and *bridge* site (Table 1, figure 6). This band gap opening can be attributed to electric field vertical to the graphene due to the molecule-graphene charge transfer, which breaks the inversion symmetry of graphene (see reference [1]).[1]

Figure 4 gives the influence of different adsorption sites of TMA/graphene on STM contrast simulation. In upper panel of figure 4, we plotted the topographic STM simulation images of different sites defined in figure 1. STM images shows a remarkable atomic resolution of the surface. An extended analysis of STM contrast along two horizontal black line directions A (neighbor of TMA) and B (underneath of TMA) is performed in figure 4(middle panel) which the position of molecule and graphene is reversed. The existed periodicity between maxima and minima along line A and B for isolated graphene have a little bit difference. However, the periodicity for graphene of neighbor and underneath of TMA molecule zone, i.e. 8-16 $\mathring{A}$ modifies to weak peaks and includes significant deformations over different sites adsorption. In middle panel of figure 4, 2D profile (maxima and minima peak) at line A displays $C - C$ bond corrugation related to neighbor of TMA molecule for all sites placed with larger amplitude from B line profile, moreover this separation for bridge and center sites is more pronounced

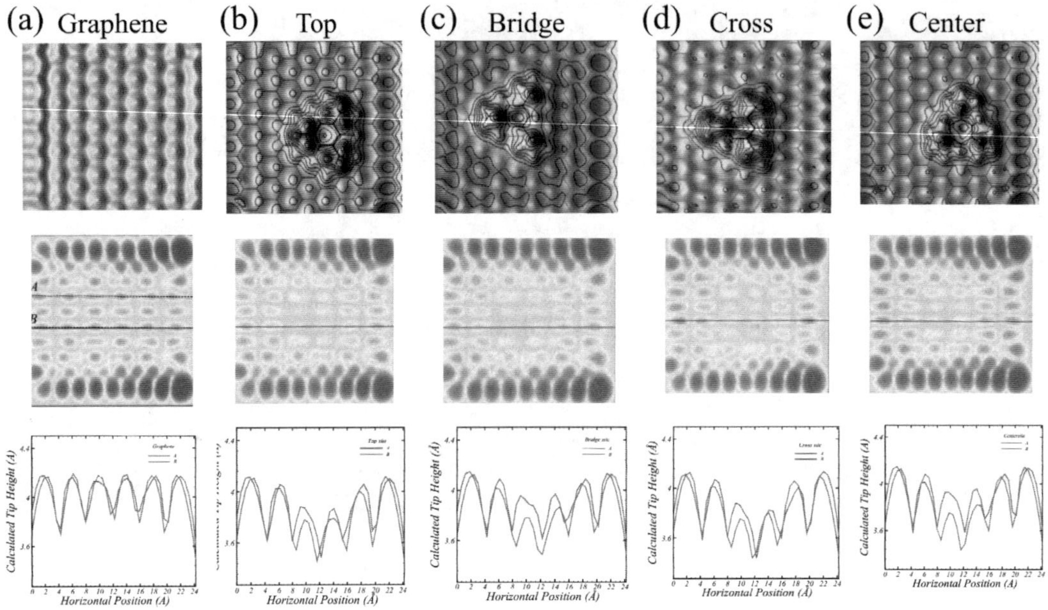

Figure 4. *Top panel: STM simulation images for different site adsorption of TMA molecule on graphene; left to right organized as top, bridge, cross and center sites as depicted in figure 1, middle panel: analysis of 2D simulation profiles; last panel: variation of tip distance for black horizontal A and B line, lower panel: particular of the middle panels with definition of differential tip hight,* $\Delta h = h_A - h_B$.

in comparison with two other sites. As a consequence, surface deformation in neighbor region of TMA for top and cross is more intense related to bridge and center sites. In other words, large perturbation term over two more stable, i.e. cross and top adsorption sites exists in TB-hamiltonian which breaks the symmetry of surface for neighboring effect and result in decreasing of degeneracy degrees or enhanced band gap opening over these two sites relative to instable sites.

3. Conclusions

The effects of adsorbing trimesic acid molecule on the electronic structure of graphene were systematically examined by first principle calculations. Different adsorption sites related to the position of central mass of TMA regard to graphene, affects electronic properties of system such as adsorption energy, band structure and charge transfer. We found that molecular adsorption TMA/graphene lead to band gap opening, while the magnitude of band gap depends on the adsorption site. This trend was acquired: the most stable configuration of molecule lead to the biggest band gap opening. We further analyzed the charge transfer mechanism for different adsorption sites and discovered that they play a significantly role in controlling the band gap. The intrinsic lateral charge transport characteristics has been affected by geometry of TMA adsorption on graphene.

4. Acknowledgments

This work was supported by the Natural Sciences and Engineering Research Council of Canada (NSERC) and the Ministère du Développement économique, de l'Innovation et de l'Exportation (MDEIE) through the PSR-SIIRI program. Computational resources were provided by Calcul Québec and Compute Canada.

5. References

[1] Shayeganfar F and Rochefort A 2014 *Langmuir* **30** 9707
[2] Macleod J M, Chaouch Z B, Perepichka D F and Rosei F 2013 *Langmuir* **29** 7318
[3] Philp D, Stoddart J F 1996 *Angew. Chem., Int. Ed. Engl.* **35** 1154
[4] Whitesides G M, Grzybowski B 2002 *Science* **295** 2418
[5] Silly F J 2012 *Phys. Chem.* C **116** 10029
[6] Liang H, Sun W, Jin X, Li H, Li J, Hu X, Teo B K, Wu K 2011 *Angew. Chem. Int. Ed.* **50** 7562
[7] Klyatskaya S, Klappenberger F, Schlickum U, Khne D, Marschall M, Reichert J, Decker R, Krenner W, Zoppellaro G, Brune H, Barth J V, Ruben M 2011 *Adv. Funct. Mater.* **21** 1230
[8] Yang Y, Wang C, 2009 *Chem. Soc. Rev.* **38** 2576
[9] Stepanow S, Lin N, Payer D, Schlickum U, Klappenberger F, Zoppellaro G, Ruben M, Brune H, Barth J V, Kern K 2007 *Angew. Chem. Int. Ed.* **46** 710
[10] Winkler, S.; Kaplan, D. L. 2000 *Rev. Mol. Biotechnol.* **74** 85
[11] Olby R, McCarty M, Maddox B, Paabo S, Chakravarti A, Little P, Bell J I, Kemp M, Ball P, Seeman N C 1953 *Nature* **171** 737
[12] Israelachvili J N, Mitchell D J, Ninham B W T 1976 *J. Chem. Soc., Faraday Trans.* **72** 1525
[13] Silien C, Risnen M T, Buck M 2010 *Small* **6** 391
[14] Rochefort A and Wuest J D 2009 *Longmuir* **25** 210

Journal of Physics: Conference Series **640** (2015) 012029 doi:10.1088/1742-6596/640/1/012029

Investigation of high pressure phase transition and electronic properties of Lutetium Nitride

Sanjay Kumar Singh[1] and U P Verma
School of Studies in Physics, Jiwaji University, Gwalior-474011,M.P, India

E-mail: sanjayfizix@gmail.com

Abstract. In the present manuscript we have investigated the structural, electronic and phase transition properties of the heaviest lanthanide lutetium nitride (LuN) compound using an *ab initio* calculations based on the density functional theory with Perdrew, Burke and Ernzerhof generalized gradient approximation (PBE-GGA) and Engel-Vosko(EV)-GGA as implemented in WIEN2k code. The basic ground state properties *viz*., lattice constants (a), bulk modulus (B_0) and its pressure derivative (B_0') and total energy (E_0) are calculated. The calculated values of lattice constant is 4.76 Å which is in good agreement with experimental value a= 4.76 Å and other theoretical value. The relative stabilities of LuN at high pressures in the NaCl (B1), CsCl (B2), zinc blende (B3) and body centred tetragonal (BCT) structures are analysed. At compressed volumes, this compound is found to favour the CsCl phase rather the body centred tetragonal phase and zinc blende as observed in other lanthanum pnictides, which has been predicted by the total energy minimization. Under compression LuN undergoes a transition from NaCl to CsCl type structure at around 250.81 GPa with a volume collapse of 3.75%. To see the effect of functional we have also computed the band structure in B1 and B2 structure. Obtained result on band structure shows that LuN are semimetal by GGA while depicts semiconducting behaviour by EV-GGA. It shows metallic nature in B2 phase.

1. Introduction

Lutetium (Lu) is the last and heaviest member of the lanthanide series with completely filled 4*f* sub-shells. Most of the lanthanide compounds crystallize in NaCl (B1) structure with space group *Fm-3m* (no. 225) under ambient conditions [1,2]. Lutetium mono nitride (LuN) crystallize in B1 structure. Most attractive applications of lanthanide nitrides (LnNs) are as fire resistive materials as well as protective materials in reactor. Lutetium nitride (LuN) is one of the interesting compounds of LnNs family of materials and attracted attention of researchers due to their unusual structural and electronic properties. It is important to study the pressure and temperature behaviour and physical properties of. lutetium mono nitride (LuN). Because of the great applications and interesting electronic and elastic properties of the LuN has been investigated in the recent past with great interest [3-6]. The majority of these compounds undergo first order structural phase transition from the NaCl to CsCl structure at high pressure. Niwa et al. [2] have successfully synthesized Ln nitrides (Ln=Ce, Pr, Gd, Lu) at high pressure and high temperature using the diamond anvil cell and laser heating system.

[1] Corresponding author

Larson et al. [3] have computed electronic structures of the LnNs in the B1 structure using the LSDA+U (local spin density approximation with Hubbard-U corrections) scheme within the density functional theory approach. Recently, Yang et al. [4] studied elastic properties and hardness of lanthanide nitrides using first principles. Painter et. al. [5] evaluated theoretical lattice constants and cohesive energy of LaN, GdN and LuN compound using the VASP package which are also good agreement with experiential value. Very recently, Yang et.al. [6] investigated the structural stability of LuN using first principles and predicted that it transform from B1 to B2 structure at 220 GPa.

2. Methodology of computations

The calculations were performed using an *ab-initio* full potential linear augmented plane wave plus local orbitals (FP-LAPW + lo) approach [7] within the framework of density functional theory (DFT) [8] as implanted in the WIEN2k package [9]. In this approach the GGA-PBE [10] and EV-GGA [11] is chosen for the exchange-correlation functional energy optimization for calculating the equilibrium properties. In this method, an orbital dependent potential is introduced for the chosen set of electron states which is present in 4*f* states of lanthanides. In this method the wave function charge density and potential are expanded by spherical harmonic function inside non overlapping spheres surrounding the atomic sites (muffin-tin (MT) spheres) and by a plane-wave (PW) basis set in the remaining space of the unit cell (interstitial region). The convergence parameter $R_{MT}K_{max}$, which controls the size of basis set in the calculations, was set to 7. The converged value of R_{MT} of Lu and N atoms are 1.51, 1.3 a.u in case of LuN. The valance wave functions inside MT spheres were expanded up to lmax=10, while the charge density was Fourier expanded up to Gmax=12 a.u^{-1}. The integrals over the first Brillouin zone (BZ) are performed in mesh of 10×10×10 for all phases in the irreducible BZ using the Monkhost-Pack special k-points approach [12]. Both the PW cut-off and the number of k-points were optimized to ensure total energy convergence. The self-consistent calculations are converged up to an accuracy of 10^{-4} Ry in the total energy of the system. The set of unit cell volume and total energy are fitted to the Murnaghan equation of state [13] to obtain equilibrium unit cell volume, optimized value of total energy, bulk modulus and pressure derivative of bulk modulus.

3. Results and discussions

3.1 Structural and phase transition properties

To investigate the ground state properties of LuN we have first optimized the value of R_{MT} and k-points. The converged value of R_{MT} are for Lu (1.51 a.u.) and N (1.3 a.u.) in LuN. The optimized value of k-points are 1000. The equilibrium structural properties *viz.* lattice parameter (a), bulk modulus (B_0) and its first order pressure derivative (B_0') and total energy (E_0) are obtained by fitting the sets of the total energy and corresponding unit cell volume in the Murnaghan's equation of state. The energy *vs.* volume curves for LuN in four different phases have been shown in Figure 1. From figure 1 it is clear that (B1) phase is most stable phase with minimum energy $E_{min.}$ = -29271.97 Ry for LuN corresponding to equilibrium volume V_0 = 182.57 a.u^3 for LuN. Calculated ground state lattice parameter (a), bulk modulus (B_0), its first order pressure derivative (B_0') obtained using GGA approximation are shown in Table 1 with other available data. From the Table 1 it is clear that the calculated equilibrium lattice constant a= 4.76 Å for LuN are very close to the experimental lattice parameters (a=4.766 for LuN [1] and other theoretical reported results [3-5].

The obtained values of B_0 is 161.10 for LuN. Our calculated lattice parameter, bulk modulus and its pressure derivatives for LuN are also in good agreement with the other theoretical and available experimental data [1-2]. From Figure 1, it is also evident that the optimization curve for B2 and BCT phase are very near to each other and optimization curve for B1 phase crosses these two phase curves at higher pressure points. This indicates that upon compression the B1 phase transforms into B2 phase. The pressure value for these transitions can be estimated by calculating the value of enthalpy (H) as a function of the applied pressure as

$$H = E + PV \quad (1)$$

Here E is the total energy of the system at pressure (P) and volume (V). The enthalpy versus pressure curve for LuN is given in Figure 2. From Figure 2, the transition pressure for LuN is found to be 250.81 GPa. The calculated transition pressure for LuN is higher than the earlier reported in literature [6]. In order to confirm these structural transitions in LuN we have also analyzed the relative volume as a function of pressure and calculated the volume collapse for B1 to B2 transitions as 3.75% for LuN.

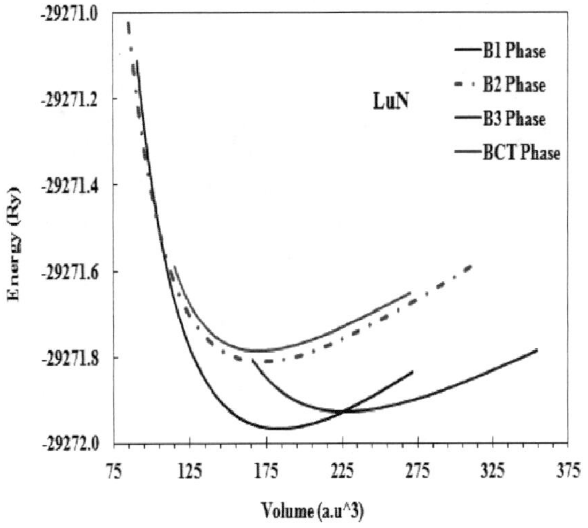

Figure 1. Energy versus volume curves of LuN.

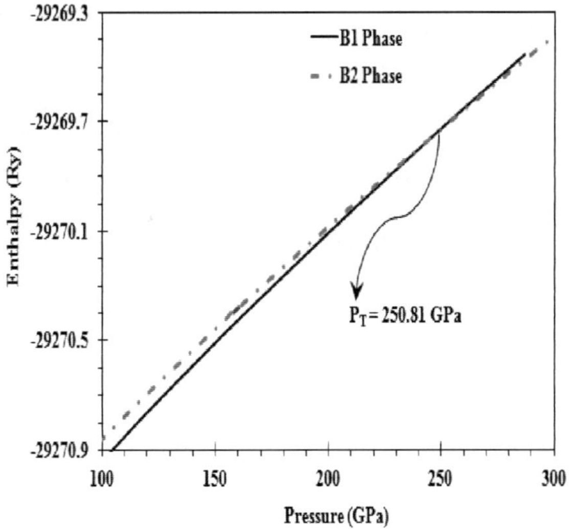

Figure 2. Enthalpy versus pressure curves for LuN.

TABLE 1. Lattice parameter (a in Å), bulk modulus (B_0 in GPa) and its pressure derivative (B_0'), phase transition pressure (P_T in GPa), the % volume collapse at P_T ($\Delta V(P_T)/V(0)$) for LuN.

Material	Phase	Space Group	a	B_0	B_0'	E_0	Reference
LuN	B1	*Fm-3m*	4.76	161.10	3.73	-29271.97	Present
			4.76	-	-	-	Expt.[1]
			4.74	-	-	-	Expt.[2]
			4.87	170	-	-	Others [3]
			4.83	183.0			Others [4]
			4.76				Others [5]
	B2	*Pm-3m*	2.94	142.68	4.02	-29271.81	Present
	B3	*F-43m*	5.12	120.52	4.32	-29271.93	Present

Material	Phase	Space Group	c/a	B_0	B_0'	E_0	Reference
LuN	BCT	*P4/mmm*	0.88	169.72	4.73	-29271.79	Present

3.2 Electronic properties of LuN

The GGA and EV-GGA approach has been used to compute the electronic structure in B1 phase at the equilibrium lattice constant which are presented in figure 3 and at high pressure B2 phase in figure 4 respectively. It can be seen from Figure 3 that there is no band gap at the Fermi level. Top of the valence band is matched to zero of the Fermi level (E_F). It is observed that top three bands in valence band region are from N *p* orbitals lying up to -3.28 eV while all fourteen completely occupied Lu 4*f* energy bands hybridized with N *p, s* bands. One of the empty band from conduction bands dips slightly at X point, as a result, LuN exhibit a semi metallic character by GGA approach while EV-GGA approach LuN shows Γ-X indirect energy band gap of 0.14 eV , Γ- Γ direct band gap is 3.4 eV which confirms the semiconducting behaviour at ambient conditions. This result is consistent with th other theoretical works [3].

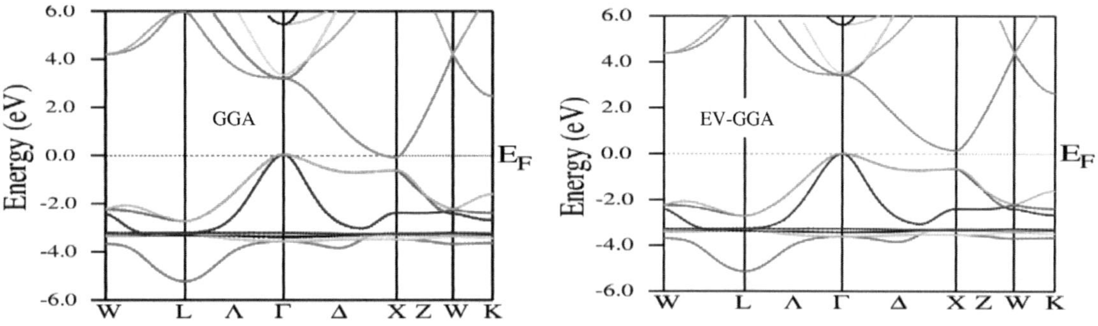

Figure 3. Band structure of LuN in B1 phase with GGA and EV-GGA.

To understand change in the nature of bands, high-pressure band structure of LuN in the B2 phase at their corresponding metallization pressures (or reduced volume) is given in Figure 4. In Lu, all the *f*-states are occupied and are present 3-4 eV below the Fermi energy and thus make up the valence band. The conduction band states are instead constituted from the anion states and the Lu *d*-states. The upper valence bands which lie above this band are mainly due to 'p'-like states of the N-atom with the top occurring at the *Γ* point. It is clear from figure 4 that at the point, the top of the valence band crosses Fermi level and at the *Γ* point the bottom of the conduction band lies down near the Fermi level. Hence, it is metallic in nature.

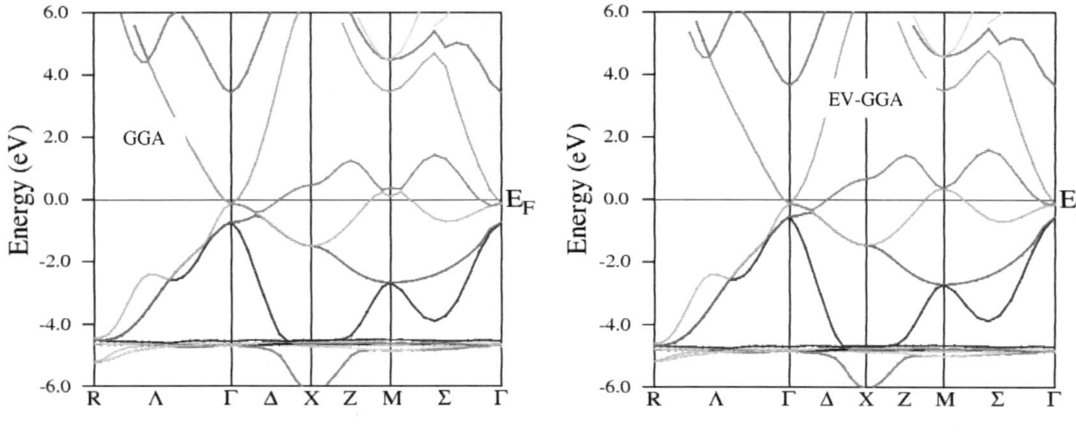

Figure 4. Band structure of LuN in B2 phase with GGA and EV- GGA.

4. Conclusions

The present paper deals with the computations of structural and phase transition properties of LuN compound GGA approach. The computed values are explained well and show good agreement with experimental data. LuN shows first-order phase transition from B1 to B2 phase under compression. The electronic structure has been computed in B1 and B2 phases which predict semi-metallic nature using GGA while small band gap semiconductor.

Acknowledgments

One of the author (SKS) is very grateful to Science and Engineering Research Board (SERB)/ Department of Science and Technology (DST), New Delhi (India) for providing international travel grant to present the work.

References

[1] Klemm W and Winkelmann G 1956 *Z. Anorg. Allg. Chem.* **288** 87
[2] Niwa K, Hasegawa M and Yagi T 2009 *J. Alloy. Compds*, **477** 493
[3] Larson P and Lambrecht W R L 2007 *Phys. Rev. B* **75** 045114
[4] Yang J, Gao F, Wang H, Gou H, Hao X and Li Z 2010 *Mate. Chem. Phys.* **119** 499
[5] Painter G S, Averill F W, Becher P F, Shibata N, Benthem K V and Pennycook S J 2008 *Phys. Rev. B* **78** 214206
[6] Yang X C, Zhang E J, Ma H Y and Xiao J P 2013 *Advanced Mate. Research*, **690** 559
[7] Hohenberg P and Kohn W 1964 *Phys. Rev. B* **136** 864
[8] Kohn W and Sham L J 1965 *Phys. Rev.* **140** A1133
[9] Blaha P, Schwarz K, Madsen G K H, Kvasnicka D and Luitz J 2001 WIEN2k_11.1 (ed.) K Schwarz Techn. Universitat Wien Austria
[10] Perdew J P, Burke S and Ernzerhof M 1996 *Phys. Rev. Lett.* **77** 3865
[11] Engel E and Vosko S H 1993 *Phys. Rev. B* **47** 13164
[12] Monkhorst H J and Pack J D 1976 *Phys. Rev. B* **13** 5188
[13] Murnaghan F D 1944 *Proc. Nat. Acad. Sci. U.S.A.* **30** 244

Lattice-switch Monte Carlo: the fcc–bcc problem

T L Underwood and G J Ackland

School of Physics and Astronomy, SUPA, The University of Edinburgh, Edinburgh, EH9 3JZ, UK

Abstract. Lattice-switch Monte Carlo is an efficient method for calculating the free energy difference between two solid phases, or a solid and a fluid phase. Here, we provide a brief introduction to the method, and list its applications since its inception. We then describe a lattice switch for the fcc and bcc phases based on the Bain orientation relationship. Finally, we present preliminary results regarding our application of the method to the fcc and bcc phases in the Lennard-Jones system. Our initial calculations reveal that the bcc phase is unstable, quickly degenerating into some as yet undetermined metastable solid phase. This renders conventional lattice-switch Monte Carlo intractable for this phase. Possible solutions to this problem are discussed.

1. Introduction

A common problem which occurs in condensed matter physics is as follows: for a given substance, which of two candidate phases is preferred at a given, say, temperature T and pressure P? This problem amounts to evaluating the free energy difference between the two phases; the preferred phase has the lower free energy. Unfortunately, calculating the free energy difference between two phases to a sufficient accuracy to solve this problem can be difficult. This is the case for a plethora of systems of practical interest, and is by no means limited to 'realistic' models of particle interactions. For instance, while the hard-sphere solid is an archetype of a 'simple' system, until relatively recently there was contention regarding whether the equilibrium phase was fcc or hcp [1]. A similar situation also existed for the Lennard-Jones solid at low temperatures and pressures [2].

The method which resolved the aforementioned hard-sphere and Lennard-Jones disputes is *lattice-switch Monte Carlo* (LSMC) [1, 2]. LSMC allows the free energy difference between two phases to be calculated efficiently. Furthermore it is 'exact' in the sense that it relies upon no approximations other than those present in the model of particle interactions it is used in conjunction with. LSMC has been applied to a wide range of systems since its inception – as summarised in Table 1[1]. This reflects the generality of the method; it can in principle be applied to any pair of phases, and any model of particle interactions. This feature of LSMC, in combination with its supposedly superior computational efficiency compared to alternative methods, makes it an attractive prospect for *ab initio* applications. However, it should be noted that claims of its superiority have proved contentious, and remain somewhat of an open question [4, 5].

[1] The incarnation of LSMC which can treat solid–*fluid* free energy differences [3] is commonly referred to as *phase-switch Monte Carlo.*

Table 1. Applications of LSMC since its inception. The notation $A \leftrightarrow B$ signifies that LSMC has been used to determine the free energy difference between phases A and B.

Phases considered	Interparticle potential	References
fcc ↔ hcp	Hard sphere	[1, 2, 4]
Various fcc and hcp phases with stacking faults	Hard sphere	[4, 6]
fcc ↔ fluid	Hard sphere	[3]
fcc ↔ hcp	Lennard-Jones	[7]
square ↔ triangular	Core-softened	[8]
fcc ↔ fluid	Lennard-Jones	[9, 10]
Various close-packed polydisperse binary phases	Hard sphere	[11]
fcc ↔ hcp (both polydisperse)	Hard sphere	[12]
fcc ↔ fluid, bcc↔fluid	r-12	[13]
fcc ↔ fluid (both polydisperse)	r-12	[14, 15, 16]
triangular ↔ triangular 'soliton staircase'	r-12	[5]

The layout of this work is as follows. In the following section we provide a brief introduction to LSMC. Detailed accounts of the method can be found in the references. In Sec. 3 we consider the application of LSMC to the fcc and bcc phases of the Lennard-Jones solid. We first describe a lattice switch between these phases, and conclude by presenting the results of our initial investigations.

2. Lattice-switch Monte Carlo: an introduction

Consider the Gibbs free energy difference $\Delta F \equiv F_1 - F_2$ between two crystalline phases 1 and 2, where F_α denotes the free energy of phase α. It can be shown that

$$\Delta F = \beta^{-1} \ln \left(p_2 / p_1 \right), \tag{1}$$

where β denotes the thermodynamic beta, and p_α denotes the probability of the system being in phase α at thermodynamic equilibrium – assuming that the system is constrained to be in either phase 1 or phase 2. The above equation can be exploited to calculate ΔF theoretically: extract p_2/p_1, the time the system spends in phase 2 relative to phase 1, from an NPT simulation of the system, and substitute this quantity into the above equation. However, this approach is intractable for simulations utilising 'realistic' particle dynamics if the typical time taken for the system to transition between the two phases is very long, in which case a reasonable estimate of p_2/p_1 cannot be deduced in a reasonable simulation time. It may even be the case that, regardless of the phase in which the simulation is initialised, the system *never* transitions to the 'other' phase during the course of the simulation. This stems from the fact that, at equilibrium, the most probable states comprise two 'islands of stability' in phase space: one within phase 1 and the other within phase 2. However, these two islands are separated by an *entropic barrier*: a region of phase space comprised of states which are very improbable at equilibrium. Hence to transition between the phases the system must traverse the entropic barrier, the success of which is very unlikely, and hence a rare occurrence.

The traditional implementation of Metropolis Monte Carlo [17] belongs to the aforementioned class of simulations which utilise 'realistic' particle dynamics. In this approach (for an NPT simulation) at each time-step a trial state of the system is generated from the current state by either altering the position of one of the particles, i.e., a 'particle move' is attempted, or altering the dimensions of the simulation box, i.e., a 'volume move' is attempted. The trial state is then accepted or rejected as the new state of the system for the next time-step according to the Metropolis algorithm [17]. The end result is that for a long simulation the states are sampled from the probability distribution corresponding to thermodynamic equilibrium. However, the important properties of the Metropolis algorithm do not rely upon the mechanism to generate trial states just described; one has considerable freedom with regards to how trial states are generated. The prospect therefore exists of generating states in a manner such that the system traverses a path in phase space which allows ΔF to be accurately calculated in relatively few time-steps. Such a path would involve frequent transitions between both phases by 'jumping over' the entropic barrier.

This is what is done in LSMC; a new type of move, a *lattice switch*, is introduced in order to supplement the aforementioned particle and volume moves. Intuitively, the system is in phase α if the positions of the particles approximately form the crystal lattice characteristic of α at the current volume. With this in mind, if the system is in phase α, one can express the position $\mathbf{r}_i$ of particle i as

$$\mathbf{r}_i = \mathbf{R}_{\alpha,i} + \mathbf{u}_i, \tag{2}$$

where $\mathbf{R}_{\alpha,i}$ denotes the position of the α crystal lattice site which is closest to i, and $\mathbf{u}_i$ denotes the displacement of i from that lattice site. The trial state σ' generated by a lattice switch from a state σ in phase 1 shares the same set of particle displacements $\{\mathbf{u}_i\}$ as σ, but the underlying set of lattice vectors for σ' is $\{\mathbf{R}_{2,i}\}$ instead of $\{\mathbf{R}_{1,i}\}$; the underlying lattice is 'switched' from σ to σ'. In other words, the lattice switch amounts to the following transformation for all i:

$$\mathbf{r}_i = \mathbf{R}_{1,i} + \mathbf{u}_i \rightarrow \mathbf{R}_{2,i} + \mathbf{u}_i. \tag{3}$$

Hence every time a lattice switch is accepted the system transitions directly to the 'other' phase, bypassing the entropic barrier.

This is, however, only half of the story. One might expect that by regularly attempting lattice switches the system will frequently transition between phases, allowing ΔF to be efficiently evaluated as described above. However, if one does this with Metropolis Monte Carlo, one finds that lattice switches are too rarely accepted for this approach to be useful. The reason for this is that, for the states visited during a typical simulation, σ' is almost always of a much higher energy than σ, and hence lattice switches will almost always be rejected by the Metropolis algorithm. Crucially, there exist states for which σ and σ' are of comparable energy; from such states switches have a reasonable chance of success. We refer to these states as *gateway states*, since they provide the key to jumping between both phases. Unfortunately, these states are almost never visited dduring a Metropolis Monte Carlo simulation. To encourage successful lattice switches, we therefore use *multicanonical Monte Carlo* [18, 19, 20] – which provides a means of sampling selected states more (or less) frequently than is the case at equilibrium, while still allowing equilibrium properties of the system to be evaluated – to more frequently visit gateway states. The result is that lattice switches are accepted reasonably often, and both phases are explored in a reasonable simulation time. To elaborate, we introduce a quantity $\mathcal{M}$ which characterises how 'gateway-like' a state is, with $\mathcal{M} = 0$ corresponding to 'perfectly gateway-like', and $|\mathcal{M}| \gg 0$ corresponding to 'very un-gateway-like'. The specific definition of the quantity $\mathcal{M}$ depends on the system under consideration. For solid–solid free energy differences in soft-potential systems, the following definition of $\mathcal{M}$ has been used:

$$\mathcal{M}(\{\mathbf{u}_i\}) = E(\{\mathbf{R}_{1,i} + \mathbf{u}_i\}) - E(\{\mathbf{R}_{2,i} + \mathbf{u}_i\}), \tag{4}$$

where $E(\{\mathbf{r}_i\})$ denotes the energy of the state with particle positions $\{\mathbf{r}_i\}$. The first term on the right-hand side is the energy associated with the displacements $\{\mathbf{u}_i\}$ for phase 1, and the second term is the analogous quantity for phase 2. Note that $\mathcal{M}(\{\mathbf{u}_i\}) = 0$ if the energies associated with $\{\mathbf{u}_i\}$ in both phases are identical. In this case the energy cost of a lattice switch from either phase is zero, and hence the states associated with $\{\mathbf{u}_i\}$ in both phases are gateway states. Defining a macrostate as a collection of all states with the same $\mathcal{M}$, we then sample all macrostates in our multicanonical simulation with equal probability. The result is that the gateway-like macrostates are frequently visited.

3. The fcc–bcc transition in the Lennard-Jones solid

We now turn to the problem of using LSMC to evaluate the free energy difference between the fcc and bcc phases in the Lennard-Jones system. Our motivation behind this is twofold. Firstly, the fcc–bcc transition is of profound importance to metallurgy. Our ultimate aim is to apply LSMC to this transition using more realistic models of metals than the Lennard-Jones model, such as the embedded atom model [21], or even *ab initio* models. However, given that there has yet to be a LSMC study of the fcc–bcc transition, it is sensible to 'tread carefully' and first study the fcc–bcc transition using the simpler, and better understood, Lennard-Jones model before proceeding to more uncharted waters. Secondly, there has been speculation that there is a region in the phase diagram of the Lennard-Jones system at high temperatures and pressures, below the melting curve, where the bcc phase will be preferred over the fcc [22]. It would be interesting to test this hypothesis, the confirmation of which would have far-reaching consequences given the widespread use of the Lennard-Jones model to describe real systems.

3.1. The fcc–bcc lattice switch

After deciding to apply LSMC to a certain system, the first problem one encounters is the choice of lattice switch. A lattice switch is a one-to-one mapping of particle positions in one phase to another phase. Hence it is necessary that the supercells used to represent both phases have the same number of particles. The Bain orientation relationship (see, e.g., Ref. [23]) provides a means for achieving this for the case of the fcc and bcc phases. To elaborate, both fcc and bcc crystals can be recast as bct crystals: the bcc crystal is equivalent to a bct crystal in which the bct unit cell has equal dimensions, i.e. $a = b = c$; the fcc crystal is equivalent to a bct crystal in which $a = b = c/\sqrt{2}$. This is illustrated in Fig. 1. Thus by tiling N_a, N_b and N_c bct unit cells corresponding to fcc or bcc along the a-, b- and c-directions, one can construct a bcc and a fcc supercell which both contain the same number of lattice sites $2N_aN_bN_c$. Specifically, the positions of the lattice sites in one of the supercells are given by $(n_a a, n_b b, n_c c)$ and $(n_a a + a/2, n_b b + b/2, n_c c + c/2)$ where $n_a = 0, 1, \ldots, (N_a - 1)$ and similarly for n_b and n_c.

We have just described how to construct an fcc and a bcc supercell with the same number of lattice sites. For the NVT ensemble we are interested in fcc and bcc phases with the same density ρ, and the supercells should reflect this. It can be shown that $a_{\text{bcc}} = 2^{1/3}\rho^{-1/3}$, $a_{\text{fcc}} = 2^{1/6}\rho^{-1/3}$ and $c_{\text{fcc}} = 2^{2/3}\rho^{-1/3}$. Hence for fcc and bcc crystals of equal density it is necessary that $a_{\text{fcc}} = 2^{-1/6}a_{\text{bcc}}$ and $c_{\text{fcc}} = 2^{1/3}a_{\text{bcc}} = 2^{1/3}c_{\text{bcc}}$. Therefore, starting from the bcc supercell, if one applies the transformation

$$a \to 2^{-1/6}a, \quad b \to 2^{-1/6}b, \quad c \to 2^{1/3}c \tag{5}$$

to the lattice site positions and the dimensions of the supercell, then the end result is an fcc supercell with the same density. This corresponds to stretching the bcc supercell in the c-direction, while simultaneously compressing the supercell in the a- and b-directions to preserve its volume. Taking this idea further, if one makes the aforementioned transformation, but keeps the displacements $\{\mathbf{u}_i\}$ of the particles in the supercell unchanged, then one has performed a density-preserving lattice switch from bcc to fcc. Obviously the converse operation is a density-preserving lattice switch from fcc to bcc. In an NPT ensemble it may be the case that, say,

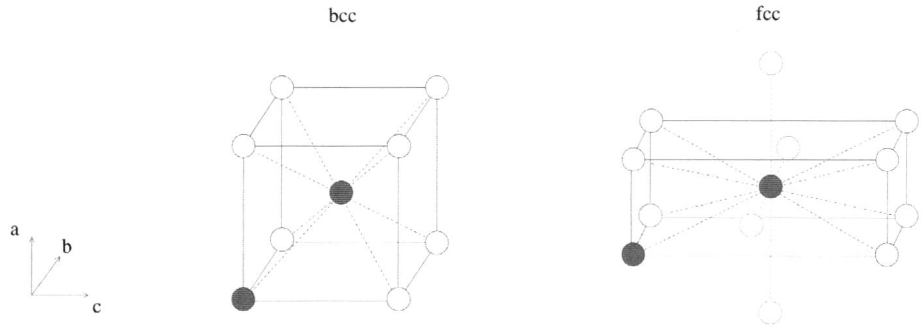

Figure 1. (Colour online) Schematic diagram illustrating the bct representations of the bcc and fcc crystal structures. Each unit cell contains two particles, which are represented by grey circles. Image particles are represented by white circles. Dashed lines connect the particle at the centre of each unit cell to its nearest neighbours. In the fcc case, four of the twelve nearest neighbours lie beyond the edges of the unit cell. These neighbours and the corresponding dashed lines are coloured red.

the bcc phase is of a lower density than the fcc phase, in which case a switch which increases the density upon transforming from bcc to fcc, and correspondingly lowers the density upon transforming from fcc to bcc, may yield a more efficient simulation. The above discussion can be easily adapted to treat such non-density-preserving lattice switches.

3.2. Results of initial investigations: instability of the bcc phase

Finally we turn to our simulations, the methodology of which closely resembles that of Ref. [7]. Before performing a LSMC simulation to calculate ΔF to a high degree of accuracy, one must optimise the step size used in the particle and volume moves used to generate trial states each time step. To do this it is sufficient to use Metropolis Monte Carlo simulations locked into one of the phases, with a small system size. Furthermore, such simulations act as a 'sanity check' before more accurate simulations are undertaken. It was during such simulations that we noticed that the bcc phase would quickly degenerate into some other – as yet unidentified – metastable phase. The same was not observed to occur for the fcc phase. This is illustrated in Fig. 2, which shows the radial distribution functions (RDFs) at the end of NVT Metropolis Monte Carlo simulations of systems of 250 particles at $\rho \approx 1.1$ and $\beta = 3.333$ initialised in the bcc and fcc phases, where we are using reduced units as described in Ref. [7]. Each simulation comprised 1×10^7 particle moves, and repeats of the simulations yielded very similar RDFs to those shown in the figure. Note that the peaks in the fcc RDF are in excellent agreement with those of the perfect fcc crystal; however, the same cannot be said for the bcc RDF.

The fact that the bcc phase is so short-lived makes it impossible to apply LSMC as it stands to determine the free energy between fcc and bcc in the Lennard-Jones system. A similar problem was described in Ref. [8] for two-dimensional core-softened systems. It should be borne in mind that if one considers *any* two phases, at least one of them will be metastable, and hence will destabilise given a long enough simulation time; we require that the two phases under consideration do not destabilise before the simulation time required to determine ΔF to the desired accuracy is reached. With regards to the fcc–bcc problem, the following question comes to mind: is there a way in which the system can be kept in the bcc phase for long enough to gather decent statistics relevant to calculating ΔF, or indeed any other property? One might think that a particle move mechanism which constrains the system to remain within the phase under consideration is a valid means of preventing the bcc phase from destabilising. However, 'hard wall' constraints on the particle positions can lead to 'drift' in the centre-of-mass of the

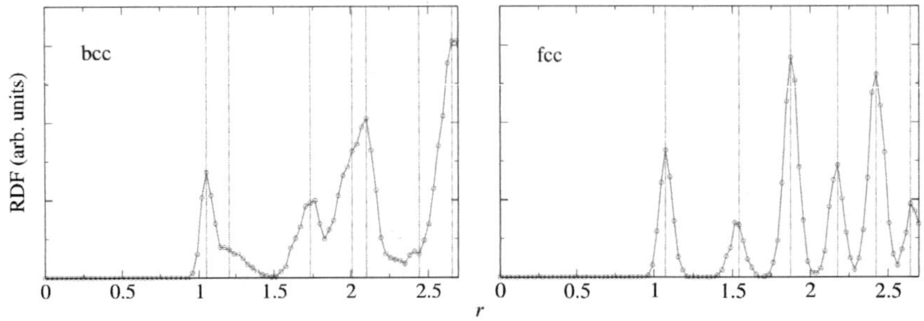

Figure 2. (Colour online) The radial distribution functions obtained at the completion of simulations of the Lennard-Jones system initialised in the bcc and fcc phases – as described in the main text. The red curve and circles in the left(right) panel is the RDF corresponding to the simulation initialised in the bcc(fcc) phase. r denotes the distance from a particle, in reduced units. The dashed, black vertical lines indicate the locations of the 'bins' of the RDF histogram which have non-zero ordinates for a perfect bcc or fcc crystal. In other words, in each panel the nth line from the left corresponds to the nth nearest neighbour shell in the analogous perfect crystal.

simulation, which may invalidate the final results [2]. An alternative approach is to softly 'tether' particles to their lattice sites in the multicanonical simulation through judicious choice of the weight function. A similar idea is used in applying LSMC to fluids [3], and may be worth exploring as a means of addressing instability in solids.

Acknowledgments

This work was supported by the Engineering and Physical Sciences Research Council.

References

[1] Bruce A D, Wilding N B and Ackland G J 1997 *Phys. Rev. Lett.* **79**(16) 3002–3005
[2] Bruce A D, Jackson A N, Ackland G J and Wilding N B 2000 *Phys. Rev. E* **61**(1) 906–919
[3] Wilding N B and Bruce A D 2000 *Phys. Rev. Lett.* **85**(24) 5138–5141
[4] Pronk S and Frenkel D 1999 *J. Chem. Phys.* **110** 4589–4592
[5] Wilms D, Wilding N B and Binder K 2012 *Phys. Rev. E* **85**(5) 056703
[6] Mau S C and Huse D A 1999 *Phys. Rev. E* **59**(4) 4396–4401
[7] Jackson A N, Bruce A D and Ackland G J 2002 *Phys. Rev. E* **65**(3) 036710
[8] Wilding N B and Magee J E 2002 *Phys. Rev. E* **66**(3) 031509
[9] Errington J R 2004 *J. Chem. Phys.* **120** 3130–3141
[10] McNeil-Watson G C and Wilding N B 2006 *J. Chem. Phys.* **124** 064504
[11] Jackson A N and Ackland G J 2007 *Phys. Rev. E* **76**(6) 066703
[12] Yang M and Ma H 2008 *J. Chem. Phys.* **128** 134510
[13] Wilding N B 2009 *Mol. Phys.* **107** 295–299
[14] Wilding N B 2009 *J. Chem. Phys.* **130** 104103
[15] Sollich P and Wilding N B 2010 *Phys. Rev. Lett.* **104**(11) 118302
[16] Wilding N B and Sollich P 2010 *J. Chem. Phys.* **133** 224102
[17] Metropolis N, Rosenbluth A W, Rosenbluth M N, Teller A H and Teller E 1953 *J. Chem. Phys.* **21** 1087–1092
[18] Berg B A and Neuhaus T 1991 *Phys. Lett. B* **267** 249–253
[19] Berg B and Neuhaus T 1992 *Phys. Rev. Lett.* **68**(1) 9–12
[20] Smith G R and Bruce A D 1995 *J. Phys. A: Math. Gen.* **28** 6623
[21] Daw M and Baskes M 1984 *Phys. Rev. B* **29**(12) 6443–6453
[22] Rahman A and Jacucci G 1984 *Nuovo Cimento D* **4** 357–381
[23] Mittemeijer E J 2010 *Fundamentals of materials science* (Berlin; London: Springer)

Development of the SSPH Method for Real-Space Electronic Structure Calculations

Yasunari Zempo, and Soichiro Sugimoto

Computer and Information Sciences, Hosei University,
3-7-2 Kajino, Koganei 184-8584 Japan

E-mail: zempo@hosei.ac.jp

Abstract. We have applied symmetric smoothed particle hydrodynamics (SSPH) to real-space electronic structure calculations. The electronic structure of a simple atom such as hydrogen, helium and lithium is calculated. The results of SSPH are compared to that of the higher-order finite difference method (FD), and are in good agreement with that of FD. Our results indicate that SSPH can be applicable to the electronic structure calculations that require high accuracy.

1. Introduction

The real-space approach is widely used for electronic structure calculation based on density functional theory [1, 2, 3]. In particular, it is used for large systems including many atoms, because a real-space mesh is suitable for large-scale parallel computing. The high-order finite-difference method (FD) is frequently employed in the calculation on a regular mesh. However, it has equal resolution everywhere, which seems to be an inefficient distribution of computation points and to be expensive. If non-uniform distribution of computation points can be used, we can significantly reduce computational costs.

Smoothed Particle Hydrodynamics (SPH) is a typical meshfree particle method in which the system is represented by a finite set of arbitrarily distributed particles without using any mesh. SPH is widely applied to hydrodynamic problems which deal with complex shapes, large deformations and free surfaces. Recently, the SPH based method has been applied to solve non-hydrodynamic partial differential equations such as the wave equation, the diffusion equation, Maxwell's equations and Poisson's equation [4, 5, 6, 7]. On the other hand, there are few studies for the electronic structure calculation [8].

The SPH based method may lead to an efficient real-space technique because of arbitrarily distributed computation points. However, it is generally known that standard SPH method has low accuracy [9]. In considering the application to practical electronic structure calculation, obtaining the almost equal accuracy to finite-difference method by using meshfree particle method is one of the problems. As one of the improved techniques, Symmetric Smoothed Particle Hydrodynamics (SSPH) [10, 11] have been proposed. This method successfully increases the accuracy of standard SPH using Taylor series expansion.

In this paper, we have employed SSPH since it is the simplest method in the higher-order series expansion. The purpose of this paper is evaluation of SSPH for the electronic structure calculation. For simplicity, we have applied SSPH to the Kohn-Sham equation [12, 13] to solve a simple atom such as hydrogen, helium, and lithium. In the evaluation, we focus on the relative error of the eigenvalue. The result is compared to that of FD.

The paper is organized as follows. The formulation of SSPH is in the next section. The numerical tests are shown in section.3. Finally, we summarize the conclusions and discussion in the last section.

2. Method

In SPH, the wave function is approximated using integral form, and the system is represented by a finite number of particles. The concept of the integral representation of a wave function $\psi(\mathbf{r})$ starts from the following identity. The delta function is replaced by a kernel function $W(\mathbf{r}, h)$.

$$\psi(\mathbf{r}) = \int_\Omega \psi(\mathbf{r}')\delta(\mathbf{r}-\mathbf{r}')\, d\mathbf{r}' \simeq \int_\Omega \psi(\mathbf{r}')W(\mathbf{r}-\mathbf{r}', h)\, d\mathbf{r}', \tag{1}$$

where h is the smoothing length. As the kernel function, we adopt the Wendland function [14].

$$W(r,h) = \begin{cases} \dfrac{3r}{4h}\left(1-\dfrac{1}{2}\dfrac{r}{h}\right)^5 \left\{2\left(\dfrac{r}{h}\right)^2 + \dfrac{5r}{2h} + 1\right\} & \left(0 \le \dfrac{r}{h} \le 2\right) \\ 0 & \left(2 \le \dfrac{r}{h}\right) \end{cases} \tag{2}$$

SSPH corrects the approximation of SPH using the Talor series expansion. Using the Taylor series expansion of $\psi(\mathbf{r})$, the right-hand side of Eq.(1) and its moments can be rewritten as follows.

$$\int_\Omega \psi(\mathbf{r}')W(\mathbf{r}-\mathbf{r}', h)\, d\mathbf{r}' \simeq \sum_{l=0}^{m} \left(\int_\Omega (\mathbf{r}-\mathbf{r}_i)^{k+l}\, W(\mathbf{r}-\mathbf{r}_\mathbf{i}, h) d\mathbf{r}\right) \left(\frac{1}{l!}\frac{\partial^l \psi}{\partial \mathbf{r}^l}\right)\Bigg|_{\mathbf{r}_i}, \quad (k = 0, 1, \ldots, m) \tag{3}$$

This expansion can be expressed by the simultaneous linear equation $P = KD$, where the integral Eq.(3) is replaced by the summation of the particles. As the small volume of the j th particles around the particle i is expressed ΔV_j,

$$\begin{aligned} P &= \begin{pmatrix} P_k \end{pmatrix}, & P_k &= \sum_j^N (\mathbf{r}_j - \mathbf{r}_i)^k\, \psi(\mathbf{r}_j) W_{ij} \Delta V_j \\ K &= \begin{pmatrix} K_{kl} \end{pmatrix}, & K_{kl} &= \int_\Omega (\mathbf{r}-\mathbf{r}_\mathbf{i})^k\, (\mathbf{r}-\mathbf{r}_i)^l\, W(\mathbf{r}-\mathbf{r}_\mathbf{i}, h) d\mathbf{r} \\ D &= \begin{pmatrix} D_l \end{pmatrix}, & D_l &= \left(\frac{1}{l!}\frac{\partial^l \psi}{\partial \mathbf{r}^l}\right)\Bigg|_{\mathbf{r}_i}, \end{aligned} \tag{4}$$

where $W_{ij} = W(\mathbf{r}_j - \mathbf{r}_i, h)$.

The electronic structure calculation is performed within the framework of LDA [12, 13]. Taking account of $D_0 = \psi(\mathbf{r}_i)$ and $D_2 = \frac{1}{2}\frac{\partial^2 \psi(\mathbf{r})}{\partial \mathbf{r}^2}\big|_{\mathbf{r}_i}$, the Kohn-Sham equation at the position of the i th particle in SSPH can be written as

$$\left(V(\mathbf{r}_i) \quad 0 \; -1 \; \ldots \; 0\right) D = \left(\varepsilon \quad 0 \quad 0 \; \ldots \; 0\right) D, \tag{5}$$

where $V(\mathbf{r}) = V_{ec}(\mathbf{r}) + V_H(\mathbf{r}) + V_{XC}(\mathbf{r})$. As an ionic potential $V_{ec}(\mathbf{r})$ we use, $V_H(\mathbf{r})$ is the Hartree potential, and $V_{XC}(\mathbf{r})$ is the exchange-correlation potential. From the electronic charge density $\rho(\mathbf{r}) = \sum_i^{occ} |\psi_i(\mathbf{r})|^2$, the Hartree and exchange-correlation potential is determined. In SSPH, the particle position is arbitrarily distributed. However, we have to calculate K^{-1} in each particle position, and to be careful to keep the matrix K regular.

3. Results and discussion

In the expression of the SSPH discretization, we solve Eq.(5) for simple atoms such as hydrogen, helium, and lithium. To solve the electronic structure of the atom, there is various conventional computational programs (AT) with high accuracy. Thus, we use it for the evaluation of the calculation of SSPH together with that FD. The SSPH particles are placed on the same regular mesh point, because the results are compared with those of FD. The calculation system was a cube 16 a.u. on a side. The particles of SSPH are regularly distributed on the same spacing points with the distance $\Delta x = 0.2$ a.u..

Table 1. $1s$ orbital energies.

	SSPH	FD	AT
H	−0.230	−0.230	−0.260
He	−0.543	−0.547	−0.570
Li ($2s$)	−0.094	−0.965	−0.128
($1s$)	−1.722	−1.728	−1.922

Using the $1s$ orbital energy, the results of SSPH, FD, and AT is listed on Table.1. As for Lithium, the $2s$ orbital energy was also calculated. The results are graphically shown in Fig.1. The results of SSPH and FD are also compared with those of AT. The results of SSPH are in good agreement with those of FD. However, there are some errors between SSPH and AT. We adopted that the number of particles of SSPH was the same that of FD. Also there is no correction on the boundary of the system. This basically causes the difference between SSPH and AT.

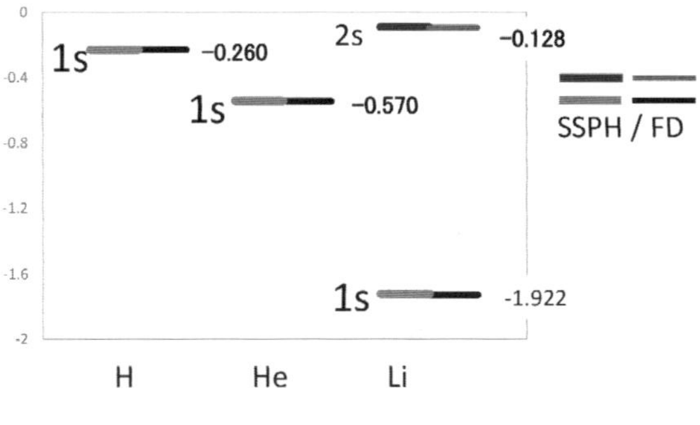

Figure 1. Comparison of $1s$ orbital energies. Left hand sides and the other side show the results of SSPH, and those of FD, respectively.

Similar to the procedure as expressed in Eq.(5), we can easily derive the SSPH form of the Poisson equation. We solved the Poisson equation $\nabla^2 V_H(\mathbf{r}) = -4\pi\rho(\mathbf{r})$, using SSPH. The calculated charge profiles are shown in Fig.2 and Fig.3. The calculated Hartree energy is in good

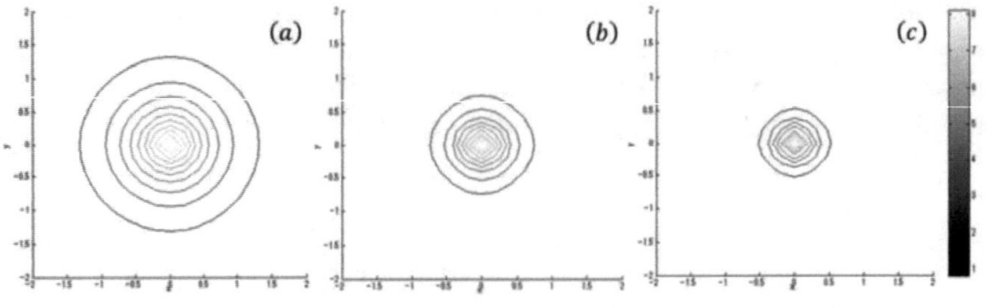

Figure 2. Charge density profiles of (a)H, (b)He, and (c)Li.

agreement in the region far from the core $V_H(r) \simeq Z_n/r$. In the vicinity of the core, the number of the particles is too few to describe the steep change of $V_H(r)$. Figure 3 shows the calculated Hartree energy profile of He as a function of distance from the core. In the core region, there is a large difference. The errors of SSPH are mainly caused by this difficulty.

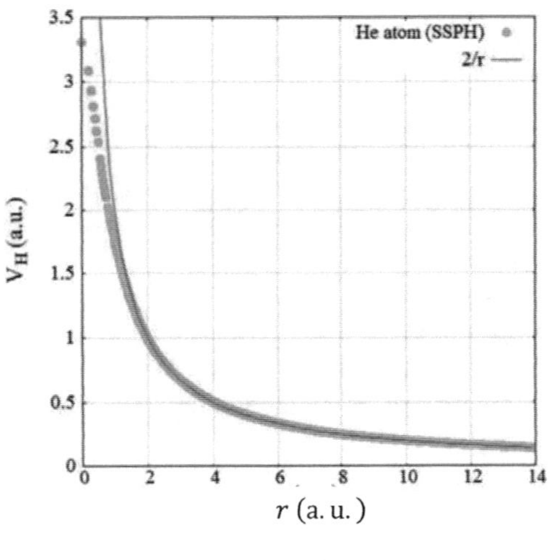

Figure 3. Calculated Hartree energy of He. The dotted line shows the result of SSPH, which can be compared to the solid line of $2/r$.

4. Summary

We presented SSPH as one of the discretization techniques in real-space electronic structure calculations. SSPH was applied to real-space electronic structure calculations of a simple atom such as hydrogen, helium and lithium. The accuracy was similar to that of FD. If the number of the particles of SSPH is enough to describe the steep change of the field function, SSPH provides the fairly good accuracy. Our results indicate that SSPH can be applicable to the electronic structure calculations that require high accuracy.

Acknowledgments
The authors would like to express their gratitude to Professor S. S. Kano for useful discussions and encouragement. This work was partially supported by JSPS Grants-in-Aid for

Scientific Research (C) Grant number 25390158, Sumitomo Chemical Co., Ltd. and Simulatio Corporation.

References

[1] Chelikowsky J, Troullier N, Wu K and Saad Y 1994 *Phys. Rev.* B**50**, 11355
[2] Beck T L 2000 *Rev. Mod. Phys.* **72**, 1041
[3] Zempo Y, Akino N, Ishida M, Ishitobi M and Kurita Y 2008 *J. Phys.: Condens. Matter.* **20**, 064231
[4] Laguana P 1995 *Astrophys. J.* **439**, 89
[5] Stranex T and Wheaton S 2011 *Comput. Method. Appl. Math.* B**200**, 392
[6] Tascano E, Balasi G D and Tortorici A 2012 *Appl. Math. Comput.* **218**, 8906
[7] Ala G, Francomano E, Tortorici A, Tascano E and Viola F 2006 *J. Comput. Appl. Math.* **191**, 194
[8] Sugimoto S and Zempo Y 2014 *J. Phys.: Conf. Ser.* **510**, 012037
[9] Lucy L B 1977 *Astron. J.* **439**, 1013
[10] Batra R C and Zhang G M 2008 *Compt. Mech.* **41**, 527
[11] Zhang G M and Batra R C 2009 *Compt. Mech.* **43**, 321
[12] Hohenberg P and Kohn W 1964 *Phys. Rev.* **136**, B864
[13] Kohn W and Sham L. J 1965 *Phys. Rev.* **140**, A1133
[14] Wendland H 1995 *Adv. Comput. Math.* **4**, 389

Toward Estimating Current Densities in Magnetohydrodynamic Generators

V. A. Bokil[1], **N. L. Gibson**[1], **D. A. McGregor**[1,2] **and C. R. Woodside**[2]

[1] Oregon State University, Corvallis, OR 97331-4605, USA
[2] National Energy Technology Laboratory, Albany, OR 97321-2198, USA

E-mail: mcgregod@math.oregonstate.edu

Abstract. We investigate the idea of reconstructing current densities in a magnetohydrodynamic (MHD) generator channel from external magnetic flux density measurements in order to determine the existence and location of damaging arcs. We model the induced fields, which are usually neglected in low magnetic Reynold's number flows, using a natural fixed point iteration. Further we present a sensitivity analysis of induced fields to current density profiles in a 3D, yet simplified model.

1. Introduction

Direct power extraction via magnetohydrodynamic (MHD) principles offers a potential step improvement in thermal efficiencies over energy systems utilizing traditional turbomachinery [1]. This is principally due to the lack of moving parts in an MHD generator, as the temperature limits of the moving parts tend to limit cycle temperatures in traditional combustion driven systems. It was established that a major weakness toward commercialization of MHD power generation was the durability of the current collectors on the walls of the generator (electrodes). The electrodes must withstand harsh conditions, and the most damaging and perhaps most difficult to predict phenomenon experienced in the generator was arcing. Consider the example of an oxifuel kerosene MHD Generator with a water-cooled, copper channel. The combustion product will be at approximately 2500 K while the channel will be kept at a temperature near 500 K. This large difference in temperature causes a thermal boundary layer to form in the plasma, where the bulk flow will be much hotter than a thin layer near the edge of this channel. As the plasma is thermally ionized, the conductivity will drop in this boundary layer. Large arcs of high current density will then form near the electrodes as the current which is forced through the electrodes will have to "jump the conductivity gap."

In the arc state we expect the current densities in the channel to be many orders of magnitude larger than in the diffuse state. Given these large differences in current density, the induced magnetic fields are measurably different near the arc. Therefore properties of the current density may be inferred via measurements of the induced fields. The idea of reconstructing current densities from external magnetic flux density measurements has been successfully applied to fuel cells and vacuum arc remelters [2]. The standard approach to this problem is to apply the Biot-Savart law and solve a system of *integral equations.* This formulation unfortunately requires many assumptions on the geometry and the model parameters. Instead, one can formulate the inversion by way of a simulation-based parameter estimation. This technique requires the simulation of a forward problem whose inputs are parameterized explicitly. One then matches the solution of the model to measured data by minimizing the discrepency between data and simulation using non-linear optimization techniques in the parameter space.

It is our goal here to provide a proof of concept for inversion, via sensitivity analysis, that current densities inside the channel of a magnetohydrodynamic generator (MHDG) can indeed be estimated from external measurements of the induced magnetic fields. This is in comparison to using measurements of the applied field which will most often already be known. It is worth noting that the induced fields are expected to be significantly smaller than the applied field of the generator, which will be a practical issue in the design of experiments.

2. Modeling

A complete description of the bulk flow of a ionized gas in the presence of a magnetic field is governed by a coupling of the Euler Equations to the Maxwell Equations [3]. It is important to observe that many of these parameters, e.g. electrical conductivity, can reasonably be assumed to be constitutive and are truly dependent upon other state parameters, e.g. gas temperature. This implies that the coupling between systems in the equations below may be stronger than is apparent. The following system of partial differential equations can be found in [4]. We use the convention that for a vector field $\mathbf{f}$, the function $\sqrt{\mathbf{f}\cdot\mathbf{f}}$ is referred to as f.

$$\rho\left(\partial_t + \mathbf{u}\cdot\nabla\right)\mathbf{u} = \mathbf{J}\times\mathbf{B} - \nabla p \qquad \text{(Conservation of Momentum)} \quad (2.1a)$$

$$\rho\partial_t\mathcal{E} + \rho\mathbf{u}\cdot\nabla\left(\frac{\mathbf{u}\cdot\mathbf{u}}{2} + U\right) = -\nabla\cdot(\mathbf{u}p) - \nabla\cdot\left(\mu^{-1}\mathbf{E}\times\mathbf{B}\right) \qquad \text{(Conservation of Energy)} \quad (2.1b)$$

$$(\partial_t + \mathbf{u}\cdot\nabla)\rho = -\rho\nabla\cdot\mathbf{u} \qquad \text{(Conservation of Mass)} \quad (2.1c)$$

$$\nabla\times\mu^{-1}\mathbf{B} = \mathbf{J} \qquad \text{(Ampere's Law)} \quad (2.1d)$$

$$\partial_t\mathbf{B} = -\nabla\times\mathbf{E} \qquad \text{(Faraday's Law)} \quad (2.1e)$$

$$\mathbf{J} = \sigma(\mathbf{E} + \mathbf{u}\times\mathbf{B}) + \frac{\beta}{B}\mathbf{J}\times\mathbf{B} \qquad \text{(Ohm's Law with Hall effect)} \quad (2.1f)$$

$$\nabla\cdot\mathbf{B} = 0 \qquad \text{(Gauss' Law for Magnetism)} \quad (2.1g)$$

where, the energy $\mathcal{E}$ is defined as $\mathcal{E} = \dfrac{\mathbf{u}\cdot\mathbf{u}}{2} + U + \dfrac{\epsilon\mathbf{E}\cdot\mathbf{E}}{2} + \dfrac{\mathbf{B}\cdot\mathbf{B}}{2\mu}$. Taking the divergence of Ampere's law gives us the *continuity equation* $\nabla\cdot\mathbf{J} = 0$. The variables in the Hall-MHD system are defined in Table 1.

Table 1. State variables for the Hall-MHD Model (2.1).

$\mathbf{u}$	plasma velocity	U	thermal energy of the plasma
p	pressure	$\mathbf{B}$	magnetic flux density
$\mathbf{E}$	electric field	$\mathbf{J}$	current density
β	Hall parameter	μ	magnetic permeability
σ	electrical conductivity	ρ	plasma density

We make two assumptions which are reasonable for the applications of interest [5].

Assumption 1 : The magnetic field is the sum of applied and induced fields $\mathbf{B} = \mathbf{b}_0 + \mathbf{b}$ where the applied field $\mathbf{b}_0 = (0, 0, b_0)^T$ is constant. In addition we assume that the applied field is of much greater intensity than the induced fields, e.g. $b \ll |b_0|$. The magnetic Reynolds number ($R_m \propto \mu\sigma u$) is known to be much less than one, due to the low conductivity and magnetic permeability, in the case MHD generators and is proportional to the ratio $\frac{b}{|b_0|}$.

Assumption 2 : The generator is in equilibrium, namely that all time derivatives are 0. An argument can be made that if the generator is run for a long time it will reach an equilibrium state. However, we have little reason to suspect that arcs are not a temporal effect. We make this assumption as a first attempt at understanding the problem and hope to lift this assumption in future work.

By applying these two assumptions to the above model we end up with the following system of equations, in which $\eta = \sigma^{-1}$ and where Poynting's Theorem is used to simplify the right-hand side of (2.1b)

$$\begin{cases} \rho\mathbf{u}\cdot\nabla\mathbf{u} - \nabla p = \mathbf{J}\times\mathbf{b}_0 \\ \rho\mathbf{u}\cdot\nabla\left(\frac{u^2}{2}+T\right) + \nabla\cdot(\mathbf{u}p) = \mathbf{J}\cdot\mathbf{b}_0 \\ \nabla\cdot(\rho\mathbf{u}) = 0 \end{cases} \quad \text{Aerostatic System} \tag{2.2}$$

$$\begin{cases} \nabla\times\mathbf{b} = \mu\mathbf{J} \\ \nabla\cdot\mathbf{b} = 0 \end{cases} \quad \text{Magnetostatic System} \tag{2.3}$$

$$\begin{cases} \nabla\times\eta\mathbf{J} = \nabla\times(\mathbf{u}\times\mathbf{b}_0) + \beta\nabla\times(\eta\mathbf{J}\times\frac{\mathbf{b}_0}{b_0}) \\ \nabla\cdot\mathbf{J} = 0 \end{cases} \quad \text{Current density System} \tag{2.4}$$

We first note that this formulation implies a natural iterative solution technique for the problem, namely that the Aerostatic and magnetostatic systems could each be solved separately assuming the solutions to the others were given. One could then update $\mathbf{J}$ using Ohm's law as a natural fixed point iteration, and then repeat the larger iteration loop. Second, the two assumptions above lead to a situation in which the induced magnetic fields respond to the system state but do not influence it. From this standpoint, the question of whether the induced fields are good candidates for providing observability of the current density, reduces to a question of the sensitivity of the Magnetostatic system to different features of $\mathbf{J}$. We focus on this question for the remainder of this paper.

3. A Numerical Method for the Magnetostatic System

The magnetostatic problem is the classical div-curl system. There are a number of ways to solve for the induced magnetic flux density given $\mathbf{J}$. It follows from the Helmholtz decomposition [6] that since $\mathbf{b}$ is divergence free it must be the curl of some vector potential $\mathbf{a}$. We will assume the *Coulomb gauge*, namely that

$$\mathbf{b} = \nabla\times\mathbf{a} \quad \text{where} \quad \nabla\cdot\mathbf{a} = 0. \tag{3.1}$$

With this choice established we can now rewrite the magnetostatic system as the saddle point problem: Find $(\mathbf{a}, \lambda)$ satisfying

$$\begin{cases} \nabla\times\nabla\times\mathbf{a} + \nabla\lambda = \mu\mathbf{J} \\ \nabla\cdot\mathbf{a} = 0 \end{cases} \tag{3.2}$$

In this formulation we have introduced a *Lagrange multiplier* λ that can be interpreted as a non-physical magnetic pressure. Since, $\nabla\cdot\mathbf{J} = 0$, assuming that $\mathbf{J}$ is smooth, we can show that $\lambda = 0$ in its domain. This ensures that system (3.2) is weakly consistent with the strong formulation of the magnetostatic problem.

We now define function spaces in which we will seek variational solutions to the saddle point problem.

$$\mathbf{H}(\text{curl}, G) = \{\mathbf{c}\in\mathbf{L}^2(G) : \nabla\times\mathbf{c}\in\mathbf{L}^2(G)\} \tag{3.3}$$

$$\mathbf{H}(\text{div}, G) = \{\mathbf{c}\in\mathbf{L}^2(G) : \nabla\cdot\mathbf{c}\in L^2(G)\} \tag{3.4}$$

$$H^1(G) = \{\psi\in L^2(G) : \nabla\psi\in\mathbf{L}^2(G)\} \tag{3.5}$$

$$\mathbf{V} = \{\mathbf{c} \in \mathbf{H}(\mathrm{curl}, G) : \mathbf{c} \times \mathbf{n} = 0 \text{ on } \partial G\} \tag{3.6}$$
$$W = \{\psi \in H^1(G) : \psi = 0 \text{ on } \partial G\} \tag{3.7}$$

We can then pose (3.2) variationally on a smooth domain G as follows

$$\text{Find } (\mathbf{a}, \lambda) \in \mathbf{V} \times W : \quad \begin{cases} \int_G \nabla \times \mathbf{a} \cdot \nabla \times \mathbf{c} + \int_G \nabla\lambda \cdot \mathbf{c} = \int_G \mu \mathbf{J} \cdot \mathbf{c} \\ \int_G \mathbf{a} \cdot \nabla\varphi = 0 \end{cases}, \quad \forall (\mathbf{c}, \varphi) \in \mathbf{V} \times W. \tag{3.8}$$

Under the assumption that $\mathbf{J}$ is in the dual of $\mathbf{V}$, the well-posedness of the variational problem (3.8) can be proved in the framework of the Babuska-Brezzi theory for saddle-point problems [6]. To avoid contamination of the solution due to reflections generated by Dirichlet type boundary conditions for the field $\mathbf{a}$, we will make the computational domain large enough so that our measurements should be far from boundary effects.

3.1. Mimetic Finite Difference Method

To discretize the continuum problem we will use a Mimetic Finite Difference (MFD) method introduced in [6]. The MFD method is, in essence, a generalization of Yee scheme type staggered difference methods to very general geometries. While in our experiments (shown below) we have used Cartesian grids, the nature of these methods is such that they can be easily adapted to the more general geometries which we expect to encounter in practical meshings of a MHD generator channel and casing.

As the details of this method can be found in [6], we will provide here only a brief overview of the method. First consider a given mesh whose nodes we will call $\mathcal{T}_h$, whose edges we will call $\mathcal{E}_h$ and whose faces we will call $\mathcal{F}_h$. On each of these structures there is a natural association with functions H^1, $\mathbf{H}(\nabla\times)$ and $\mathbf{H}(\nabla\cdot)$ respectively. Thus on each structure we assume the following degrees of freedom

$$f \in H^1(G), \quad f_h = (f(\mathbf{x}_i) : \mathbf{x}_i \in \mathcal{T}_h), \tag{3.9}$$
$$\mathbf{v} \in H(\mathrm{curl}, G), \quad \mathbf{v}_h = \left(\frac{1}{|e_i|}\int_{e_i} \tau_i \cdot \mathbf{v} : e_i \in \mathcal{E}_h \text{ and } \tau_i \text{ tangent to } e_i\right), \tag{3.10}$$
$$\mathbf{w} \in H(\mathrm{div}, G), \quad \mathbf{w}_h = \left(\frac{1}{|f_i|}\int_{f_i} \mathbf{n}_i \cdot w : f_i \in \mathcal{F}_h \text{ and } n_i \text{ normal to } f_i\right). \tag{3.11}$$

Given that we will be working with a lowest order method it also natural to then associate edge and face degrees of freedom with midpoints of edges and barycenters of faces as these represent a sufficiently accurate quadrature. We construct, for each set, a mass matrix $\mathbb{Q}_\mathcal{T}, \mathbb{Q}_\mathcal{E}, \mathbb{Q}_\mathcal{F}$ which produces an $\mathcal{O}(h)$ approximation of the L^2 inner product for a form defined on each topology. For example,

$$\mathbf{u}_h, \mathbf{v}_h \in \mathcal{E}_h : \mathbf{v}_h^T \mathbb{Q}_\mathcal{E} \mathbf{u}_h \approx \int_G \mathbf{u} \cdot \mathbf{v} + \mathcal{O}(h). \tag{3.12}$$

These matrices are constructed by producing the matrix on each cell of the discretization and then assembling them in a standard way. One can also construct discrete operators which satisfy analogies of the Fundamental Theorem of Calculus, Stokes Theorem, and Divergence Theorem, as follows

$$\mathcal{GRAD}_h : \mathcal{T}_h \to \mathcal{E}_h, \mathcal{GRAD}_h f_h = \left(\frac{f(\mathbf{x}_i) - f(\mathbf{x}_j)}{|e_k|} : \text{ for } \mathbf{x}_i, \mathbf{x}_j \text{ the endpoints of } e_k \in \mathcal{E}_h \right) \quad (3.13)$$

$$\mathcal{CURL}_h : \mathcal{E}_h \to \mathcal{F}_h, \mathcal{CURL}_h \mathbf{v}_h = \left(\frac{1}{|f_i|} \sum_{e_j \in f_i} \sigma_{ij} |e_j| \mathbf{v}(e_j) : f_i \in \mathcal{F}_h, \sigma_{ij} = \pm 1 \right). \quad (3.14)$$

Given our assumptions about the continuum problem we will seek $\mathbf{a}_h \in \mathcal{E}_h$ and $\lambda_h \in \mathcal{T}_h$ given $\mathbf{j}_h \in \mathcal{E}_h$. By accounting carefully for where the image of each differential operator lies, one can construct a discrete system as follows

$$\begin{bmatrix} \mathcal{CURL}_h^T \mathbb{Q}_{\mathcal{F}} \mathcal{CURL}_h & \mathbb{Q}_{\mathcal{E}} \mathcal{GRAD}_h \\ -\mathcal{GRAD}_h^T \mathbb{Q}_{\mathcal{E}} & \end{bmatrix} \begin{bmatrix} \mathbf{a}_h \\ \lambda_h \end{bmatrix} = \begin{bmatrix} \mathbb{Q}_{\mathcal{E}} \mathbf{j}_h \\ 0 \end{bmatrix}. \quad (3.15)$$

With the discrete forward problem posed, we now describe the computation of sensitivities.

4. Sensitivity Analysis

In order to determine the viability of the response of the induced magnetic flux density to changes in the current density we will perform a sensitivity analysis. There are number of features of current densities which are suspected to occur in MHD Generator channels which we would like to be able to detect.

Total Current Given that the current is extracted from the channel by the load applied across the electrodes and will be a design parameter for the generator.

Current Density Experimental evidence from legacy MHD research suggests that the destructive macro-arcs which form at electrodes will much denser current profiles than the diffuse state. As our long term goal is to detect the location of the arcs inside the generator and facilitate an understanding of their dynamics, this is perhaps the most critical parameter for sensitivity.

Direction of Current Density It is known from earlier MHD work that the Hall effect causes a tilt to the current density, pointing the current slightly in the direction of the fluid flow [7]. Sensitivity to this parameter would allow one to estimate the magnitude of the Hall effect near the sensor.

Given these three features, we have developed the following parameterized current profile

$$\mathbf{j}(x, y, z; J_m, s, \theta) = \mathbf{v} \frac{J_m}{\sqrt{2\pi s^2}} \exp\left(\frac{1}{2s^2} \left| \left(\mathbb{I} - \mathbf{v}\mathbf{v}^T \right) \begin{bmatrix} x \\ y \\ z \end{bmatrix} \right|^2 \right), \quad \mathbf{v} = \begin{bmatrix} \cos\theta \\ \sin\theta \\ 0 \end{bmatrix}. \quad (4.1)$$

This current density is a Gaussian around the line passing through the origin pointing in the direction $(\cos\theta, \sin\theta, 0)^T$. The parameter J_m (A/m^2) controls the total current in the system, the parameter s (m) controls the spread of the density profile, and θ controls the tilting of the arc due to the Hall effect. A significant feature of this formulation is that the profile is naturally divergence free as all variation happens orthogonally to the direction the vector field is pointing.

To perform our sensitivity analysis we fix two parameters and vary the third. We compute actual magnetic flux density values instead of derivatives in order to additionally inform the necessary specifications of measurement equipment. We assume our domain is $[-0.5, 0.5]^3$, the magnetic permeability is constant and on the order of 10^{-6} (which is on the order of magnitude of air at STP) and we measure the magnetic flux density at $(0, 0.015, 0.015)$ and $(.25, 0.015, 0.015)$.

The center result is to estimate the magnitude of fields very close to the arc, while the short distance away is to demonstrate the effect of measuring outside the channel near where sensors would be placed. The sensitivity results are depicted in Figure 1. Figure 1(a) shows high sensitivity to J_m, while Figure 1(b) shows an increasing sensitivity to s in the limit toward smaller diameter, i.e., dense arcs. Finally, Figure 1(c) shows a significant lack of sensitivity to angle at a short distance from the arc.

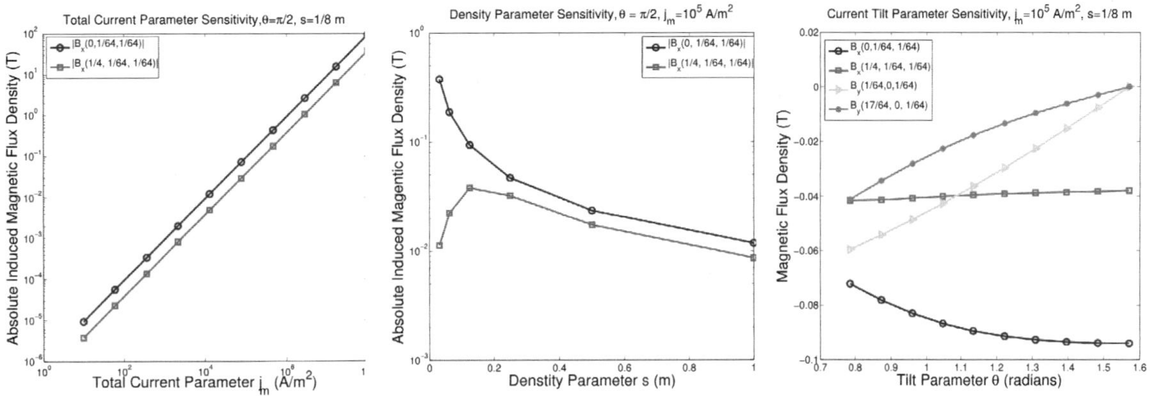

(a) (b) (c)

Figure 1. Analysis of the sensitivity of the induced magnetic flux density to the three current density parameters: J_m, s, and θ.

5. Conclusions

In this paper we have shown that the standard modeling assumptions for the bulk flow of plasma in an MHD generator channel lead to a system in which the induced magnetic flux density responds to the physical system state but does not influence the system. We have created a current density profile which includes several important features of arcs which we wish to be able to detect, namely total current, the direction of the arc, and the width of the density of the arc. Our numerical experiments have confirmed that the magnetic flux density is sensitive to these three features, which is a necessary condition for inversion.

Acknowledgments

This work was supported by the US Department of Energy, National Energy Technology Laboratory grant RES1100426/015.

References

[1] Woodside C R 2012 Direct power extraction with oxy-combustion: An overview of magnetohydrodynamic research activites at the NETL-Regional University Alliance Pittsburgh Coal Conference

[2] Woodside C R and King P 2009 Characterizing arc motion and distribution during vacuum arc remelting *TMS2009 International Symposium on Liquid Metal Processing and Casting. Santa Fe: TMS Publications* vol 75

[3] Davidson P 2001 *An Introduction to Magnetohydrodynamics* (Cambridge University Press)

[4] Rosa R 1987 *Magnetohydrodynamic energy conversion* 2nd ed (McGraw-Hill)

[5] Rosa R J, Krueger C H and Shioda S 1991 *IEEE Trans. Plasma Sci.* **19** 1180–1190

[6] Lipnikov K, Manzini G, Brezzi F and Buff A 2011 *J. Comp. Phys.* **230** 305–328

[7] Huba J D 2003 *Hall Magnetohydrodynamics: A Tutorial* vol LNP 615 (Springer)

Journal of Physics: Conference Series **640** (2015) 012033 doi:10.1088/1742-6596/640/1/012033

PIC Algorithm with Multiple Poisson Equation Solves During One Time Step

Junxue Ren[1,2], Trenton Godar[1], James Menart[1], Sudhakar Mahalingam[3], Yongjun Choi[3], John Loverich[3], and Peter H. Stoltz[3]
[1]*Wright State University, Dayton, OH 45435, USA*
[2]*Beihang University, Beijing, 100191, P.R.China*
[3]*Tech-X Corporation, Boulder, CO 80303, USA*

Email: rjx_buaa@sina.com

Abstract: In order to reduce the overall computational time of a PIC (particle-in-cell) computer simulation, an attempt was made to utilize larger time step sizes by implementing multiple solutions of Poisson's equation within one time step. The hope was this would make the PIC simulation stable at larger time steps than an explicit technique can use, and using larger time steps would reduce the overall computational time, even though the computational time per time step would increase. A three-dimensional PIC code that tracks electrons and ions throughout a three-dimensional Cartesian computational domain is used to perform this study. The results of altering the number of times Poisson's equation is solved during a single time step are presented. Also, the size of the time that can be used and still maintain a stable solution is surveyed. The results indicate that using multiple Poisson solves during one time step provides some ability to use larger time steps in PIC simulations, but the increase in time step size is not significant and the overall simulation run time is not reduced

1. Introduction

Basically, there are three types of models used in plasma modeling: fluid models, fully kinetic models (or particle-in-cell), and hybrid models. The advantage of the particle-in-cell (PIC) method is its physical fidelity and the detailed results that it produces. The disadvantage of PIC methods, especially the explicit technique, is very long computational time. It is desired to find a technique to reduce these computational times. This was the motivation for this investigation.

In order to decrease the computation time for the PIC method, the implicit PIC technique has been under development for a number of years. Implicit methods in the most general sense fall under one of two categories, the direct implicit method [2] or the implicit moment method [3]. The direct implicit method is a predictor-corrector type of method which requires an iterative solution to be fully implicit. The implicit moment method aims to use fluid assumptions in order to estimate spatial properties, such as the electric field and the magnetic field, at a future time. Though the goal of the implicit and semi-implicit methods is to increase the size of the time step, while keeping the simulation stable, they are generally still restricted by the need for the particle to cross less than one cell in a single time step [4]. This is a restrictive limitation in sheath regions of high density plasmas where the relevant length scale is on the order of the Debye length.

In this paper, using multiple Poisson equation solves within one time step is investigated. The authors have not been able to find this technique documented in the literature and thus have undertaken this work to test this idea. Because we have not been able to find this technique documented in the literature, we will give it a name. We will call this method of PIC solution, the multiple Poisson solves technique. The ultimate goal of testing the multiple Poisson solves technique is to decrease the overall computational time of the PIC simulation by increasing the maximum stable time step size that can be used in the simulation. Of course, the time step has to be large enough to overcome the extra computational time for solving Poisson's equation several times for each time step.

2. Mathematical Model

2.1. Charged particle kinetics equations

Two important equations for describing the motions of charged particles in an electric field are the Newton-Lorentz equation and Poisson's equation. The Newton-Lorentz equation is

$$m\frac{d\vec{v}}{dt} = \vec{F} = q\left(\vec{E} + \vec{v} \times \vec{B}\right), \tag{1}$$

where m is the mass of the particle, $\vec{v}$ is the velocity, t is the time, $\vec{F}$ is the force, q is the particle's charge, $\vec{E}$ is the electric field, and $\vec{B}$ is the magnetic field. While the electric field vector can change both the speed of the particle and its direction of travel, the magnetic field can only change the direction of the particle's travel and is unable to increase or decrease its speed.

The electric field value is determined from the gradient of the electric potential, $\vec{\Phi}$, using

$$\vec{E} = -\nabla\vec{\Phi}. \tag{2}$$

The electric potential is found using Poisson's equation,

$$-\frac{\rho}{\varepsilon_0} = \nabla^2\Phi, \tag{3}$$

where ρ is the charge density and ε_0 is the permittivity of free space. The charge density is found using the particle number densities of ions, n_i, and electrons, n_e,

$$\rho = q\left(n_i - n_e\right). \tag{4}$$

The three-dimensional PIC code used in this work includes the solution of Poisson's equation to determine the electric fields throughout the computational domain that are caused by the boundary conditions and the charged particles. Also included is the solution of the Newton-Lorentz equation of motion to obtain the path of all ions and electrons when traveling through these electric fields. While the developed PIC code can handle magnetic fields, no magnetic fields are included in the simulations performed for this paper. Full coupling between the particle motion calculations (equation 1) and the electric field calculation (equations 2, 3, and 4) is maintained. The details of the solving process of equations (1) through (4) can be found in reference [5] and are not given here.

2.2. Explicit Stability Criteria

The explicit PIC algorithm is subject to a number of stability criteria [6] which limit its performance capabilities. These criteria define limits of the numerical parameters used, which include the maximum time step size, maximum grid spacing, and minimum number of computational particles per cell. These limiting values typically depend on characteristics of the plasma such as plasma frequency, Debye length, and electron velocity (electron temperature). The criteria which apply to the explicit PIC technique are:

$$n_{cell} \geq 20, \tag{5}$$

$$v_{p,x}\Delta t \leq \Delta x, v_{p,y}\Delta t \leq \Delta y, v_{p,z}\Delta t \leq \Delta z \tag{6}$$

$$\omega_{pe}\Delta t \leq 2 \tag{7}$$

and

$$\frac{\lambda_D}{0.3} \leq \min(\Delta x, \Delta y, \Delta z). \tag{8}$$

In equations (5) through (8), n_{cell} is the number of computational particles per cell for one species, $v_{p,x}$, $v_{p,y}$ and $v_{p,z}$ are the particle velocities in each of the three Cartesian directions, Δx, Δy and Δz are the control volume sizes in each of the three Cartesian directions, ω_{pe} is the plasma frequency, and λ_D is the Debye length.

2.3. Multiple Poisson Solves Technique

The idea behind the multiple Poisson solves technique is to prevent unrealistic charge separation during one time step. This technique is classified as semi-implicit because it evaluates the electric fields a number of times during the course of a single time step. In the multiple Poisson solves technique some fraction of the total number of computational particles are moved using the Newton-Lorentz equation, then Poisson's equation is solved using the new positions of the particles, then another fraction of the computational particles are moved using the Newton-Lorentz equation, then Poisson's equation is solved again, and so on. This process is repeated within a given time step until all computational particles have been moved. Figure 1 shows the flow chart for the multiple Poisson solves technique, and, for comparison purposes, the flow chart of the explicit technique. The difference between the two techniques can be seen from this figure. This figure shows that multiple Poisson solves can be done during the movement of the electrons and during the movement of the ions. For the results presented in this work, multiple solves are only done during the electron moves. Multiple solves during the ion moves have been tried and shown to have little effect. The reason is that ions move such a small distance relative to the electrons for a given time step.

An important input parameter used in the multiple Poisson's solve technique is Npps. This stands for the number of particles per solve. This number indicates how many computational particles are moved between Poisson solves. The lower this number the more Poisson solves per time step. The number of Poisson equation solves per time step can be obtained by dividing the total number of computational electrons by Npps.

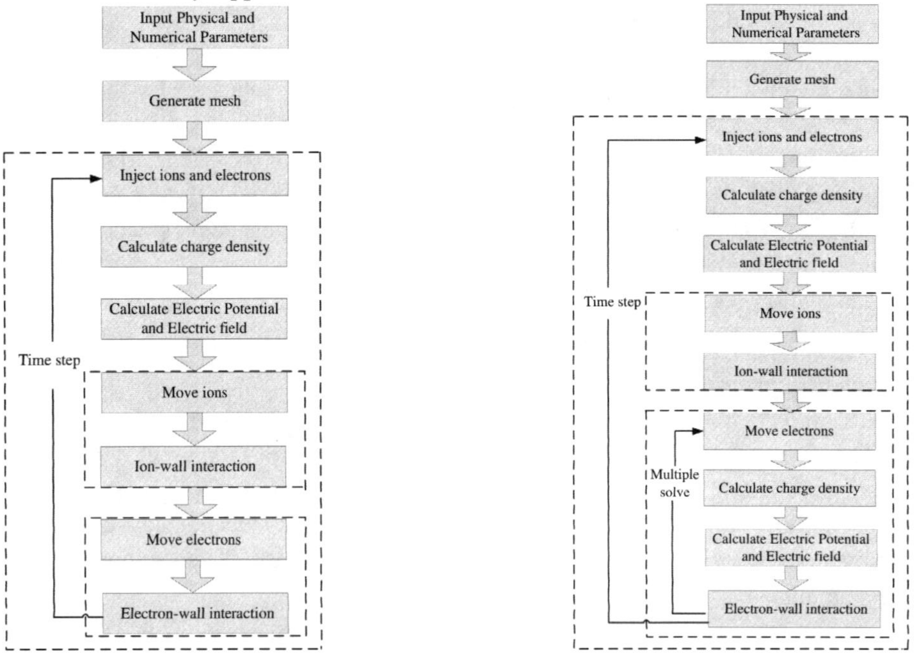

(a) Explicit Technique (b) Multiple Poisson Solves Technique

Figure 1. Flow charts of the explicit PIC technique and multiple Poisson solves PIC technique.

3. Results

3.1. Geometry and Physical Conditions

The geometry used to test the multiple Poisson solves technique is a simple cubical box with the wall at x = 0 set to 3 volts and the wall at x = 10mm set to 6 volts. The walls at y =0, y = 10mm, z = 0, and z = 10mm are dielectric walls that are specular reflectors of all particles incident upon them. This particular problem could have be simulated in two dimensions, but the computer program developed to perform this work is a full three-dimensional simulation. Validation results for this newly developed computer program are given in reference [5].

Only electrons and positively charged hydrogen ions are considered in this analysis and these particles are injected uniformly throughout the cubical box at equal rates of 2.5×10^{14} particles/second. Thus the plasma in the box is a low density plasma with no neutral particles and only consists of electrons and positive ions. The initial velocity distribution of the electrons and ions are uniform and the magnitudes of these velocities are 100,000 m/s and 1,000 m/s, respectively. The total number of the particles increases as the time increases, until the number of injected particles is equal to the rate at which particles are lost to the absorbing 3 and 6 volt walls.

The cell size used in all simulations is $0.5\times0.5\times0.5$mm; thus 8000 cells are used to cover the cubical computational domain. The weighting of the computation particles is 500 real particles per computational particle. The time step size is varied depending on the particular study being undertaken. The total simulation time to reach steady state is about 1×10^{-5} seconds which is 10,000 time steps when a time step size of 1×10^{-9} seconds is used.

3.2. Stable Multiple Poisson Solves Technique Results

Figure 2 shows electric potential profiles across different planes of the cubical computational domain that intersect the z-axis at different locations. All of the results are basically the same and the figures show that the electric potential in the bulk of the plasma is 6.4 volts. The sheaths that form in the vicinity of the two walls with voltages of 3 volts and 6 volts are clearly visible. The explicit PIC technique and analytical results give a bulk plasma potential of 6.3 volts. Thus the multiple Poisson solves technique increases the plasma potential by 0.1 volts. The results shown in Figure 2 were obtained with the multiple Poisson solves technique using a time step size of 1×10^{-9} seconds, a Npps of 30,000, and a total number of time steps of 10,000. A Npps value of 30,000 for this case means Poisson's equation was solved 27 times per time step when the solution reached its steady state number of particles. There are about 810,000 particles in the cubical box when the particle number reaches steady state. Poisson's equation was solved fewer times per time step as the electrons were building up to their steady state value.

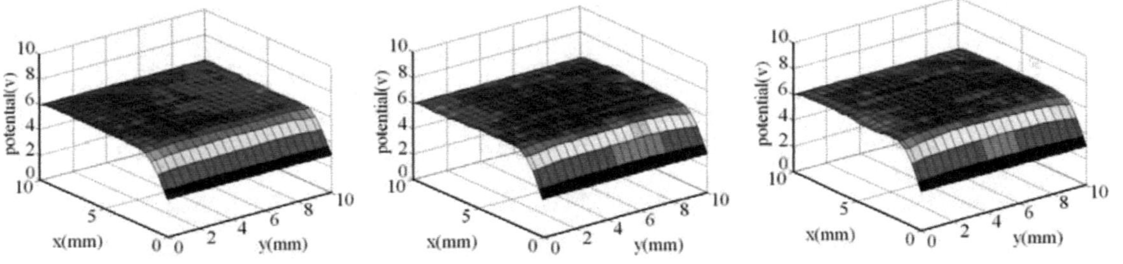

(a) z = 0 mm (b) z = 2.25cm (c) z = 7.25cm

Figure 2. Electric potential profiles at different slices using the multiple Poisson solves technique with a Npps value of 30,000 and a time step size of $1x10^{-9}$ seconds.

3.3. Effect of Time Step and Npps

Figure 3 gives the time history of the total number of electrons in the computational domain for different computational techniques. Figure 3(a) gives the time history of the number electrons in the computational domain using the explicit technique with a number of different time steps. Figures 3(b) and 3(c) provide total number of electron history plots using the multiple Poisson solves technique with different Npps values and times steps of 2×10^{-9} and 4×10^{-9} seconds. Note that the explicit technique results are also shown on these graphs. It can be seen from these three graphs that the explicit technique results are stable when the time step size is less than or equal to 1×10^{-9} seconds, but it is unstable when time steps greater than or equal to 2×10^{-9} seconds are used. The multiple Poisson solves technique is stable for time steps up to and including 2×10^{-9} seconds when the Npps values are 5,000, 10,000 and 30,000. If Npps is 50,000, 70,000 or 90,000 the multiple Poisson solves technique is unstable at time steps of 2×10^{-9}. For a time step of 4×10^{-9} seconds all techniques are unstable. The flat lines seen in Figure 3(c) do not indicate stability, but are flat because particles are leaving the chamber rapidly and the effects of the sheaths are being bypassed. These results clearly indicate that the multiple Poisson solves technique enhances stability at slightly larger time steps if the Npps value is chosen correctly. However, these results also indicate that the multiple Poisson solves technique produces slightly different numbers of electrons in the computational domain than the explicit technique.

Figure 4 gives the electric potentials at the z-direction mid-plane, for different time step sizes, using the multiple solves technique. Reasonable electric potential profiles are seen for time steps of 1 $\times 10^{-9}$ and 2×10^{-9} seconds, but greatly enlarged values are seen with a time step of 4×10^{-9} seconds. The reason for this is the simulation results are unstable with a time step of 4×10^{-9} seconds, and this is obvious if one watches a video of the voltage profiles changing as a function of time.

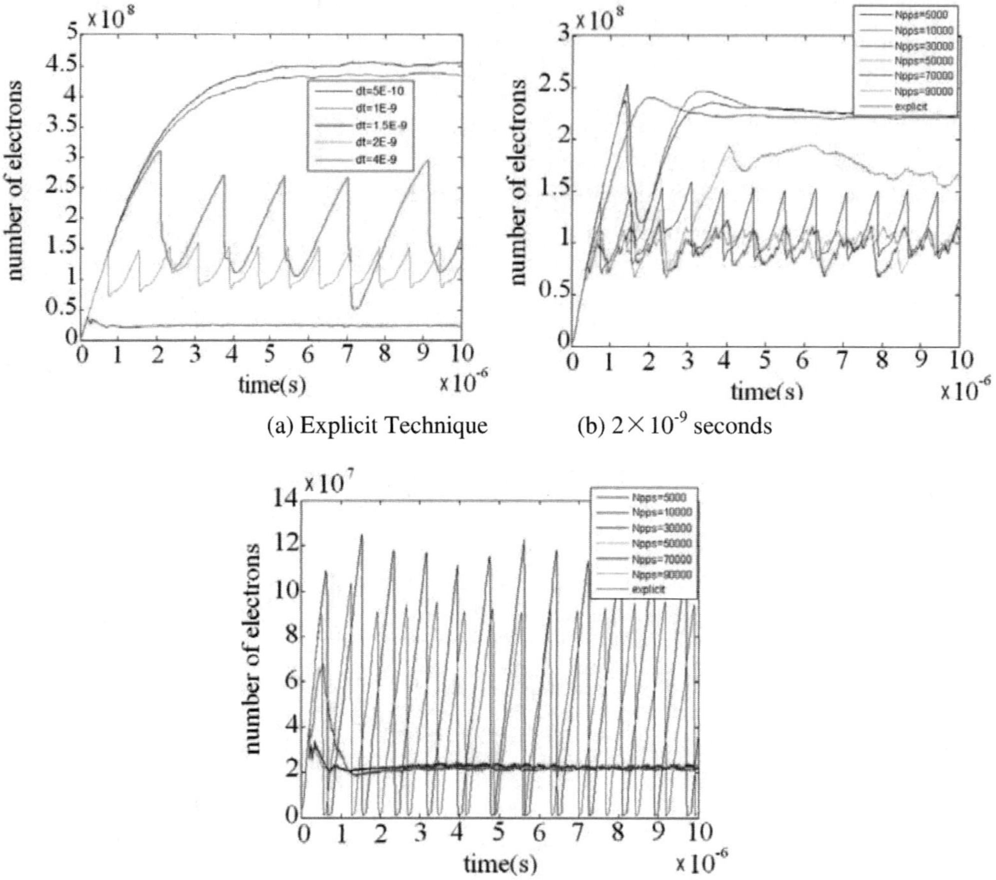

(a) Explicit Technique (b) 2×10^{-9} seconds

(c) 4×10^{-9} seconds

Figure 3. Time history plots of the total number of electrons in the computational domain for (a) the explicit technique using different time step sizes,(b) the multiple Poisson solves technique with a time step size of 2×10^{-9} seconds using different Npps values, and (c) the multiple Poisson solves technique with a time step size of 4 $\times 10^{-9}$ seconds using different Npps values.

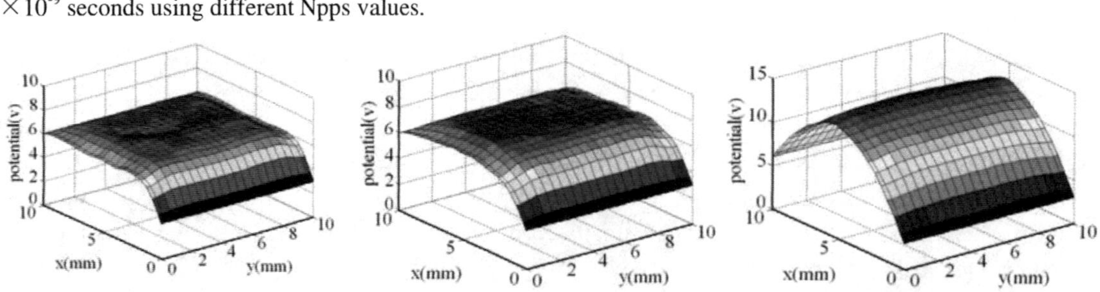

(a) 1×10^{-9} seconds (b) 2×10^{-9} seconds (c) 4×10^{-9} seconds

Figure 4. Electrical potential surface plots at the mid-plane of the z-direction using different time steps with the multiple Poisson solves technique.

Figure 5 gives the electron density as a function of the time step size and Npps. The black line is the explicit result and these results converge when the time step size is less than 1×10^{-9} seconds. When the time step size is bigger than 1×10^{-9} seconds, the results with the multiple Poisson solves technique are different than the explicit technique results; however, decreasing the time step size causes the multiple Poisson solves technique results to converge with the explicit technique results.

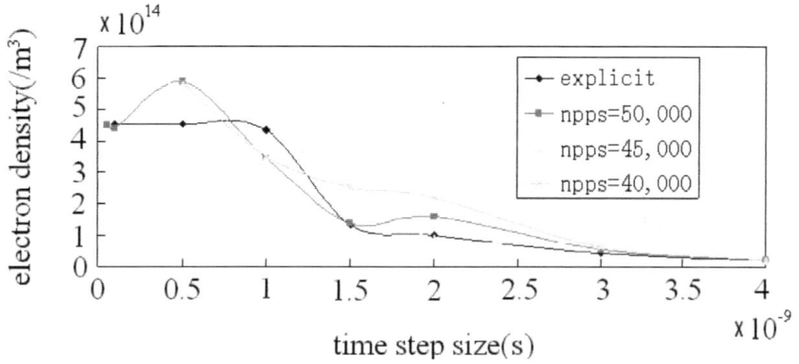

Figure 5. Electron density as a function of the time step size and Npps.

3.4. Computational Times

Table 1 shows the comparison of computational times for different time steps and different values of Npps. The first line in this Table provides times for the explicit technique. The cases were run on a computer equipped with an AMD Phenom II x6 1045 T 2.7 GHz processor. The least computational time with the multiple Poisson solves technique producing a stable result is 3.75 hours using at time step size of 2×10^{-9} seconds and a Npps of 40,000. However, this time is still larger than the time of 1.91 hours obtained with the explicit technique using a time step size of 1×10^{-9} seconds. Thus the goal of decreasing the computation time with the multiple Poisson solves technique has not been obtained.

Table 1. Computational times in hours using different time step sizes and different Npps.

Npps \ Δt	1×10^{-9}s	1.5×10^{-9}s	2×10^{-9}s	3×10^{-9}s
Explicit	1.91	0.80	0.47	0.28*
5,000	120.22	47.40	35.67	5.96*
10,000	59.64	25.76	16.41	3.02*
30,000	19.49	9.59	6.28	1.42*
40,000	19.85	10.04	3.75	0.94*
45,000	18.61	8.18	1.80*	0.79*
50,000	12.59	3.15*	2.56*	0.72*
70,000	5.11*	2.09*	1.04*	0.48*
90,000	3.32*	1.66*	0.93*	0.44*

*unstable results

4. Conclusions

In this work results of the effects of altering the number of times Poisson's equation is solved during one time step are presented. The results include a study of the effect of time step and the number of times the Poisson equation is solved in a single time step via the parameter Npps (number of particles per solve). The results indicate that small increases in the time step size over that which can be used with a purely explicit technique can be obtained; however, overall computational times are not decreased because of the extra time required to perform multiple Poisson solves in a single time step. In addition, the results produced by the multiple Poisson solves technique are slightly different than those produced by the explicit technique, except for the cases where very small time steps are used This work was able to find no advantages of using the multiple Poisson solves technique over a purely explicit PIC technique.

Acknowledgments

The authors would like to thank the National Aeronautics and Space Administration (NASA), Glenn Research Center, for funding this project under the 2011 Phase-II Small Business Technology Transfer Research (STTR) program. We would also like to thank our technical monitors Dr. Jon Van Noord and George Soulas.

References

[1] Mahalingam S 2007 *Particle Based Plasma Simulation for an Ion Engine Discharge Chamber* (Dayton, PhD:Wright State University).
[2] Friedman A, Landon A B, and Cohen B I 1981 *Comments Plasma Phys.Controlled Fusion* **6** 225-236.
[3] Mason R J 1981 *J. Comput. Phys.* **41** 233-244.
[4] Parker S E, Friedman A, Ray S L, *et al.* 1993 *J. Comput. Phys.* **107**, 388-402.
[5] Godar T J, Ren J X, and Menart J A 2014 *Particle In Cell (PIC) Algorithm Advancement for Plasma Modelling of an Ion Thruster Discharge Chamber*, AIAA Region III Student Conference, Cleveland, OH.
[6] Birdsall C K, Langdon A B 1991 *Plasma Physics Via Computer Simulation.* IOP Publishing Ltd, New York.

Computational investigation of flow control by means of tubercles on Darrieus wind turbine blades

K Sevinç[1] G Özdamar[2] U Şentürk[2] and A Özdamar[2]
[1]: FIGES Inc., Tepekule Is Merkezi, No:40/510, 35010 Bayraklı, Izmir, Turkey
[2]: Ege University Faculty of Engineering Mechanical Engineering Department, 35040 Bornova, Izmir, Turkey

E-mail: aydogan.ozdamar@ege.edu.tr

Abstract. This work presents the current status of the computational study of the boundary layer control of a vertical axis wind turbine blade by modifying the blade geometry for use in wind energy conversion. The control method is a passive method which comprises the implementation of the tubercle geometry of a humpback whale flipper onto the leading edge of the blades. The baseline design is an H-type, three-bladed Darrieus turbine with a NACA 0015 cross-section. Finite-volume based software ANSYS Fluent was used in the simulations. Using the optimum control parameters for a NACA 634-021 profile given by Johari et al. (2006), turbine blades were modified. Three dimensional, unsteady, turbulent simulations for the blade were conducted to look for a possible improvement on the performance. The flow structure on the blades was investigated and flow phenomena such as separation and stall were examined to understand their impact on the overall performance. For a tip speed ratio of 2.12, good agreement was obtained in the validation of the baseline model with a relative error in time-averaged power coefficient of 1.05%. Modified turbine simulations with a less expensive but less accurate turbulence model yielded a decrease in power coefficient. Results are shown comparatively.

1. Introduction

Boundary layer control is a major concern in aeronautical research which aims to improve performance and decrease costs in device operation. Mostly, the control is applied to alter the fluid flow on the walls which are the streamlined lifting surfaces. The useful force that is created by the fluid on these walls is heavily influenced by the parameter known as the angle of attack. Each airfoil section has its own critical angle of attack value above which the phenomenon called stall occurs where flow separates from walls and sharp decrease in the useful force is observed. The boundary layer control methods briefly aim to delay stall or increase the critical angle of attack if possible. The methods are classified into two: (1) Passive control methods that don't require any energy input such as vortex generators, fixed slats, leading edge tubercles etc., (2) Active control methods where a certain amount of energy is required in application such as, suction, blowing, synthetic jets, plasma actuators etc. In particular, tubercles (figure 1, 2, 3, 4) take part in vortex generation by the excitation of flow that results in an increase in the lift in the post-stall region [1].

Journal of Physics: Conference Series **640** (2015) 012034 doi:10.1088/1742-6596/640/1/012034

Figure1. Tubercles on humpback whale flippers [1].

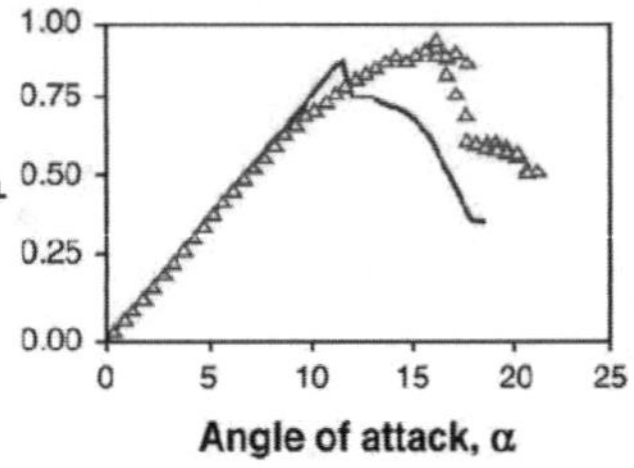

Figure 2. Photographs by W.W. Rossiter Experimental lift curves [2].

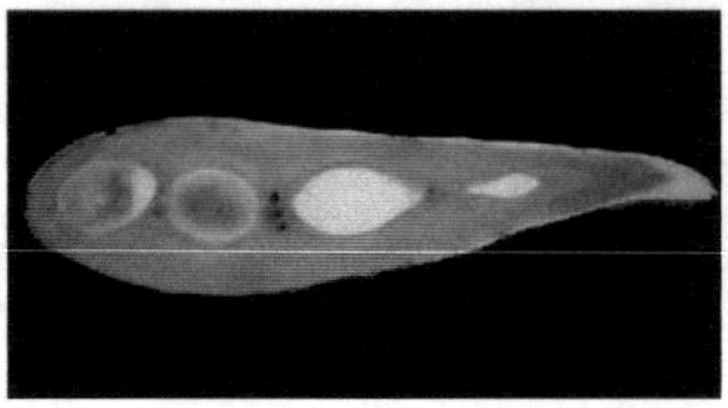

Figure 3. Tubercles on humpback whale flippers [1].

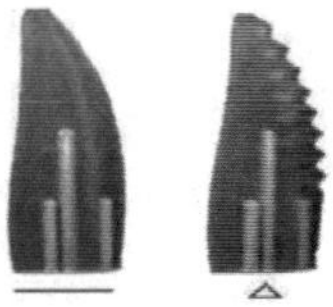

Figure 4. Photographs by W.W. Rossiter Experimental lift curves [2].

Wind turbine performance relies on blade aerodynamics where similar flow characteristics are observed. Modern horizontal axis wind turbines have twisted blades to account for the change in the angle of attack at design conditions. However, turbines often operate at off-design conditions where stall occurs and causes sharp decrease in performance. In order to improve the energy efficiency of the machinery, efforts are made to control the boundary layer on the airfoil cross-section [3, 4, 5]. Flow computations based on numerical methods (finite volume, finite element methods etc.) have become standard tools in turbomachinery design. Howell et al. [6] reported the performance of a vertical axis wind turbine found from computational fluid dynamics simulations. Authors studied the problem in 2D and 3D with a transient sliding mesh model and compared the results with experiments. Trivellato and Castelli [7] presented the influence of the Courant-Friedrichs-Lewy (CFL) number on the computed torque of a 2D vertical axis wind turbine model and proposed a relation for time step selection. In this work, the results of a computational study is reported which investigates the effect of the leading edge tubercles on the performance of a three-bladed, vertical axis wind turbine. Numerical research was conducted using the commercial code ANSYS Fluent 14.5. In order to validate the methodology, a baseline design was modeled and simulated based on the experimental performance testing given by Bravo et al. [8]. The baseline turbine blades had constant cross sections with a NACA 0015 profile. This procedure was followed by the implementation of the leading edge tubercles on the turbine. Geometric parameters of the tubercle geometry were determined by following the experimental research conducted by Johari et al. [9] on a NACA 634-021 profile. Current study focuses on a certain operating point of the turbine with a tip speed ratio of 2.12. Qualitative and quantitative results derived from CFD simulations are shown and discussed.

2. Turbine details

The wind turbine under consideration is a vertical axis, three-bladed Darrieus turbine which is also known as the H-type rotor. The velocity triangle obtained from a two-dimensional analysis at any cross section of the blades is given in figure 5. Here, **V** is the free stream velocity; **U** is the circumferential velocity at radius (R) and ω is the relative velocity as seen by an observer moving with the blades. The tip speed ratio of the turbine is defined as

$$\lambda = \frac{|\mathbf{U}|}{|\mathbf{V}|} = \frac{\Omega R}{V} \tag{1}$$

where Ω stands for the angular velocity. From the geometry, the angle of attack (α) is found to be a dependent of the blade azimuthal position (θ) and tip speed ratio as

$$\alpha = \tan^{-1}\left(\frac{\sin\theta}{\cos\theta + \lambda}\right) \tag{2}$$

which indicates that the angle of attack itself is a time-dependent parameter since $\theta = \Omega t$.
Pressure differences between the suction and pressure sides of the blade create the useful moment (M) which is non-dimensionalized as a torque coefficient

$$C_M = \frac{M}{\rho V^2 L R^2} \tag{3}$$

and the power coefficient which reflects the energy efficiency of the blades is defined as

$$C_P = \frac{M\omega}{\rho V^3 L R} \tag{4}$$

For a vertical axis wind turbine, C_M and C_P are both time-dependent parameters. For convenience, their time-averaged values are used to indicate the performance as follows (overbars will be omitted after Eq. 5):

$$\bar{C}_{M,P} = \frac{1}{T}\int_0^T C_{M,P}\,dt \tag{5}$$

The baseline model in this study has constant NACA 0015 cross-section with a chord (c) of 0.4 m, blade span (*L*) of 3 m and radius (*R*) of 1.25 m. figure 5. shows the wind tunnel performance testing results reproduced from Bravo et al. [8] for the case with a free stream velocity of 8 m/s.

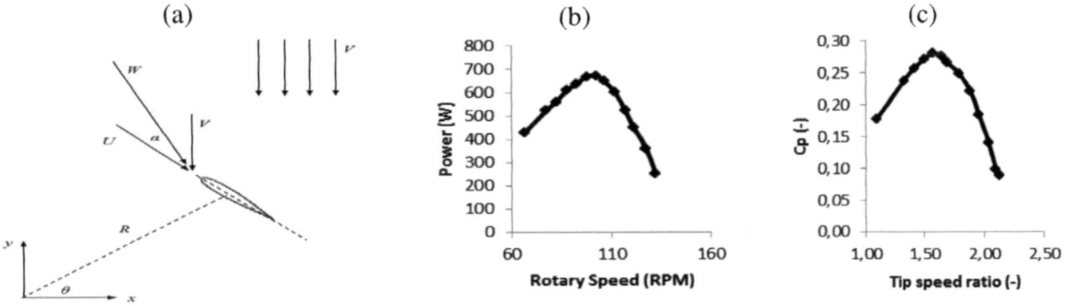

Figure 5. (a) The velocity triangle on a blade cross section (b, c) Baseline turbine performance curves reproduced from Bravo et al. [8].

Minimum and maximum values of the tip speed ratio are estimated from the original plot as 1.09 and 2.12 respectively. For these cases, angle of attack variations based on Equation 2 are plotted in figure

3. Here, dotted lines show the static stall angle as 12° for a NACA 0015 airfoil which clearly demonstrate that the turbine operates in a stalled condition mostly.

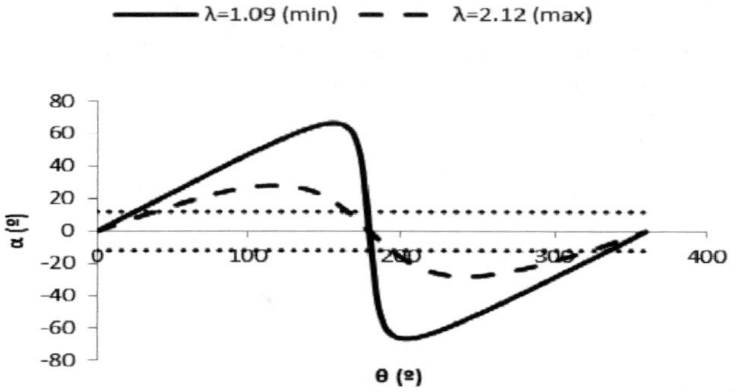

Figure 6. Angle of attack variations for tip speed ratio extrema.

In order to look for a possible improvement of the energy efficiency of this turbine, the influence of the leading edge tubercles is investigated. A suitable geometry is obtained from Johari et al. [9] where authors conducted parametric research of leading edge tubercles on a NACA 634-021 profile. Here, case 8M was chosen since it yielded better lift characteristics in the post-stall region while having almost the same drag compared to other tubercle cases. It should be emphasized that since the airfoil types are essentially different, the exact same post-stall behavior is not expected. Nevertheless, it clearly provides information on geometric parameters and implementation. figure 7 shows the extruded view of the blades before and after modification.

(a)

(b)

Figure 7. Extruded view of the blades (a) Baseline model (b) Modified model

3. Computational study

Under the assumptions of three-dimensional, unsteady, viscous and incompressible flow, the velocity and the pressure fields were calculated by using commercial CFD code ANSYS Fluent 14.5. To accelerate convergence, initial fields were calculated using the Multiple Reference Frames Model (MRFM) which is basically a steady-state approach with different reference frames attached to zones that have relative motion with respect to each other. Transient simulations were run using the Sliding Mesh Model (SMM). The governing equations for the SMM are as follows [10]:

$$\frac{\partial \rho}{\partial t} + \nabla \rho(\mathbf{V} - \mathbf{U}) = 0 \tag{6}$$

$$\frac{\partial(\rho \mathbf{V})}{\partial t} + \nabla \cdot [\rho(\mathbf{V} - \mathbf{U})] = -\nabla P + \nabla \cdot \overline{\overline{\tau}} + F \tag{7}$$

Here, ρ is the fluid density, P is the pressure field, F stands for the body forces and $\overline{\overline{\tau}}$ is the stress tensor $\overline{\overline{\tau}} = \mu[\nabla \mathbf{V} + (\nabla \mathbf{V})^T - (2/3)\mathrm{I}\nabla \cdot \mathbf{V}]$. The problem domain was made up of two zones (rotating

and stationary) with overlapping interfaces. The stationary zone represents the wind tunnel having a cross section of 9 m×9 m as in Bravo et al. [8] with extended downstream yielding a total length of 22.5 m. Rotating zone enclosed the three blades of the turbine. Any other parts such as the rotating shaft, connecting elements etc. were excluded in the model.
Standard air properties were used (ρ=1.225 kg/m^3 and μ=1.7894e-05 kg/(ms)) resulting in a Reynolds number based on chord ($Re_c = \rho Vc/\mu$) of ~219000. Velocity inlet (8 m/s) and pressure outlet (0 Pa gauge static pressure) boundary conditions were defined and no-slip wall boundary condition was assigned on the blades and the tunnel walls. For the validation of the baseline model which is relatively simpler in terms of geometry, *k-ω* SST turbulence model was used with default model constants. Hybrid grids were used with inflation layers on solid walls to resolve the boundary layers where the first cell height, growth rate and total number of inflation layers were ~0.1 mm, 1.2 and 18, respectively. Due to the grid quality considerations, baseline grid had approximately 9 million cells. Minimum and average orthogonal qualities were approximately 0.19 and 0.87 for the baseline blade while 0.004 and 0.82 for the modified blades. Using the same grid properties on the modified model yielded an increase in grid size and computation time. For this reason, the turbulence model was switched to k-ε realizable with default model constants. Pressure-based solver was used with the PISO scheme for pressure-velocity coupling. Second-order upwind schemes for momentum, turbulent kinetic energy and specific dissipation rate equations and a second order implicit discretization for transient terms were used. In addition to residual monitoring, the moment coefficient on each blade was monitored and recorded at each time step. Time step size was defined as 1e-03 s and calculated fields were recorded once in every 15° turn.

4. Results and discussion

Instantaneous torque on a single blade and total torque on the blades are shown as a function of dimensionless time (t*=t/T) in figure 8 comparative evaluation of curves depict that the baseline model is superior in general.

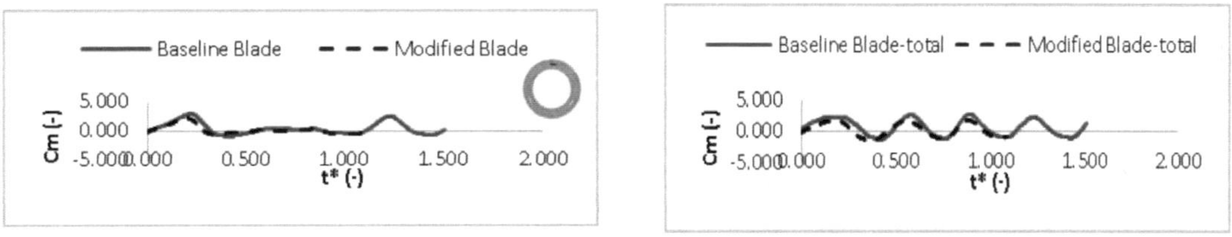

Figure 8. Torque coefficient versus dimensionless time. The initial blade position is shown on the plot.

A closer look on the blade tips are given in figure 9. These figures are also in line with the plot in figure 6 in terms of the angle of attack so that the blade in figure 9a is in the unstalled region.

(a) (b)

Figure 9. Comparative surface streamlines of two blades. The blade positions are shown on the figures where the flow is from left to right. (a) low angle of attack, (b) high angle of attack.

Time-averaged C_P values for both blade types are presented in table 1 for an overall performance comparison. Results show that, for the investigated tip speed ratio, CFD results are in good agreement with the experiments given by Bravo et al. [8]. However, the turbine with leading edge tubercles yielded a decrease of 54% in C_P yielding a value of 4.0%. This result is discussed as follows: In the simulations for the baseline geometry, *k-ω* SST turbulence model was used. After an iterative mesh refinement study, maximum and average y^+ values obtained on the blades were 10.4 and 2.9 respectively. Aiming for a low y^+ value with the grid quality mentioned above ensued a good estimate for the time-averaged power coefficient. The modified geometry on the other hand was modeled with the k-ε realizable turbulence model where maximum and average y^+ values obtained on the blades were 320.9 and 14.9 respectively. In the recent years *k-ω* SST turbulence model gained popularity due to its superiority compared to other one-equation and two-equation models especially for strongly separated airfoil flows. It's believed that more accurate performance estimation would be achieved with *k-ω* SST turbulence model for the modified blade. However, due to the grid size, current study exploited the k-ε realizable turbulence model for the modified blade.

Table 1. Power coefficient comparison.

	Bravo et al. [8]	**Baseline (CFD)**	**Modified (CFD)**
C_P (%)	8.8	8.7	4.0

5. Conclusions

The numerical investigation of the influence of leading edge tubercles on vertical axis wind turbines has been presented in this study. Simulations were run for a single tip speed ratio for the baseline and modified blade forms. Post-processing tools were exploited to gain a better understanding of the flow around the blades. An overall conclusion would be that this turbine type is prone to stalled operation and boundary layer control would be a way to improve performance. Current study is confined in obtaining a solution field for the modified geometry with the k-ε turbulence model. Results for this tip speed ratio yielded a decrease in power coefficient about 54%. Further studies will likely to involve research on the use of *k-ω* SST turbulence model, grid sensitivity analysis and performance estimation for several other tip speed ratios.

6. References

[1] Fish F E and Lauder G V 2006 Passive And Active Flow Control By Swimming Fishes And Mammals *Annual Rev. of Fluid Mechanics* **38**.1 193-224.

[2] Miklosovic D, Murray M, Howle L and Fish E 2004 Leading-edge Tubercles Delay Stall on Humpback Whale (Megaptera Novaeangliae) Flippers *Physics of Fluids* **16**.5 L 39.

[3] Johnson S J, Van Dam C P and Berg D E 2008 Active Load Control Techniques for Wind Turbines. *Sandia Report*: SAND 2008-4809.

[4] Potočar E, Branko Š, Marko H and Matjaž E 2012 Control of Separation Flow over a Wind Turbine Blade with Plasma Actuators Strojniški Vestnik – *Journal of Mechanical Engineering* **58**.1 37-45.

[5] Pechlivanoglou G 2013 Passive and active flow control solutions for wind turbine blades PhD Thesis, Berlin Technical University.

[6] Howell R, Ning Q, Jonathan E and Naveed D 2010 Wind Tunnel and Numerical Study of a Small Vertical Axis Wind Turbine *Renewable Energy* **35**.2 412-22.

[7] Trivellato F and M. Raciti C 2014 On the Courant–Friedrichs–Lewy Criterion of Rotating Grids in 2D Vertical-axis Wind Turbine Analysis *Renewable Energy* **62** 53-62.

[8] Bravo R, Tullis S and Ziada S 2007 Performance testing of a small scale vertical axis wind turbine, *Canadian Conf. on Appl. Mechanics*.

[9] Johari H, Henoch C, Custodio D and Levshin A 2007 Effects of Leading-Edge Protuberances on Airfoil Performance *AIAA Journal* **45**.11 2634-642.

[10] ANSYS Fluent 14.5 Theory Guide, ANSYS Inc. 2010.

Journal of Physics: Conference Series **640** (2015) 012035 doi:10.1088/1742-6596/640/1/012035

Numerical Simulation of Cavitation Characteristics for Pump-jet Propeller

Yao Shi, Guang Pan, Qiaogao Huang, and Xiaoxu Du
School of Marine Science and Technology, Northwestern Polytechnical University, China

E-mail: sy880408@163.com

Abstract. With $k-\varepsilon$ turbulent model, non-cavitating performance of a pump-jet propeller was obtained by calculating RANS equations. The comparison between calculation results and experiment data shown that the numerical model and method was reliable. The cavitating hydrodynamic performance of it was calculated and analyzed with mixture homogeneous flow cavitation model based on Rayleigh-Plesset equations and sliding mesh. The effects of different inlet velocity ratio, cavitation number and flow velocity on cavitation characteristics of pump-jet were studied. When the cavitation occurred on the blades, the propeller thrust and torque decreased significantly, thereby causing open water efficiency reduced 15%. For the same cavitation number, as the inlet velocity ratio decreased, the pump-jet propeller blade cavitation phenomenon was more obvious. While for the same ratio, the smaller the number of cavitation, cavitation phenomenon was more remarkable. The more significant was that while the cavitation number was greater than a certain value, the blade cavitation phenomenon disappeared.

1. Introduction

Pump-jet propeller is a new type of underwater propulsion system, its main characteristic is to use a single rotor to boost and the application of deceleration ducted which could make the propeller working in a low velocity environment and improve the propeller cavitation performance. However, with the development of the submarine body to the direction of large-scale and high speed, it makes the increased probability of cavitation of jet propulsion pump. Cavitation would not only reduce the propulsion performance, produce cavitation erosion, cause the vibration of underwater navigation body, but also produce the enough cavitation noise to expose itself. Therefore, understanding and mastering the accurate forecasting technology of pump-jet cavitation becomes more and more important.

At present, the study of cavitating flows has two main methods-experiment and numerical analysis. Experimental research can accurately reflect the features of flow field variation, but it needs a lot of manpower, financial resource, and some complex conditions can't be studied; however, numerical analysis has advantages of full information, the unlimited objective condition, the cavitating flows analysis plays an increasingly important role.

Up till now, there are less public reports about the pump-jet cavitating flows, mainly concentrated in the aspect of propeller and hydraulic machinery. Liu Zhanyi[1] forecast the hydrodynamic performance of spray pump propeller based on CFD technology, the simulation results were similar with the test results, but it didn't involve the phenomenon of cavitating flows. Shi WeiDong[2] studied the

cavitation characteristics of axial flow pump under different tip clearance by using CFX software and the results agreed well with the test results. Yang Qiongfang[3] obtained E779A propeller cavitation by using Sauer cavitation model and modified shear stress transport turbulence model. Ji Bin[4-5] simulated the unsteady cavitation of propeller flow of three dimensional airfoils and propeller by using equilibrium flow cavitation model and turbulence modification model based on filter, the length and velocity field of cavitating flows and pressure coefficient agreed well with the experimental values.
Due to the cavitating flows involved quality transmission, pump-jet flow field which had the characteristic of unsteady and high Reynolds number, the numerical simulation of pump-jet cavitating flow field is difficult. Based on theory of viscous flow and multiphase flow, this thesis solved the pump-jet three-dimensional full channel cavitating steady flow field by using homogeneous multiphase model, turbulent model and sliding grid technique; studied the pump-jet performance under different speed ratio, cavitation number and inflow speed, the relevant experimental results matched well with the paper's results, and what's more, it provides a certain technical support for research and design of pump-jet in the future.

2. Numerical Computation Method

2.1. Flow Filed Basic Control Equation

Mixing multiphase flow model to solve the navier-stokes equation which based on viscous flow theory value was the major method. Gas-liquid mixture two-phase flow model had considered the interaction and alternate-slip velocity of flow field in the process of phase change, and lead into the cavitation model by phase change rate. Assuming that the density of gas and liquid two-phase flow of flow field were constant, treated the gas- liquid two-phase flow as unified mixed flow with a changeable density, its density is a function of a gas phase volume fraction. On the one hand, the navier-stokes equation was used to solve the physical quantities of unified mixed flow, the physical quantities could be expressed in average volume of gas phase and liquid phase. On the other hand, the introduction of gas phase transmission equation could solve the gas phase volume fraction. Looked the mixed flow as a uniform fluid, the control equation was[7]:

(1) Mixed phase continuity equation:

$$\begin{cases} \dfrac{\partial \rho}{\partial t} + \nabla \cdot (\rho \boldsymbol{u}) = 0 \\ \dfrac{\partial}{\partial t}(\rho f) + \nabla(\rho f \boldsymbol{u}) = m_{fg} \\ \dfrac{1}{\rho} = \dfrac{f}{\rho_p} + \dfrac{1-f}{\rho_q} \end{cases} \qquad (1)$$

(2) Momentum equation of mixed phase flow:

$$\frac{\partial(\rho \boldsymbol{u})}{\partial t} + \nabla \cdot (\rho \boldsymbol{u} \cdot \boldsymbol{u}) = -\nabla p + \frac{1}{3}\nabla\left[(\mu + \mu_t)\nabla \cdot \boldsymbol{u}\right] + \nabla \cdot \left[(\mu + \mu_t)\nabla \boldsymbol{u}\right]$$

In equation: ρ is density of mixed phase flow, ρ_p is density of cavity phase, ρ_q is density of flow phase, f is mass component of cavity phase; $\boldsymbol{u}$ is velocity vector of mixed phase, m_{fg} is interphase transport quality when cavity is born and collapsed, μ_t is Mixed viscous coefficient.

2.2. Cavitation Model

The mass transfers between cavity and water lead to the generation and condensation of cavitation, cavity generation and condensation rate is controlled by the following Rayleigh-Plesset equation [8-9]:

$$R_B \frac{d^2 R_B}{dt^2} + \frac{3}{2}\left(\frac{dR_B}{dt}\right)^2 + \frac{2\sigma}{\rho_q R_B} = \frac{p_v - p}{\rho_q}$$

$$\rho = \alpha_p \rho_p + \alpha_q \rho_q$$

$$m_{fg} = 4F\rho_p \pi R_B^2 \left(\frac{2|p_v - p|}{3\rho_q} \right)^{1/2} \operatorname{sgn}(p_v - p)$$

In equation: α_p is volume fraction of cavity phase, α_q is volume fraction of flow phase, R_B is cavity radius, p_v is pressure inside the cavity, p is pressure around the cavity, σ is surface tension factor, F is experience coefficient of cavity generation and condensation.

In this paper, a standard $k-\varepsilon$ turbulence model was adopted to close the equation, the model is widely used in engineer and suitable for solving high Reynolds number flow.

3. Computation Model and Grid

3.1 Computation Model

In this paper, the simulation model of pump-jet propeller is shown in figure 1. The propeller has 11 pieces of rotor blade, 9 pieces of stator vane, the rotors are in the former of the model, the stators are in the behind, rotors rotate clockwise (from the front looking towards back).Considering the clearance flow between the rotors and the inner surface of propeller, the minimum clearance of tip clearance flow is 1mm.

Figure 1. Model of pump-jet propeller

In order to better simulate the flow in the pump-jet, the front inner-axis of pump was designed as a half ellipsoid fair water cap, short axis connected smoothly with the front inner-axis of pump. At the same time, to improve the accuracy of numerical simulation, this paper adopted full channel model for numerical simulation.

According to the nature of the physical problems in this paper, the computational domains could be divided into three parts: rotor domain, stator domain and flow field calculation domain. Among them, the rotor domain was rotation calculation domain, the others were static domains. The paper simulated the interaction between rotor and static domains by the sliding grid technology.

3.2 Computational Domain and Mesh Generation

The performance test of pump-jet is to measure the open water characteristic, in order to make the numerical simulation in conformity with the test condition; this paper puts the pump-jet model into a cylindrical computational domain of same coaxial line with pump-jet, just shown in figure 2. Set D is the diameter of the pump-jet rotor, the distances between the pump model and the entrance boundary of the computational domain and the exit boundary are respectively *4D* and *6D*, the diameter is *10D*.

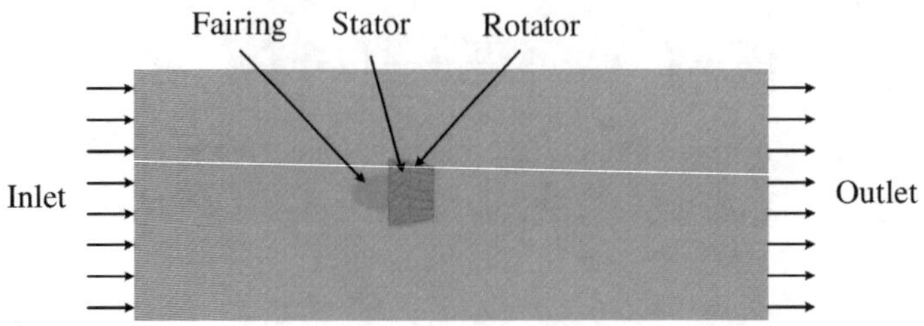

Figure 2. Computational domain and boundary

Use the blocking grid technology to generate high quality structured grid. In the surface of pump vane, use the O type mesh for encryption to accurately simulate the boundary layer. The whole calculation domain can be divided into a total of about 1.5 million cells, the rotor field has 748902 units, the stator domain has 457254 units. Figure 3 (a) and (b) respectively shows the mesh of blades and the mesh of rotator.

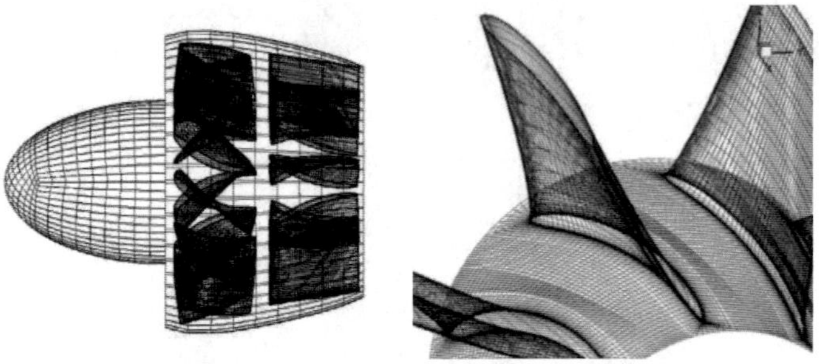

(a) Mesh of blades (b) Mesh of rotator

Figure 3. Mesh of computational domain

3.3 Boundary Condition

To complete the numerical simulation of pump-jet propeller cavitation flow field, we need to set up the reasonable boundary conditions, the specific settings are as follows: the inlet boundary is speed entrance, the turbulence intensity is 5% of default parameter, the cavity volume fraction of entrance is 0, the volume fraction of liquid is 1; the reference pressure is 0Pa, the liquid temperature is 25℃, the saturated vapor pressure is 3574pa, the average diameter of cavity is 2×10^{-6} m; the outlet boundary set as the export pressure, the cavitation number σ is controlled by pressure:

$$\sigma=\left(p_{out}-p_v\right)\left(0.5\rho v_\infty^2\right)^{-1/2}$$

In equation: p_{out} is export pressure, p_v is saturated vapor pressure. Set pipe wall as no-slip wall, the area near wall uses the reinforced wall function model. The cylinder boundary of computational domain meet the requirement that wall can freely slide. The coupled analysis of dynamic-static effect between rotation domain and stationary domain can be realized by setting the Frozen Rotor interface implementation.

4. Research of Pump-jet Cavitation Characteristic

For the convenience of the analysis of calculation results, define the related physical quantities as dimensionless variables which shown in table 1.

Table 1. Define of dimensionless variable

Variable	Define
inlet speed ratio	$J=v\left(nD\right)^{-1/2}$
rotor thrust coefficient	$K_{T_t}=T_t\left(\rho n^2D^4\right)^{-1/2}$

rotor torque coefficient	$K_{Q_t} = Q_t\left(\rho n^2 D^5\right)^{-1/2}$
stator duct system thrust coefficient	$K_{T_s} = T_s\left(\rho n^2 D^4\right)^{-1/2}$
stator duct system torque coefficient	$K_{Q_s} = Q_s\left(\rho n^2 D^5\right)^{-1/2}$
total thrust coefficient	$K_T = K_{T_t} + K_{T_s}$
total torque coefficient	$K_Q = K_{Q_t}$
open water efficiency	$\eta = JK_T\left(2\pi K_Q\right)^{-1/2}$

In table, v is the far field flow velocity, n is the rotor speed, D is the rotor diameter, ρ is the fluid density, T_t and Q_t are respectively the thrust and torque of rotor, T_s and Q_s are respectively the thrust and torque of stator duct system.
This paper calculated the open water performance of pump-jet propeller with different cavitation numbers, different inflow velocity ratio and different inflow velocity, and analyzed the influence law of open water performance on the above factors.

4.1 Calculation Result Analysis with Different Inlet Velocity Ratio
From the velocity ratio definition, control it by changing the speed of pump-jet and inlet speed. Fig.4 is the graph of cavitation and non-cavitation open water efficiency at different inlet velocity ratio of $1.6730 \sim 3.5222$, $\sigma = 1.4721$ and $v = 25.72m/s$.

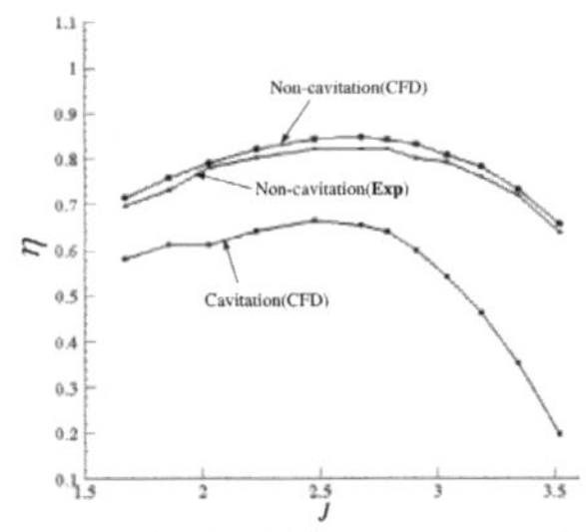

Figure 4. Efficiency at different inlet velocity ratio

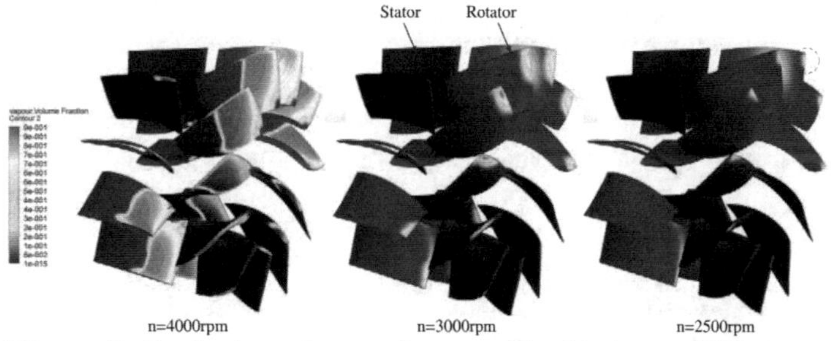

Figure 5. Cavitation of pump-jet propeller blades at different inlet velocity ratio

As can be seen from figure 4, with the increase of inlet speed ratio, the propeller cavitation and non-cavitation open water efficiency showed a trend of increasing first and then decreasing, and reached the maximum when the velocity ratio was 2.4786. When the propeller cavitation didn't happen, the propeller efficiency was above 70% and had a good agreement with the test results[11]. The test value was slightly lower than the numerical simulation results, this might be caused that the simulation conditions were not in accordance with the test conditions. The open water efficiency of propeller cavitation is lower 15% above than the efficiency of propeller non-cavitation. Thus, improving the resistance to cavitation performance of propeller blades in the propeller design process is important. In addition, the results of numerical simulation show that the torque coefficient of stator and rotor system is almost the same, the maximum error is only 1.96%, it shows that the balance feature of pump-jet is very good.
Figure 5 is the cavitation cloud of the pump-jet propeller blade under three kinds of typical working conditions. In the figure, the red area is air, the blue area is water (the same with below). Can be seen from the diagram, at low speed, the cavitation occurs only in the root of the blade tip; as the speed

increases gradually, the cavitation areas gradually expand from the blade tip root to the blade suction surface, when $n = 4000rpm$, almost half of the suction surface happens cavitation.

After the cavitation, the part of propeller blade worked in the gas, it made the thrust and torque reduced, so was the open water efficiency of propeller The open water efficiency is a comprehensive evaluation of ratio, thrust and torque coefficient. Due to the extent of thrust and torque, the open water efficiency showed the trend of increasing and later reducing. It is also well explained the change laws of propeller open water efficiency in Fig.4.

4.2 Calculation Result Analysis with Different Cavitation Number

From the cavitation number definition, we can change the inlet flow speed and outlet pressure to control the cavitation number. In Fig.6, there are the pump-jet propeller open water efficiency curves when the flow velocity is $v = 25.72m/s$, the cavitation number- σ is $1.4721 \sim 8.8869$.

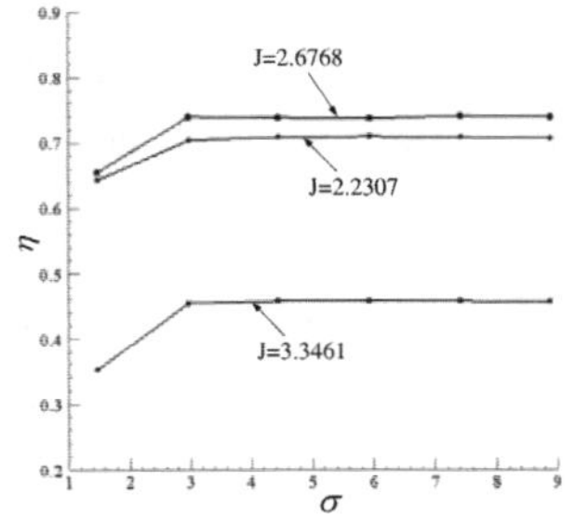

Figure 6. Efficiency at different cavitation number

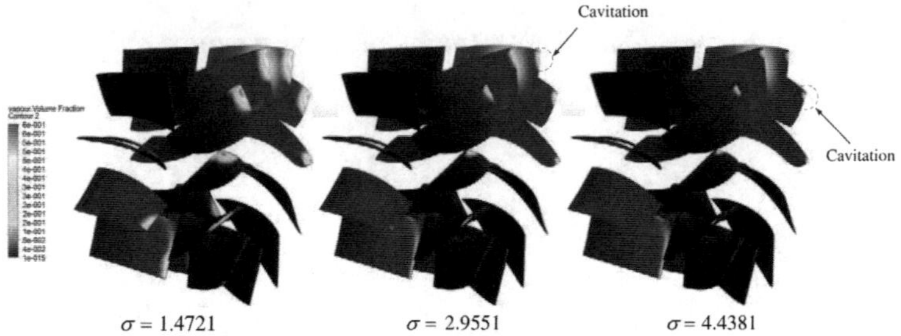

Figure 7. Cavitation of pump-jet propeller blades at different cavitation number

As can be seen from Figure 6, with the increase of cavitation number, the propeller open water efficiency increased firstly and then kept a basic invariable feature under different inlet speed ratio. When the cavitation number was 1.4721, the pump-jet efficiency was minimum; when the cavitation number was greater than 3, with the increase of cavitation number, the propeller open water efficiency remained unchanged. Analyzed its reason, the smaller cavitation number meant the outlet pressure was low, at this point, the rotation areas of the pump-jet propulsion pruned to cavitation; the cavitation number increased, so was the outlet pressure, it made the difference between the evaporation pressure with the local fluid pressure was bigger, so the cavitation would disappear, then the pump-jet propeller open water efficiency remained unchanged. Actual calculation results show that when the cavitation number is bigger than 5.921, the propeller blade cavitation phenomenon exists not any more.

Figure 7 is the cavitation cloud of the surface of the propeller blade with three typical cavitation numbers. From the diagram, the cavitation number is bigger, the propeller cavitation area on the surface of the blade is smaller. When the cavitation number is 4.4381, the cavitation phenomenon occurs only in the tip of the blade. Therefore, with the loss of the cavitation number, the cavitation happens near the clearance area at first, and gradually increases, at the same time, it expands along the radial to the blade suction surface.

5. Conclusion

In this paper, a pump-jet propeller as the research object, its performance has been studied with different inlet speed ratio, cavitation number and inflow velocity, and the grid independence has also been checked, the following conclusions are obtained by analysis:

(1)The numerical simulation of cavitation flow field can better predict the area where the cavitation occurs, and the cavitation phenomenon occurs in the root of the blade tip at first and gradually develops to the suction side and it happens in the suction surface before the pressure surface.

(2)Under the same cavitation number, with the increase of inlet speed ratio, the propeller open water efficiency shows a trend of increasing firstly and decreasing later, and the inlet speed ratio is smaller, the cavitation phenomenon is more obvious.

(3)Under the same inlet speed ratio, with the increase of cavitation number, the propeller open water efficiency increases firstly and then keeps a basic invariable feature. The cavitation number is smaller, the cavitation phenomenon is more obvious; when the cavitation number is greater than a specific value, the blade cavitation disappears.
(4)Under the same rotation speed, with the increase of inflow velocity, the propeller open water efficiency shows a trend of monotonically decreasing.

Acknowledgment
This study was undertaken an part of a project supported by Ministry of education " program for New Century Excellent talents "(NCET-09-0074) , National Natural Science Foundation of China Projects(51279165), and National Defense Basic Research Project(B2720132014), China, which is gratefully acknowledged by the authors.

6. References:
[1] LIU Zhanyi, SONG Baowei, HUANG Qiaogao, et al. Applying CFD Technique to Calculating Successfully Hydrodynamic Performance of Water Jet Pump. 2010 *Journal of Northwestern Polytechnical University*, 28(5):724-729.
[2] SHI Weidong, LI Tongtong, ZHANG Desheng, et al. Effect of tip clearance on the cavitation and flow field of axial flow pumps.2013 *Journal of Huazhong University*, 41(4):21-25.
[3] YANG Qiongfang, WANG Yongsheng, ZHANG Zhihong. Assessment of the Improved Cavitation Model and Modified Turbulence Model for Ship Propeller Cavitation Simulation. 2012 Journal of Mechanical Engineering, 48(9):178-185.
[4] JI Bin, LUO Xianwu, WU Yulin, et al. Numerical analysis of unsteady cavitating turbulent flow and shedding horse-shoe vortex structure around a twisted hydrofoil. 2013 International Journal of Multiphase Flow 51:33-43.
[5] JI Bin, LUO Xianwu, PENG Xiaoxing. Unsteady Numerical Simulation of Cavitating Turbulent Flow Around a Highly Skewed Model Marine Propeller. 2011 *Journal of Fluids Engineering*, 133, 1-8.
[6] JI Bin, LUO Xianwu, WU Yulin, et al. Partially-Averaged Navier–Stokes method with modified k–e model for cavitating flow around a marine propeller in a non-uniform wake. 2012 *International Journal of Heat and Mass Transfer*, 55:6582-6588.
[7] Wang Xianfu. Cavitating and Supercavitating Flows Theory and Applications[M]. Beijing: National Defence Industry Press, 2009.
[8] Brennen C E. Cavitation and Bubble Dynamics[M] . Oxford: Oxford University Press, 1994: 50-79.
[9] Johann F G. Centrifugal Pumps[M]. New York: Springer Berlin Heidelberg, 2008:257-333.
[10] The Test Report about Hydrodynamic Characteristics of Pump-jet Propeller [R]. CSSRC, 2009. (in Chinese)

CFD simulation of hydrogen deflagration in a vented room

I C Tolias[1,2], A G Venetsanos[1], N C Markatos[2] and C T Kiranoudis[2]
[1]Environmental Research Laboratory, National Center for Scientific Research Demokritos, Agia Paraskevi, 15310, Greece
[2]National Technical University of Athens, School of Chemical Engineering, Heroon Polytechniou 9, Zografou, 15780, Greece

E-mail: tolias@ipta.demokritos.gr

Abstract. In the present work, CFD simulations of hydrogen deflagration in a real scale vented room are performed. Two ignition points were simulated: at the wall opposite to the vent (back ignition) and at the center of the chamber (center ignition). The overpressure time series and flame front velocities are compared with the experimental results. The combustion model is based on the turbulent flame speed concept. The turbulent flame speed is calculated based on a modification of Yakhot's equation, in order to account for all the main physical mechanisms which appear in hydrogen deflagrations. Special attention is given to the simulation of Rayleigh-Taylor instability. This instability occurs at the vent area and results in sudden explosion of the mixture that has been pushed outside the chamber at the initial stage of the explosion. The importance of this external explosion to the generated overpressures inside the chamber is highlighted. The agreement between experimental and computational results is satisfactory in both back ignition and center ignition cases.

1. Introduction

Hydrogen is a very promising alternative fuel which is expected to play a significant role in the near future. However, significant safety issues are associated with it. In the case of an accidental release, hydrogen mixes with air and can form a flammable mixture over a wide range of concentrations. An accidental release in a closed space can have catastrophic consequences in the case of an explosion. The confined space will lead to the development of much higher overpressure compared to a similar explosion in an open space. The venting of the hot products out of the closed space is the final protection measure in order to mitigate the damages of the explosion. Even thought there are guidelines for the minimum size of the vent for a given space, there are not reliable and commonly accepted [1]. As a result experiments and simulations for confined hydrogen explosions have to be conducted, in order to gain a better insight on the phenomenon.

An experiment revealing the complexity of vented hydrogen explosions was carried out by Cooper et al. [2]. Four characteristic overpressure peaks were identified: The first one took place because of the opening of the vent, the second one was caused by the external explosion of the fuel which has been pushed outside the room during the initial stages of the explosion, the third one by the reduction of the combustion rate when the flame touches the walls and the last one by the instabilities which enhance the combustion process because of flame-acoustic interaction. A recent analysis of the peak overpressures in vented explosion has been made by Chao et al. in [3], where the effects of fuel, enclosure size, ignition location, vent size and obstacles were examined. Special attention to external explosion has been paid by several authors such as Harrison and Eyre (1987) [4], Jiang et al. (2005) [5], Ferrara et al. (2008) [6], and Proust and Leprette (2010) [7].

In the past years, the increase of computational power has rendered Computational Fluid Dynamics (CFD) as a very attractive methodology for risk assessment of hydrogen applications. Previous CFD studies about hydrogen deflagration in vented spaces have been performed by Baraldi et al. (2010) [8], Bauwens et al. (2011) [9] and Keenan et al. (2014) [10]. In [8] a comparison of CFD models has been made for hydrogen deflagration in a 0.95 m^3 vented vessel. In [9] and [10] hydrogen deflagration in a 64 m^3 room was simulated. In both works special attention to the modelling of the Rayleigh-Taylor instability was paid. The instability was modelled through an additional transport equation which takes into account the generation and the suppression rate of the instability.

2. Experiment overview

Experiments of hydrogen deflagration in real-scale vented room were performed by FM-Global [9]. The top view of the room with its dimensions can be seen in figure 1. The room has a volume of 64 m^3 (4.6 x 4.6 x 3.0 m) and a square vent of 5.4 m^2 is located at a vertical wall. The room is filled with homogeneous hydrogen-air mixture of 18% v/v. Two cases for the ignition position were studied: one at 0.25 m from the center of the wall opposite to the vent (back ignition) and one at the center of the room (central ignition). The pressure was measured at the center of the right wall (point P1 in figure 1). The experimental pressure was filtered with an 80 Hz low pass filter.

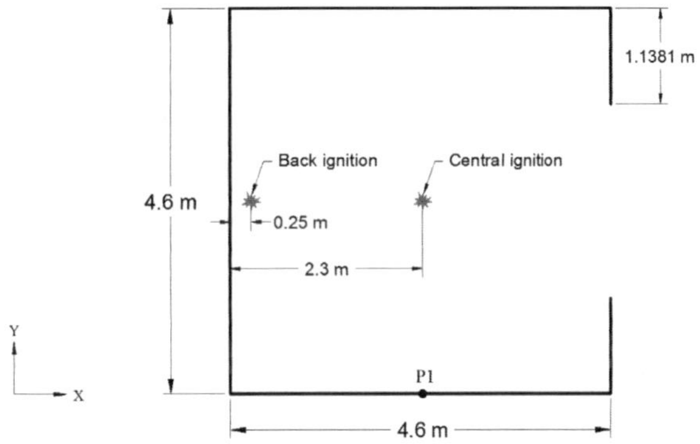

Figure 1. Top view of the experimental facility.

3. Mathematical methodology

3.1. Governing equations

For the CFD simulations the ADREA_HF code was used [11]. The model used solves the space-averaged Navier-Stokes equations along with the energy equation and the conservation equation of each of the mass fraction of the species that take part in the combustion process. Turbulence is modelled using the RNG LES. The set of main equations that was used are presented in [12].

The implemented in ADREA_HF code combustion model [12], originally proposed in [13,14], is based on turbulent flame speed concept. In the utilized model, the calculation of the turbulent flame speed is based on Yakhot's equation for premixed turbulent combustion [15]:

$$S_t = \Xi_k \cdot \Xi_{lp} \cdot \Xi_f \cdot \Xi_{RT} \cdot S_u \cdot \exp\left(\frac{u'}{S_t}\right)^2 \quad (1)$$

where S_u is the laminar burning velocity (constant, equal to 0.64 m/s in our simulations) and u' the subgrid scale velocity. Yakhot's equation has been derived based on the renormalization group theory

Journal of Physics: Conference Series **640** (2015) 012036 doi:10.1088/1742-6596/640/1/012036

and at its original form does not include the Ξ factors. These factors have been included in the equation in order to take into account the mechanisms which can increase the turbulence burning velocity in deflagrations.

One mechanism which leads to an acceleration of the flame front, is the turbulence generated by the flame front itself. As described in [16], the source of this turbulence is the thermal expansion of the combustion products. The upper limit for a flame wrinkling factor due to the turbulence generated by the flame front itself can be estimated by the equation $\Xi_k^{\max} = (E-1)/\sqrt{3}$ where E is the expansion coefficient. The wrinkling factor, Ξ_k, gradually increases from the value of one at the ignition point to the maximum value of $\Xi_k^{\max}$ at the critical radius where the turbulence is fully developed. This critical radius is approximately equal to $R_0 = 1.0 - 1.2\,m$ for near stoichiometric hydrogen-air mixtures. The formula that was used for the wrinkling factor which takes under consideration these parameters is:

$$\Xi_k = 1 + \left(\psi \cdot \Xi_k^{\max} - 1\right) \cdot \left[1 - \exp\left(-R/R_0\right)\right] \tag{2}$$

where R is the distance from the ignition point and ψ is a model constant which represents the level at which the maximum value $\Xi_k^{\max}$ is reached ($0 \le \psi \le 1$). This constant is suggested to be equal to 0.5 for near stoichiometric mixtures and equal to 1 for lean hydrogen-air mixtures. In the present study we set $R_0 = 1.0$, $E = 5.2$ and $\psi = 0.90$.

Another mechanism which leads to acceleration of the flame front is the preferential diffusion which causes an increment of the burning velocity. Based on the work of Zimont and Lipatnikov [17], a correction factor Ξ_{lp} for the laminar burning velocity can be calculated for a given mixture composition. Based on [17], the value of this factor for 18% hydrogen-air mixture composition should be 2.08. This value is very close to the value 1.96 of the ratio of the initial flame speed measured in the experiment to the theoretical flame speed [9]. The value of 1.96 for Ξ_{lp} was chosen in the present simulations.

The fractal structure of the flame front increases flame surface and thus flame speed. The Ξ_f coefficient takes under account this phenomenon. Details about this coefficient can be found in [18].

As the combustion develops inside the enclosure, unburned mixture is pushed outside through the vent. When the flame reaches the vent, this mixture will start burning leading to the external explosion. This explosion seems to have much more strength than combustion models predict. The main reasons that have been proposed for that behavior are: 1) the creation of extensive turbulence in the area outside the vent, because of the jet-like flow and 2) the development of Rayleigh-Taylor instability due to the acceleration of the less dense products through the vent to the more dense unburned mixture. Both reasons potentially lead to an augmentation of the reaction rate which leads to a violent external explosion. Thus, the models should be improved. In this respect a Ξ_{RT} factor has been added to the combustion model in order to reproduce the reaction rate enhancement due to the Rayleigh-Taylor (RT) instability. Its value is constant at the area immediately outside the vent where the RT instability develops and equal to one elsewhere. A similar modelling approach has been used in a large scale vented methane explosion in [19].

3.2. Numerical details

ADREA-HF uses the finite volume method on a staggered Cartesian grid. For the discretization of the convective terms in the momentum equations a second order accurate bounded central scheme was used while in the conservation equations of species and energy a second order accurate bounded linear upwind scheme. For the time advancement, the second order accurate Crank-Nicolson numerical scheme was chosen. The time step is automatically adapted according to prescribed error bands and the desired Courant–Friedrichs–Lewy (CFL) number which maximum value was set equal to 0.8.

The computational domain was extended in all directions outside the room in order to minimize the effect of the implied boundary conditions. Its total size was 42.3 x 60.0 x 30.0 m. A more extended domain was examined and showed zero impact on the results. The total number of active computational cells was approximately 707,000. The length of the computational cells was almost uniform and equal to 0.1 m inside the room and at the area outside the vent. In all exit planes (lateral, front, back and top) the non-reflecting type boundary condition for the normal velocities was chosen, while for the parallel to the exit planes' velocity components, zero-gradient boundary condition was applied. As initial conditions, a stagnant flow field with no turbulence was specified.

4. Results and discussion

In figure 2 the pressure inside and outside the room with and without the inclusion of RT instability modelling for the back ignition case is shown. The outside pressure was monitored at the distance of 1.2 m from the center of the vent. No experimental data are available for the outside pressure. We observe that external explosion strongly affects the pressure inside the room. The enhancement of the combustion rate at the area outside the vent, which was modelled with Ξ_{RT} factor, leads to the increase of the external pressure. This results also in the increase of the internal pressure, due to the suppression of the venting process. Modelling of the RT instability is crucial for the accurate prediction of the maximum overpressure inside the enclosure. When $\Xi_{RT} = 1.0$ (no RT modelling) the external pressure is lower and the combustion model fails to reproduce the maximum pressure value inside the room. In figure 2 the experimental and simulated flame speed velocity normalized by the laminar flame speed (which is equal to $E \cdot S_u = 3.3$ m/s) are also shown. The agreement up to a distance of 3.6 m from the ignition point is excellent. After that point we observe that the predicted flame speed is higher than the experimental.

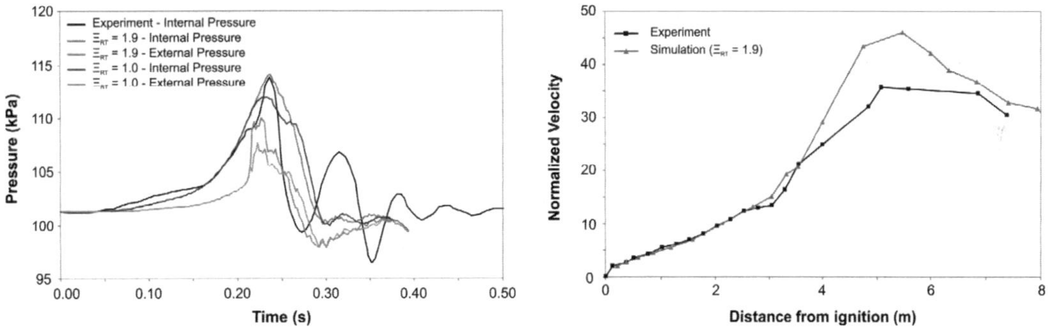

Figure 2. Pressure inside and outside the room for two different values of Ξ_{RT} (left) and flame velocity normalized by laminar flame speed (right) for back ignition case.

Figure 3 shows the hydrogen mass fraction (the grid that was used is also presented), the flow velocity and the subgrid scale velocity for back ignition at t=0.222 sec. We can clearly see the unburned mixture that has been pushed outside the vent. The mixing with the ambient air is very low. This behavior has been observed also in a smaller scale experiment in [7]. The unburned mixture starts burning as the flame front exits the vent. Velocity contours reveal the increase of the flow speed after the vent and the strong turbulence that the flame front experience. Consequently, a violent external explosion occurs.

In figure 4, the pressure inside and outside the room, with and without the inclusion of RT instability modelling for the central ignition case is shown. We observe that very good agreement with the experimental pressure was achieved by using the same value of the Ξ_{RT} factor (equal to 1.9) as in the back ignition case. This is an important finding because it implies that the enhancement of the combustion rate outside the vent weakly depends on the ignition position. Different ignition point

Journal of Physics: Conference Series **640** (2015) 012036 doi:10.1088/1742-6596/640/1/012036

implies different flow characteristics. Figure 5 shows the hydrogen mass fraction, the flow velocity and the subgrid scale velocity at the time of the external explosion. Comparing with the case of back ignition (figure 3), we notice that the value of velocity (both resolved and sub grid scale) is smaller. The smaller value is justified by the fact that for the central ignition case the flame does not have the time to accelerate much and it reaches the vent with a smaller velocity, because it is closer to it compared to the back ignition case. Even though velocity differs significantly, the Ξ_{RT} value equal to 1.9 gives very good results for both cases. This leads to the conclusion that simple models and correlations for the external explosion may exist. Of course, simulations of more experiments need to be carried out.

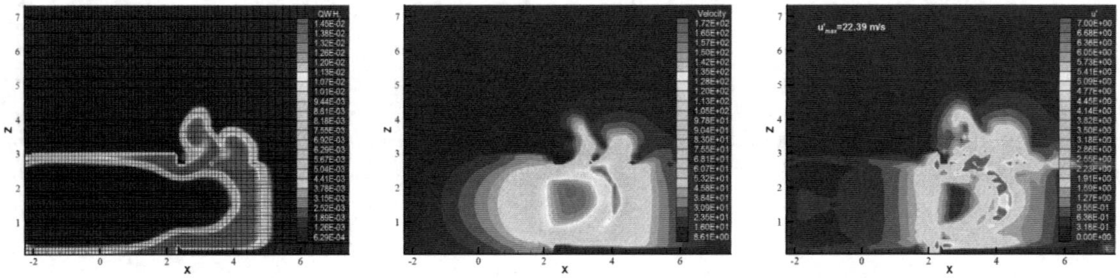

Figure 3. Hydrogen mass fraction (left), velocity (middle) and subgrid scale velocity (right) contours at t=0.222 seconds (back ignition).

In figure 4 the overpressure with $\Xi_{RT} = 1.0$ model is also shown. As in the previous case, the external pressure is lower compared to the $\Xi_{RT} = 1.9$ case. This leads to the underestimation of the maximum pressure value inside the room. In the same figure the experimental and simulated flame speed velocity normalized by laminar flame speed are presented. The agreement up to a distance of 2.5 m from the ignition point is satisfactory. After that point some discrepancies between the experimental and the predicted flame speed are observed.

In figure 4 we observe also that a second peak exists in the experimental pressure which the numerical model fails to predict. This peak is related to the combustion of the unburned fuel that has been trapped inside the enclosure. Structural response of the room, which interacts with the acoustics, and flame-acoustic interaction enhance the combustion process at this stage giving rise to strong pressure oscillations. A modelling approach of this phenomenon needs to be developed for the accurate prediction of this second pressure peak.

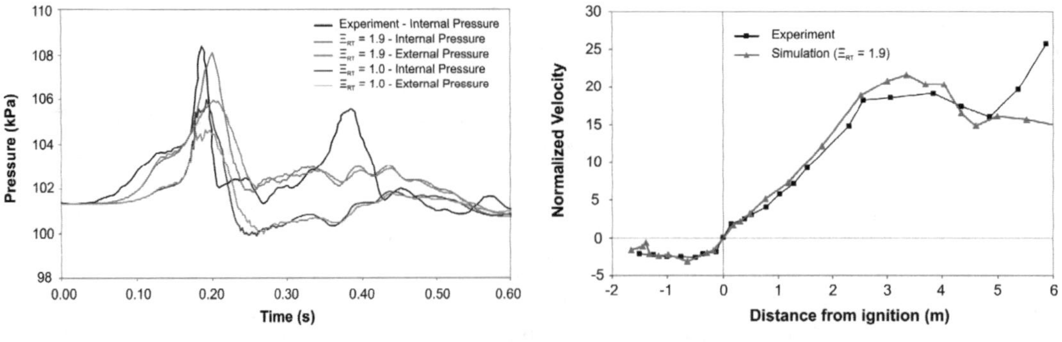

Figure 4. Pressure inside and outside the room for two different values of Ξ_{RT} (left) and flame velocity normalized by laminar flame speed (right) for central ignition case.

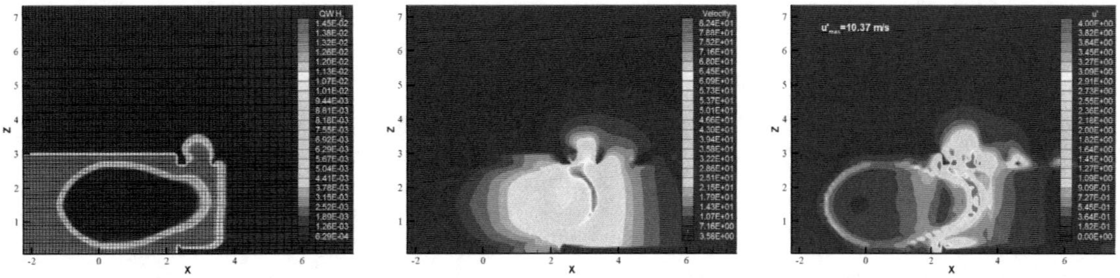

Figure 5. Hydrogen mass fraction (left), velocity (middle) and subgrid scale velocity (right) contours at t=0.175 seconds (central ignition).

5. Conclusions

Hydrogen deflagration in a real scale vented room was simulated. Two cases were examined, a back ignition and a central ignition case. A multi-phenomenon combustion model was used which takes into account the main mechanisms of the deflagration process such as turbulence, turbulence generated by the flame front, preferential diffusion and fractal geometry. A simple extension of this model is proposed in order to take into account the enhancement of the combustion rate at the area outside the vent, due to the Rayleigh-Taylor instability. The model revealed very good agreement with the experimental results. The maximum pressure inside the room was predicted accurately only by the modified model. Despite the different flow characteristics between the two cases, the same value of the Ξ_{RT} factor (equal to 1.9) was found to be appropriate for both cases. This shows that simple models and correlations regarding this factor may be derived.

Acknowledgments

This work has been supported by the project HyIndoor ("Pre-normative research on safe indoor use of fuel cells and hydrogen systems", Grant agreement no: 278534). The first author would also like to acknowledge the "IKY Fellowships of Excellence for Postgraduate Studies in Greece - Siemens Program" for the financial support.

References

[1] Bauwens C R and Dorofeev S B 2013 *5th Int. Conference on Hydrogen Safety* Brussels
[2] Cooper M G, Fairweather M and Tite J P 1986 *Combust. Flame* **65** 1–14
[3] Chao J, Bauwens C R and Dorofeev S B 2011 *Proc. Combust. Inst.* **33** 2367–74
[4] Harrison A J and Eyre J A 1987 *Combust. Sci. Technol.* **52** 91–106
[5] Jiang X, Fan B, Ye J and G. Dong 2005 *J. Loss Prev. Process Ind.* **18** 21–6
[6] Ferrara G et al. 2008 *J. Hazard. Mater.* **155** 358–68
[7] Proust C and Leprette E 2010 *Process Saf. Prog.* **29** 231–5
[8] Baraldi D et al. 2010 *Int. J. Hydrogen Energy* **35** 12381–90
[9] Bauwens C R, Chaffee J and Dorofeev S B 2011 *Int. J. Hydrogen Energy* **36** 2329–36
[10] Keenan J J, Makarov D V and Molkov V V 2014 *Int. J. Hydrogen Energy* **39** 20467–73
[11] Venetsanos A G, Papanikolaou E and Bartzis J G 2010 *Int. J. Hydrogen Energy* **35** 3908–18
[12] Tolias I C, Venetsanos A G, Markatos N and Kiranoudis C T 2014 *Int. J. Hydrogen Energy* **39** 20538–46
[13] Molkov V, Makarov D and Schneider H 2006 *J. Phys. D. Appl. Phys* **39** 4366–76
[14] Molkov V, Verbecke F and Makarov D 2008 *Combust. Sci. Technol.* **180** 796–808
[15] Yakhot V 1988 *Combust. Sci. Technol.* **60** 191–214
[16] Karlovitz B, Denniston D W and Wells F E 1951 *J. Chem. Phys.* **19** 541
[17] Zimont V and Lipatnikov A 1995 *Chem. Phys. Rep* **14** 993–1025
[18] Molkov V V 2012 *Fundamentals of Hydrogen Safety Engineering* www.bookboon.com
[19] Molkov V, Makarov D and Puttock J 2006 *J. Loss Prev. Process Ind.* **19** 121–9

Journal of Physics: Conference Series **640** (2015) 012037 doi:10.1088/1742-6596/640/1/012037

Quantum Matter: Topology and Correlations driving new materials, phases, and phenomena

Nandini Trivedi
Department of Physics, The Ohio State University, 191 W. Woodruff Avenue, Columbus, OH 43210, U.S.A.

E-mail: trivedi.15@osu.edu

Abstract. I review the classification of quantum matter as a function of Coulomb interactions U and spin-orbit coupling λ in units of the bandwidth. The region where all three of these energy scales are comparable, as in the 4d and 5d transition metal oxides and dichalcogenides is the most promising for discovering new materials and new phenomena. I discuss what it means to extract a model from a material and the hierarchy of computational methods for single and multi-orbital systems. As specific examples I describe half metallic high-T_c ferromagnetism in the double perovskite Sr_2FeMoO_6 in which the kinetic energy of Mo (d^1) electrons drives ferromagnetic exchange between the Fe (d^5) core spins. I will also discuss predictions for materials with a d^4 electronic configuration with strong spin-orbit coupling and show that there is a large parameter space where instead of a J=0 band insulator one surprisingly finds ground states with orbitally entangled ferromagnetic states. I will conclude with a broad outlook for new discoveries in this arena.

1. Introduction

It is evident that over the past decades, the discovery, understanding and applications of new solid-state materials have played a crucial role in modern technology. However, devices today based on silicon have hit a bottleneck in terms of the amount of information that can be packed in nano-dimensions and still avoid the complications due to heating [1]. For the design of the next generation of multifunctional devices, we need new paradigms, new principles and new classes of materials. The question then arises: which specific materials do we investigate? In this review article, we start by giving a big picture for guiding exploration of novel materials and then focus on two stories that highlight half metallic conduction in a high-T_c ferromagnet and novel ferromagnetism in strongly spin-orbit coupled systems.

If we classify quantum matter in the plane described by spin-orbit coupling (SOC) λ and Coulomb correlations U relative to the bandwidth W (Figure 1), we have the traditional metals and insulators at weak U and λ shown in the bottom left quadrant. These would include metals such as aluminum, insulators such as silicon that have ruled our electronics applications, and the newly emerging semimetal graphene with Dirac linear dispersion [2-3]. At large U, in the bottom right quadrant, are the transition metal oxides with partially filled 3d shells that have dominated materials research in the past decades, leading to such spectacular phenomena as high-T_c superconductivity [4-5] and colossal magnetoresistance (CMR) [6-7]. Even though high Tc superconductivity in the cuprate family was discovered almost 25 years ago the problem still remains unsolved. This is because strong correlations make density functional theory (DFT) approaches inapplicable and theoretical predictions unreliable, making the feedback loop between theory and experiments difficult. Once electrons interact strongly and correlations dominate, the movement of one electron requires the adjustment of all the other electrons and the electronic structure, which is essentially a map of how a single electron behaves in a solid, becomes an inadequate description of the problem. From a materials point of view, oxides are notorious for problems of segregation and inhomogeneity across various length scales such that controlling the composition and homogeneity of oxides over entire range of carrier concentration is highly challenging.

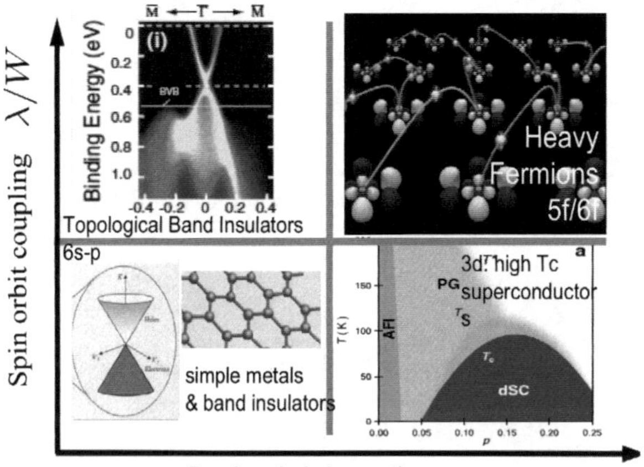

Figure 1: Map of quantum matter as a function of spin-orbit coupling λ and Coulomb correlations U in units of the bandwidth W. At weak U and weak λ are the band metals and insulators; bottom right at stronger U are the 3d high-T_C superconductors and colossal magnetoresistive manganites; at the top left at weak U but strong λ are the 6s and 6p topological band insulators; at the top right at strong U and strong λ are the 5f and 6f heavy fermion systems.

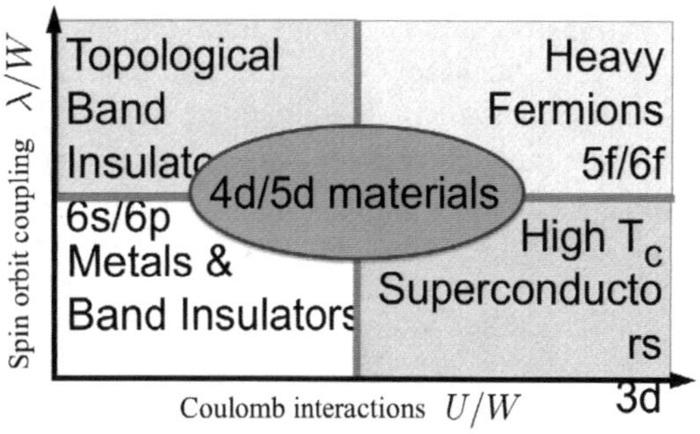

Figure 2: The unchartered territory for new materials, new phases and new phenomena is in the middle where all the energy scales Coulomb interactions U, spin-orbit coupling λ, and bandwidth W become comparable. We expect the 4d and 5d transition metal oxides and dichalcogenides to straddle the sweet region where all the energy scales are comparable making them ideal for controlling their properties using chemical tuning, strain, magnetic and electric fields to achieve the desired functionalities.

It is important to note that for multi-orbital systems, we are simplifying the situation by denoting the interactions by a single parameter U which should more realistically be broken down into intra-orbital and inter-orbital (Hund's) interactions. However, for the present discussion a single U axis suffices to bring out the qualitative aspects of the discussion. Continuing on, in the large λ - small U quadrant (Figure 1 top left) we have the prediction [8-10] and subsequent detection [11-12] of topological band insulators (TBI) in s- and p- band materials. The topological properties of these materials are best preserved in the absence of broken time reversal symmetry. One of the primary difficulties in developing useful applications based on TBI has been the inability to pin the chemical potential within the band gap which not only mixes signals from the topological surface states and the bulk bands in measuring their properties but is also a hindrance for many applications.

The uncharted territory for new materials and new physics is in the middle involving 4d and 5d transition metals where all energy scales, W, U, and λ are comparable as emphasized in Figure 2. We expect new phases and new phenomena to emerge from the close interplay of these energy scales. In this context the transition metal oxides (TMO) [13-16] and transition metal dichalcogenides (TMDCs) [17-19] with comparable correlations, SOC and bandwidth W, are emerging as the dream materials both in terms of their fundamental properties and potential applications [20]. The TMDCs have layered structures similar to graphene that makes them easier to grow compared to the oxide based materials. In addition they can have can have considerably richer properties compared to graphene given both the tunable SOC and correlations. Both of the classes of materials, the TMO and TMDC materials with heavy transition metal atoms and with moderate correlations have the potential to combine the properties of the second (top left) and fourth (bottom right) quadrants shown in Figures 1 and 2. They bring together the advantages of topologically-protected surface states arising from the large SOC, as well as the magnetism from correlations. There is the potential for using strain to modulate the orbitals and through that the magnetism via the spin-orbit coupling. Given their moderate correlations, they are theoretically much more tractable and allow for a quick iterative cycle between experiment and theory.

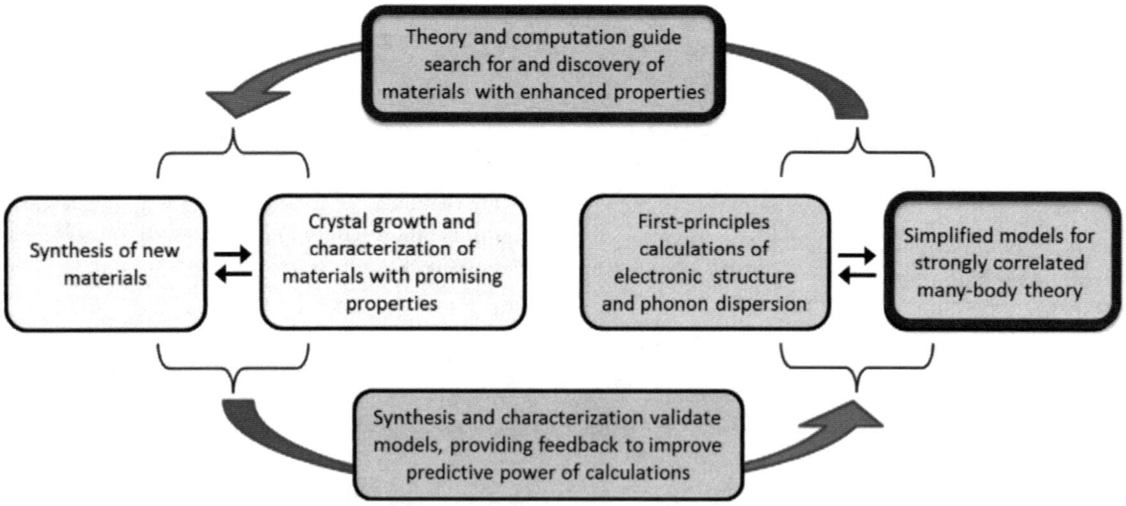

Figure 3: The interaction between theory and experiment generating both "Materials Inspired" and "Computation/Theory Inspired" routes to accelerated materials design. The former starts with the synthesis of new materials (including chemical modifications of known materials), the observation of phenomena yielding puzzles, followed by theory, modeling and computation providing insights. In the latter route, theory and computation provide “universal” phase diagrams in multi-parameter space of bandwidth W, interactions and SOC, and the challenge is to find target materials that potentially describe phenomena at the “interesting” points in the phase diagram. This then propels theory into new unexplored regimes to make new predications and guide the next loop in materials synthesis.

2. Hierarchy of computational complexity and extracting a meaningful model

The first step in modelling a material is to use the known atoms and crystal structure and obtain the electronic structure using density functional theory (DFT). At the next level of complexity we can enhance the tightbinding electronic structure by including spin orbit coupling. From this we can often extract a tightbinding Hamiltonian by a downfolding procedure focusing on a few target bands. For example, in the case of graphene, we can obtain a tightbinding Hamiltonian with a single orbital on a honeycomb lattice that describes the electronic structure at low energies near the chemical potential. Density functional methods include Coulomb correlations within an effective one-electron picture. Several packages are available for calculating the electronic structure given the types of atoms and the crystal structure [21]. The topological properties must be elucidated by a topological band analysis of the associated wave functions [22-24]. Once correlations are included, even in a single orbital model, the complexity increases enormously. At the simplest level we obtain a one-band Hubbard model [25]. To treat the onsite interaction [Eq. (2)], one approach is to embed a single site or a cluster of sites in a fermionic bath and solve the problem self-consistently, known as dynamical mean field theory (DMFT). The "dynamical" in DMFT indicates that quantum fluctuations, even though at the level of a single site or a cluster of sites, are included [26]. Even the solution of the single site problem requires quantum Monte Carlo (QMC) methods [27,28] for its solution, though other solvers using exact diagonalization and perturbation theory are also prevalent. There has been recent progress in solving the impurity model by using the hybridization expansion version of the continuous time quantum Monte Carlo (CT-HYB) [29] implemented in the TRIQS code [30] for rotationally invariant interaction using conserved quantities to speed up the calculations.

Single site DMFT largely avoids the sign problem, which is one of its great virtues. The sign problem however begins to appear in more sophisticated versions like cluster DMFT or in dynamical cluster approximation (DCA), and also in real space approaches like determinantal quantum Monte Carlo (DQMC). There are however difficulties of going to low temperatures because of the fermion sign problem and accessing the superconducting state and difficulties of obtaining the real frequency spectral function and density of states even at higher temperatures in the pseudogap region because of uncertainties associated with analytic continuation.

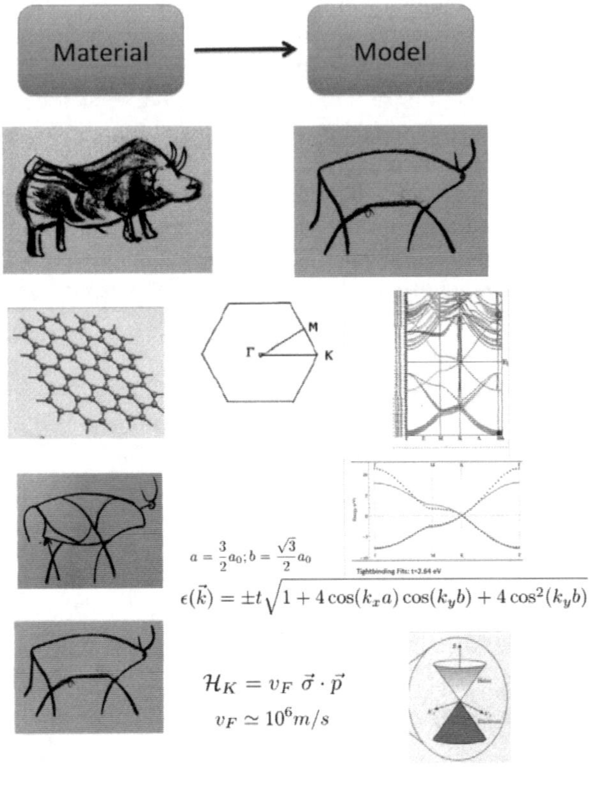

Figure 4: Using Picasso "bull" pictures to illustrate the successive extraction of the model in the context of graphene. Graphene is a honey comb lattice of carbon atoms; each C consists of a filled helium core plus 4 valence electrons $2s^2\ 2p^2$ occupying (s, p_x,p_y,p_z) orbitals. The honeycomb lattice is represented as a triangular lattice with 2 atoms in the unit cell i.e. with 8 electrons and 8 orbitals in the unit cell. As the unit cells are brought together to make a solid, the orbitals overlap and form 8 bands. In general, C also has unoccupied d-orbitals that form bands. This electronic structure is the "higher level" model for graphene. At the next level of simplification, if the physics we are considering is only sensitive to states close to the Fermi energy, we can extract a tight-binding model with a single band. This band is primarily of p_z character, with some mixing of other orbitals. We obtain an excellent fit to the actual single band with a single hopping parameter t=2.64 eV. If necessary the quality of the fit can be improved with additional hopping parameters. Now if one is interested in physics on the scale of few 100 meV then it is possible to focus on just the linear dispersion near the Dirac points and simplify the Hamiltonian to $H = \pm v_F \vec{\sigma} \bullet \vec{p}$. This brings out the beauty of graphene, its topological nature, the presence of two Dirac points with linear dispersion at K and K' in the Brillouin zone with opposite Berry curvature.

As discussed above, there has been considerable progress using dynamical mean field theory and its cluster- based generalizations, which however still requires DQMC for the solution of the impurity or cluster problem as well as maximum entropy methods for extracting dynamical information. It is important in this context to bring up complementary variational methods that are based on a wave function and are constructed by using some intuition for the phenomenon being proposed. The variational estimate provides an upper bound to the energy and can be creatively used to get information about moments of spectral functions. Most importantly, because the absolute square of the wave function is sampled, they do not suffer from the sign problem. In the context of the Hubbard model, projected wave functions have provided understanding about the origin of two energy scales, a pairing scale vs. a coherence scale, with completely different doping dependencies, and the broadening of the spectral function with underdoping [31,32].

As we move into the uncharted territory of 4d and 5d transition metal oxides and dichalcogenides, we must include the complexity of multi-orbitals, the crystal field splittings, the intra- and inter-orbital correlations and spin-orbital coupling. How do we begin to make sense of such complexity? Clearly, the problem must first be investigated using simple test cases with exact diagonalization, effective models and simple mean field theory first to build understanding and identify directions followed by multi-orbital generalizations of DMFT thereby paving the way for a full QMC analysis.

<u>Iterated cycle between theory, computation and experiments:</u> Several conceptual steps are necessary to go from a material to a model as best illustrated by Picasso [33] in his sketches of the "bull" going sequentially from a realistic rendition ("the material") to a simplified sketch capturing the bare essentials ("the model"). The degree of simplification can depend on the questions that are being explored. In other words there need not be a single model for a given material but depending on the nature of the questions, for example, whether charge degrees are important or spin, the range of energies being explored and other factors one can have a series of models. This too is best illustrated by Picasso's bull pictures shown below in the context of graphene in Figures 4. To summarize then, the first step in modeling a new material is to start with the known elements and structure and obtain the full electronic structure from first principles DFT methods to identify the important orbitals. This can be followed by an *ab initio* downfolding procedure to obtain an effective low energy tight-binding Hamiltonian, with further reduction toward an effective magnetic Hamiltonian. The effective Hamiltonians can then be used to calculate phase diagrams, identifying phases and their transport and dynamical susceptibilities using mean field theory and QMC methods.

<u>Interplay of Spin-Charge-Orbital degrees (Fig 5 top):</u> In half-filled electronic systems, strong Coulomb correlations U can open a Mott gap in the density of states. By favoring single occupancy of electrons on a site, the strong U also leads to the formation of local moments. The low energy sector is dominated by spin interactions that typically gives rise to antiferromagnetic ground states in half filled systems but also competing ferromagnetic interactions based on Goodenough-Kanamori rules in multi-orbital systems. In the presence of spin-orbit coupling acting in the e_g or t_{2g} sectors, the resulting magnetic interactions become directional, and can lead to Mott-assisted insulators. A new tuning knob now emerges—lattice strain—which can manipulate the orbitals and thereby the spin degrees.

We can see the vast panorama of possibilities in material synthesis and design that is available by tuning SOC and interaction parameters (both intra-orbital and inter-orbital), lattice structure and the valency of the transition metal ion. In order to keep the rest of the discussion somewhat focused, we next demonstrate such a multi-pronged theoretical and computational approach applied to two problems based on the double perovskite structure.

Double

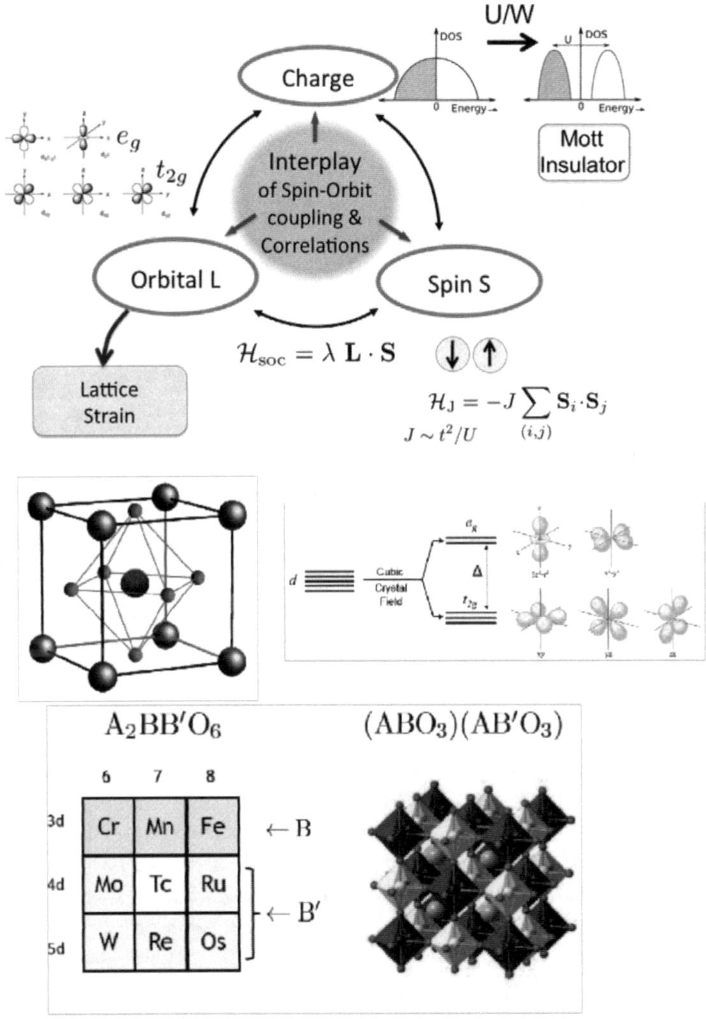

Figure 5: (top) Illustrating the charge, spin and orbital coupling in d orbital systems. (middle) The perovskite structure ABO_3 with the transition metal ion B (blue) in the center of an oxygen (red) octahedral cage. The lowering of symmetry from spherical splits the 5-fold degenerate d orbitals into a 2-fold degenerate e_g and a 3-fold degenerate t_{2g} orbitals. (bottom) A double perovskite structure with two transition metal ions B, B' expands the possibilities of new materials with new properties.

3A. Perovskites with half metallicity and high-T_c Ferrimagnetism

Sr_2FeMoO_6 (SFMO) is an important example of a material that exhibits a fully polarized metallic ground state and a ferromagnetic T_c of around 425K [34-36]. This combination of properties has the potential for an enormous technological impact in spintronic applications such as tunneling magnetoresistance devices. What is the fundamental mechanism underlying such behavior?

In SFMO Fe^{3+} is in a $3d^5$ electronic configuration and because of the large Hund's coupling the five electrons completely fill the $d\uparrow$ orbitals and behave as an $S=5/2$ core spin. Mo^{5+} is in a $4d^1$ electronic configuration. This Mo electron in the t_{2g} orbital could be either up or down in principle. However, it chooses to be down because the electron can gain kinetic energy by delocalizing via the Fe $d\downarrow$ orbitals that are empty. In addition to previous mean field theories [37-40] we use exact diagonalization (ED) to solve the quantum mechanics of ''fast'' itinerant electrons moving in the background of ''slow'' *S*=5/2 spins, for which we use a $T \neq 0$ classical Monte Carlo (MC) simulation [41-43].

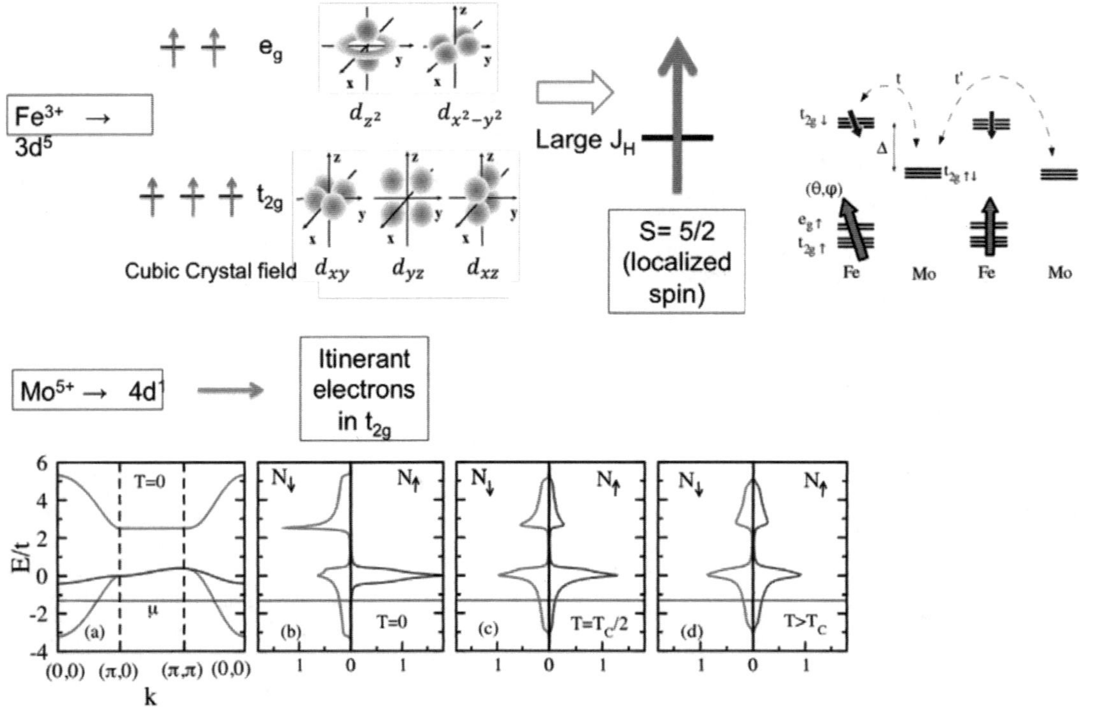

Figure 6: (top) The quantum itinerant Mo electron moving in the fluctuating magnetic field of the Fe core spins treated classically. (bottom) The spin-resolved (red: ↓; blue ↑) electronic structure of the Mo electron and the density of states found to be strongly temperature dependent from half metallic at T=0 to unpolarized above Tc.

Using standard ED-MC techniques, supplemented by *T*=0 variational calculations, we obtain the results shown in Figs. 7 and 8. We show in Fig. 7 the band structure and the spin-resolved density of states (DOS) $N_\sigma(E)$ for the ferromagnetic ground state of SFMO with n=0.33 electrons per unit cell per plane. These results are consistent with photoemission experiments [44]. We consider three different characterizations of magnetization: the core Fe spin magnetization, the itinerant Mo spin magnetization and the itinerant spin polarization at the Fermi surface. We find that the conduction electrons, with magnetization $M_{el}(T=0)=1\mu_B$ (per unit cell), are polarized opposite to the Fe spins, with core spin magnetization $M_S(0)=5\mu_B$. The half-metallic ground state, with $N_\uparrow(0)=0$ and $N_\downarrow(0)\neq 0$, has a net magnetization $M(0)=M_S(0)-M_{el}(0)=4\mu_B$. We see from Fig. 8 that $M_S(T), M_{el}(T)$ and hence the total magnetization $M(T)=M_S(T)-M_{el}(T)$ all have essentially the same T dependence. Another quantity of great interest is the conduction electron polarization $P(T)$ at E_f which determines the tunneling magnetoresistance. This is defined below:

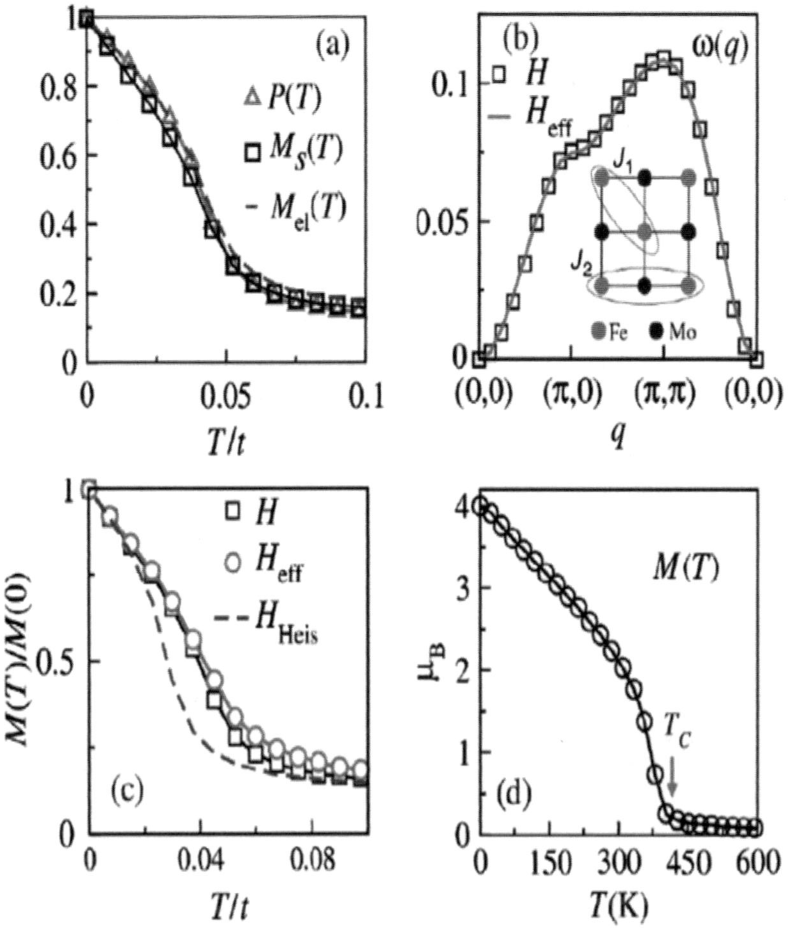

Figure 7: The core spin magnetization, $M_S(T)$ conduction electron magnetization $M_{el}(T)$ and polarization $P(T)$ at E_f calculated using a combination of exact diagonalization for the itinerant electrons moving in a spin texture of the core electrons obtained by classical Monte Carlo. (b) Spin-wave dispersion of the full Hamiltonian H, obtained by exact diagonalization, compared with that of the effective Hamiltonian H_{eff}. Inset: Fe-Mo lattice showing the nearest-neighbor J_1 and next nearest- neighbor J_2interactions between Fe-Fe ions described by H_{eff}. (c) The normalized magnetization for three Hamiltonians: (1) the exact H with itinerant spins moving in a background of fluctuating core spins (black squares), (2) the effective magnetic Hamiltonian for double perovskites (red circles), and (3) the Heisenberg model as a test case (blue dashes). These results are obtained on 8^3 systems with error bars no larger than the symbol size. (d) Magnetization $M(T)$from 3D simulations of H_{eff} on 16^3 systems; T_C is obtained by from finite scaling

$P(T) = \dfrac{N_\downarrow(0) - N_\uparrow(0)}{N_\downarrow(0) + N_\uparrow(0)}$ measured at E_f. In Fig. 8, we also see that $P(T)$ follows the $M_S(T)$.

Effective Spin Hamiltonian: The results of Fig. 8 imply that the core spin magnetization $M_S(T)$ and the electronic polarization $P(T)$ are linked. This is an important observation, since the former is much easier to measure. With this motivation, we derive an effective Hamiltonian H_{eff} for the core spins by generalizing the two-site Anderson-Hasegawa [45] analysis for manganites to double perovskites. To derive H_{eff}, we find the exact solution of the full H for one electron in two unit cells. The Hilbert space has three states per unit cell: Fe $t_{2g\downarrow}$ and Mo $t_{2g\uparrow,\downarrow}$ leading to a 6 x 6 matrix for two unit cells. We analytically find its lowest eigenvalue as a function of the angle $(\theta_i - \theta_j)$ between core spins and find the nearest-neighbor (***J***$_1$) and next-nearest-neighbor (***J***$_2$) interaction energies; see the inset in Fig. 8(b). Expressing these in terms of $S_i \cdot S_j$ where each S_i is a unit vector, we obtain the effective

Hamiltonian $H_{eff} = -J_1 \sum_{\langle i,j \rangle} F_1(S_i \cdot S_j) - J_2 \sum_{\langle\langle i,j \rangle\rangle} F_2(S_i \cdot S_j)$ where the functions $F_1(x) = 8\sqrt{2+\sqrt{2+2x}}$ and $F_2(x) = (5+\sqrt{5})\sqrt{6+2\sqrt{3+2x}}$. Our two-unit cell analysis gives explicit expressions for J_1 and J_2, both of which are ferromagnetic, with their scale set by the kinetic energy t of delocalization. We emphasize that the double square-root form of H_{eff} is quite different from the (single square-root) Anderson-Hasegawa model. Next, we need to understand how we can use H_{eff} going beyond the simple two-unit cell derivation. Specifically: (i) how can we relate J_1 and J_2 to t, t', Δ and the filling?

(ii) To what extent does H_{eff} capture the essential physics of the full Hamiltonian H? To validate the effective model H_{eff}, we show in Figure 8 that it reproduces the magnetization $M(T)$ of the full Hamiltonian H over the entire range of temperatures. In contrast, the Heisenberg model H_{Heis} gives quite different results, except in the $T \to 0$ limit with small spin deviations.

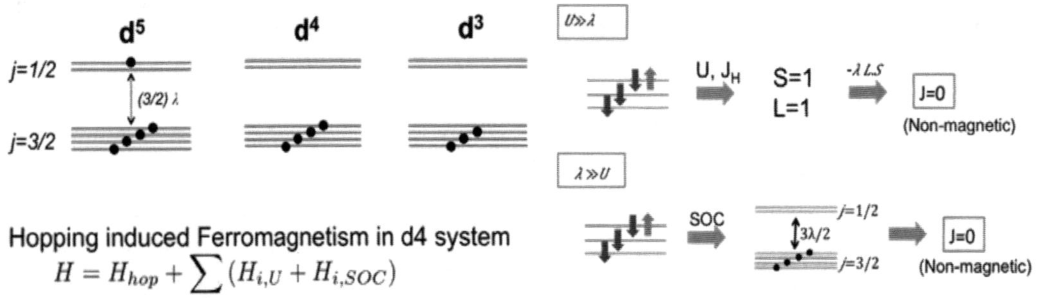

$$H = H_{hop} + \sum_i (H_{i,U} + H_{i,SOC})$$

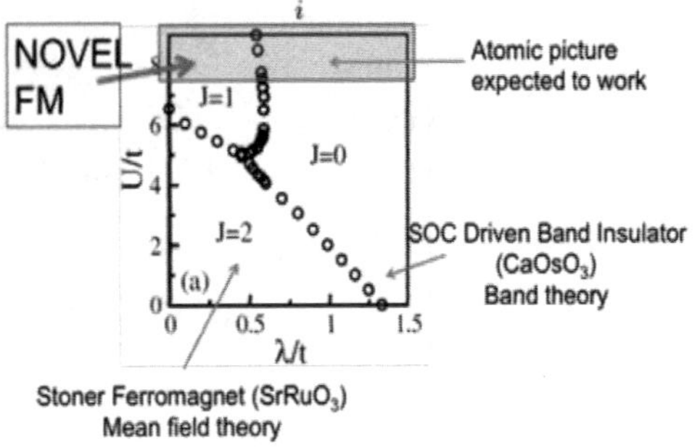

Figure 8: (top) In the presence of strong SOC the t_{2g} orbitals split into a 4-fold degenerate j=3/2 and 2-fold degenerate j=1/2 orbitals. The atom is in a J=0 state for 4 electrons. The same conclusion of a non-magnetic atomic state is reached in the large U limit. (bottom) Phase diagram for 2 atoms shows non-trivial magnetic phases even though there are no moments on the atoms.

Thus, we cannot describe the $M(T)$ for the DP's using a Heisenberg model, while H_{eff} provides an excellent description of the ED-MC result.

The classical H_{eff} can be easily simulated on large 3D lattices, unlike the full H, and the results are shown in Fig. 8. We note the linear drop in $M(T)$ at low T, due to classical spin waves, followed by a rapid suppression of M at the phase transition. Our $M(T)$ results are in qualitative agreement with recent experiments [46]. We estimate T_c in the infinite volume limit using the finite size scaling of results obtained on L^3 systems with $L = 8,12,16$. For $t'/t = 0.1$ and $\Delta/t = 2.5$ we find $T_c = 0.14t$. Comparing this to $T_c^{\exp} = 420$ K for pure SFMO, we obtain $t = 0.27$ eV consistent with Ref. [36].

3B: Novel Ferromagnetism in d^4 electronic configuration

The interplay of strong interactions and spin-orbit coupling (SOC) naturally combines in the 5d transition metal materials, which hold the potential for hosting new phases of matter with entangled spin, orbital and charge degrees of freedom. Already there are many predictions for exotic topological matter, for example the topological Mott insulators and Weyl semi-metals. Recent experiments demonstrating that Sr_2IrO_4 is an unusual Mott insulator with a half filled *j*=1/2 band resulting from strong SOC [47] have prompted the search for Weyl semi-metals in iridium pyrochlores [48]. We emphasize that most of the focus in this field to date has been on iridium based materials with a d^5 electronic configuration. In the limit of strong spin orbit coupling the t_{2g} orbitals are split into *j*=1/2 and *j*=3/2 multiplets. For the d^5 configuration, all of the *j*=3/2 states are filled and the only low-energy degree of freedom is a single hole in the *j*=1/2 manifold which is sufficient to understand most phenomena. This picture is often given in terms of non-interacting atomic levels, but it also holds in the strongly interacting limit for the d^5 configuration.

The physics is dramatically different for other fillings. Mott insulators with $d^1$and d^2 configuration have been shown to exhibit exotic magnetic phases [49, 50] in the presence of large SOC. In the d^3 case, SOC is quenched in a cubic environment [51] and the problem reduces to a conventional spin-only model. This leaves only the d^4 case, which has been largely ignored because it is expected that large SOC and strong interactions give rise to a non-magnetic state in the atomic limit [49], though see [52]. We show [53] that, contrary to the above expectation, the d^4 configuration has a rich magnetic phase diagram as a function of SOC and Hubbard *U*. In particular, at large *U* there is a quantum phase transition from the expected non-magnetic insulator of local *J*=0 singlets to a ferromagnetic insulator with a nonzero local moment as SOC is reduced. This moment arises from super-exchange induced mixing, at a single site, of magnetic states largely from the higher energy triplet with the singlet. Our result provides an interesting counterexample to the commonly held notion that Mott insulators have well defined local moments that cannot be affected by small perturbations (compared to the interaction scale *U*).

To begin, we perform exact diagonalization on a two-site system with minimal assumptions. From here we can extract the effective magnetic Hamiltonian and then generalize to a lattice. The lattice problem can then be analyzed by various methods such as mean field theory and variational methods. For the two-site problem the number of states required for exact diagonalization are 12 choose 8 which is approximately 500 states arising from the 3 (t_{2g} orbitals) x 2 (sites) x 2(spin) and 8 electrons. As a function of λ and *U* there are three different magnetic states as shown in Figure 9: (i) a non-magnetic state *J*=0 in the large λ limit, (ii) a ferromagnet with *J*=2 for small λ and moderate *U*, and (iii) a ferromagnet with *J*=1 at large values of U and small λ. The magnetic phases can be understood easily in limiting cases. The *J*=0 state at large λ and small *U* corresponds to a band insulator with a completely filled *j*=3/2 manifold. The post-perovskite material $NaIrO_3$ and perovskites $BaOsO_3$ and $CaOsO_3$ are believed to be in such a state [54,55,56]. With increasing *U*, this band insulator smoothly crosses over into a *J*=0 Mott insulator (as defined by real-space localization of charge, and a single-particle excitation gap set by the large interaction energy scale). In the limit of small λ and moderate *U*, the *J*=2 ferromagnet is essentially the Stoner ferromagnet seen in $SrRuO_3$. The most interesting phase is the *J*=1 ferromagnet at large *U* and small λ, which has been often overlooked because the naive atomic limit analysis predicts a nonmagnetic insulator. However, its existence can be understood as follows. For large U, each site *i* has a well defined $|L_i| = 1$ and $|S_i| = 1$. The superexchange interaction is ferromagnetic because of Hund's coupling, and that therefore stabilizes the total *S*=2 state for the two-site problem. The total orbital moment is however not maximized and a total *L*=1 state is realized; upon including SOC in the $L-S$ coupling scheme yields the observed *J*=1 state. In sharp contrast, an atomic limit analysis gives $J_i = 0$ singlet states at each site in accordance with Hund's third rule with the caveat that $L_i = -1$ for t_{2g} orbitals.

We propose candidate materials from the double perovskite family, with cubic crystal field, which can be tuned across the magnetic transition by chemical substitution and/or pressure. In the double perovskite $A_2BB'O_6$ where A is an alkaline earth element while B and B' are two different transition metal ions, ordered in a 3D checkerboard pattern, if we choose the B sites to have completely filled shells, the bandwidth is suppressed, giving rise to a Mott insulator. When combined with a 4d or a 5d element on the B' site, we have the ideal model system where large SOC competes with ferromagnetic exchange. Of particular interest is La_2ZnRuO_6 which is an insulator with Ru in the d^4 configuration. Two different samples grown by two different groups have shown very different magnetic states. One group has found a ferromagnetic state with T_C=165K [57], while the other found a non-magnetic state [58]. We believe La_2ZnRuO_6 is very close to the phase boundary and small differences in the lattice parameter may be the origin of this discrepancy. A resonant x-ray scattering study under pressure will be an ideal experiment to observe the phase transition. Another closely related material is La_2MgRuO_6 [57], which is also a promising candidate.

4. Conclusions

At this stage we are just beginning to scratch the problem of multiorbitals with intra (Hubbard U) and inter (Hund's) orbital interactions and spin-orbit coupling λ. We expect to see the growth of paradigmatic multi-orbital Hamiltonians and effective magnetic Hamiltonians with directional interactions that harbor non-trivial magnetic Mott insulators and metallic states. With the corresponding materials growth from the transition metal oxide and dichalcogenide families, we expect to see a rich dialog between theory and experiment.

5. Acknowledgments: I would like to thank my collaborators Oinam Nganba Meetei, Onur Erten and Mohit Randeria for fruitful collaborations and discussions and the support from the Center of Emergent Materials CEM, an NSF MRSEC, under Grants DMR-0820414 and DMR-1420451.

References:

[1] V V Zhirnov, R K Cavin, J A Hutchby and G I Bourianoff *Proc. IEEE* **91**, 1934–39 (2003).
[2] A K Geim and K S Novoselov *Nat. Mat.* **6**, 183–191 (2007).
[3] K S Novoselov, D Jiang, F Schedin, T J Booth, V V Khotkevich, S V Morozov and A K Geim *Proc. Nat. Acad. Sciences* **102**, 10451–3 (2005).
[4] P A Lee, N Nagaosa and X G Wen *Rev. Mod. Phys.* **78**, 17–85 (2006).
[5] J Orenstein and A J Millis *Science* **288**, 468–474 (2000).
[6] M B Salamon and M Jaime *Rev. Mod. Phys.* **73,** 583–628 (2001).
[7] E Dagotto *New J. Phys.* **7,** 67/1–28 (2005).
[8] C L Kane *Nat. Phys.* **4,** 348–9 (2008).
[9] C L Kane and E J Mele *Phys. Rev. Lett.* **95**, 226801/1–4 (2005).
[10] C L Kane and E J Mele *Phys. Rev. Lett.* **95**, 146802/1–4 (2005).
[11] M Z Hasan and C L Kane *Rev. Mod. Phys.* **82**, 3045–67 (2010).
[12] X–L Qi and S–C Zhang *Rev. Mod. Phys.* **83**, 1057–1110 (2011).
[13] D Pesin and L Balents *Nat. Phys.* **6**, 376–81 (2010).
[14] G Chen, R Pereira and L Balents *Phys. Rev. B* **82**, 174440/1–25 (2010).
[15] G Chen and L Balents *Phys. Rev. B* **84**, 094420/1–13 (2011).
[16] X Wan, A M Turner, A Vishwanath and S Y Savrasov *Phys. Rev. B* **83**, 205101/1–9 (2011).
[17] J A Wilson and A D Yoffe *Adv. Phys.* **28**, 193–335 (1969).
[18] J A Wilson, F J Di Salvo and S Mahajan *Adv. Phys.* **24**, 117–201 (1975).
[19] K Rossnagel *J. Phys: Cond. Mat.* **23**, 213001/1–24 (2011).
[20] Q H Wang, K Kalantar–Zadeh, A Kis, J N Coleman and M S Strano *Nat. Nanotechnology* **7**, 699–712 (2012).
[21] Software packages: Wien2K www.wien2K.at; Quantum Espresso www.quantum–espresso.org
[22] D Xiao, M C Chand and Q Niu *Rev. Mod. Phys.* **82**, 1959 (2010).
[23] M V Berry *Proc. R. Soc. Lond.* **392**, 45 (1984).
[24] T Fukui, Y Hatsugai and H Suzuki *J. Phys. Soc. Japan* **74**, 1674 (2005).

[25] M Imada, A Fujimori and Y Tokura *Rev. Mod. Phys.* **70**, 1039 (1998).
[26] A Georges, G Kotliar, W Krauth and M J Rozenberg *Rev. Mod. Phys.* **68**, 13 (1996).
[27] R Blankenbecler, D J Scalapino and R L Sugar *Phys. Rev. D* **24**, 2278 (1981).
[28] T Paiva, Y–L Loh, M Randeria, R T Scalettar and N Trivedi *Phys. Rev. Lett.* **107**, 086401 (2011).
[29] P Werner, A Comanac, L de' Medici, M Troyer and A J Millis *Phys. Rev. Lett.* **97**, 076405 (2006).
[30] M Ferrero and O Parcollet "TRIQS: a Toolbox for Research in Interacting Quantum Systems." http://ipht.cea.fr/triqs
[31] A Paramekanti, M Randeria and N Trivedi *Phys. Rev. Lett.* **87**, 217002 (2001).
[32] A Paramekanti, M Randeria and N Trivedi *Phys. Rev. B* **70**, 054504 (2004).
[33] Inspired by Picasso "Bull" pictures.
[34] For a recent review, see D Serrate, J M De Teresa and M R Ibarra *J. Phys. Cond. Mat.* **19**, 023201 (2007).
[35] K–I Kobayashi, T Kimura, H Sawada, K Terakura and Y Tokura *Nature* **395**, 677 (1998).
[36] D D Sarma, P Mahadevan, T Saha–Dasgupta, S Ray and A Kumar *Phys. Rev. Lett.* **85**, 2549 (2000).
[37] A Chattopadhyay and A J Millis *Phys. Rev. B* **64**, 024424 (2001).
[38] J L Alonso, L A Fernández, F Guinea, F Lesmes and V Martín–Mayor *Phys. Rev. B* **67**, 214423 (2003).
[39] L Brey, M J Calderón, S Das Sarma and F Guinea *Phys. Rev. B* **74**, 094429 (2006).
[40] P Sanyal and P Majumdar *Phys. Rev. B* **80**, 054411 (2009).
[41] O Erten, O Nganba Meetei, A Mukherjee, M Randeria, N Trivedi and P Woodward *Phys. Rev. Lett.* **107**, 257201 (2011).
[42] O N Meetei, O Erten, A Mukherjee, M Randeria, N Trivedi and P Woodward *Phys. Rev. B* **87**, 165104 (2013).
[43] O Erten, O N Meetei, A Mukherjee, M Randeria, N Trivedi and P Woodward *Phys. Rev. B* **87**, 165105 (2013).
[44] T Saitoh, M Nakatake, A Kakizaki, H Nakajima, O Morimoto, S Xu, Y Moritomo, N Hamada and Y Aiura *Phys. Rev. B* **66**, 035112 (2002).
[45] P W Anderson and H Hasegawa *Phys. Rev.* **100**, 675 (1955).
[46] A J Hauser, R E A Williams, R A Ricciardo, A Genc, M Dixit, J M Lucy, P M Woodward, H L Fraser and F Y Yang *Phys. Rev. B* **83**, 014407 (2011).
[47] B J Kim, H Ohsumi, T Komesu, S Sakai, T Morita, H Takagi and T Arima *Science* **323**, 1329 (2009).
[48] K Ueda, J Fujioka, Y Takahashi, T Suzuki, S Ishiwata, Y Taguchi and Y Tokura *Phys. Rev. Lett.* **109**, 136402 (2012).
[49] G Chen, R Pereira and L Balents *Phys. Rev. B* **82**, 174440 (2010).
[50] G Chen and L Balents *Phys. Rev. B* **84**, 094420 (2011).
[51] O N Meetei, O Erten, M Randeria, N Trivedi and P Woodward *Phys. Rev. Lett.* **110**, 087203 (2013).
[52] G Khaliullin *Phys. Rev. Lett.* **111**, 197201 (2013).
[53] O N Meetei, W S Cole, M Randeria and N Trivedi *Phys. Rev. B* **91**, 054412 (2015).
[54] L Du, X Sheng, H Weng and X Dai *Europhys. Lett.* **101**, 27003 (2013).
[55] L Du, L Huang and X Dai *Eur. Phys. J. B* **86**, 94 (2013).
[56] I I Mazin and D J Singh *Phys. Rev. B* **56**, 2556 (1997).
[57] R I Dass, J–Q Yan and J B Goodenough *Phys. Rev. B* **69**, 094416 (2004).
[58] K Yoshii, N Ikeda and M Mizumaki *Phys. Stat. Sol.*(a) **203**, 2812–17 (2006).

Numerical Studies of the Quantum Adiabatic Algorithm

A. P. Young[1] and Itay Hen[2]
[1]Department of Physics, University of California, Santa Cruz, California 95064
[2]Information Sciences Institute, University of Southern California, Marina del Rey, California 90292

E-mail: `petery@ucsc.edu`, `itayhen@isi.edu`

Abstract. Quantum annealers promise to solve practical optimization problems potentially faster than conventional classical computers. One of the major ongoing debates in this context pertains to their robustness against the decohering effects of finite temperature and interactions with the environment. We argue that even in an ideal setting of very low temperatures and in the absence of a decohering environment, quantum annealers do not necessarily perform better than classical heuristic solvers. Here, we numerically study the performance of the quantum adiabatic algorithm (QAA) on a variety of constraint satisfaction problems and a spin glass problem by studying the size dependence of the minimum energy gap during the evolution of the QAA. We do so by employing Quantum Monte Carlo schemes as these allow us to study these problems at much larger scales than exact methods would allow. We find that in all cases a quantum phase transition occurs and the minimum gap decreases exponentially with system size, leading to an exponentially large running time for the QAA. Based on these and other results, we briefly discuss potential modifications to the QAA that may improve the scaling of the minimum gap, leading to faster quantum adiabatic algorithms.

1. Introduction

In the standard paradigm for quantum computing, called the "gate" model, the quantum state of the qubits is acted on by a series of discrete unitary transformations [1, 2]. A general quantum state of N qubits is a linear combination of 2^N basis states, and the unitary transformations act in parallel on all of them. Thus, *if phase coherence can be maintained*, these gate operations act in a *massively parallel* manner. Unfortunately, in order to get information out one must do a measurement, in which one observes just one state, with the appropriate probability, so a huge amount of information is lost at this point. It might therefore seem that the potential advantage of quantum parallelism is completely lost in practice. However, by doing some clever preprocessing *before* the measurement there are some problems for which quantum parallelism can produce spectacular gains. The best known, and most important, is Shor's algorithm [3] for factoring large integers. While the best classical algorithm for factoring an integer with n bits takes a time proportional to the exponential of (a fractional power of) n (exponential scaling) Shor's algorithm take a time proportional to a (fairly low) power of n (polynomial scaling). Thus Shor's algorithm would be much more efficient than a classical one for large n. Unfortunately, there are huge experimental problems in maintaining phase coherence of more than a very small number of qubits for the time necessary to do a computation, and these are not yet solved.

Nonetheless, there is no doubt that, if they could be solved, a large *quantum speedup* would be obtained for certain problems by exploiting *quantum parallelism.*

Quantum parallelism arises because a quantum state can be a linear superposition of basis states (e.g. a single spin can be simultaneously "up" and "down"). However, this is only *one* of the ways quantum state differs from a classical state. Another difference is that a quantum system can *tunnel* through energy barriers to get to regions of configuration space that would be impossible classically. Consequently, an alternative paradigm for quantum computing has emerged which aims to take advantage of tunneling (rather than quantum parallelism). This approach is called *quantum annealing* [4, 5] (QA) and is very closely related to the *quantum adiabatic algorithm* (QAA) proposed by Farhi et al. [6]. These methods aim to solve optimization problems, in which one has to find the minimum of a function of many variables (which we will call the "energy") with constraints. Such problems occur widely in the sciences, engineering and industry. Practical problems are difficult because no configuration of the variables simultaneously minimizes all terms in the Hamiltonian, a concept known as "frustration" [7, 8] in the spin glass community, so one cannot find the ground state by repeatedly cycling through each of the variables and minimizing with respect to them one at a time. This "greedy" algorithm finds a local minimum but not, in general, the absolute minimum. For the hardest optimization problems, there are no known efficient algorithms to solve them in sub-exponential time. For these, the time τ to find the global minimum increases exponentially with the system size on a classical computer, i.e.

$$\tau \propto \exp(\mu N)\,. \tag{1}$$

The question of whether a quantum computer can do better than this using QA/QAA is one of the most intriguing open questions in the field to date, as there is no proof that a quantum computer exploiting quantum tunneling could solve optimization problems more efficiently than a classical computer. On the other hand, there is no proof that it could not. Hence, in practice one has to resort to numerics to try to answer this question. While it is probably too optimistic to expect that a quantum computer could convert the exponential dependence in N in Eq. (1) to a power-law (polynomial) dependence, it would still be valuable if a quantum computer could reduce substantially the value of the coefficient μ since one would then be able to study larger sizes than those possible on a classical computer.

A general purpose *classical* method to avoid getting stuck in the nearest local minimum, using physics ideas, was proposed many years ago [9] and is called simulated annealing (SA). In this approach one simulates, using classical Monte Carlo methods [10], the Hamiltonian at a non-zero temperature T, which means that there is some probability to make a move which *increases* the energy, and so the system can escape from a local minimum. During the course of the simulation the temperature is gradually decreased to zero and one expects that the system will end up in the global minimum if the temperature reduction is done slowly enough.

Quantum annealing (QA) has a similar motivation to that of simulated (i.e. classical, or thermal) annealing (SA), but rather than including thermal fluctuations whose strength tends to zero at the end of the run, one adds a term to the Hamiltonian which does not commute with the "problem" Hamiltonian (the one that we are finding the ground state of), which thereby induces quantum fluctuations. The amount of this extra Hamiltonian, and hence the strength of the quantum fluctuations, is decreased to zero during the run. QA is often implemented using Quantum Monte Carlo simulations in which the time to perform the calculation is taken to be the annealing time (number of Monte Carlo sweeps) [5, 11] rather that being obtained from correlations in the "imaginary" time direction of the simulation. Following Ref. [11] we shall call this approach simulated quantum annealing (SQA).

The quantum adiabatic algorithm (QAA) shares many similarities with QA. However, it is designed to makes use of the adiabatic theorem of quantum mechanics [12] as follows. We denote the problem Hamiltonian, whose ground state we want, by $\mathcal{H}_P$. Typically this will be a function of bits, $z_i = 0$ or 1, $(i = 1, 2, \cdots N)$, or equivalently Ising spins $\sigma_i^z = \pm 1$. One example would be a spin glass Hamiltonian where

$$\mathcal{H}_P = -\sum_{\langle i,j \rangle} J_{ij} \sigma_i^z \sigma_j^z - \sum_i h_i \sigma_i^z \, , \tag{2}$$

in which the two-spin interactions J_{ij} can have either sign at random. In the spin glass context the fields h_i are often set to zero. To induce quantum fluctuations, we add a non-commuting "driver" Hamiltonian $\mathcal{H}_D$, which, in the simplest case, is just a transverse field which takes the same value on each site, i.e.

$$\mathcal{H}_D = -\Gamma \sum_i \sigma_i^x \, , \tag{3}$$

where σ_i^x and σ_i^z are now Pauli spin operators. Typically we shall set the parameter Γ to unity. The interactions between the qubits are adjusted so that they have the following Hamiltonian,

$$\mathcal{H}(s) = s(t)\mathcal{H}_P + [\, 1 - s(t) \,]\mathcal{H}_D \, , \tag{4}$$

where $s(t)$ is a parameter varying smoothly with time, from 0 at $t = 0$ to 1 at the end of the algorithm, $t = \mathcal{T}$, so the initial Hamiltonian is $\mathcal{H}_D$ and the final Hamiltonian is $\mathcal{H}_P$. The qubits are initially prepared with $\sigma_i^x = 1$, so the system starts in its ground state. If the time evolution in Eq. (4) is done slowly enough, the adiabatic theorem of Quantum Mechanics [12] ensures that the system will stay close to the ground state of the instantaneous Hamiltonian throughout the evolution, so that one finally obtains a state close to the ground state of $\mathcal{H}_P$. At this point, measuring the state will give the solution of the original problem with high probability. The crucial question, which will be the focus of this article, is how does the time to maintain adiabicity, and hence to solve the problem with high probability, vary with system size N.

A bottleneck of the QAA occurs when the gap between the ground state and the first excited state becomes very small, because it is hard to maintain adiabatic evolution in this case. An upper bound for the run time needed to find the ground state of $\mathcal{H}_P$ is given in terms of the eigenstates $\{|n\rangle\}$ and eigenvalues $\{E_n\}$ of the Hamiltonian $\mathcal{H}(s)$, by [13, 14]

$$\mathcal{T} \gg \hbar \, \frac{|\mathrm{max}_s V_{10}(s)|}{(\Delta E_{\mathrm{min}})^2} \, , \tag{5}$$

where ΔE_{min} is the minimum of the first excitation gap $\Delta E_{\mathrm{min}} = \mathrm{min}_s \Delta E$ with $\Delta E = E_1 - E_0$, and $V_{n0} = \langle 0|\mathrm{d}\mathcal{H}/\mathrm{d}s|n\rangle$. Typically, matrix elements of $\mathcal{H}$ scale as a low polynomial of the system size N, and the question of whether the running time depends polynomially or exponentially with N therefore depends on how the minimum gap ΔE_{min} scales with N. This means that if the gap becomes exponentially small at any point in the evolution, then the computation requires an exponential amount of time, rendering the QAA inefficient. The dependence of the minimum gap on the system size for a given problem is therefore a central issue in determining the running time of the QAA. The minimum gap will vary for different samples of a given size, so some averaging over instances of the same size is required. Here, we will determine performance of the QAA by looking at the "typical" minimum gap.

The study of QA/QAA is not just of theoretical interest, as there are now experiments on real hardware with several hundred qubits. A Vancouver-based company D-Wave has produced a machine, D-Wave One, with 128 superconducting qubits to implement the QAA, and recently has come out with bigger version, D-Wave Two, having 512 qubits [15, 16]. Thus one can compare the running time of real quantum hardware with that of numerical simulations, both classical and simulated quantum annealing, for non-trivial optimization problems [11, 17]. Since the D-Wave machine is affected by the temperature being non-zero and by non-thermal noise, one can also investigate expderimentally whether quantum tunneling of a large number of qubits is destroyed by these effects, in which case the machine is effectively doing thermal annealing rather than quantum annealing. Since the system has an energy gap, the hope is that noise is not fatal to quantum annealing, as it is to the gate model of quantum computing [18, 19]. The qubits in the D-Wave machine are not claimed to maintain coherence during the run, so we shall describe this machine as a "quantum annealer" rather than a quantum computer, and use the latter term for devices which do maintain coherence.

Two types of questions need to be answered about QA/QAA:

(a) Suppose we consider an ideal situation in which the temperature is zero and also there isn't any non-thermal noise. In these ideal circumstances, does the time to solve a hard optimization problem with QA/QAA scale better with size than for a classical algorithm such as simulated annealing? Assuming that scaling is exponential, this is equivalent to asking if the coefficient μ in Eq. (1) is smaller for the quantum algorithm than for a classical algorithm.

(b) Now include the effects of non-zero temperature and non-thermal noise which are present in the D-Wave machine. Do these destroy quantum tunneling and render the machine effectively a thermal annealer? If not, does the run time on the D-Wave machine scale better with size than that of a classical simulation?

Concerning (b), some recent work [11, 20, 17] indicates that the D-Wave machine is doing quantum annealing, though it does not appear that the scaling with size is better than that of an optimized classical algorithm. Experiments with an even larger number of qubits, e.g. 2048, would be helpful in elucidating the asymptotic performance of the D-Wave hardware.

Here we will summarize our own work [21, 22, 23, 24, 25, 26, 27] which focuses on the ideal situation in (a) above. We shall consider several optimization problems, and compare the scaling with size of the run time of the QAA with that of a heuristic classical algorithm similar to simulated annealing. We will find that the basic implementation of the QAA which we use is not better than the classical algorithm for any of the problems studied. We will also consider how one might improve this implementation of the QAA to get better results.

2. Numerical results

2.1. Method

To find the minimum gap for a specific instance of a specific problem, we perform quantum Monte Carlo simulations for a range of s values (s being the interpolating adiabatic parameter) that bracket the minimum gap. The specific method we use in this study is known as the stochastic series expansion (SSE) algorithm [28, 29] which involves a Taylor series expansion of the partition function $\mathrm{Tr}[\mathrm{e}^{-\beta\hat{H}}]$ and uses a discrete representation of continuous imaginary time. This discretization however does not introduce errors into the algorithm as is the case in the alternative path-integral formulation, where one usually performs a Trotter-type discretization of imaginary time, see e.g. Refs. [30, 21], though formulations in continuous imaginary time also exist [31, 32]. Here β is the inverse temperature $1/T$ (in our units $k_B = 1$).

The SSE algorithm has several properties that are very useful in addressing the problems we focus on in this study. Firstly, it works in continuous imaginary time as discussed above. Secondly, it allows not only local updates of system configurations but also global cluster updates, which in most cases prove to be more efficient than single-spin-flip updates. These global updates are achieved by dividing the configurations of the system produced by the QMC into clusters and then flipping a fraction of them within each sweep of the simulation [33]. An important bonus of cluster updates is the existence of "improved estimators" for determining time-dependent correlation functions, for which the signal to noise is much better than with conventional measurements.

In addition, we speed up equilibration by implementing "parallel tempering" [34], where simulations for different values of s are run in parallel and spin configurations with adjacent values of s are swapped with a probability satisfying the detailed balance condition. Traditionally, parallel tempering is performed for systems at different temperatures, but here the parameter s plays the role of (inverse) temperature.

The gap of the system for a given instance and a given s value is extracted by analyzing measurements of (imaginary) time-dependent correlation functions of the type

$$C_A(\tau) = \langle \mathcal{A}(\tau)\mathcal{A}(0)\rangle - \langle A\rangle^2 , \tag{6}$$

where the operator $\mathcal{A}$ is some measurable physical quantity. In the low temperature limit, $\Delta E \ll T$ where $\Delta E = E_1 - E_0$, the system is in its ground state so the imaginary-time correlation function is given by

$$C_A(\tau) = \sum_{m=1} |\langle 0|\mathcal{A}|m\rangle|^2 \left(e^{-\Delta E_m \tau} + e^{-\Delta E_m(\beta-\tau)}\right) , \tag{7}$$

where $\Delta E_m = E_m - E_0$. At long times, τ, the correlation function is dominated by the smallest gap, $\Delta E \equiv \Delta E_1$, (as long as the matrix element $|\langle 0|\mathcal{A}|1\rangle|^2$ is nonzero). On a log-linear plot $C_A(\tau)$ then has a region where it is a straight line whose slope is the negative of the gap. This can therefore be extracted by linear fitting. The difficulty is that the straight-line region is at long times where the signal is very small and is easily dominated by the Monte Carlo noise. However, the signal to noise can be dramatically improved by optimizing the choice of the operator $\mathcal{A}$, as discussed in Ref. [25].

As we shall see, a couple of the problems we analyze have global bit-flip symmetry in which case, a subtle difficulty arises in the computation of the gap. Eigenstates of the Hamiltonian are either even or odd under bit-flip symmetry (in particular, the ground state is even). In the $s \to 1$ limit, states occur in even-odd pairs with an exponentially small gap (see Fig. 1 for an illustration). Therefore, the quantity of interest is the gap to the first even state. To do that, we consider correlation functions of even quantities, so there are only matrix elements between states of the same parity. However, the lowest odd level becomes very close to the ground state near where the gap to the first even excited state has a minimum, hence this lowest odd state becomes thermally populated, with the result that odd-odd gaps are present in the data as well. We have eliminated these undesired contributions by projecting out the symmetric subspace of the Hamiltonian: In standard quantum Monte Carlo simulations one imposes periodic boundary conditions in imaginary time. To project out the symmetric subspace one imposes, instead, free boundary conditions. The properties of the symmetric subspace can then be obtained, for $\beta \to \infty$, by measurements far from the boundaries. We have incorporated this idea into our QMC scheme, and used this modified algorithm in the simulations of problems with global bit-flip symmetry (the reader is referred to Ref. [23] for a more detailed discussion of this issue).

For each of the studied s values we extract the gap and interpolate the minimum value using a simple quadratic fit. We determine the typical minimum gap for different sizes by considering about 50 instances for each size and extracting the minimum gap for each of them. We take the *median* value of the minimum gap among the different instances for a given size as a measure of the "typical" minimum gap for that size.

2.2. SAT problems

In Ref. [22], we studied three optimization problems which had previously been suggested [35, 36, 37] as good potential candidates for detailed investigation. These are of the "constraint satisfaction" (SAT) type in which there are N bits and M "clauses", each clause being a logical condition on a small number of the bits. The problem is to decide whether there is an assignment of the N bits which satisfies all of M clauses. Each bit is represented in the Hamiltonian by

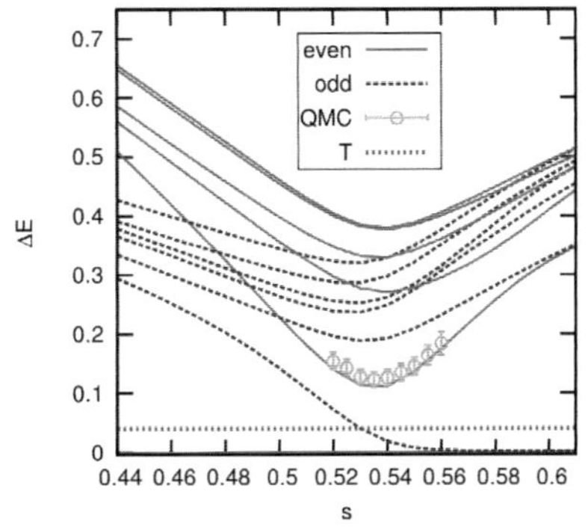

Figure 1. (Color online) Energy gaps to even (solid, red) and odd (dashed, blue) excited states for an $N = 16$ instance of the locked 2-in-4 problem, which has bit-flip symmetry as discussed in the text. The dotted line shows a characteristic value of another important energy scale in the problem, temperature. In the region where the gap to the first even state has a minimum, the gap to the first odd state becomes very small and is inevitably thermally populated. Hence, odd-odd gaps appear in this region as well as even-even gaps. This is the reason why we use a non-standard Monte Carlo algorithm for this problem which projects out the symmetric subspace, so only even-even gaps are present in the data. The figure also shows the gap obtained from the even-subspace projected QMC in vicinity of the minimum. It agrees with exact diagonalization within the error bars.

the z-component of a Pauli matrix, σ_i^z, where i labels the spin. Each clause is converted to an energy function which depends on the spins associated with the clause, such that the energy is zero if the clause is satisfied and is a positive integer if it is not. The general structure of the problem Hamiltonian $\mathcal{H}_p$ is therefore

$$\mathcal{H}_P = \sum_{a=1}^{M} \mathcal{H}_a \,, \tag{8}$$

where a is the clause index and $\mathcal{H}_a$ is the energy associated with the clause and is a function of the spins σ_i^z belonging to it.

Clearly, it is easy to satisfy all clauses if the ratio $\alpha \equiv M/N$ is small enough. In fact, one expects an exponentially large number of satisfying assignments in this region. Conversely, if M/N is very large, with high probability there will be a conflict between different clauses. Hence there is a "satisfiability transition" at some value α_s where the number of satisfying assignments goes to zero. It is particularly hard to solve satisfiability problems close to the transition [38], so we will work in this region. Furthermore, when studying the efficiency of the QAA numerically [6, 30, 21], it is convenient to consider instances with a unique satisfying assignment (USA), i.e. the ground state of the problem Hamiltonian is non-degenerate. This also means that we must take a value of α very close to α_s.

We studied three different SAT problems which go under the arcane (for the physics community) names of "locked 1-in-3 SAT", "locked 2-in-4 SAT", and "3-reg-3-XORSAT". The first two problems are "locked" problems – a term first introduced by Zdeborová and Mézard [35, 36] for problems with instances having the following two properties: (i) every variable is in at least two clauses, and (ii) one can not get from one satisfying assignment to another by flipping a single bit. These locked problems have several properties that make them eminently suitable as benchmarks: They are analytically "simple", but are computationally hard. Also, fluctuations between instances are smaller than with "unlocked" problems. In the

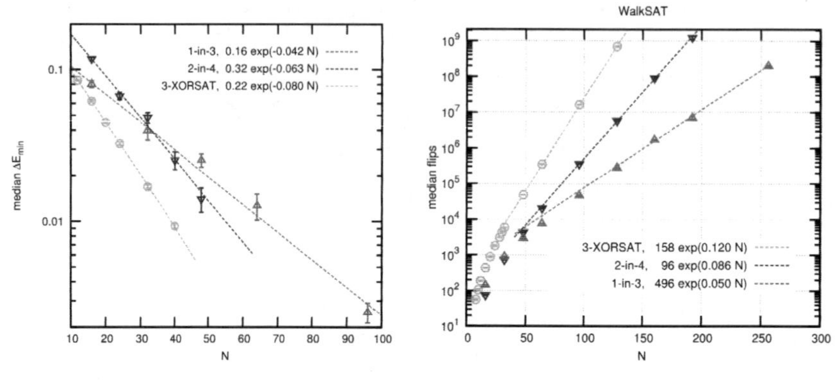

Figure 2. (Color online) Left: Median gap on a log-linear scale for the locked 1-in-3 (red), locked 2-in-4 (blue) and 3-reg-3-XORSAT (green) problems. The straight-line fits are fairly reasonable, indicating that the complexity for these problems is indeed exponential. Right: Median number of flips to solution of the WalkSAT algorithm applied to the same three problems also on a log-linear scale.

model	μ(QAA)	μ(WalkSAT)	Ratio
locked 1-in-3 SAT	0.084(3) [22]	0.0505(5) [40]	1.66
locked 2-in-4 SAT	0.126(5) [22]	0.0858(8) [40]	1.47
3-reg-3-XORSAT	0.159(2) [23]	0.1198(20) [40]	1.32

Table 1. Values of μ, the coefficient of N in the exponential running time expression in Eq. (1), for the Quantum Adiabatic Algorithm (QAA) and for the analogous classical algorithm WalkSAT, as well as the ratios between them, for three SAT problems.

1-in-3 SAT problem each clause consists of three bits chosen randomly, and the clause is satisfied if one of the bits is one and the others are zero. Similarly, in locked 2-in-4 instances, a clause has four bits, and is satisfied if two are zero and two are one. The reader is referred to Ref. [22] for a more detailed description of the problems. For our purposes here we only need to know that these are three models with which we tested the performance of the QAA. The "3-reg-3-XORSAT" is an example of a p-spin model with $p = 3$ of the type that has been studied in the spin glass community (see, e.g., Ref. [39]) The latter model was studied for small sizes, up to $N = 24$, using diagonalization by Jörg *et al.* [37] while Refs. [22] and [23] extended the range of sizes up to $N = 40$ by Quantum Monte Carlo simulations, obtaining consistent results. For all three models we find [22, 23] that the minimum gap decreases exponentially with N (see left pane of Fig. 2).

As discussed in Sec. 1, the computation time is proportional to $1/\Delta E_{\rm min}^2$ (neglecting N dependence of matrix elements) and since we find [22, 23] that $\Delta E_{\rm min} \sim \exp(-c\,N)$, the running time can be written in the form of Eq. (1) with $\mu = 2\,c$.

It is interesting to compare the efficiency of the QAA with that of a classical algorithm. In Ref. [40] it was argued that a reasonable classical algorithm to compare with QAA is the heuristic local search algorithm known as WalkSAT [41], which is similar in spirit to simulated annealing in that both make moves which reduce the "energy", but also sometimes make moves which increase it to avoid being trapped in the nearest local minimum. In the WalkSAT algorithm, the running time is proportional to the number of "bit flips" the algorithm makes (for more details, the reader is referred to Ref [40]). These results are summarized in Fig. 2 (right pane). Writing the median number of flips as $N_{\rm flips} \propto e^{\mu N}$, we compare in Table 1 [22] the values of μ defined by Eq. (1) for the QAA and WalkSAT.

As Table 1 indicates, the exponent coefficients obtained with WalkSAT are somewhat smaller than those of the QAA, suggesting that the latter algorithm, in the specific way it was implemented in Ref. [22], is slightly less efficient than its corresponding classical one for these three problems. It is also evident from the table that the harder the problem is for WalkSAT, the harder it also is for QAA.

Previous work [42, 37, 43] had argued that a first order quantum phase transition occurs for a broad class of random SAT problems, and indeed our results are consistent with the gap being a minimum *at* a first order quantum phase transition.

2.3. A spin glass problem

SAT problems are standard fare for computer scientists working on optimization, but are rather unfamiliar to physicists. We have therefore also studied [23] a problem in the physics domain, a spin glass. A spin glass model should have both disorder and "frustration". Most spin glass models [44] achieve this by having the spins on a regular lattice and choosing the interactions between neighbors to be random with a significant probability of having either sign. Here, by contrast, we put the spins on the vertices of a regular random graph of coordination number (degree) equal to three, rather than a regular lattice, and make all the interactions equal and antiferromagnetic. Locally the graph looks like a tree but, since there are no free ends, there are large loops with a typical size of order $\ln N$. Disorder and frustration come entirely from these large loops. Disorder arises because the connections between the sites are chosen at random so the structure of the loops is random. Frustration arises because half the loops, on average, have an odd number of links and these are frustrated. Because of these odd-size loops, it is not possible to form a regular antiferromagnetic state and the system is actually a spin glass [45]. The problem Hamiltonian is very simple:

$$\mathcal{H}_P = \sum_c \sigma^z_{i_1,c}\sigma^z_{i_2,c}, \tag{9}$$

where c runs over the $3N/2$ links, and (i_1, c) and (i_2, c) refer to the two sites on a link. Again we choose samples (instances) for which $\mathcal{H}_P$ has a unique ground state, apart from the global symmetry of flipping all the σ^z_i.

For small s the system is dominated by the transverse field and equilibration is fast; the system is in a quantum paramagnetic phase. As s increases, the system goes through a quantum spin glass phase transition at a critical value $s_c \simeq 0.36$, while for still larger values of s the system is in a quantum spin glass phase. We were not able to equilibrate the system all the way in the spin glass phase up to the end point $s = 1$. However, even considering just the part of the spin glass phase which could be equilibrated ($s_c = 0.36 < s < 0.5$) we found that the minimum gap decreases exponentially with size, as shown in Fig. 3 (left pane).

Interestingly, if we just consider data right in the vicinity of the critical point, the minimum gap in this region only seems to decrease with a power of the system size [23], at least for the range of sizes that could be studied.

We see that the bottleneck for the QAA occurs at a different place for the SAT problems and the spin glass. For SAT, the gap becomes very small *at* the (first order) quantum phase transition, while for the spin glass it becomes very small *beyond* the (second order) transition in the spin glass phase.

3. Modifications to the simplest approach

The numerical studies described above have made it clear that a simple application of the QAA, i.e., an equal-weight transverse-field driver Hamiltonian and a linear interpolating schedule, does not lead to quantum speedups, even in the ideal setting of zero-temperature and in the absence

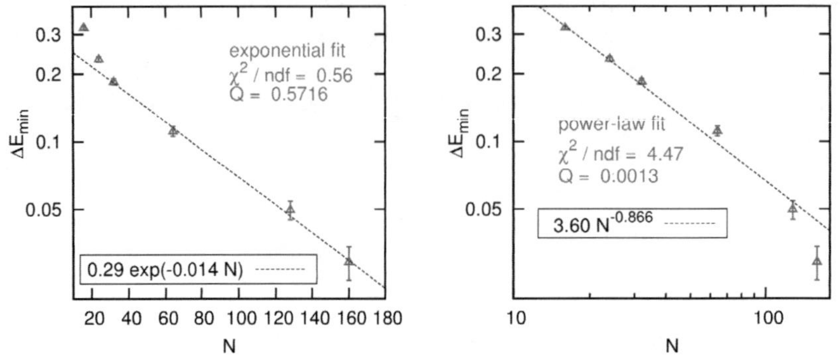

Figure 3. (Color online) Left panel is the median minimum gap on a log-lin scale for the spin glass problem. The quality of the straight-line fit is good, ($Q = 0.57$), indicating that the minimum gap decreases exponentially with N. (The two smallest sizes are omitted from the fit since they seem to be affected by corrections to scaling.) A power-law fit works less well as shown in the right hand panel.

of coupling to a decohering environment. One could then ask whether other, perhaps more appropriate, implementations of the QAA could lead to a better complexity scaling and what is the effort that would be involved in finding such variants of the QAA.

There are many ways in which one could generalize and expand on the QAA approach taken here to possibly greatly increase the success probability of the algorithm, or equivalently, shorten its required runtime. To mention a few examples, it is now known that a carefully chosen annealing schedule could improve the error-scaling of the algorithm [46] and/or yield a better scaling of the runtime with respect to the minimum gap [47]. Other studies show that taking different paths in 'Hamiltonian space' between the initial and final Hamiltonians may also prove beneficial [31, 48]. Finally, it has been shown that in some cases, thermal effects (i.e., nonzero temperature) that are normally thought of having destructive effects, populating unwanted excited states, may in fact assist in the annealing process by providing an additional mechanism for traversing energy barriers [49].

The quantum adiabatic algorithm studied here did not utilize any a priori knowledge about the SAT or spin-glass instances being solved in the construction of the algorithm (as both the driver Hamiltonian and the annealing schedule had been determined independently of those beforehand). It is well known that for some problems, the utilization of such knowledge in the implementation of the QAA may prove advantageous. One notable example is the oracular problem of searching for a marked item in an unstructured database addressed by Roland and Cerf [47] later extended to other problems [26, 27], which showed that tailoring the driver Hamiltonian as well as the adiabatic schedule to the problem at hand (which for these problems translated to an explicit calculation of the gap as a function of the adiabatic parameter) may lead to quantum speedups.

In satisfiability problems, one can not compute the value or location of the minimum gap beforehand as these vary between instances and doing so for each instance is likely to be as difficult as solving the problem itself. However, it is plausible to assume that one could utilize the fact that these Hamiltonians are "frustration-free" [50], i.e., that the solutions of the total Hamiltonian are also solutions of the individual clauses, to gain a favorable scaling of the gap. One approach would be to consider for each instance the class of problem Hamiltonians

$$\mathcal{H}_p = \sum_{a=1}^{M} c_a \mathcal{H}_a \,, \tag{10}$$

where c_a are positive real-valued weights. That the solutions of the original satisfiability problem are also solutions of the individual clauses ensures that for any given set of instances $\{\mathcal{H}_a\}$, the ground states of the entire class of problem Hamiltonians are solutions of the original problem as well, and one therefore has the freedom to choose these positive weights as deemed fit. The above is also true for the family of driver Hamiltonians

$$\mathcal{H}_D = -\sum_i \Gamma_i \sigma_i^x \,, \tag{11}$$

where different values for the nonzero weights Γ_i may also be considered.

An appropriate choice of values for the weights c_a and/or Γ_i could be sought by 'blind' trial and error, in which case a polynomial number of QAAs with different weight sets is attempted and the configuration of weights producing the shortest runtime (or largest minimum gap) is then picked out. A presumably more efficient method would be to adjust the weight values adaptively after running the QAA and examining the obtained bit configuration and the clauses it violated. Assigning larger weights to the violated clauses will render them more unlikely to be violated in the next round, thereby presumably also making the minimum gap larger. It should be noted that this type of techniques can be straightforwardly tested using the QMC techniques used here, however it should be emphasized that these are worthy of implementing only if the computational effort involved in adjusting the weights turns out to be smaller than the computational gain associated with the growth of the minimum gap.

4. Conclusions

We have studied the efficiency of the QAA in its most basic implementation, when applied to several optimization problems, ignoring the effects of finite temperature and non-thermal noise which occur in experiments on the D-Wave hardware. We used Quantum Monte Carlo simulations to compute the minimum gap and then took the run time to be proportional to one over the square of the minimum gap. For each problem we found that the QAA was not more efficient than a classical heuristic solver, WalkSAT, in the sense that the coefficient μ in Eq. (1) was a little larger for the QAA than for WalkSAT. We also discussed possible modifications to the QAA that might improve the situation.

Acknowledgments

We thank Eddie Farhi and Vadim Smelyanskiy for many stimulating discussions about quantum annealing. This work is supported in part by the National Science Foundation under Grant No. DMR-DMR-1207036.

References

[1] M. A. Nielsen and I. L. Chuang, *Quantum Computation and Quantum Information.* Cambridge, England: Cambridge University Press, 2000.

[2] N. D. Mermin, *Quantum Computer Science.* Cambridge: Cambridge University Press, 2007.

[3] P. W. Shor, "Algorithms for quantum computing: discrete logarithms and factoring," in *Proc. 35th Symp. on Foundations of Computer Science* (S. Goldwasser, ed.), (Los Alamitos CA), p. 124, IEEE Computer Society Press, 1994.

[4] T. Kadowaki and H. Nishimori, "Quantum annealing in the transverse Ising model," *Phys. Rev. E*, vol. 58, p. 5355, 1998.

[5] G. Santoro, R. Martoňák, E. Tosatti, and R. Car, "Theory of quantum annealing of an Ising spin glass," *Science*, vol. 295, p. 2427, 2002.

[6] E. Farhi, J. Goldstone, S. Gutmann, J. Lapan, A. Lundgren, and D. Preda, "A quantum adiabatic evolution algorithm applied to random instances of an NP-complete problem," *Science*, vol. 292, p. 472, 2001. A longer version of the paper appeared in arXiv:quant-ph/0104129.

[7] G. Toulouse, "Theory of the frustration effect in spin glasses," *Commun. Phys.*, vol. 2, p. 115, 1977.

[8] K. Binder and A. P. Young, "Spin glasses: Experimental facts, theoretical concepts and open questions," *Rev. Mod. Phys.*, vol. 58, p. 801, 1986.

[9] S. Kirkpatrick, C. D. Gelatt, Jr., and M. P. Vecchi, "Optimization by simulated annealing," *Science*, vol. 220, p. 671, 1983.

[10] M. E. J. Newman and G. T. Barkema, *Monte Carlo Methods in Statistical Physics*. New York, USA: Oxford University Press Inc., 1999.

[11] S. Boixo et. al., "Quantum annealing with more than one hundred qubits," *Nature Phys.*, vol. 10, p. 2018, 2014.

[12] A. Messiah, *Quantum Mechanics*. Amsterdam: North-Holland, 1962.

[13] G. H. Wannier, "Probability of violation of the Ehrenfest principle," *Physics*, vol. 1, p. 251, 1965.

[14] E. Farhi, J. Goldstone, and S. Gutmann, "Quantum adiabatic algorithms versus simulated annealing." (arXiv:quant-ph/0201031), 2002.

[15] M. W. Johnson *et al.*, "Quantum annealing with manufactured spins," *Nature*, vol. 473, pp. 194–198, 2011.

[16] A. J. Berkley *et al.*, "Tunneling spectroscopy using a probe qubit," *Phys. Rev. B*, vol. 87, p. 020502(R), 2013.

[17] T. F. Rønnow, Z. Wang, J. Job, S. Boixo, S. V. Isakov, D. Wecker, J. M. Martinis, D. A. Lidar, and M. Troyer, "Defining and detecting quantum speedup," *Science*, vol. 345, no. 6195, pp. 420–424, 2014.

[18] A. M. Childs, E. Farhi, and J. Preskill, "Robustness of adiabatic quantum computation," *Phys. Rev. A*, vol. 65, p. 012322, 2001.

[19] M. Amin, D. V. Averin, and J. A. Nesteroff, "Decoherence in adiabatic quantum computation," *Phys. Rev. A*, vol. 79, p. 022107, 2009. (arXiv:0708.0384).

[20] L. Wang et. al., "Comment on "Classical signature of quantum annealing"." (unpublished), 2013.

[21] A. P. Young, S. Knysh, and V. Smelyanskiy, "First order phase transition in the Quantum Adiabatic Algorithm," *Phys. Rev. Lett.*, vol. 104, p. 020502, 2010.

[22] I. Hen and A. P. Young, "Exponential complexity of the quantum adiabatic algorithm for certain satisfiability problems," *Phys. Rev. E.*, vol. 84, p. 061152, 2011.

[23] E. Farhi, D. Gosset, I. Hen, A. Sandvik, P. Shor, A. P. Young, and F. Zamponi, "The performance of the quantum adiabatic algorithm on random instances of two optimization problems on regular hypergraphs," *Phys. Rev. A*, vol. 86, p. 052334, 2012.

[24] I. Hen and A. P. Young, "Solving the graph-isomorphism problem with a quantum annealer," *Phys. Rev. A.*, vol. 86, p. 042310, 2012.

[25] I. Hen, "Excitation gap from optimized correlation functions in quantum monte carlo simulations," *Phys. Rev. E.*, vol. 85, p. 036705, 2012.

[26] I. Hen, "Continuous-time quantum algorithms for unstructured problems," *J. Phys. A*, vol. 47, p. 045305, 2014.

[27] I. Hen, "How fast can quantum annealers count?," *J. Phys. A*, vol. 47, p. 235304, 2014.

[28] A. Sandvik, "Stochastic series expansion method with operator-loop update," *Phys. Rev. B*, vol. 59, p. R14157, 1999.

[29] A. Sandvik, "A generalization of Handscomb's quantum Monte Carlo scheme — Application to the 1-D Hubbard model," *J. Phys, A*, vol. 25, p. 3667, 1992.

[30] A. P. Young, S. Knysh, and V. Smelyanskiy, "Size dependence of the minimum excitation gap in the Quantum Adiabatic Algorithm," *Phys. Rev. Lett.*, vol. 101, p. 170503, 2008.

[31] E. Farhi, J. Goldstone, D. Gosset, S. Gutmann, H. B. Meyer, and P. Shor, "Quantum adiabatic algorithms, small gaps, and different paths," *Quant. Inf. Comp.*, vol. 11, p. 181, 2009. (arXiv:0909.4766).

[32] F. Krzakala, A. Rosso, G. Semerjian, and F. Zamponi, "On the path integral epresentation for quantum spin models and its application to the quantum cavity method and to Monte Carlo simulations," *Phys. Rev. B*, vol. 78, p. 134428, 2008.

[33] A. Sandvik, "Stochastic series expansion method for quantum Ising models with arbitrary interactions," *Phys. Rev. E*, vol. 68, p. 056701, 2003.

[34] K. Hukushima and K. Nemoto, "Exchange Monte Carlo method and application to spin glass simulations," *J. Phys. Soc. Japan*, vol. 65, p. 1604, 1996.

[35] L. Zdeborová and M. Mézard, "Locked constraint satisfaction problems," *Phys. Rev. Lett.*, vol. 101, p. 078702, 2008.

[36] L. Zdeborová and M. Mézard, "Constraint satisfaction problems with isolated solutions are hard," *J. Stat. Mech.*, vol. 12, p. P12004, 2008.

[37] T. Jörg, F. Krzakala, G. Semerjian, and F. Zamponi, "First-order transitions and the performance of quantum algorithms in random optimization problems," *Phys. Rev. Lett.*, vol. 104, p. 207206, 2010. (arXiv:0911.3438).

[38] S. Kirkpatrick and B. Selman, "Critical behavior in the satisfiability of random Boolean expressions," *Science*, vol. 264, p. 1297, 1994.

[39] C. K. Thomas and H. G. Katzgraber, "Optimizing glassy p-spin models," *Phys. Rev. E*, vol. 83, p. 046709, 2011.

[40] M. Guidetti and A. P. Young, "Complexity of several constraint satisfaction problems using the heuristic, classical, algorithm, WalkSAT," *Phys. Rev. E*, vol. 84, p. 011102, 2011. (arXiv:1102.5152).

[41] For information about WALKSAT see `http://www.cs.rochester.edu/~kautz/walksat/`.

[42] T. Jörg, F. Krzakala, J. Kurchan, and A. C. Maggs, "Simple glass models and their quantum annealing," *Phys. Rev. Lett.*, vol. 101, p. 147204, 2008.

[43] T. Jörg, F. Krzakala, J. Kurchan, A. C. Maggs, and J. Pujos, "Energy gaps in quantum first-order mean-field-like transitions: The problems that quantum annealing cannot solve," *Europhys. Lett.*, vol. 89, p. 40004, 2010. (arXiv:0912.4865).

[44] S. F. Edwards and P. W. Anderson, "Theory of spin glasses," *J. Phys. F*, vol. 5, p. 965, 1975.

[45] Y. Matsuda, H. Nishimori, and L. Zdeborova, "Random-field p-spin-glass model on regular random graphs," *J. Phys. A: Math. and Theor.*, vol. 44, p. 185002, 2007.

[46] N. Wiebe and N. S. Babcock, "Improved error-scaling for adiabatic quantum evolutions," *New Journal of Physics*, vol. 14, no. 1, p. 013024, 2012.

[47] J. Roland and N. J. Cerf, "Quantum search by local adiabatic evolution," *Phys. Rev. A*, vol. 65, p. 042308, 2002.

[48] A. Perdomo-Ortiz, S. E. Venegas-Andraca, and A. Aspuru-Guzik, "A study of heuristic guesses for adiabatic quantum computation," *Quantum Information Processing*, vol. 10, p. 33–52, mar 2010. n/a.

[49] N. Dickson *et al.*, "Thermally assisted quantum annealing of a 16-qubit problem," *Nature Comm.*, vol. 4, p. 1903, 2013.

[50] S. Bravyi and B. Terhal, "Complexity of stoquastic frustration-free hamiltonians," *SIAM J. Comput.*, vol. 39, pp. 1462–1485, Nov. 2009.

Lanczos steps to improve variational wave functions

Federico Becca[1], Wen-Jun Hu[2], Yasir Iqbal[3], Alberto Parola[4], Didier Poilblanc[5], and Sandro Sorella[1]

[1] Democritos National Simulation Center, Istituto Officina dei Materiali del CNR and SISSA-International School for Advanced Studies, Via Bonomea 265, I-34136 Trieste, Italy
[2] Department of Physics and Astronomy, California State University, Northridge, California 91330, USA
[3] Institute for Theoretical Physics and Astrophysics, Julius-Maximilian's University of Würzburg, Am Hubland, D-97074, Würzburg, Germany
[4] Dipartimento di Scienza e Alta Tecnologia, Università dell'Insubria, Via Valleggio 11, I-22100 Como, Italy
[5] Laboratoire de Physique Théorique UMR-5152, CNRS and Université de Toulouse, F-31062 Toulouse, France

E-mail: becca@sissa.it

Abstract. Gutzwiller-projected fermionic states can be efficiently implemented within quantum Monte Carlo calculations to define extremely accurate variational wave functions for Heisenberg models on frustrated two-dimensional lattices, not only for the ground state but also for low-energy excitations. The application of few Lanczos steps on top of these states further improves their accuracy, allowing calculations on large clusters. In addition, by computing both the energy and its variance, it is possible to obtain reliable estimations of exact results. Here, we report the cases of the frustrated Heisenberg models on square and Kagome lattices.

1. Introduction

Obtaining accurate results in strongly-correlated systems remains an extremely difficult task in systems where the sign-problem prevents one to use numerically exact quantum Monte Carlo methods. [1] In the recent past, a large variety of numerical approaches have been proposed and devised to study simple models of interacting electrons on the lattice. Among the most promising ones, we mention post density-matrix renormalization group (DMRG) techniques, based upon tensor networks. [2, 3] As far as insulating phases are concerned, one important issue is to describe Mott insulators, in which the insulating character is driven by the strong electron-electron interaction. When lowering the temperature, the general expectation is that some symmetry-breaking phenomena take place, leading for example to magnetic or Peierls (i.e., dimerized) phases. In two spatial dimensions, a continuous symmetry cannot be spontaneously broken at finite temperatures; still, the ground state may possess magnetic long-range order. In this regard, the existence of quantum spin liquids, i.e., states that do not break any symmetry down to zero temperature, in two dimensions represents a fascinating problem. The key feature to impede long-range ordering is given by the presence of frustrating interaction, namely the presence of competing super-exchange couplings that strongly enhance quantum fluctuations. The simplest example that captures the low-energy physics of Mott insulators and may give rise

Journal of Physics: Conference Series **640** (2015) 012039 doi:10.1088/1742-6596/640/1/012039

to spin-liquid ground states is the frustrated Heisenberg model:

$$\mathcal{H} = J_1 \sum_{\langle ij \rangle} \mathbf{S}_i \cdot \mathbf{S}_j + J_2 \sum_{\langle\langle ij \rangle\rangle} \mathbf{S}_i \cdot \mathbf{S}_j, \tag{1}$$

where both J_1 and J_2 are positive and $\mathbf{S}_i$ are spin-1/2 operators at each lattice site i; $\langle ij \rangle$ and $\langle\langle ij \rangle\rangle$ denote sum over nearest-neighbor and next-nearest-neighbor sites, respectively. Here, we consider both the square and Kagome lattices with N sites: $N = L \times L$ for the square lattice and $N = 3 \times L \times L$ for the Kagome lattice. All energies are given in units of J_1.

The ground state properties of these two models have been widely debated in the past twenty years, with contradictory results. Recent DMRG calculations suggest that the ground state on the Kagome lattice for $J_2 = 0$ is a gapped Z_2 spin liquid, [4] while the ground state of the $J_1{-}J_2$ model on the square lattice is a gapless spin liquid for $0.45 < J_2/J_1 < 0.5$ and a (gapped) valence-bond solid for $0.5 < J_2/J_1 < 0.6$. [5] Here, we present a variational Monte Carlo method to assess ground-state properties of these models. The starting variational wave functions can be substantially improved by the application of few Lanczos steps, which can be easily affordable for rather large clusters (i.e., few hundreds sites). Moreover, the calculation of both the energy and its variance makes it possible to perform a zero-variance extrapolation to estimate exact energies, not only for the ground state but also for low-energy excitations. [6, 7, 8]

2. Variational wave functions

The variational wave functions are defined through the mean-field Hamiltonian for the Abrikosov-fermion representation of the spin-1/2 operators: [9]

$$\mathcal{H}_{MF} = \sum_{i,j,\sigma} (t_{i,j} + \mu \delta_{ij}) c^\dagger_{i,\sigma} c_{j,\sigma} + \sum_{i,j} (\eta_{i,j} + \zeta \delta_{ij})(c^\dagger_{i,\uparrow} c^\dagger_{j,\downarrow} + c^\dagger_{j,\uparrow} c^\dagger_{i,\downarrow}) + h.c., \tag{2}$$

where for each bond (i,j) there are hopping $t_{i,j}$ and/or pairing $\eta_{i,j}$ terms; on-site terms, i.e., a real chemical potential μ and/or a complex pairing ζ, may be also considered. Given any eigenstate $|\Psi_{MF}\rangle$ of the mean-field Hamiltonian (2), a physical state for the spin model can be obtained by a projection onto the subspace with one fermion per site:

$$|\Psi_v\rangle = \mathcal{P}_G |\Psi_{MF}\rangle, \tag{3}$$

where $\mathcal{P}_G = \prod_i (n_{i,\uparrow} - n_{i,\downarrow})^2$ is the Gutzwiller projector, $n_{i,\sigma} = c^\dagger_{i,\sigma} c_{i,\sigma}$ being the local density. The flexibility of this approach is given by the fact that *different* spin liquids, having for example $U(1)$ or Z_2 gauge structure and gapped or gapless spinon spectrum, may be obtained by changing the pattern of the $t_{i,j}$'s and the $\eta_{i,j}$'s. [10] Moreover, dimerized or chiral states can be also easily obtained within this approach. [11, 12]

In the following, we will consider the square lattice with $J_2/J_1 = 0.5$ and the Kagome lattice with both $J_2 = 0$ and $J_2 = 0.25$, corresponding to cases where the high frustration may stabilize fully symmetric spin liquids. For the square lattice, we will consider a state that is obtained by taking a real pairing η_{xy} (with d_{xy} symmetry) on top of the $U(1)$ state with nearest-neighbor hopping t and real pairing $\eta_{x^2-y^2}$ (with $d_{x^2-y^2}$ symmetry). The d_{xy} term is responsible for breaking the $U(1)$ gauge symmetry down to Z_2. Restricting this coupling along the $(\pm 2, \pm 2)$ bonds implies four Dirac points in the mean-field spectrum. [6] For the Kagome lattice, we will consider the $U(1)$ state that is described by nearest-neighbor real hoppings such to have 0 magnetic flux through triangles and π flux through hexagons. [13] At the mean-field level, there are gapless excitations at two Dirac points. A slight improvement of the energy can be achieved by considering a next-nearest-neighbor real hopping. [11] A gapped wave function, corresponding

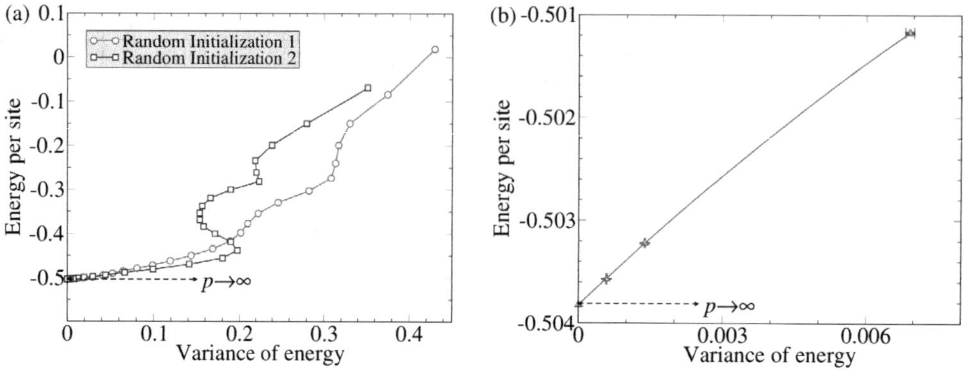

Figure 1. The energy per site versus the energy variance per site for the J_1-J_2 Heisenberg model on the square lattice with $J_2/J_1 = 0.5$ and $L = 6$. Exact Lanczos diagonalizations starting from two random initializations (a) and quantum Monte Carlo calculations starting from the best variational state (3) (b) are reported. Points correspond to different values of p.

to a Z_2 spin liquid (labelled by $Z_2[0,\pi]\beta$) can be obtained by adding on-site and next-nearest-neighbor pairing terms. [14] A gapless spin liquid with a large Fermi surface (labeled by uniform RVB) is obtained when no magnetic fluxes are present.

Within this formalism, it is straightforward to construct not only an Ansatz for the ground state but also for low-energy excitations. In this respect, it is useful to consider a particle-hole transformation for the down electrons on the mean-field Hamiltonian (2), i.e., $c^\dagger_{i,\downarrow} \to c_{i,\downarrow}$, such that the transformed Hamiltonian conserves the total number of particles. Then, the ground state is obtained by filling the lowest N orbitals, with suitable boundary conditions (either periodic or anti-periodic) in order to have a unique mean-field state. Spin excitations can be obtained by creating the appropriate Bogoliubov quasi-particles (spinons) and possibly switching boundary condition. In particular, it is easy to construct a unique and real wave function containing four spinon excitations with total momentum $k = (0,0)$, while generic $S = 1$ excitations would require complex wave functions. Therefore, we will consider the case of a $S = 2$ excitation with momentum $k = (0,0)$, for both the square and the Kagome lattices. By computing separately the energies of the $S = 0$ and $S = 2$ states, the spin gap Δ_2 is studied as a function of the cluster size.

3. Lanczos step procedure

In order to systematically improve the variational wave functions, we can apply a number p of Lanczos steps to the starting variational wave function:

$$|\Psi_p\rangle = \left(1 + \sum_{m=1}^{p} \alpha_m \mathcal{H}^m\right) |\Psi_v\rangle, \tag{4}$$

where α_m are p additional variational parameters. Clearly, whenever $|\Psi_v\rangle$ is not orthogonal to the exact ground state, $|\Psi_p\rangle$ converges to it for large p. Unfortunately, on large sizes, only few steps can be efficiently afforded: here, we consider the case with $p = 1$ and $p = 2$ ($p = 0$ corresponds to the original variational wave function). Furthermore, an estimate of the exact energy may be obtained by the variance extrapolation. Indeed, for accurate variational states $|\Psi_p\rangle$ with energy E_p and variance σ_p^2, it is easy to prove that

$$E_p \approx E_{\rm ex} + {\rm const} \times \sigma_p^2, \tag{5}$$

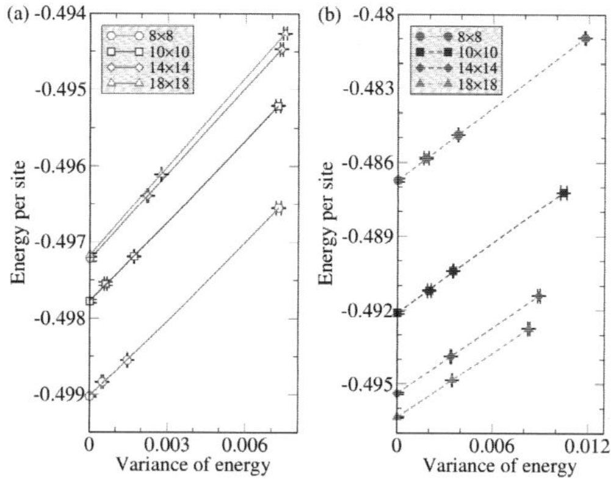

Figure 2. Energies per site for $S = 0$ (a) and $S = 2$ (b) versus the energy variance per site for the Heisenberg model on the square lattice with $J_2/J_1 = 0.5$. The initial state is the best variational Ansatz (3). The variance extrapolated results are also shown.

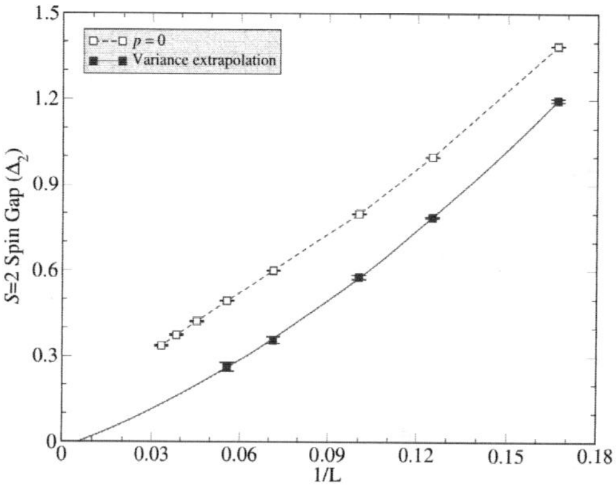

Figure 3. The size scaling of the $S = 2$ spin gap for the best variational wave function and the zero-variance extrapolation on the square lattice with $J_2/J_1 = 0.5$. The thermodynamic extrapolation gives $\Delta_2 = -0.04(5)$.

where

$$E_p = \frac{\langle\Psi_p|\mathcal{H}|\Psi_p\rangle}{N}, \tag{6}$$

$$\sigma_p^2 = \frac{\langle\Psi_p|\mathcal{H}^2|\Psi_p\rangle - \langle\Psi_p|\mathcal{H}|\Psi_p\rangle^2}{N}, \tag{7}$$

are the energy and variance per site, respectively. Therefore, the exact energy $E_{\rm ex}$ may be extracted by fitting E_p versus σ_p^2 and performing a zero-variance extrapolation.

It should be emphasized that the Lanczos step procedure of Eq. (4) is not size consistent if p is not increased with the number of sites N. Indeed, both the energy and variance improvements with respect to the original state $|\Psi_v\rangle$ vanish for $N \to \infty$ and fixed p. Nevertheless, it is remarkable that a sizable improvement is obtained even for rather large clusters with few hundred sites. By contrast, the zero-variance extrapolation remains size consistent, as shown below.

4. Monte Carlo sampling

By using the variational Monte Carlo technique, the properties of the Gutzwiller-projected wave function can be easily assessed. Indeed, expectation values of any operator $\mathcal{O}$, including $\mathcal{H}$, are given by:

$$\frac{\langle\Psi_p|\mathcal{O}|\Psi_p\rangle}{\langle\Psi_p|\Psi_p\rangle} = \sum_x \frac{\langle\Psi_p|x\rangle\langle x|\mathcal{O}|\Psi_p\rangle}{\langle\Psi_p|\Psi_p\rangle} = \sum_x \frac{|\langle\Psi_p|x\rangle|^2}{\langle\Psi_p|\Psi_p\rangle}\frac{\langle x|\mathcal{O}|\Psi_p\rangle}{\langle x|\Psi_p\rangle}, \tag{8}$$

where $\{|x\rangle\}$ represents a complete and orthogonal basis. By interpreting $P(x) = \langle\Psi_p|x\rangle|^2/\langle\Psi_p|\Psi_p\rangle$ as a probability distribution, a Markov chain can be constructed to sample Eq. (8). In particular, the Gutzwiller projector is automatically implemented by taking $\{|x\rangle\}$ written in terms of electron configurations in real space with only singly-occupied sites. Within the fermionic representation, an efficient Metropolis algorithm can be implemented with local electron moves, e.g., spin flips, $|x\rangle \to |x'\rangle$. For $p = 0$, whenever few electrons are moved, the computational cost of $P(x')/P(x)$ only requires $O(1)$ operations, while the updating when a new configuration is accepted requires $O(N^2)$ operations. In order to have uncorrelated electron configurations, N moves must be done, so that the variational algorithm scales like $O(N^3)$. When the first Lanczos step is implemented, the only extra cost is that $P(x')/P(x)$ scales as $O(N)$, while the second Lanczos step gives $O(N^2)$; in both cases the updating algorithm remains $O(N^2)$. Therefore, up to $p = 2$, the variational calculations have the same computational effort as the $p = 0$ case. In some cases, in order to have a stable simulation for p Lanczos steps, it is important to adopt a regularization scheme to avoid vanishingly small determinants. Here, we consider to sample configurations $|x\rangle$ such that

$$\sum_{x' \neq x} \left| \frac{\langle x|\mathcal{H}|x'\rangle\langle x'|\Psi_v\rangle}{\langle x|\Psi_v\rangle} \right| < \frac{N}{\epsilon}. \tag{9}$$

A finite ϵ represents a restriction of the Hilbert space and, therefore, slightly raises the variational energy. On the other hand, a finite ϵ makes the simulation more stable, since it eliminates large statistical fluctuations. Therefore, one must find a compromise between these two effects and fix a value that optimizes the results: empirically, we find that the best choice is given by ϵ ranging from 10^{-6} to 10^{-5}.

5. Results

We start by showing how the Lanczos step procedure and the variance extrapolation work for a case where exact diagonalizations are also available, namely the Heisenberg model on the square lattice with $L = 6$ and $J_2/J_1 = 0.5$. In Fig. 1, we report the calculations for the ground state in two different cases, either by starting from random initial states (i.e., what is usually done in the Lanczos method) or by initializing with the best variational wave function (3). While in the former cases, many Lanczos steps are needed to reach a good energy per site and, therefore, also the linear regime of Eq. (5), the latter one gives excellent results even with two steps, the zero-variance extrapolation being exact within the statistical error. We would like to stress that abrupt reductions of the energy, with almost constant variance, or even non-monotonic behaviors, can appear when the initial wave function has large overlaps with excited states.

By using the best Gutzwiller projected state, we can extend the Lanczos step procedure to larger systems, where exact diagonalizations are not possible. In Fig. 2, we report the calculations for both the $S = 0$ ground state and the $S = 2$ excitation. Calculations are performed up to $L = 10$ with $p = 0$, 1, and 2, and up to $L = 18$ with $p = 0$ and 1. Even though it is visible that the energy/variance gain reduces when increasing L, the slope of the fit remains similar, allowing a size-consistent zero-variance extrapolation. From these results, the $S = 2$ gap is shown in Fig. 3. We expect a vanishing gap in the thermodynamic limit for the variational calculations with $p = 0$, since the Z_2 gauge structure is not expected to alter the mean-field properties. [10] However, our numerical results up to $L = 30$ cannot definitively confirm this fact. Most importantly, the zero-variance extrapolation for the gap suggests that the frustrated Heisenberg model on the square lattice is gapless for $J_2/J_1 = 0.5$.

Let us move to the Kagome lattice. In Fig. 4, we report the variance extrapolation for a small lattice with 48 sites (i.e., $L = 4$) for both $J_2 = 0$ and $J_2/J_1 = 0.25$. For the former case, no matter what is the initial state (gapless with Dirac points or a large Fermi surface, or even gapped), the extrapolated energy is the same, within the statistical errors. Notice, however, that the actual values of the initial energies and variances are quite different for these three cases. By contrast, for the case with $J_2 = 0.25$, a rather different behavior is found when starting from the $U(1)$ Dirac state or the Z_2 gapped one. Remarkably, although the latter one gives a slightly better variational energy for $p = 0$, the Lanczos extrapolation performs much better for the Dirac state, where a smooth fit is possible. Instead, for the $Z_2[0,\pi]\beta$ state, the second Lanczos step makes a sensible energy gain, while the variance does not improve. This behavior is reminiscent of what is seen when the initial wave function has some large overlap with excited

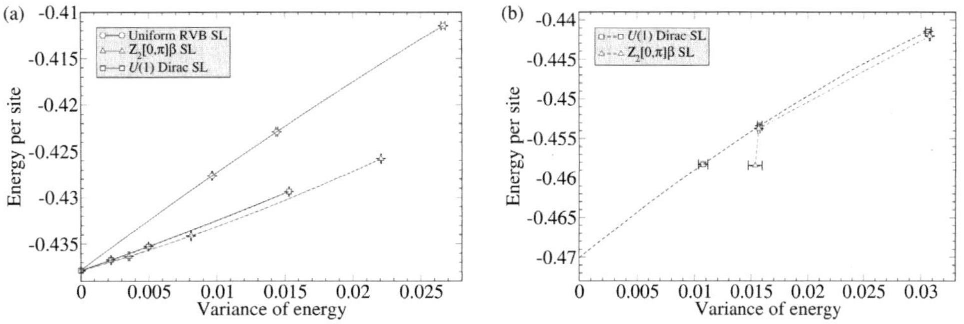

Figure 4. Energies per site for the $S = 0$ states for the Heisenberg model on the Kagome lattice on 48 sites (i.e., $L = 4$) for $J_2 = 0$ (a) and $J_2/J_1 = 0.25$ (b). Both the $U(1)$ Dirac spin liquid and the gapped $Z_2[0, \pi]\beta$ (for which the next-nearest-neighbor pairing η is fixed to 1) spin liquid are reported; for $J_2 = 0$, the uniform RVB state with vanishing magnetic fluxes is also reported.

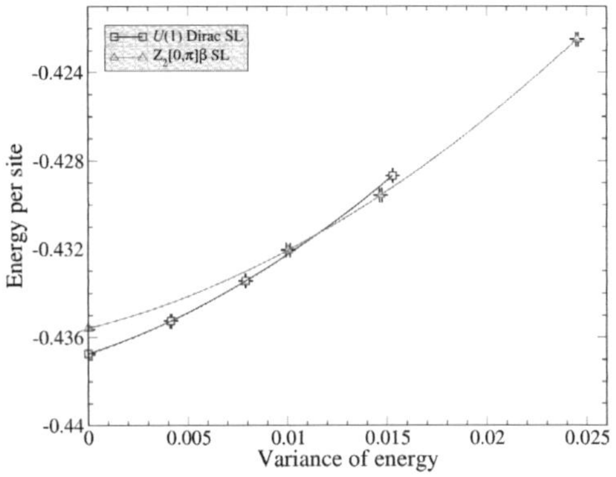

Figure 5. The energy per site versus the variance per site on the Kagome lattice with $L = 8$ and $J_2 = 0$ for the $U(1)$ Dirac spin liquid and the gapped $Z_2[0, \pi]\beta$ spin liquid.

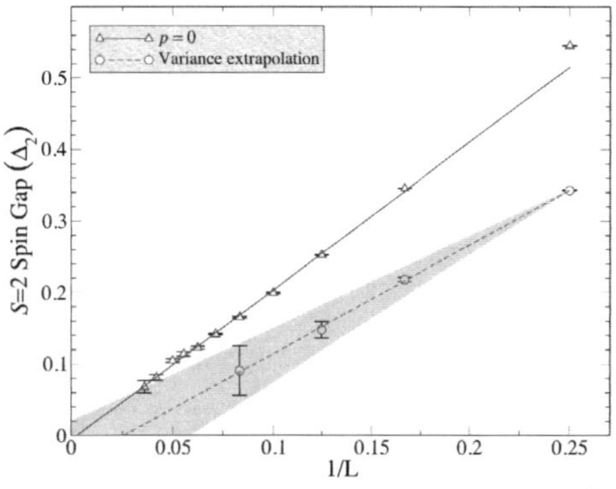

Figure 6. The size scaling of the $S = 2$ spin gap for the $U(1)$ Dirac state and the zero-variance extrapolation on the Kagome lattice with $J_2 = 0$. The thermodynamic extrapolation gives $\Delta_2 = -0.04(6)$.

states (see Fig. 1 for the square lattice). The evidence that the $Z_2[0, \pi]\beta$ state is not a good approximation even when $J_2 = 0$ is reported in Fig. 5. For $L = 8$, the Lanczos extrapolation of the $U(1)$ Dirac spin liquid gives a better energy than the one obtained from the Z_2 state.

By performing a size-scaling extrapolation of the gap by using the $U(1)$ Dirac state, we obtain the results shown in Fig. 6. First of all, it is clear that the Gutzwiller projector does not open a spin gap on top of the Dirac state: this is a non-trivial outcome, given the fact that a $U(1)$ gauge structure may give rise to strong interactions among spinons that may completely change the mean-field picture. Most importantly, also after the Lanczos step extrapolation, the system is compatible with a gapless spectrum.

6. Conclusions

In summary, we have shown that extremely accurate calculations for the energy per site are possible in frustrated two-dimensional lattice by using variational wave functions that are built from Abrikosov fermions. Few Lanczos steps can be applied to the pure variational state, giving a remarkable energy gain, also for clusters with few hundreds sites. Moreover, performing a zero-variance extrapolation, it is possible to obtain accurate values of the exact energies, not only for the ground state but also for excited states.

References

[1] Kaul R K, Melko R G and Sandvik A W 2013 *Annu. Rev. Con. Mat. Phys.* **4** 179
[2] Schollwoek U 2011 *Annals of Physics* **326** 96
[3] Orus R 2014 *Annals of Physics* **349** 117
[4] Yan S, Huse D A and White S R 2011 *Science* **332** 1173
[5] Gong S S, Zhu W, Sheng D N, Motrunich O I and Fisher M P A *Phys. Rev. Lett.* **113** 027201
[6] Hu W J, Becca F, Parola A and Sorella S 2013 *Phys. Rev. B* **88** 060402
[7] Iqbal Y, Becca F, Sorella S and Poilblanc D 2013 *Phys. Rev. B* **87** 060405
[8] Iqbal Y, Poilblanc D and Becca F 2014 *Phys. Rev. B* **89** 020407
[9] Baskaran G and Anderson P W 1988 *Phys. Rev. B* **37** 580
[10] Wen X G 2002 *Phys. Rev. B* **65** 165113
[11] Iqbal Y, Becca F and Poilblanc D 2011 *Phys. Rev. B* **83** 100404
[12] Bieri S, Messio L, Bernu B and Lhuillier C 2014 *Preprint* arXiv:1411.1622
[13] Ran Y, Hermele M, Lee P A and Wen X G 2007 *Phys. Rev. Lett.* **98** 117205
[14] Lu Y M, Ran Y and Lee P A 2011 *Phys. Rev. B* **83** 224413

Uni10: an open-source library for tensor network algorithms

Ying-Jer Kao[1,2], Yun-Da Hsieh[1], Pochung Chen[3]

[1] Department of Physics , National Taiwan University, Taipei, 10607, Taiwan

[2] Center for Advanced Study in Theoretical Science, National Taiwan University, Taipei, 10607, Taiwan

[3] Department of Physics, National Tsing Hua University, Hsin-Chu, 30013, Taiwan

E-mail: `yjkao@phys.ntu.edu.tw`

Abstract. We present an object-oriented open-source library for developing tensor network algorithms written in C++ called Uni10. With Uni10, users can build a symmetric tensor from a collection of bonds, while the bonds are constructed from a list of quantum numbers associated with different quantum states. It is easy to label and permute the indices of the tensors and access a block associated with a particular quantum number. Furthermore a network class is used to describe arbitrary tensor network structure and to perform network contractions efficiently. We give an overview of the basic structure of the library and the hierarchy of the classes. We present examples of the construction of a spin-1 Heisenberg Hamiltonian and the implementation of the tensor renormalization group algorithm to illustrate the basic usage of the library. The library described here is particularly well suited to explore and fast prototype novel tensor network algorithms and to implement highly efficient codes for existing algorithms.

1. Introduction

Tensor network (TN) algorithms[1] have emerged in recent years as a promising tool to study the physics of quantum many-body systems. The most well-known example is the density matrix renormalization group (DMRG) algorithm for quasi-one-dimensional systems[2], which variationally optimizes the matrix product state (MPS)[3]. DMRG has been successfully used to study frustrated quantum spin systems[4], and topological ordered systems[5]. Borrowing the concepts of entanglement from the quantum information science, proposals to extend MPS to higher dimensions have also been put forward. Two-dimensional generalizations such as projective entangled-pair states (PEPS)[6, 7] and the multi-scale entanglement renormalization ansatz (MERA) [8, 9], have been used to study many-body physics in two dimensions. Recently, ideas from TNs have also been used to study systems other than the condensed matter systems such as the holographic principles of AdS/CFT, and loop quantum gravity[1]. In short, TN algorithms have become an important tool to study non-perturbative aspects of many strongly interacting systems. From a broader perspective, tensors are widely used in different branches of sciences and engineering, such as pattern and image recognition, signal processing, machine learning and computational neuroscience to represent very complicated multidimensional data with multiple aspects [10]. It is then nature to use TNs to represent these multidimensional data and to use tensor decomposition algorithms to find the low-dimension and low-rank tensors that best approximate the complicated data.

To motive the need for a tensor network library, consider the contraction of a 2D tensor network. Currently, there exist many different schemes, such as the corner transfer matrix method (CTM)[11, 12], the tensor renormalization group (TRG)[13], plaquette renormalization[14], high-order tensor

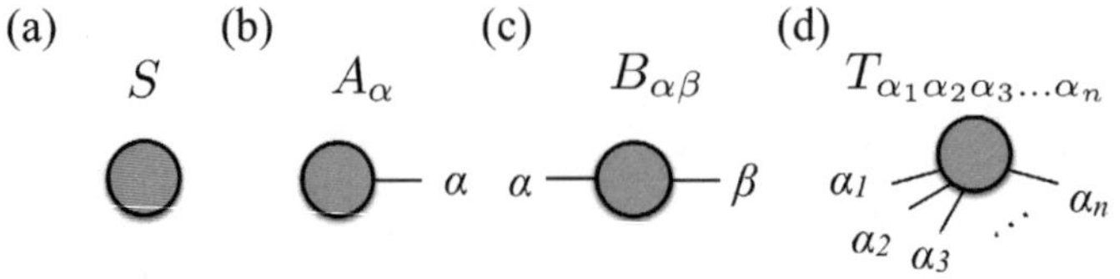

Figure 1: (Color online.) Tensor network diagrams: (a) scalar, (b) vector, (c) matrix and (d) rank-n tensor

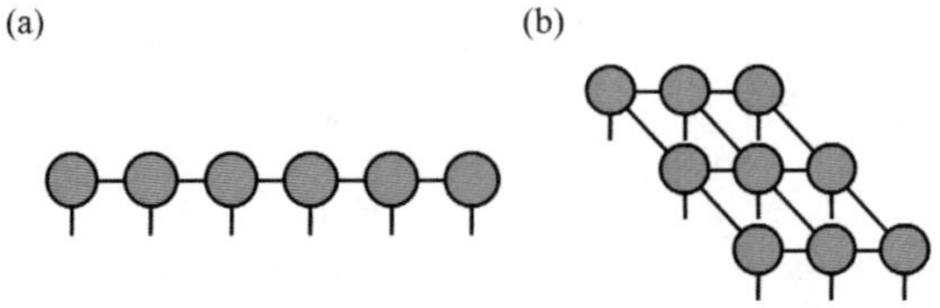

Figure 2: (Color online.) Two examples of tensor network diagrams: (a) Matrix Product State (MPS) for 6 sites with open boundary conditions; (b) Projected Entangled Pair State (PEPS) for a 3×3 lattice with open boundary conditions.

renormalization group (HOTRG)[15], and many others. All these algorithms involves contracting tensors, i.e., summing over common indices between tensors. However, programming tensor network algorithms is tedious and prone to errors. The task of keeping track of tensor indices while performing contraction of a complicated tensor network can be daunting. It is hence desirable to have a platform that provides the bookkeeping and optimization to speed up the software development .

In this paper, we describe a fully open-source implementation of a library designed for the development of the tensor network algorithms called the *Universal Tensor Network library* or **Uni10**. This software distinguishes itself from other available software solutions by providing the following advantages:

- Fully implemented in C++.
- Aimed toward applications in tensor network algorithms.
- Provides basic tensor operations with an easy-to-use interface.
- Provides a class `Network` to process and store the graphical representations of the networks.
- Provides an engine to construct and analyze the contraction tree for a given network.
- Provides a collection of Python wrappers called **pyUni10** which interact with the compiled C++ library to take advantage of the Python language for better code readability and faster prototyping, without the sacrifice of speed.
- Provides behind-the-scene optimization and acceleration.

The paper is organized as follows: Sec. 2 gives a brief introduction to the tensor networks and their graphical representation; Sec. 3 summarizes the key components of the Uni10 library and their usage; Sec. 4 gives a sample code implementing the TRG method using pyUni10; we conclude in Sec. 5.

2. Tensors, tensor networks, and tensor network diagrams

Tensors can be regarded as the multi-dimensional generalizations of matrices. For a given tensor $\mathbf{T}$, the rank is the number of indices. Therefore, a rank-0 tensor is a scalar (S) , a rank-1 tensor is a vector (A_α), and a rank-2 tensor is a matrix ($B_{\alpha\beta}$). The number of all possible values for an index is the dimension of the index. We can perform index contraction for a set of tensors by summing over all of the possible

Journal of Physics: Conference Series **640** (2015) 012040 doi:10.1088/1742-6596/640/1/012040

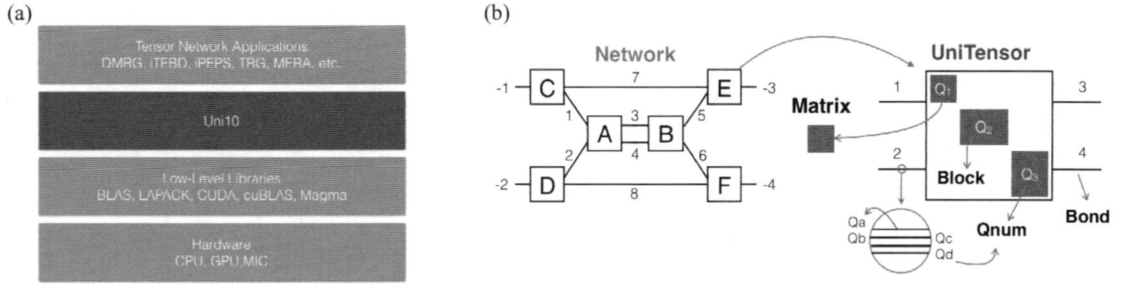

Figure 3: (a) Uni10 serves as a middleware between the tensor network applications and the low-level numerical libraries. (b) Hierarchy of the uni10 classes.

values for the common indices. For example, a matrix multiplication can be regarded the contraction of two rank-2 tensors

$$T_{\alpha\beta} = \sum_{\gamma=1}^{\chi} A_{\alpha\gamma} B_{\gamma\beta}, \tag{1}$$

where χ is the dimension of the index γ. A more complicated example is to contract four tensors to give a complex number,

$$G = \sum_{\alpha\beta\gamma\delta\xi\eta\zeta} A_{\alpha\beta\gamma\delta} B_{\beta\gamma\delta\xi} C_{\xi\eta\zeta\delta} D_{\eta\zeta\alpha}. \tag{2}$$

It is convenient to introduce a graphical representation of tensors and TNs using the tensor diagrams (Fig. 1). Tensors are represented by a circle or a rectangle, and indices are represented by lines emerging from these shapes. A TN is thus represented by a set of shapes interconnected by lines. The internal lines connecting tensors correspond to contracted indices, and external lines correspond to open indices in the TN. Figure 2 shows the graphical representation for MPS and PEPS respectively.

3. Uni10

3.1. Overview

The key object in Uni10 is the Abelian symmetric tensor class `UniTensor` together with auxiliary classes for quantum numbers `Qnum`, blocks `Block`, and bonds `Bond`. In addition, the library provides a network class `Network` to store and process the details of the graphical representations of the networks. Table 1 lists the major classes in Uni10, and Fig 3(b) shows the hierarchy of the different classes.

For a given network, Uni10 provides an engine construct and analyze the pair-wise contraction tree. A heuristic algorithm is implemented to search for an optimal pairwise contraction order based on the computation and memory constraints of the system. The library also provides a collection of wrappers for Python called **pyUni10** to take advantage of the for Python language that is designed for better code readability, and faster prototyping.

In the following, we will use the pyUni10 module to demonstrate some features of Uni10. The following `import` statement should be inserted at the beginning of all the Python codes,

```
from pyUni10 import *
```

3.2. Symmetric Tensors

In order to define a symmetric tensor, one must first construct an object that carries a quantum number. In Uni10, a quantum number object is created through the `Qnum` class. To create a `Qnum` object named `q10` to represent a U(1) quantum number $U_1 = 1$ with even parity one can use the code:

```
q10 = Qnum(1, PRT_EVEN)
```

Table 1: Major Uni10 classes

Class	Description
`UniTensor`	A symmetric tensor
`Bond`	A bond with a list of quantum numbers $\{q_1, q_2, \ldots\}$
`Qnum`	A composite $U(1) \times Z_2 \times Z_2^F$ quantum number
`Matrix`	A matrix
`Block`	A map from a quantum number to a matrix
`Network`	A network

While to create a quantum number object `q_11` with $U_1 = -1$ with odd parity one can use the code:

```
q_11 = Qnum(-1, PRT_ODD)
```

Using the quantum number objects, one can proceed to define a bond which contains these quantum numbers. In Uni10, two types of bonds are defined: the *incoming* and *outgoing* bonds. This can be indicated using `BD_IN` and `BD_OUT` constants. To define an incoming bond of a three-fold degenerate quantum number 0 the following code can be used:

```
Q=Qnum(0)
Qlist=[Q,Q,Q]
Bi=Bond(BD_IN,Qlist)
```

In cases where none of the symmetries is used, Uni10 provides a simple interface to create a non-symmetric bond. To create a bond with dimension four the following code can be used:

```
Bo=Bond(BD_OUT,4)
```

Where a default quantum number is assigned to all the quantum states. When no symmetries are used, there is no need to distinguish between the incoming and outgoing bonds. However, the incoming bonds are always used as the column indices for the internal storage of Uni10, and the specification is preserved in the API.

A tensor is created using these bonds as shown in the code below:

```
A=UniTensor([Bi,Bi,Bo,Bo])
```

In Uni10, each bond is associated with an integer label. A default set of labels will be assigned during the tensor creation but the user can set the labels in any way that is convenient. In Uni10, the tensor operations are overloaded, for example, the expression in Eq. (2) is represented in Uni10 as

```
G= A*B*C*D
```

Here bonds with the same label will be summed. A more efficient way to contract tensors is using the `Network` class, which will be described in details in Sec. 3.4.

3.3. Example: Heisenberg Hamiltonian

Here we demonstrate how to construct a two-site Heisenberg Hamiltonian for $S = 1$,

$$H = \mathbf{S}_1 \cdot \mathbf{S}_2. \tag{3}$$

First we create incoming and outgoing bonds with dimension 3,

```
bdi = uni10.Bond(uni10.BD_IN, 3 )
bdo = uni10.Bond(uni10.BD_OUT, 3 )
```

Next we create a tensor named `H` with two incoming and two outgoing bonds of dimension 3 that will be used to store the Hamiltonian of the Heisenberg model,

```
H = UniTensor([bdi, bdi, bdo, bdo])
```

We can fill in the elements of `H` by putting in the elements directly,

```
heisenberg_s1=[\
1, 0, 0, 0, 0, 0, 0, 0, 0,\
0, 0, 0, 1, 0, 0, 0, 0, 0,\
0, 0,-1, 0, 1, 0, 0, 0, 0,\
0, 1, 0, 0, 0, 0, 0, 0, 0,\
0, 0, 1, 0, 0, 0, 1, 0, 0,\
0, 0, 0, 0, 0, 0, 0, 1, 0,\
0, 0, 0, 0, 1, 0,-1, 0, 0,\
0, 0, 0, 0, 0, 1, 0, 0, 0,\
0, 0, 0, 0, 0, 0, 0, 0, 1\
]
H.setRawElem(heisenberg_s1)
```

Next, we show how to build the same Hamiltonian in the U(1) symmetric basis. This can be achieved following the steps described above by specifying the quantum numbers on the bonds. First, create `Qnum` objects holding U(1) quantum numbers 0, 1, and -1,

```
q0 = uni10.Qnum(0)
q1 = uni10.Qnum(1)
q_1 = uni10.Qnum(-1)
```

Then create incoming and outgoing bonds carrying these quantum numbers,

```
bdi = uni10.Bond(uni10.BD_IN, [q1, q0, q_1] )
bdo = uni10.Bond(uni10.BD_OUT, [q1, q0, q_1] )
```

Finally, create a tensor `H_U1` from the incoming and outgoing bonds, and fill in the elements,

```
H_U1 = uni10.UniTensor([bdi, bdi, bdo, bdo], "H_U1")
H_U1.setRawElem(heisenberg_s1)
```

The `setRawElem` method will fill in Hamiltonian tensor according to the local basis on the bonds, but the tensor is stored in the symmetric basis, i.e., in terms of blocks. Uni10 provides a compact way to output the contents of the tensor with the **`print`** command,

```
print H_U1
```

The output includes the type, the quantum numbers and their degeneracies of each bond, the blocks corresponding to the total quantum numbers formed by the individual bonds, and the elements in each block,

```
**************** H_U1 ****************
             _____________
            |             |
      0____|3            3|____2
            |             |
      1____|3            3|____3
            |             |
            |_____________|

================BONDS===============
IN : (U1 = 1, P = 0, 0)|1, (U1 = 0, P = 0, 0)|1, (U1 = -1, P = 0, 0)|1, Dim = 3
IN : (U1 = 1, P = 0, 0)|1, (U1 = 0, P = 0, 0)|1, (U1 = -1, P = 0, 0)|1, Dim = 3
OUT: (U1 = 1, P = 0, 0)|1, (U1 = 0, P = 0, 0)|1, (U1 = -1, P = 0, 0)|1, Dim = 3
OUT: (U1 = 1, P = 0, 0)|1, (U1 = 0, P = 0, 0)|1, (U1 = -1, P = 0, 0)|1, Dim = 3

===============BLOCKS===============
--- (U1 = -2, P = 0, 0): 1 x 1 = 1
  1.000
--- (U1 = -1, P = 0, 0): 2 x 2 = 4
  0.000  1.000
```

```
  1.000  0.000
--- (U1 = 0, P = 0, 0): 3 x 3 = 9
 -1.000  1.000  0.000
  1.000  0.000  1.000
  0.000  1.000 -1.000
--- (U1 = 1, P = 0, 0): 2 x 2 = 4
  0.000  1.000
  1.000  0.000
--- (U1 = 2, P = 0, 0): 1 x 1 = 1
  1.000
Total elemNum: 19
***************** END ****************

The number of the blocks = 5
(U1 = -2, P = 0, 0) 1 x 1 = 1
  1.000
(U1 = -1, P = 0, 0) 2 x 2 = 4
  0.000  1.000
  1.000  0.000
(U1 = 0, P = 0, 0) 3 x 3 = 9
 -1.000  1.000  0.000
  1.000  0.000  1.000
  0.000  1.000 -1.000
(U1 = 1, P = 0, 0) 2 x 2 = 4
  0.000  1.000
  1.000  0.000
(U1 = 2, P = 0, 0) 1 x 1 = 1
  1.000
```

3.4. Network

TNs are collections of tensors and it is important to store the information about the connection between tensors. An important feature of Uni10 is the `Network` class. Figure 5(a) shows a typical TN with three tensors W, WT, and H. The labels indicates how these tensors are connected. Two tensors connected through an internal line will share a common label. This TN can be represented by a text file with entries showing the tensors and the labels for each tensors [Fig. 5(b)]. Uni10 provides a simple interface to create a network named `net` by reading in the network file,

```
net = uni10.Network("network.net")
```

Put tensors to `net`,

```
net.putTensor("H", H_U1)
net.putTensor("W", W)
net.putTensor("WT", WT)
```

Finally, perform contractions inside the tensor network,

```
T=net.launch()
```

the resulting tensor `T` has an incoming bond labeled -1, and an outgoing bond labeled -2, as indicated in the `network.net`.

(a)

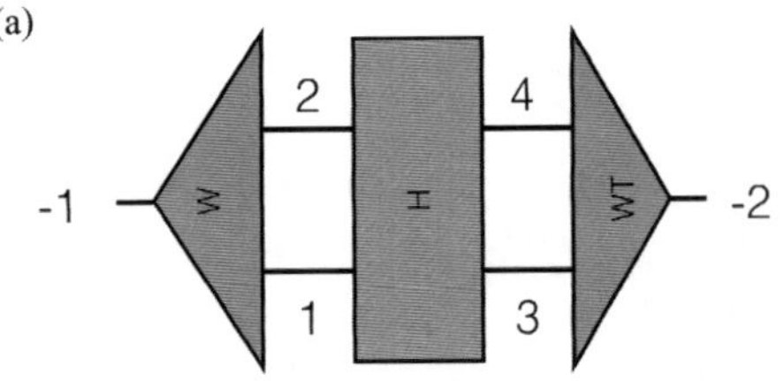

(b)

```
H: 1 2; 3 4
W: -1; 1 2
WT: 3 4; -2
TOUT: -1; -2
ORDER:
```

Figure 4: (Color online.) (a) A network with three tensors. (b) The network file `network.net` representing the connection between the tensors.

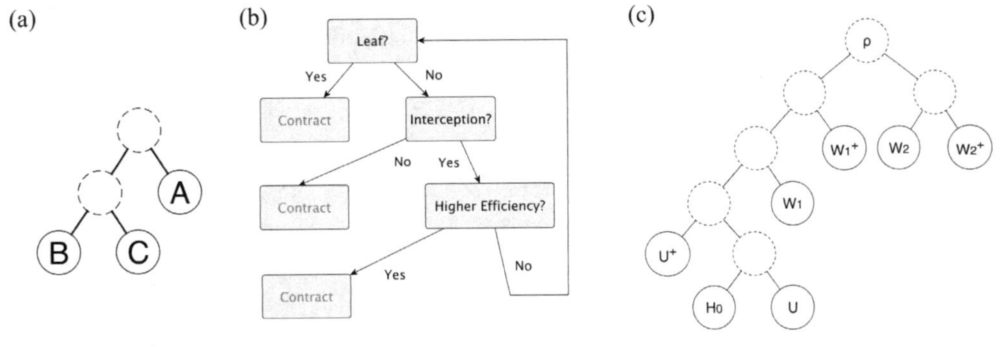

Figure 5: (Color online.) (a) Pairwise contraction of three tensors. (b) Contraction algorithm. (c) Pairwise contraction tree.

3.5. Network Contraction Algorithm

Finding the optimal pairwise contraction order of a general TN is NP-hard[16], and search algorithms are available to find the optimal order in some special cases[17]. Here, we describe a network contraction algorithm which aims to find a locally optimized pair-wise contractions order to contract the whole tensor network. The algorithm has two major features: First, it is fast, and the time complexity is of order $O(\log N)$ to $O(N)$, depending on the balance of the contraction tree. Second, the algorithm uses heuristics to optimize the contraction order and users can provide initial guess to improve the optimization process.

We measure the efficiency of a pairwise contraction of two tensors A and B by defining the contraction efficiency ϵ in terms of the numbers of tensor elements $N(A)$, $N(B)$ and $N(A*B)$,

$$\epsilon(A,B) \equiv (N(A)+N(B))/N(A*B). \tag{4}$$

It is clear that $\epsilon(A,B)=\epsilon(B,A)$. For a given pairwise contraction of A and B, we insert a third tensor C by using a local optimization algorithm to determine the most efficient way for the three tensor to contract. We compare $\epsilon(A,B)$, $\epsilon(B,C)$ and $\epsilon(C,A)$ and contract first the pair with the largest efficiency, followed by the contraction of the resulting tensor with the remaining tensor [Fig. 5(a)]. This procedure is repeated until all the tensors are inserted into the network [Fig. 5(b)]. For a given tensor network, a binary tree for the contraction order will be generated [Fig. 5(c)]. Since the algorithm can only find the local optimal contraction order, the final contraction order may not be optimal. To overcome this, users can propose a contraction order in the network file to serve as the initial guess for the search algorithm.

4. Sample Code: TRG

Here we present a code sample to demonstrate how to compute the magnetization along the z-axis $\langle\sigma_z\rangle$ using TRG for the transverse Ising model. In the following example, the tensors `GaL` and `GbL` of the ground state wave function are obtained through some update scheme, such as iTEBD. The detailed description of the TRG algorithm can be found in Ref. [13]. The goal of the algorithm is to compute the expectation value of an operator $\hat{O}$,

$$\langle O\rangle = \frac{\langle\Psi|\hat{O}|\Psi\rangle}{\langle\Psi|\Psi\rangle} \tag{5}$$

for a given TNS wave function in the two sublattice form,

$$|\Psi\rangle = \sum_{s_i,s_j} \mathrm{Tr} \prod_{i\in A, j\in B} \Gamma^A_{u_i d_i l_i r_i}[s_i]\Gamma^B_{u_j d_j l_j r_j}[s_j]|s_i s_j\rangle. \tag{6}$$

To compute the norm (denominator) and numerator of Eq. (5), we first define two functions `makeT`, and `makeImpurity` to form double tensors and impurity tensors[13] from Γ^A and Γ^B,

```
def makeT(GL):
    GLT = copy.copy(GL)
    GLT.transpose()
    GL.setLabel([0, 1, 2, 3, 4])
    GLT.setLabel([-1, -2, -3, -4, 0])
    T = GLT * GL
    for i in xrange(1, 5):
        T.combineBond([-i, i])
    return T

def makeImpurity(GL, Ob):
    GLT = copy.copy(GL)
    GLT.transpose()
    Ob.setLabel([0, 5])
    GL.setLabel([5, 1, 2, 3, 4])
    GLT.setLabel([-1, -2, -3, -4, 0])
    I = Ob * GL
    I = GLT * I
    for i in xrange(1, 5):
        I.combineBond([-i, i])
    return I
```

The function `trgSVD` performs the singular value decomposition for the A and B sublattices, and perform truncation,

```
def trgSVD(wch, T, chi):
    if wch % 2 == 0:
        T.permute([-4, -3, -1, -2], 2)
    else:
        T.permute([-1, -4, -2, -3], 2)

    svd = T.getBlock().svd()
    chi = chi if chi < svd[1].row() else svd[1].row()
    bdi_chi = uni10.Bond(uni10.BD_IN, chi)
    bdo_chi = uni10.Bond(uni10.BD_OUT, chi)
    S0 = uni10.UniTensor([T.bond(0), T.bond(1), bdo_chi])
    S1 = uni10.UniTensor([bdi_chi, T.bond(2), T.bond(3)])
    svd[1].resize(chi, chi)
    for i in xrange(chi):
        svd[1][i] = np.sqrt(svd[1][i])
    S0.putBlock(svd[0].resize(svd[0].row(), chi) * svd[1])
    S1.putBlock(svd[1] * svd[2].resize(chi, svd[2].col()))
    return S0, S1
```

The function `trgContract` performs the contraction of four tensors given the TN described in `TRG_net`,

```
def trgContract(Ss, TRG_net):
    for i in xrange(4):
        TRG_net.putTensor(i, Ss[i])
    return TRG_net.launch()
```

The function `updateAll` performs the renormalization of the tensors,

```
def updateAll(Ts, Imps, chi, TRG_net):
    S = []
    I = []
    for r in xrange(2):
        S.append(trgSVD(r, Ts[r%len(Ts)], chi))
```

```
    for r in xrange(4):
        I.append(trgSVD(r, Imps[r], chi))
    a, b, c, d = 0, 1, 2, 3
    Ts = [trgContract([S[a][0], S[b][0], S[a][1], S[b][1]], TRG_net)]
    maxel = max([Ts[0][i] for i in xrange(Ts[0].elemNum())])
    Ts[0] *= (1.0 / maxel)

    Imps=[trgContract([S[a][0],I[b][0],I[a][1],S[b][1]],TRG_net)*(1.0/maxel),\
            trgContract([S[a][0],S[b][0],I[c][1],I[b][1]],TRG_net)*(1.0/maxel),\
            trgContract([I[c][0],S[b][0],S[a][1],I[d][1]],TRG_net)*(1.0/maxel),\
            trgContract([I[a][0],I[d][0],S[a][1],S[b][1]],TRG_net)*(1.0/maxel)]

    return Ts, Imps
```

The function `trgExpectation` compute the expectation value using the impurity tensors,

```
def trgExpectation(Ts, exp_net):
    for step in xrange(4):
        exp_net.putTensor(step, Ts[step % len(Ts)])
    return exp_net.launch()[0]
```

The main function reads in the tensor networks (Fig. 6),

```
TRG_net = uni10.Network("TRG.net")
exp_net = uni10.Network("expectation.net")
```

Construct the σ_z operator,

```
bdi = uni10.Bond(uni10.BD_IN, 2)
bdo = uni10.Bond(uni10.BD_OUT, 2)
Mz = uni10.UniTensor([bdi, bdo], "Mz")
Mz.setElem([1, 0, 0, -1])
```

Prepare the double tensors and impurity tensors,

```
Ts = [makeT(GaL), makeT(GbL)]
Imps = [makeT(GaL), makeT(GbL), makeImpurity(GaL, Mz), makeT(GbL)]
```

Finally, perform the renormalization steps,

```
for i in xrange(8):
    Ts, Imps = updateAll(Ts, Imps, chi2, TRG_net)
M=trgExpectation(Imps, exp_net)
norm=trgExpectation(Ts, exp_net)
```

For brevity, the code segments associated with initialization is omitted. The complete pyUni10 source code for the transverse Ising model can be found at the Uni10 website [18].

5. Conclusions

We have presented a new, open-source C++ library for the tensor network algorithms. This library is suitable for implementing a wide range of TN algorithms. In this work we have described the basic structure of the library, and how to construct a spin-1 Heisenberg Hamiltonian. For more in-depth documentation, and more elaborate examples using pyUni10 API listed in Table 1, the reader should visit the Uni10 website at `http://www.uni10.org`. Further optimization of Uni10 using speed accelerators such as Graphic Processing Unit (GPU) and Many Integrated Core (MIC) architectures and Matlab wrappers will be available in future releases.

Acknowledgments

We thank Tama Ma and Sukhbinder Singh for their help in the development of Uni10. We acknowledge inspiring discussions with A. Sandvik, G. Evenbly, G. Vidal, and Ian McCulloch. The support

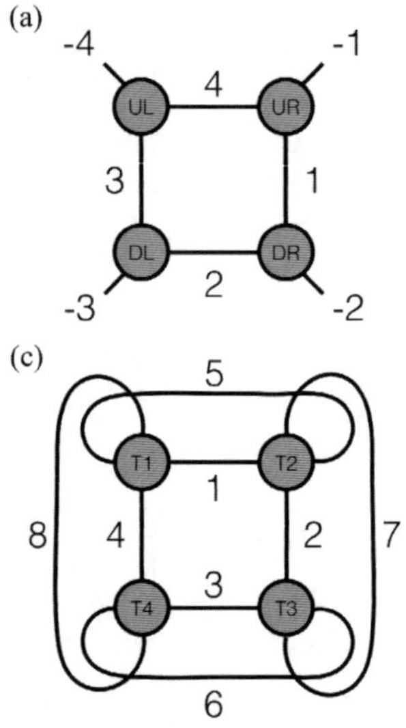

(b)

```
UR: 4, 1; -1
DR: 1, 2; -2
DL: -3; 3, 2
UL: -4; 4, 3
TOUT: -1, -2, -3, -4
```

(d)

```
T1: 8, 1, 4, 5
T2: 7, 5, 2, 1
T3: 2, 6, 7, 3
T4: 4, 3, 8, 6
TOUT:
```

Figure 6: (Color online.) (a) A TN for tensor renormalization in the TRG algorithm. (b) Network file `TRG.net`. (c) A TN for expectation values. (d) Network file `expectation.net`.

from MOST in Taiwan through Grants No. 100-2112-M-002-013-MY3, 100-2923-M-004-002 -MY3, 102-2112-M-002-003-MY3, 101-2112-M-007-010-MY3 as well as the support from National Taiwan University Grant No. 101R891004 are acknowledged.

Appendix

Here we list the commands available in pyUni10 (v0.9.1). Full documentation, including API and other examples, is available at Uni10 Website [18].

Table 1: pyUni10 global constants, functions, classes and class methods.

Constants	
BD_IN	Defines an incoming bond
BD_OUT	Defines an outcoming bond
PRT_EVEN	Defines particle parity even in a bosonic system
PRT_ODD	Defines particle parity odd in a bosonic system
PRTF_EVEN	Defines particle parity even in a fermionic system
PRTF_ODD	Defines particle parity odd in a fermioninc system
Functions	
contract()	Performs tensor contraction
otimes()	Performs tensor product
combine()	Combines bonds
takeExp()	Returns the exponential of a matrix
Qnum class	
Qnum()	Creates a quantum number
QnumF()	Creates a fermionic quantum number
Qnum.U1()	Obtains the U(1) quantum number
Qnum.assign()	Assigns a quantum number to Qnum
Qnum.assignF()	Assigns a fermonic quantum number to Qnum
Qnum.prt()	Obtains the particle number parity
Qnum.prtF()	Obtains the fermionic number parity
Bond class	
Bond()	Creates a bond
Bond.Qlist()	Returns the Qnum objects in a Bond
Bond.assign()	Assigns Qnum to Bond
Bond.change()	Changes the type of Bond

`Bond.combine()`	Combine bonds
`Bond.degeneracy()`	Returns the degeneracy of quantum number
`Bond.dim()`	Returns total dimension of `Bond`
`Bond.type()`	Returns type of `Bond`
`Matrix` class	
`Matrix()`	Creates a matrix
`Matrix.col()`	Returns number of columns of `Matrix`
`Matrix.eigh()`	Returns the eigenvalues and eigenvectors
`Matrix.elemNum()`	Returns the total number of elements
`Matrix.getElem()`	Returns the reference to matrix elements
`Matrix.identity()`	Sets `Matrix` to 1 on the diagonal and 0 otherwise
`Matrix.isDiag()`	Returns whether `Matrix` is diagonal
`Matrix.load()`	Loads elements of `Matrix` from a file
`Matrix.norm()`	Returns the L^2 norm of `Matrix`
`Matrix.orthoRand()`	Generates an orthogonal basis with random elements and assigns to `Matrix`
`Matrix.randomize()`	Sets the elements of `Matrix` with random number in $[0, 1)$
`Matrix.resize()`	Resizes `Matrix`
`Matrix.row()`	Returns number of rows of `Matrix`
`Matrix.save()`	Saves `Matrix` to a file
`Matrix.setElem()`	Sets the elements of `Matrix`
`Matrix.set_zero()`	Sets all the elements in `Matrix` to zero
`Matrix.sum()`	Returns the sum of all elements
`Matrix.svd()`	Performs singular value decomposition of `Matrix`
`Matrix.trace()`	Returns the trace of `Matrix`
`Matrix.transpose()`	Returns the transpose of `Matrix`
`UniTensor` class	
`UniTensor()`	Creates a symmetric tensor
`UniTensor.assign()`	Restructures `UniTensor` with new bonds
`UniTensor.blockNum()`	Returns the total number of blocks
`UniTensor.blockQnum()`	Returns the quantum number of a block in `UniTensor`
`UniTensor.bond()`	Returns a bond in `UniTensor`
`UniTensor.bondNum()`	Returns the total number of bonds
`UniTensor.combineBond()`	Combines bonds
`UniTensor.elemCmp()`	Compares the elements of two tensors
`UniTensor.elemNum()`	Returns the total number of elements
`UniTensor.getBlock()`	Returns the block of a given quantum number
`UniTensor.getBlocks()`	Returns the map from quantum numbers to blocks
`UniTensor.getElem()`	Returns the reference of the elements
`UniTensor.getName()`	Returns the name of `UniTensor`
`UniTensor.getRawElem()`	Returns the raw elements of `UniTensor` with row(column) basis defined by the incoming (outgoing) bonds
`UniTensor.identity()`	Sets blocks to identity matrix.
`UniTensor.inBondNum()`	Returns the number of incoming bonds in `UniTensor`
`UniTensor.label()`	Returns label(s) of `UniTensor`
`UniTensor.orthoRand()`	Randomly generates orthogonal bases and assigns to blocks
`UniTensor.partialTrace()`	Takes partial trace
`UniTensor.permute()`	Permutes the order of bonds
`UniTensor.printRawElem()`	Prints raw elements of `UniTensor`
`UniTensor.putBlock()`	Puts matrix into block
`UniTensor.randomize()`	Sets the elements of `UnTensor` with random number in $[0, 1)$
`UniTensor.save()`	Saves `UniTensor` to file
`UniTensor.setElem()`	Assigns the elements to `UniTensor`, replacing the originals
`UniTensor.setLabel()`	Assigns a new label of a bond
`UniTensor.setName()`	Sets the name of `UniTensor`
`UniTensor.setRawElem()`	Assigns raw elements to `UniTensor`
`UniTensor.set_zero()`	Sets all elements to zero
`UniTensor.similar()`	Checks if the bonds of two tensors are the same
`UniTensor.trace()`	Takes the trace

`UniTensor.transpose()`	Transposes all blocks associated with quantum numbers
`Network` class	
`Network()`	Creates a network
`Network.launch()`	Performs the full contraction of `Network`
`Network.profile()`	Returns the total memory usage of `Network`
`Network.putTensor()`	Put tensors into `Network`
`Network.putTensorT()`	Put transposed tensors into into `Network`

References

[1] Orús R 2014 *The European Physical Journal B* **87** 280–18
[2] White S R 1993 *Phys. Rev. B* **48**(14) 10345–10356
[3] Schollwöck U 2011 *Annals of Physics* **326** 96–192
[4] Depenbrock S, McCulloch I P and Schollwöck U 2012 *Phys. Rev. Lett.* **109** 067201
[5] Zaletel M P, Mong R S K and Pollmann F 2013 *Phys. Rev. Lett.* **110** 236801
[6] Verstraete F, Murg V and Cirac J 2008 *Advances in Physics* **57** 143–224
[7] Verstraete F and Cirac J I 2004 Renormalization algorithms for quantum-many body systems in two and higher dimensions (*Preprint* `cond-mat/0407066`)
[8] Vidal G 2007 *Phys. Rev. Lett.* **99** 220405
[9] Evenbly G and Vidal G 2009 *Phys. Rev. B* **79**(14) 144108
[10] Cichocki A 2014 Era of Big Data Processing: A New Approach via Tensor Networks and Tensor Decompositions (*Preprint* `arXiv:1403.2048`)
[11] Jordan J, Orus R, Vidal G, Verstraete F and Cirac J I 2008 *Phys. Rev. Lett.* **101** 250602
[12] Orús R and Vidal G 2009 *Physical Review B* **80** 094403
[13] Jiang H C, Weng Z Y and Xiang T 2008 *Phys. Rev. Lett.* **101** 090603
[14] Wang L, Kao Y J and Sandvik A W 2011 *Physical Review E* **83** 056703
[15] Xie Z Y, Chen J, Qin M P, Zhu J W, Yang L P and Xiang T 2012 *Physical Review B* **86** 045139
[16] Lam C C, P S and Rephael W 1997 *Parallel Processing Letters* **07** 157–168
[17] Pfeifer R N C, Haegeman J and Verstraete F 2014 *Physical Review E* **90** 033315
[18] URL `http://www.uni10.org`

Journal of Physics: Conference Series **640** (2015) 012041 doi:10.1088/1742-6596/640/1/012041

Numerical studies of various Néel-VBS transitions in SU(N) anti-ferromagnets

Ribhu K. Kaul[1] **and Matthew S. Block**[2]

[1]Department of Physics & Astronomy, University of Kentucky, Lexington KY 40506-0055

E-mail: rkk@pa.uky.edu

[2]Department of Physics & Astronomy, California State University, Sacramento CA 95819

E-mail: matthew.block@csus.edu

Abstract. In this manuscript we review recent developments in the numerical simulations of bipartite SU(N) spin models by quantum Monte Carlo (QMC) methods. We provide an account of a large family of newly discovered sign-problem free spin models which can be simulated in their ground states on large lattices, containing $O(10^5)$ spins, using the stochastic series expansion method with efficient loop algorithms. One of the most important applications so far of these Hamiltonians are to unbiased studies of quantum criticality between Néel and valence bond phases in two dimensions – a summary of this body of work is provided. The article concludes with an overview of the current status of and outlook for future studies of the "designer" Hamiltonians.

1. Overview

The study of ground states of lattice models of quantum spins has become a major field in condensed matter physics [1]. Despite their simplicity these models can be extremely hard to study theoretically, due in part to the rich variety of ground states that they can host. Recent years have seen a dramatic increase in the use of numerical methods to study the ground states of quantum spin models. Of particular interest are "unbiased" numerical methods, which solve for physical properties of model systems with numerical errors that can be controlled and estimated in a reliable fashion. Quantum Monte Carlo occupies a special place among the unbiased methods because when applicable, it is the only technique that is able to access the large systems sizes required for reliable extrapolation to the thermodynamic limit, especially in the proximity of critical phenomena [2].

The simplest ground state that can arise in a quantum spin system is a magnetic ground state, where the expectation value of the spin on a site is finite, causing the spin to effectively "point" in a certain direction, thus spontaneously breaking the global symmetry associated with the rotation of spins. Such magnetic states are well known to arise in the low-energy space of classical spin models and their appearance in quantum models can be understood from semi-classical arguments. The focus of recent numerical studies of spin models has been in large part on the nature of the *non-magnetic phases* that arise at $T = 0$ due to quantum fluctuations and the quantum phase transitions separating magnetic and non-magnetic phases. The non-magnetic phases may either break another symmetry, most often a lattice translational symmetry in which case they are called "solids" or be completely symmetric in both the lattice and spin symmetries in which case they are called "liquids." The central questions of interest to numerical studies are: What is the precise characterization of the non-magnetic phases, both of the "solid" and "liquid" type that are found in simple microscopic models? How can the phase transitions between the magnetic and non-magnetic phases observed in numerical simulations be understood in terms of long wavelength quantum field theories?

While these general questions have received copious attention from a multitude of complementary approaches in different contexts over the last three decades [2, 3, 4, 5], in this review we will outline how new "designer" Hamiltonians with SU(N) symmetry have contributed to answers concerning phase transitions between SU(N) symmetry breaking magnetic phases and lattice translational symmetry breaking valence-bond solid states, for which an exotic direct continuous transition – "deconfined criticality" – has been proposed [6, 7, 8]. For a fuller appreciation of this short review, a familiarity with the deconfined theory is recommended. In the interest of space this is not provided here, the interested reader is encouraged to refer to the original literature. In Section 2 we briefly review the discovery of a large family of "designer" SU(N) models that do not suffer from the sign problem. In Sec 3 we describe the phase diagrams obtained for the Hamiltonians that have been studied so far and the nature of the critical points that arise between different phases. Finally, in Section 4 we conclude with an outlook and directions for future work.

2. SU(N) Hamiltonians, loop models and the sign problem

In order to use unbiased quantum Monte-Carlo techniques efficiently, one needs to identify models that do not suffer from the sign problem. It is well known that Hamiltonians that satisfy Marshall's sign condition are also sign-problem free. So a basic questions of central importance to the numerical simulations of quantum spin models is: What is the family of spin models that satisfies Marshall's sign criteria?

In order to make this question concrete, we specialize our considerations to bipartite models with a specific representation of SU(N) symmetry, originally introduced to condensed matter physics by Affleck [9]. This realization of SU(N) symmetry requires spins on one sub-lattice to transform in the fundamental representation of SU(N) and spins on the other sub-lattice to transform in the conjugate to fundamental representation. We note here for $N = 2$, since the fundamental and conjugate to fundamental representations are identical, this realization of SU(2) symmetry gives rise to the familiar Heisenberg-like models.

The standard quantum-classical mapping allows us to rewrite the quantum statistical mechanics of d-dimensional Hamiltonians as a classical statistical mechanics problem in $d+1$-dimensions, where the extra dimension is of extent $\beta = 1/T$. The stochastic series expansion (SSE) is an elegant and well-documented method to execute this step [10]. When carried out and as discussed in detail [11], the Affleck SU(N) spin models on bipartite lattices can be mapped to *oriented tightly packed loop models with N colors* in one higher dimension. In order to carry out Monte Carlo sampling we require these configurations to have positive weights. In the language of loop models it is straightforward to systematically write down all possible interactions that keep the weights of the loop configurations positive. When the quantum-classical mapping is run backwards, the interactions in the classical loop model correspond to terms in quantum Hamiltonians that are Marshall positive. This picture allows one to systematically write down a large class of SU(N) spin Hamiltonians that are Marshall positive. We note parenthetically that the Marshall positivity of these "designer" Hamiltonians is not obvious when viewed directly in the spin language, and that the "designer" models include all previously known Marshall positive spin models as particular cases.

As a concrete example of the results, let us discuss the familiar $N = 2$ case. Here the spins $\vec{S}$ on either sub-lattice can be written in terms of Pauli matrices in the usual way. Previously known sign-problem-free Hamiltonians include (with $J_1, J_2, Q > 0$) [12],

$$\begin{aligned} H_{J_1} &= J_1 \vec{S}_1^A \cdot \vec{S}_1^B \\ H_{J_2} &= -J_2 \vec{S}_1^A \cdot \vec{S}_2^A \\ H_Q &= -Q\left(\vec{S}_1^A \cdot \vec{S}_1^B - \frac{1}{4}\right)\left(\vec{S}_2^A \cdot \vec{S}_2^B - \frac{1}{4}\right) \end{aligned} \tag{1}$$

where the superscript A, B indicates the sublattice the spin lives on and the subscript indicates the different spins on a given sublattice. Thus the J_1 interaction is defined on two spins, one of which lives on the A sublattice and the other on the B sublattice, the J_2 interaction is defined on two spins on the same sublattice, and the Q interaction is defined on four spins, two on the A sublattice and two on the B sublattice. As an application of the new strategy sketched above, a new linearly independent four-spin interaction can be shown to be Marshall positive,

$$\begin{aligned} H_R &= R\left(\vec{S}_1^A \cdot \vec{S}_2^A - \frac{1}{4}\right)\left(\vec{S}_1^B \cdot \vec{S}_2^B - \frac{1}{4}\right) \\ &- R\left(\vec{S}_1^A \cdot \vec{S}_1^B - \frac{1}{4}\right)\left(\vec{S}_2^A \cdot \vec{S}_2^B - \frac{1}{4}\right) - R\left(\vec{S}_1^A \cdot \vec{S}_2^B - \frac{1}{4}\right)\left(\vec{S}_1^A \cdot \vec{S}_2^B - \frac{1}{4}\right) \end{aligned} \tag{2}$$

It is straightforward to prove the Marshall positivity of the above interaction by direct evaluation of its matrix elements when $R > 0$, even though its positivity is not apparent from a naive inspection of the term.

We note following the strategy described in Ref. [11], all the Marshall-positive interactions involving an arbitrary large number of A and B spins and with any N (of SU(N)) can be systematically written down. Also from the way the interactions are written above it is clear that they can be written on any bipartite lattice in any dimension, and can be made to preserve the lattice symmetry through an appropriate summation over the entire lattice.

Detailed reviews and pedagogical introductions to the kind of QMC algorithms that are used to simulate the spin models introduced in this Section may be found in [10, 2].

3. Phase Diagrams & Quantum Criticality

The only two phases found so far in the numerical simulations of the bipartite SU(N) sign-free models discussed in Sec. 2 are Néel and "valence-bond" phases (VBS). By "valence-bond" phases we mean phases that are smoothly connected to a product state of two-site valence bond coverings (see Fig. 1 for cartoons of such coverings). The numerical study of the Néel and VBS phases of the Affleck SU(N) models was initiated early on [13, 14]. However, the most intriguing aspect of the phase diagram, the critical point, was first accessed with the introduction of the four-site Q interaction 1 for the $N = 2$ case [12]. While the pioneering study was carried out for $N = 2$ on the square lattice, subsequent work has carried out the study for a large range of N and for a variety of bipartite lattices and interactions.

In order to summarize the studies in "conceptual" rather than historical order, we need one technical result from the deconfined theory. The deconfined theory predicts that for bipartite SU(N) magnets in two dimensions, the Néel-VBS critical point is described by a nc-$\mathbb{CP}^{N-1}$ critical field theory, only for certain "columnar" VBS states. In actual lattice realization, whether in numerical simulations or real materials, a perturbation to the nc-$\mathbb{CP}^{N-1}$ theory, λ_q (for the experts, it is the fugacity of q-monopoles in the gauge field) is present. The central difference between the various bipartite lattices is the integer q, which can be intuitively understood as the minimum degree of degeneracy of the VBS phase on the particular bipartite lattice under consideration. In order for the deconfined critical point to exist λ_q must be *irrelevant* at the

Journal of Physics: Conference Series **640** (2015) 012041 doi:10.1088/1742-6596/640/1/012041

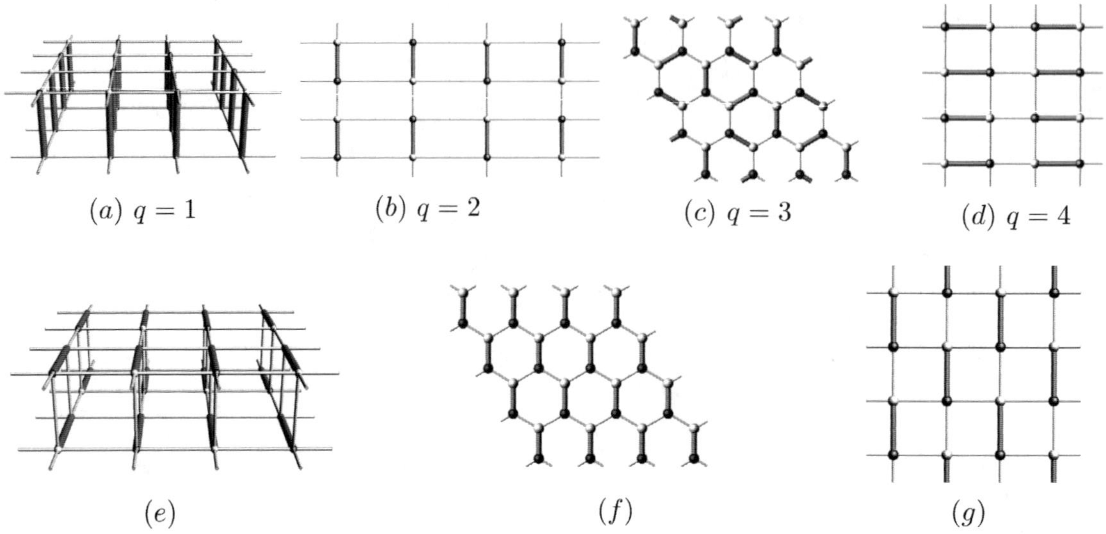

Figure 1. Summary of valence-bond patterns that have been accessed in numerical simulations of two-dimensional bipartite SU(N) designer Hamiltonians. The quantum phase transition into each of these phases from SU(N) symmetry breaking Néel phases is the subject of this review. The thick red bonds when put in an SU(N) singlet state provide a cartoon state for the symmetry breaking. (a-d) are the "columnar" patterns for which the theory of a continuous deconfined criticality can hold for sufficiently large N. The value of q is the degeneracy of the VBS coverings in these states. (e) is a four-fold degenerate VBS state on a bilayer to which there is a first order transition. (f,g) are three- and four-fold staggered VBS states into which there are also first order transitions.

monopole-free fixed point. If it is relevant one expects a first order transition or a continuous transition in a universality class different from the nc-$\mathbb{CP}^{N-1}$ universality.

Let us begin with the case of the bilayer lattice. The simplest non-magnetic state in the bilayer system is a non-degenerate ($q = 1$) rung singlet state (see Fig. 1(a)). For $N = 2$ there is beautiful evidence that supports the theoretical expectation [15] that there is a continuous critical point between Néel and rung-singlet state in the $2 + 1$-dimensional O(3) universality class [16]. For $N \geq 4$ numerical simulations have found evidence for a first order transition, which is expected from Landau theory [17]. The case $N = 3$ appears to have either a very weakly first order or a continuous transition. The field theoretical scenario explaining a possible continuous $N = 3$ transition has been nicely summarized in Ref. [18]. Other numerical work has studied the Néel-VBS transition on the rectangular ($q = 2$) [19], honeycomb ($q = 3$) [19, 20, 21] and square ($q = 4$) [12, 22, 23, 24] lattices for a range of N by using a judicious choice of the designer couplings defined in Section 2. See Fig. 1(b-d) for cartoons of the VBS states. Both first order transitions for small N and continuous transitions for large N have been identified. The critical value of N at which the transition turns continuous decreases as q increases, as expected theoretically. The appearance of continuous and first-order transitions occur in a systematic way that can be attributed to whether the lattice anisotropy in the form of λ_q is relevant or irrelevant at the nc-$\mathbb{CP}^{N-1}$ fixed point. The results may be summarized in Table 1, which shows whether the deconfined critical point between an SU(N) magnet and q-degenerate VBS state (when it is the minimum degeneracy on a particular lattice) is stable. We note that in the bilayer model

$N=\infty, 1/N$	I	I	I	I	$\dots$	I	nc-$\mathbb{CP}^{N-1}$
$\dots$							
$N=10$	R	I	I	I		I	nc-$\mathbb{CP}^9$
$N=9$	R	I	I	I		I	nc-$\mathbb{CP}^8$
$N=8$	R	I	I	I		I	nc-$\mathbb{CP}^7$
$N=7$	R	I	I	I		I	nc-$\mathbb{CP}^6$
$N=6$	R	I	I	I		I	nc-$\mathbb{CP}^5$
$N=5$	R	I	I	I		I	nc-$\mathbb{CP}^4$
$N=4$	R	I	I	I		I	nc-$\mathbb{CP}^3$
$N=3$	R	R	I	I		I	nc-$\mathbb{CP}^2$
$N=2$	R	R	I	I		I	nc-$\mathbb{CP}^1$
$N=1$	R	R	R	I		I	XY
$N=0$	R	R	R	R		R	photon
	$q=1$	$q=2$	$q=3$	$q=4$	$\dots$	$q=\infty$	

Table 1. Table showing the inferred relevance (R) or irrelevance (I) of q-monopoles at the nc-$\mathbb{CP}^{N-1}$ fixed point, which various studies summarized in the text have allowed us to complete. Numerical simulations of the Néel-VBS transition in the models discussed here only allow studies for $N \geq 2$. The entries with R correspond to an unstable fixed point, and I to a stable fixed point allowing for deconfined criticality. At some currently unknown critical value of $N > 10$, the $q = 1$ case switches from R to I. Adapted from Ref. [19]

($q = 1$) the R is expected to turn to I at a finite value of N – analytic estimates from large-N theories suggest this happens around $N \approx 25$ [25]. However, such large values of N have not been accessed numerically yet. For the values of N for which there is a continuous transition, the critical exponents should not depend on the bipartite lattice geometry, since λ_q is irrelevant if there is a continuous transition. Numerical measurements for the anomalous dimension of the Néel and VBS fields are consistent with this expectation and are consistent with results from the analytic $1/N$ expansion as shown in Fig. 2

Having discussed the cases of a "columnar" VBS state to which the theory of deconfined criticality applies, it is also of interest to numerically study the Néel-VBS transition in situations where the deconfined criticality does *not* apply, as a non-trivial check on the theory. As we shall see in the examples below there are a number of different reasons why this can happen:

(1) In the bilayer geometry, consider the phase transition between a four-fold degenerate columnar VBS (c-VBS), see Fig. 1(e), and the SU(N) Néel state. The c-VBS on the bilayer breaks exactly the same symmetries as the single-layer c-VBS state (see Fig. 1(d)) and thus one might conclude naively based on "Landau" theory that they are described by the same critical phenomena. However, it is well known that in the bilayer geometry the Berry phases that are crucial for deconfined criticality cancel between the layers, eliminating the possibility of an exotic continuous transition. Consistent with this expectation, numerical studies have found that the c-VBS-Néel transition in the bilayer is first order [17].

(2) Back to the single-layer case, one can tune designer coupling to favor spontaneous symmetry breaking different from the "columnar" states for which the original proposal of deconfined criticality was made [6]. Indeed, studies have been carried out for the transition between Néel and "staggered" VBS for SU(2), on the square lattice by tuning a "designer" six-spin interaction [26] and on the honeycomb by tuning a four-spin interaction [27]. The theory of a continuous deconfined critical point does not generalize to the staggered VBS case, and in

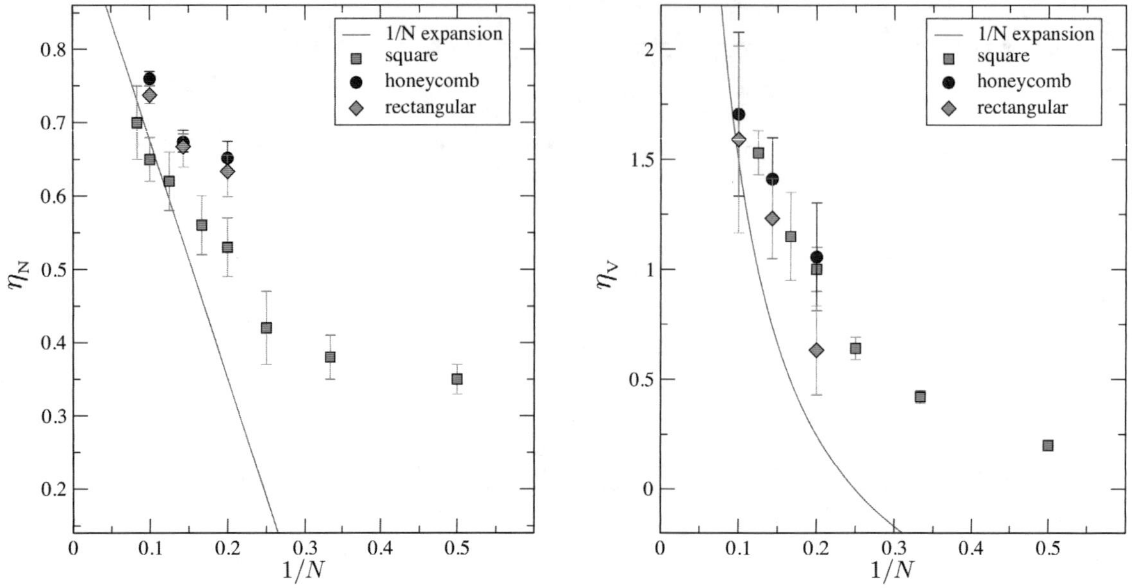

Figure 2. Comparison of anomalous dimensions of Néel and VBS operators in the case of continuous transitions for $q = 2, 3$ and 4. (a) Anomalous dimension of the Néel order parameter as a function of $1/N$. (b) Anomalous dimension of the VBS order parameter as a function of $1/N$. The gray squares are the results of a square lattice study ($q = 4$) [22, 23]. The blue circles are results from the honeycomb lattice ($q = 3$) and the green diamonds are results from the rectangular lattice ($q = 2$). The red line is the $1/N$ expansion ($\eta_N = 1 - 32\pi^2/N, 1 + \eta_V = \delta_1 N$ with $\delta_1 \approx 0.2492$). The universality of the exponents with respect to q is a direct consequence of the irrelevance of λ_q. The agreement of the exponents with the $1/N$ computation is strong evidence for the emergence of the nc-$\mathbb{CP}^{N-1}$ universality at the Néel-VBS transition. Adapted from Ref. [19].

the absence of any plausible alternative one expects a restoration of "Landau" theory and hence a first-order transition. Consistently, both studies find clear evidence for first-order transitions between the Néel and staggered VBS phases.

(3) There have been two studies of the Néel-VBS transition in designer Hamiltonians in *three-dimensional* bipartite SU(N) systems thus far. Both studies, which are on cubic lattices but with distinct Hamiltonians, find first-order transitions between Néel and columnar VBS and no evidence for any new intervening phases [28, 29]. The absence of a direct continuous transition in $3+1$ dimensions is again consistent with the deconfined criticality scenario since various aspects of the $\mathbb{CP}^{N-1}$ field theory are specific to $2+1$ dimensions and are known to be invalid in $3+1$ dimensions.

4. Outlook

In the previous section, we have summarized how the deconfined criticality scenario and the $\mathbb{CP}^{N-1}$ universality, convincingly explain unbiased numerical studies of the SU(N) designer Hamiltonians in a variety of studies carried out by different groups, which have probed various distinct aspects of the deconfined criticality scenario.

One concern is that the transition could be weakly first order and the numerical studies, which are necessarily limited to finite size systems, may not have accessed sizes large enough to detect a first-order transition [30]. If this is indeed the case, the transition must be so weakly first order that the correlation length exceeds the sizes of the largest lattices on which many of the designer Hamiltonians have been simulated, since no direct evidence for first-order behavior has been found in the cases where a continuous transition has been claimed. This makes the discussion of the presence of such a weakly first-order transition somewhat academic.

Nevertheless, there are well established corrections to scaling observed in the numerical studies of the Néel-VBS transition in the designer Hamiltonians [31, 32, 24, 21]. The origin of these corrections to the asymptotic behavior is not completely understood. Three options might be considered: (a) the transition is described completely by deconfined criticality; the corrections to scaling are just the usual corrections to asymptotic behavior that arise from deviants from the true fixed point due to irrelevant operators, finite size, etc. and will vanish in the thermodynamic limit; (b) the transition is weakly first order and no non-compact $\mathbb{CP}^{N-1}$ fixed point exists; the deviation from scaling is claimed as incipient first order behavior; (c) there is something fundamentally new in the scaling behavior of deconfined critical points that needs to be understood; the "corrections" are part of the true asymptotic behavior and their existence can be understood from a field theoretic argument that has so far been overlooked.

Currently, it is not possible to rule out any of the three options categorically. Given the large body of evidence presented in Section 3 that appears consistent with the deconfined scenario and the lack of direct evidence for a first-order transition, Occum's razor suggests that (a) is the most likely explanation. The corrections to scaling arising due to an irrelevant operator with a small exponent [33]. As a theoretical possibility, option (c) is the most exciting and is an interesting area for further field theoretic work [34, 35]. Likewise, some positive theoretical reasoning that supports the existence of a first-order transition and explains why it is so weak, or direct evidence numerical evidence for a first-order transition, would strengthen the case for option (b).

It is also possible that the transition is first order for small-N and becomes continuous only for some finite value of $N > 2$. Numerical studies on the designer Hamiltonians do not find a dramatic difference in the simulations between $N = 2$ and $N > 2$ for the cases where a continuous transition is found. It would be of interest to extend the studies of the $N = 2$ nc-$\mathbb{CP}^{N-1}$ field theory on a lattice, to $N > 2$, which have not been carried out [30, 36].

Beyond the study of deconfined criticality, an as yet unanswered question is what other phases can be accessed in the family of sign-free SU(N) spin models presented in Sec. 2. The only two phases found so far in these models are Néel and VBS phases, with the VBS phases being the simplest possible, i.e., they are each connected without phase transition to a cartoon state that is simply a direct product of two-site SU(N) singlets. There is no evidence for plaquette VBS states in the designer models, despite many approximate studies favoring such a state in similar models. A related unanswered question is whether the SU(N) "designer" Hamiltonians can host a spin liquid, and if so how to design such models. We note that Marshall positive models with simpler U(1) symmetries have been shown convincingly to host Z_2 spin liquids [37, 38], but no such model with SU(N) symmetry is known yet.

Finally, studies of the designer models with internal symmetries different from SU(N), studies of their phase diagrams with quenched disorder, and studies of their spectral properties and doping with a small concentration of holes all provide exciting directions for future research.

Acknowledgments
The authors would like to acknowledge Jon D'Emidio, Roger Melko, and Anders Sandvik for collaboration on related work and NSF DMR-1056536 for partial financial support.

References
[1] Balents L 2010 *Nature* **464** 199–208 URL http://dx.doi.org/10.1038/nature08917

[2] Kaul R K, Melko R G and Sandvik A W 2013 *Annu. Rev. Cond. Matt. Phys* **4** 179 URL http://www.annualreviews.org/doi/abs/10.1146/annurev-conmatphys-030212-184215
[3] Stoudenmire E and White S R 2012 *Annual Review of Condensed Matter Physics* **3** 111–128 URL http://www.annualreviews.org/doi/abs/10.1146/annurev-conmatphys-020911-125018
[4] Sachdev S 1999 *Quantum Phase Transitions* (Cambridge University Press)
[5] Laflorencie N and Poilblanc D 2004 Simulations of pure and doped low-dimensional spin-1/2 gapped systems *Quantum Magnetism* (*Lecture Notes in Physics* vol 645) (Springer Berlin Heidelberg) pp 227–252 URL http://dx.doi.org/10.1007/BFb0119595
[6] Senthil T, Vishwanath A, Balents L, Sachdev S and Fisher M P A 2004 *Science* **303** 1490–1494 (*Preprint* http://www.sciencemag.org/content/303/5663/1490.full.pdf) URL http://www.sciencemag.org/content/303/5663/1490.abstract
[7] Senthil T, Balents L, Sachdev S, Vishwanath A and Fisher M 2004 *Phys. Rev. B* **70**(14) 144407 URL http://link.aps.org/doi/10.1103/PhysRevB.70.144407
[8] Senthil T, Balents L, Sachdev S, Vishwanath A and Fisher M P A 2005 *Journal of the Physics Society of Japan* URL http://journals.jps.jp/doi/abs/10.1143/JPSJS.74S.1
[9] Affleck I 1985 *Phys. Rev. Lett.* **54**(10) 966–969 URL http://link.aps.org/doi/10.1103/PhysRevLett.54.966
[10] Sandvik A W 2010 *AIP Conf. Proc.* **1297** 135 URL http://scitation.aip.org/content/aip/proceeding/aipcp/10.1063/1.3518900
[11] Kaul R K 2014 Designing marshall positive su(n) quantum spin hamiltonians URL http://arxiv.org/abs/1403.5678
[12] Sandvik A W 2007 *Phys. Rev. Lett.* **98**(22) 227202 URL http://link.aps.org/doi/10.1103/PhysRevLett.98.227202
[13] Santoro G, Sorella S, Guidoni L, Parola A and Tosatti E 1999 *Phys. Rev. Lett.* **83**(15) 3065–3068 URL http://link.aps.org/doi/10.1103/PhysRevLett.83.3065
[14] Harada K, Kawashima N and Troyer M 2003 *Phys. Rev. Lett.* **90** 117203 URL http://journals.aps.org/prl/abstract/10.1103/PhysRevLett.90.117203
[15] Chakravarty S, Halperin B and Nelson D 1988 *Phys. Rev. Lett.* **60**(11) 1057–1060 URL http://link.aps.org/doi/10.1103/PhysRevLett.60.1057
[16] Wang L, Beach K and Sandvik A 2006 *Phys. Rev. B* **73**(1) 014431 URL http://link.aps.org/doi/10.1103/PhysRevB.73.014431
[17] Kaul R K 2012 *Phys. Rev. B* **85**(18) 180411 URL http://link.aps.org/doi/10.1103/PhysRevB.85.180411
[18] Nahum A, Chalker J, Serna P, Ortuño M and Somoza A 2013 *Phys. Rev. B* **88**(13) 134411 URL http://link.aps.org/doi/10.1103/PhysRevB.88.134411
[19] Block M S, Melko R G and Kaul R K 2013 *Phys. Rev. Lett.* **111**(13) 137202 URL http://link.aps.org/doi/10.1103/PhysRevLett.111.137202
[20] Pujari S, Damle K and Alet F 2013 *Phys. Rev. Lett.* **111**(8) 087203 URL http://link.aps.org/doi/10.1103/PhysRevLett.111.087203
[21] Harada K, Suzuki T, Okubo T, Matsuo H, Lou J, Watanabe H, Todo S and Kawashima N 2013 *Phys. Rev. B* **88**(22) 220408 URL http://link.aps.org/doi/10.1103/PhysRevB.88.220408
[22] Lou J, Sandvik A and Kawashima N 2009 *Phys. Rev. B* **80**(18) 180414 URL http://link.aps.org/doi/10.1103/PhysRevB.80.180414
[23] Kaul R and Sandvik A 2012 *Phys. Rev. Lett.* **108**(13) 137201 URL http://link.aps.org/doi/10.1103/PhysRevLett.108.137201
[24] Kaul R K 2011 *Phys. Rev. B* **84** 054407 URL http://journals.aps.org/prb/abstract/10.1103/PhysRevB.84.054407
[25] Murthy G and Sachdev S 1990 *Nucl. Phys. B* **344** 557
[26] Sen A and Sandvik A W 2010 *Phys. Rev. B* **82**(17) 174428 URL http://link.aps.org/doi/10.1103/PhysRevB.82.174428
[27] Banerjee A, Damle K and Paramekanti A 2011 *Phys. Rev. B* **83**(13) 134419 URL http://link.aps.org/doi/10.1103/PhysRevB.83.134419
[28] Beach K and Sandvik A 2007 *Phys. Rev. Lett.* **99**(4) 047202 URL http://link.aps.org/doi/10.1103/PhysRevLett.99.047202
[29] Block M S and Kaul R K 2012 *Phys. Rev. B* **86**(13) 134408 URL http://link.aps.org/doi/10.1103/PhysRevB.86.134408
[30] Kuklov A, Matsumoto M, Prokof'ev N, Svistunov B and Troyer M 2008 *Phys. Rev. Lett.* **101**(5) 050405 URL http://link.aps.org/doi/10.1103/PhysRevLett.101.050405
[31] Sandvik A 2010 *Phys. Rev. Lett.* **104**(17) 177201 URL http://link.aps.org/doi/10.1103/PhysRevLett.104.177201

[32] Banerjee A, Damle K and Alet F 2010 *Phys. Rev. B* **82**(15) 155139 URL http://link.aps.org/doi/10.1103/PhysRevB.82.155139
[33] Bartosch L 2013 *Phys. Rev. B* **88**(19) 195140 URL http://link.aps.org/doi/10.1103/PhysRevB.88.195140
[34] Sandvik A, Kotov V and Sushkov O 2011 *Phys. Rev. Lett.* **106**(20) 207203 URL http://link.aps.org/doi/10.1103/PhysRevLett.106.207203
[35] Nogueira F and Sudbø A 2012 *Phys. Rev. B* **86**(4) 045121 URL http://link.aps.org/doi/10.1103/PhysRevB.86.045121
[36] Motrunich O I and Vishwanath A 2008 *http://arxiv.org/abs/0805.1494*
[37] Isakov S V, Hastings M B and Melko R G 2011 *Nat Phys* **7**(10) 772–775 URL http://www.nature.com/nphys/journal/v7/n10/abs/nphys2036.html
[38] Isakov S, Kim Y and Paramekanti A 2006 *Phys. Rev. Lett.* **97**(20) 207204 URL http://link.aps.org/doi/10.1103/PhysRevLett.97.207204

Competition between the Haldane insulator, superfluid and supersolid phases in the one-dimensional Bosonic Hubbard Model

G G Batrouni[1,2,3,4], V G Rousseau[5], R T Scalettar[6] and B Grémaud[3,4,7,8]

[1] INLN, Université de Nice–Sophia Antipolis, CNRS; 1361 route des Lucioles, 06560 Valbonne, France
[2] Institut Universitaire de France, 103, Boulevard Saint-Michel, 75005 Paris, France
[3] Centre for Quantum Technologies, National University of Singapore, 3 Science Drive 2, Singapore 117543, Singapore
[4] Merlion MajuLab, CNRS-UNS-NUS-NTU International Joint Research Unit UMI 3654, Singapore
[5] Department of Physics and Astronomy, Louisiana State University, Bâton Rouge, Louisiana 70803, USA
[6] Physics Department, University of California, Davis, California 95616, USA
[7] Laboratoire Kastler Brossel, Ecole Normale Supérieure CNRS, UPMC; 4 Place Jussieu, 75005 Paris, France
[8] Department of Physics, National University of Singapore, 2 Science Drive 3, Singapore 117542, Singapore

Abstract. We use quantum Monte Carlo (QMC), density matrix renormalization group (DMRG) and time evolution block decimation (TEBD) to study the competition between several exotic quantum phases in the one-dimensional extended bosonic Hubbard model and map its phase diagram. Of particular interest is the Haldane insulating phase which we show to be present only for density $\rho = 1$. We also show that the supersolid phase is present for a wide range of parameters including commensurate densities.

1. Introduction

The bosonic Hubbard model [1] has attracted enormous attention and has been exploited to understand many physical phenomena such as adsorption of bosonic atoms on surfaces[2], the effect of disorder on superfluids and the appearance of the compressible Bose glass phase [1, 3], localization of flux lines and superconducting vortices[4], universal conductivity in thin disordered films[5, 6, 7] and quantum phase transitions between strongly correlated exotic phases. In addition, in the hardcore limit, the BHM can be mapped onto Heisenberg spin models and thus offers the opportunity to study magnetic phenomena under various conditions. Study of the BHM intensified with the experimental realization of Bose-Einstein condensates and the ability to load them in optical lattices [8]. Under experimentally realizable conditions, these systems are described by the BHM and its extensions [9] with highly tunable parameters and in one, two and three dimensions.

In its simplest form, which has only on-site contact interactions, the ground state of the BHM exhibits two phases [1]. At integer filling and strong repulsion, the particles cannot hop between sites and the system is an incompressible Mott insulator (MI) which is replaced by a superfluid (SF) phase at weak coupling. At incommensurate fillings, the system is always SF. Extending this model with the addition of longer range interactions or anisotropic hopping terms leads to new exotic phases. For example, extensive quantum Monte Carlo (QMC) simulations have shown that a strong enough near neighbor repulsion can lead to insulating incompressible density wave order (CDW) at integer and half odd integer fillings. Doping these phases can lead to phase separation or to supersolid (SS) phases [10, 11, 12, 13, 14, 15, 16, 17].

At a density $\rho = 1$ and large contact repulsion, the BHM is in the MI phase. Quantum fluctuations in this phase allow bosons to jump to neighboring sites giving occasional vacant sites and multiply occupied sites. If one assumes that at most doubly occupied sites are present, then the system exhibits empty, singly and doubly occupied sites and can be mapped onto the $S_z = 0, \pm 1$ states of a spin-1 Heisenberg system. Exploiting this mapping, it was in fact shown [18, 19] that this model (with near and next near neighbor interactions) has a Haldane insulating (HI) phase [20, 21] which is a gapped phase characterized by a non-local (string) order parameter. The phase diagram at $\rho = 1$ was also mapped [22].

However, several important questions remain open. Does the HI exist for other integer fillings of the system or is it a special property of the unit filling case? The SS phase found in one dimension [15] was obtained by doping a CDW phase: Does this phase also exist for commensurate fillings in one dimension for parameter choices similar to those in two [23] and three dimensions [24]? If the SS phase exists for commensurate fillings, where is it situated in the phase diagram relative to the CDW, MI and HI phases? Here, we use DMRG (from the ALPS library[25]), QMC and TEBD to answer these questions. The QMC is done with periodic boundary conditions and thus allows us to sample different winding number sectors and therefore to calculate the superfluid density (see below). DMRG and TEBD calculations are performed with open bounday conditions which allows us to gauge the finite size effects compared to QMC. In addition, it is much easier to measure excited state energies and excitation spectra using DMRG and TEBD than QMC. All calculations reported here were done in the canonical ensemble.

2. The Model

The one dimensional extended BHM we shall study is described by the Hamiltonian,

$$H = -t\sum_i (a_i^\dagger a_{i+1} + a_{i+1}^\dagger a_i) + \frac{U}{2}\sum_i n_i\,(n_i - 1) + V\sum_i n_i n_{i+1}. \qquad (1)$$

The sum over i extends over the L sites of the lattice, periodic boundary conditions were used in the QMC and open conditions with the DMRG and TEBD. The hopping parameter, t, is put equal to unity and sets the energy scale, a_i ($a_i^\dagger$) destroys (creates) a boson on site i, $n_i = a_i^\dagger a_i$ is the number operator on site i, U and V are the onsite and near neighbor interaction parameters.

Several quantities are needed to characterize the phase diagram. The superfluid density is given by [26]

$$\rho_s = \frac{\langle W^2 \rangle}{2td\beta L^{d-2}}, \qquad (2)$$

where W is the winding number of the boson world lines, d is the dimensionality and β the inverse temperature. The CDW order parameter, $S(k=\pi)$, the one-particle Green function and the momentum distribution, n_k, are given by

$$S(k) = \frac{1}{L}\sum_{r=0}^{L-1} \mathrm{e}^{ikr}\langle n_0 n_r \rangle, \quad G(r) = \langle a_0^\dagger a_r \rangle, \quad n_k = \frac{1}{L}\sum_{r=0}^{L-1} \mathrm{e}^{ikr} G(r). \qquad (3)$$

Table 1. Order parameters characterizing various phases.

	ρ_s	$S(\pi)$	Δ_c	Δ_n	$\mathcal{O}_p(L_{max})$	$\mathcal{O}_s(L_{max})$
MI	0	0	$\neq 0$	$= \Delta_c$	$\neq 0$	0
CDW	0	$\neq 0$	$\neq 0$	$\neq 0$	$\neq 0$	$\neq 0$
SF	$\neq 0$	0	0	0	0	0
HI	0	0	$\neq 0$	$\neq 0$	0	$\neq 0$
SS	$\neq 0$	$\neq 0$	0	0	$\neq 0$	$\neq 0$

The charge gap is given by,

$$\Delta_c(n) = \mu(n) - \mu(n-1) = E_0(n+1) + E_0(n-1) - 2E_0(n), \tag{4}$$

where $\mu(n) = E_0(n+1) - E_0(n)$ and $E_0(n)$ is the ground state energy of the system with n particles and is obtained both with QMC and DMRG. The neutral gap, Δ_n, is obtained using DMRG by targeting the lowest excitation with the same number of bosons. The various phases are summarized in Table 1 where, defining $\delta n_j = n_j - \rho$, the two non-local order parameters which are important to characterize the HI are the string ($\mathcal{O}_s$) and parity ($\mathcal{O}_p$) order parameters,

$$\mathcal{O}_s(|i-j| \to \infty) = \langle \delta n_i e^{i\pi \sum_{k=i}^{j} \delta n_k} \delta n_j \rangle, \;\; \mathcal{O}_p(|i-j| \to \infty) = \langle e^{i\pi \sum_{k=i}^{j} \delta n_k} \rangle. \tag{5}$$

3. Results

We calculated the order parameters for a wide range of parameters using the Stochastic Green Function (SGF) QMC algorithm [27], DMRG and TEBD. To characterize the phases and map out the phase diagram, some parameters are fixed while others are varied. For example, in Fig.1 we fix $V/U = 3/4$ and $\rho = 1$ and calculate several physical quantities as t/U is varied. In the right panel, we show (a) the CDW $S(k = \pi)$ and superfluid (ρ_s) order parameters, (b) the string ($\mathcal{O}_s$) and parity ($\mathcal{O}_p$) order parameters and (c) the neutral (Δ_n) and charge (Δ_c) gaps. $S(\pi)$ vanishes at $t/U \approx 0.22$ beyond which ρ_s appears to be nonvanishing. However, finite size studies show that $\rho_s \to 0$ as $L \to \infty$ indicating that the zone $0.22 < t/U < 0.4$ is not superfluid. At the same time, $\mathcal{O}_s(t/U > 0.22) \neq 0$ and the gaps are nonvanishing too. This indicates that the transition at $t/U = 0.22$ is from the CDW phase to the gapped Haldane insulating phase. As t/U increases further (not shown) the system will leave the HI and enter the SF phase. Note that in the $\rho = 1$ CDW phase $\Delta_c > \Delta_n$ and that Δ_n vanishes at the CDW-HI transition but Δ_c remains nonzero. The meaning of the two unequal gaps can be seen in the center and right panels of Fig.1 showing the dispersion, $\omega(k)$. The charge gap is given by $\Delta_c = \omega(k \to 0)$ while Δ_n is the *smallest* excitation gap at any k. The smallest gap here is at $k = \pi$ thus giving the neutral gap $\Delta_n = \omega(k = \pi)$. So, it is possible for Δ_c and Δ_n to be unequal and for one of them to vanish but not the other. At the CDW-HI Δ_n vanishes due to the loss of order at the ordering vector $k^* = \pi$.

The question of what happens at other fillings, especially commensurate ones, is very interesting. For example, does the HI appear at other integer fillings? We address this question in Fig.2 where we show in the left panel the same quantities as in Fig.1. These two figures exhibit some marked differences. The QMC results, Fig.2(a), show that ρ_s becomes nonzero before $S(\pi)$ vanishes indicating the simultaneous presence of long range density order (CDW)

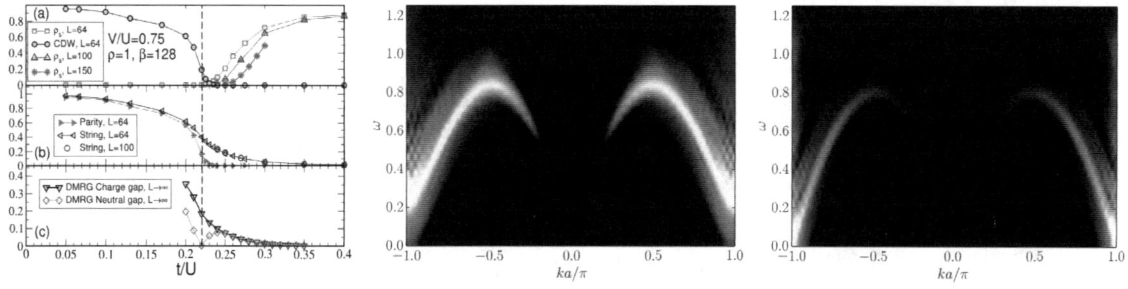

Figure 1. Left: Order parameters (a,b) and gaps (c) versus t/U at fixed $V/U = 3/4$. Center: excitation spectrum, $\omega(k)$ at $t/U = 0.2$ (CDW). Right: excitation spectrum, $\omega(k)$ at $t/U = 0.22$ (CDW-HI transition). All are at fixed $\rho = 1$.

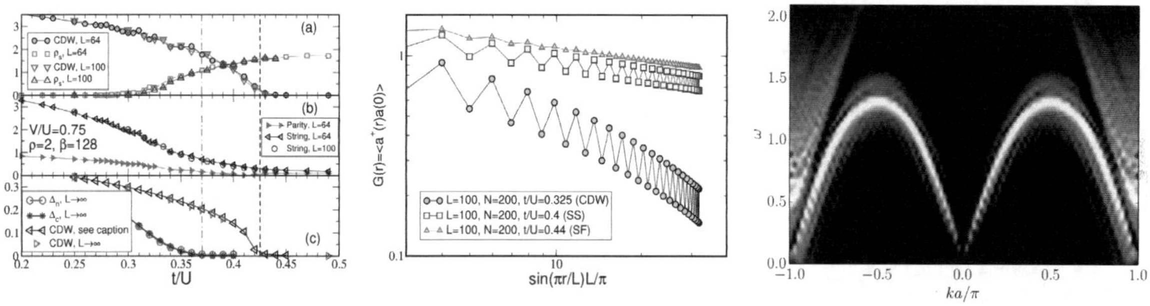

Figure 2. Left: Order parameters (a,b) and gaps (c) versus t/U at fixed $V/U = 3/4$. Center: $G(r)$ in for $t/U = 0.325, 0.4, 0.44$. Right: excitation spectrum, $\omega(k)$ at $t/U = 0.38$, in the SS phase. All are at fixed $\rho = 2$.

and superfluidity. This is confirmed by DMRG in panel (c) which shows $\Delta_n = \Delta_c$ vanish together at $t/U = 0.37$ while the CDW order is still finite. Therefore, this region of coexistence, $0.37 < t/U < 0.425$, corresponds to the supersolid (SS) phase: No HI is apparent at $\rho = 2$. The center panel of Fig.2 shows that the one-body Green function decays as a power, $G(r) \propto r^{-1/2K}$, in the SS and SF phases, which resolves the disagreement between various bozonization results concerning the SS phase. We found $K = 2.24$ for $t/U = 0.4$ (SS) and $K = 2.5$ for $t/U = 0.44$ (SF). In the CDW phase, $G(r)$ decays exponentially. The right panel of Fig.2 shows the excitation spectrum in the SS phase. Note that $\Delta_c = \omega(k \to 0) \to 0$, indicating gapless excitations. The same is seen at $k = \pi$; in this phase $\Delta_c = \Delta_n = 0$.

By repeating such measurements at many fillings, we map out the phase diagram in the $(\mu/U, t/U)$ plane at fixed $V/U = 3/4$ and confirm the absence of the HI at all fillings except $\rho = 1$. Figure 3 shows that the $\rho = 1/2$ CDW lobe is almost entirely surrounded by SF, all lobes at $\rho > 3/2$ are surrounded by SS and the region of the $\rho = 1$ lobe which sticks out of the SS is the HI phase. The transition from the $\rho = 3/2$ CDW lobe goes directly to the SF phase without passing through SS. We verified that this behavior persists at other values of the ratio V/U. It is interesting to note that the HI phase is present only at $\rho = 1$ and that the SS phase is present over a large region in the phase diagram. It is particularly interesting that the SS phase exists at *commensurate* fillings.

Journal of Physics: Conference Series **640** (2015) 012042 doi:10.1088/1742-6596/640/1/012042

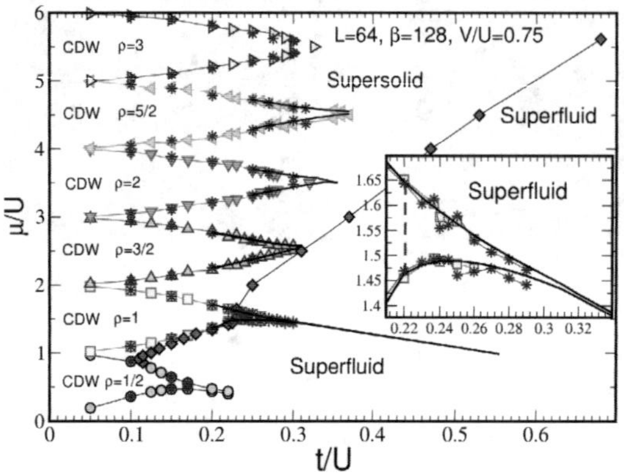

Figure 3. The phase diagram at fixed ratio $V/U = 3/4$. The inset is a zoom on the tip of the $\rho = 1$ lobe. All symbols represent results from QMC simulations for $L = 128$ (stars) and $L = 64$ (all other symbols). The solid black lines near the lobe tips are DMRG results with $L = 192$. The end points of the lobes are obtained by studying the finite size dependence of Δ_n using DMRG except for $\rho = 1$ which is obtained using QMC by extrapolating $\mathcal{O}_s$ to the thermodynamic limit.

Another informative view is to show the phase diagram at fixed density. This is given in the left panel of Fig.4. The open squares are the results of Ref. [22], all other symbols are results of

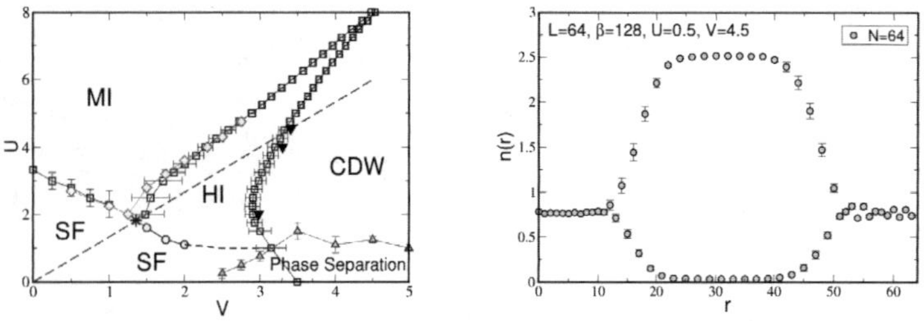

Figure 4. Left: The phase diagram in the (U, V) plane for $\rho = 1$. Right: The density profile displaying phase separation.

our QMC simulations. Our results confirm those of Ref. [22] where the error bars are small and improve them where the error bars are large and we have determined part of the SF-HI boundary. The straight line, $U = 4V/3$, shows where the measurements leading to Fig.1 were taken. The star on this line is the tip of the $\rho = 1$ lobe in Fig.3. In addition, we have determined the boundary of a phase separation (PS) region whose presence had not been suspected in the literature. Identifying this PS is accomplished by examining the density profiles. The right panel of Fig.4 shows such a profile exhibiting an island of SS region in a sea of SF.

In Fig.3 we showed that, for $V/U = 3/4$, the HI appears only at $\rho = 1$. To explore if the HI can appear for other ratios, we map in Fig.5 the phase diagram in the (U, V) plane

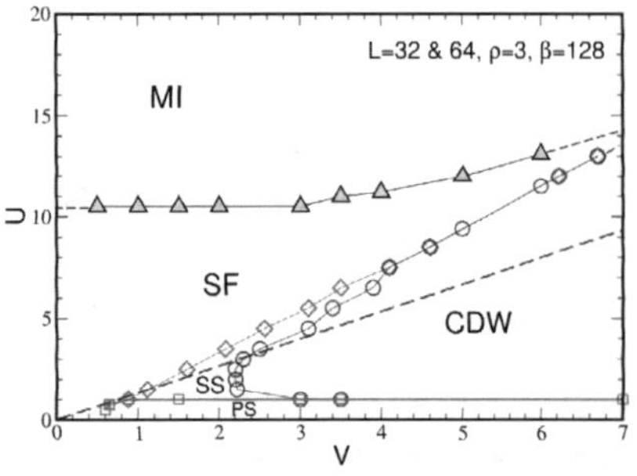

Figure 5. Applying the same methods as for $\rho = 1$, the $\rho = 3$ phase diagram is mapped out. Unlike for $\rho = 1$, there is no HI here but there is SS in addition to the MI, SF, CDW and PS regions.

for $\rho = 3$. This figure shares some general features with Fig.4 but there are important differences: (a) No HI phase appears anywhere in the phase diagram but instead (b) a supersolid phase appears at the tip of the CDW region.

4. Conclusions

We used QMC simulations and DMRG/TEBD calculations to determine the phase diagram of the extended bosonic Hubbard model in one dimenion. Of special interest is the HI phase and where it appears in the phase diagram. The HI is a gapped insulating phase characterized by a nonlocal order parameter; it is expected to appear in the Heisenberg spin model for odd integer values of the spin. Deep in the MI lobes, one can map the BHM onto the Heisenberg model with $S = \rho$ and might then expect the HI to appear for odd values of ρ. We have determined the phase diagram in the $\mu/U, t/U)$ plane for $V/U = 3/4$ and confirmed that the HI is indeed present for $\rho = 1$ but we found that it is absent at all other densities. This absence is again confirmed by mapping the $\rho = 3$ phase diagram in the (U, V) plane where it is shown that a SS phase appears but no HI. Furthermore, we showed that the phase diagrams at fixed filling exhibit a region of phase separation between SS and SF phases in the small-U large-V regions.

5. Acknowledgements

This work was supported by: the CNRS-UC Davis EPOCAL joint research grant; by NSF-PIF-1005503 grant; by the France-Singapore Merlion program (PHC Egide and FermiCold 2.01.09); by grant DOE DE-NA0001842-0. The Centre for Quantum Technologies is a Research Centre of Excellence funded by the Ministry of Education and National Research Foundation of Singapore.

References

[1] Fisher M P A, Weichman P B, Grinstein G and Fisher D S 1989 *Phys. Rev.* B**40** 546
[2] Zimanyi G T, Crowell P A, Scalettar R T, Batrouni G G 1994 *Phys. Rev.* B **50** 6515
[3] Batrouni G G, Scalettar R T and Zimanyi G T 1991 *Phys. Rev. Lett.* **66** 3144
[4] Nelson D R and Vinokur V M 1992 *Phys. Rev. Lett.* **68** 2398; Nelson D R and Vinokur V M 1993 *Phys. Rev.* B**48** 13060
[5] Cha M-C, Fisher M P A, Girvin S M, Wallin M and Young A P 1991 *Phys. Rev.* B**44** 6883
[6] Batrouni G G, Larson B, Scalettar R T, Tobochnik J and Wang J 1993 *Phys. Rev.* B**48** 9628
[7] Chen K, Liu L, Deng Y, Pollet L and Prokofev N 2014 *Phys. Rev. Lett.* **112** 030402
[8] Greiner M, Mandel O, Esslinger T, Hänsch T W and Bloch I 2002 Nature **415** 39
[9] Jaksch D, Briegel H J, Cirac J I, Gardiner C W and Zoller P 1999 *Phys. Rev. Lett.* **82** 1975
[10] Batrouni G G, Scalettar R T, Zimanyi G T and Kampf A P 1995 *Phys. Rev. Lett.* **74** 2527
[11] Batrouni G G and Scalettar R T 2000 *Phys. Rev. Lett.* **84** 1599
[12] Wessel S and Troyer M 2005 *Phys. Rev. Lett.* **95**, 127205
[13] Sengupta P, Pryadko L P, Alet F, Troyer M and Schmid G 2006 *Phys. Rev. Lett.* **94** 207202
[14] van Otterlo A, Wagenblast K-H, Baltin R, Bruder C, Fazio R and Schön G 2005 *Phys. Rev.* B **52** 16176
[15] Batrouni G G, Hébert F and Scalettar R T 2006 *Phys. Rev. Lett.* **97** 087209
[16] Pollet L, Picon J D, Büchler H P and Troyer M 2010 *Phys. Rev. Lett.* **104** 125302
[17] Capogrosso-Sansone B, Trefzger C, Lewenstein M, Zoller P and Pupillo G 2010 *Phys. Rev. Lett.* **104** 125301
[18] Dalla Torre E G, Berg E and Altman E 2006 *Phys. Rev. Lett.* **97** 260401
[19] Berg E, Dalla Torre E G, Giamarchi T and Altman E 2008 *Phys. Rev.* B**77** 245119
[20] Haldane F D M 1983 *Phys. Lett.* **93A** 464; Haldane F D M 1983 *Phys. Rev. Lett.* **50** 1153
[21] den Nijs M and Rommelse K 1989 *Phys. Rev.* B**40** 4709
[22] Rossini D and Fazio R 2012 *New J. Phys.* **14** 065012
[23] Ohgoe T, Suzuki T and Kawashima N 2012 *Phys. Rev.* B**86** 054520
[24] Ohgoe T, Suzuki T and Kawashima N 2012 *Phys. Rev. Lett.* **108** 185302
[25] Bauer B. et al. (ALPS collaboration) 2011 *J. Stat. Mech.* **P05001**.
[26] Pollock E L and Ceperley D M 1984 *Phys. Rev.* B**30** 2555
[27] Rousseau V G 2008 *Phys. Rev.* E**77** 056705; Rousseau V G 2008 **E78** 056707; Rousseau V G and Galanakis D arXiv:1209.0946.

XXVI IUPAP Conference on Computational Physics (CCP2014) IOP Publishing
Journal of Physics: Conference Series **640** (2015) 012043 doi:10.1088/1742-6596/640/1/012043

1D valence bond solids in a magnetic field

Adam Iaizzi and Anders W. Sandvik

Department of Physics, Boston University, 590 Commonwealth Avenue, Boston, Massachusetts 02215, USA

Abstract. A Valence bond solid (VBS) is a nonmagnetic, long-range ordered state of a quantum spin system where local spin singlets are formed in some regular pattern. We here study the competition between VBS order and a fully polarized ferromagnetic state as function of an external magnetic field in a one-dimensional extended Heisenberg model—the J-Q_2 model—using stochastic series expansion (SSE) quantum Monte Carlo simulations with directed loop updates. We discuss the ground state phase diagram.

1. Introduction

A valence bond solid (VBS) is a long-range nonmagnetic ordered state that can appear in certain quantum spin systems. In a VBS (also called the spin-Peierls state) spins spontaneously pair up to form singlets in some regular pattern, breaking lattice translational symmetry but retaining spin rotation symmetry. Recent innovations in models and simulation techniques have enabled large-scale numerical studies of this type of ground state and associated quantum phase transitions. In one dimension, these studies have found evidence for the fractionalization of triplons into deconfined spinons carrying spin $S = 1/2$ in the VBS phase [1]. Since lattice translational symmetry is discrete, the VBS transition is allowed in one-dimensional systems at zero temperature. We here add an external magnetic field and study its effect on the VBS.

The J-Q model is at its core a sign-problem-free system for studying valence bond solids, associated quantum phase transitions, and other related phenomena using quantum Monte Carlo (QMC) methods [2]. It supplements a standard Heisenberg J exchange term with a multispin interaction composed of a product of two (or more) singlet projection operators,

$$P_{ij} = \frac{1}{4} - \vec{S}_i \cdot \vec{S}_j. \tag{1}$$

Several variants of the J-Q model have been studied extensively in two dimensions [2–7]. The J-Q_2 model has a Q term with two singlet projection operators; its one-dimensional realization can be written as

$$H \equiv -J \sum_i P_{i,i+1} - Q \sum_i P_{i,i+1} P_{i+2,i+3} + h \sum_i S_i^z \tag{2}$$

where $J, Q > 0$ indicate antiferromagnetic interactions. In this paper we will fix $J = 1$ and use the dimensionless parameters $q \equiv Q/J$ and $h \equiv h_{bare}/J$. At zero temperature and zero field the 1D J-Q_2 model produces a VBS when q exceeds $q_c = (Q/J)_c = 0.84831$ [1].

2. Methods

The primary computational tools we are using in this work are exact diagonalization of the Hamiltonian for small systems and a stochastic series expansion (SSE) QMC method for larger systems. Our QMC simulations are based on the directed loop algorithm for the anisotropic Heisenberg model in an external field [8], with added procedures to account for the Q term. Strictly speaking, this is a finite temperature method, but since any finite size system will have a finite excitation gap, we can reach the zero temperature regime by using a temperature much smaller than the gap.

3. Results

We begin by showing in Fig. 1 a schematic phase diagram based on previous work and our own calculations (discussed in more detail below). Along the h-axis, with $q = 0$, we have the standard Heisenberg chain in an external field, which undergoes a transition to a fully polarized ferromagnet at $h = h_c$. Along the q-axis for $h = 0$ we have the previously studied zero-field J-Q_2 chain, which for $q < q_c$ has a quasi-Néel ordered state with spin correlations decaying with distance r as $1/r$ (up to logarithmic corrections) [9], and at $q = q_c$ undergoes a phase transition to a VBS [1,10]. This transition is of the same type (similar to a Kosterlitz-Thouless transition) as in the the well-studied frustrated J_1-J_2 chain.

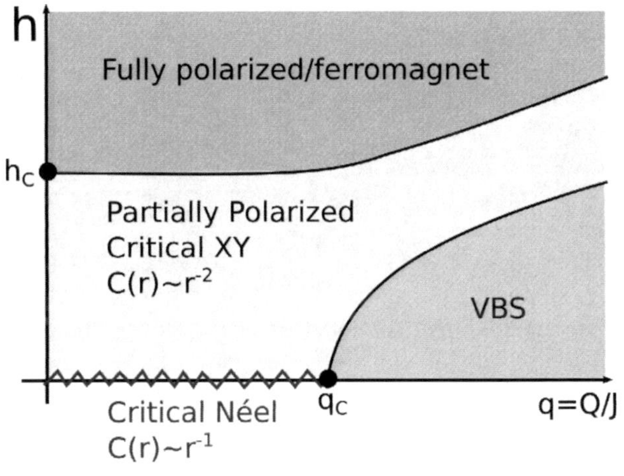

Figure 1. Schematic preliminary phase diagram of the 1D JQh model in the plane of the coupling ratio q and magnetic field h.

The area off of the axes of Fig. 1 has not been studied previously. We believe there are three different phases: a fully polarized state, a partially polarized state with critical XY spin correlations (like the Heisenberg chain in a magnetic field) and a VBS state. The critical magnetic field, $h_c(q)$, can be quantitatively extracted from data such as those shown in Fig. 5. In some cases we see clear magnetization jumps, i.e. a first-order transition, but it remains an open question whether the transition is first-order along the whole line, or only above some minimum value of q.

The q_c line extends out from the known zero-field transition point; the rest of the line depends on how the VBS is destroyed by the magnetic field. The elementary excitations of the $h = 0$ VBS in one dimension are pairs of deconfined spinons. Each pair of spinons arises from breaking one of the singlet bonds, creating a triplet which fractionalizes into two independently propagating domain walls between the two possible VBS ordering patterns. We believe that an arbitrarily low density of such deconfined spinons will destroy the VBS state (using arguments analogous to the standard ones for breaking of a discrete symmetry by temperature in one dimension, e.g., in the Ising chain). Thus, the magnetic field required to destroy the VBS should be simply related to the spin gap.

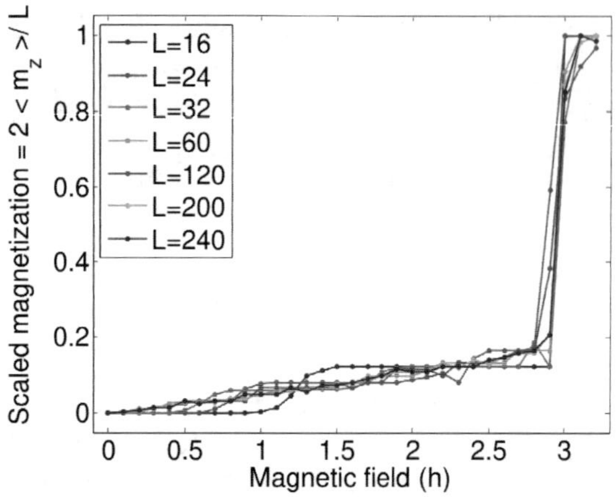

Figure 2. Scaled magnetization plotted as a function of applied magnetic field with $q = 1.2$ for a wide range of sizes. Computed using QMC with periodic boundary conditions.

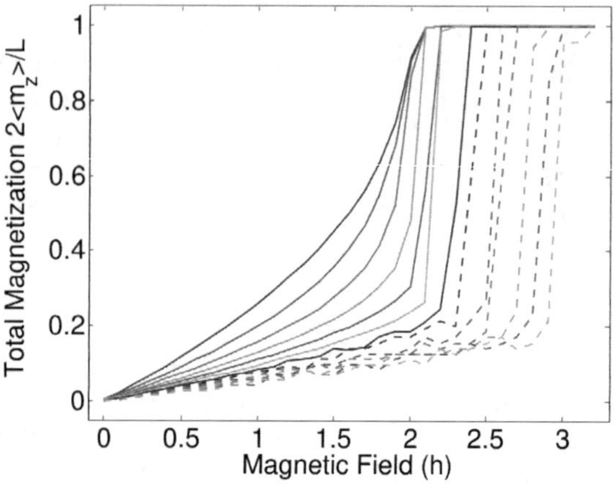

Figure 3. Scaled magnetization plotted as a function of applied magnetic field for a range of different values of $q = Q/J$. From the left (solid blue), $q = 0.0, 0.1, 0.2 ... 1.2$. Computed using QMC with $L = 140$ and open boundary conditions.

Starting in the VBS phase ($q = 1.2$) and increasing the magnetic field we observe a jump in the magnetization. In Fig. 2 we plot the scaled magnetization,

$$\langle m \rangle = \frac{2}{L} \left| \sum S_i^z \right| \tag{3}$$

for $0 \leq h \leq 3.2$ for system sizes from $L = 16$ to $L = 240$. At every size we find the same behavior: roughly linear response until $m \approx 1/6$ followed by a sharp jump to the fully polarized state. This resembles the so-called metamagnetic transition demonstrated in the anisotropic J_1-J_2 chain [11–14].

In Fig. 3 we observe the magnetization jump developing from $q = 0$, the Heisenberg limit. For $q = 0$, the ground state becomes fully polarized at $h \approx 2$. There is no jump and $m(q, h)$ should be continuous in the thermodynamic limit. For small values of $q > 0$, $m(h)$ becomes steeper, but based on this data we cannot determine if there is a jump in the magnetization until larger values of q. As q is increased the jump becomes more distinct and the maximum magnetization before the jump asymptotically approaches $m_c(q) \to 1/6$. Fig. 3 was generated using open boundary conditions in an attempt to 'pin' the structure of the ground state right before the magnetization jump. No significant differences were observed between results generated with open and periodic boundary conditions.

To get a better idea of the behavior around this phase transition, we have solved small chains using exact diagonalization. In Fig. 4 we plot ground state energy of an 18 site chain for each magnetization sector, $E_0(h, m) = E_{min}(q, m) - hm$, for a selection of magnetic fields. For each magnetic field, the black dot shows which value of m is the ground state. For $h = 0$, we can see that there is a slight change in curvature at the high magnetization states; this is key to the transition because it allows the fully polarized state to 'outrun' intermediate polarizations when h is increased. A similar change in curvature was observed in the metamagnetic transition in the J_1-J_2 model [12]. This change in curvature appears only for $q > 0$; it is not a feature of the Heisenberg model. For $h = 2.7$, the ground state has a magnetization of $m = 0.22$; at just a slightly higher field, $h = 2.8$, the ground state is fully polarized. The ground state jumps from $m = 0.22$ to $m = 1$ without passing though any of the intermediate states.

Journal of Physics: Conference Series **640** (2015) 012043 doi:10.1088/1742-6596/640/1/012043

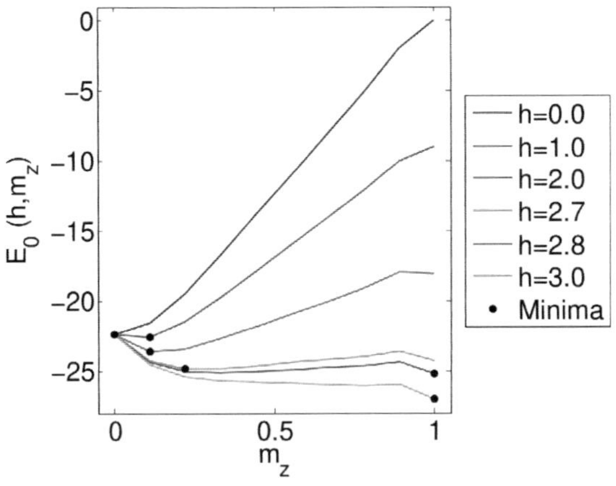

Figure 4. Ground state energy as a function of magnetization for different values of h showing either side of the magnetization jump. Computed using exact diagonalization for $L = 18$ and $q = 1.2$ with periodic boundary conditions.

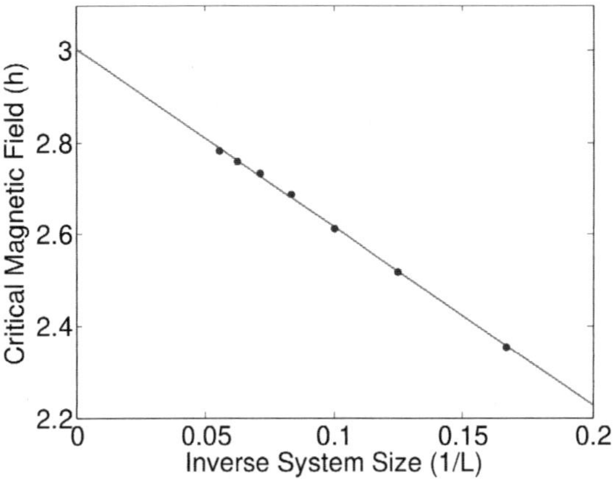

Figure 5. Finite size scaling: critical magnetic field for $q = 1.2$ plotted against $1/L$. Plotting against inverse system size we can extract h_c in the thermodynamic limit by taking the y-intercept of a linear fit. We find $h_c(L \to \infty) \approx 3$ with $R^2 < 0.999$. Computed using exact diagonalization with periodic boundary conditions.

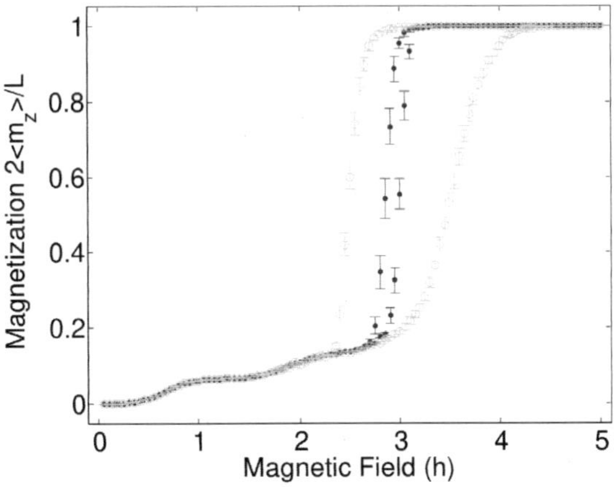

Figure 6. Simulation time hysteresis for a $L = 30$ chain with $q = 1.2$ and $\beta = 8$. ● corresponds to a field velocity of $v_1 = 2.5 \times 10^{-7}$ in units of magnetic field per Monte Carlo sweep, ○ has a faster velocity of $v_2 = 10^{-5}$. Computed using QMC data with periodic boundary conditions.

We can also use exact diagonalization to examine the critical magnetic field; in Fig. 5 we plot $h_c(q = 1.2, 1/L)$ for $L = 6$ to $L = 18$. From this we obtain a good linear fit which predicts that $h_c(L \to \infty) = 3$ with $R^2 > 0.999$, which is consistent with our QMC data. By repeating this process over a mesh of q values we find a line consistent with the qualitative picture presented in Fig. 1.

In Fig. 6, we present QMC data for $q = 1.2$ and $L = 30$. The system exhibits hysteresis when the magnetic field is ramped up and back down, a hallmark behavior in QMC simulations of a first-order phase transition. For clarity, only two different sweep velocities are shown here; a finer mesh of sweep velocities revealed a consistent positive relationship between velocity and the size of the hysteresis effect.

4. Discussion

The magnetization jumps and hysteresis lead us to believe that the J-Q_2 model produces a first-order transition from a partially magnetized to a fully polarized state. The mechanism for this transition is currently unknown and is the subject of an ongoing investigation. We suspect it may bear resemblance to the metamagnetic transition.

We draw these preliminary conclusions from limited data. At low temperature and high values of q and h our QMC method encounters difficulty reaching equilibrium, especially around the magnetization jump. This often causes simulations to become stuck in metastable magnetization states and produce poor estimates of statistical error. As a result, the magnetization curves in Figs. 2 and 3 are not very smooth.

We are currently developing a replica exchange method (also known as "quantum parallel tempering") we hope will ameliorate this problem. Replica exchange was developed for simulating classical systems with slow relaxation times [15] and has since been adapted for use with quantum systems [16]. Despite equilibration issues, we are reasonably confident in our preliminary results and the conclusions we have drawn from them, since repeated independent simulations produce data that match to within small statistical noise.

This project is part of a larger plan to study the J-Q-h model in two dimensions. The simulation methods developed for one-dimensional systems should work with few modifications in two dimensions.

Acknowledgments

We would like to thank Kedar Damle for valuable discussions facilitated by the APS-IUSSTF Physics PhD Student Visitation Program. This work was supported by the NSF under grants No. DMR-1104708 and DMR-1410126.

References

[1] Tang Y and Sandvik A W 2011 *Physical Review Letters* **107** 157201
[2] Sandvik A 2007 *Phys. Rev. Lett.* **98** 227202
[3] Sandvik A W 2010 *Phys. Rev. Lett.* **104**(17) 177201
[4] Lou J, Sandvik A W and Kawashima N 2009 *Phys. Rev. B* **80**(18) 180414
[5] Jin S and Sandvik A W 2013 *Phys. Rev. B* **87**(18) 180404
[6] Kaul R K, Melko R G and Sandvik A W 2013 *Annu. Rev. Condens. Matter Phys.* **4** 179
[7] Kaul R K 2014 *arXiv:cond-mat/1403.5678*
[8] Syljuåsen O F and Sandvik A W 2002 *Phys. Rev. E* **66**
[9] Singh R R P, Fisher M E and Shankar R 1989 *Phys. Rev. B* **39**(4) 2562
[10] Sanyal S, Banerjee A and Damle K 2011 *Phys. Rev. B* **84**(23) 235129
[11] Gerhardt C, Mütter K H and Kröger H 1998 *Phys. Rev. B* **57**(18) 11504
[12] Hirata S 1999 *arXiv:cond-mat/9912066*
[13] Aligia A 2000 *Phys. Rev. B* **63**(1) 014402
[14] Dmitriev D V and Krivnov V Y 2006 *Phys. Rev. B* **73**(2) 024402
[15] Hukushima K and Nemoto K 1996 *Journal of the Physical Society of Japan* **65** 1604
[16] Sengupta P, Sandvik A W and Campbell D K 2002 *Phys. Rev. B* **65**(15) 155113

Matrix-Product-State Algorithm for Finite Fractional Quantum Hall Systems

Zhao Liu and R. N. Bhatt

Department of Electrical Engineering, Princeton University, Princeton, New Jersey 08544, USA

E-mail: zhaol@princeton.edu

Abstract. Exact diagonalization is a powerful tool to study fractional quantum Hall (FQH) systems. However, its capability is limited by the exponentially increasing computational cost. In order to overcome this difficulty, density-matrix-renormalization-group (DMRG) algorithms were developed for much larger system sizes. Very recently, it was realized that some model FQH states have exact matrix-product-state (MPS) representation. Motivated by this, here we report a MPS code, which is closely related to, but different from traditional DMRG language, for finite FQH systems on the cylinder geometry. By representing the many-body Hamiltonian as a matrix-product-operator (MPO) and using single-site update and density matrix correction, we show that our code can efficiently search the ground state of various FQH systems. We also compare the performance of our code with traditional DMRG. The possible generalization of our code to infinite FQH systems and other physical systems is also discussed.

1. Introduction

Topologically ordered phases of matter attract a great deal of interest currently in condensed matter physics. Fractional quantum Hall (FQH) states [1, 2], as one of the most well-known topological states, provide examples of some of the most exotic features such as excitations with a fraction of the electron charge that obey anyonic statistics [3, 4] and are essential resources in topological quantum computation [5].

Exact diagonalization (ED) in small systems has traditionally been the central numerical technique of theoretical research of FQH states. Despite its considerable success in studying some of the robust FQH states like the $\nu = 1/3$ Laughlin state [6], the largest system size that ED can reach is seriously limited by the exponential growth of the Hilbert space. This weakens its capability in understanding some more complex FQH systems, for example those with non-Abelian anyons and Landau level mixing. Therefore, new algorithms, like density-matrix-renormalization-group (DMRG) [7], were applied to FQH systems and the computable system size was increased by almost a factor of two compared with ED [8, 9, 10, 11, 12, 13].

Very recently, it was realized that some model FQH states and their quasihole excitations have exact matrix-product-state (MPS) representation [14, 15, 16, 17]. Motivated by this, and considering that the DMRG algorithm can be apparently formulated in the MPS language [18], here we report a MPS code for finite FQH systems on the cylinder geometry, which is different from the one developed for infinite systems [19]. We will show the structure of our code, explain how to use it to search the FQH ground states, compare the performance of our code with ED and traditional two-site DMRG algorithm, and discuss possible generalizations.

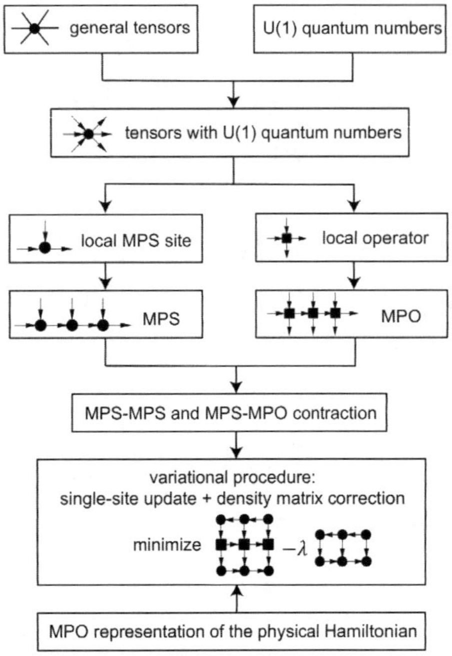

Figure 1. The structure of our MPS code. The most basic part is located on the top, based on which we can implement the whole MPS algorithm step by step to search the ground state and ground energy. The arrows on the bonds of tensors represent $U(1)$ currents. The concepts of $U(1)$-symmetric tensors and their graphical representation are explained in Ref. [20]. The details of the variational procedure can be found in Ref. [18].

2. Code structure

We show the structure of our code in Fig. 1. The most basic part is the implementation of general tensors and $U(1)$ quantum numbers. Based on that, we can construct tensors that conserve $U(1)$ quantum numbers [20]. Each site in the MPS (MPO) is a special case of this kind of tensor with three (four) indices. Then we can implement the chains of these sites, namely MPS and MPO, and the contraction between them. Finally, with the MPO representation of the physical Hamiltonian as an input, we can do variational procedure (single-site update [18] and density matrix correction [21]) to minimize the expectation value of the energy λ.

3. Solve the FQH problem

For a FQH system with N_e electrons and N_s flux in a single Landau level on the cylinder of circumference L (in units of the magnetic length), any translational invariant two-body Hamiltonian can be written as

$$H = \sum_{l}\sum_{m>0}\sum_{n>0} V_{m,n} c^\dagger_{m+n+l} c^\dagger_{n+l} c_{m+2n+l} c_l + h.c. + \sum_{l}\sum_{m>0} V_{m,0} c^\dagger_{m+l} c^\dagger_l c_{m+l} c_l, \tag{1}$$

where $c^\dagger_l$ (c_l) creates (annihilates) an electron on the Landau level orbital $l = 0, 1, ..., N_s - 1$, and m and n determine the range of H. The total electron number $N_e = \sum_{l=0}^{N_s-1} c^\dagger_l c_l$ and total quasi-momentum $K = \sum_{l=0}^{N_s-1} l c^\dagger_l c_l$ are $U(1)$ good quantum numbers.

In our code, the procedure to search for the ground state of H in a fixed (N_e, K) sector is as follows:

- Construct an initial MPS state $|\Psi_0\rangle$ (either a random or a special state) with fixed (N_e, K).
- Construct a MPO representation of H, which can be generated by the finite state automaton [22]. There are two choices when applying the finite state automaton method. The simplest one is to keep all $V_{m,n}$ satifying $|V_{m,n}|$ larger than some threshold (for example 10^{-12}) and $m, n \leq \Lambda$, where Λ is the truncation of the interaction range. The bond dimension

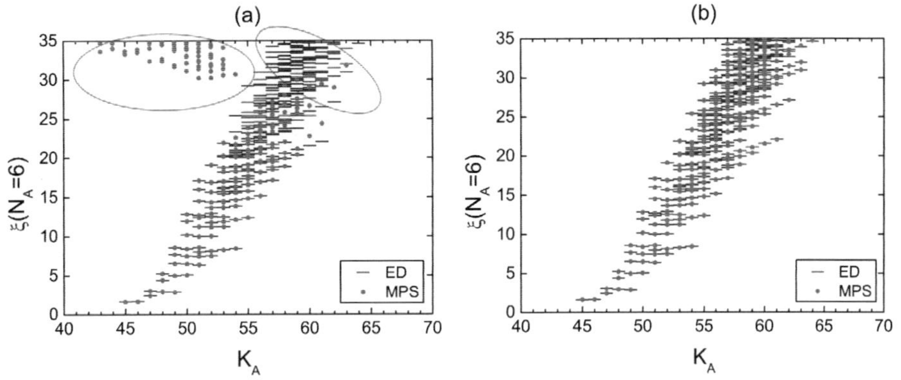

Figure 2. The orbital-cut entanglement spectra obtained by ED (blue dashes) and MPS algorithm (red dots) for $N_e = 12$ electrons at $\nu = 1/3$ with $L = 20$. $\Lambda = 12, \varepsilon = 10^{-6}, \eta = 10^{-10}$ in (a) and $\Lambda = 12, \varepsilon = 10^{-10}, \eta = 10^{-14}$ in (b). The difference from the ED results, which are caused by nonzero ε and η, are indicated by the green circles in (a).

of the MPO, D_{MPO}, obtained in this way is proportional to Λ^2 [19]. The other choice is numerically much more efficient. For each $n \leq \Lambda$, we first select all $V_{m,n}$ satisfying $|V_{m,n}|$ larger than some threshold (for example 10^{-12}), and then approximate them by an exponential expansion $C_n A_n^{m-1} B_n$, with an error $\varepsilon_{m,n} = |V_{m,n} - C_n A_n^{m-1} B_n|$. Here C_n, A_n and B_n are real $1 \times d_n$, $d_n \times d_n$ and $d_n \times 1$ matrices, respectively, whose optimal values can be found by the state space representation method in the control theory [23]. This exponential expansion of $V_{m,n}$ can reduce the bond dimension of MPO by a factor of two compared with the first choice and is especially suitable for long-range interactions [23, 24]. We find that typically $\Lambda \approx 10$ and $\varepsilon \approx 10^{-8}$ are enough for a good representation of H.

- Do the variational procedure with the MPO representation of H for the initial MPS state $|\Psi_0\rangle$. In the density matrix correction $\widetilde{\rho} = \rho + \alpha\Delta\rho$ [21], α should be a small number (we choose $\alpha = 10^{-5}$ in the following). Eigenstates of the corrected reduced density matrix $\widetilde{\rho}$ with eigenvalues larger than η are kept. The smallest and largest allowed number of kept states can also be set. When the energy goes up, the density matrix correction is switched off. After the energy converges, we get a candidate of the ground state of H.
- Try different values of α and η to make sure we are not trapped in local energy minimum.

4. Comparison with exact diagonalization

In this section, we study a system of $N_e = 12$ electrons at filling $\nu = 1/3$ with $N_s = 3N_e - 2 = 34$ and $L = 20$, which is easily accessible for ED. By comparing the MPS results with ED, we demonstrate that there are two error sources in our MPS code: one is the quality of the MPO representation of the Hamiltonian, the other is the density matrix truncation.

We use Haldane's $\mathcal{V}_1$ pseudopotential [25] as the Hamiltonian, for which

$$V_{m,n} = -\frac{16\pi^2\sqrt{2\pi}}{L^3} m(m+2n) e^{-\frac{2\pi^2}{L^2}[(m+n)^2+n^2]}. \tag{2}$$

The ground state in the $(N_e, K = 3N_e(N_e - 1)/2)$ sector is the exact Laughlin state with exactly zero ground energy. In Fig. 2, we use the orbital-cut entanglement spectrum [26], which is a fingerprint of the topological order in FQH states, to analyze the error sources of our MPS algorithm. We observe two kinds of difference between the entanglement spectra obtained by

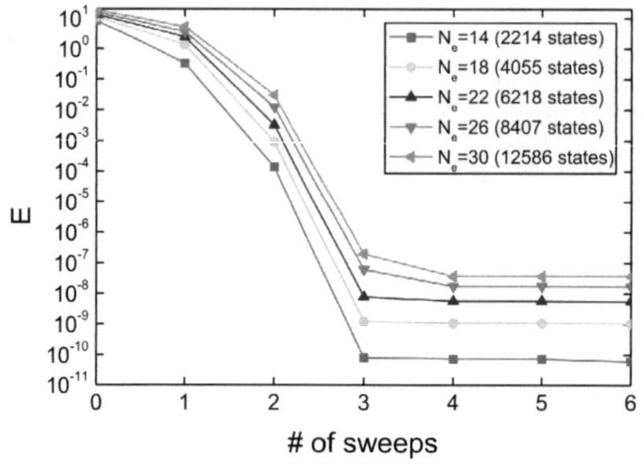

Figure 3. The ground energy at $\nu = 1/3$ with $\Lambda = 12, \varepsilon = 10^{-8}, \eta = 10^{-13}$ versus the number of sweeps obtained our MPS algorithm for various system sizes. We consider square samples, namely the circumference of the cylinder $L = \sqrt{2\pi N_s}$. The maximal number of kept states is given in the bracket for each system size.

ED and MPS algorithm, as indicated by the green circles in Fig. 2(a). The extra levels in the left circle are caused by the relatively poor quality (large ε) of the MPO representation of the Hamiltonian, and the missing levels in the right circle are caused by the relatively large density matrix truncation error η. After improving the quality of the MPO representation (reduce ε) and increase the accuracy of the density matrix truncation (reduce η), we find the difference between MPS and ED results essentially disappears, as shown in Fig. 2(b).

5. Performance of our MPS code

We now apply our MPS code to systems at $\nu = 1/3$ with sizes beyond the ED limit. Again, we choose the interaction between electrons as $\mathcal{V}_1$ pseudopotential, so the ground state in the $(N_e, K = 3N_e(N_e - 1)/2)$ sector is the exact Laughlin state with exactly zero ground energy.

In Fig. 3, we fix the accuracy of the MPO (Λ and ε) and the density matrix truncation (η), and study the convergence of the ground energy on various square samples with $L = \sqrt{2\pi N_s}$. We choose the root configuration at $\nu = 1/3$ [27] as the initial state $|\Psi_0\rangle$. The number of kept states in the density matrix truncation is controlled by η and changes during the sweeps. This is a little different from the usual DMRG algorithm where a fixed number of kept states is usually set at the beginning. However, in our MPS code, we can still track the number of kept states in each sweep and select the maximal one, which is also shown in Fig. 3. With the increase of the system size from 14 electrons to 30 electrons, the entanglement in the system grows due to the increase of circumference L. Thus the computational cost of the MPS simulation also increases, reflected by the fact that the maximal number of kept states grows by a factor of 6. For the largest system size (30 electrons), our computation takes 19 days by 15 CPU cores on a computer cluster with 512GB memory. The final energy that we obtain is very close to the theoretical value 0 for each system size, but grows from roughly 10^{-11} for 14 electrons to 10^{-8} for 30 electrons. This is because higher accuracy is needed for larger system size and cylinder circumference.

We also want to compare our MPS code with the traditional two-site DMRG code. We consider 20 electrons at $\nu = 1/3$ with $\mathcal{V}_1$ interaction and study the convergence of the ground energy for different density matrix truncations η (Fig. 4). The number of kept states in the two-site DMRG algorithm is set to be equal with the maximal number of kept states in the MPS sweeps. With the decrease of η, the ground energy for both of the MPS algorithm and

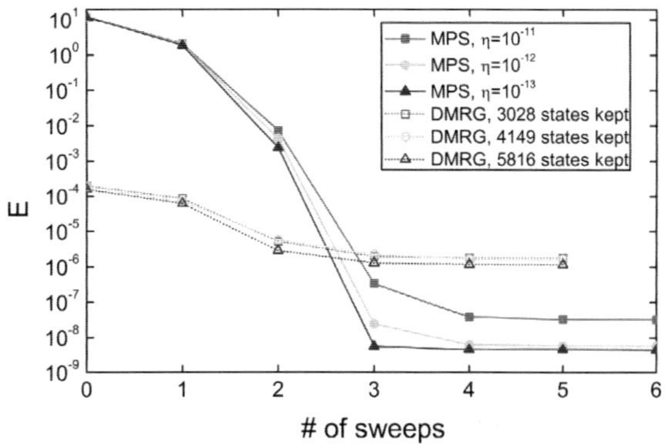

Figure 4. The ground energy of $N_e = 20$ electrons at $\nu = 1/3$ with $L = 20, \Lambda = 12, \varepsilon = 10^{-8}$ versus the number of sweeps for our MPS algorithm and traditional two-site DMRG algorithm. The number of kept states in the DMRG algorithm is set to be equal with the maximal number of kept states in the MPS sweeps.

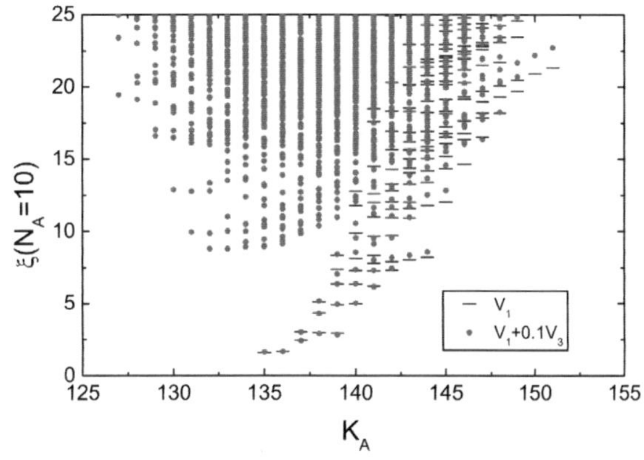

Figure 5. The orbital-cut entanglement spectra obtained by MPS algorithm for pure $\mathcal{V}_1$ (green dashes) and $\mathcal{V}_1 + 0.1\mathcal{V}_3$ interactions (red dots) for $N_e = 20$ electrons at $\nu = 1/3$ with $L = 20, \Lambda = 12, \varepsilon = 10^{-8}, \eta = 10^{-12}$.

two-site DMRG algorithm goes closer to 0. However, the MPS algorithm can reach lower energy than two-site DMRG.

When the interaction goes beyond the pure $\mathcal{V}_1$ pseudopotential, the ground state at $\nu = 1/3$ is no longer the exact Laughlin state with exactly zero ground energy. To achieve this, we use a combination of Haldane's $\mathcal{V}_1$ and $\mathcal{V}_3$ pseudopotentials with $\mathcal{V}_3 = 0.1\mathcal{V}_1$. Considering the strength of $\mathcal{V}_3$ is much smaller than $\mathcal{V}_1$, we expect that the ground state at $\nu = 1/3$ is still in the Laughlin phase, although not the exact Laughlin state. Because $\mathcal{V}_3$ pseudopotential is a longer-range interaction than $\mathcal{V}_1$ pseudopotential, the bond dimension of the MPO is larger than that of pure $\mathcal{V}_1$ pseudopotential. We calculate the ground-state entanglement spectrum of 20 electrons at filling $\nu = 1/3$, and find that the low-lying part indeed matches that of the exact Laughlin state, although generic levels appear in the high energy region (Fig. 5).

6. Discussion

In this work, we here reported a MPS code for finite FQH systems on the cylinder geometry. By comparing with ED and traditional two-site DMRG, we show its capability of searching for the FQH ground states. Compared with the MPS code for infinite FQH systems [19], our code is more suitable to study the physics in finite systems, such as edge effects.

There are several possible directions for the future work. We can generalize our code to the infinite cylinder, as in Ref. [19], where a multi-site update was used. It would be interesting to see the performance of single-site update and density matrix correction in the that case. We can also go beyond short-range Haldane's pseudopotential to deal with some long-range Hamiltonians, such as dipole and Coulomb interactions. However, the bond dimension of MPO increases fast with the interaction range. Therefore, we will need to truncate the interaction differently and study results as a function of the truncation length. Finally, because the only system-dependent part of our code is the MPO representation of the physical Hamiltonian, it can readily be used in other many-body systems such as spin chains and lattice models (such as fractional Chern insulators) [28, 29].

Acknowledgments

We thank M. P. Zaletel, M. C. Banuls, and F. D. M. Haldane for the useful discussion, and Sonika Johri for the DMRG calculation shown in Figure 4. This work was supported by the Department of Energy, Office of Basic Energy Sciences through Grant No. DE-SC0002140.

References

[1] R. B. Laughlin, Phys. Rev. Lett. **50**, 1395 (1983).
[2] G. Moore, and N. Read, Nucl. Phys. B **360**, 362 (1991).
[3] J. M. Leinaas and J. Myrheim, Nuovo Cimento Soc. Ital. Fis. **37B**, 1 (1977).
[4] D. Arovas, J. R. Schrieffer, and F. Wilczek, Phys. Rev. Lett. **53**, 722 (1984).
[5] C. Nayak, S. H. Simon, A. Stern, M. Freedman, and S. Das Sarma, Rev. Mod. Phys. **80**, 1083 (2008).
[6] F. D. M. Haldane Phys. Rev. Lett. **51**, 605 (1983).
[7] S. R. White, Phys. Rev. Lett. **69**, 2863 (1992); Phys. Rev. B **48**, 10345 (1993).
[8] N. Shibata and D. Yoshioka, Phys. Rev. Lett. **86**, 5755 (2001).
[9] E. J. Bergholtz and A. Karlhede, arXiv:cond-mat/0304517.
[10] A. E. Feiguin, E. Rezayi, C. Nayak, and S. Das Sarma, Phys. Rev. Lett. **100**, 166803 (2008).
[11] D. L. Kovrizhin, Phys. Rev. B **81**, 125130 (2010).
[12] J. Zhao, D. N. Sheng and F. D. M. Haldane Phys. Rev. B **83**, 195135 (2011).
[13] Z.-X. Hu, Z. Papic, S. Johri, R. N. Bhatt, and P. Schmitteckert, Phys. Lett. A **376**, 2157 (2012).
[14] M. P. Zaletel and R. S. K. Mong, Phys. Rev. B 86, 245305 (2012).
[15] B. Estienne, Z. Papic, N. Regnault, B. A. Bernevig, Phys. Rev. B **87**, 161112(R) (2013).
[16] B. Estienne, N. Regnault, B. A. Bernevig, arXiv: 1311.2936.
[17] Y.-L. Wu, B. Estienne, N. Regnault, B. A. Bernevig, Phys. Rev. Lett. **113**, 116801 (2014).
[18] U. Schollwoeck, Ann. Phys. (Amsterdam) **326**, 96 (2011).
[19] M. P. Zaletel, R. S. K. Mong, and F. Pollmann, Phys. Rev. Lett. **110**, 236801 (2013).
[20] S. Singh, R. N. C. Pfeifer, and G. Vidal, Phys. Rev. B **83**, 115125 (2011).
[21] S. R. White, Phys. Rev. B **72**, 180403(R) (2005).
[22] G. M. Crosswhite and D. Bacon, Phys. Rev. A **78**, 012356 (2008).
[23] M. P. Zaletel, R. S. K. Mong, F. Pollmann, E. H. Rezayi, arXiv: 1410.3861.
[24] G. M. Crosswhite, A. C. Doherty, and G. Vidal, Phys. Rev. B **78**, 035116 (2008).
[25] F. D. M. Haldane, Phys. Rev. Lett. **51**, 605 (1983).
[26] H. Li and F. D. M. Haldane, Phys. Rev. Lett. **101**, 010504 (2008).
[27] B. A. Bernevig, F. D. M. Haldane, Phys. Rev. Lett. **100**, 246802 (2008).
[28] Z. Liu, D. L. Kovrizhin, E. J. Bergholtz, Phys. Rev. B **88**, 081106(R) (2013).
[29] A. G. Grushin, J. Motruk, M. P. Zaletel, F. Pollmann, arXiv: 1407.6985.

Mott glass phase in a diluted bilayer Heisenberg quantum antiferromagnet

Nv-Sen Ma[1], Anders W. Sandvik[2], and Dao-Xin Yao[1]

[1]State Key Laboratory of Optoelectronic Materials and Technologies, School of Physics and Engineering, Sun Yat-sen University, Guangzhou 510275, People's Republic of China

[2]Department of Physics, Boston University, 590 Commonwealth Avenue, Boston, Massachusetts 02215, USA

E-mail: sandvik@bu.edu, yaodaox@mail.sysu.edu.cn

Abstract. We use quantum Monte Carlo simulations to study a dimer-diluted $S = 1/2$ Heisenberg model on a bilayer square lattice with intralayer interaction J_1 and interlayer interaction J_2. Below the classical percolation threshold p_c, the system has three phases reachable by tuning the interaction ratio $g = J_2/J_1$: a Néel ordered phase, a gapless quantum glass phase, and a gapped quantum paramagnetic phase. We present the ground-state phase diagram in the plane of dilution p and interaction ratio g. The quantum glass phase is certified to be of the gapless Mott glass type, having a uniform susceptibility vanishing at zero temperature T and following a stretched exponential form at $T > 0$; $\chi_u \sim \exp(-b/T^{\alpha})$ with $\alpha < 1$. At the phase transition point from Néel ordered to Mott glass, we find that the critical exponents are different from those of the clean system described by the standard $O(3)$ universality class in 2+1 dimensions.

1. Introduction

The effects of disorder (randomness) on the critical behavior represent an important aspect of quantum phase transitions (QPTs). Disorder brings with it a plethora of profound phenomena beyond those present in clean systems [1, 2], with many open questions of broad interest in quantum many-body physics. Many different kinds of disorders can be considered, e.g., random dilution of the degrees of freedom or random variations in the coupling constants. Among the possible disorder channels in quantum spin systems, random dilution has attracted special attention because of the possibilities to achieve it experimentally by substitution of magnetic ions by non-magnetic ones, e.g., replacing Cu bu Zn or Mg in the high-T_c cuprates [3]. The case is also of particular interest from a theoretical point of view as there is simple notion of classical "geometrical" fluctuations which interplay with the quantum fluctuations [4]. Thus, a multicritical point can appear if those two fluctuations diverge together, as a classical percolation point is approached while at the same time tuning the quantum fluctuations. A prototypical example is the relatively well understood transverse-field Ising model with dimensionality $d > 1$ [4, 5]. Many works have also shown that a multicritical point can be realized in the diluted bilayer Heisenberg model with different intralayer interaction J_1 and interlayer interaction J_2, if the dilution is accomplished by random removal of entire dimers (instead of diluting individual spins, which leaves behind "dangling" spins which lead to different physics) [6, 7]. The bilayer model is illustrated in Fig. 1 and is further studied in this paper.

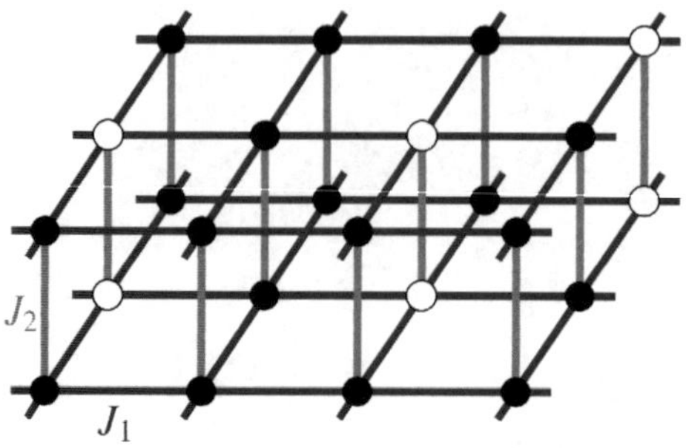

Figure 1. The bilayer Heisenberg model with intraplane coupling J_1 (blue bonds) and interplane coupling J_2 (red vertical bonds). The open circles stand for removed dimers (i.e., the two spins are removed, not just the coupling between them). By tuning the coupling ratio $g = J_2/J_1$ a QPT can be reached for any dilution fraction $p < p_c$, with $p_c \approx 0.407$ [8] being the classical percolation point.

According to previous studies, at the classical percolation point, $p_c \approx 0.407$, the percolating cluster remains ordered until the quantum critical point located at $g_c^* = 0.118(6)$ [9, 6]. In this paper, we study the QPT when $p < p_c$ and obtain the phase diagram at $T = 0$ (shown in Fig. 2) using the efficient stochastic series expansion (SSE) quantum Monte Carlo (QMC) method [10]. While this criticality has been studied previously [6, 7], we here re-examine it in light of recent studies [11, 12] of other, similar quantum phase transitions where no changes in the critical exponents were found in the presence of disorder—in violation of the Harris criterion [13] for the relevance of disorder. In addition, we study "quantum glass" (Griffiths) behavior away from the critical curve. Such phases come in two main variants in random quantum spin systems and boson systems: the Bose glass (BG) and Mott glass (MG) [14]. Both of them are gapless, with the BG having a non-zero uniform susceptibility χ_u (compressibility in Boson language, where the magnetization of a spin system is mapped onto the density of the Boson system) at $T = 0$ while the MG has a vanishing χ_u at $T = 0$ [15]. It is believed that the MG can only exist in random systems with particle-hole symmetry [16], though recently evidence has been found of similar behavior also without this symmetry (an MG or and effective MG with extremely small essentially undetectable $T = 0$ compressibility) [17]. In our previous work, the MG phase was investigated in a two-dimensional square-lattice dimerized Heisenberg model with three different nearest-neighbor couplings [12] and the suscetibility was found to follow a stretched exponential form. We here find the same characteristic form in the diluted bilayer.

2. Model and critical behavior

The Hamiltonian of the $S = 1/2$ spin model illustrated in Fig. 1 is give by

$$H = J_1 \sum_{a=1,2} \sum_{\langle ij \rangle} \mathbf{S}_{i,a} \cdot \mathbf{S}_{j,a} + J_2 \sum_i \mathbf{S}_{i,1} \cdot \mathbf{S}_{i,2}, \qquad (1)$$

where $a = 1$ (2) stands for the upper (lower) plane of the bilayer system. Dimers of spins, one in the upper layer and one in the lower layer (coupled by the intra-dimer coupling J_2), are removed at random with probability $p < p_c$. We normalize by setting the intra-plane interaction $J_1 = 1$. In the clean system, $p = 0$, the system goes through a QPT between Néel to quantum disordered phase at the critical point located at $g_c(p = 0) = 2.525(2)$ [18]. The critical exponents are the same as the classical $3D$ Heisenberg model, i.e., the 3D O(3) universality class. When $g = 0$, the model turns into two separate diluted $2D$ square lattice Heisenberg and at the classical

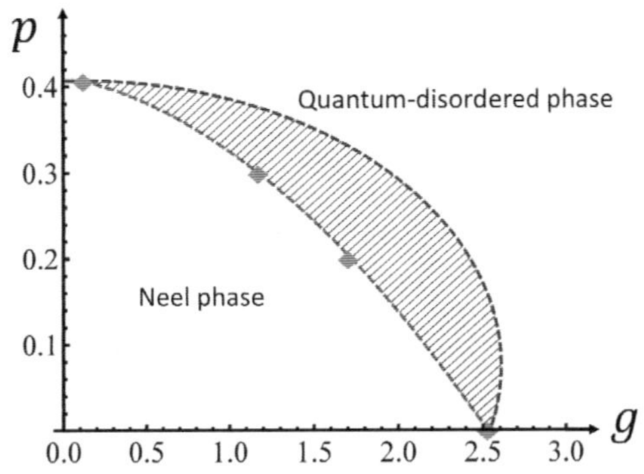

Figure 2. Phase diagram of the diluted bilayer Heisenberg model. The shaded area stands for the MG phase existing between the Néel and quantum-disordered phases.

percolation point $p_c \approx 0.407$ the percolating cluster remains ordered at $T = 0$. This order persists in the coupled layers as well up to $g_c(p = p_c) \approx 0.11$, which is a multi-critical point with simultaneous critical geometrical (percolation) fluctuations and critical quantum fluctuations of the spins on those classically critical clusters.

We here consider the system between the special points $g_c(p = 0)$ and $g_c(p = p_c)$, on the critical curve as well as for larger g, in the shaded glass region in the phase diagram of Fig. 2. This is a preliminary report of results of SSE QMC calculations on lattices as large as $N = 2 \times 64 \times 64$ sites, averaged over 100 dilution configurations in each case. For a given disorder realization we identify the largest cluster and only use it in the simulations. We take advantage of the "β-doubling" technique [19] to quickly equilibrate the system to the very low temperatures needed to study the ground state.

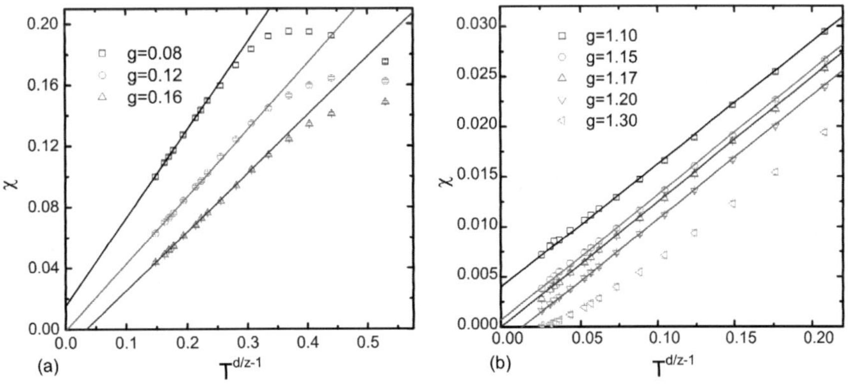

Figure 3. Analysis of the uniform susceptibility delivering the dynamic exponent by adjusting the value of z for the best linear behavior of χ_u versus $T^{2/z-1}$ for coupling ratios g close to the QPT. In (a) $p = p_c = 0.407$ while in (b) $p = 0.3$. At p_c the fitted line passes through $(0, 0)$ when $g \approx 0.12$ and $z \approx 1.36$. For $p = 0.3$ we observe linear behavior when $z \approx 1.14$ and locating the zero intercept (indicating the QPT) gives $g_c \approx 1.17$.

3. Dynamic exponent

Close to a quantum-critical point, the uniform susceptibility in the thermodynamic limit should have the following characteristic form [20]:

$$\chi_u = a + b/T^{d/z-1}, \quad (2)$$

where a and b are constants, with $a = 0$ at g_c, and d is the dimensionality which at p_c in the cluster fractal dimensionality, $d = 91/48$ [8], and for $p < p_c$ $d = 2$. We calculate χ_u in the simulations using the standard expression

$$\chi_u = \frac{\beta}{N}\left\langle \left(\sum_{i=1}^{N} S_i^z\right)^2 \right\rangle. \quad (3)$$

To test the form (2) we follow the T-dependence for different system sizes and only report results down to the T for which good convergence with L has been achieved.

From Eq. (2), we can see that χ has a linear relation with $T^{d/z-1}$. If z is not known, we can use it as an adjustable parameter and optimize it for linear scaling. In addition, when a line fitted to the data has vanishing intercept a the g value will be the critical one. Results close to the multicritical point (p_c,g_c) are shown in Fig. 3(a). We here use $z = 1.36$ as obtained in Ref. [9] and see that the critical $g_c \approx 0.12$, which is consistent with the previous results. This procedure has not previously been considered for $p < p_c$ and we show results at $p = 0.3$ in Fig. 3(b), where we use $d = 2$ and adjust z to achieve a linear behavior. We again observe remarkably good scaling for several g values, but now with $z = 1.14(2)$. The critical point is at $g_c \approx 1.17$ for $p = 0.3$. Note that a cross-over away from the linear behavior can be seen for $g > g_c(p)$ but for $g < g_c$ we do not observe any cross-overs at the temperatures studied. At p_c it has been speculated that the scaling may hold all the way down to $T = 0$ because the percolating cluster has vanishing stiffness (which sets the scale of the cross-over) [9], but for $p < p_c$ there must eventually be a cross-over because the stiffness is non-zero.

We have also studied the critical behavior at $p = 0.2$ and find the critical point $g_c \approx 1.70$ with $z = 1.08(3)$. Considering the error bars, the z-value is not very different from the value obtained at $p = 0.3$, and the most likely scenario is that z is the same but different from 1 for all $p < p_c$, which would agree with Ref. [7], where the critical exponents were found not not change with p in the classical model. In that study, a scaling correction was used because the fitting curve deviates significantly from the pure power-law behavior. However, in our case the power-law fitting curves are of high quality with $\chi^2 \approx 1$ for $g = 1.17$ in Fig. 3. The dynamic exponent is also quite different from the random-coupling square lattice Heisenberg model, where $z = 1$ always appears to hold [12]. This implies that the two kinds of disordered systems have quite different low-energy physics systems, which would not be expected at first sight and for which there is no theoretical explanation as of yet.

The Harris criterion can normally be used to determine the relevance of disorder at classical phase transitions. It says that the correlation length exponent ν in the clean system has to fulfill $\nu \geq 2/d$ if disorder is irrelevant. Otherwise, the critical exponents will change when disorder is introduced in order to satisfy the inequality [13]. Although it has been thought to be applicable to QPTs, several disordered systems have shown violations of the Harris criterion. [12, 11] The dimensionality d for a QPT in Harris criterion is the number of dimensions which have disorder. Thus, in this model $d = 2$ and disorder is expected to be relevant to the clean bilayer antiferromagnet which has $\nu \approx 0.705 < 1$, as we have found here (unlike the system in Ref. [12]). Although we have not yet computed ν, we note that the change in z indicates a new universality class.

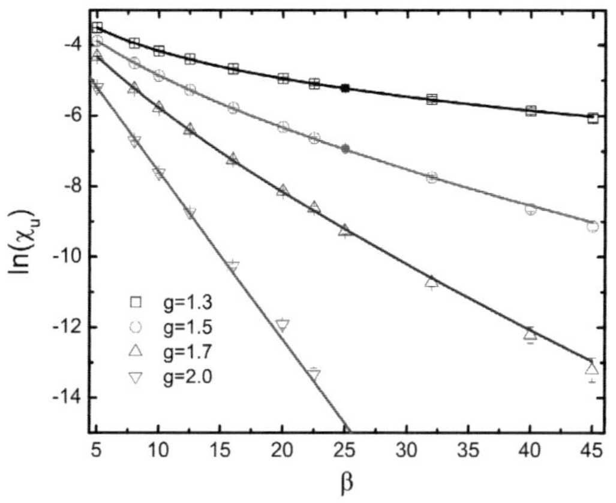

Figure 4. Inverse-temperature dependence of the logarithm of the magnetic susceptibility at $p = 0.3$ and g larger than the critical value. When $g = 1.3, 1.5$ and 1.7, the behaviors (fitted curves) follow the form (5) with $\alpha = 0.21(1)$, 0.58(1), and 0.71(2), respectively. When $g = 2$, the linear fitting illustrates the standard decay of the form (4) with a gap Δ.

4. Mott-Glass Phase

When disorder is introduced in, a new phase, often called a "quantum glass" should emerge. The uniform susceptibility χ_u should approach a nonzero value at low T in Néel phase, while in the gapped quantum-disordered phase

$$\chi_u \sim \exp\left(-\frac{\Delta}{T}\right), \tag{4}$$

with the spin gap Δ. If the system is in the gapless quantum glass phase of the Mott type, χ_u vanishes at zero temperature, and the following form has been found empirically:

$$\chi_u \sim \exp\left(-\frac{b}{T^\alpha}\right), \tag{5}$$

where $\alpha = 1/2$ in a diluted $S = 1$ system [21, 22] and $0 < \alpha < 1$ in a disordered $S = 1/2$ Heisenberg square lattice. [12] Accordingly, the temperature dependence of $\ln(\chi_u)$ is a useful quantity to distinguish different ground states and test the existence of the MG state. The results for $p = 0.3$ when $g > g_c$ are presented in Fig. 4. It can be observed that χ_u follows the form (5) with $0 < \alpha < 1$ when $g > g_c$ but not far away from the critical point. This implies that the MG phase exists in this bond diluted bilayer Heisenberg antiferromagnet. The form (5) also holds $p = 0.2$.

5. Summary

We have studied the critical behavior of a dimer-diluted bilayer $S = 1/2$ Heisenberg model below the classical percolation point using the SSE Monte Carlo simulation method. Based on our calculations, we obtain the phase diagram shown in Fig. 2. For $p = 0.3$, the QPT at zero temperature occurs at $g \approx 1.17$ and the uniform susceptibility changes with temperature by following the power law form $\chi = a + b/T^{d/z-1}$, with $d = 2$. Below the percolation threshold, the dynamic exponent $z = 1.18(3)$ was found. It is different from the clean case, where $z = 1$, and consistent with the Harris criterion, according to which the disorder of bond dilution is relevant to the system and the new critical exponents different from the clean case could appear. We have also calculated z for other disorder strengths and found no significant changes with p, though the claculation becomes difficult for $p < 2$ due to slow approach (with the system size) to the asymptotic critical scaling behavior. From the analysis of critical points, we find the gapless MG phase with vanishing χ_u above the critical g_c for $p < p_c$, where a stretched exponential relationship of χ versus T holds.

Acknowledgments

We would like to thank Wenan Guo for stimulating discussions. NSM and DXY acknowledge support from National Basic Research Program of China (2012CB821400), NSFC-11275279, RFDPHE of China (20110171110026), Fundamental Research Funds for the Central Universities of China, and NCET-11-0547. The work of AWS was supported by the NSF under Grants No. DMR-1104708 and DMR-1410126.

References

[1] J. P. Attfield, A. L. Kharlanov, and J.A.McAllister, *Nature* **394**, 157 (1998).
[2] B.Huchestein, *Rev. Mod. Phys.* **67**, 357 (1995).
[3] O.P. Vajk *et al.*, *Science* **295**, 1691 (2002).
[4] T. Senthil and S. Sachdev, *Phys. Rev. Lett* **77**, 5392 (1996).
[5] R. R.dos Santos, *J.Phys.*C **15**,3141 (1982).
[6] O. P. Vajk and M. Greven, *Phys. Rev. Lett.* **89**, 177202 (2002).
[7] R. Sknepnek, T. Vojta, and M. Vojta, *Phys. Rev. Lett.* **93**, 097201 (2004).
[8] D. Stauffer and A. Aharony, *Introduction to Percolation Theory* (Taylor and Francis, London, 1991).
[9] A. W. Sandvik, *Phys. Rev. Lett.* **96**, 207201 (2006).
[10] A. W. Sandvik, *Phys. Rev.* B **56**, 11678 (1997), *ibid* **59**, R14157(R) (1999).
[11] D. X. Yao, J. Gustafsson, E. W. Carlson, and A. W. Sandvik,*Phys. Rev.* B **82**, 172409 (2010).
[12] N. Ma, A. W. Sandvik and D. Yao, *Phy. Rev.* B **90**, 104425 (2014).
[13] A. B. Harris, *J. Phys.* C **7**, 1671 (1974).
[14] K. G. Balabanyan, N. Prokofe'v, and B. Svistunov, *Phys. Rev. Lett.* **95**,055701 (2005).
[15] T. Giarmarchi, P. Le Doussal, and E. Orignac, *Phys. Rev.* B **64**, 245119 (2001).
[16] N. Prokof'ev and B. Svistunov, *Phys. Rev. Lett.* **92**, 015703 (2004).
[17] Y. Wang, W. Guo, and A. W. Sandvik, *Phys. Rev. Lett.* 114, 105303 (2015).
[18] L. Wang, K. S. D. Beach, and A. W. Sandvik, *Phy. Rev.* B **73**, 014431 (2006).
[19] A. W. Sandvik, *Phy. Rev.* B **66**, 024418 (2002).
[20] A. V. Chubukov, S. Sachdev, and J. Ye, *Phy. Rev.* B **49**, 11919 (1994).
[21] T. Roscilde and S. Haas, *Phys. Rev. Lett.* **99**, 047205 (2007).
[22] R. Yu, T. Roscilde, and S. Haas, *Phys. Rev. Lett.* **94**, 197204 (2005); Phys. Rev. B **73**, 064406 (2006)

Entropy and thermopower in the 2D t-J model

W O Putikka

Physics Department, The Ohio State University, 1760 University Dr, Mansfield, OH 44906, USA

E-mail: putikka.1@osu.edu

Abstract. The entropy of the two-dimensional t-J model is investigated using its 12th order high temperature series. A direct Padé extrapolation of the entropy series doesn't converge well for temperatures below $T \sim J$. The series coefficients are exact polynomials so the series convergence can be improved by modifying the series that is extrapolated. By subtracting a scaled version of the series for the entropy of the Heisenberg antiferromagnet from the t-J entropy series the low temperature convergence is greatly improved. Using this technique results are obtained for the full range of electron densities and temperatures. The electron density is an adjustable parameter in the series coefficients allowing the density dependence of the entropy and the density derivative at fixed temperature $\partial S/\partial n|_T$ to be determined accurately. The density derivative depends strongly on temperature, unlike noninteracting models. The density derivative is also an approximation to the experimentally measured thermopower.

1. Introduction

The two-dimensional t-J model has been studied for many years as a model for the copper oxide planes found in high temperature superconductors. Despite considerable effort [1] the basic thermodynamic properties of this model are still not well understood. The calculation reported here investigates the entropy by means of its high temperature series calculated to 12th order in inverse temperature β. In principle, knowing the full temperature dependence of the entropy and the value of the free energy at one temperature is sufficient to fully determine the free energy and thus all the thermodynamic properties of the model. This more general calculation is left for a future publication. In this paper our attention is restricted to the entropy per site S and its density derivative at fixed temperature $\partial S/\partial n|_T$.

The t-J model Hamiltonian is given by

$$H = -t \sum_{\sigma\langle ij\rangle} \left(c^\dagger_{i\sigma} c_{j\sigma} + c^\dagger_{j\sigma} c_{i\sigma} \right) + J \sum_{\langle ij\rangle} \vec{S}_i \cdot \vec{S}j, \tag{1}$$

along with the constraint of no double occupancy. The high temperature series for the Helmholtz free energy per site is calculated by a linked cluster expansion, taking the form

$$\beta F = \sum_{i=0}^{N} f_i (\beta J)^i, \tag{2}$$

where the $f_i(n, t/J)$ are exact polynomials in terms of the electron density n and the ratio of the model coupling constants t/J. For the calculation reported here $N = 12$ and we fix $t/J = 2.5$.

The series for the entropy per site is also calculated in terms of the f_i as

$$S = k_{\rm B} \sum_{i=0}^{N} (i-1) f_i (\beta J)^i \tag{3}$$

and the energy per site is

$$\beta E = \sum_{i=0}^{N} i f_i (\beta J)^i. \tag{4}$$

2. Details of the Calculation

To reach low temperatures, the series expansion for the entropy needs to be extrapolated by Padé approximants. With the available 12th order series the direct extrapolation of the t-J entropy only converges for $T \gtrsim J$. A low temperature scale remains unresolved in this calculation. To improve the convergence we can try to extract this low temperature scale by forming the series for

$$\Delta S(n,T) = S_{tJ}(n,T) - n^*_{AF} S_{AF}(J^*,T) - \alpha^* S_{TB}(n,T), \tag{5}$$

where S_{AF} is the entropy of the Heisenberg antiferromagnet (found by setting $n = 1$ in the t-J entropy), S_{TB} is the tight-binding model entropy and the starred parameters are adjusted to produce a function ΔS with a simple temperature dependence that extrapolates well to low temperatures. A least squares fit of the [6/6] Padé to a fourth order polynomial is calculated to extend ΔS to $T = 0$. For $n \geq 0.55$ a reasonable ΔS can be found by setting $n^*_{AF} = n$, $\alpha^* = 0$ and adjusting J^*. To extract S_{tJ} we need to know $S_{AF}(T)$ which is determined by

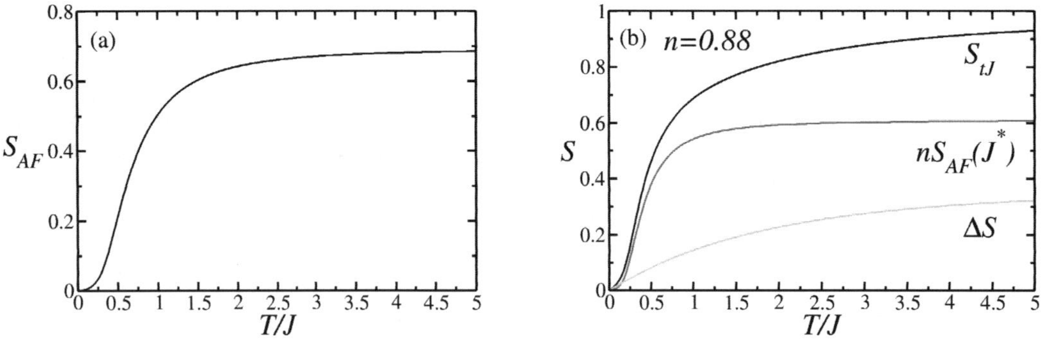

Figure 1. a) Heisenberg model entropy vs. temperature. b) An example of the fitting procedure described in the text for the t-J model entropy at $n = 0.88$. The t-J entropy is the sum $S_{tJ} = nS_{AF}(J^*) + \Delta S$, with $J^* = 0.638173J$.

using a combination of high precision Monte Carlo data [2, 3] and series expansion results for $E_{AF}(T)$. The entropy is found from the energy by using the thermodynamic relations $E = F + TS$ and $F_{AF}(T) = E_{0AF} - \int_0^T S_{AF}(T')dT'$. The integal is evaluated by the trapezoidal rule with step size $\Delta T = 10^{-5}J$ and the temperature dependence of the energy is given by $\tilde{E}_{AF}(T) = E_{AF}(T) - E_{0AF}$ with ground state energy $E_{0AF} = -0.669437J$ [4]. Using these expressions the temperature dependence of the entropy $S_{AF}(T)$ is determined iteratively by

$$S_{AF}(j\Delta T) = \frac{2\tilde{E}_{AF}(j\Delta T)}{(2j-1)\Delta T} + \frac{2}{(2j-1)} \sum_{i=1}^{j-1} S_{AF}(i\Delta T), \tag{6}$$

where j runs from 1 to 500,000 to cover the range $0 \leq T \leq 5J$ and $\tilde{E}_{AF}(0) = S_{AF}(0) = 0$. Fig(1a) shows the result for the temperature dependence of $S_{AF}(T)$. Using the relation above between $F_{AF}(T)$, $S_{AF}(T)$ and the value $F_{AF}(5J) = -3.5043614J$ determined from the series expansion the ground state energy found by integrating $S_{AF}(T)$ given by Eq(6) is $E^{est}_{0AF} = -0.66947J$ and $E^{est}_{0AF} - E_{0AF} = -3.3 \times 10^{-5}J$, comparable to the error estimates in the Monte Carlo data [2, 3].

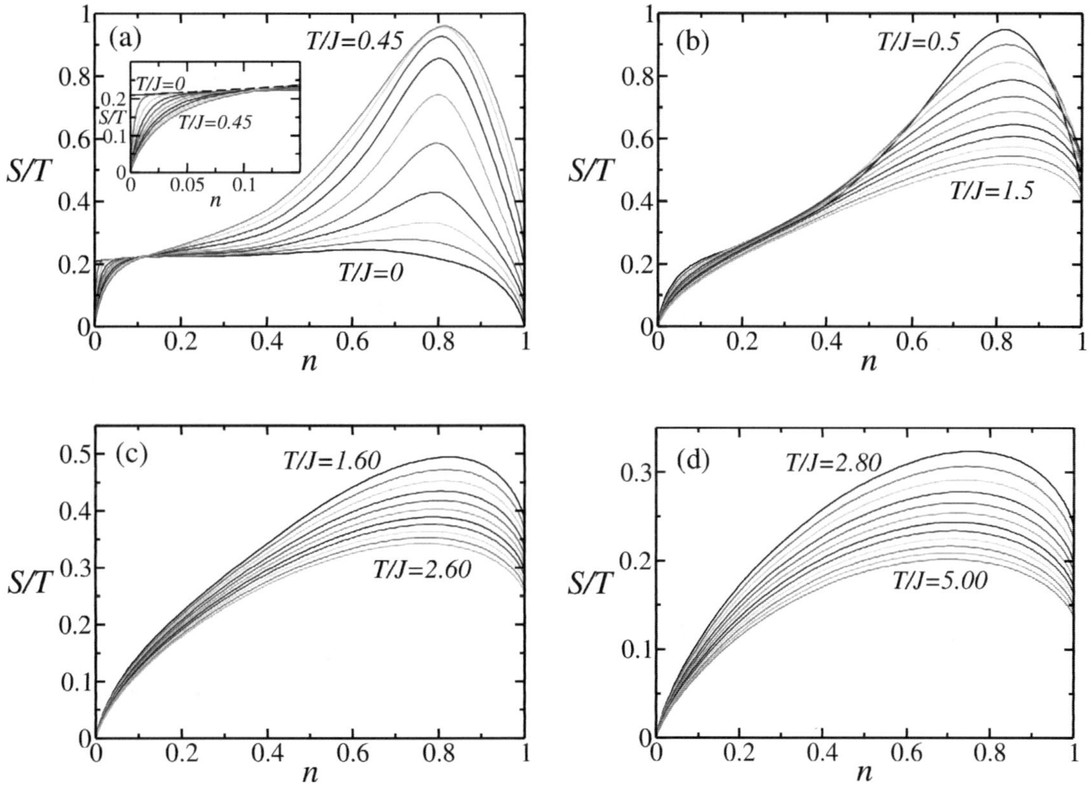

Figure 2. A selection of data plotted as S/T vs. n. a) $0 \leq T/J \leq 0.45$ in steps of 0.05. The dashed line in the inset is the tight-binding model γ. b) $0.5 \leq T/J \leq 1.5$ in steps of 0.1. c) $1.6 \leq T/J \leq 2.6$ in steps of 0.1. d) $2.8 \leq T/J \leq 5$ in steps of 0.2.

The temperature dependence of the Heisenberg entropy $S_{AF}(T)$ can be used as a known function to reverse the ΔS calculation and extract $S_{tJ}(n, T)$. Fig(1b) gives an example of this procedure for $n = 0.88$. Two temperature scales emerge from this analysis: a high temperature scale $T \sim t$ and a low temperature scale $T \sim J^*$. For $n < 0.55$ the t-J entropy crosses over to the tight-binding model entropy and all the starred parameters in Eq(5) are required. A complete discussion of this more complicated fitting procedure at low n is deferred to a future publication, though here we note to order n the series coefficients for the t-J model free energy are exactly the same as the tight-binding model free energy so as $n \to 0$ we have $S_{tJ} \to S_{TB}$ at all temperatures.

3. Results

The extrapolation in temperature to find $S_{tJ}(n, T)$ is done separately for each n. For $0.10 \leq n \leq 0.92$ the spacing of densities is $\Delta n = 0.01$, while tighter spacing is used for the highest and lowest density ranges. Data were accumulated for 226 densities, with 5,000 temperature values for each density with a uniform spacing of $\Delta T = 0.001J$. A selection of the entropy data is shown in Fig(2), with S_{tJ}/T plotted as a function of n for a range of temperatures.

From the entropy data an interesting quantity to calculate is the density derivative at fixed temperature $\partial S/\partial n|_T$. This derivative has been investigated in earlier calculations [1], but here we have greatly improved density resolution allowing a more detailed investigation than before. The entropy initially grows upon doping away from half filling and at high temperatures the configurational entropy goes through a maximum at $n = 2/3$. For noninteracting systems the configurational entropy maximum remains true for all temperatures. For example, the maximum entropy for the tight-binding model is found at $n = 1$ for all temperatures. Even for hard core bosons the entropy maximum remains fixed at $n = 0.5$ for all temperatures. The entropy maximum for the t-J model is strongly temperature dependent as shown in Fig(3). As the temperature decreases from $T \sim 5J$ to $T \sim J$ the entropy maximum moves up to

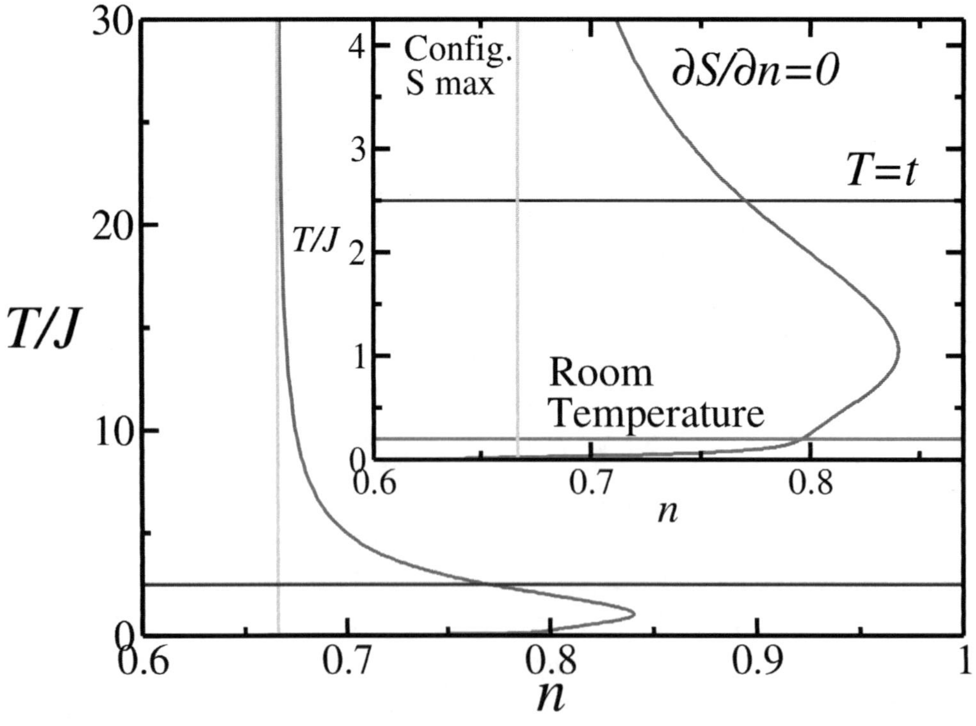

Figure 3. Red line: $\partial S/\partial n = 0$ on a temperature vs. density plot. Green line: configurational entropy maximum at $n = 2/3$, independent of temperature. Blue line: approximate temperature scale $T = t$ for the deviation of $\partial S/\partial n = 0$ from the configurational entropy maximum. Magenta line: $T = 0.2J$, approximately room temperature where $\partial S/\partial n = 0$ at $n = 0.796$.

$n = 0.84$. In this temperature range the interactions in the t-J model are decreasing the

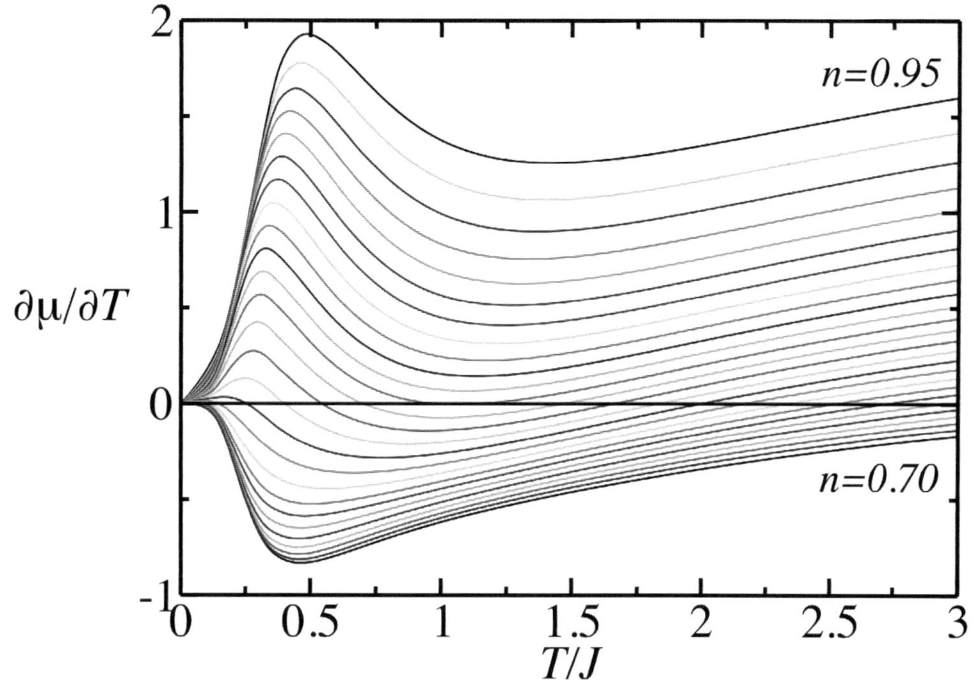

Figure 4. Plot of $\partial\mu/\partial T$ vs. temperature for the density range $0.7 \leq n \leq 0.95$. The spacing of the densities is 0.01.

entropy at higher densities than would be expected from the configurational entropy. This is an indication of correlations developing in this density and temperature range. The temperatures are too high for these to be antiferromagnetic correlations and the density and temperature ranges don't match the pseudogap found in high temperature superconductors. For $T \lesssim J$ the entropy maximum reverses and moves back to lower densities, with the temperature dependence decreasing sharply down to $n = 0.8$ and a low temperature tail below $T = 0.2J$ extending down to $n = 0.62$. Shastry [5] has shown that $\partial S/\partial n$ is an approximation for the thermopower. In experiments [6, 7] on cuprate superconductors the thermopower is observed to change sign as a function of doping. In particular the thermopower is zero at room temperature for 23% doping in many cuprate superconductors [6, 7]. For $T = 0.2J$ (corresponding to room temperature) the t-J model entropy maximum is at 20.4% doping.

Further information on the correlations developing at lower temperatures in the t-J model can be found by considering the full temperature and density dependence of $\partial S/\partial n$. By the equality of the mixed partial derivatives of the free energy we also have

$$\left.\frac{\partial \mu}{\partial T}\right|_n = -\left.\frac{\partial S}{\partial n}\right|_T, \tag{7}$$

where μ is the chemical potential. The relation in Eq(7) provides another means to interpret the derivative. Fig(4) shows $\partial\mu/\partial T$ as a function of temperature for the range of densities where $\partial S/\partial n = 0$. The low temperature positive peak in $\partial\mu/\partial T$ for $n \gtrsim 0.8$ is due to the attractive antiferromagnetic interaction between oppositely oriented spins decreasing the chemical potential at low temperatures. It costs less energy to add an electron to the system when the temperature falls below the effective spin interaction energy. For $n \lesssim 0.8$ in Fig(4) the opposite happens: a fairly broad negative peak in $\partial\mu/\partial T$ develops. Below this peak it costs more energy to add an electron to the system than for temperatures above the peak. For the full range of densities shown in Fig(4) the high temperature limit of $\partial\mu/\partial T$ is positive.

4. Conclusions

The strong temperature dependence of $\partial S/\partial n = 0$ shows the interactions in the t-J model are producing competing correlations. For $n \gtrsim 0.84$ antiferromagnetic fluctuations are dominant at low temperatures. For $n \lesssim 0.8$ the dominant fluctuations at low temperature are d-wave pair fluctuations as shown in Ref. 8. For densities $0.8 \lesssim n \lesssim 0.84$ there is a crossover between the two dominant fluctuations. The different fluctuations also give very different temperature dependences to $\partial\mu/\partial T$. The antiferromagnetic fluctuations reduce the energy needed to add an electron at low temperatures while the d-wave pair fluctuations increase the energy needed to add an electron at low temperatures.

Acknowledgments

The author acknowledges the support of the Ohio Supercomputer Center.

References

[1] Jaklič J and Prelovšek P 2000 *Adv. in Phys.* **49** 1
[2] Sandvik A 2010 *AIP Conf. Proc.* **1297** 135
[3] Sandvik A 2011 *Preprint* arXiv:1101.3281
[4] Sandvik A W 1997 *Phys. Rev.* B **56** 11678
[5] Peterson M R and Shastry B S 2010 *Phys. Rev.* B **82** 195105
[6] Obertelli S, Cooper J R and Tallon J L 1992 *Phys. Rev.* B **46** 14928
[7] Honma T and Hor P H 2008 *Phys. Rev.* B **77** 184520
[8] Putikka W O and Luchini M U 2006 *Phys. Rev. Lett.* **96** 247001

Deciding the fate of the false Mott transition in two dimensions by exact quantum Monte Carlo methods

D. Rost[1,2] and N. Blümer[1]

[1] Institute of Physics, Johannes Gutenberg University, Mainz, Germany
[2] Grad. School Materials Science in Mainz, Johannes Gutenberg University, Mainz, Germany

E-mail: rostda@uni-mainz.de

Abstract. We present an algorithm for the computation of unbiased Green functions and self-energies for quantum lattice models, free from systematic errors and valid in the thermodynamic limit. The method combines direct lattice simulations using the Blankenbecler-Scalapino-Sugar quantum Monte Carlo (BSS-QMC) approach with controlled multigrid extrapolation techniques. We show that the half-filled Hubbard model is insulating at low temperatures even in the weak-coupling regime; the previously claimed Mott transition at intermediate coupling does not exist.

1. Introduction

Numerous studies of the Hubbard model [1] have greatly enhanced our understanding of strongly correlated electron systems within the last four decades. Specialized methods, namely the semi-analytic Bethe ansatz and the density matrix renormalization group (DMRG) [2], yield reliable high-precision results (only) in the case of one spatial dimension. Conversely, the dynamical mean-field theory (DMFT) [3, 4] provides deep insight in the limit of high dimensionality. However, the intermediate regime of two (and three) dimensions is less well understood, e.g., with respect to pseudogap physics [5–7] and high-T_c superconductivity. In this paper, we will address another open question: the nature of the Mott metal-insulator transition (MIT) [8,9] of the half-filled Hubbard model in two dimensions ($D = 2$).

As indicated by shaded regions in Fig. 1, the Hubbard model is insulating at half filling (1 electron per site) and sufficiently strong coupling, in any dimension, while it is metallic at weak coupling – as long as the translational symmetry is not broken, e.g., by magnetic order. For $D = 3$, one finds antiferromagnetic (AF) ordering [10], which implies insulating behavior, below the Néel temperature T_N (gray dash-dotted line). Such long-range order is ruled out by the Mermin-Wagner theorem in $D = 2$ (at temperature $T > 0$). Single-site DMFT predicts a MIT with critical interactions $2.3 \lesssim U_c \lesssim 2.9$ (pink dotted line). This transition line is shifted to weaker interactions within cluster DMFT (CDMFT), which includes short-range correlations on small clusters (violet solid line) [11]. However, the low-T metallic phase implied by the CDMFT result is incompatible with several earlier studies highlighting effects of "short-range antiferromagnetism" [7,12,13]. To clarify the situation, we apply the Blankenbecler-Scalapino-Sugar quantum Monte Carlo (BSS-QMC) [14] approach with controlled multigrid extrapolation techniques to selected points in the phase diagram (green crosses). We will show that these points are separated by a MIT line (or narrow crossover) and conclude that the $2D$ Mott physics is quite similar to the $3D$ case [15], in spite of the Mermin-Wagner theorem.

Journal of Physics: Conference Series **640** (2015) 012047 doi:10.1088/1742-6596/640/1/012047

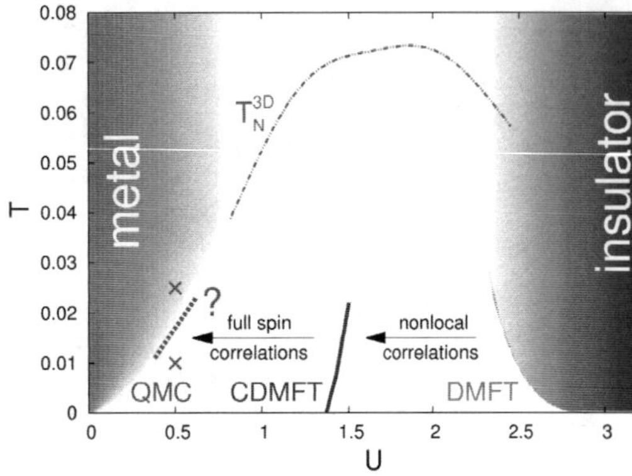

Figure 1. Schematic representation of the MIT in the Hubbard model at half filling on a square lattice. Green crosses: parameters selected for our unbiased QMC study. Gray dash-dotted line: Néel temperature for the cubic lattice (rescaled).

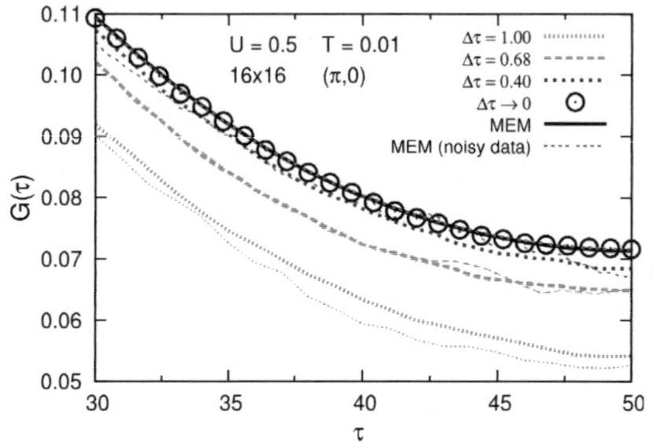

Figure 2. Extrapolation of BSS-QMC Green functions to $\Delta\tau \to 0$. Analytic continuation (MEM, black line) regularizes the results of pointwise extrapolations (symbols) and yields stable results (gray line) even in the case of noisy raw data (thin color lines).

2. Model and Methods

The single-band Hubbard Hamiltonian is given by

$$\hat{H} = -\,t \sum_{\langle ij\rangle,\sigma} \hat{c}^{\dagger}_{i\sigma}\hat{c}_{j\sigma} + U\sum_{i}\left(\hat{n}_{i\uparrow}-\frac{1}{2}\right)\left(\hat{n}_{i\downarrow}-\frac{1}{2}\right), \tag{1}$$

with number operators $\hat{n}_{i\sigma} = \hat{c}^{\dagger}_{i\sigma}\hat{c}_{i\sigma}$, next-neighbor hopping t, and local interactions U. In this paper, we set $t = 0.25$ and consider finite-size clusters with periodic boundary conditions.

The BSS-QMC algorithm is based on a Trotter-Suzuki decomposition of the partition function

$$Z = \mathrm{Tr}\left(e^{-\beta(H_t+H_U)}\right) \approx Z_{\Delta\tau} = \mathrm{Tr}\left(\prod_{l=0}^{\Lambda} e^{-\Delta\tau H_t}e^{-\Delta\tau H_U}\right), \tag{2}$$

where H_U (H_t) corresponds to the interaction (kinetic) term in (1), $\beta = 1/T$ is the inverse temperature ($k_b \equiv 1$) and Λ denotes the number of time slices ($\Lambda = \beta/\Delta\tau$). A discrete Hubbard-Stratonovich transformation simplifies the interaction term, which is quartic in the fermionic operators, to a quadratic form and a coupling to an auxiliary Ising field h. As usual for quadratic Hamiltonians, the fermionic degrees of freedom can be integrated out and the partition function is expressed by determinants of matrices:

$$Z_{\Delta\tau} = \sum\nolimits_{\{h\}} \det\left[M_{\uparrow}^{\{h\}}\right]\det\left[M_{\downarrow}^{\{h\}}\right]. \tag{3}$$

The sum in (3), and finally the computation of all observables, is performed by Monte Carlo methods. The main results of the BSS-QMC algorithm are Green functions $G_{ij}(\tau_l)$ on a discrete imaginary-time grid $\{\tau_l \in [0,\beta]\}$ for each pair of real-space coordinates $\{i,j\}$ on the given lattice. The involved matrix operations for the calculation of the determinants in (3) lead to a scaling of $\mathcal{O}(\Lambda\, N_c^3)$, which restricts the algorithm to relatively small clusters ($N_c \lesssim 400$).

To arrive at reliable conclusions regarding the Mott transition using BSS-QMC, one has first to eliminate all systematic errors (finite-size as well as Trotter errors) and, secondly, calculate the self-energy on the Matsubara axis, $\Sigma(i\omega_n)$, from the imaginary-time Green function in a stable manner. These steps are described in the following.

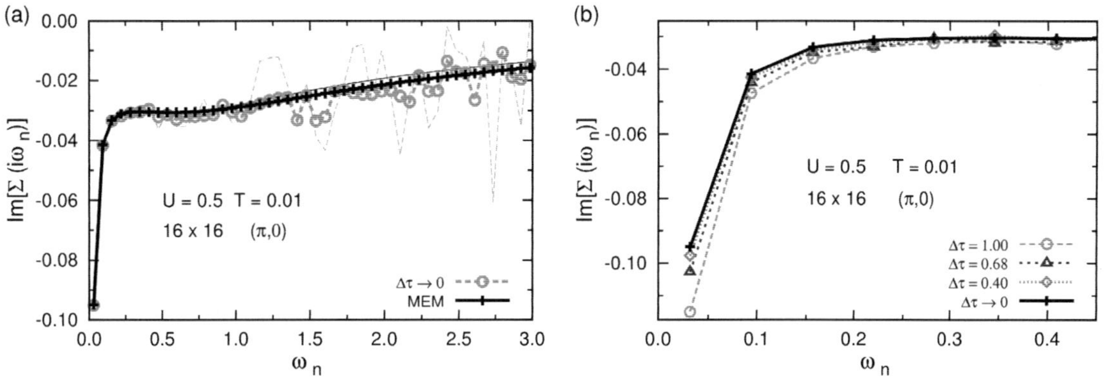

Figure 3. (a) Influence of statistical noise in the raw data $G(\tau)$ on the self-energy $\Sigma(i\omega_n)$ and stabilization by MEM regularization. Thin lines: noisy data, thick lines and symbols: high precision results. (b) Impact of Trotter discretization $\Delta\tau$ on the Matsubara self-energy.

3. Elimination of systematic errors

As the elimination of Trotter errors is well established [16–19], we will sketch the scheme only briefly and focus on the problem specific and physically more relevant finite-size analysis and present a stable method to calculate unbiased Green functions, even if the raw data are noisy.

Trotter bias – The raw BSS-QMC Green functions (thick colored lines in Fig. 2) do not only live on different imaginary-time grids $\{\tau_i\}$, depending on the chosen discretization $\Delta\tau$, but are also shifted with respect to each other (and to the exact solution). After aligning them on a common fine grid [20] the Trotter errors can be eliminated with high precision by piecewise extrapolations of $\Delta\tau \to 0$ (circles) [19]. However, some fluctuations remain inevitably, especially when using noisy raw data (using a factor of 10 less QMC sweeps; thin colored lines). These can be greatly reduced by regularization via a maximum entropy method (MEM), which computes corresponding spectral functions via the inversion of

$$G(\tau) = -\int_{-\infty}^{\infty} d\omega\, A(\omega)\, \frac{e^{-\omega\tau}}{1+e^{-\omega\beta}} \;. \tag{4}$$

Note that the intermediate spectra $A(\omega)$ are used only for producing continuous and smooth Green functions $G(\tau)$ via (4); in this combination, the procedure is stable. After this step, the results based on high- or low-precision data (thick black line vs. gray line) are hardly distinguishable and can both be used for stable computations of self-energies.

Self-energy – These quasi-continuous $G(\tau)$ can be reliably Fourier transformed to the Matsubara axis:

$$G(i\omega_n) = \int_0^{\beta} d\tau\, G(\tau)\, e^{-i\omega_n\tau} \;. \tag{5}$$

A quantity of great interest for the analysis of the Mott transition is the imaginary part of the momentum-resolved self-energy, which is connected to the BSS-QMC Green function G and the non-interacting Green function $\mathcal{G}$ via a Dyson equation [21]

$$\Sigma_{\mathbf{k}}(i\omega_n) = \mathcal{G}_{\mathbf{k}}^{-1}(i\omega_n) - G_{\mathbf{k}}^{-1}(i\omega_n) = i\omega_n + \mu - \epsilon_{\mathbf{k}} - G_{\mathbf{k}}^{-1} \;, \tag{6}$$

where μ is the chemical potential and $\epsilon_{\mathbf{k}}$ the dispersion of the noninteracting problem. As shown in Fig. 3 (b), the MEM regularizing procedure leads to estimates (crosses and black line) of the imaginary part of the self-energy (the real part vanishes at $\mathbf{k} = (\pi, 0)$ due to particle-hole

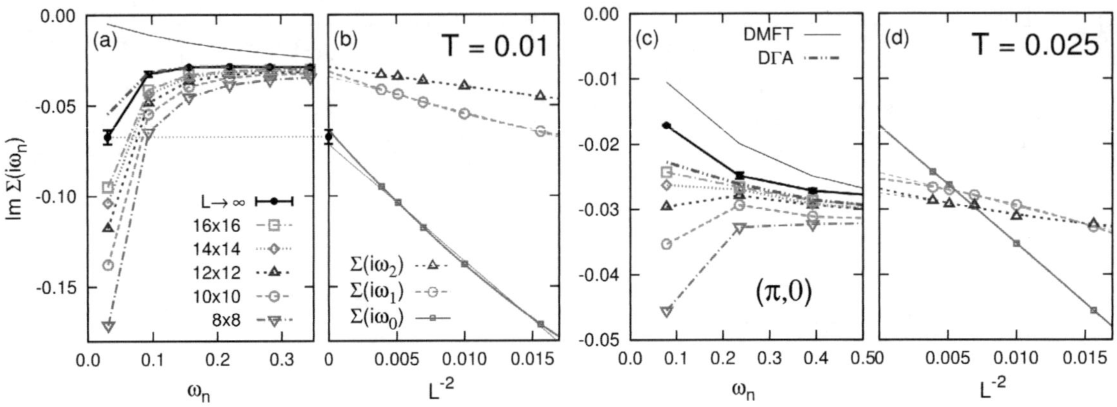

Figure 4. Finite-size scaling of Matsubara self-energy at $U = 0.5$, $\mathbf{k} = (\pi, 0)$. Finite-size BSS-QMC data (open symbols and broken colored lines), extrapolated BSS-QMC results in the thermodynamic limit (circles and black bold solid line), and DΓA data (gray line) versus Matsubara frequency ω_n at $T = 0.01$ (a) and $T = 0.025$ (c); also shown are momentum-independent single-site DMFT results (thin black lines). (b)+(d) Finite-size BSS-QMC (symbols) data for the first three Matsubara frequencies versus inverse system size plus extrapolations in linear order in L^{-2} (thin lines) and quadratic order (thick lines).

symmetry) which are smooth even at higher frequencies. In contrast, the direct results (green circles) fluctuate significantly even when based on high-precision raw data (and even more so for noisy raw data; thin dashed green lines). Note that both methods agree very well at the lowest Matsubara frequencies, where the QMC predictions are most reliable.

In Fig. 3 (b) the influence of the Trotter discretization is shown. The Trotter biased self-energies (broken colored lines) deviate from the exact result (black solid line) for large values of $\Delta\tau$, in particular at the lowest frequencies. However, in the parameter regime explored in this study, we can afford small enough values of $\Delta\tau$ so that the elimination of Trotter errors is less essential than the finite-size extrapolation to be discussed below.

Finite-size extrapolation of the self-energy – Using the methods described above, unbiased and stable estimates of the Matsubara self-energy have been obtained for square $L \times L$ clusters with sizes ranging from 8×8 to 16×16 (with a computational effort that varies as $N_c^3 = L^6$, i.e., by a factor of $2^6 = 64$ in this range). Results at the "anti-nodal" [22] momentum point $\mathbf{k} = (\pi, 0)$ [throughout the paper, unit lattice spacing $a = 1$ is assumed] are shown (colored symbols and broken lines) in Fig. 4(a) for the low temperature $T = 0.01$ and in Fig. 4(c) for the elevated temperature $T = 0.025$, respectively. Evidently, the finite-size effects are enormous: while the smallest systems (8×8, triangles) show insulating behavior, i.e., a strong enhancement of $\Sigma(i\omega_n)$ towards small frequencies, at both temperatures, this tendency is reduced with increasing L at $T = 0.01$ [Fig. 4(a)] and completely eliminated at $T = 0.025$ [Fig. 4(c)]. Obviously, careful extrapolations are needed for reliable predictions in the thermodynamic limit.

These are, indeed, possible, as shown in Fig. 4(b) for $T = 0.01$ and in Fig. 4(d) for $T = 0.025$, respectively, for the three lowest Matsubara frequencies (where the finite-size effects are largest): as a function of L^{-2}, the finite-size results (symbols) can be fitted with second-order polynomials (thick lines) with high precision; as these fit functions have small curvatures, linear extrapolations (thin lines) to the thermodynamic limit (i.e., $L^{-2} \to 0$) deviate only slightly from the quadratic ones. We use these deviations as error bars and the arithmetic average of both extrapolation procedures as final result, as indicated for $\Sigma(i\omega_0)$ by a black symbol in Fig. 4(b).

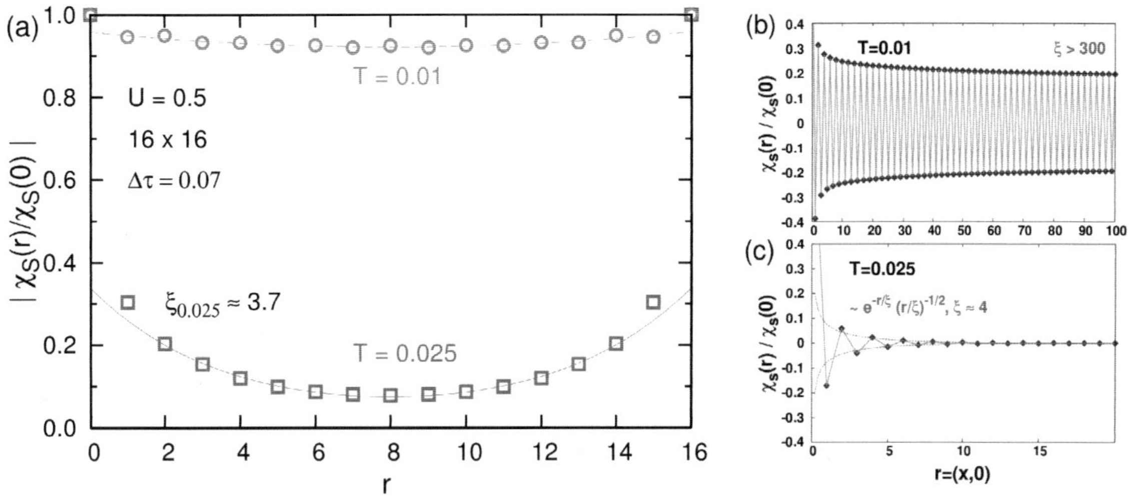

Figure 5. Real-space spin correlation function. (a) The correlation length for $T = 0.025$ (red squares) from BSS-QMC calculations for a 16×16 lattice is estimated by an exponential fit (red solid line). For $T = 0.01$ (green circles) ξ is much larger than the extend of the lattice. *Right panel:* The corresponding estimations from DΓA data [25] for $T = 0.01$ (b) and $T = 0.025$ (c) confirm the BSS-QMC results.

4. Mott physics and correlation lengths

These final QMC results [black symbols and lines in Fig. 4(a) and Fig. 4(c)] show that the character of the system changes drastically between the selected phase points (crosses in Fig. 1): while the QMC self-energy indicates a metallic phase at $T = 0.025$, very similar to the DMFT solution [thin black line in Fig. 4(c)], the low-temperature phase is clearly insulating - and completely unlike the paramagnetic DMFT solution [thin black line in Fig. 4(a)]. In contrast, the QMC results are strikingly similar to the DΓA predictions (gray dash-dotted lines), especially at low T. The dynamical vertex approximation ($D\Gamma A$) [23, 24] is a diagrammatic extension of DMFT which works directly in the thermodynamic limit (without any need for finite-size extrapolations); for the full set of $D\Gamma A$ results see [25]. QMC and DΓA together show convincingly that the suspected phase transition (dotted line in Fig. 1) is, indeed, present.

The question remains how exactly this metal-insulator transition (or crossover) is related to magnetic properties. After all, long-range AF order is excluded, in $D = 2$, by the Mermin-Wagner theorem. However, as shown in Fig. 5(a), the spin correlations decay much more slowly at $T = 0.01$ (circles) than at $T = 0.025$ (squares). More precisely, the correlation length at the elevated temperature can be estimated (using 16×16 QMC data) to $\xi \approx 3.7$ while it is clearly too large ($\xi \gg 30$) at $T = 0.01$ to allow precise determination using QMC. For this observable, DΓA predictions, shown in Fig. 5(b) and Fig. 5(c) are more reliable, yielding correlation lengths of $\xi \approx 4$ at the higher temperature, in excellent agreement with QMC, and $\xi > 300$ at $T = 0.01$, also consistent with QMC. Using DΓA we could also show that the temperature dependence of ξ changes its character at the Mott transition (see [25]).

Conclusion and acknowledgements

In spite of its simplifications, the Hubbard model remains a challenging target for theorists, especially in the interesting case of two spatial dimensions. All available methods have certain limitations, many of which are of particular concern in the nonperturbative intermediate-coupling regime at low temperatures and in the presence of longer-range correlations. Therefore, a single method can hardly yield authoritative results; some predictions may even be completely off (such as the CDMFT transition line in Fig. 1).

In the study presented in this work (with focus on the QMC methodology; for the full story, see [25]), we have applied two methods (an unbiased variant of BSS-QMC as well as the dynamical vertex approximation) with completely different characteristics. Only the near-perfect agreement between both sets of results makes the predictions truly authoritative (and validates technical choices on either side). We have settled one important question at half filling, namely the character of the Mott metal-insulator transition on the square lattice: it is driven by exponentially long-ranged [25] antiferromagnetic correlations, which act similarly to the AF long-range order in the cubic case, and is **not** connected to a quantum-critical point at $U > 0$.

The same methodology may be useful in other parameter ranges, e.g., for frustrated or anisotropic Hubbard models, doped systems, or multi-band models. It might also be worthwhile to compute transport properties, within bubble approximation or beyond, in order to observe the metal-insulator transition even more directly (than via the self-energy).

We acknowledge support from the research unit FOR 1346 of the German Research Foundation (DFG) and the graduate school GSC 266.

References

[1] Hubbard J 1959 *Phys. Rev. Lett.* **3** 77–78
[2] Schollwöck U 2005 *Rev. Mod. Phys.* **77** 259
[3] Metzner W and Vollhardt D 1989 *Phys. Rev. Lett.* **62** 324–327
[4] Georges A, Kotliar G, Krauth W and Rozenberg M J 1996 *Rev. Mod. Phys.* **68** 13–125
[5] Lee P A and Wen X-G 2006 *Rev. Mod. Phys.* **78** 17
[6] Armitage N, Fournier P and Greene R 2010 *Rev. Mod. Phys.* **82** 2421
[7] Rost D, Gorelik E V, Assaad F F and Blümer N 2012 *Phys. Rev.* B **86** 155109
[8] Mott N F 1968 *Rev. Mod. Phys.* **40** 677
[9] Gebhard F 1997 *The Mott Metal-Insulator Transition* (Springer Berlin)
[10] Staudt R, Dzierzawa M and Muramatsu A 2000 *Eur. Phys. J. B* **17** 411
[11] Park H, Haule K and Kotliar G 2008 *Phys. Rev. Lett.* **101** 186403
[12] Gorelik E V, Rost D, Paiva T, Scalettar R, Klümper A and Blümer N 2012 *Phys. Rev.* A **85** 061602
[13] Chang C-C, Scalettar R T, Gorelik E V and Blümer N 2013 *Phys. Rev.* B **88** 195121
[14] Blankenbecler R, Scalapino D J, and Sugar R L 1981 *Phys. Rev.* D **24** 2278
[15] Fuchs S, Gull E, Troyer M, Jarrell M and Pruschke T 2011 *Phys. Rev.* B **83** 235113
[16] Blümer N 2007 *Phys. Rev.* B **76** 205120
[17] Blümer N 2008 Preprint *arXiv*:08011222
[18] Gorelik E V and Blümer N 2009 *Phys. Rev.* A **80** 051602(R)
[19] Rost D, Assaad F F and Blümer N 2013 *Phys. Rev.* E **87** 053305
[20] This nontrivial task is achieved using reference models and spline interpolation of differences between measured and reference Green functions [7, 16, 17, 27, 30].
[21] Fetter A L and Walecka J D 1971 *Quantum Theory of Many-Particle Systems* McGraw-Hill
[22] In the context of high-T_c and pseudogap physics, one distinguishes "nodal" from "anti-nodal" points on the noninteracting Fermi surface. The observed dichotomy in the pseudogap phase can be viewed as "momentum-selective" Mott physics [26], in analogy to orbital-selective Mott phases [27–29] observed in systems with inequivalent bands.
[23] Toschi A, Katanin A A and Held K 2007 *Phys. Rev.* B **75** 045118
[24] Held K, Katanin A A and Toschi A 2008 *Prog. Theor. Phys. Suppl.* **176** 117
[25] Schäfer T, Geles F, Rost D, Rohringer G, Arrigoni E, Held K, Blümer N, Aichhorn M and Toschi A 2014 *Phys. Rev.* B **91** 125109
[26] Ferrero M, Cornaglia P, De Leo L, Parcollet O, Kotliar G and Georges A 2009 *Phys. Rev.* B **80** 064501
[27] Knecht C, Blümer N and van Dongen P G V 2005 *Phys. Rev.* B **72** 081103(R)
[28] Jakobi E, Blümer N and van Dongen P G V 2009 *Phys. Rev.* B **80** 115109
[29] Jakobi E, Blümer N and van Dongen P G V 2013 *Phys. Rev.* B **87** 205135
[30] Blümer N and Kalinowski E 2005 *Phys. Rev.* B **71** 195102

Topological properties of a Valence-Bond-Solid

Hui Shao[1], Wenan Guo [1,2,*] and Anders W. Sandvik[3,†]

[1] Department of Physics, Beijing Normal University, Beijing 100875, China
[2] State Key Laboratory of Theoretical Physics, Institute of Theoretical Physics, Chinese Academy of Science, Beijing 100190, China
[3] Department of Physics, Boston University, 590 Commonwealth Avenue, Boston, Massachusetts 02215, USA

E-mail: *waguo@bnu.edu.cn, †sandvik@bu.edu

Abstract. We present a projector quantum Monte Carlo study of the topological properties of the valence-bond-solid ground state in the J-Q_3 spin model on the square lattice. The winding number is a topological number counting the number of domain walls in the system and is a good quantum number in the thermodynamic limit. We study the finite-size behavior and obtain the domain wall energy density for a topological nontrivial valence-bond-solid state.

1. Introduction

While topological order has mainly been discussed in the context of exotic states such as quantum spin liquids [1, 2], similar topological quantum numbers can also arise in systems with long-range order. The simplest example of a topological quantum number is the winding number of a dimer model [3]. In the classical close-packed dimer model on the square lattice (as an example of a simple bipartite lattice), a winding number $W = (w_x, w_y)$ can be defined by assigning a directionality (arrow) from sublattice A to sublattice B for each dimer. Superimposing any such configuration of arrows onto two reference configurations (conventionally one with the dimers forming horizontal or vertical columns), closed loops are formed and the winding numbers correspond to the total x and y currents normalized by the system length L [3].

In the quantum dimer model (QDM), the classical dimer configurations are the basis states. For $S = 1/2$ quantum spins on a bipartite lattice, any total spin singlet can be expanded in Valence-bond states (VBs) with positive-defined expansion coefficients. The VBs are dimers with the added spin-singlet internal structure and the configurations can again be classified according to the winding number [4]. Since the Hamiltonian of the QDM contains only local operators that have no matrix element between states of different winding numbers, the Hilbert space can be divided into different winding number sectors, i.e., the topological winding number is a good quantum number. However, for general VB states, longer bonds are introduced, which makes the bond operators in the Hamiltonian non-local. For finite size systems, the winding number can be changed if bonds of lengths exceeding 1/4 of the system length appear. Therefore, the winding number can only be conserved in the thermodynamic limit.

To investigate the emergent conserved winding number in an $S = 1/2$ spin system, we here study the J-Q_3 Model on the square lattice [5]. The Hamiltonian of the model is

$$H = -J\sum_{\langle ij\rangle} C_{ij} - Q_3 \sum_{\langle ij,kl,mn\rangle} C_{ij}C_{kl}C_{mn} \tag{1}$$

with $C_{ij} = \frac{1}{4} - \mathbf{S}_i \cdot \mathbf{S}_j$ recognized as a singlet projector. The J term of H is the standard Heisenberg exchange and the Q_3 terms have the singlet projectors arranged in columns, as shown in Fig. 1(a). This model hosts a columnar valence-bond solid (VBS) ground state when Q_3/J is sufficiently large [6]. Ignoring quantum fluctuations, the VBS ground state is sketched as an ordered VBS breaking the Z_4 symmetry of the Hamiltonian, as shown in Fig. 1(b). As previously studied in the context of QDMs [7], a VBS with domain walls on a periodic lattice corresponds to a non-zero winding number—a case of $W = (1, 0)$ is shown in Fig. 1(c).

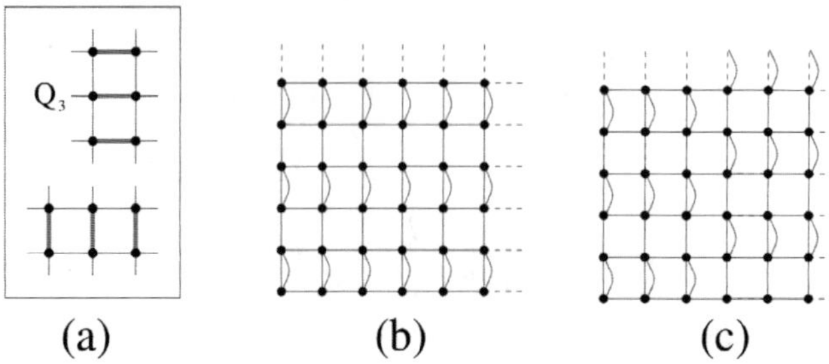

Figure 1. Illustrations of (a) the Q_3 terms, (b) the columnar VBS state, and (c) a VBS state with winding number $W = (1, 0)$ (on a periodic lattice).

In this paper, we consider the pure Q_3 model without any J term to study the VBS ground state. We will demonstrate explicitly, using projector quantum Monte Carlo (PQMC) calculations, that there is an emergent topological quantum number (a winding number) associated with the VBS order in the thermodynamic limit, while in a finite system the winding number is not conserved. In this case the topological sectors correspond to different types of domain walls, which are unstable (non-conserved) in finite systems but become stable in the thermodynamic limit.

2. Projector Quantum Monte Carlo method

The model with $Q_3 > 0$ in (1) does not have sign problems. We thus take advantage of the PQMC method with efficient loop update algorithms [8, 9, 10] to study the model. The method is based on applying a high power of the Hamiltonian to a trial state $|\psi_0\rangle$,

$$|\psi_m\rangle = (-H)^m |\psi_0\rangle. \quad (2)$$

For sufficient large m, $|\psi_m\rangle$ approaches the ground state $|0\rangle$ of the model. As the ground state $|0\rangle$ of a bipartite J-Q_3 model is guaranteed to be a total spin singlet, it is particularly convenient to use a trial state expanded in the VB basis $|V_r\rangle$ in the singlet sector. Then only the singlet sector is considered from the outset, leading to faster convergence with m.

The energy of the projected state can be calculated in the following way [11]

$$E = \frac{\langle N|H|\psi_m\rangle}{\langle N|\psi_m\rangle} = -\frac{\langle N|(-H)(-H)^m|\psi_0\rangle}{\langle N|(-H)^m|\psi_0\rangle}, \quad (3)$$

where $|N\rangle$ is one of the Néel states in the z-spin basis, which has equal overlap with all VBs.

By writing $(-H)^m$ as a sum over all possible strings of the individual Q_3 terms in Eq. (1), the above expression is evaluated by implementing importance sampling of the operator strings with the efficient loop algorithm [9, 10]. The implementation of the algorithm to the J-Q_3 Hamiltonian has been described briefly in [12]. In the end, the energy E is calculated as $E = -\langle n_d + n_f/2\rangle$ with n_f the number of bond flips and n_d the number of diagonal operations.

3. Results

We chose a basis state $|\psi_0\rangle = |V_r\rangle$ as the trial state and consider the cases that $|V_r\rangle$ in different winding number sectors W. The detailed definition of the winding number of a VB state can be found, e.g., in Refs. [3, 2]. In the PQMC simulations, besides the energy, we also sample the probability, $P(W)$, of a projected state in the topological sector $W = (w_x, w_y)$. This is done by calculating the winding number of each projected VBs $P_k|V_r\rangle$, with P_k a operator string with length m generated in the MC processes.

In the case that the trial state is a VBs in the winding number sector $W = (0,0)$, the ground columnar VBS state will be projected out quickly, i.e., within a small m/N, which can be defined as a projecting "time" (closely related to imaginary time [13]). This is indicated by the convergence of the ground state energy density $e_0(L)$ for a system with linear size L.

Now turn to the cases that the trial state is in the nontrivial winding number sector $W \neq (0,0)$. For small systems, the columnar VBS state is again projected out after some projection time m/N. This is indicated by the convergence of the energy density $e(L)$ to the ground state value $e_0(L)$. Meanwhile, the probability $P(W)$ decreases (to 0 for $L \to \infty$). However, as the system size increases, the projection time m/N needs to grow as well in order for the ground state to be obtained. In the thermodynamic limit, we expect that the system will stay in the sector with the initial winding number W, and then it is also plausible that the energy of the system will converges to a value $e_W > e_0$ corresponding to the lowest excited state within the sector W.

To demonstrate such behavior, we introduce the energy density $e_W(L)$ of states in the winding number sector W, which is obtained by only sampling those states in the sector W. Figure 2 shows the "time evolution" of the probability $P(W)$ for the system staying in the original winding number sector $W = (0,0), W = (1,0)$, and $W = (2,0)$ as a function of m/L^2, respectively, for a system with linear size $L = 96$ (lower panel). The corresponding energy density $e_W(L)$ converges to the values which are higher than $e_0(L)$, if $W \neq 0$ (upper panel).

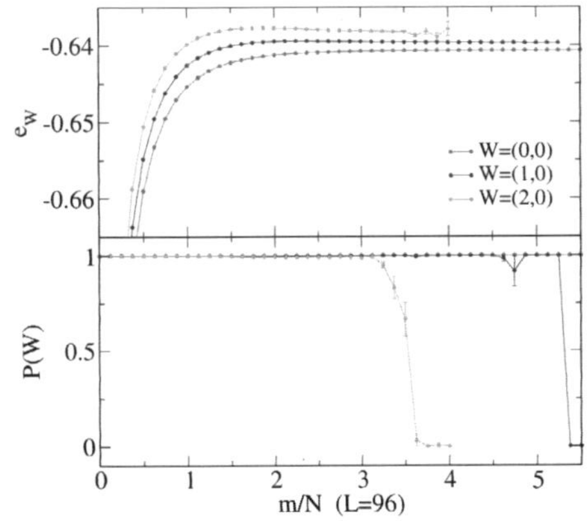

Figure 2. The probability $P(W)$ and the energy density $e_W(L)$ as functions of time m/N (projector power rescaled by the system volume).

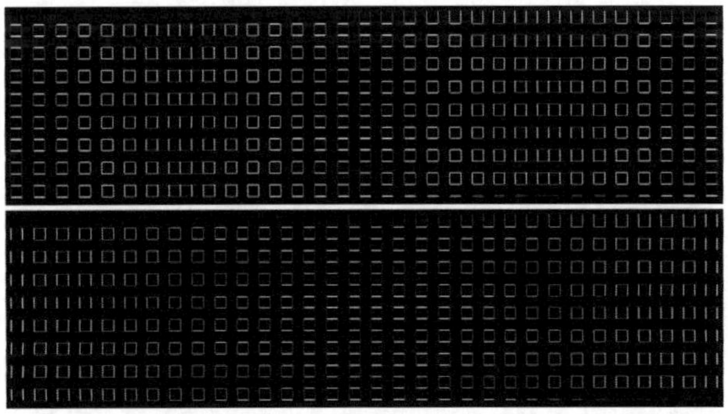

Figure 3. Snapshots of $\langle B_\alpha(\mathbf{r})\rangle$ for a periodic system with winding number $W = (1,0)$, in which a 2π domain wall (four separate $\pi/2$ domain walls) is formed (upper panel) and for an open system with appropriate boundary conditions in which a π domain wall is forced (lower panel).

We now study the reason of the energy gap between a system in a nontrivial topological sector and in the $W = 0$ ground state. It is well known that the ground state of the J-Q_3 is the columnar VBS. The VBS state can be detected by the columnar VBS order parameter, which is defined by the operators

$$D_x = \frac{1}{N}\sum_{x,y} B_{\hat{x}}(x,y)(-1)^x, \qquad D_y = \frac{1}{N}\sum_{x,y} B_{\hat{y}}(x,y)(-1)^y. \tag{4}$$

where the dimer operator B_α is used:

$$B_\alpha(\mathbf{r}) = \mathbf{S}(\mathbf{r}) \cdot \mathbf{S}(\mathbf{r}+\alpha) \tag{5}$$

where $\alpha = \hat{x}, \hat{y}$ denotes the lattice unit vector in the two directions. In a columnar VBS, either D_x or D_y has a nonzero expectation value.

However, if a state bears a nontrivial winding number, the lattice translational symmetry with period two is also broken. As an example, Fig. 3 (upper panel) shows a snapshot of the bond configuration $\langle B_\alpha(\mathbf{r})\rangle$ for a projected state with winding number $W = (1,0)$. To describe such a state, it is useful to define the local order parameters for the VBS state. Consider the case that there is still translational invariant VBS with period two along one axis of the square lattice, say y axis. The one-dimensional local x and y order parameters as a function of the x coordinate are defined [14] as follows:

$$D_x(x) = [\langle B_{\hat{x}}(x,y)\rangle - \frac{1}{2}\langle B_{\hat{x}}(x-1,y)\rangle - \frac{1}{2}\langle B_{\hat{x}}(x+1,y)\rangle](-1)^x, \tag{6}$$

$$D_y(x) = [\langle B_{\hat{y}}(x,y)\rangle - \langle B_{\hat{y}}(x,y+1)\rangle](-1)^y. \tag{7}$$

These two local order parameters are independent of the y coordinate. Following theoretical expectations [15], The VBS angle $\theta(x)$ can also be defined [14],

$$\theta(x) = \mathrm{atan}\left[\frac{D_y(x) + D_y(x+1)}{2D_x(x)}\right], \tag{8}$$

such that $\theta = 0$ and $\theta = \pi$ for a fully x or y oriented VBS order, respectively.

To understand the energy gap between system in an nontrivial topological sector and in the $W = 0$ ground state, we first set asymmetric open boundary conditions on a $L \times L$ cylindrical lattice, where the x-direction boundaries have been modified. By removing Q_3 terms with vertical bonds closest to the left edge on every even row and closest to the right edge on every odd row, we obtain a state with forced VBS angle $\theta(0) = \pi/2$ and $\theta(L) = -\pi/2$. A snapshot of the bond strength $\langle B_\alpha(\mathbf{r})\rangle$ for the state projected out from simple trial state with a corresponding domain wall is shown in Fig. 3 (lower panel). More properties of the projected state are shown in Fig. 4 (right), where the two local order parameters and the VBS angle are plotted as functions of x coordinate and various system sizes L. In this case, the local order parameter $D_x(x)$ is maximized in the center and goes to zero at boundaries; $D_y(x)$ is about 0.75 at the left boundary, but gradually changes to -0.75 at the right boundary. The VBS angle $\theta(x)$ changes from $\pi/2$ to $-\pi/2$. Clearly a domain wall is formed, as expected. According to the total change of the VBS angle, we define this as a π domain wall (which consist of two $\pi/2$ domain walls).

The energy density of the state with π domain wall $e_\pi(L)$ can be calculated as described above. The resulting $e_\pi(L)$ is larger than the ground state energy density $e_0(L)$ of a $L \times L$ cylindrical lattice with symmetric boundary modification (i.e., forcing a VBS with no domain wall). The energy difference is entirely due to the presence of the domain wall. We thus define the domain wall energy per unit length as

$$\kappa_\pi(L) = [e_\pi(L) - e_0(L)]L \tag{9}$$

As shown in Fig. 5, when system size goes to infinity, the domain wall energy per unit length converge to a constant κ_π, which is estimated as 0.434(9), as listed in Table 1.

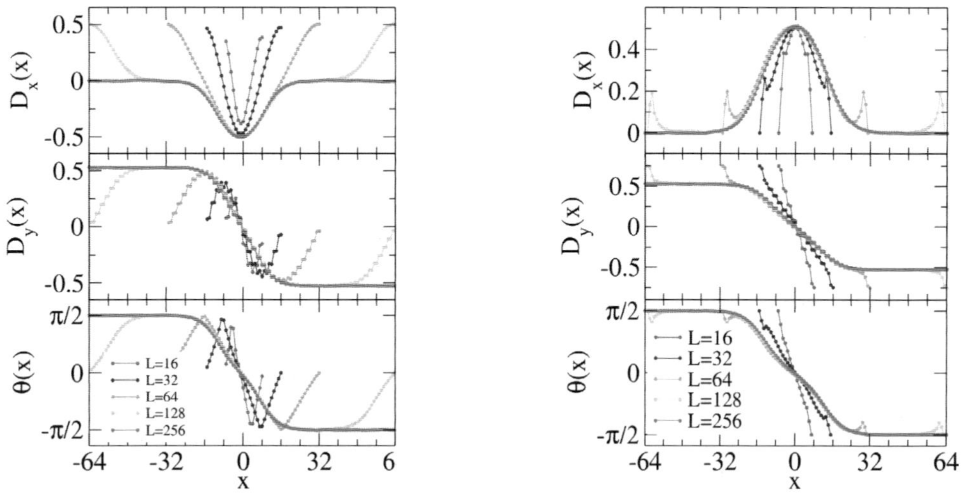

Figure 4. The local order parameters $D_x(x)$, $D_y(x)$ and the VBS angle $\theta(x)$ for projected states in the topological sector $W = (1, 0)$ with periodic boundaries (left) and for the states with a π domain wall (right) in a system with asymmetric open boundaries.

Similarly, a $\pi/2$ domain wall can realized by setting asymmetric open boundary conditions forcing horizontal and vertical bond patterns at the edges. The energy difference to $e_0(L)$ multiplied by the system size again converges to a constant when system size L tends to infinity (see Fig. 5). The constant is thus identified as the domain wall energy $\kappa_{\pi/2}$ for the $\pi/2$ domain wall with the estimated value being a half of the π domain wall energy, as listed in Tab. 1.

We now return to systems with periodic boundaries. In the case of large enough system, the projected energy eigenstate is still in the winding number sector W of the initial state. Take the $W = (1, 0)$ case as an example. Figure 4 (left panel) shows the local order parameters and VBS angel as functions of the x coordinate. $D_x(x)$ has a minimum in the center and maximums at the left and right sides, while $D_y(x)$ is almost zero at the edges gradually grows to a maximum of 0.5, then decays to a minimum about -0.5, and finally returns to 0. As the result, the VBS angle $\theta(x)$ changes from 0 to $\pi/2$, then to $-\pi/2$, and back to 0. Domain walls are clearly formed here and their dynamics in a large system is very slow and practically locked in to their locations in the initial state. According to the total change of the VBS angle we define this as a 2π domain wall (which can be regarded as four elementary $\pi/2$ domain walls). A finite-size scaling analysis of the energy density gap to the ground state ($W = 0$) $e_W(L) - e_0(L)$ multiplied by the system size L is shown in Fig. 5. Clearly, the scaled gap converges to a constant, which can be understood as the 2π domain wall energy per unit length $\kappa_{2\pi}$. The estimated $\kappa_{2\pi}$ is indeed twice that of the π domain wall, as seen in Table 1.

We further calculated the energy density gap $e_W(L) - e_0(L)$ from the projected state in the winding number sector $W = (2, 0)$ to the ground state for several system sizes. The local order parameters and VBS angles as functions of the x-direction shows that there is a 4π domain wall. Again the energy gap multiplied by the system size converges to a constant, which is the 4π domain wall energy. The estimated domain wall energy is 0.205(9), which is about four times of the π domain wall energy.

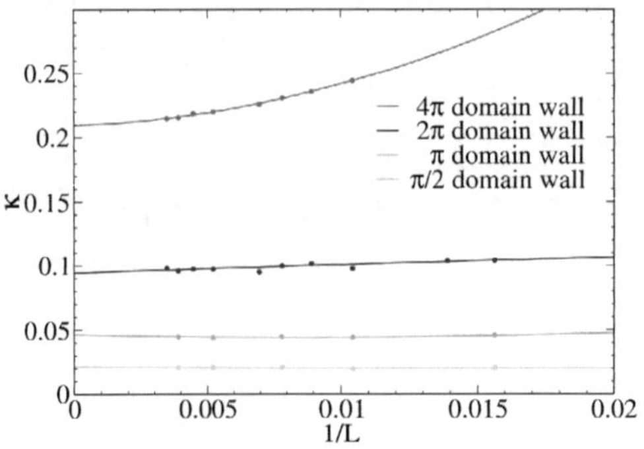

Figure 5. Finite-size scaling analysis of the domain wall energies. The curves are fitted polynomials, used to extrapolate the energies listed in Table 1.

domain wall	$\pi/2$	π	2π	4π
m/N	6	6	4	3
κ	0.0208(5)	0.0434(9)	0.093(9)	0.205(9)

Table 1. Domain wall energy per unit length for various domain walls. m/N is the projection "time" at which the energy is calculated in each case.

4. Summary and Conclusions

Using a ground-state projector QMC method in the VB basis, we have studied projected states in various winding number sectors of the J-Q_3 model on the square lattice. We showed that the projected state stays in the winding number sector of the trial state when system size goes to infinity. The energy of the states in a nontrivial winding number sector have gaps to the VBS ground state in the $W = (0, 0)$ sector, due to the presence of domain walls (the winding number counting these domain walls). Such domain walls are stable only for infinite size, thus, the winding number is an emergent quantum number in the thermodynamic limit.

Acknowledgments

This research was supported by the NSFC under Grant No. 11175018 (WG) and by the NSF under Grants No. DMR-1104708, PHY-1211284 and DMR-1410126 (AWS). SH gratefully acknowledges support from the organization committee of the CCP2014. WG would like to thank Boston University's Condensed Matter Theory Visitors program.

References

[1] P. W. Anderson, Science **235**, 1196 (1987).
[2] Y. Tang, A. W. Sandvik, and C. L. Henley, Phys. Rev. B **84**, 174427 (2011).
[3] D. S. Rokhsar and S. A. Kivelson, Phys. Rev. Lett. **61**, 2376 (1988).
[4] N. Bonesteel, Phys. Rev. B **40**, 8954 (1989).
[5] A. W. Sandvik, Phys. Rev. Lett. **98**, 227202 (2007).
[6] J. Lou, A. W. Sandvik, and N. Kawashima, Phys. Rev. B **80**, 180414(R) (2009).
[7] S. Papanikolaou, K. S. Raman, and E. Fradkin, Phys. Rev. B **75**, 094406 (2007).
[8] A. W. Sandvik, and H. G. Evertz, Phys. Rev. B **82**, 024407 (2010).
[9] H. G. Evertz, G. Lana, and M. Marcu, Phys. Rev. Lett. **70**, 875 (1993).
[10] H. G. Evertz, Adv. Phys. **52**, 1 (2003).
[11] A. W. Sandvik, Phys. Rev. Lett. **95**, 207203 (2005).
[12] A. W. Sandvik, AIP Conf. Proc. **1297**, 135 (2010).
[13] C.-W. Liu, A. Polkovnikov, and A. W. Sandvik, Phys. Rev. B **87**, 174302 (2013).
[14] A. W. Sandvik, Phys. Rev. B **85**, 134407 (2012).
[15] M. Levin and T. Senthil, Phys. Rev. B **70**, 220403 (2004).

Recent progress in simulations of gauge theories on the lattice

Luigi Del Debbio
Higgs Centre for Theoretical Physics, The University of Edinburgh, Edinburgh, United Kingdom.
E-mail: `luigi.del.debbio@ed.ac.uk`

Abstract. There has been considerable progress in simulating gauge theories coupled to fermions in the last decade. As a result, Monte Carlo simulations are now capable of taking the fermion determinant properly into account when generating gauge configurations even for light fermion masses. We briefly summarize the physical intuitions behind this progress, and discuss the impact on the range of problems that can currently be adressed using a computational approach to evaluate the path integral of gauge theories defined on discretized Euclidean spacetime lattices. We focus in particular on recent result for QCD at the physical point, and on new nonperturbative models for physics beyond the Standard Model.

1. Introduction

At the International Symposium on Lattice Field Theory in Berlin in 2001, a panel discussion was dedicated to the cost of quantum chromodynamics (QCD) simulations with dynamical fermions. The computational cost of simulating light dynamical fermions was identified as the main obstacle towards robust studies of quantum field theories in the nonperturbative regime by means of numerical simulations. The computing power required for such simulations was summarized by A. Ukawa in the plot taken from Ref. [1], and reproduced in Fig. 1, which became known as the *Berlin wall* plot. It is clear, even before entering into any technical discussion, that the computational effort diverges as the fermion mass is reduced. The actual scaling law proposed at the time was:

$$\text{cost} \propto m_\pi^{-6}\, L^5\, a^{-7}\,, \tag{1}$$

where m_π is the pion mass, which in QCD is proportional to the square root of the fermion mass, L is the physical size of the lattice, and a is the lattice spacing.

In this talk, I will briefly summarize the basic principles underlying Monte Carlo simulations of gauge theories, focussing on two main points: the importance of dynamical fermions, and the difficulties that arise when trying to reach the chiral limit; the latter are clearly exposed by the scaling reported in Eq. 1. It is interesting to discuss some of the solutions that have been found to this problem, as they are based on a deeper understanding of the underlying physics. I will then briefly recall recent results in QCD that have an important impact on flavor phenomenology. The final section of this talk is devoted to the possibility of studying new models of nonperturbative dynamics beyond the Standard Model by numerical simulations. A discussion of the identification of strongly-interacting fixed points in four-dimensional gauge theories using Monte Carlo data is presented.

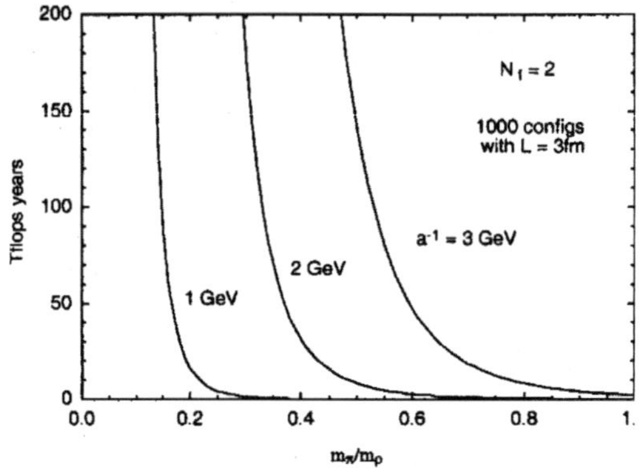

Figure 1. *Berlin wall* plot, illustrating the cost of numerical simulations as a function of the fermion mass and the lattice spacing. The curves show the cost in TFlops years of generating 1000 independent gauge configurations as a function of the fermion mass. The three curves correspond to three different values of the lattice spacing a, while the lattice linear size is kept constant. Plot from Ref. [1]

2. Simulating dynamical fermions

Gauge theories can be defined in a discretized Euclidean space starting from the path integral:

$$Z = \int \mathcal{D}U \mathcal{D}\psi \mathcal{D}\bar{\psi}\, e^{-S[\psi,\bar{\psi},U]}. \tag{2}$$

The integration variables in the path integral are the fermion fields $\psi(x)$ and $\bar{\psi}(x)$, defined on the sites of the lattice, and the link variables $U(x,\mu)$ that describe the degrees of freedom associated with the gauge field. The integration over the link variables is the Haar measure over the gauge group. In a finite discretized volume, Eq. (2) is a well-defined multidimensional integral.

The dynamics of the theory is encoded in the action S, which determines the statistical weight of each field configuration. Different choices of the lattice action S are currently used by practitioners in the field; in general these actions differ by lattice artefacts, i.e. by operators that are suppressed by some power of the lattice spacing a. For our purposes we do not need to describe the lattice action in detail, we only assume that some discretized version of the Dirac operator $D(x,y;U)$ exists, so that:

$$S = \sum_{x,y} \bar{\psi}(x)\, D(x,y;U)\, \psi(y) + S_g[U]\,. \tag{3}$$

Note that the Dirac operator depends on the gauge link variables, and encodes the interaction of the fermions with the gauge field. The fermion mass is one of the parameters that enter the definition of the Dirac operator. We also assume that we have a discretized version of the pure gauge action S_g, which we do not need to specify. For instance QCD, the theory of strong interactions, is defined as an SU(3) gauge theory coupled to three flavors of light fermions in the fundamental representation of the gauge group.

The action is quadratic in the Grassmann variables $\bar{\psi}$ and ψ, and therefore the integration over the fermionic fields can be done analytically, yielding:

$$Z = \int \mathcal{D}U e^{-S_g[U]} \left(\det D[U]\right)^{n_f}, \tag{4}$$

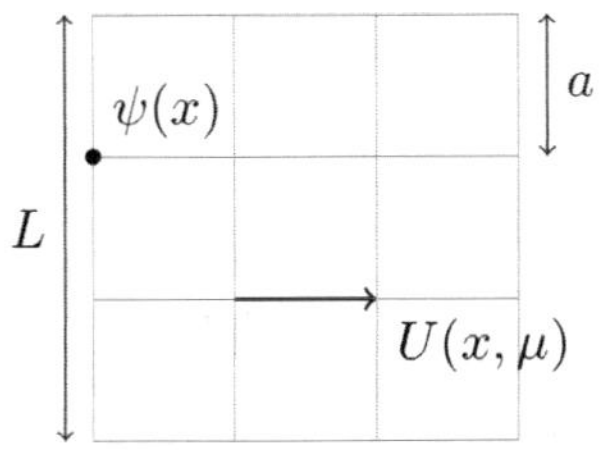

Figure 2. Field variables in a lattice formulation of gauge theories. The fermionic fields are defined on the sites of the lattice, the gauge fields are associated to the links. The linear size of the system in physical units is L, while the lattice spacing is a. The number of sites in each direction is $N = L/a$. The discretization of spacetime acts as a regulator for the high-energy divergences, the inverse lattice spacing a^{-1} being the UV cutoff.

where n_f is the number of fermionic species in the theory, which we assumed to be degenerate in this example. Eq. 4 shows that the probability distribution of the gauge link variables depends on the fermionic determinant, as well as on the gauge part of the action. Ignoring the contribution of the fermionic determinant defines the so-called *quenched approximation*, where the link variables are distributed according to the value of the gauge action only. Clearly such an approximation violates unitarity, and introduces systematic errors whose size is unacceptable for the precision studies in contemporary phenomenology. *Dynamical fermions* is the name used to identify those simulations that take fully into account the contribution of the Dirac determinant.

In order to understand the cost of simulating dynamical fermions as shown in Fig. 1, it is useful to recall how the fermionic determinant is usually computed. Introducing complex bosonic fields, the determinant can be written as:

$$\det D[U]^2 = \int \mathcal{D}\phi\mathcal{D}\phi^* \exp\left[-\sum_{x,y\in\Lambda} \left|D^{-1}(x,y;U)\phi(y)\right|^2\right]. \tag{5}$$

It is clear from the latter equation that the cost of computing the fermionic determinant is dictated by the cost of inverting the Dirac operator D, i.e. by the cost of solving:

$$\sum_y D(x,y)\psi(y) = \eta(x), \tag{6}$$

for a given source $\eta(x)$. The Dirac operator is a large, sparse matrix, and Eq. 6 can be solved using one of several iterative Krylov space solvers, see e.g. Ref. [2]. The inversion of D becomes more computationally expensive as the condition number of the Dirac operator increases, *i.e.* the small eigenvalues of the Dirac operator are responsible for the cost of the simulations summarised in the *Berlin wall* plot. We shall refer to the eigenstates corresponding to these low eigenvalues as *low modes* or *low eigenstates* of the Dirac operator. Clearly any IR regulator, e.g. the fermion mass or the finite volume of the box, will prevent a substantial increase of the condition number of D; as summarized by the scaling law above, the problem arises only when trying to reach the physical regime of small masses and large volumes.

A very successful way to improve this scaling behaviour, based on the factorization of the Dirac determinant, was proposed shortly after the Berlin conference in Ref. [3]. In this talk, we will focus on a complementary approach, which relies on understanding the dynamical properties of the Dirac low eigenmodes. It is important to point out here that the number of low modes of the Dirac operator is driven by the dynamics of the system under study. In a theory where chiral symmetry is spontaneously broken, as is the case in QCD, the order parameter of the

Journal of Physics: Conference Series **640** (2015) 012049 doi:10.1088/1742-6596/640/1/012049

broken symmetry, is directly related to the density of zero modes of the massless Dirac operator through the Banks-Casher relation [4]:

$$\langle\bar{\psi}\psi\rangle = \pi \lim_{\lambda\to 0} \lim_{V\to\infty} \rho(\lambda, 0)\,, \tag{7}$$

where $\rho(\lambda, m)$ denotes the density of modes with eigenvalue λ, and m is the fermion mass that appears in the Dirac operator . A non-vanishing chiral condensate in the left-hand size of Eq. 7 implies a non-vanishing density on the right-hand side. As shown in Ref. [5], integrating the Banks-Casher relation yields the number of low modes with eigenvalue λ such that $|\lambda| < M$:

$$N_{|\lambda|<M} = \frac{2}{\pi} M \langle\bar{\psi}\psi\rangle V + O(M^2)\,. \tag{8}$$

It is clear from this equation that, as the fermion mass goes to zero, the number of low modes of D increases linearly with the physical volume of the system.

3. Deflated inverters

The problem of the large condition number for the Dirac operator can be alleviated using deflation. Deflation techniques are well known to practitioners in lattice QCD, see e.g. Refs. [6, 7, 8, 9, 10]; they are based on the idea of projecting out the low modes of the Dirac operator. The Dirac operator restricted to the orthogonal complement of the space spun by the low modes has a better condition number, and hence can be inverted at a lower cost. More specifically, let us denote by $\mathcal{S}$ the subspace spun by N orthogonal vectors:

$$\mathcal{S} = \mathrm{span}\{\phi_1, \dots, \phi_N\}\,, \tag{9}$$

then the projector on $\mathcal{S}$ can be defined as:

$$P\psi = \sum_{k=1}^{N} \phi_k (\phi_k, \psi)\,. \tag{10}$$

The little Dirac operator is defined to be the restriction of the Dirac operator to the subspace $\mathcal{S}$:

$$PDP\psi = \sum_{k,l} \phi_k A_{kl} (\phi_l, \psi), \quad A_{kl} = (\phi_k, D\phi_l)\,. \tag{11}$$

If P is effective at projecting on the low eigenmodes of the Dirac operator, then the condition number of the Dirac operator in the orthogonal subspace is clearly improved. As the number of low modes scales with the physical volume, a naive deflation procedure has a cost $O(V^2)$, and becomes rapidly unaffordable as large volumes are used.

Multi-level algorithms have been proposed a long time ago in order to improve the scaling of the cost with the size of the system [12]. These ideas were developed and applied to QCD with Wilson fermions in Refs. [14, 15, 16, 17]. Unfortunately it is not possible to present here a full account of the work done in this area. A recent account of this work was presented at the Lattice conference this year. Instead here we shall summarize briefly one particular approach, proposed by Lüscher in Ref. [5].

Lüscher's idea is based on an empirical property of low eigenmodes called *local coherence*; it refers to the fact that the low eigenmodes of the Dirac operator, when restricted to a small physical volume, are well approximated by a relatively small number of independent vectors. Local coherence suggests a new recipe for constructing deflation subspaces. Here we follow closely the notation introduced in Ref. [5].

The lattice is divided into non-overlapping rectangular blocks, denoted by Λ. In each block, a local deflation subspace can be defined by choosing a set of orthonormal fields ϕ_l^Λ, where $l = 1, \ldots, N_s$. The total number of fields introduced in this way is $N_b N_s$, where N_b denotes the number of blocks. The block fields are defined starting from N_s global fields ψ, projecting them to the blocks:

$$\phi_l^\Lambda(x) = \begin{cases} \psi_l(x), & \text{if } x \in \Lambda, \\ 0 & \text{otherwise} \end{cases}, \tag{12}$$

and orthonormalizing the resulting fields. They can be used to build a projector as explained above, and for any given normalized field ψ, the accuracy of the approximation provided by the deflation subspace can be monitored by introducing the deficit:

$$\epsilon = ||\,(1 - P)\,\psi||^2\,. \tag{13}$$

Given a set of low eigenmodes of $D^\dagger D$, a domain-decomposed subspace can be constructed from a smaller subset of these modes chosen randomly. Local coherence implies that the deficit for *all* the low modes is small, independently of whether they were included in the the subset used to define the projector. In the numerical experiments presented in Ref. [5], 12 eigenmodes out of 48 were chosen at random to construct the subspace; the remaining 36 modes were well approximated within this subspace with deficits of the order of 3% to 6%, using blocks of size ≈ 0.3 fm.

In practice an efficient deflation subspace can be obtained starting from global fields that are only approximate eigenmodes of the Dirac operator. These can be computed at small cost using some relaxation procedure, which yields a set of N_s global fields such that:

$$||D\psi_l|| \leq M||\psi||\,, \tag{14}$$

where M is the size of the largest eigenvalue that we want to project out using deflation. The important point here is that the number of modes computed does not depend on the volume of the lattice.

Recent numerical investigations using Wilson fermions show that this approach overcomes the problem of the proliferation of low modes with volume, and delivers a signifcant improvement in the inversion of the Dirac operator for small fermion masses. An eloquent example of the improved performance, as reported in Ref. [18], is shown in Fig. 3.

A similar procedure has been recently implemented for domain-wall fermions in Ref. [11]. Domain wall fermions are a five-dimensional formulation of lattice fermions which preserves some kind of chiral symmetry at finite lattice spacing [19]. The five-dimensional formulation requires to solve a number of new problems; the implementation in Ref. [11] has successfully addressed these issues, and has delivered a significant gain in the inversion of the Dirac operator for chiral fermions.

In all these cases, it is interesting to note that the algorithmic progress has been driven by the understanding of the underlying dynamics, and its relation to the cost of numerical simulations.

4. QCD at the physical point

As a result of the recent algorithmic improvements, simulations of lattice QCD are now performed on large volumes and physical light quark masses. These simulations allow for an unprecedented control of systematic errors, and high statistical accuracy. Current lattice results provide numerous nonperturbative inputs for precision studies in the flavor sector of the Standard Model.

It is not possible to cover the whole spectrum of QCD results in this talk; instead we will have to make a selection, and refer the reader to the proceedings of the annual Lattice conference for

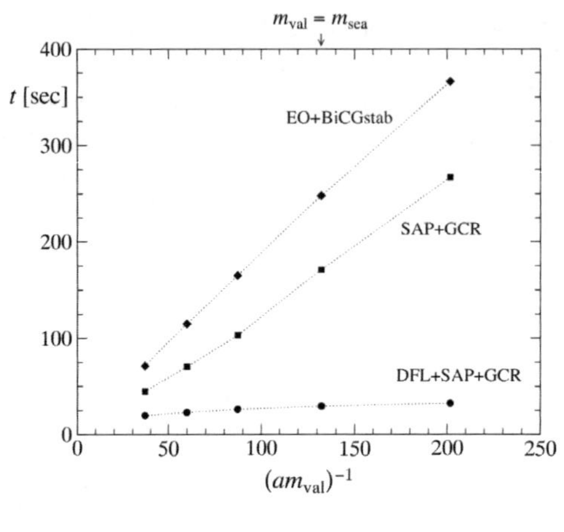

Figure 3. Cost of the inversion of the Dirac operator using deflation as discussed in the text. Figure from Ref. [18].

a more complete survey of current results. Several new results have been reviewed in parallel sessions at this conference, they illustrate well the breadth of the topics that are currently under study:

- nonperturbative QCD contributions to the anomalous magnetic moment of the muon;
- isospin breaking and QED effects;
- strongly-interacting matter;
- structure of nucleons;
- nonperturbative inputs for flavor physics phenomenology;
- theories of composite dark matter;
- supersymmetry.

The calculation of the hadronic contributions to the anomalous magnetic moment of the muon were review by T. Izubuchi. This computation is important in view of the planned experiments aiming at measuring $g-2$. Several results have appeared this year, see e.g. Refs. [20, 21, 22, 23, 24, 25], and further references therein.

As a consequence of the precision of current simulations, isospin and hence QED effects have become a significant source of systematic error; first studies have appeared that address the formulation of lattice QCD in the presence of isospin breaking, and coupled to a U(1) electromagnetic field. Interesting results have appeared recently [26, 27].

The Equation of State of QCD describes the phase transition that strongly-interacting matter undergoes as the temperature of the system is increased. Recent results have been reported this year in Refs. [28, 29].

Current lattice simulations also allow nucleons to be studied from first principles. Computations of the nucleon vector and axial form factors, of the nucleon axial charge, and quark momentum fraction can now be computed with quarks at the physical point, with an accuracy at around 10% [30]. See also Ref. [31] for an interesting new approach.

An up-to-date summary of lattice results for flavor physics is provided by the FLAG collaboration in Ref. [32], and was summarized at this conference in a parallel session by A. El-Khadra. The combination of new lattice gauge configurations generated with dynamical light quarks, and several complementary formulations for the heavy quark sector, yield a varierty of results with high accuracy. A review of results was presented at this conference, see e.g. [33, 34].

Novel work on the possibility of constructing models of composite dark matter was reported in one of the parallel sessions. The idea behind these models is that Dark Matter is made of the hadrons of a new strongly-interacting sector. Investigating numerically the nuclear spectroscopy of these theories allows the computation of the nonperturbative parameters that describe the dynamics of these systems. Ultimately these inputs are crucial for the construction of a predictive and constrained theory. The results of these studies for an SU(2) gauge theory with two Dirac fermions in the fundamental representation, using Wilson fermions, have appeared in Refs. [35, 36]. Similar directions have been investigated in Refs. [37, 38].

Finally, recent progress in formulating supersymmetric theories on a discrtized space-time have been reviewed by S. Catterall. An excellent review of the theoretical issues can be found in Ref. [39].

Hopefully this partial report shall help the reader to get a snapshot of the recent directions in which the field is moving. The list of references should also provide an entry point to the contemporary literature on these topics.

5. IR fixed points for composite Higgs models

The discovery of the Higgs boson at the LHC is an impressive verification of the Standard Model. The first measurements of the couplings of the Higgs boson show that the Brout-Englert-Higgs (BEH) mechanism provides an excellent description of electroweak symmetry breaking. However the Higgs sector of the Standard Model lagrangian does not provide any insight on the dynamical origin of the symmetry breaking, which is simply parametrized by the Higgs potential. The measured value of the mass of the Higgs requires new physics beyond the Standard Model to explain the stabilization of the electroweak scale. Traditional solutions like minimal supersymmetry need to be finely tuned in order to be compatible with current observations.

Assuming that the Higgs boson is a composite particle provides an appealing explanation of the mechanism of electroweak symmetry breaking; indeed it would be very much in line with the BCS explanation of superconductivity, which in turn inspired the BEH mechanism. In composite scenarios the nonperturbative dynamics of some new sector beyond the Stantard Model is advocated in order to explain the origin of the electroweak scale, in analogy to the appearance of an hadronic scale in QCD. The problem with these kind of models is that computations in the nonperturbative regime of quantum field theories are difficult, and hence quantitative predictions do not come easy. The naive idea that the new sector could be a rescaled version of QCD is already ruled out by LHC data, and therefore some new dynamics needs to be found if these scenarios have any chance of describing experimental results.

Strongly-interacting dynamics that is qualitatively different from QCD could be obtained by deforming theories that possess an infrared (IR) fixed point [40, 41, 42]. This kind of fixed point is usually thought to be generated by the quantum effects of a sufficiently large number of massless fermions, hence these problems could only be addressed as a result of the algorithmic progresses discussed above. Numerical simulations of gauge theories regulated on a space-time lattice provide a tool to study the nonperturbative regime of these theories from first principles. Hence they have a great potential to become a valuable instrument for addressing the questions discussed above. In the rest of this article, we will discuss some of the methods that have been developed to identify the existence of fixed points, and to compute the associated anomalous dimensions, using lattice simulations.

The behaviour of couplings in the neighbourhood of a fixed point is described by the so-called linearized renormalization group (RG) equations. The RG equations describe the change of the couplings in a theory as the energy scale is varied: the couplings are said to *flow* in parameter space. This flow is described by the beta functions:

$$\mu \frac{d}{d\mu} \hat{g}_k = \beta_k(\hat{g}) ,$$

where $\hat{g}$ are dimensionless couplings, and μ is the energy scale. The fixed points are identified by the zeroes of the beta functions, and the equations in the neighbourhood of a fixed point can be linearized:

$$\beta_k(\hat{g}^*) = 0 \implies \beta_k(\hat{g}) = \left.\frac{\partial \beta_k}{\partial \hat{g}_j}\right|_* (\hat{g}_j - \hat{g}_j^*) .$$

Working in a basis in the space of couplings where the matrix $\left.\frac{\partial \beta_k}{\partial \hat{g}_j}\right|_*$ is diagonal, the linearized evolution is easily solved:

$$\hat{g}_k(\mu) = \hat{g}_{k,0} \left(\frac{\mu}{\Lambda_{\rm UV}}\right)^{-y_k} \implies \boxed{\Lambda_{\rm IR} = \hat{g}_0^{1/y_k} \Lambda_{\rm UV}} , \tag{15}$$

where y_k are the eigenvalues of the linearized evolution matrix, and are related to the anomalous dimensions of operators at the fixed point. In Eq. 15 $\hat{g}_{k,0}$ is the value of the coupling at the UV scale $\Lambda_{\rm UV}$, while $g_k(\mu)$ is the value at the scale μ. The IR scale $\Lambda_{\rm IR}$ is defined as the value of the energy scale where the coupling becomes of order 1. It is clear that a small value of the exponent y_k leads naurally to a large separation between the UV and the IR scale, which otherwise would require a fine-tuning of the coupling at the UV scale g_{k0}. Hence we see explicitly that the existence of a fixed point with small exponents y_k can be used to generate a hierarchy of scales without fine-tuning.

Theories that have an IR fixed point are said to be inside the *conformal window.* Analytical results about the conformal window are difficult to derive in the absence of further symmetries, like e.g. supersymmetry. The expectation for gauge theories coupled to fermions, usually based on some kind of approximations, is that these theories develop a fixed point as the number of fermions n_f is increased, and hence there is a range in n_f for which there is an IR fixed point before asymptotic freedom is lost. The conjectured conformal window for SU(N) theories coupled to fermions in different representations of the gauge group are shown in Fig. 4.

Phenomenologically viable models can be built by deformations of theories inside the conformal window, or by finding theories close to the edge of the conformal window. Since the existence of an IR fixed point has some unambiguous signatures, we focus on the possibility of finding robust evidence that a given theory is indeed inside the conformal window. This entails a characterization of the fixed point obtained from a determination of the anomalous dimensions.

The existence of a IR fixed point has its theoretical signatures. Here we shall concentrate on the scaling of the spectrum of the theory. Indeed if the theory becomes scale invariant at large distances, it cannot develop a mass gap, and therefore *all* the masses of the states in the spectrum are expected to scale to zero as the fermion mass is sent to zero, i.e. as the explicit source of breaking of scale invariance is removed. More precisely, the masses scale to zero as a power of the fermion mass, and the scaling exponent is determined by the critical dimension of the mass operator in the theory [44, 45, 46]:

$$M_H \propto m^{1/(1+\gamma^*)} , \tag{16}$$

where M_H is the mass of the physical states, m is the fermion mass, and γ^* is the anomalous dimension of the mass at the fixed point. Clearly this behaviour is at odds with the one expected

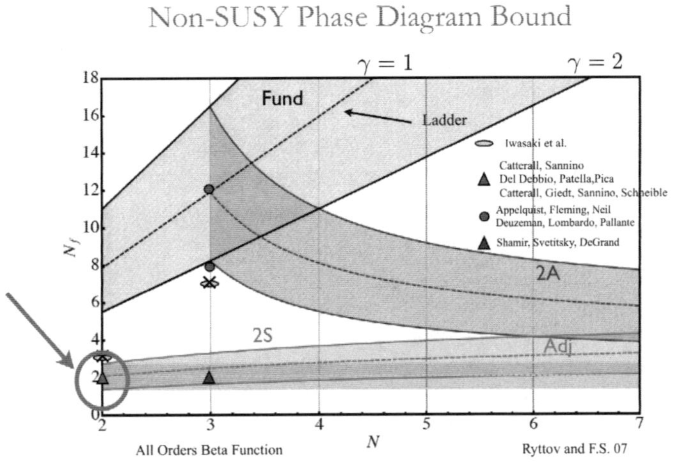

Figure 4. Conformal windows for SU(N) theories coupled to fermions. The coordinate on the horizontal axis labels the gauge group, while the one on the vertical axis represents the number of fermion families. Above the upper bound of each band, the theory is no longer asymptotically free. Different colours correspond to different representations of the fermions. The arrow indicates the SU(2) theory with two adjoint fermions for which we report results below. The phase diagram sketched here is taken from Ref. [43].

in a theory outside the conformal window, where chiral symmetry breaking leads to massless pions, but other states remain massive even when the fermion mass goes to zero. The latter is the behaviour that is expected e.g. in QCD.

Figure 5 shows the spectrum for a number of states computed from Monte Carlo simulations of an SU(2) gauge theory with two Dirac fermions in the adjoint representation [47]. Although the scaling of the masses to zero as a function of the fermion mass is encouraging, a more quantitative analysis is needed in order to conlcude that the theory possesses a fixed point. Systematic errors due to the finite value of the fermion mass, and the finite volume of the system, could obscure the physics that instead we would like to highlight.

A more quantitative description is obtained by looking at ratios of masses. The upper part of Fig. 6 shows the ratio of the mass of the vector particle M_V to the mass of the pseudoscalar one M_{PS}. This ratio remains constant as the mass of the fermions is decreased, as expected from the scaling argument presented above, and in sharp contrast with the expectation in a theory where chiral symmetry is spontaneously broken. In the latter case, the pseudoscalar particle becomes massless in the chiral limit because it is a Goldstone boson, while the rest of the spectrum remains finite, and hence the ratio in the plot should diverge. The lower part of the figure shows the ratio of the pseudoscalar mass to the square root of the string tension. Once again the qualitative behaviour is consistent with the existence of an IR fixed point with a single relevant operator which we identify as the fermion mass. A careful analysys shows that finite volume effects are sizeable for the smallest fermion masses that have been simulated so far [48]. Finite-volume effects are clearly visible when comparing the results obtained on lattices of increasing volume in the upper plot. As expected these effects are only relevant for light fermion masses: this is the regime where signatures of IR-conformality can be found, and it is crucial that all sources of systematic error are studied in detail.

Evidence for the existence of a fixed point can be reinforced by combining multiple analyses,

Journal of Physics: Conference Series **640** (2015) 012049 doi:10.1088/1742-6596/640/1/012049

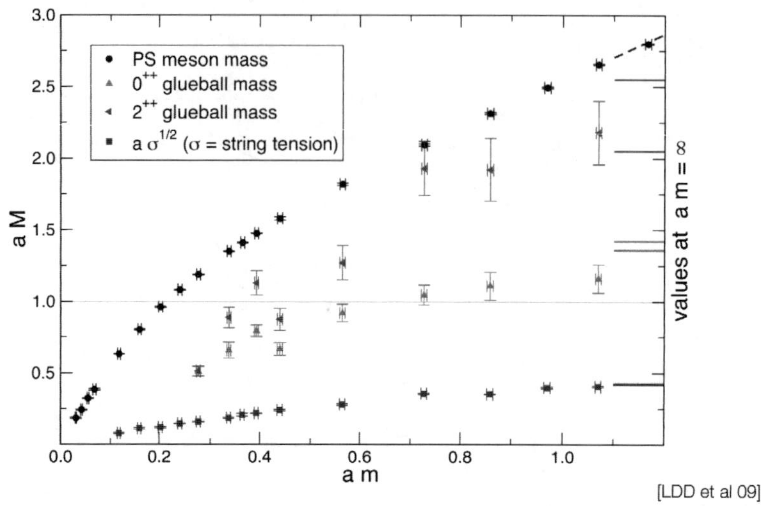

Figure 5. Scaling of the spectrum as a function of the fermion mass. The fermion mass is defined as the PCAC mass, and hence the point at $m = 0$ is the point where explicit symmetry breaking is absent. The masses of all the states scale to zero at the same rate as predicted for theories with an IR fixed point.

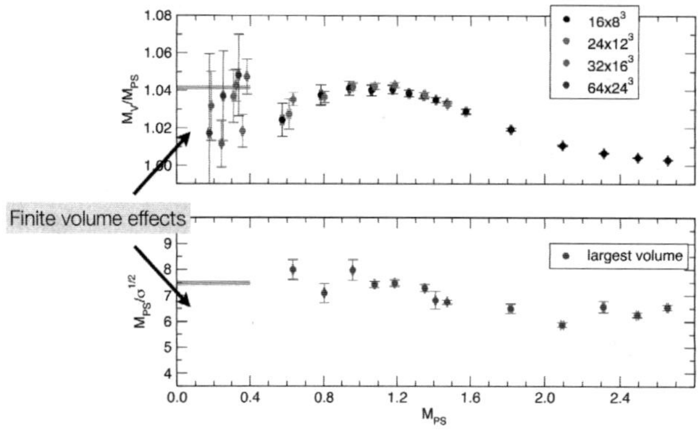

Figure 6. Ratios of masses as a function of the fermion mass. Different colours indicate results from lattices with different volumes. All results are obtained at a single value of the lattice spacing a.

e.g. by looking simultaneously at the scaling of several quantities. In Ref. [49] scaling relations for dimensionful quantities including matrix elements of operators were established, which can be tested in numerical studies. Also the study of the density of eigenvalues of the Dirac operators in Ref. [50] has provided a determination of the mass anomalous dimension for SU(2) with adjoint fermions. The simulations for this model are all consistent with the existence of an IR fixed point, and mass anomalous dimension $\gamma^* = 0.37(2)$, which is related to the critical exponent by $y_m = 1 + \gamma^*$. This is so far the strongest numerical evidence in favour of the existence of an IR fixed point.

Recent investigations have suggested novel theories as candidates for building BSM scenarios. All these theories are at the edge of the conformal window, and could be characterized by multiple scales, with large separations between them. Some theories, like the the SU(3) gauge theory with sextet fermions [60] have a light scalar state in the spectrum, which could be a candidate for a composite Higgs boson. On the other hand, the recent studies in Ref. [57] have found for the first time a theory with a large mass anomalous dimension. The interested reader can find a compilation of recent results in Refs. [51, 52, 53, 54, 55, 56, 57, 58, 59, 60, 61]. While this list is not exhaustive, it should provide a suitable overview of the current problems in this field.

6. Outlook

There has been a remarkable progress in numerical simulations of lattice field theories in the last decade, which has led to the possibility of simulating light dynamical fermions efficiently on current hardware. From the algorithmic point of view, we have witnessed a very impressive improvement, partly driven by a better understanding of the physical origin of the critical slowing down of the simulation codes for light fermion masses. The performance of the current algorithms has opened up two major research directions.

On the one hand simulations of QCD are now performed at physical quark masses, and hence some major sources of systematic errors have been removed from lattice inputs in flavor physics. As a consequence of this increase in accuracy, new questions need to be addressed, like e.g. the impact of electrodynamics and isospin breaking. There has been theoretical work in this direction, and first numerical results have started to appear.

On the other hand, simulations with dynamical fermions have made possible the numerical study of theories beyond QCD. Some tantalizing results have emerged from Monte Carlo simulations that have identified strongly-interacting theories as candidates to describe new sectors beyond the Standard Model. The challenge in the near future will be to scrutinize these theories even more closely, in order to draw robust conclusions that may be useful for model building.It will be interesting to follow the developments in this area as the LHC restarts its experiments in 2015.

References

[1] A. Ukawa [CP-PACS and JLQCD Collaborations], Nucl. Phys. Proc. Suppl. **106** (2002) 195.
[2] Y. Saad. *Iterative Methods for Sparse Linear Systems.* Society for Industrial and Applied Mathematics, Philadelphia, PA, USA, 2nd edition, 2003.
[3] M. Hasenbusch, Phys. Lett. B **519** (2001) 177 [hep-lat/0107019].
[4] T. Banks and A. Casher, Nucl. Phys. B **169** (1980) 103.
[5] M. Luscher, JHEP **0707** (2007) 081 [arXiv:0706.2298 [hep-lat]].
[6] R. G. Edwards, U. M. Heller and R. Narayanan, Phys. Rev. D **59** (1999) 094510 [hep-lat/9811030].
[7] H. Neff, N. Eicker, T. Lippert, J. W. Negele and K. Schilling, Phys. Rev. D **64** (2001) 114509 [hep-lat/0106016].
[8] R. B. Morgan and W. Wilcox, Nucl. Phys. Proc. Suppl. **106** (2002) 1067 [hep-lat/0109009].
[9] D. Darnell, R. B. Morgan and W. Wilcox, Nucl. Phys. Proc. Suppl. **129** (2004) 856 [hep-lat/0309068].
[10] Dean Darnell, Ronald B. Morgan, and Walter Wilcox. *Linear Algebra and its Applications*, 429(10):2415 – 2434, 2008.
[11] P. A. Boyle, arXiv:1402.2585 [hep-lat].
[12] A. Brandt, *Math. Comp.* **31**:330 – 390, 1977.
[13] M. Brezina, R. Falgout, S. MacLachlan, T. Manteuffel, S. McCormick, J. Ruge, *SIAM Journal on Scientific Computing* **25**(6):1896 – 1920, 2004.
[14] J. Brannick, R. C. Brower, M. A. Clark, J. C. Osborn and C. Rebbi, Phys. Rev. Lett. **100** (2008) 041601 [arXiv:0707.4018 [hep-lat]].
[15] R. Babich, J. Brannick, R. C. Brower, M. A. Clark, S. D. Cohen, J. C. Osborn and C. Rebbi, PoS LAT **2009** (2009) 031 [arXiv:0912.2186 [hep-lat]].
[16] R. Babich, J. Brannick, R. C. Brower, M. A. Clark, T. A. Manteuffel, S. F. McCormick, J. C. Osborn and C. Rebbi, Phys. Rev. Lett. **105** (2010) 201602 [arXiv:1005.3043 [hep-lat]].
[17] A. Frommer, K. Kahl, S. Krieg, B. Leder and M. Rottmann, SIAM J. Sci. Comput. **36** (2014) A1581 [arXiv:1303.1377 [hep-lat]].
[18] M. Luscher, arXiv:1002.4232 [hep-lat].

[19] D. B. Kaplan, Phys. Lett. B **288** (1992) 342 [hep-lat/9206013].
[20] T. Blum, S. Chowdhury, M. Hayakawa and T. Izubuchi, arXiv:1407.2923 [hep-lat].
[21] B. Chakraborty, C. T. H. Davies, G. C. Donald, R. J. Dowdall, J. Koponen, G. P. Lepage and T. Teubner, arXiv:1403.1778 [hep-lat].
[22] M. Benayoun, J. Bijnens, T. Blum, I. Caprini, G. Colangelo, H. Czy, A. Denig and C. A. Dominguez *et al.*, arXiv:1407.4021 [hep-ph].
[23] A. Francis, V. Guelpers, G. Herdoiza, H. Horch, B. Jaeger, H. B. Meyer and H. Wittig, arXiv:1410.7491 [hep-lat].
[24] M. Della Morte, A. Francis, G. Herdoiza, H. Horch, B. Jger, A. Jttner, H. Meyer and H. Wittig, PoS LATTICE **2014** (2014) 162 [arXiv:1411.1206 [hep-lat]].
[25] A. Francis, V. Glpers, B. Jger, H. Meyer, G. von Hippel and H. Wittig, PoS LATTICE **2014** (2014) 128 [arXiv:1411.7592 [hep-lat]].
[26] S. Borsanyi, S. Durr, Z. Fodor, C. Hoelbling, S. D. Katz, S. Krieg, L. Lellouch and T. Lippert *et al.*, arXiv:1406.4088 [hep-lat].
[27] R. Zhou and S. Gottlieb, PoS LATTICE **2014** (2014) 024 [arXiv:1411.4115 [hep-lat]].
[28] S. Borsanyi, Z. Fodor, C. Hoelbling, S. D. Katz, S. Krieg, C. Ratti and K. K. Szabo, arXiv:1410.7917 [hep-lat].
[29] A. Bazavov *et al.* [HotQCD Collaboration], Phys. Rev. D **90** (2014) 9, 094503 [arXiv:1407.6387 [hep-lat]].
[30] S. Syritsyn, T. Blum, M. Engelhardt, J. Green, T. Izubuchi, C. Jung, S. Krieg and M. Lin *et al.*, arXiv:1412.3175 [hep-lat].
[31] A. J. Chambers, R. Horsley, Y. Nakamura, H. Perlt, D. Pleiter, P. E. L. Rakow, G. Schierholz and A. Schiller *et al.*, Phys. Rev. D **90** (2014) 014510 [arXiv:1405.3019 [hep-lat]].
[32] S. Aoki, Y. Aoki, C. Bernard, T. Blum, G. Colangelo, M. Della Morte, S. Drr and A. X. El Khadra *et al.*, Eur. Phys. J. C **74** (2014) 9, 2890 [arXiv:1310.8555 [hep-lat]].
[33] A. X. El-Khadra, PoS LATTICE **2013** (2014) 001 [arXiv:1403.5252 [hep-lat]].
[34] N. H. Christ, J. M. Flynn, T. Izubuchi, T. Kawanai, C. Lehner, A. Soni, R. S. Van de Water and O. Witzel, arXiv:1404.4670 [hep-lat].
[35] W. Detmold, M. McCullough and A. Pochinsky, arXiv:1406.2276 [hep-ph].
[36] W. Detmold, M. McCullough and A. Pochinsky, arXiv:1406.4116 [hep-lat].
[37] A. Hietanen, R. Lewis, C. Pica and F. Sannino, JHEP **1407** (2014) 116 [arXiv:1404.2794 [hep-lat]].
[38] T. Appelquist *et al.* [Lattice Strong Dynamics (LSD) Collaboration], Phys. Rev. D **89** (2014) 094508 [arXiv:1402.6656 [hep-lat]].
[39] S. Catterall, D. B. Kaplan and M. Unsal, Phys. Rept. **484**, 71 (2009) [arXiv:0903.4881 [hep-lat]].
[40] M. J. Strassler, hep-th/0309122.
[41] D. K. Hong, S. D. H. Hsu and F. Sannino, Phys. Lett. B **597**, 89 (2004) [hep-ph/0406200].
[42] M. A. Luty and T. Okui, JHEP **0609**, 070 (2006) [hep-ph/0409274].
[43] D. D. Dietrich and F. Sannino, Phys. Rev. D **75** (2007) 085018 [hep-ph/0611341].
[44] T. DeGrand and A. Hasenfratz, Phys. Rev. D **80** (2009) 034506 [arXiv:0906.1976 [hep-lat]].
[45] L. Del Debbio and R. Zwicky, Phys. Rev. D **82** (2010) 014502 [arXiv:1005.2371 [hep-ph]].
[46] L. Del Debbio and R. Zwicky, Phys. Rev. D **89** (2014) 014503 [arXiv:1306.4038 [hep-ph]].
[47] L. Del Debbio, B. Lucini, A. Patella, C. Pica and A. Rago, Phys. Rev. D **80** (2009) 074507 [arXiv:0907.3896 [hep-lat]].
[48] F. Bursa, L. Del Debbio, D. Henty, E. Kerrane, B. Lucini, A. Patella, C. Pica and T. Pickup *et al.*, Phys. Rev. D **84** (2011) 034506 [arXiv:1104.4301 [hep-lat]].
[49] L. Del Debbio and R. Zwicky, Phys. Lett. B **700** (2011) 217 [arXiv:1009.2894 [hep-ph]].
[50] A. Patella, Phys. Rev. D **86** (2012) 025006 [arXiv:1204.4432 [hep-lat]].
[51] J. Rantaharju, T. Karavirta, V. Leino, T. Rantalaiho, K. Rummukainen and K. Tuominen, arXiv:1411.4879 [hep-lat].
[52] M. P. Lombardo, K. Miura, T. J. N. da Silva and E. Pallante, arXiv:1410.0298 [hep-lat].
[53] T. DeGrand, Y. Shamir and B. Svetitsky, Phys. Rev. D **88** (2013) 5, 054505 [arXiv:1307.2425].
[54] T. DeGrand, Y. Liu, E. T. Neil, Y. Shamir and B. Svetitsky, arXiv:1412.4851 [hep-lat].
[55] Y. Aoki *et al.* [the LatKMI Collaboration], Phys. Rev. D **89** (2014) 111502 [arXiv:1403.5000 [hep-lat]].
[56] Y. Aoki *et al.* [LatKMI Collaboration], Phys. Rev. Lett. **111** (2013) 16, 162001 [arXiv:1305.6006 [hep-lat]].
[57] A. Athenodorou, E. Bennett, G. Bergner and B. Lucini, arXiv:1412.5994 [hep-lat].
[58] T. Appelquist *et al.* [LSD Collaboration], Phys. Rev. D **90** (2014) 114502 [arXiv:1405.4752 [hep-lat]].
[59] T. Appelquist, R. C. Brower, M. I. Buchoff, M. Cheng, G. T. Fleming, J. Kiskis, M. F. Lin and E. T. Neil *et al.*, Phys. Rev. Lett. **112** (2014) 111601 [arXiv:1311.4889 [hep-ph]].
[60] Z. Fodor, K. Holland, J. Kuti, D. Nogradi, C. Schroeder and C. H. Wong, Phys. Lett. B **718** (2012) 657 [arXiv:1209.0391 [hep-lat]].
[61] J. Kuti, PoS LATTICE **2013** (2014) 004.

Journal of Physics: Conference Series **640** (2015) 012050 doi:10.1088/1742-6596/640/1/012050

Supersymmetry on a Lattice

Simon Catterall

Department of Physics, Syracuse University, Syracuse, NY 13244, US

E-mail: smcatterall@gmail.com

Abstract. Attempts to formulate supersymmetric field theories on discrete spacetime lattices have a long history. However, until recently most of these efforts have failed. In this talk I will review some of the new ideas that have finally allowed a solution to this problem for certain supersymmetric theories. I will focus my attention, in particular, on N=4 super Yang-Mills which forms one of the pillars of the AdSCFT correspondence connecting gravitational theories in anti-deSitter space to gauge theories living on the boundary of that space. It is also integral to the scattering amplitudes program and is under intensive study using conformal bootstrap techniques. I will review the theoretical construction of the model and describe what is known analytically about the lattice theory. Following on from this I will present preliminary results stemming from the first large scale lattice study of N=4 Yang-Mills. Our results are consistent with the existence of a single deconfined phase for the theory in which the static fermion potential shows only a Coulomb-like behavior. Unlike QCD there is no confining, chirally broken phase even at strong coupling.

1. Introduction

The problem of formulating supersymmetric theories on lattices has a long history going back to the earliest days of lattice gauge theory. However, after initial efforts failed to produce useful supersymmetric lattice actions the topic languished for many years. Indeed a folklore developed that supersymmetry and the lattice were mutually incompatible. However, recently, the problem has been re-examined using new tools and ideas such as topological twisting, orbifold projection and deconstruction and a class of lattice models have been constructed which maintain one or more supersymmetries exactly at non-zero lattice spacing. For a general review of these approaches see [1] and references therein.

While in low dimensions there are many continuum supersymmetric theories that can be discretized this way, in four dimensions there appears to be a unique solution to the various constraints: N=4 super Yang-Mills (SYM). The availability of a supersymmetric lattice construction for this theory is clearly very exciting from the point of view of exploring the connection between gauge theories and string/gravitational theories and is complementary to approaches based on the conformal bootstrap or large N integrability considerations.[1]

First, let me explain why discretization of supersymmetric theories resisted solution for so long. The central problem is that naive discretizations of continuum supersymmetric theories break supersymmetry completely and radiative effects lead to a profusion of relevant supersymmetry breaking counter terms in the renormalized lattice action. The coefficients to

[1] Additional work on N=4 Yang-Mills which does not rely on these formulations can be found in [2, 3, 4, 5, 6].

these counter terms must then be carefully fine tuned as the lattice spacing is sent to zero in order to arrive at a supersymmetric theory in the continuum limit.

In most cases this is both unnatural and practically impossible particularly if the theory contains scalar fields. Of course, one might have expected problems; the supersymmetry algebra is an extension of the Poincare algebra which is explicitly broken on the lattice. Specifically, there are no infinitesimal translation generators on a discrete spacetime so that the algebra $\{Q, \overline{Q}\} = \gamma^\mu p_\mu$ is broken at the classical level.

One option is to arrange for supersymmetry to emerge as an accidental symmetry in the continuum limit in a manner analogous to the way Poincaré symmetry is restored in lattice QCD for vanishing lattice spacing. In the latter case the existence of a subgroup of discrete translational and rotational symmetries in the lattice action protect the theory from developing relevant counter terms which violate the full continuum symmetry which then re-emerges automatically in the continuum limit. Ideally one would like to arrange for something similar to happen in the case of supersymmetry.

In the case of N=1 SYM chiral symmetry can enforce just such a condition the only relevant susy breaking operator is the gaugino mass which is prohibited by chiral symmetry. If one employs a lattice action which respects chiral symmetry such as domain wall fermions one can hope that supersymmetry, while broken at non-zero lattice spacing, should be recovered without fine tuning in the continuum limit. Unfortunately, this nice state of affairs does not persist to more interesting models such as super QCD or models with extended supersymmetry containing scalar fields.

2. Twisting and the lattice

In the last five years or so this problem has been revisited using new theoretical tools and ideas and a set of lattice models have been constructed which retain exactly some of the continuum supersymmetry at non-zero lattice spacing. The basic idea is to use this subalgebra to constrain the effective lattice action and protect the theory from the dangerous susy violating counter terms. The class of theories which can be treated this way is quite restrictive: the continuum theories need to possess multiples of 2^D supercharges but nevertheless this set includes many interesting theories.

These theories were first constructed using orbifold methods [7] but afterwards it was realized that the same lattice theories could be derived using ideas based on topological twisting [8, 9, 10] The basic idea of twisting goes back to Witten in his seminal paper on topological field theory [11]. In our context the idea is to decompose the fields of the theory in terms of representations not of the original (Euclidean) rotational symmetry $SO_{\rm rot}(D)$ but a *twisted* rotational symmetry which is the diagonal subgroup of this symmetry and a $SO_{\rm R}(D)$ subgroup of the R-symmetry of the theory.

$$SO_{\rm twist}(D) = {\rm diag}\left(SO_{\rm rot}(D) \times SO_{\rm R}(D)\right). \tag{1}$$

To be explicit consider the case where the number of supersymmetries is 2^D. The supercharges of the theory transform under Lorentz L and flavor/R symmetries F as

$$q^i_\alpha \to L_{\alpha\beta} F_{ij} q_{\beta j}. \tag{2}$$

Under the twisted diagonal symmetry with $F = L$ they hence transform like a matrix $q \to LqL^T$ of size $2^{D/2} \times 2^{D/2}$. This matrix can be expanded on products of Dirac gamma matrices

$$q = QI + Q_\mu \gamma_\mu + Q_{\mu\nu}\gamma_\mu\gamma_\nu + \ldots \tag{3}$$

with the various supercharges now transforming as antisymmetric tensors or p-forms under the twisted symmetry. It is easy to show from the original supersymmetry algebra that the scalar supercharge Q is nilpotent and that the momentum is Q-exact. This latter fact makes it plausible that the entire energy-momentum tensor is also Q-exact and hence that the twisted theory has the structure of a co-homological topological field theory. Indeed we will see that the action for twisted $N = 4$ Yang-Mills can be written as the sum of Q-exact and Q-closed pieces.

It is important at this point to recognize that the transformation to twisted variables corresponds to a simple change of variables if one lives only in flat space. The merit of this change of variables is that the resultant theory is more suitable for discretization. A true topological field theory only results when the scalar charge is treated as a BRST charge and attention is restricted to states annihilated by this charge. In the language of the supersymmetric parent theory such a restriction corresponds to a projection to the vacua of the theory. It is not employed in these lattice constructions.

When one discretizes the theory one tries to preserve just the subalgebra $Q^2 = 0$ which then guarantees that the lattice theory is invariant under Q. Of course one must simultaneously require that the theory not suffer from doubling problems and, in the case of Yang-Mills theories retains exact gauge invariance. Remarkably, both of these additional constraints may be satisfied if the discretization is chosen carefully.

The absence of fermion doubling is intimately linked to the fact that the fermions, like the supercharges, are treated as a set of antisymmetric tensors in the twisted reformulation of the supersymmetric theory. For the case of 2^D supersymmetries it is possible to show that the original action describing $2^{D/2}$ degenerate Majorana fermions, can be re-written in an entirely geometrical way requiring only the (gauged) exterior derivative d and its adjoint $d^\dagger$.

Indeed, the twisted fermions satisfy the Kähler-Dirac equation which takes the form [12]

$$(d - d^\dagger)\Psi = 0 \tag{4}$$

where $\Psi = (\eta, \psi_\mu, \chi_{\mu\nu}, \ldots)$ is the collection of antisymmetric tensor fermion fields. This equation, unlike its spinorial cousin can be discretized in such a way that the discrete solutions map one-to-one with those of the continuum equation, as was shown many years ago by Rabin [13] and thereby evades fermion doubling problems. In practice this necessitates assigning p-form fermions to p-cells on the lattice and replacing the exterior derivative and its adjoint by forward and backward difference operators respectively. The well-known staggered fermion formulation of lattice QCD can be related to this prescription for discretization of the Kähler-Dirac equation. This structure for the fermionic action when combined with the exact Q susy and gauge invariance will then dictate where the bosons of the lattice theory are located. To understand how this works in more detail let me first describe the bosonic sector of the continuum twisted Yang-Mills theory.

The gauge fields are singlets under the R-symmetry and so transform as vectors again under twisted rotations. The scalars however typically transform as scalars under the original rotations but vectors under the R-symmetry. Thus they also transform as vectors under twisted rotations. In practice they may be combined with the gauge fields to yield complexified connections in the twisted theory with the scalar fields being encoded in the imaginary parts of the gauge field. Notice that while the bosonic degrees of freedom are composed of complex connections the final action is only invariant under the usual U(N) gauge symmetry corresponding to the real part of this complex connection. In the lattice theory the complex gauge fields are then placed on corresponding links in the lattice. Unlike lattice QCD they are taken to live in the algebra of $GL(N, C)$. The associated measure is simply the flat measure. This procedure does not violate gauge invariance since the full measure involves contributions from both the complex boson and its complex conjugate and the Jacobians that result from these two measures after performing a gauge transformation cancel against reach other.

3. Lattice N=4 Yang-Mills

In four dimensions the constraint that the target theory possess 16 supercharges singles out a single theory N=4 SYM. The continuum twist of N=4 that is the starting point of the construction was first written down by Marcus in 1995 [14] although it now plays a important role in the Geometric-Langlands program and is hence sometimes called the GL-twist [15] . This four dimensional twisted theory is most compactly expressed as the dimensional reduction of a five dimensional theory in which the ten (one gauge field and six scalars) bosonic fields are realized as the components of a complexified five dimensional gauge field while the 16 twisted fermions naturally span a real Kähler-Dirac field in five dimensions. Unlike the case of say lattice QCD we are not free to use any lattice in our construction; the lattice should possess a set of links that correspond to the different components of the twisted fields. This suggests a lattice whose unit cell contains *five* equivalent basis vectors on which the boson fields reside.

Such a four dimensional lattice exists - the A_4^* lattice and is constructed from the set of five basis vectors v_a pointing out from the center of a four dimensional equilateral simplex out to its vertices together with their inverses $-v_a$. It is the four dimensional analog of the 3D bcc lattice and the two dimensional triangular lattice (A_2^*). Complexified Wilson gauge link variables U_a are placed on these links together with their Q-superpartners ψ_a. Another 10 fermions are associated with additional face links $v_a + v_b$ with $a > b$. Finally, the exact scalar supersymmetry implies the existence of a single fermion for every lattice site.

The lattice action corresponds to a discretization of the Marcus twist on this A_4^* lattice and is given below

$$S = \sum_{\mathbf{n}} \mathrm{Tr}\left[\mathcal{F}_{ab}^{\dagger}(\mathbf{n})\mathcal{F}_{ab}(\mathbf{n}) + \frac{1}{2}\left(\overline{\mathcal{D}}_a^{(-)}\mathcal{A}_a(\mathbf{n})\right)^2 - \chi_{ab}(\mathbf{n})\mathcal{D}_{[a}^{(+)}\psi_{b]}(\mathbf{n}) - \eta(\mathbf{n})\overline{\mathcal{D}}_a^{(-)}\psi_a(\mathbf{n})\right] + S_{\mathrm{cl}} \quad (5)$$

where

$$S_{\mathrm{cl}} = -\frac{1}{4}\epsilon_{abcde}\chi_{de}(\mathbf{n} + \boldsymbol{\mu}_a + \boldsymbol{\mu}_b + \boldsymbol{\mu}_c)\overline{\mathcal{D}}_c^{(-)}\chi_{ab}(\mathbf{n} + \boldsymbol{\mu}_c),$$

and the lattice field strength is given by

$$\mathcal{F}_{ab}(\mathbf{n}) = \mathcal{A}_a(\mathbf{n})\mathcal{A}_b(\mathbf{n} + \boldsymbol{\mu}_a) - \mathcal{A}_b(\mathbf{n})\mathcal{A}_a(\mathbf{n} + \boldsymbol{\mu}_b).$$

It is invariant under lattice gauge transformations and a global permutation symmetry S^5 and can be proven free of fermion doubling problems as discussed before. Furthermore it can be shown to reduce to the Marcus twist of N=4 in the naive continuum limit. It is also invariant under the lattice Q-symmetry:

$$\begin{aligned} Q\,\mathcal{A}_a &= \psi_a & Q\,\psi_a &= 0 \\ Q\,\chi_{ab} &= -\overline{\mathcal{F}}_{ab} & Q\,\overline{\mathcal{A}}_a &= 0 \\ Q\,\eta &= d & Q\,d &= 0, \end{aligned}$$

To this lattice action we need to add two additional terms which break the exact Q-symmetry. The first of these is given below:

$$S_1 = \mu^2 \sum_{x,a}\left[\frac{1}{N}\mathrm{Tr}\left(\overline{\mathcal{A}}_a(x)\mathcal{A}_a(x)\right) - I\right]^2. \quad (6)$$

This term ensures that the lattice gauge field has the expansion

$$\mathcal{A}_a(x) = I + A_a(x) + \ldots\ldots \quad (7)$$

The unit matrix that appears in this expression is necessary for the lattice action to target the correct twisted four dimensional Yang-Mills theory in the naive continuum limit. It represents a vacuum expectation value for the $U(1)$ component of the scalar field.[2] The second term that is needed for practical simulations is

$$S_2 = \kappa \sum_{x,a,b} |\det P_{ab} - 1|^2 \tag{8}$$

This term targets the $U(1)$ sector of the theory and is needed to suppress strong lattice artifacts that arise in the $U(1)$ sector at strong coupling. It has the effect of enforcing an approximate projection of the theory down to $SU(N)$. These terms break Q-supersymmetry but any new counterterms induced by these terms must possess couplings which vanish in the limit $\mu^2, \kappa \to 0$. Thus they do not induce additional fine tuning problems. With these additions the lattice theory may be simulated using the standard algorithms of lattice QCD - see the recent paper [17].

The lattice constructions I have described play a role similar to lattice QCD in providing a non-perturbative definition of the corresponding continuum supersymmetric field theory. But at least in the case of sixteen supercharges they also have potential value as tools to explore the connection between gauge theory and string and supergravity models. Of course the most obvious example of the latter is the AdS-CFT correspondence which in its weakest form tell us that type IIB string theory should be dual to maximally supersymmetric large N Yang-Mills theory at strong coupling. The constructions described here may provide a new window on this strongly coupled theory both conceptually and practically through the use of Monte Carlo simulation.

Furthermore these dualities are conjectured to persist for type II string theory containing Dp-branes and (p+1)-dimensional Yang-Mills with 16 supercharges. The simplest of these dualities corresponds to large N Yang-Mills quantum mechanics compactified on a thermal circle and IIa supergravity at low temperature. Quite a bit of numerical work both on and off the lattice has already been attempted for this system yielding encouraging agreement between the thermodynamics of the Yang-Mills system and the corresponding black hole solution in supergravity[18].

4. Analytical results

The presence of an exact supersymmetry at non zero lattice spacing leads to some remarkable conclusions. In [19] a perturbative calculation was carried out with the conclusion that the (divergent part of) the one loop diagrams of the lattice theory matched those in the continuum twisted theory. This has an immediate consequence; the beta function of the lattice theory vanishes and no fine tuning is needed to target the continuum theory at the one loop level. Furthermore, and perhaps even more remarkably, one can prove that the classical moduli space of the theory survives to all orders of perturbation theory (for $\mu^2 = \kappa = 0$). Thus all scalar fields remain massless as a consequence of the single lattice supersymmetry.

On the question of restoration of the other (twisted) supersymmetries there has also been progress; in [20] it was proved that restoration of a set of discrete R symmetries is sufficient to ensure that all supersymmetries are inherited in the continuum limit. And in [21] a RG argument is given that only a single marginal coupling needs to be fine tuned to achieve this. This is comparable to the tuning needed for Wilson fermions in simulations of lattice QCD and shows that the new formulations have all but solved the traditional fine tuning problems of lattice supersymmetry in this case.

These are powerful results but to understand the physical content of the theory eg its scaling dimensions requires techniques that can be used at strong coupling and for generic observables.

[2] See [16] for a similar non-compact formulation.

Journal of Physics: Conference Series **640** (2015) 012050 doi:10.1088/1742-6596/640/1/012050

This has motivated the development of numerical codes that reveal the structure of the theory for any value of the gauge coupling and for arbitrary numbers of colors [17, 22]. We turn to this next.

5. Technical aspects of simulations

These mainly relate to the practical problems of simulating these theories. Since we cannot represent anticommuting numbers on a computer one typically integrates out the fermionic degrees of freedom before devising a numerical algorithm to sample the path integral. This procedure generates the Pfaffian or determinant of a large sparse matrix whose size is given by the total number of degrees of freedom. In general, expectation values must be weighted by this Pfaffian in the path integral. However in many cases the Pfaffian is complex and hence carries a non-trivial phase factor. Monte Carlo algorithms require that the weight be be real, positive definite so typically this phase is dropped in the simulation. In principle, this can be corrected for when measuring expectation values by a well known re-weighting procedure. However, this technique will fail if the phase fluctuates too rapidly: this is the well known sign problem of finite density QCD simulations.

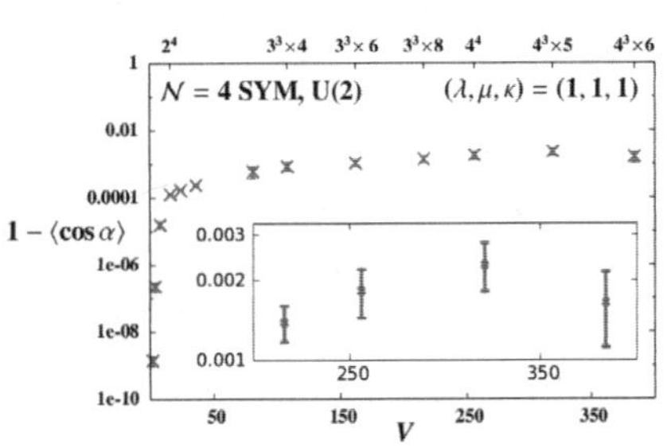

Figure 1. Pfaffian vs volume

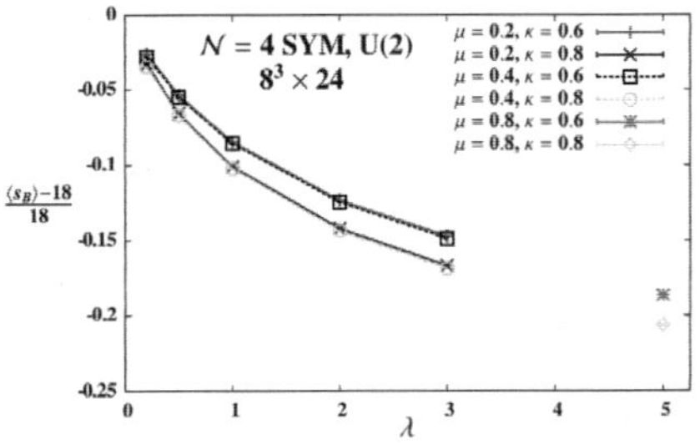

Figure 2. Average bosonic action vs coupling

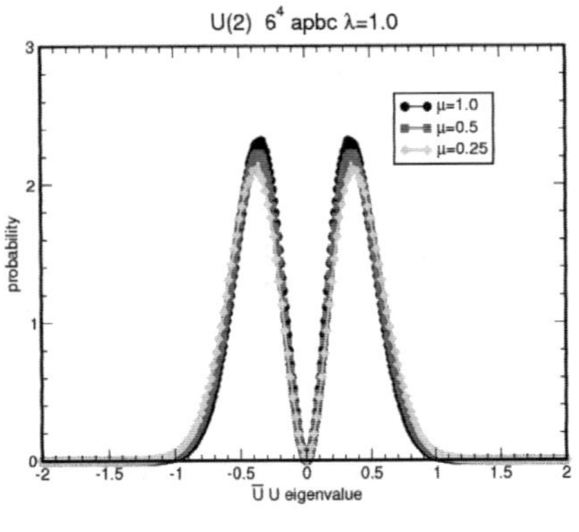

Figure 3. Distribution of scalar eigenvalues for several μ^2 and $\lambda = 1$

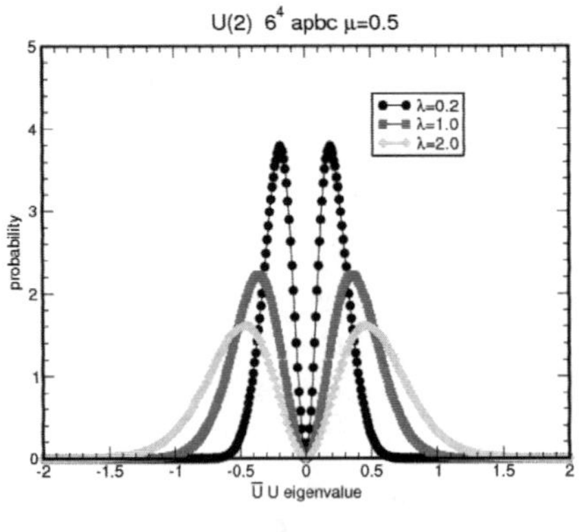

Figure 4. Distribution of scalar eigenvalues for $\mu = 0.5$ with varying λ

A priori it is not known whether this sign problem will make simulation of these supersymmetric lattice systems impossible. Luckily, preliminary numerical work has been rather encouraging in this respect at least for the sixteen supercharge models - see figure 1 and results in [23, 24]. There is no hint in the numerical data of the extensive growth of the phase with system volume that would be expected in any system experiencing a sign problem. It is an open theoretical question as to why this is so; there is no symmetry that guarantees positivity in the N=4 fermion operator and so the observed numerical result appears to be dynamical in origin; the phase quenched ensemble simply never visit regions of the configuration space where the Pfaffian changes sign. In figure 2 we also show results for the expectation value of the bosonic action. This is related to a Q Ward identity and hence can be calculated exactly. The results show that in the limit $\mu^2 \to 0$ and $\kappa \to 0$ this limit is approached smoothly so that this limit is thought to be well under control. Again, these results were obtained in the phase quenched ensemble and the agreement with the exact result serves as another indicator that the system does not suffer from a sign problem.

There is one further obstruction that potentially may ruin attempts at numerical simulation; the presence of a non-trivial moduli space. The vacua of the continuum and lattice models consists of constant commuting matrices. In particular the magnitude of the scalar fields can a priori diverge along so-called flat directions. One might worry that this effect might destabilize the numerical simulations. We find that this is not the case – figures 3 and 4 show the distribution of eigenvalues of the scalar fields extracted from $\mathcal{A}_a^\dagger \mathcal{A}_a$ as μ^2 and the 't Hooft coupling λ are varied. One sees that the distributions are bounded for all gauge couplings and independent of μ^2 in the limit in which that parameter is sent to zero. The reason for this is simple; in the Monte Carlo we integrate over a set of fermion zero modes that arise on the flat directions. The presence of these zero modes yields a strong suppression of the flat directions in the path integral. In the infinite volume continuum theory these modes are not integrated since they are non-normalizable and the distribution of scalars is indeed flat. In a finite volume we must integrate them and they provide a natural stabilization mechanism.

To understand the possible phases of the lattice theory we have computed the string tension which can be extracted from the static potential [24, 25]. The string tension is shown in figure 5 and shows a value that is consistent with zero out to 't Hooft couplings $\lambda \sim 2.5$. Complementary to this we find that the static potential *can* be well fit with a Coulomb potential as is shown in figure 6. The bold line indicates the perturbative prediction.

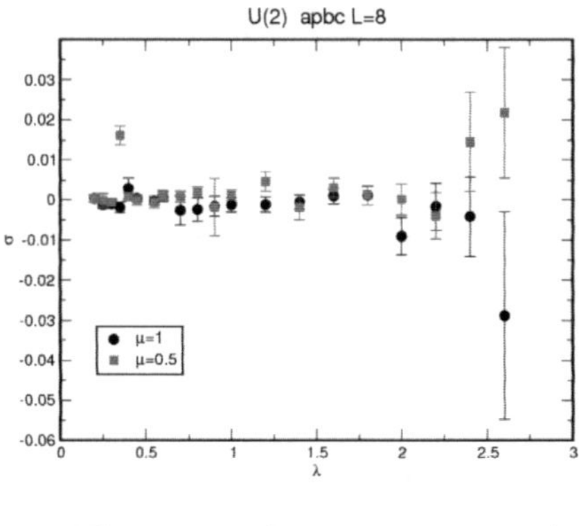

Figure 5. String tension vs lambda

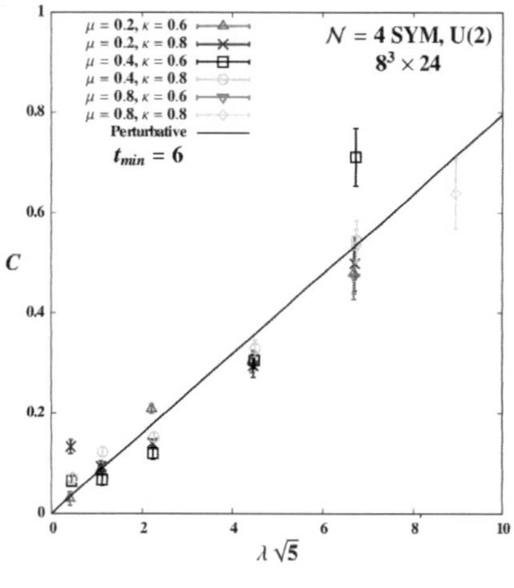

Figure 6. Coulomb coefficient vs coupling

6. Summary

In this talk I have described a new class of lattice gauge theories which retain an exact supersymmetry at non zero lattice spacing. The presentation has focused on N=4 Yang-Mills which is arguably the most interesting of these theories. This lattice theory exhibits several remarkable properties; the scalars are naturally massless to all orders in lattice perturbation theory, the beta function is zero at one loop and one has to tune a single log divergent operator to take a continuum limit in which all the symmetries are restored. Large scale simulations are underway and preliminary results favor a single phase system with no chiral symmetry breaking or confinement for arbitrarily strong coupling. We are currently focusing on methods to measure scaling dimensions of operators in this would be conformal field theory in an effort to make contact with other continuum approaches. The lattice has the potential to provide a well-defined non-perturbative regulator for this theory valid for any number of colors and coupling. As such it may give us a tool to explore quantum and string effects in gravity.

Acknowledgments
I would like to acknowledge my collaborators Poul Damgaard, Tom Degrand, Joel Giedt, Anosh Joseph, David Schaich and Aarti Veernala as well as numerous important conversations over the years with David B. Kaplan and Mithat Ünsal. Numerical simulations have been performed using USQCD facilities at Fermilab. The author received partial support from the DOE under grants SC0009998 and SC0008669.

7. References
[1] Catterall S, Kaplan D B and Unsal M 2009 *Phys. Rept.* **484** 71–130 (*Preprint* 0903.4881)
[2] Hanada M, Matsuura S and Sugino F 2011 *Prog. Theor. Phys.* **126** 597–611 (*Preprint* 1004.5513)
[3] Honda M, Ishiki G, Nishimura J and Tsuchiya A 2011 *PoS* **Lattice 2011** 244 (*Preprint* 1112.4274)
[4] Ishii T, Ishiki G, Shimasaki S and Tsuchiya A 2008 *Phys. Rev.* **D78** 106001 (*Preprint* 0807.2352)
[5] Ishiki G, Kim S W, Nishimura J and Tsuchiya A 2009 *Phys. Rev. Lett.* **102** 111601 (*Preprint* 0810.2884)
[6] Ishiki G, Kim S W, Nishimura J and Tsuchiya A 2009 *JHEP* **0909** 029 (*Preprint* 0907.1488)
[7] Kaplan D B and Unsal M 2005 *JHEP* **0509** 042 (*Preprint* hep-lat/0503039)
[8] Catterall S 2008 *JHEP* **0801** 048 (*Preprint* 0712.2532)
[9] Damgaard P H and Matsuura S 2008 *Phys. Lett.* **B661** 52–56 (*Preprint* 0801.2936)
[10] Unsal M 2006 *JHEP* **0610** 089 (*Preprint* hep-th/0603046)
[11] Witten E 1988 *Commun.Math.Phys.* **117** 353
[12] Banks T, Dothan Y and Horn D 1982 *Phys. Lett.* **B117** 413
[13] Rabin J M 1982 *Nucl. Phys.* **B201** 315
[14] Marcus N 1995 *Nucl. Phys.* **B452** 331–345 (*Preprint* hep-th/9506002)
[15] Kapustin A and Witten E 2007 *Commun. Num. Theor. Phys.* **1** 1–236 (*Preprint* hep-th/0604151)
[16] Palumbo F 1990 *Phys. Lett.* **B244** 55–57
[17] Schaich D and Degrand T 2014 (*Preprint* 1410.6971)
[18] Catterall S, Joseph A and Wiseman T 2010 *JHEP* **1012** 022 (*Preprint* 1008.4964)
[19] Catterall S, Dzienkowski E, Giedt J, Joseph A and Wells R 2011 *JHEP* **1104** 074 (*Preprint* 1102.1725)
[20] Catterall S, Giedt J and Joseph A 2013 *JHEP* **1310** 166 (*Preprint* 1306.3891)
[21] Catterall S and Giedt J 2014 *accepted for publication JHEP* (*Preprint* 1408.7067)
[22] Catterall S and Joseph A 2012 *Comput. Phys. Commun.* **183** 1336–1353 (*Preprint* 1108.1503)
[23] Catterall S, Galvez R, Joseph A and Mehta D 2012 *JHEP* **1201** 108 (*Preprint* 1112.3588)
[24] Catterall S, Schaich D, Damgaard P H, DeGrand T and Giedt J 2014 *Phys.Rev.* **D90** 065013 (*Preprint* 1405.0644)
[25] Catterall S, Damgaard P H, Degrand T, Galvez R and Mehta D 2012 *JHEP* **1211** 072 (*Preprint* 1209.5285)

Semimetal-Insulator transition without a condensate

Venkitesh Ayyar
Duke University, Durham NC, USA.
E-mail: vpa@phy.duke.edu

Abstract. We study a lattice field theory model containing two flavors of massless staggered fermions with an onsite four-fermion interaction. The symmetry of the model forbids non-zero fermion bilinear order parameters that can generate a fermion mass. At weak couplings, we expect a massless fermion (Semimetal) phase. At strong couplings, we can argue for the existence of massive fermions (Insulator phase) without the formation of any fermion bilinear condensate. Using Monte Carlo calculations in three space-time dimensions, we find evidence for a direct second order phase transition between the two phases.

1. Introduction

It is well known that relativistic four-fermion field theories in three dimensions can contain strongly interacting second order fixed points [1, 2] . The main motivation for their study is to understand new mechanisms for fermion mass generation that may be realized in nature. The conventional mechanism of fermion mass generation is via spontaneous symmetry breaking of chiral symmetries which is signalled through a non-zero fermion bilinear condensate. In this work, we explore a more exotic mechanism of fermion mass generation without such condensates.

We study a simple lattice four-fermion model in three Euclidean dimensions. This model is a limiting case of a Yukawa model. In the weak-coupling regime, since the coupling is irrelevant, we have a massless phase. In the strong coupling regime, we can argue for the existence of a massive phase. Interestingly, all fermion bilinear condensates are found to vanish in both phases. This exotic massive phase with zero condensates has been studied before [3]. However, a first principles Monte Carlo calculation to study the transition between these phases in 3D had not been done before. In this work, we perform such a calculation and show that these two phases are separated by a single second-order phase transition. A detailed description of this work has been published in [4].

2. Model

We study a simple four-fermion model containing two flavors of staggered fermions whose Euclidean action is given by

$$S = \sum_{i=1,2} \sum_{x,y} \overline{\psi}_{x,i} \, M_{x,y} \, \psi_{y,i} - U \sum_{x} \left\{ \overline{\psi}_{x,1} \psi_{x,1} \overline{\psi}_{x,2} \psi_{x,2} \right\} \tag{1}$$

where $\overline{\psi}_{x,i}, \psi_{x,i}, i = 1,2$ are four independent Grassmann valued fields. The matrix M is the well known staggered fermion matrix given by

$$M_{x,y} = \sum_{\hat{\alpha}} \frac{\eta_{x,\hat{\alpha}}}{2} \left[\delta_{x,y+\hat{\alpha}} - \delta_{x,y-\hat{\alpha}} \right] \tag{2}$$

where $x \equiv (x_1, x_2, x_3)$ denotes a lattice site on a 3 dimensional cubic lattice and $\hat{\alpha} = \hat{1}, \hat{2}, \hat{3}$ represent unit lattice vectors in the three directions. The staggered fermion phase factors are defined as usual: $\eta_{x,\hat{1}} = 1, , \eta_{x,\hat{2}} = (-1)^{x_1}$, and $\eta_{x,\hat{3}} = (-1)^{x_1+x_2}$. We will study cubical lattices of equal size L in each direction with anti-periodic boundary conditions.

It can be shown that, in addition to the usual space-time lattice transformations (translation, axis reversal, rotation [5, 6]) , the action is symmetric under internal $SU(4)$ transformations given by :

$$\begin{pmatrix} \psi_{x_e,1} \\ \overline{\psi}_{x_e,1} \\ \psi_{x_e,2} \\ \overline{\psi}_{x_e,2} \end{pmatrix} \rightarrow V \begin{pmatrix} \psi_{x_e,1} \\ \overline{\psi}_{x_e,1} \\ \psi_{x_e,2} \\ \overline{\psi}_{x_e,2} \end{pmatrix} \text{ and } \begin{pmatrix} \psi_{x_o,1} \\ \overline{\psi}_{x_o,1} \\ \psi_{x_o,2} \\ \overline{\psi}_{x_o,2} \end{pmatrix} \rightarrow V^* \begin{pmatrix} \psi_{x_o,1} \\ \overline{\psi}_{x_o,1} \\ \psi_{x_o,2} \\ \overline{\psi}_{x_o,2} \end{pmatrix} \tag{3}$$

where x_e and x_o refer to even and odd lattice sites resepectively and V is a $SU(4)$ matrix in the fundamental representation.

The observables we wish to measure are the susceptibilities :

$$\chi_1 = \frac{1}{2L^3} \sum_{x,y,x\neq y} \langle \overline{\psi}_{x,1} \psi_{x,1} \overline{\psi}_{y,1} \psi_{y,1} \rangle. \tag{4}$$

$$\chi_2 = \frac{1}{2L^3} \sum_{x,y,x\neq y} \langle \overline{\psi}_{x,1} \psi_{x,1} \overline{\psi}_{y,2} \psi_{y,2} \rangle. \tag{5}$$

where expectation values are defined as

$$\left\langle \mathscr{O} \right\rangle = \frac{1}{Z} \int [d\overline{\psi}\, d\psi]\, \mathscr{O}\, \mathrm{e}^{-S(\overline{\psi},\psi)} \tag{6}$$

with Z being the partition function. The presence of a condensate can be inferred when these susceptibilities diverge as L^3 for large values of L. A constant susceptibility at large L would signal a zero condensate. Using the above $SU(4)$ symmetry, it can be shown that all other fermion bilinear susceptibilities can be expressed in terms of χ_1 and χ_2. Another observable that we measure is the local four-point condensate defined by

$$\rho_m = \frac{1}{L^3} \sum_x \langle \overline{\psi}_{x,1} \psi_{x,1} \overline{\psi}_{x,2} \psi_{x,2} \rangle. \tag{7}$$

3. The Fermion Bag Approach

The traditional method for studying four-fermion field theories is by introducing an auxiliary field to convert the four-fermion term into a fermion bilinear. In this work we use an alternative Monte Carlo approach introduced a few years ago, called the fermion bag approach [7]. A review of the fermion bag approach can be found in [8].

In the Fermion bag approach, by defining a binary lattice field $n_x = 0, 1$, we divide the lattice sites into monomers ($n_x = 1$) and free sites ($n_x = 0$). Calling each such distribution a configuration $[n]$, (Figure 1) we can express our observables as a sum over these configurations. For example, the partition function can be written in the strong and weak coupling limits as follows:

$$Z = \sum_{[n]} U^k \prod_{\mathscr{B}} \Big(\mathrm{Det}(W_{\mathscr{B}})\Big)^2 \tag{8}$$

$$Z = \Big(\mathrm{Det}(M)\Big) \sum_{[n]} U^k \Big(\mathrm{Det}(G)\Big)^2, \tag{9}$$

Figure 1. An example of a monomer configuration $[n]$ showing free fermion bags on a two dimensional lattice. The blue circles denote the monomer sites.

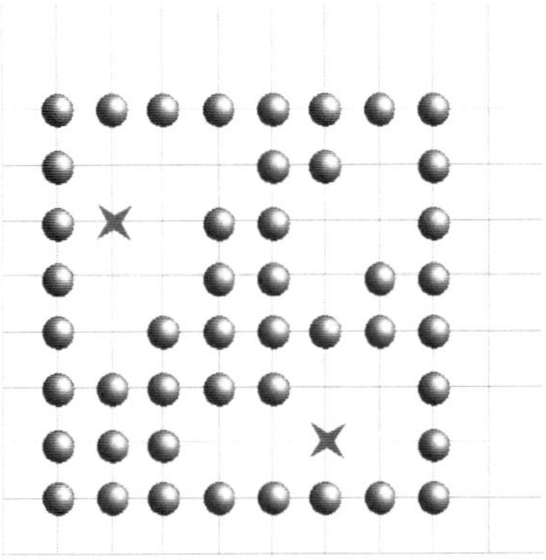

Figure 2. An example of a monomer configuration $[n]$ giving a zero contribution for the fermionic operator. The crosses denote the source points x and y.

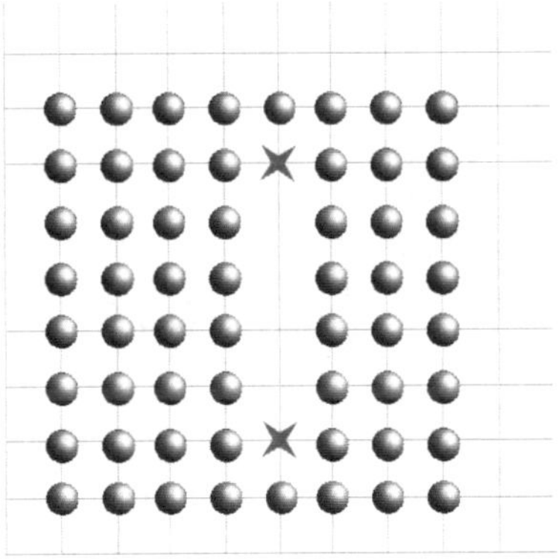

Figure 3. The leading contribution with one fermion bag, at strong coupling. The crosses denote the source points x and y.

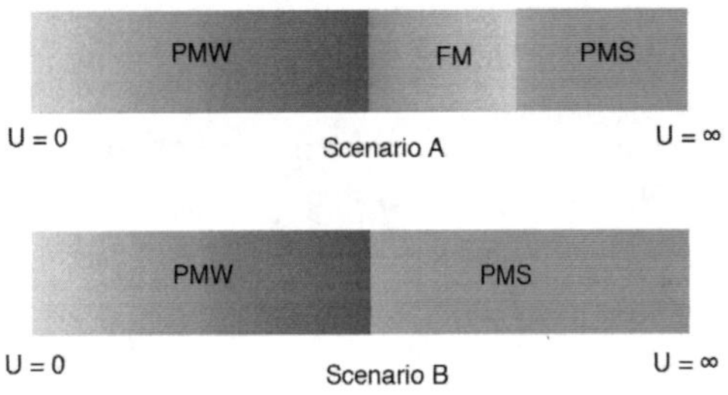

Figure 4. The two possible phase diagrams for our model based on previous studies. Our work is consistent with scenario B with a second order transition between the PMW phase and the PMS phase.

where k represents the number of monomers in the configuration, M is the free staggered fermion matrix defined in (2), $W_{\mathscr{B}}$ represents a free staggered fermion matrix connecting the sites within the bag $\mathscr{B}$, while G represents a $k \times k$ free staggered propagator matrix connecting monomer sites.

There are two equivalent ways to look at fermion bags :

1 In the weak coupling limit, we can think of fermions as living on the monomer sites and being able to hop on to other monomer sites through the free fermion propogators. The collection of all monomer sites can be thought of as a fermion bag.

2 In the strong coupling limit, we can think of fermions as living on the free sites and being able to hop on to other free sites due to the free action. Each group of connected free sites can be thought

of as forming a fermion bag.

3.1. Phases

Using the Fermion bag approach, it is easy to visualize the different phases of the model in the strong and weak coupling limits. In 3D, since the four-fermion coupling is irrelevant, we must have a massless phase at weak couplings. To understand the behavior at strong couplings, let us write the fermionic correlator in the form:

$$\langle \psi^1_x \overline{\psi^1_y} \rangle \;=\; \frac{1}{Z}\sum_{[n]} U^{N_m} \prod_{\mathscr{B}} (\det(W_B))^2\, {W^{-1}}_{\mathscr{B};x,y} \tag{10}$$

where ${W^{-1}}_{\mathscr{B};x,y}$ is the inverse of the Dirac operator within the free fermion bag $\mathscr{B}$ containing the points x and y. For large U, the leading contribution will come from configurations with very few free sites and hence the free fermion bags will be small. Since the free action only connects nearest-neighbors, fermions cannot hop between different bags. Hence if the sites x and y where the fermion sources are inserted belong to different bags, (Fig 2) then (10) requires the fermionic correlator to be zero. Hence, the leading contributions will come from configurations having just one fermion bag containing both points x and y (Fig 3). Since large fermion bags are exponentially supressed, the correlator decays exponentially, implying a massive phase at large couplings. Similarly, using a slightly different argument we can show that the bosonic correlator also decays exponentially at large couplings [4]. This implies that the condensate is zero.

As mentioned in the Introduction, such exotic mechanisms of fermion mass generation have appeared in literature in the context of Yukawa models. Mean field calculations have shown a phase transition from a massless phase (referred to as PMW or weak paramagnetic phase) to an intermediate massive phase with a fermion bilinear chiral condensate (referred to as FM phase) and another transition from this intermediate phase to a new massive phase with zero fermion bilinear condensates (referred to as PMS phase)[9]. Interestingly, other mean field calculations [10, 11] for 2 flavors in 3D give a single first-order phase transtion from the PMW to the PMS phases. These two scenarios are shown in Fig 4. In our work, we find a single second-order transition from the PMW to PMS phases.

4. Analysis and results

We have collected data for cubical lattices with sizes ranging from 12^3 upto 28^3, using three different Monte-Carlo algorithms. A detailed description of our algorithms and analysis can be found in [4].

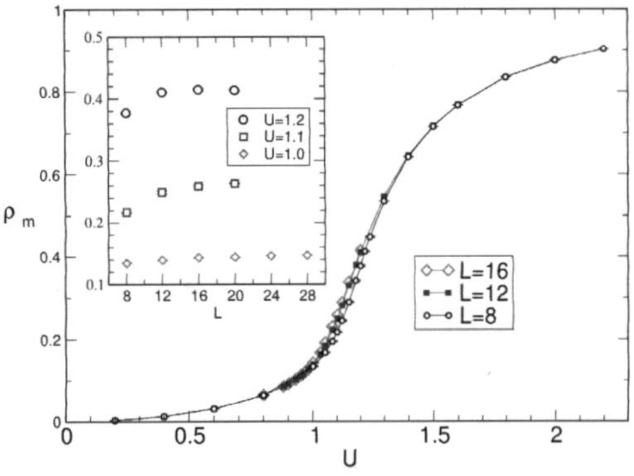

Figure 5. Variation of the average monomer density ρ_m as a function of coupling U for lattices of size 8,12,16.

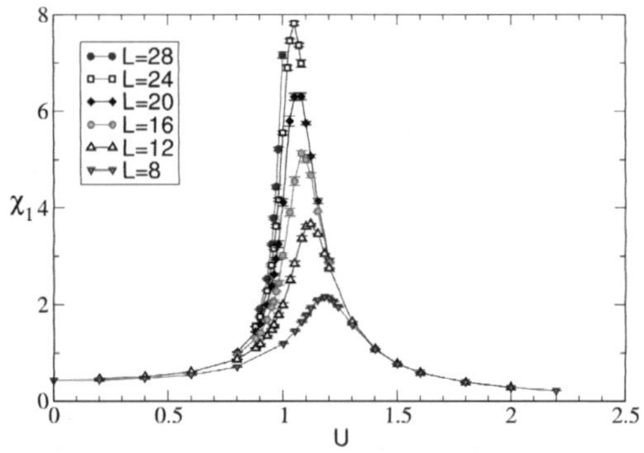

Figure 6. Variation of the susceptibility χ_1 as a function of coupling U.

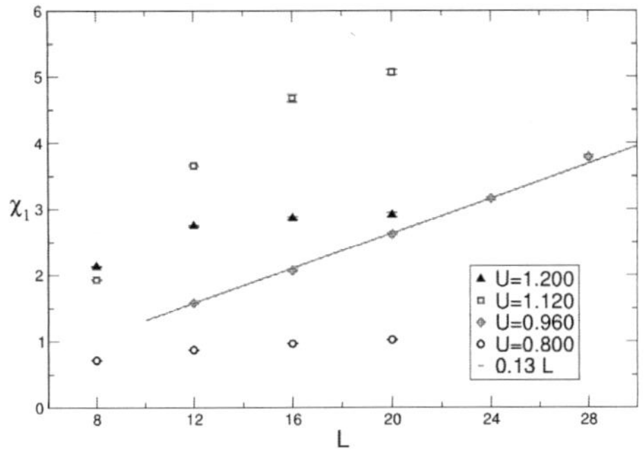

Figure 7. Variation of the susceptibility χ_1 as a function of lattice size L. Note the linear rise of the susceptibility for $U = 0.96$.

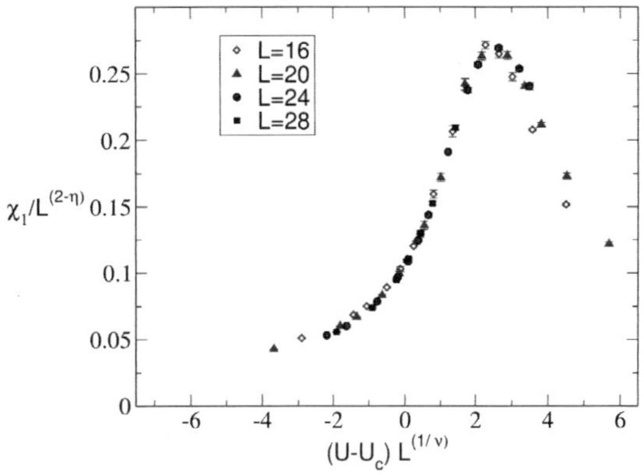

Figure 8. Evidence for a second order phase transition

Looking at the behavior of the four-point condensate ρ_m as a function of the coupling U in Figure 5 it is clear that it is a smooth function, increasing from 0 for small couplings, rising sharply near $U \sim 1$ and approaching 1 for large couplings. The thermodynamic limit is reached by L=16. Figure 6 shows the behavior of the susceptibility χ_1 as a function of the coupling U for various values of the lattice size L. The susceptibility is a smooth function of U, reaching a maximum for $U \sim 1$. If we look at the behavior of χ_1 as a function of L for different values of U in Figure 7, it can be seen that the susceptibility saturates with lattice size for both small and large couplings. The lack of an L^3 divergence implies the absence of a condensate. However near $U = 0.96$, the susceptibility increases linearly with L, which is consistent with a second order critical point. The susceptibility χ_2 also shows a similar behavior. All this points to a single phase-transition in the region close to $U_c = 0.96$.

We have used scaling relations of both susceptibilities and performed a combined fit to extract the critical exponents. We present two such fits in table 1.

Fit	U_c	ν	η	χ^2
1	0.957(1)	0.95(5)	0.940(9)	1.1
2	0.958(1)	1.24(2)	0.884(1)	1.9

Table 1. Critical exponents obtained using two different fits

Thus, although the critical point seems to be constrained well enough, the exponents ν and η are not. This suggests the need for data on larger lattices to constrain the exponents.

From the theory of second-order phase transitions, near a second-order critical point, we know that the susceptibility should vary continuously i.e.

$$\chi \sim L^{2-\eta} f\left((U - U_c)L^{\frac{1}{\nu}}\right) \tag{11}$$

Hence, a plot of $\frac{\chi}{L^{2-\eta}}$ vs $(U - U_c)L^{\frac{1}{\nu}}$ should be a smooth function. Figure 8 shows that this is indeed true. This points to a second-order phase transition between the PMW phase and the PMS phase.

5. Conclusions and Future work

We have studied a lattice model in three Euclidean dimensions in which fermions can acquire a mass, but without a fermion bilinear condensate. The transition from massless to massive phase seems second order. This suggests the possibility of defining a continuum quantum field theory at the critical point.

It would be interesting to ask if there is a Semimetal-Insulator transition belonging to this universality class.

To obtain the precise values of the critical exponents, we plan to extend this calculation to larger lattices. We also plan to extend this work to 4D to explore the existence of a similar phase transition there.

Acknowledgments
Work done in collaboration with Shailesh Chandrasekharan. The material presented here is based upon work supported by the U.S. Department of Energy, Office of Science, Nuclear Physics program under Award Number DE-FG02-05ER41368. This research was done using resources provided by the Open Science Grid, which is supported by the National Science Foundation and the U.S. Department of Energy's Office of Science.

References
[1] B. Rosenstein, B. Warr, and S. Park,
Dynamical symmetry breaking in four-fermion interaction models,
Phys.Rept. 205, 59 (1991).

[2] S. Hands, A. Kocic, and J. B. Kogut,
Four-Fermi theories in fewer than four dimensions,
Annals Phys. 224, 29 (1993).

[3] A. Hasenfratz, W.-q. Liu, and T. Neuhaus,
Phase Structure and Critical points in a scalar-fermion Model,
Phys.Lett. B236, 339 (1990).

[4] V. Ayyar and S. Chandrasekharan,
Massive fermions without fermion bilinear condensates,
arXiv:1410.6474, Oct 2014.

[5] C. van den Doel and J. Smit,
Dynamical symmetry breaking in two flavor SU(N) and SO(N) lattice gauge theories ,
Nucl.Phys. B228, 122 (1983).

[6] M. F. Golterman and J. Smit,
Self-energy and flavor interpretation of staggered fermions ,
Nucl.Phys. B245, 61 (1984).

[7] S. Chandrasekharan,
Fermion bag approach to lattice field theories ,
Phys. Rev. D 82, 025007 (2010).

[8] S. Chandrasekharan,
Fermion bag approach to Fermion sign problems,
The European Physical Journal A 49, 1 (2013), ISSN 1434-6001.

[9] I.-H. Lee, J. Shigemitsu, and R. E. Shrock,
Study of different lattice formulations of a Yukawa model with a real scalar field ,
Nucl.Phys. B334, 265 (1990).

[10] M. A. Stephanov and M. Tsypin,
Phase diagram for the lattice scalar-fermion model,
Phys.Lett. B236, 344 (1990).

[11] M. A. Stephanov and M. Tsypin,
The Phase structure of the U(1) Higgs-fermion lattice theory ,
Phys.Lett. B242, 432 (1990).

Electromagnetic effects on the light hadron spectrum

S Basak[1], A Bazavov[2], C Bernard[3], C DeTar[4], E Freeland[5], J Foley[4], Steven Gottlieb[6], U M Heller[7], J Komijani[3], J Laiho[8], L Levkova[4], R Li[4], J Osborn[9], R L Sugar[10], A Torok[4], D Toussaint[11], R S Van de Water[12] and R Zhou[12] [MILC Collaboration]

[1] NISER, Bhubaneswar, Orissa 751005, India
[2] Department of Physics and Astronomy, University of Iowa, Iowa City, IA 52240, USA
[3] Department of Physics, Washington University, St. Louis, MO 63130, USA
[4] Department of Physics and Astronomy, University of Utah, Salt Lake City, UT 84112, USA
[5] Liberal Arts Department, School of the Art Institute of Chicago, Chicago, IL, USA
[6] Department of Physics, Indiana University, Bloomington, IN 47405, USA
[7] American Physical Society, One Research Road, Ridge, NY 11961, USA
[8] Department of Physics, Syracuse University, Syracuse, NY 13244, USA
[9] ALCF, Argonne National Laboratory, Argonne, IL 60439, USA
[10] Physics Department, University of California, Santa Barbara, CA 93106, USA
[11] Physics Department, University of Arizona, Tucson, AZ 85721, USA
[12] Theoretical Physics Department, Fermilab, Batavia, IL 60510, USA

E-mail: sg@indiana.edu

Abstract. For some time, the MILC Collaboration has been studying electromagnetic effects on light mesons. These calculations use fully dynamical QCD, but only quenched photons, which suffices to NLO in χPT. That is, the sea quarks are electrically neutral, while the valence quarks carry charge. For the photons we use the non-compact formalism. We have new results with lattice spacing as small as 0.045 fm and a large range of volumes. We consider how well chiral perturbation theory describes these results and the implications for light quark masses.

1. Introduction

The MILC collaboration has been studying electromagnetic and isospin-violating effects in pions and kaons for a number of years [1–4]. These effects are crucial for determining the three light-quark masses. The u, d, and s quark masses are both fundamental parameters of the standard model of elementary particle physics and important for phenomenology. The uncertainty in the size of the electromagnetic contributions to the pion and kaon masses is a major part of the error on each of the masses and the greatest source of error on the m_u/m_d mass ratio [5].

The electromagnetic error in m_u/m_d depends on the error in our estimate of $(M^2_{K^+} - M^2_{K^0})^\gamma$ where γ indicates the electromagnetic contribution to the difference in the squares of the charged and neutral kaon masses. (There is also a difference coming from the quark masses.) In 1960, Dashen [6] showed that at leading order electromagnetic mass splittings for the mesons are mass independent, *i.e.*, $(M^2_{K^+} - M^2_{K^0})^\gamma = (M^2_{\pi^+} - M^2_{\pi^0})^\gamma$. We can parameterize the higher order effects via a parameter ϵ defined by $(M^2_{K^+} - M^2_{K^0})^\gamma = (1+\epsilon)(M^2_{\pi^+} - M^2_{\pi^0})^{\rm exp}$. This definition of ϵ, which uses the experimental mass difference on the RHS, is slightly different from what we have used before, in which $(M^2_{\pi^+} - M^2_{\pi^0})^\gamma$ appears on the RHS.

Table 1. Parameters of the (2+1)-flavor asqtad ensembles used in this study. Volumes marked with $*$ are currently used in the finite volume studies, but not in the full analysis. The quark masses m_l' and m_s' are the light and strange dynamical masses used in the runs. The number of configurations listed as '132+52' for the $a \approx 0.12$ fm, $48^3 \times 64$ ensemble gives values for two independent streams, the first in single precision, and the second in double. At the moment, we treat them as separate data, and do not average the results.

$\approx a$[fm]	Volume	β	m_l'/m_s'	# configs.	L (fm)	$m_\pi L$
0.12	$12^3 \times 64^*$	6.76	0.01/0.05	1000	1.4	2.7
	$16^3 \times 64^*$	6.76	0.01/0.05	1303	1.8	3.6
	$20^3 \times 64$	6.76	0.01/0.05	2254	2.3	4.5
	$28^3 \times 64$	6.76	0.01/0.05	274	3.2	6.3
	$40^3 \times 64^*$	6.76	0.01/0.05	115	4.6	9.0
	$48^3 \times 64^*$	6.76	0.01/0.05	132+52	5.4	10.8
	$20^3 \times 64$	6.76	0.007/0.05	1261	2.3	3.8
	$24^3 \times 64$	6.76	0.005/0.05	2099	2.7	3.8
0.09	$28^3 \times 96$	7.09	0.0062/0.031	1930	2.3	4.1
	$40^3 \times 96$	7.08	0.0031/0.031	1015	3.3	4.2
0.06	$48^3 \times 144$	7.47	0.0036/0.018	670	2.8	4.5
	$56^3 \times 144$	7.465	0.0025/0.018	433	3.3	4.4
	$64^3 \times 144$	7.46	0.0018/0.018	826	3.7	4.3
0.045	$64^3 \times 192$	7.46	0.0028/0.014	861	2.8	4.6

Our calculation is done with dynamical QCD but with quenched, noncompact QED. The analysis of Bijnens and Danielsson [7] shows that quenched QED is sufficient for a controlled calculation of ϵ at NLO in $SU(3)$ chiral perturbation theory.

In table 1, we list the ensembles of gauge configurations used for the results reported here. The two larger volume $a \approx 0.06$ fm and the $a \approx 0.045$ fm ensembles were not included in our results presented at Lattice 2014 [8]. The $40^3 \times 64$ and $48^3 \times 64$ ensembles with $\beta = 6.76$ were generated after Lattice 2014 and first discussed at this conference, but they are included in Ref. [8].

2. Finite volume effects

Because the photon is massless, finite volume effects are an important issue and a source of uncertainty in our earlier results. For $a \approx 0.12$ fm, we had volumes of $20^3 \times 64$ and $28^3 \times 64$ and found small finite volume effects compared to what was seen in Ref. [9]. Electromagnetic finite volume effects have been calculated by Hayakawa and Uno [10] using chiral perturbation theory. They found rather large effects; however, there is not a unique choice of how to treat finite-volume zero modes for the $U(1)$ field in the noncompact formalism. For the temporal modes A_0 in Coulomb gauge, all modes for 3-momentum $\vec{k} = 0$ must be dropped. That is, $A_0(\vec{0}, k_0) = 0$ for all k_0. For the spatial modes, the action density is $[(\partial_0 A_i)^2 + (\partial_j A_i)^2)]/2$, so the only mode that must be dropped is that with $(\vec{k}, k_0) = (\vec{0}, 0)$. Hayakawa and Uno dropped all A_i modes with $\vec{k} = 0$. However, we have only dropped modes with both $\vec{k} = 0$ and $k_0 = 0$. In our case, the finite size effects are smaller, although the magnitude of the effect does depend on T/L where T (L) is the temporal (spatial) extent of the lattice. Figure 1 compares the finite volume effects for the two choices.

In order to test the prediction, we added four new volumes this year with $L = 12$, 16, 40, and 48. In Fig. 2, we show our results for six values of L. We plot electromagnetic splittings of squared masses $\Delta M^2_{xy} \equiv M^2_{xy} - M^2_{x'y'}$, where the first meson is constructed from valence quarks x and y with charges q_x, q_y and masses m_x, m_y, and the second meson is made from quarks x'

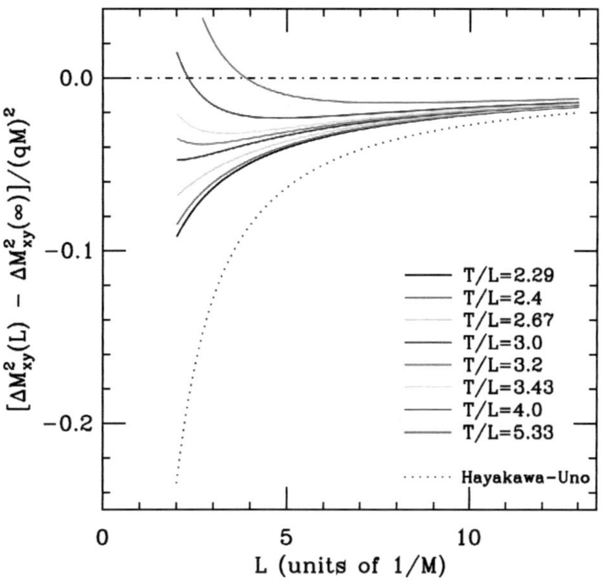

Figure 1. Comparison of electromagnetic finite volume (FV) effects in our scheme with that in Ref. [10]. q and M are the charge and mass of the meson, respectively. The values of T/L correspond to aspect ratios used in our ensembles, with 4.0 and 5.33 corresponding to the two smallest volumes in our FV study. Our FV effects are a factor of 2–3 smaller in most of the relevant range.

and y' with the same masses but with the quark charges set to zero. To compare results from different ensembles, we multiply the squared-mass difference by r_1^2 where r_1 is a length scale determined from the static potential. Each curve is a fit with a single parameter, the infinite volume limit. In other words, the shape of the curve, but not its height, is determined by the theory. The horizontal lines show the infinite volume limit for two mass combinations labeled 'pion' and 'kaon.' It is now clear why we saw such a small difference between $L = 20$ and 28. The slope of each curve is low in that region. It is also worth noting that for the 'pion,' the sign of the effect changes within the range of our new calculation. With our improved understanding of the finite volume effects, we can more accurately adjust our results to the infinite volume limit with a smaller residual error. This reduces our errors on ϵ and m_u/m_d, as we show later.

3. Chiral fit and extrapolation

We have calculated meson masses for a large number of valence-quark mass and charge combinations. On some ensembles, we include charges larger than the physical values. On others, we calculated all combinations of charges $\pm 2/3e$, $\pm 1/3e$, and 0. In the first step of the analysis, values are corrected for finite-volume effects. The corrections are 7–10% for pions (mesons with two light quarks) and 10–18% for kaons (mesons with one light quark and one whose mass is near the strange quark mass). Because of the high level of correlation among the many meson masses, we thin the data set and sometimes have to consider uncorrelated fits. A typical fit might include 150 points.

Once the chiral fit for uncharged sea quarks is determined, we can set valence and sea quark masses equal, set m_s to its correct value and take the continuum limit. We then set the sea quark charges to their physical value using the NLO chiral logs. The last adjustment is very small for the kaon and vanishes identically for the pion. Figure 3 shows a small subset of the data for the uncorrelated fit that determines our central value. Note how much smaller the electromagnetic effects are for the neutral mesons (right).

There is a complication involving the neutral pion. The physical pion propagator involves

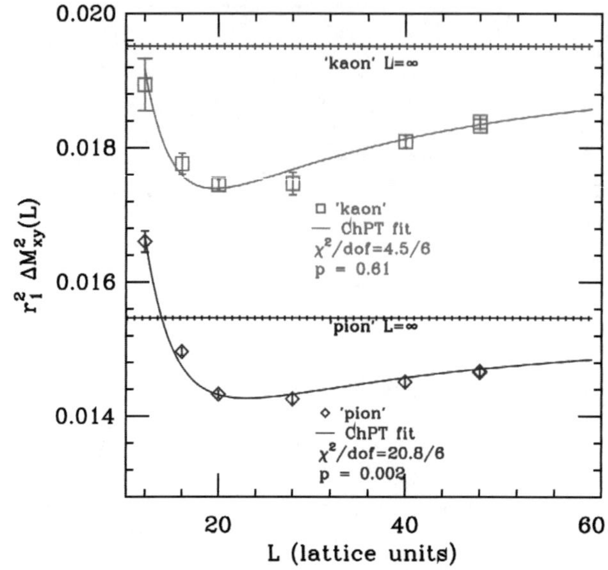

Figure 2. Finite volume effects at $a \approx 0.12$ fm and $am'_l = 0.01, am'_s = 0.05$ as a function of spatial lattice length L for two different meson masses: a unitary 'pion' (blue) with degenerate valence masses $m_x = m_y = m'_l$, and a 'kaon' (red) with valence masses $m_x = m'_l$ and $am_y = 0.04$, close to the physical strange quark mass. The fit lines are to the FV form from SχPT, and have one free parameter each, the infinite volume value (shown by horizontal solid lines with dotted lines for errors).

disconnected diagrams in which the quark and anti-quark annihilate. These contributions are difficult to calculate accurately because they are noisy. We drop the disconected diagrams and use the RMS average mass of $u\bar{u}$ and $d\bar{d}$ mesons. We denote this state as "π^0." However, electromagnetic contributions to neutral mesons vanish in the chiral limit, so both the true $(M^2_{\pi^0})^\gamma$ and our $(M^2_{\text{“}\pi^0\text{”}})^\gamma$ are small.

In the last stage of the analysis, we subtract our results for the "π^0" and K^0 from the corresponding charged meson to arrive at the physical electromagnetic mass differences. This is shown as the purple lines on the LHS of Fig. 3. The vertical lines represent the appropriate sum of quark masses for the charge-averaged pion and kaon masses. The horizontal line is the experimental value of the pion splitting. Thus, the distance between the horizontal line and the intersection of the purple line with the vertical, dashed-dotted physical pion line is an indication of the size of the systematic error. Looking at the ratio of the experimental result for the pion splitting to our kaon splitting, we get $\epsilon = 1.02(4)$. Alternatively, we may use our result for the electromagnetic pion splitting, and we find $\epsilon = 0.84(5)$. We have recently added the newer ensembles with $a \approx 0.06$ and 0.045 fm to our analysis. In this case, we drop the $a \approx 0.12$ fm ensembles and can get good correlated fits when the data is thinned. One such fit is shown in Fig. 4. In this case, we see that our result for the pion splitting is a bit low. This results in $\epsilon = 0.97(5)$. If we take the experimental value of the pion splitting, our result is $\epsilon = 0.83(4)$.

4. Results and future plans

Our current (preliminary) result for ϵ does not include the fit shown in Fig. 4, but it does include other variations on the fits shown in Fig. 3 in which different subsets of our data are included. Our result is [8]

$$\epsilon = 0.84(5)_{\text{stat}}(18)_{a^2}(6)_{\text{FV}} \ . \tag{1}$$

The first error is statistical, the second from variations in the continuum extropolation, and the third is our (hopefully conservative) estimate of the residual finite volume error that may

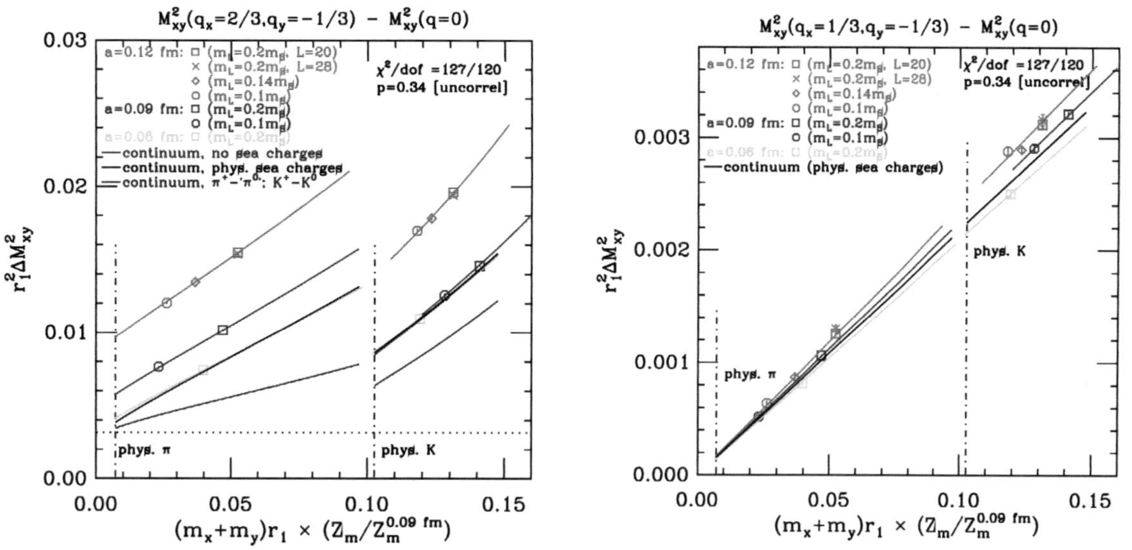

Figure 3. Central fit to the EM splitting ΔM^2_{xy} *vs.* the sum of the valence-quark masses. Only a small subset of the partially quenched data set included in the fit is shown: the points for $a \approx 0.09$ fm and ≈ 0.06 fm, as well as the 'pion' (x-axis value < 0.07) points for $a \approx 0.12$ fm, have unitary values of the valence masses, while the 'kaon' (x-axis value > 0.1) points for $a \approx 0.12$ fm have $m_x = m_l'$ but $m_y = 0.8m_s'$, which is closer to the physical strange mass than m_s' itself. The data have been corrected for finite volume effects using NLO SχPT. The red, blue, and green curves correspond to the three lattice spacings. The black and purple curves are extrapolations, see text.

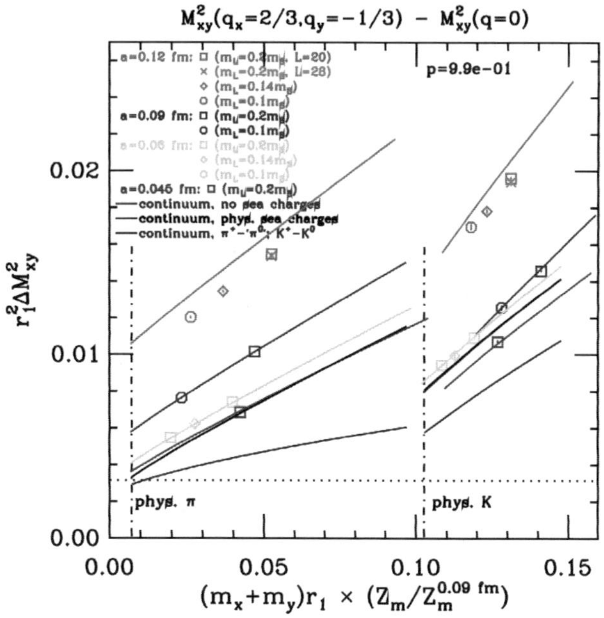

Figure 4. Similar to Fig. 3 except that $a \approx 0.12$ fm ensembles are not included in the fit and the correlations are included. Only a subset of the fitted data is shown.

remain after our correction based on the NLO formula. Previously, we have determined m_u/m_d by using our pseudoscalar mass results on our asqtad ensembles. However, we have recently been generating highly improved staggered quark (HISQ) ensembles for which the chiral extrapolation is much better controlled. Using our new value of ϵ in Eq. (1) with the HISQ light meson masses [11] gives a preliminary value for the ratio

$$m_u/m_d = 0.4482(48)_{\rm stat}({}^{+\ 21}_{-115})_{a^2}(1)_{{\rm FV}_{\rm QCD}}(165)_{\rm EM}, \tag{2}$$

where here "EM" denotes all errors from electromagnetism, while "$\mathrm{FV_{QCD}}$" refers to finite-volume effects in the pure QCD calculation. The electromagnetic error has been reduced by more than a factor of two from our previous result [5].

We plan on several future improvements. First, we will continue to analyze the ensembles that were not in our prior work. We will also repeat the analysis of quenched electromagnetic effects on the HISQ ensembles. The HISQ ensembles have several advantages. They should have smaller discretiziation errors, and the chiral extrapolation errors should be much smaller, as we have ensembles tuned to the physical light quark mass. Finally, the finite volume errors should be reduced as the HISQ lattices are larger than those for asqtad. We are also working on a fully dynamical $SU(3) \times U(1)$ code to allow us to include charged sea-quark mass effects [12] . This will make possible controlled calculations of many additional quantities. We have also calculated electromagnetic effects on the baryon spectrum, but did not have time to report on that.

Acknowledgments
The spectrum running was done on computers at the National Center for Supercomputing Applications, Indiana University, the Texas Advanced Computing Center (TACC), and the National Institute for Computational Science (NICS). Configurations were generated with resources provided by the USQCD Collaboration, the Argonne Leadership Computing Facility, and the National Energy Research Scientific Computing Center, which are funded by the Office of Science of the U.S. Department of Energy; and with resources provided by the National Center for Atmospheric Research, NICS, the Pittsburgh Supercomputer Center, the San Diego Supercomputer Center, and TACC, which are funded through the National Science Foundation's XSEDE Program. This work was supported in part by the U.S. DOE and the NSF.

References

[1] Basak S *et al.* (MILC Collaboration) 2008 *PoS* **LATTICE2008** 127 (*Preprint* 0812.4486)
[2] Torok A, Basak S, Bazavov A, Bernard C, DeTar C *et al.* 2010 *PoS* **LATTICE2010** 127
[3] Basak S *et al.* (MILC Collaboration) 2012 *PoS* **LATTICE2012** 137 (*Preprint* 1210.8157)
[4] Basak S *et al.* (MILC) 2013 *PoS* **CD12** 030 (*Preprint* 1301.7137)
[5] Bazavov A, Toussaint D, Bernard C, Laiho J, DeTar C *et al.* 2010 *Rev.Mod.Phys.* **82** 1349–1417 (*Preprint* 0903.3598)
[6] Dashen R F 1969 *Phys.Rev.* **183** 1245–1260
[7] Bijnens J and Danielsson N 2007 *Phys.Rev.* **D75** 014505 (*Preprint* hep-lat/0610127)
[8] Basak S *et al.* (MILC Collaboration) 2014 *PoS* **LATTICE2014** (*Preprint* 1409.7139)
[9] Portelli A, Durr S, Fodor Z, Frison J, Hoelbling C *et al.* 2011 *PoS* **LATTICE2011** 136 (*Preprint* 1201.2787)
[10] Hayakawa M and Uno S 2008 *Prog.Theor.Phys.* **120** 413–441 (*Preprint* 0804.2044)
[11] Bazavov A *et al.* (Fermilab Lattice, MILC) 2014 *Phys.Rev.* **D90** 074509 (*Preprint* 1407.3772)
[12] Zhou R and Gottlieb S 2014 *PoS* **LATTICE2014** 024 (*Preprint* 1411.4115)

From quarks to hadrons and back - spectral and bulk phenomena of strongly interacting matter

Stefan Krieg
Budapest-Marseille-Wuppertal and Wuppertal-Budapest collaborations
Department of Physics, University of Wuppertal, Gaußstr. 20, D-42119, Germany
IAS, Jülich Supercomputing Centre, Forschungszentrum Jülich, Jülich, D-52425, Germany

Abstract. Recent results on the proton-neutron mass splitting, relevant for the stability of ordinary matter, and on the properties of matter in under extreme conditions as created by heavy ion experiments, obtained through numerical simulations of Lattice Quantum Chromo- and Electrodynamics are discussed.

1. Introduction

Computing, from first principles, the hadron masses to percent accuracy [1], is only possible through simulations of Lattice Quantum Chromodynamics (QCD). With the advent of the present class of Pflop Machines and novel simulation algorithms, we now can proceed to compute per-mil effects in the particle spectrum, i.e., the proton-neutron mass difference. This difference is due to a subtle cancellation of already small effects (due to the mass difference of the up- and down-quarks and the presence of electromagnetic interactions). In the first part of these proceedings, we report on a project [2, 3] to compute this and other mass differences using simulations of Lattice Quantum Chromo- and Electrodynamics and discuss the new simulation methods and the highly efficient code employed.

In the case of the proton and the neutron, quarks and gluons are confined to the hadron. If we, however, increase the temperature of the system sufficiently, both particles will 'melt' and quarks and gluons behave as free particles ('quark-gluon-plasma'). This transition is described by the Equation of State (EoS) of QCD [4]. In the second half of these proceedings, we discuss a full result for the EoS [5] *excluding* as well as an ongoing project (e.g. [6, 7]) aimed at calculating the EoS *including* the effects of a dynamical charm quark, which becomes relevant for temperatures larger than 300-400 MeV.

2. Mass splittings from simulations of Lattice Quantum Electro- and Chromodynamics

Lattice Quantum Chromodynamics (LQCD) has matured significantly over the last decade. By now, calculations of many properties of strongly interacting matter are available. Simpler quantities such as proton and neutron masses are routinely computed to percent precision in the continuum limit [1, 8, 9, 10]. Properties of nucleons computed on the lattice, such as the charge radius of the proton, agree with experiment (see, e.g., [11, 12, 13], for a recent review see [14]). In this particular case, however, the level of precision reached does not allow one to solve the

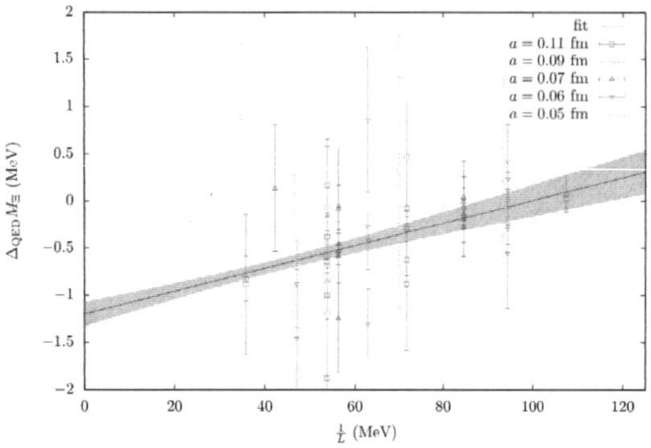

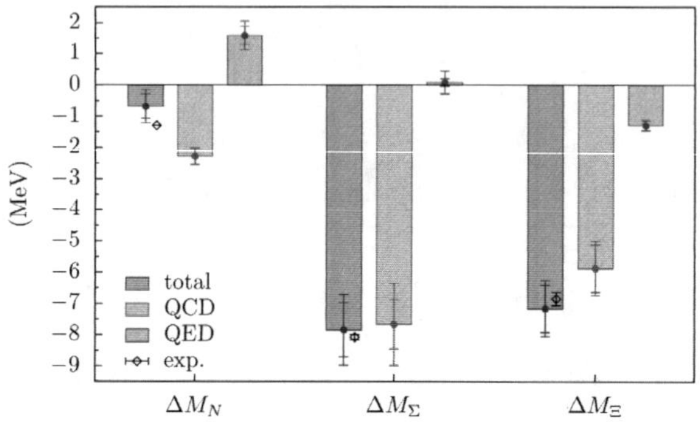

Figure 1. *Left:* Finite size effects due to QED. Shown is the dependence of the QED part of the Xi splitting on the inverse box size L. *Right:* Final result for the mass splittings. In this calculation [2], the achieved precision for the proton-neutron mass splitting (ΔM_N) excludes zero by slightly more than one sigma.

"proton radius puzzle" [15] for which better control of systematics, such as excited state effects, is required.

For hadron mass splittings, permill precision has been reached already [2, 3]. These calculations calculate splittings in hadron masses due to the violations of the (approximate) isospin symmetry in nature, i.e., due to mass and electromagnetic charge differences between the different quark flavors. In [2] these small effects were included by simulating lattice QCD+QED in the electro-quenched approximation, where the sea quarks were uncharged (and, as a matter of fact, up and down quarks were mass-degenerate). The most challenging problem faced by such calculations including electro-magnetic effects are finite-size effects. This is illustrated on the l.h.s. of Figure 1. The most significant caveat of this calculation is the fact that the signal is of the same order as the leading source of error due to the uncharged sea quarks.

This can only be remedied by a fully dynamical calculation, including dynamical QED and non-degenerate quark masses. Such a calculation was presented in [3]. For it to be successful, a range of issues had to be addressed. The hadron propagators suffer from unphysical contributions that are proportional to the quark charges e_q, the signal being of order e_q^2. In the case of the aforementioned quenched QED calculation with uncharged sea quarks, one could average propagators with negative and positive charges, thereby eliminating the $O(e_q)$ contributions to the results. This is, however, not possible in an unquenched simulation. The approach taken was, therefore, to simulate at larger than physical charges, where the noise is much smaller, and to use a set of ensembles with uncharged quarks to interpolate to the physical point. It also turns out that the previously used method to subtract problematic zero modes of QED (which is permissible since these are of zero measure in the continuum) violated reflection positivity. As a consequence there are no mass plateaus and results depend on the ratio of the temporal over spatial box size (T/L). By subtracting more modes (also of zero measure), however, this problem could be solved. As mentioned above, finite size effects are large when QED is included in the calculation. For the target precision aimed at in this project, since the finite size deviations dominate the signal, the first to terms of the $1/L$ expansion were calculated analytically and shown to be universal. The non-universal coefficient of the $1/L^3$ term (which depends, among others, on the charge radius) was then fitted. In order to be certain that this setup is correct, dedicated pure QED simulations were performed to check the finite volume correction terms. A new "fourier accelerated" HMC was used to update the QED fields, since the generic method suffered from extremely long autocorrelations. The propagator calculations were sped up using

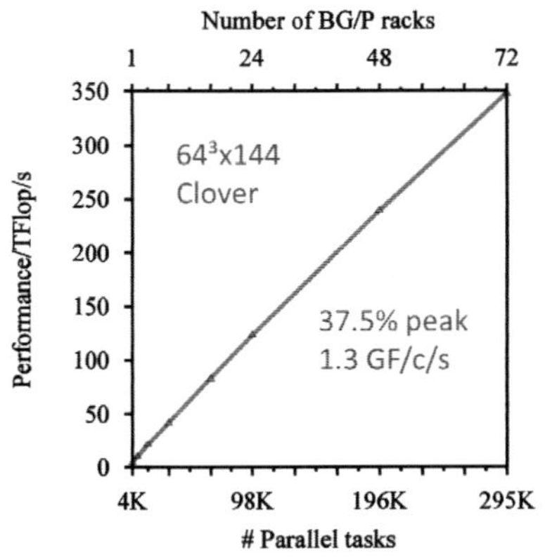

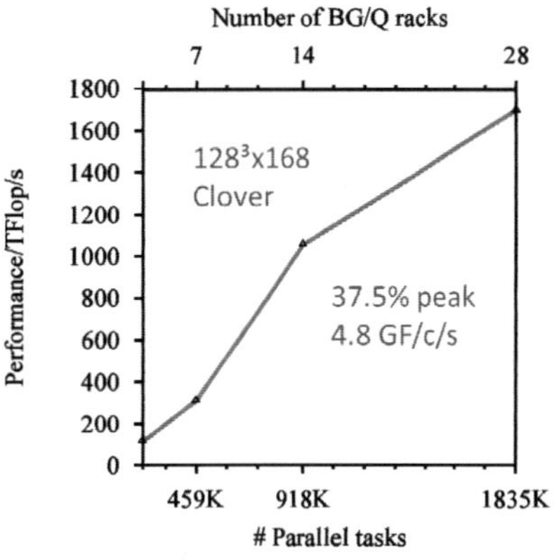

Figure 2. *Left:* Perfect strong scaling as seen on the (now obsolete) Blue Gene/P. Up to 300,000 parallel threads could be run at a sustained performance of about 350 TFlop/s. *Right:* Strong scaling on the Blue Gene/Q. On this architecture and the lattice size chosen for the strong scaling analysis, a sweet spot at about 1,000,000 parallel threads can be seen. At 1,800,000 parallel threads we achieve, for this particular lattice volume, a sustained performance of about 1.7 PFlop/s.

a new multilevel-method [16, 17] combined with "all mode averaging" [18]. Finally, the analysis combined a Kolmogorov-Smirnov analysis for the fit intervals with the histogram method of [1] and was crosschecked independently. The final results [3] now exclude zero for the proton-neutron mass splitting on the five sigma level and make predictions for other mass splittings that have not yet been measured or where the experimental uncertainties are significantly larger.

3. Performance

Our simulation software has been ported to a range of different architectures, such as clusters (e.g., "Juropa" at JSC, Jülich) or highly scalable architectures such as Blue Gene/Q, Cray XE6, or Cray XC40 (e.g., "Hermit" and "Hornet" at HLRS, Stuttgart or "Juqueen" at JSC, Jülich). In Figure 2 we show the performance obtained on the Blue Gene/Q architecture. Our code scales well up to 1.8 million hardware threads or the full 28 Racks of "Juqueen". We have hand-tuned both the parallel and serial parts of the code. The performance of the multi-shift CG used in the RHMC [19] algorithm achives a sustained performance of 3.2 GFlop/core/s (CG is more efficient, BiCGstab similar) and the considerably more complicated multi-level solver [16, 17] reaches 2.0 GFlop/core/s (but converges, compared to e/o preconditioned BiCGstab, orders of magnitude faster [16]).

4. Full result for the $N_f = 2 + 1$ equation of state

The rapid transition from the quark-gluon-plasma 'phase'[1] to the hadronic phase in the early universe and the QCD phase diagram are subjects of intense study in present heavy-ion experiments (LHC@CERN, RHIC@BNL, and the upcoming FAIR@GSI). This transition can be studied in a systematic way in Lattice QCD (for recent reviews see, e.g., [21, 22, 23, 10]). The equation of state (EoS) of QCD, (i.e, the pressure p, energy density ϵ, trace anomaly $I = \epsilon - 3p$, entropy $s = (\epsilon + p)/T$, and the speed of sound $c_s^2 = dp/d\epsilon$ as functions of the temperature) has been determined by several lattice groups, however, a full result has only recently become

[1] Since this transition is a cross-over [20], this use of the term 'phase' is somewhat abusive, and indicates only the dominant degrees of freedom.

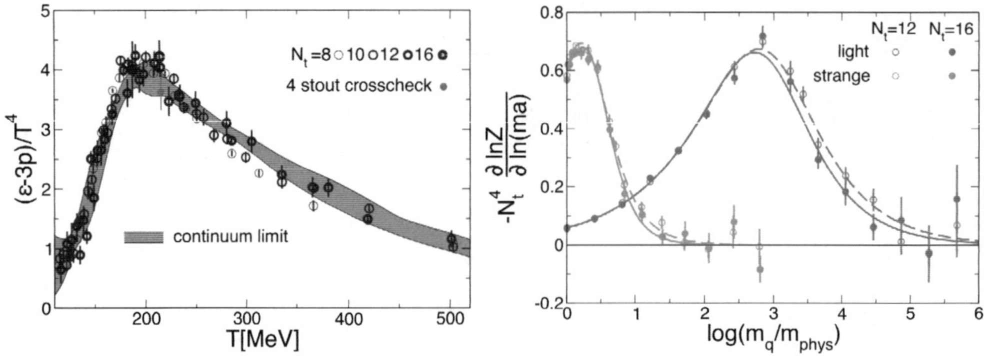

Figure 3. *Left:* The trace anomaly as a function of the temperature. The continuum extrapolated result with total errors is given by the shaded band. Also shown is a cross-check point computed in the continuum limit with our new and improved lattice action at $T = 214$ MeV, indicated by a smaller filled red point, which serves as a crosscheck on the peak's height. *Right:* Setting the overall scale of the pressure: integration from the infinitely large mass region down to the physical point using a range of dedicated ensembles and time extents up to $N_t = 16$; the sum of the areas under the curves gives p/T^4.

available: reference [4] constitutes a full result at three characteristic temperatures, which we have now extended to the full temperature range and made available electronically [5] .

Our calculation is based on a tree-level Symanzik improved gauge action with 2-step stout-link improved staggered fermions. The precise definition of the action can be found in ref. [24], its advantageous scaling properties are studied in ref. [25, 26, 27].

We include two different scale setting procedures in our 'histogram' method [1] used to estimate systematical errors, along with a range of other fit methods, each of which is an in principle completely valid approach. We then calculated the goodness of fit Q and weights based on the Akaike information criterion AICc [28, 29] and looked at the unweighted or weighted (based on Q or AICc) distribution of the results. The median is the central value, whereas the central region containing 68% of all the possible methods gives an estimate on the systematic uncertainties. This procedure provides very conservative errors. Here, we used four basic types of continuum extrapolation methods, two continuum extrapolation ranges, seven ways to determine the subtraction term at T=0, and the aforementioned two scale procedures. Finally, we included eight options to determine the final trace anomaly by choosing among various spline functions, giving altogether 4·2·7·2·8=896 methods. Note that using either an AICc or Q based distribution changed the result only by a fraction of the systematic uncertainty. Furthermore, the unweighted distribution always delivered consistent results within systematical errors.

The continuum extrapolated trace anomaly is shown in Figure 3.

5. Update on the $N_f = 2 + 1 + 1$ Equation of State

So far, the equation of state is known only in 2+1 flavor QCD. The contribution from the sea charm quarks most likely matter at least for $T > 300 - 400$ MeV (for an illustration, see Figure 4).

The $N_f = 2+1$ lattice results of the previous section agree with the HRG at low temperatures and are correct for small to medium temperatures, and, as is shown in Figure 4, at temperatures of about 1 GeV perturbative results become sufficiently precise. Therefore, we need to calculate the EoS with a dynamical charm only for the remaining temperatures in the region of

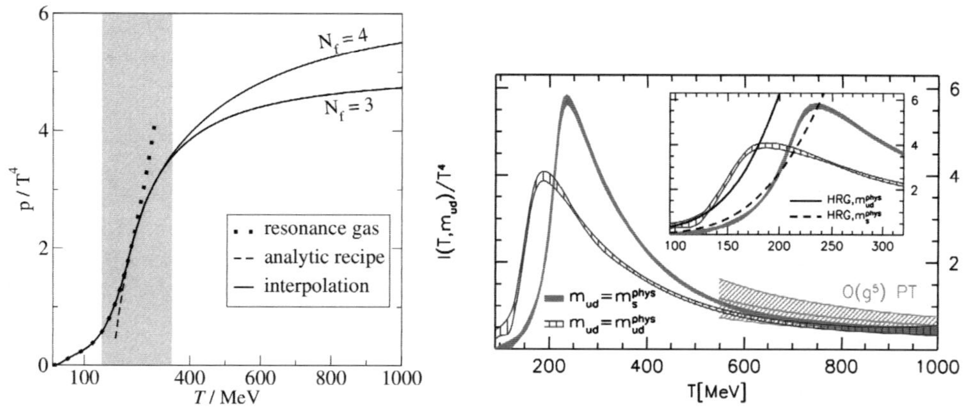

Figure 4. *Left:* Laine and Schroeder's perturbative estimate of the effect of the charm in the QCD equation of state [30]. *Right:* Wuppertal-Budapest [4] and perturbative (up to $O(g^5)$) results for the equation of state.

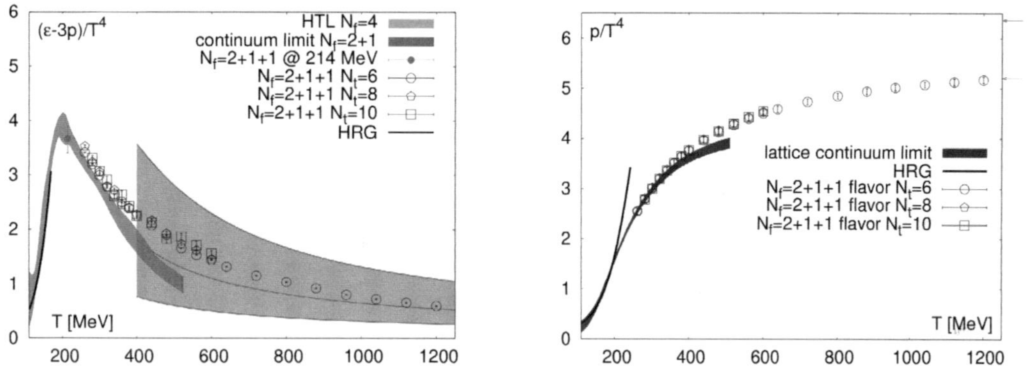

Figure 5. *Left:* Preliminary results for the charmed EoS. For comparison, we show the HRG result, the $N_f = 2 + 1$ band, and, at high Temperatures, the HTL result [31], where the central line marks the HTL expectation for the EoS with the band resulting from (large) variations of the renormalization scale. *Right:* Preliminary result for the pressure, errors indicate the Stefan-Boltzmann value. All errors are statistical only.

approximately 300 MeV $< T <$ 1000 MeV.

We are using a new lattice action for these calculations. More precisely, where our $N_f = 2+1$ calculation used an action with 2 levels of stout gauge link averaging in the coupling of the fermions to the gauge fields, we increased this to 4 levels with a smearing parameter of $\rho = 0.125$ (for further details see [5]). The crosscheck point shown in Figure 3 was computed using this new action. Since it perfectly agrees with the $N_f = 2 + 1$ results, even though it was computed using a dynamical charm, we can be certain that at temperatures at and below $T = 214$ MeV, we can rely on the $N_f = 2 + 1$ results.

Our preliminary results are shown in Figure 5, all errors are statistical only. Our results span a region of temperatures from $T = 214$ MeV up to $T = 1.2$ GeV. At the low end we make contact to the $N_f = 2 + 1$ equation of state, and at large temperatures to the HTL result. We thus cover the full region of temperatures, from low temperatures, where the HRG gives reliable results, to high temperatures, where we make contact with perturbation theory. Our present set of data points will be extended to $N_t = 12$, in order to allow for a controlled continuum extrapolation.

6. Conclusions

The precision of Lattice QCD results at finite temperature has increased significantly over the last years. We discussed how permill level precision could be achieved for the particle spectrum of QCD, presented a full result (all sources of uncertainties controlled) for the $N_f = 2+1$ EoS [5], and shown how to include a dynamical charm quark for the $N_f = 2+1+1$ EoS and presented preliminary results.

Acknowledgments

Computations were performed on JUQUEEN at Forschungszentrum Jülich, HERMIT at HLRS, Stuttgart, and on the QPACE machine and on GPU clusters [32] at University of Wuppertal. We acknowledge PRACE for awarding us resources on JUQUEEN at Forschungszentrum Jülich. This work was partially supported by the DFG Grant SFB/TRR 55 and ERC no. 208740.

References

[1] Durr S, Fodor Z, Frison J, Hoelbling C, Hoffmann R *et al.* 2008 *Science* **322** 1224–1227 (*Preprint* 0906.3599)
[2] Borsanyi S, Drr S, Fodor Z, Frison J, Hoelbling C *et al.* 2013 *Phys.Rev.Lett.* **111** 252001 (*Preprint* 1306.2287)
[3] Borsanyi S, Durr S, Fodor Z, Hoelbling C, Katz S *et al.* 2014 (*Preprint* 1406.4088)
[4] Borsanyi S, Endrodi G, Fodor Z, Jakovac A, Katz S D *et al.* 2010 *JHEP* **1011** 077 (*Preprint* 1007.2580)
[5] Borsanyi S, Fodor Z, Hoelbling C, Katz S D, Krieg S *et al.* 2014 *Phys.Lett.* **B730** 99–104 (*Preprint* 1309.5258)
[6] Ratti C, Borsanyi S, Endrodi G, Fodor Z, Katz S D *et al.* 2013 *Nucl.Phys.* **A904-905** 869c–872c
[7] Borsanyi S, Fodor Z, Hoelbling C, Katz S D, Krieg S *et al.* 2014 (*Preprint* 1410.7917)
[8] Fodor Z and Hoelbling C 2012 *Rev.Mod.Phys.* **84** 449 (*Preprint* 1203.4789)
[9] Kronfeld A S 2012 *Ann.Rev.Nucl.Part.Sci.* **62** 265–284 (*Preprint* 1203.1204)
[10] Hoelbling C 2014 (*Preprint* 1410.3403)
[11] Green J, Engelhardt M, Krieg S, Negele J, Pochinsky A *et al.* 2014 *Phys.Lett.* **B734** 290–295 (*Preprint* 1209.1687)
[12] Green J, Negele J, Pochinsky A, Syritsyn S, Engelhardt M *et al.* 2012 *Phys.Rev.* **D86** 114509 (*Preprint* 1206.4527)
[13] Green J, Negele J, Pochinsky A, Syritsyn S, Engelhardt M *et al.* 2014 *Phys.Rev.* **D90** 074507 (*Preprint* 1404.4029)
[14] Constantinou M 2014 *PoS* **LATTICE2014** 001 (*Preprint* 1411.0078)
[15] Pohl R, Antognini A, Nez F, Amaro F D, Biraben F *et al.* 2010 *Nature* **466** 213–216
[16] Frommer A, Kahl K, Krieg S, Leder B and Rottmann M 2013 (*Preprint* 1307.6101)
[17] Frommer A, Kahl K, Krieg S, Leder B and Rottmann M 2014 *SIAM J. Sci. Comput.,* **36** A1581–A1608 (*Preprint* 1303.1377)
[18] Blum T, Izubuchi T and Shintani E 2013 *Phys.Rev.* **D88** 094503 (*Preprint* 1208.4349)
[19] Clark M and Kennedy A 2007 *Phys.Rev.Lett.* **98** 051601 (*Preprint* hep-lat/0608015)
[20] Aoki Y, Endrodi G, Fodor Z, Katz S and Szabo K 2006 *Nature* **443** 675–678 (*Preprint* hep-lat/0611014)
[21] Fodor Z and Katz S 2009 (*Preprint* 0908.3341)
[22] Philipsen O 2013 *Prog.Part.Nucl.Phys.* **70** 55–107 (*Preprint* 1207.5999)
[23] Petreczky P 2012 *PoS* **ConfinementX** 028 (*Preprint* 1301.6188)
[24] Aoki Y, Fodor Z, Katz S and Szabo K 2006 *JHEP* **0601** 089 (*Preprint* hep-lat/0510084)
[25] Aoki Y, Fodor Z, Katz S and Szabo K 2006 *Phys.Lett.* **B643** 46–54 (*Preprint* hep-lat/0609068)
[26] Aoki Y, Borsanyi S, Durr S, Fodor Z, Katz S D *et al.* 2009 *JHEP* **0906** 088 (*Preprint* 0903.4155)
[27] Borsanyi S *et al.* 2010 *JHEP* **1009** 073 (*Preprint* 1005.3508)
[28] Akaike H 1974 *IEEE Transactions on Automatic Control* **19** 716–723
[29] Hurvich C and Tsai C L 1989 *Biometrica* **76** 297–307
[30] Laine M and Schroder Y 2006 *Phys.Rev.* **D73** 085009 (*Preprint* hep-ph/0603048)
[31] Andersen J O, Leganger L E, Strickland M and Su N 2011 *JHEP* **1108** 053 (*Preprint* 1103.2528)
[32] Egri G I, Fodor Z, Hoelbling C, Katz S D, Nogradi D *et al.* 2007 *Comput.Phys.Commun.* **177** 631–639 (*Preprint* hep-lat/0611022)

Journal of Physics: Conference Series **640** (2015) 012054 doi:10.1088/1742-6596/640/1/012054

Nucleon Structure on a Lattice at the Physical Point

Sergey Syritsyn

RIKEN/BNL Research Center, Brookhaven National Laboratory, Upton, NY, 11973, USA

E-mail: ssyritsyn@quark.phy.bnl.gov

Abstract. We report initial nucleon structure results computed on lattices with 2+1 dynamical Möbius domain wall fermions at the physical point generated by the RBC and UKQCD collaborations. At this stage, we evaluate only connected quark contributions. In particular, we discuss the nucleon vector and axial-vector form factors, nucleon axial charge and the isovector quark momentum fraction. From currently available statistics, we estimate the stochastic accuracy of the determination of g_A and $\langle x\rangle_{u-d}$ to be around 10%, and we expect to reduce that to 5% within the next year. To reduce the computational cost of our calculations, we extensively use acceleration techniques such as low-eigenmode deflation and all-mode-averaging (AMA). We present a method for choosing optimal AMA parameters.

1. Introduction

Exploring nucleon structure in lattice QCD at the physical point became possible in the recent few years, and many collaborations have started calculations that no longer require chiral extrapolations. This is a major step forward because it eliminates one of the major sources of systematic uncertainty that made difficult validating methods and achieving high precision in lattice QCD.

Nucleon structure calculations at the physical point are extremely demanding. This usually justifies choosing an affordable fermion action such as Wilson [1] or twisted mass [2], even though they do not respect chiral symmetry of QCD. However, a number of nuclear and high energy physics problems, e.g., proton decay and neutron oscillations, require chiral symmetry for computing corresponding nucleon matrix elements, and one has to use substantially more expensive domain wall or overlap fermion actions. Studying nucleon structure, the nucleon vector and axial vector form factors in particular, with these actions is therefore a necessary stepping stone for developing methods to facilitate calculations that will preserve chiral symmetry.

In this report, we show our initial results from computing nucleon structure directly at the physical point using gauge configurations generated with the $N_f = 2+1$ dynamical domain wall fermion action [3]. We use various improvement techniques, which we discuss in Sec. 2, and present initial, low-statistics results for the nucleon form factors, axial charge and quark momentum fraction in Sec. 3.

2. Methodology

For this initial study with chiral quarks at the physical point, we use one ensemble of gauge configurations generated by the RBC/UKQCD collaborations with the inverse lattice spacing $a^{-1} = 1.729(4)$ GeV and the pion mass $m_\pi = 139.2(4)$ MeV [3]. This ensemble is generated with the Iwasaki gauge action and the Möbius Domain Wall fermion (MDWF) action for $N_f = 2+1$

Journal of Physics: Conference Series **640** (2015) 012054 doi:10.1088/1742-6596/640/1/012054

quarks. The Möbius formulation [4] of chiral fermions allows one to shorten the fifth ("flavor") dimension of the traditional domain wall fermion action without increasing the quark residual mass. The lattice size is $48^3 \times 96$, corresponding to $m_\pi L = 3.86$, which should be sufficient to substantially suppress finite volume effects, as known from the meson sector.

The main challenge in computing hadron observables with domain wall-like actions is the cost of calculating light quark propagators, especially at the physical point. We use two methods to accelerate our calculations: acceleration of the conjugate gradient algorithm (CG) with low-eigenmode deflation and improved stochastic sampling, or *All-mode Averaging (AMA)* [5]. The idea behind AMA is to compute cheap approximate samples with large statistics and then correct for any potential bias by comparing a subset of approximate samples to exact solutions,

$$\langle\mathcal{O}\rangle_{\rm imp} = \langle\mathcal{O}_{\rm approx}\rangle_{N_{\rm approx}} + \langle\Delta\mathcal{O}\rangle_{N_{\rm exact}}\,, \quad \Delta\mathcal{O} = \mathcal{O}_{\rm exact} - \mathcal{O}_{\rm approx}\,, \tag{1}$$

$$\left(\delta\mathcal{O}_{\rm imp}\right)^2 = \frac{1}{N_{\rm approx}}{\rm Var}\{\mathcal{O}_{\rm approx}\} + \frac{1}{N_{\rm exact}}{\rm Var}\{\Delta\mathcal{O}\} \tag{2}$$

One usually takes a large number of approximate samples at different locations on the same lattice configuration, exploiting the translational invariance of the QCD ensemble average, while for $\Delta\mathcal{O}$ smaller statistics is sufficient to estimate the difference arising from the approximation. We obtain approximate samples using quark propagators computed with truncated CG. Low-eigenmode deflation is applied in computing both exact and approximate samples. In addition to accelerating the CG convergence for the exact samples, deflation helps to reduce the bias of approximate samples and make them cheaper by substantially reducing the number $n_{\rm CG}$ of CG iterations required for good approximation.

We compute $N_{EV} = 500$ lowest eigenmodes for deflation using ARPACK with $n = 200$ Chebyshev polynomial acceleration. With the range of deflated eigenvalues $\lambda_{\rm max}/\lambda_{\rm min} \approx 10^2$ (see Fig. 1), the convergence rate of CG is accelerated by a factor of ≈ 10, which was confirmed by direct tests. The number of eigenvalues we can use is limited by the total amount of memory available to a single job. In fact, using $(3\ldots4)\times$ more eigenvectors, which would span the spectrum between the light and strange quark masses, is expected to improve the AMA efficiency significantly [6].

In order to apply the AMA effectively, one has to choose the optimal approximation (number of truncated CG steps $n_{\rm CG}$) and ratio of exact and approximate samples $N_{\rm approx}/N_{\rm exact}$. Such optimization should take into account the change in the precision of the approximate samples, as well as their total cost. We use the product of their stochastic variance and the cost required for their computation, normalized by the cost of a single exact sample, as the figure of merit:

$${\rm Cost}_{\rm imp}\cdot{\rm Var}\{\mathcal{O}_{\rm imp}\} \sim \left(1 + \frac{n_{\rm CG}^{\rm approx}}{n_{\rm CG}^{\rm exact}}\cdot\frac{N_{\rm approx}}{N_{\rm exact}}\right)\cdot\left({\rm Var}\{\Delta\mathcal{O}\} + \frac{N_{\rm exact}}{N_{\rm approx}}{\rm Var}\{\mathcal{O}_{\rm approx}\}\right). \tag{3}$$

Estimates of this FOM for varying combinations of $n_{\rm CG}$ and $N_{\rm approx}/N_{\rm exact}$ are shown in Fig. 2 for the nucleon three-point functions yielding the axial charge and the charge radius[1]. From that, we derive optimal $N_{\rm approx}/N_{\rm exact} = 32$ and $n_{\rm CG}^{\rm approx} = 400$, which reduce stochastic variance by a factor of $\times(2.5\ldots3)$.

Excited states in lattice nucleon matrix elements are a major source of systematic error, and one has to study the dependence of nucleon-operator three-point functions on the separation between sources and sinks $T = t_{\rm sink} - t_{\rm source}$. For this initial study, we select $T/a = 8, 9, 10, 12$ corresponding to 0.91, 1.03, 1.14, and 1.37 fm. Since the statistics are not sufficient for reliable multi-exponential fits, we resort to simpler methods such as summation (e.g., Ref. [7]) to estimate and reduce excited-state contributions at this stage.

[1] More precisely, the matrix element of the vector current with the minimal non-zero momentum transfer.

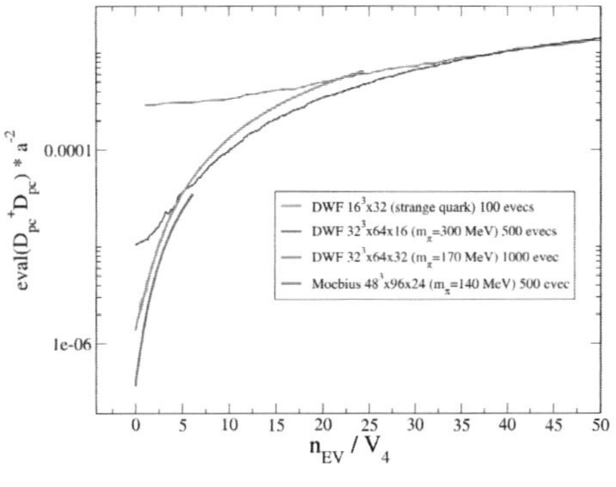

Figure 1. Deflation eigenvalues for different domain wall operators, scaled with volume and lattice cutoff (courtesy of T. Blum, T. Izubuchi, and E. Shintani). The green line corresponds to the current work.

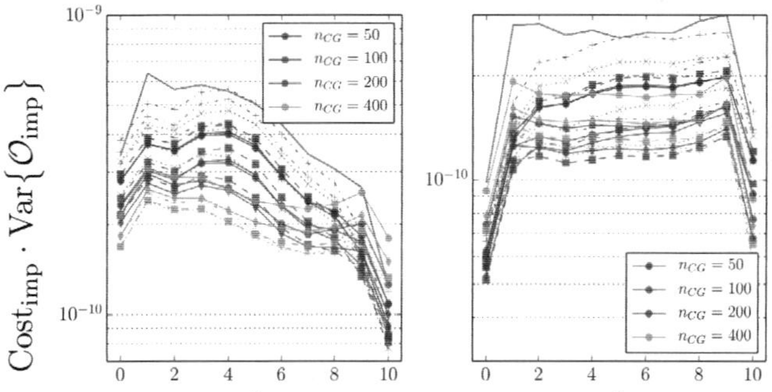

(a) $\mathrm{Im}\langle 000|\bar{q}\gamma_3\gamma_5 q|000\rangle$ (b) $\mathrm{Re}\langle 001|\bar{q}\gamma_4 q|000\rangle$

Figure 2. Figure of merit for AMA optimization (see Eq. 2.3) of the nucleon axial charge (a) and charge radius (b) three-point function plateaus. Symbols ●, ◇, ■, "x" and "+" correspond to $N_{\text{approx}}/N_{\text{exact}} = 4, 8, 16, 32, 64$, respectively, and the solid line corresponds to exact-only samples (no AMA).

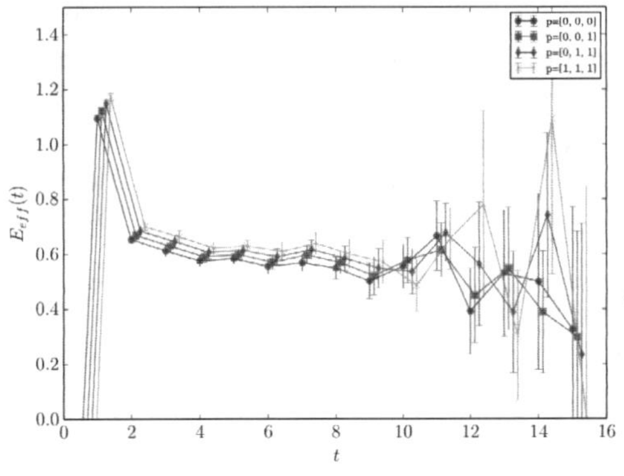

Figure 3. Plateaus for nucleon effective energies for the lowest momenta.

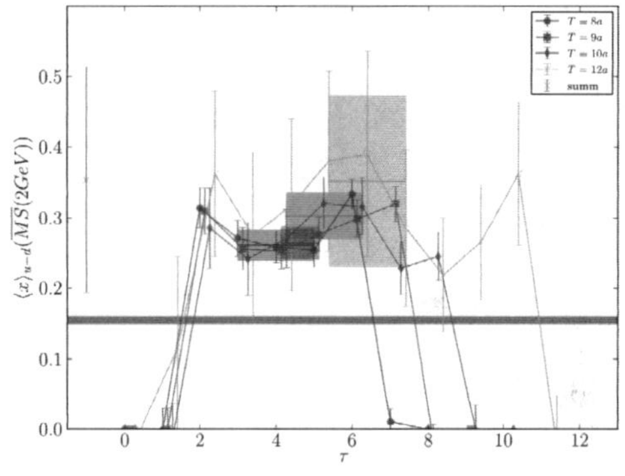

Figure 4. Plateaus of quark momentum fraction $\langle x\rangle_{u-d}$. The renormalization is taken from Ref. [9].

3. Initial Results

We have analyzed 20 gauge configurations separated by 80 MD steps to minimize autocorrelations; these configurations are spread evenly over the entire available ensemble except for the first 640 MD steps necessary for thermalization. We compute one exact and 32 approximate samples per configuration. At this initial stage, we focus on nucleon structure "benchmark quantities": the nucleon axial charge, Dirac and Pauli isovector radii, isovector magnetic moment and quark momentum fraction. As shown in Fig. 3, the effective energies $E_{\text{eff}}(t) = \log\big(C_{2\text{pt}}(t)/C_{2\text{pt}}(t+1)\big)$ become too noisy after $t = 11$, indicating that the source-sink separation $T/a = 12$ is the largest one that can be analyzed with the current statistics. In order to extract nucleon form factors from nucleon matrix elements we use standard methods that can be found elsewhere (e.g., see Refs. [8, 9, 1, 10]).

In Figs. 4 and 5 we show quark momentum fraction, vector charge, and axial charge

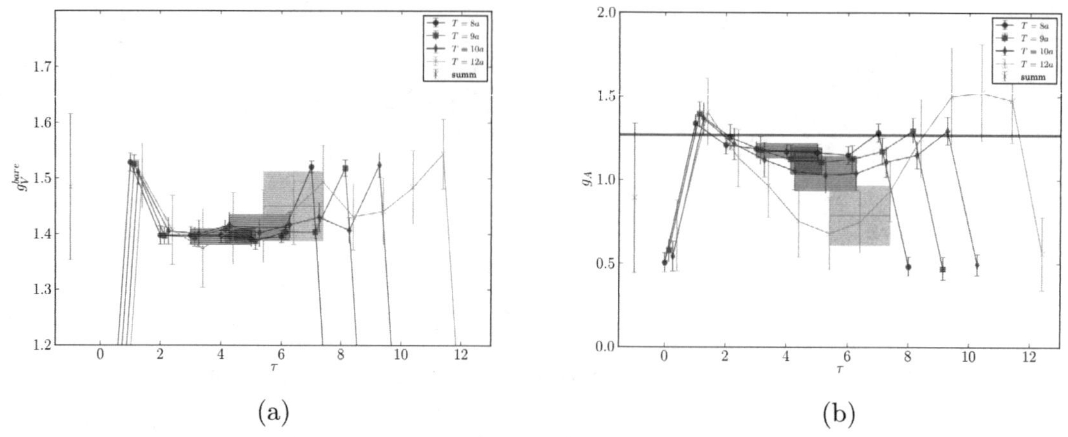

Figure 5. Plateaus of vector (a) and axial (b) charges of the nucleon. The leftmost point represents the "summation" value.

plateaus, their central plateau average values and the values obtained with the summation method [7] (applied to data with all the four separations T). Although we have not calculated renormalization factors for the momentum fraction for this ensemble, we can use the factors from earlier calculations with Domain Wall fermions with heavier pion masses and the identical lattice spacing [9] to convert bare quantities to the $\overline{\text{MS}}$ scheme at 2 GeV. The nucleon isovector quark momentum fraction is known to have substantial contributions from excited states (see, e.g. [1, 11, 12]), and lattice QCD results typically overestimate experiment by 30-60%. Our initial values for shorter separations also significantly deviate from experiment, while the longest separation $T/a = 12$ has insufficient precision to investigate whether it is excited states that cause this systematic effect. The "summation" method yields a value that is consistent with experiment by virtue of its larger statistical uncertainty.

The vector charge plateaus agree for all separations, and we use the central value at $T = 8a$ as an approximate renormalization factor for the vector and axial-vector current operators, $Z_A = Z_V = \left(g_V\big|_{T=8a}\right)^{-1}$. The value of the axial charge g_A is below the experimental value $g_A = 1.2723(23)$ [13]. We note that the deviation from the experiment is $(2\ldots3)\sigma$, and it can still be attributed to a statistical fluctuation. In addition, the plateaus indicate significant excited state contamination. Although the value from the summation method (Fig. 5(b)) agrees with experiment, the central plateau values move away from it with increasing separation T, a behavior that was also observed in other studies close to the physical point [1]. Additional statistics and careful analysis of excited states are required to understand this phenomenon.

The statistics are not yet sufficient to analyze the form factors to extract the radii reliably, especially with larger separations. Since no chiral extrapolation is needed, it is more informative to compare the form factors directly to experiment. In Figure 6 we show the isovector Dirac and Pauli form factors in the region of small momentum transfer $0 \le Q^2 \lesssim 0.6$ GeV^2, together with phenomenological fits of experimental data [14]. The Dirac form factor at small separations $T/a = 8, 9, 10$ deviates from the experiment, while the values at the largest separation $T/a = 12$, as well as the "summation" value, agree within statistics. This indicates that the deviations are likely caused by excited state contributions, in agreement with earlier findings that the isovector Dirac radius is subject to large excited state effects [1], although more statistics are necessary to make a certain conclusion. The isovector Pauli form factor is in better agreement with phenomenology, although its values are even less precise and the "summation" method

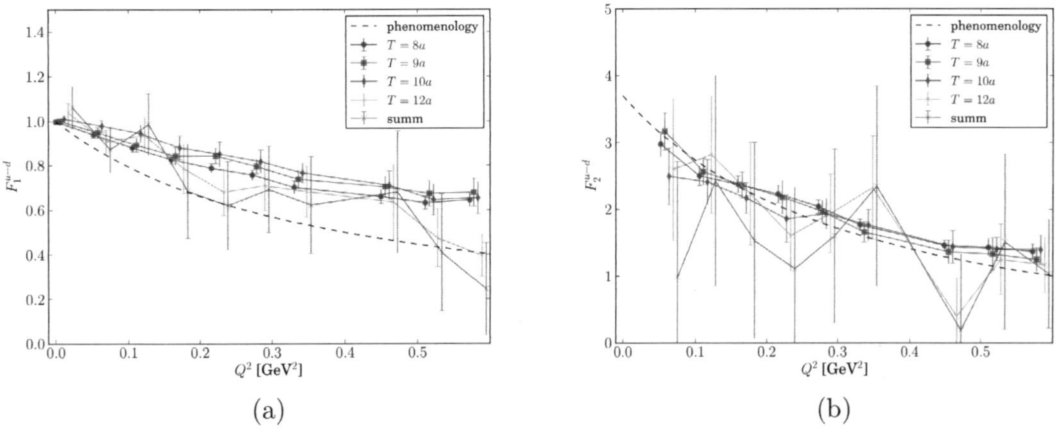

Figure 6. Vector form factors of the nucleon: isovector Dirac (a) and Pauli (b).

yields nearly 100% uncertainty.

The form factors of the axial-vector current are shown in Fig. 7. The axial form factor $G_A(Q^2)$ is interesting for two reasons. First, its forward value $G_A(0) = g_A$ is the axial charge that has been discussed above; second, its Q^2-dependence determines the axial radius of the nucleon and plays an important role in the physics of neutrino scattering and meson production. The axial radius, defined similarly to the charge radius, is typically underestimated in lattice calculations by a factor of 2 [15]. From Fig. 7(a) we see that our initial results follow the same pattern: the slope of the axial form factor is substantially smaller than experimental fits. It is not clear yet whether increasing statistics with larger source-sink separations $T/a \geq 10$ will indicate excited state contamination and suggest a solution to this problem; however, comparing $T/a = 8, 9$ and 10 shows very little dependence on the source-sink separation T/a.

The induced pseudoscalar form factor $G_P(Q^2)$ (Fig. 7(b)) is relevant for low-energy QCD dynamics. Its value is measured in meson production off nucleons and nuclei, as well as in muon capture experiments [16]. Its low-momentum behavior is believed to be governed by the pion pole $\sim (Q^2 + m_\pi^2)^{-1}$ and therefore is very sensitive to the pion mass. Calculations with the physical pion mass are especially important for this form factor. Our initial results indicate that there are substantial contributions from excited states, and as the source-sink separation increases, we observe better agreement with the phenomenology.

4. Discussion

Stochastic uncertainty of the initial results that we report indicates that substantially more statistics will be needed to compute lattice QCD "benchmark" quantities with required precision. Currently, we plan to quadruple statistics. Since, for example, our present stochastic uncertainties for the axial charge and the quark momentum fraction with separation $T = 1.14$ fm are $(\delta g_A)_{T/a=10} \approx 10\%$ and $(\delta\langle x\rangle_{u-d})_{T/a=10} \approx 11\%$, this will reduce their uncertainties down to $\approx 5\%$ and make reliable analysis of excited states possible.

Although computing resources increase steadily, and the calculation is fully feasible with the current methodology, we are actively studying approaches to both reduce the cost of computing approximate samples and to make them more precise. An obvious improvement is to simply use more eigenvectors in the CG deflation; however, this approach is limited by the size of the 5D eigenvectors (≈ 5.6 TiB per configuration). Other possible directions are using approximations to the Möbius fermion operator with shorter L_5, multigrid methods, and

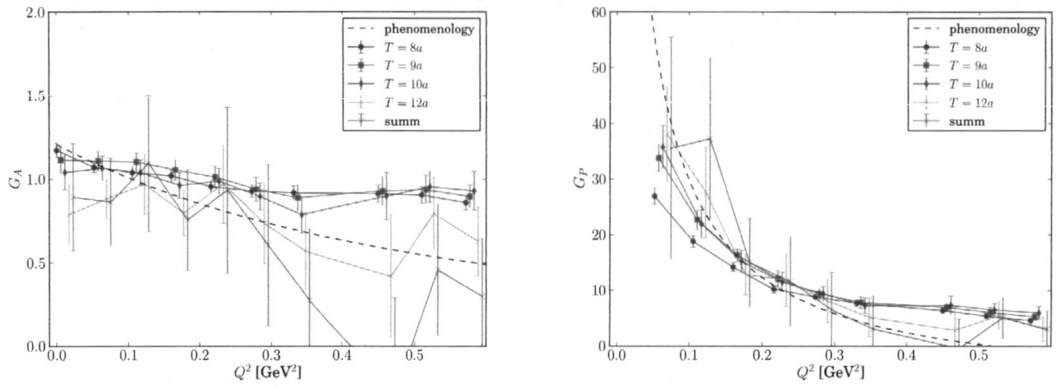

Figure 7. Axial vector form factors of the nucleon: axial (a) and induced pseudoscalar (b).

hierarchical deflation [17].

A combination of the improved computing methods and increasing computing resources will allow us to accomplish a reliable calculation of nucleon structure with chirally symmetric action in the near future.

Acknowledgments

We thank the RBC and UKQCD collaborations for providing us with the gauge configurations. Computations for this work were carried out on facilities of the USQCD Collaboration, which are funded by the Office of Science of the U.S. Department of Energy. This work was supported by RIKEN Foreign Postdoctoral Researcher Program (S.N.S.) and the U.S. Department of Energy (DOE) and the Office of Nuclear Physics under grants DE-FG02-96ER40965 and DE-SC0011090.

References

[1] Green J, Engelhardt M, Krieg S, Negele J, Pochinsky A *et al.* 2014 *Phys.Lett.* **B734** 290 (*Preprint* 1209.1687)
[2] Alexandrou C, Constantinou M, Drach V, Jansen K, Kallidonis C *et al.* 2013 *PoS* **LATTICE2013** 292 (*Preprint* 1312.2874)
[3] Blum T *et al.* (RBC Collaboration, UKQCD Collaboration) 2014 (*Preprint* 1411.7017)
[4] Brower R C, Neff H and Orginos K 2005 *Nucl.Phys.Proc.Suppl.* **140** 686–688 (*Preprint* hep-lat/0409118)
[5] Blum T, Izubuchi T and Shintani E 2013 *Phys.Rev.* **D88** 094503 (*Preprint* 1208.4349)
[6] Blum T and Izubuchi T *private communications*
[7] Mathur N, Dong S, Liu K, Mankiewicz L and Mukhopadhyay N 2000 *Phys.Rev.* **D62** 114504 (*Preprint* hep-ph/9912289)
[8] Yamazaki T *et al.* 2009 *Phys. Rev.* **D79** 114505 (*Preprint* 0904.2039)
[9] Aoki Y, Blum T, Lin H W, Ohta S, Sasaki S *et al.* 2010 *Phys.Rev.* **D82** 014501 (*Preprint* 1003.3387)
[10] Green J, Negele J, Pochinsky A, Syritsyn S, Engelhardt M *et al.* 2014 *Phys.Rev.* **D90** 074507 (*Preprint* 1404.4029)
[11] Bali G S, Collins S, Gläßle B, Göckeler M, Najjar J *et al.* 2014 (*Preprint* 1408.6850)
[12] Jäger B, Rae T, Capitani S, Della Morte M, Djukanovic D *et al.* 2013 (*Preprint* 1311.5804)
[13] Olive K *et al.* (Particle Data Group) 2014 *Chin.Phys.* **C38** 090001
[14] Kelly J J 2004 *Phys. Rev.* **C70** 068202
[15] Syritsyn S (LHP Collaboration) 2011 *AIP Conf.Proc.* **1374** 305–308
[16] Clayton S M 2010 *AIP Conf.Proc.* **1222** 407–411
[17] Boyle P A 2014 (*Preprint* 1402.2585)

Journal of Physics: Conference Series **640** (2015) 012055 doi:10.1088/1742-6596/640/1/012055

Targeting the Conformal Window: Scalars on the Lattice

Evan Weinberg[1], Rich Brower[1], Anna Hasenfratz[2], Claudio Rebbi[1], and Oliver Witzel[3†]

1 Department of Physics, Boston University, Boston, MA, USA
2 Department of Physics, University of Colorado Boulder, Boulder, CO, USA
3 Center for Computational Science, Boston University, Boston, MA, USA

E-mail: weinbe2@bu.edu

Abstract. The light Higgs boson of the Standard Model could arise as the consequence of the weakly broken conformal symmetry in a strongly interacting gauge theory. Here we present a novel idea to study the transition from conformal to confining behavior using an $SU(3)$ gauge theory with four light and eight heavy flavors. This system interpolates between the 12-flavor conformal and the 4 flavor chirally broken theory as the mass of the heavy flavors are varied. We show first results on our determination of the iso-singlet 0^{++} state.

1. Introduction

In 2012 the Higgs boson was discovered at the LHC [1, 2] adding the final missing piece to the electroweak sector of the Standard Model. In the coming years the experiments will improve their precision and resolve further properties of the Higgs particle. It is of particular interest is to learn more about the origin of the Higgs boson and the nature of electroweak symmetry breaking. The fundamental, self-interacting scalar as described by the Standard Model poses a theoretical challenge because it is not ultraviolet complete. One viable scenario is to consider the Higgs boson as a composite particle with its own underlying strongly interacting but chirally broken gauge theory [3, 4]. In such composite Higgs models the Higgs boson arises as a fermionic 0^{++} iso-singlet bound state.

QCD-like systems have a plethora of low-energy bound states and we expect to find a similar spectrum in scaled-up QCD models. Viable composite Higgs models have to be very different, predicting that the mass of any non-Goldstone states is much heavier than the iso-singlet scalar. In addition to an unusual spectrum, composite Higgs models have to satisfy other constraints, like predicting the experimentally observed Higgs decay width or the S-parameter correctly. These constraints could be met by a strongly interacting theory with a weakly broken conformal symmetry — a conjecture which can be numerically tested using lattice gauge theory.

Recent lattice results look very promising: there is increasing evidence for an infrared fixed point (IRFP) for 12 fundamental flavors [5–12] and two groups have reported on a low mass scalar in the SU(3) gauge theory: one for 8 fundamental flavors [13], the other for 2 sextet flavors [14]. These investigations are limited by an integer flavor number. *A priori* it is not obvious

† Present address: School of Physics & Astronomy, The University of Edinburgh, Edinburgh EH9 3FD, UK

that a theory with an integer fermion number will be close enough to the conformal window where the IRFP appears. Also there are many more gauge groups or fermion representations to be considered. To facilitate such a search and overcome the limitation of integer flavor numbers, we proposed to simulate a theory with $N_\ell = 4$ light and $N_h = 8$ heavy flavors [15]. Keeping the four light flavors near the chiral limit, i.e. $m_\ell \approx 0$, we use the eight heavy flavors with mass $m_h \geq m_\ell$ to interpolate between the $N_\ell = 4$ chirally broken theory and the $N_h + N_\ell = 12$ flavor mass-deformed conformal theory. Phenomenlogically it may be more interesting to simulate a model with, e.g., 2+1 flavors of sextet fermions since this has exactly three massless Goldstone bosons as needed by electroweak symmetry breaking. Our choice of 4+8 flavors is motivated by our use of the staggered formulation to discretize the fermions which has a multiple of four as natural choice. Furthermore, the above mentioned results indicate that the limiting cases of four and twelve flavors have the desired properties, implying 4+8 flavors should certainly interpolate in the relevant parameter space, although the anomalous dimension of the $SU(3)$ 12-flavor system may be too small ($\gamma_m^\star \approx 0.24$) [9, 11, 16] to satisfy phenomenological walking constraints.

In our first numerical studies presented at Lattice 2014 [17], we demonstrated that the renormalized running coupling in our setup becomes a "walking" coupling in an intermediate energy range. Having a walking regime is necessary to satisfy known phenomenological constraints. We also found that the energy range of walking increases as we lower the mass of the eight heavy flavors, i.e., as we approach the IRFP in the 12-flavor theory, confirming our expectations. Next we study the impact of a walking gauge coupling on the meson spectrum. While the computation of the (connected) meson spectrum of states like the pion, rho, or a_0 is standard, the computation of the 0^{++} sigma particle brings additional challenges because it requires the evaluation of disconnected contributions. Disconnected contributions arise from quark-anti-quark pairs coupled only via a gluon to the state of interest. The evaluation of those is numerically much more demanding. In this paper we focus on the computation of the mass of the 0^{++} meson and show preliminary results for the meson spectrum of our 4+8 flavor theory.

In the following section we introduce the setup of our simulations and provide details on the generated ensembles of gauge field configurations. Next we discuss in Section 3 the details and results of our computation of the connected and disconnected spectrum, before we finally conclude.

2. Numerical setup

We use a lattice action based on nHYP [18] smeared staggered fermions and the plaquette gauge action containing fundamental and adjoint terms [9, 19]. Gauge field ensembles are generated using the hybrid Monte Carlo (HMC) update algorithm [20] as implemented in the FUEL software package [21]. The combined choice of the gauge and fermion actions has been extensively tested in projects using various number of fermion flavors (see e.g. [9, 19]) and proven to provide numerically stable simulations for our choice of parameters. Moreover, taste breaking effects of staggered fermions are largely suppressed when nHYP smeared. So far we generated ensembles with a range of different light masses m_ℓ and heavy masses m_h, analyzing two different volumes $24^3 \times 48$ and $32^3 \times 64$ at one value of the gauge coupling ($\beta = 4.0$, $\beta_a = -\beta/4$ [22]). In this paper we focus on the 4+8 flavor ensembles summarized in the left panel of Fig. 1, a total of 21 different ensembles at five different values of the light quark mass m_ℓ and three different values of the heavy quark mass m_h. Work is still in progress to extend some of the existing runs and add new ensembles at different light or heavy quark masses. The color coding in the above mentioned plot highlights in green, orange and red those ensembles which are most useful for our determinations of the 0^{++} scalar with the colors indicating finite size effects (negligible for green and severe for red). As expected we observe finite size effects becoming significant when reducing the masses. Simulations with $m_\ell = 0.035$ turned out to have a too large light mass to

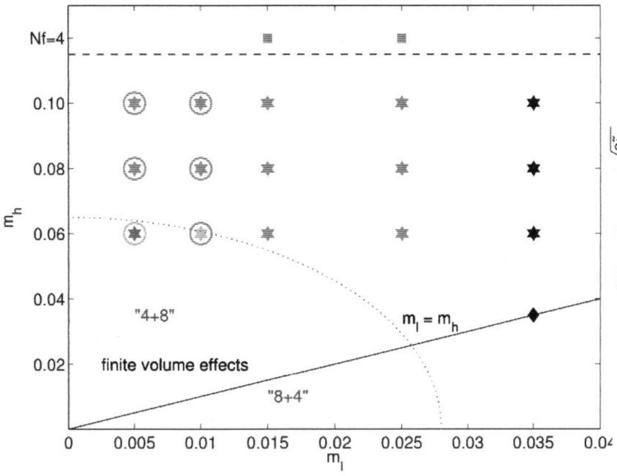

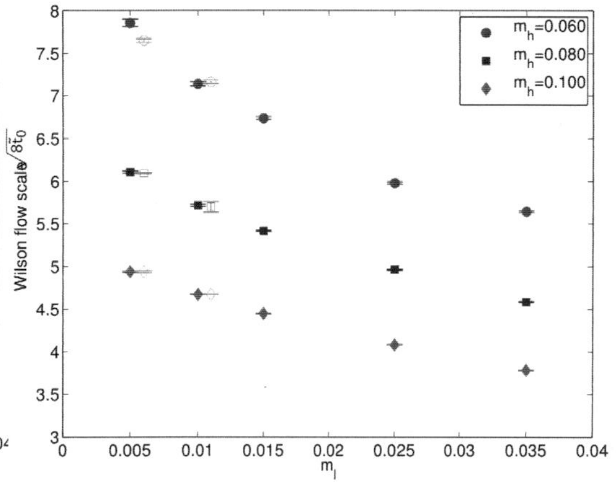

Figure 1. Left panel: light (m_ℓ) and heavy (m_h) mass values for the simulations carried out on $24^3 \times 48$ lattices (filled symbols) and $32^3 \times 64$ lattice (open circles). The colors are meant to caution about finite size effects, likely negligible for green, but of increasing importance as the color turns to orange and red. The black data points are likely to heavy and troubled by cut-off effects. Right panel: The gradient flow scale $\sqrt{8\tilde{t_0}}$ for our 21 ensembles vs. m_ℓ. Filled symbols show values determined on $24^3 \times 48$ lattices, open symbols (shown with a small horizontal offset) refer to the values measured on $32^3 \times 64$ lattices. The strong dependence on both the heavy and light fermion masses is most likely the effect of the IRFP of the 12 flavor system.

reliably extract the 0^{++} and are not considered in our analysis.

The right plot in Fig. 1 shows the lattice scale of our ensembles and how it depends on the light and heavy quark mass. We determine the lattice scale using the concept of gradient or Wilson flow [23–25] which defines the quantity t_0. In order to reduce $\mathcal{O}(a^2)$ lattice artifacts we consider the t-shift improved gradient flow by shifting the Wilson flow parameter t by a small constant τ_0 [10]. Empirically we find that $\tau_0 = 0.1$ is close to optimal and removes nearly all cut-off effects of t_0 [17]. We denote our t-shifted scale by $\tilde{t_0}$ and use the quantity $\sqrt{8\tilde{t_0}}$ to compare the lattice scales of our different ensembles as shown in the right plot of Fig. 1. Reducing cut-off effects is important since currently we only have data at one β value which does not allow us to take a proper continuum limit. As is shown in the plot, our scale has a strong, non-linear dependence on m_ℓ and changes significantly with m_h. Simulating at lower values of m_ℓ, as needed to approach the chiral limit, will therefore force us to simulate on larger volumes. As a rule of thumb, values of $\sqrt{8\tilde{t_0}} \lesssim L/5$ are typically considered to be safe from finite size effects as verified by comparing the full ($24^3 \times 48$) and open ($32^3 \times 64$) symbols. We observe a significant difference between the two volumes only for the simulation with the lightest input masses, $m_\ell = 0.005$, $m_h = 0.060$.

3. Calculating the Light flavor spectrum

As mentioned above, the iso-singlet 0^{++} meson receives special attention in models beyond the Standard Model as it could be a Higgs candidate. Its mass is considerably more difficult to measure numerically than other, non-singlet mesons because both quark-line connected and disconnected diagrams contribute to the correlator

$$C_{0^{++}}(t) \equiv \frac{N_\ell}{4} C_{\text{disc}}(t) - C_{\text{conn}}(t). \tag{1}$$

The disconnected contribution is given by

$$C_{\rm disc}(t) = \sum_{t_0} \left(\langle\bar\psi\psi\rangle(t_0) - \langle\langle\bar\psi\psi\rangle\rangle_{\rm e}\right)\left(\langle\bar\psi\psi\rangle(t_0+t) - \langle\langle\bar\psi\psi\rangle\rangle_{\rm e}\right) \quad (2)$$

where $\langle\langle\bar\psi\psi\rangle\rangle_{\rm e}$ denotes the ensemble average of the fermion condensate. Accurately measuring the ensemble average of the condensate is essential for these calculations as large errors in $\langle\langle\bar\psi\psi\rangle\rangle_{\rm e}$ can destroy the correlator of Eq. (2) at large t.

Further complicating matters, only a stochastic estimate of the matrix element $\langle\bar\psi\psi\rangle(t)$ is feasible. Care must be and is taken to ensure no bias is introduced by incorrectly combining stochastically measured quantities. We compute it on each considered configuration using N_r $U(1)$ noise sources η_i with the property

$$\lim_{N_r\to\infty} \frac{1}{N_r}\sum_i \eta_i^\dagger(\vec{x},t)\eta_i(\vec{y},t') = \delta_{\vec{x},\vec{y}}\delta_{t,t'}. \quad (3)$$

Inverting the staggered Dirac operator $D((\vec{x},t),(\vec{y},t'))$ on the noise source η_i gives

$$\phi_i(\vec{x},t) = D((\vec{x},t),(\vec{y},t'))^{-1}\eta_i(\vec{y},t'), \quad (4)$$

and we can compute the desired matrix element as

$$\langle\bar\psi\psi\rangle(t) = \lim_{N_r\to\infty}\frac{1}{N_r}\sum_i\sum_{\vec{x}}\eta_i^\dagger(\vec{x},t)\phi_i(\vec{x},t). \quad (5)$$

In order to enhance the signal, we use $N_r = 6$ noise sources spreading over the full 4-d volume of the lattice diluted in time, color, as well as even/odd in space [26]. This gives a reasonable balance between numerical costs and the signal-to-noise ratio. Our dilution pattern is consistent with the similar calculation performed in Ref. [14]. Moreover, we use an improved operator for the chiral condensate, unique to naïve and staggered fermions, which further reduces variance in our measurement [27].

The above parameter choices require 1728 inversions of a single color fermion matrix for each $24^3\times 48$ configuration, and 2304 inversions for each $32^3\times 64$ configuration. Using our full ensembles we are able to determine the disconnected correlators to a few percent precision this way and extract the mass of the corresponding 0^{++} states. For comparison, for a $24^3\times 48$ ($32^3\times 64$) lattice volume, it suffices to perform 18 (24) inversions of a single color fermion matrix to evaluate the quantities of the connected spectrum, e.g., the pion and rho meson, using wall sources to determine their mass with a precision to better than a percent on our ensembles. This illustrates the cost and difficulties of disconnected correlator calculations.

As an example we show the resulting connected and disconnected scalar correlators on the $32^3\times 64$, $m_\ell = 0.010$, $m_h = 0.060$ ensemble in the left panel of Fig. 2. Due to the much higher costs we have presently measured the disconnected correlator on only 89 configurations, with further measurements in progress.

General features of the $C_{0^{++}}$ correlators are nonetheless present: For small t the connected correlator dominates the combination and the effect of the disconnected contribution is insignificant until $t \gtrsim 8$. There the signal of $C_{\rm disc}$ is very noisy and we are not able to fit the mass reliably. As an alternative Refs. [12–14] advocate to consider the disconnected correlator only. As long as the 0^{++} state is lighter than the non-singlet a_0 state, the iso-singlet scalar mass can be extracted directly from $C_{\rm disc}$. From either correlator, values are determined by a fully correlated non-linear fit to a single 0^{++} state, with statistical errors estimated by a single-elimination jackknife. In practice, fitting to just $C_{\rm disc}$ works surprisingly well and one heuristic

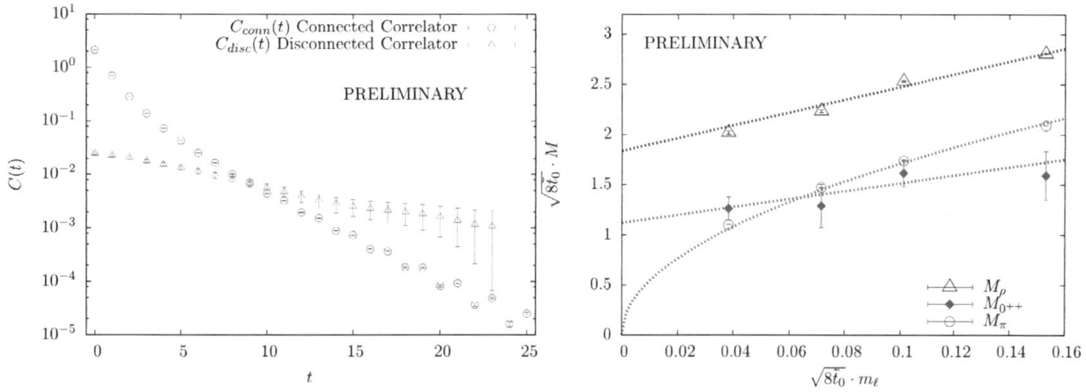

Figure 2. Left panel: Comparison of $-C_{\mathrm{conn}}(t)$ and $C_{\mathrm{disc}}(t)$ for the $32^3 \times 64$ $m_\ell = 0.010$. Right panel: The pion, rho and 0^{++} scalar light flavor spectrum for the $m_h = 0.060$ ensembles.

explanation may be that excited states in the connected and the 0^{++} correlators cancel. Using C_{disc} only we obtain values in agreement with $C_{0^{++}}$ but with smaller statistical uncertainty.

We close this section by showing in the right panel of Fig. 2 preliminary results for the light flavor pion, the rho, and the 0^{++} scalar, obtained on our 4+8 ensembles with $m_h = 0.060$ using $32^3 \times 64$ volumes for $m_\ell = 0.005$ and 0.010 and $24^3 \times 48$ for $m_\ell = 0.015$ and 0.025. The data shown are rescaled by $\sqrt{8\tilde{t}_0}$ to account for differences in the lattice scale. We observe the 0^{++} mass is lighter than the pion at large m_ℓ, crosses the pion mass, and becomes heavier toward the chiral limit. While the pion and 0^{++} may still be degenerate within two sigmas of our statistical uncertainty even at our lightest mass, our data reveal a clear trend which requires further data to be confirmed. Our preliminary results have the potential to be of great phenomenological interest and have thus far not been observed in other studies of conformal or near-conformal systems [12–14]. We have further measurements in progress, both increasing our statistics on existing ensembles and extending our analysis of the 0^{++} to additional enembles.

4. Conclusions and outlook

Exploring the phenomena of gauge-fermion systems near the conformal window is interesting in its own and may in addition reveal new "physics" which may find its application in beyond the Standard Model theories describing e.g. the Higgs boson as composite particle. We study a novel model of four light and eight heavy flavors which allows us to tune arbitrarily close to the conformal fixed point of the 12-flavor system while still being chirally broken in the infrared limit. Of special interest is determining the mass of the iso-singlet scalar which could be a Higgs candidate. This computation is in particular challenging because its measurement contains both a connected and a disconnected contribution. Using stochastic estimates for the disconnected contribution, our preliminary results reveal the very interesting case of an iso-singlet scalar which is lighter than the pion at large fermion mass, but heavier for smaller masses. Our current findings will have to be confirmed by further simulations.

Acknowledgments

Computations for this work were carried out in part on facilities of the USQCD Collaboration, which are funded by the Office of Science of the U.S. Department of Energy, on computers at the MGHPCC, in part funded by the National Science Foundation, and on computers allocated under the NSF Xsede program to the project TG-PHY120002.

We thank Boston University, Fermilab, the NSF and the U.S. DOE for providing the facilities essential for the completion of this work. R.C.B., C.R. and E.W. were supported by DOE grant DE-SC0010025. In addition, R.C.B., C.R. and O.W. acknowledge the support of NSF grant OCI-0749300. A.H. acknowledges support by the DOE grant DE-SC0010005.

References

[1] Aad G et al. (ATLAS Collaboration) 2012 Phys.Lett. **B716** 1–29 (*Preprint* 1207.7214)
[2] Chatrchyan S et al. (CMS Collaboration) 2012 Phys.Lett. **B716** 30–61 (*Preprint* 1207.7235)
[3] Yamawaki K, Bando M and Matumoto K i 1986 Phys. Rev. Lett. **56**(13) 1335–1338
[4] Appelquist T, Terning J and Wijewardhana L C R 1991 Phys. Rev. D **44**(3) 871–877
[5] Appelquist T, Fleming G, Lin M, Neil E and Schaich D 2011 Phys.Rev. **D84** 054501 (*Preprint* 1106.2148)
[6] Hasenfratz A 2012 Phys.Rev.Lett. **108** 061601 (*Preprint* 1106.5293)
[7] Aoki Y, Aoyama T, Kurachi M, Maskawa T, Nagai K i et al. 2012 Phys.Rev. **D86** 054506 (*Preprint* 1207.3060)
[8] DeGrand T 2011 Phys.Rev. **D84** 116901 (*Preprint* 1109.1237)
[9] Cheng A, Hasenfratz A, Liu Y, Petropoulos G and Schaich D 2014 Phys.Rev. **D90** 014509 (*Preprint* 1401.0195)
[10] Cheng A, Hasenfratz A, Liu Y, Petropoulos G and Schaich D 2014 JHEP **1405** 137 (*Preprint* 1404.0984)
[11] Lombardo M, Miura K, da Silva T J N and Pallante E 2014 (*Preprint* 1410.0298)
[12] Aoki Y, Aoyama T, Kurachi M, Maskawa T, Nagai K i et al. 2013 Phys.Rev.Lett. **111** 162001 (*Preprint* 1305.6006)
[13] Aoki Y et al. (the LatKMI Collaboration) 2014 Phys.Rev. **D89** 111502 (*Preprint* 1403.5000)
[14] Fodor Z, Holland K, Kuti J, Nogradi D and Wong C H 2014 PoS **LATTICE2013** 062 (*Preprint* 1401.2176)
[15] Brower R, Hasenfratz A, Rebbi C, Weinberg E and Witzel O 2014 (*Preprint* 1410.4091)
[16] Cheng A, Hasenfratz A, Petropoulos G and Schaich D 2013 JHEP **1307** 061 (*Preprint* 1301.1355)
[17] Brower R, Hasenfratz A, Rebbi C, Weinberg E and Witzel O 2014 (*Preprint* 1411.3243)
[18] Hasenfratz A, Hoffmann R and Schaefer S 2007 JHEP **0705** 029 (*Preprint* hep-lat/0702028)
[19] Cheng A, Hasenfratz A, Petropoulos G and Schaich D 2013 PoS **LATTICE2013** 088 (*Preprint* 1311.1287)
[20] Duane S, Kennedy A, Pendleton B and Roweth D 1987 Phys.Lett. **B195** 216–222
[21] Osborn J et al. Framework for unified evolution of lattices (FUEL) URL http://usqcd-software.github.io/FUEL.html
[22] Hasenbusch M and Necco S 2004 JHEP **0408** 005 (*Preprint* hep-lat/0405012)
[23] Narayanan R and Neuberger H 2006 JHEP **0603** 064 (*Preprint* hep-th/0601210)
[24] Lüscher M 2010 Commun.Math.Phys. **293** 899–919 (*Preprint* 0907.5491)
[25] Lüscher M 2010 JHEP **1008** 071 (*Preprint* 1006.4518)
[26] Foley J, Jimmy Juge K, O'Cais A, Peardon M, Ryan S M et al. 2005 Comput.Phys.Commun. **172** 145–162 (*Preprint* hep-lat/0505023)
[27] Kilcup G and Sharpe S R 1987 Nucl.Phys. **B283** 493

Modeling Galactic Rotation Curves with ultra-light scalar field dark matter

Argelia Bernal and C Soto-Campos
Universidad Autónoma del estado de Hidalgo (UAEH), Hidalgo, México.
E-mail: argelia.bernal@uaeh.edu.mx, csoto@uaeh.edu.mx

Abstract. In this work we investigate the behaviour of SFDM at small scales, specifically its viability in reproducing galactic rotation curves. We consider a scalar field with no self-interaction. In this case, the only free parameter is the mass of the scalar field. By fitting the galactic rotational curves it is found the allowed region for the mass of the scalar field. We discuss those limits with previous bounds on the mass coming from astrophysical considerations. The galactic density profiles within the SFDM is found by solving a non linear system of ordinary differential equations for the scalar and the gravitational fields. Given the appropriate boundary conditions, the system becomes an eigenvalue problem which is solved by an inner-outer interaction using an efficient computational method. This method is faster than the predecessors, and could be implemented in parallel computing.

1. Introduction

There are two well tested theories that describe all four fundamental interactions: the Standard Model of elementary particles (SM) and the general theory of relativity. The first one explains strong, weak and electromagnetic interactions while the later explains the gravitational interaction. The combination of this two theories can explain most of the astrophysical observations at cosmological distances in a consisten model that is referred as the standard model of cosmology. This standard model of cosmology predicts that 95% of the energy density of the universe is made of two new components that are not included in the SM. They are referred as the Dark Matter (DM) and the Dark Energy (DE) and its nature is one of the most intriguing and fundamental subjects in Modern Physics. (See [1] and [2] for reviews on DM and DE respectively).

Regarding dark matter, there is a plethora of candidates, most of them coming from extensions to the SM [3]. In general, the most generic DM candidate coming form particle physics is referred as Weak Interactive Massive Particle (WIMP) and is modelled as a pressureless fluid. Another viable candidate is the axion, and recently, the interest on ultra-light axion-like dark matter candidates has been increased [4]. On the context of axion-like dark matter, one of the simplest candidates is the one which does not have any coupling with the standard model. Several flavors of this simplest candidate have been studied since more than a decade ago. They have been named as Scalar Field Dark Matter (SFDM) [5, 6, 7, 12], fuzzy dark matter [8], Bose-Einstein Condensate dark matter [26] and more recently Wave dark matter [9].

The general feature of those models is that dark matter particles form a coherent state that can be described as a massive, classical scalar field, either complex or real, minimally coupled to gravity, endowed with a scalar field potential $V(\Phi)$. Different potentials have been considered. For instance, $V(\Phi) = \frac{m^2}{8\pi G\lambda^2}(\cosh(\sqrt{8\pi G}\lambda\Phi) - 1)$ [5, 6, 7], $V(\Phi) = m^2\Phi^2 + \lambda\Phi^4$ [10, 11] and the simplest one that only consider the mass term [12, 8, 9]:

$$V(\Phi) = m^2\Phi^2 . \tag{1}$$

With out the intension of to be exhaustive, a general description of the model with the simplest potential (1), to which we called from now on Ultra-Light Scalar Field Dark Matter Model (ULSFDM) is as follows: The thermal evolution of the bosonic particles and their condensation is studied in [22, 13]. Once the condensate is formed, this coherent state described by a classical scalar field with the scalar potential (1), successfully describes the dark matter evolution at cosmological scales [14, 12] that is, for a homogeneous and isotropic universe. Furthermore, the ULSFDM mimics the dynamics of linear perturbations to the homogeneous distribution of dark matter [8] [23], and recently results, shown that the ULSFDM is able of reproduce the non linear structure formation at large scales [25, 9, 23, 24].

These results indicates that ULSFDM is a robust model at cosmological scales and as viable as the more popular and well studied CDM. Even more, the ULSFDM would overcome to the CDM at smaller scales since it could not have the so called problems of CDM: the formation of singular halos and the overproduction of dwarf halos. The key piece of the ULSFDM in order to avoid the overproduction of substructure is a mass of the scalar field of the order of $m \sim 1.1 \times 10^{-23}$ [6]. This value of the mass, which is the only free parameter of the model, is in agreement with results on the cosmological evolution and on the structure formation [9]. On the other hand, regarding ULSFDM halos, there exist self-gravitating scalar field configurations that have been proposed to play this role. Such configurations do not present singularities and, for a scalar field endowed with a potential (1) with a mass of the order of $m \sim 10^{-23}$, have sizes comparable with galactic halos. There have been several works studying the viability of such configurations in reproducing different observables of galactic halos.

In this note we investigate the allowed values of the mass m in order to fit a set of observed galactic rotation curves. We restrict ourselves to consider galaxies for which the contribution of the luminous matter to the rotation curve is negligible and to model ULSFDM halos with stable self-gravitating scalar field configurations known as ground states. We found that big galaxies demand smaller values of the mass of the scalar field than those needed for small galaxies. Some galaxies will requiere masses beyond the allowed regions coming from other astrophysical considerations. This multiplicity of values for the free parameter m is an issue in modeling halos with ground state configurations and then another proposals to model ULSFDM halos should be considered. In fact, one proposal has been presented in [36], were non stable scalar fiel configurations known as excited states, have been used. However a non stable halo is not realistic. Another possibility is to use multi-state-boson-stars which are configurations where excited states coexist together with the ground state. It has been shown that there exist stable multi-state-boson-stars. [17, 18, 19]. The corresponding bounds on the mass of the scalar field considering excited states are left for a future work.

2. Ultra-light scalar field dark matter galactic halos

It is well known that self-gravitating configurations of a complex scalar field Φ with a potential given by (1) can be obtained as solutions to the coupled Einstein Klein-Gordon (EKG) equations. Such configurations, known as Boson Stars (BS), have their newtonian counterparts, the so called, Newtonian Boson Stars (NBS). Because it is a good approximation to consider galactic halos as newtonian objects, it has been proposed within the ULSFDM that such NBS, instead of relativistic BS, would play the role of galactic halos.

The non-relativistic limit of the EKG equations is the Shrödinger-Poisson system:

$$i\hbar\partial_t\psi = -\frac{\hbar^2}{2m}\nabla^2\psi + Um\psi\ , \quad (2)$$

$$\nabla^2 U = 4\pi G\rho, \quad (3)$$

where $\rho = m^2\psi\psi^*$, U is the gravitational potential and ψ is related to Φ through $\Phi = e^{-imc^2t/\bar{h}}\psi$. In fact ψ is the non relativistic part of Φ. NBS are obtained as solutions to (2-3) introducing the following ansatz

$$\psi = e^{-i\epsilon mc^2t/\bar{h}}\phi(r), \quad (4)$$

where ϵ is the energy eigenvalue, see section (3). NBS are configurations spherically symmetric, gravitationally bounded and regular every where. Therefore ULSFDM halos, modeled by NBS will have such characteristics. In particular, the tangential velocity produced by such configurations, i.e., the rotation curve produced by ULSFDM halos, can be easily computed by

$$v = \sqrt{\frac{GM(r)}{r}}, \quad (5)$$

where

$$M(r) = \int_0^r \rho dv. \quad (6)$$

It is importan to make a comment on the gravitational stability of NBS. NBS differ qualitatively on the number of nodes of its radial function ϕ. The configurations for which ϕ does not have nodes is called the ground state. For the first excited state ϕ has one node and so on. It is well known that excited states are unstable under gravitational linear perturbations [16, 27, 28, 29, 30, 31] and then, they are not suitable to model realistic halos. In contrast, the ground state is stable an therefore in this work we will consider only this state to model ULSFDM galactic halos.

Another important point to stress is that free parameters of NBS are the central value of the scalar field $\phi(0)$ and the mass m. However, while m is a fundamental parameter of the scalar field, $\phi(0)$ is a boundary condition that can be freely chosen. In fact relevant scales of observables for NBS depend on the pair $(\phi(0), m)$. For instance, in terms of these free parameters, the mass and the radius of the ground state are given by

$$M = {\phi_c}^{1/2} 2.75 \times 10^{-10}\frac{1\text{eV}}{m}M_\odot \qquad R = {\phi_c}^{-1/2} 2.43 \times 10^{-26}\frac{1\text{eV}}{m}\text{Kpc}, \quad (7)$$

where ϕ_c is a dimensionless quantity related with the value of $\phi(0)$. For Newtonian systems it is expected $\phi_c \leq 1 \times 10^{-3}$. Notice that in this regime, the boson mass m should be extremely light in order that M and R have astrophysical scales.

3. Numerical solutions for the Schrödinger-Poisson system

Taking the ansatz (4) and imposing boundary conditions of regularity and asymptotic flatness, the system (2-3) becomes a Sturm-Liouville two-point boundary value problem: given a central value for the radial function $\phi(0)$, there are discrete values $\{\epsilon_i\}$ for which solutions $\{\epsilon_i, \phi_i(r), U_i(r)\}$ satisfy the boundary conditions. Usually the shooting method is used to solve this eigenvalue problem [28, 29, 30, 18]. However we solved it using an algorithm developed in Bernstein et al [32].

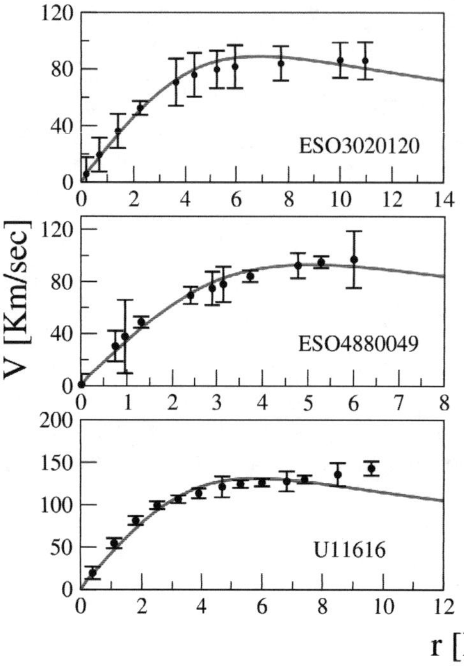

Figure 1. Best fit velocity profile for different low surface brightness galaxies.

System (2-3) can be written as:

$$\frac{\partial^2 \bar{f}}{\partial \bar{r}^2} = 2\bar{f}(\bar{u} - \bar{\epsilon}) \tag{8}$$

$$\nabla^2 \bar{u} = \frac{1}{\bar{r}^2}\bar{f}^2 \tag{9}$$

where we have re-scaled the system and we have defined $\bar{f} = \frac{\sqrt{4\pi}}{m^{3/2}} r\phi \left(\frac{m_p}{m}\right)^3$, $\bar{\epsilon} = \left(\frac{m}{m_p}\right)^4 m\epsilon$, $\bar{u} = \left(\frac{m_p}{m}\right)^4 U$.

The numerical scheme for computing solutions to (8-9) consist in: *i)* to supply an initial guess for $\bar{f}$ and to solve Eqn. (9) for $\bar{u}$. *ii)* to substitute $\bar{u}$ in (8) and to solve the eigenvalue problem for $\bar{\epsilon}$ and $\bar{f}$. *iii)* to substitute the new $\bar{f}$ in Eqn. (9) and solve it for $\bar{u}$. *iv)* to terate the previous two steps until the value of $\bar{\epsilon}$ converge sufficiently.

Equation (8) can be solved using the finite differences, while equation (9) can be solved using quadratures.

4. Comparison with observations

Once we have found a solution for the Schrödinger-Poisson system, we can compute the theoretical velocity $v^i_{theo}(\phi(0), m) = v(r_i) = \sqrt{GM(r_i)/r_i}$, with $M(r)$ as defined as eq. (6). Here we perform a single χ^2 analysis. There are two free parameters: the amplitude of the scalar field at the origin and the mass of the scalar field. Thus

$$\chi^2(\phi(0), m) = \sum_i \frac{(v^i_{theo}(\phi(0), m) - v^i_{exp})^2}{\delta v_{exp}}. \tag{10}$$

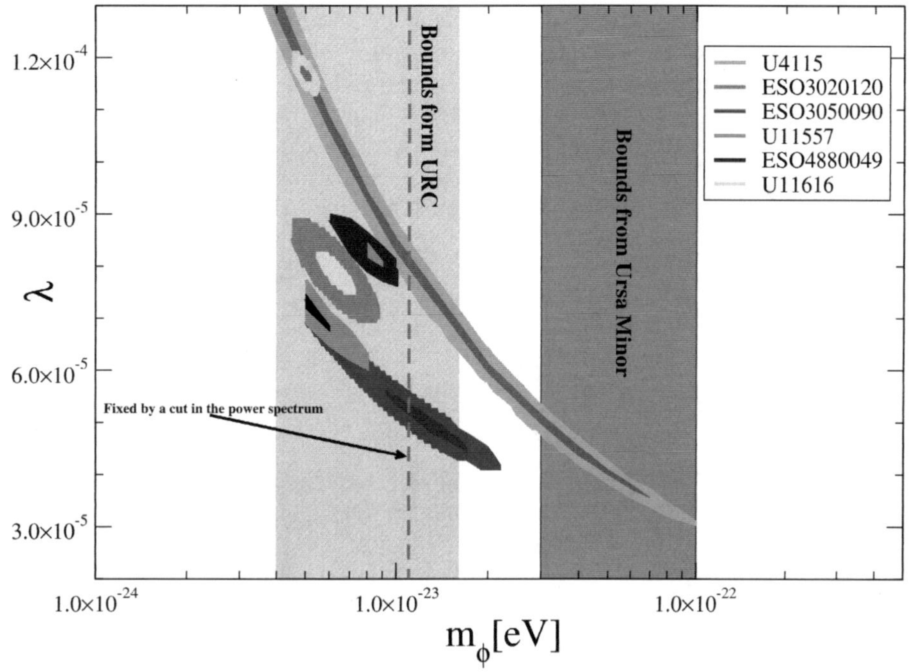

Figure 2. Allowed region for the mass of the scalar field (68 % and 90 % C.L.) for the LSB galaxies shown in Fig. 1. Dashed vertical line corresponds to $m = 1.1 \times 10^{-23}$ eV needed to have a cut in the power spectrum [6]. Other limits from the Dwarf galaxy Ursa minor and fits to the Universal Rotational Curve are also shown.

where i runs for all data points, δv_{exp} is the error in the determination of the rotational velocity and v^i_{exp} is the central value of the rotational velocity of the stars around the galaxies.

As an example we have used the data points of the galaxies ESO 3020120, ESO 4880049, ESO3050090, U11616, U11557 and U4115 taken from [33]. The best fit together with the data points are shown in Fig. 1. As can bee seen, there are values of m and $\phi(0)$ such as we have a good match between theory and observations. The allowed regions for $\phi(0)$ and m at 68% and 90% C.L. are shown in Fig. 2. There $\phi(0) \propto \lambda$.

5. Discussion

There are different constraints for the mass of the scalar field within the ULSFDM. For instance in [6], the model predicts a cut in the power spectrum that will alleviate the problems that CDM posses if $m \sim 1.1 \times 10^{-23}$ eV. Another interesting constraint was derived in [34] where in order to explain that a relic cluster of star had survived without been disrupted by the gravitational field of the halo as the observations indicate in the interior of the dwarf galaxy Ursa Minor,then $3 \times 10^{-23}\text{eV} < m < 1 \times 10^{-22}$ eV. Finally, masses between $4 \times 10^{-24}\text{eV} < m < 1.6 \times 10^{-23}$ have been derived by fitting a Universal Rotational Curve (URC) [35]. We have included those limits together with the allowed regions for the mass of the scalar field in Fig. 2.

As it can bee seen, big galaxies with high velocities requieres small masses of the scalar field, as small as those needed in [6, 35]. But in general there is a conflict with small galaxies. In particular Dwarf galaxies [34] requires bigger masses. Just the galaxy U4115, the one with the smallest radial distance and smaller velocities has compatible masses with all constraints. Is it possible to have the same mass for all galaxies? That would be a benchmark for the model, but what we see from this exercise is that it is not possible. One way to overcome this may be to perform the fit for big galaxies by using not only ground state BS but multi state boson stars [17, 18, 19]. Those configurations have bigger masses and extend longer distances, such as the mass could be lowered. Further studies are needed.

5.1. Aknowledgments

A.B. acknowledges support from CONACYT, Mexico. C. S. acknowledges partial support from UAEH.

References

[1] G. Bertone, D. Hooper and J. Silk, Phys. Rept. **405**, 279 (2005) [hep-ph/0404175].
[2] T. Padmanabhan, Phys. Rept. **380**, 235 (2003) [hep-th/0212290].
[3] M. Taoso, G. Bertone and A. Masiero, JCAP **0803**, 022 (2008) [arXiv:0711.4996 [astro-ph]].
[4] P. Arias, D. Cadamuro, M. Goodsell, J. Jaeckel, J. Redondo and A. Ringwald, JCAP **1206**, 013 (2012) [arXiv:1201.5902 [hep-ph]].
[5] T. Matos and L. A. Urena-Lopez, Class. Quant. Grav. **17**, L75 (2000) [astro-ph/0004332].
[6] T. Matos and L. A. Urena-Lopez, Phys. Rev. D **63**, 063506 (2001) [astro-ph/0006024].
[7] T. Matos and L. A. Urena-Lopez, Int. J. Mod. Phys. D **13**, 2287 (2004) [astro-ph/0406194].
[8] W. Hu, R. Barkana and A. Gruzinov, Phys. Rev. Lett. **85**, 1158 (2000) [astro-ph/0003365].
[9] H. Y. Schive, T. Chiueh and T. Broadhurst, arXiv:1406.6586 [astro-ph.GA].
[10] F. Briscese, Phys. Lett. B **696**, 315 (2011) [arXiv:1101.0028 [astro-ph.CO]].
[11] B. Li, T. Rindler-Daller and P. R. Shapiro, Phys. Rev. D **89**, 083536 (2014) [arXiv:1310.6061 [astro-ph.CO]].
[12] T. Matos, J. A. Vazquez and J. Magana, *ϕ^2 as Dark Matter*, [arXiv:0806.0683]
[13] L. A. Urena-Lopez, JCAP **0901** (2009) 014. [arXiv:0806.3093 [gr-qc]].
[14] M. S. Turner, Phys. Rev. **D28** (1983) 1243.
[15] J. F. Navarro, C. S. Frenk and S. D. M. White, Astrophys. J. **490**, 493 (1997) [astro-ph/9611107].
[16] M. Gleiser, Phys. Rev. D **38**, 2376 (1988) [Erratum-ibid. D **39**, 1258 (1989)].
[17] T. Matos and L. A. Urena-Lopez, Gen. Rel. Grav. **39**, 1279 (2007).
[18] A. Bernal, J. Barranco, D. Alic and C. Palenzuela, Phys. Rev. D **81**, 044031 (2010) [arXiv:0908.2435 [gr-qc]].
[19] L. A. Urena-Lopez and A. Bernal, Phys. Rev. D **82**, 123535 (2010) [arXiv:1008.1231 [gr-qc]].
[20] A. Bernal, "Estudio Dinámico de campos escalares autogravitantes", PhD Thesis http://www.fis.cinvestav.mx/thesis/index.php/thesis/doc_details/36-argelia-bernal-bautista
[21] Chi Wai Lai, "A Numerical Study of Boson Stars", PhD Thesis http://laplace.physics.ubc.ca/People/matt/Doc/Theses/Phd/lai.pdf
[22] A. P. Lundgren, M. Bondarescu, R. Bondarescu and J. Balakrishna, Astrophys. J. **715**, L35 (2010) [arXiv:1001.0051 [astro-ph.CO]].
[23] J. M. Torres, M. Alcubierre, A. Diez-Tejedor and D. Nunez, Phys. Rev. D **90**, 123002 (2014) [arXiv:1409.7953 [gr-qc]].
[24] D. J. E. Marsh and P. G. Ferreira, Phys. Rev. D **82**, 103528 (2010) [arXiv:1009.3501 [hep-ph]].
[25] T. P. Woo and T. Chiueh, Astrophys. J. **697**, 850 (2009) [arXiv:0806.0232 [astro-ph]].
[26] C. G. Boehmer and T. Harko, JCAP **0706**, 025 (2007) [arXiv:0705.4158 [astro-ph]].
[27] Lee, T.D., & Yang Pang, Nuclear Physics **B** 315 (1988) 477.
[28] Kusmartsev, F.V., Mielke, E.W., & Schunk, F.E., Phys. Lett. **A** 157 (1991) 465.
[29] Seidel, E., & Suen Wai-Mo, Phys. Rev. **D** 42 (1990) 384.
[30] Jetzer, P., Phys. Reports **220** (1992) 163.
[31] Bernal, A., & Guzmán, F. S., Phys. Rev.**D** 74 (2006) 063504 [astro-ph/0608523].
[32] Bernstein D. H., Giladi E., Jones K.R.W., Mod. Phys. Lett. **A**13 (1998) 2327
[33] S. S. McGaugh, V. C. Rubin and W. J. G. de Blok, Astron. J. **122**, 2381 (2001) [astro-ph/0107326].
[34] V. Lora, J. Magana, A. Bernal, F. J. Sanchez-Salcedo and E. K. Grebel, JCAP **1202**, 011 (2012) [arXiv:1110.2684 [astro-ph.GA]].
[35] A. Arbey, J. Lesgourgues and P. Salati, Phys. Rev. D **64**, 123528 (2001) [astro-ph/0105564].
[36] Sin, Sang-Jin, Phys. Rev. D **50**, 3650 (1994) [astro-ph/9205208].

Cosmological Impact of the Population III Hypernovae

Ke-Jung Chen

Department of Astronomy & Astrophysics, University of California, 1156 High St.
Santa Cruz, California 95064, USA
School of Physics and Astronomy, University of Minnesota, 116 Church St.
Minneapolis, Minnesota 55455, USA

E-mail: kchen@ucolick.org

Abstract. We investigate the stellar feedback of Pop III hypernovae by using cosmology simulations with GADGET. Recent studies of the Pop III star formation suggested that these stars might be very massive and many of them might have died as hypernovae, which are about ten times more energetic than conventional core-collapse supernovae. Such exotic supernovae might posit a strong impact on the early universe by injecting a large amount of energy and heavy elements. For investigating the cosmological impact of the first hypernovae, we use the realistic Pop III stellar evolution models including the relevant physics to simulate the formation, evolution, and hypernova explosion of a 60 $M_\odot$ Pop III star. Our results show that the gas inside the host minihalo is effectively blown out by the ramp pressure of radiation and hypernova shock. It delays the star formation inside the halo for at least 50 million years. More importantly, the pristine gas of the intergalactic medium around the halo is chemically enriched to an average metallicity of $10^{-4} - 10^{-3}$ $Z_\odot$, out to ~ 2 kpc within five hundred million years after the star dies. Such metal enrichment is able to transform the later star formation mode from Pop III to Pop II.

1. Introduction

One of the most important problems in modern cosmology is to understand how the first luminous objects such stars, accreting black holes (BHs) and galaxies, that shaped the early universe at the end of the cosmic dark ages [1, 2]. A key driver of this dramatic transformation was the gradual enrichment of the pristine universe with heavy chemical elements out of the first supernova (SN) explosions [3]. According to the modern theory of cosmological structure formation [4], the hierarchical assembly of dark matter (DM) halos provided the gravitational potential wells that allowed gas to form stars and galaxies inside them. Extending this model to the highest redshifts, one can determine the sites where the first stars, the so-called Population III (Pop III) stars, formed out of the pure hydrogen and helium gas created in the birth of the universe. Within this framework, Pop III stars are predicted to form inside of DM minihalos with total masses (DM and gas) of about 10^6 $M_\odot$ at redshifts of $z \sim 20 - 30$ [5, 6, 7].

The first stars affected the evolution of the early universe in several different ways. Massive Pop III stars were strong sources of hydrogen and helium ionizing photons that built up extensive H II, He II, and He III regions [8]. The metals forged inside Pop III stars were later dispersed into the intergalactic medium (IGM) when they died as SNe, thus quickly polluting the primordial gas such that the second generation of (Population II) stars could form. It is convenient to classify Pop III feedback mechanisms into different classes, specifically radiative feedback [9, 10], mechanical and chemical feedback due to SNe [11, 12], and X-ray feedback from accreting BH remnants [13, 14].

In the Pop III star-forming regions, metal cooling was absent because the primordial gas consisted almost exclusively of hydrogen ($\sim 76\,\%$ by mass) and helium ($\sim 24\,\%$). Molecular hydrogen was thus the dominant coolant, but owing to its quantum-mechanical structure, it was unable to cool the gas to the low temperatures typically encountered in star forming clouds today. The primordial gas, therefore, remained relatively warm, with typical temperatures of several hundred Kelvin. Hence the Jeans mass was correspondingly larger, as well, leading to the expectation that Pop III stars might have been more massive than stars formed today, with a predicted mass scale of $50\,\mathrm{M}_\odot - 100\,\mathrm{M}_\odot$ [15, 16]. Because of their high surface temperatures [17, 18], Pop III stars could effectively produce copious amounts of ionizing UV photons. Extended H II regions with size of several kpc were created before the stars died. Given the shallowness of the gravitational potential well of the host DM halos, the surrounding gas was subject to strong photo-heating, thus being able to easily escape the host minihalos [19]. This photo-evaporation suppressed further star formation inside the minihalos, thus delaying further star formation until more massive host halos emerged [20].

The character of Pop III feedback sensitively depends on the fate encountered by massive Pop III stars when they die after their short lifetime of a few million years. The fates of massive Pop III stars fall into four categories, depending upon their helium core mass at death [21]. Stars with helium cores of $2 - 8\,\mathrm{M}_\odot$ (main sequence masses $M_* = 10 - 25\,\mathrm{M}_\odot$) die in core-collapse SNe. Stars with helium cores of $8 - 35\,\mathrm{M}_\odot$ ($M_*= 25 - 90\,\mathrm{M}_\odot$) likely collapse to black holes unless they rotate very rapidly, in which case hypernovae might result. When the helium core is above $35\,\mathrm{M}_\odot$, it eventually encounters the electron/positron pair-instability. It either ejects a series of shells known as the pulsational pair-instability supernova for $M_*= 90 - 140\,\mathrm{M}_\odot$ [22], Or completely destroys the star in one cataclysmic explosion, pair-instability supernovae (PSN) for $M_*= 140 - 260\,\mathrm{M}_\odot$ [23, 24, 25]. A new type of supernova due to the GR instability in supermassive stars at $\sim 55,000\,\mathrm{M}_\odot$, which may have formed in the early universe, has now also been found in [26]. Because vast amounts of metals are ejected during a PSN explosion, even a single event could enrich about $10^7\,\mathrm{M}_\odot$ of primordial gas up to $10^{-4}\,\mathrm{Z}_\odot - 10^{-3}\,\mathrm{Z}_\odot$ [11, 12]. Even a small amount of metals could change the subsequent star formation process and might cause a transition of the stellar initial mass function (IMF) from the top-heavy mode predicted for Pop III stars to the standard IMF for later (Pop I and Pop II) generations with typical masses comparable to that of our Sun [27]. However, the results from stellar archeology [28], which studies the most metal-poor stars in the Local Group that retained the imprints from nucleosynthesis in the early universe, possibly including those from Pop III stars, in general do not support the chemical abundance pattern predicted for PSN enrichment. Theoretical PSN yields exhibit a pronounced odd-even effect resulting from a low neutron excess. In addition, the lack of any neutron capture process results in the absence of all elements heavier than the Fe peak (no r- or s-process). The Pop III core-collapse SNe and hypernovae models [29], however, can produce abundance patterns in good accord with the observation of metal poor stars. Thus, we study the Pop III hypernovae feedback using cosmology simulations with the realistic Pop III stellar models.

The structure of this paper is as follows: in Section 2, we describe our initial setup, as well as our numerical methods. The simulation results are presented in Section 3, and we conclude by discussing the broader implications in Section 4. All of the results presented in this paper use physical coordinates instead of comoving coordinates.

Journal of Physics: Conference Series **640** (2015) 012057 doi:10.1088/1742-6596/640/1/012057

2. Numerical Methodology

The major tool used for our simulations is the well-tested, massively-parallel cosmological code GADGET [30], which computes gravitational forces with a hierarchical tree algorithm and represents fluids by means of smoothed particle hydrodynamics (SPH). In order to simulate the feedback exerted by the first stars and their SNe, additional physical processes, such as cooling and chemistry of the primordial gas, radiative transfer of ionizing photons, and SN explosions, are needed and have been included in our modified version of GADGET.

Our simulations use the initial conditions based on [12], starting at $z = 100$ in a periodic box of linear size of 1 Mpc (co-moving). The ΛCDM cosmological parameters are matter density $\Omega_m = 0.3$, baryon density $\Omega_b = 0.04$, Hubble constant $H_0 = 70$ km s^{-1}Mpc^{-1}, spectral index $n_s = 1.0$, and normalization $\sigma_8 = 0.9$, based on the WMAP cosmic microwave background (CMB) measurement [31]. Our simulations marginally resolve the corresponding Jeans mass of $M_J \simeq 500\,M_\odot$. The cooling and chemistry network of the primordial gas include all relevant cooling mechanisms of primordial gas, such as H and He collisional ionization, excitation and recombination cooling, bremsstrahlung, and inverse Compton cooling; in addition, the collisional excitation cooling via H_2 and HD is also taken into account. For H_2 cooling, collisions with protons and electrons are explicitly included. We use the sink particles to mimic the star formation based on the algorithm of [32]. The important criterion for sink creation, and subsequent accretion of further SPH particles, is that the gas density exceeds a pre-specified density threshold, $n_c \sim 10^4\,\mathrm{cm}^{-3}$, but we also test for gravitational boundedness, and whether the gas is part of a converging flow, where $\nabla \cdot \vec{v} < 0$. The sink particles provide markers for the position of a Pop III star and its remnants, such as a black hole or SN, to which detailed sub-grid physics can be supplied.

When a Pop III star has formed inside the minihalo, the sink particle representing it immediately turns into a point source of ionizing photons to mimic the birth of a star. Instead of simply assuming constant rates of emission, we use the results of one-dimensional stellar evolution calculations from [29] to construct the luminosity history of the Pop III star. Indeed, luminosities exhibit considerable time variability when taking the evolution off the main sequence into account. The photons streaming from the star then establish an ionization front and build up H II regions. To trace the propagation of photons and the ionization front, we use the ray-tracing algorithm from [9]. The radiation transport is coupled to the hydrodynamics of the gas through its chemical and thermal evolution. The transfer of the H_2-dissociating photons in the Lyman–Werner (LW) band ($11.2 - 13.6$ eV) is also included.

When the star reaches the end of its lifetime, we initialize a SN blast wave by distributing the explosion energy among the gas SPH particles surrounding the sink. Because the resolution of the simulation is about 1 pc, we cannot resolve individual SNe in both mass and space. Instead, we select the particles within a region of 10 pc to share the supernova's thermal energy and metal yield. The gas within this region has mean temperatures about several million Kelvin. On this scale, the blast wave is still close to its energy-conserving phase. The explosion energy of hypernovae can be up to $10^{52} - 10^{53}$ erg, whereas a conventional core-collapse SN has about 10^{51} erg. Since we here cannot resolve the fine-grained mixing due to fluid instabilities in the early SN ejecta, developing on a scale far below 1 pc. However, we approximately model the coarse-grain mixing on cosmologically relevant timescales by applying the SPH diffusion scheme from [33] which approximately model the effect of unresolved, sub-grid turbulence as a diffusion process, linking the corresponding transport coefficients to the local physical conditions at the grid scale.

After the SN explosion, metal cooling must be considered in the cooling network. We assume that C, O, and Si are produced with solar relative abundances, which are the dominant coolants for the gas contaminated by the first hypernovae. There are two distinct temperature regimes for these species. In low temperature gas, $T < 2 \times 10^4$ K, we use a chemical network presented in [34], which follows the chemistry of C, C^+, O, O^+, Si, Si^+, and Si^{++}, supplemental to the primordial species discussed above. At high temperatures, $T \geq 2 \times 10^4$ K, due to the increasing number of ionization states, a full non-equilibrium treatment of metal chemistry becomes computationally prohibitive. Instead of directly solving the cooling network, we use the cooling rate table based on [35], which gives effective cooling rates for hydrogen and helium line cooling, as well as bremsstrahlung, at different metallicities. Dust cooling is not included because it would only become important at densities much higher than what is reached in our simulations.

We choose a $60\,M_\odot$ Pop III star from the library of [29] as the target of study. This star evolves in total for about 3.7 Myr before it dies as a hypernova with an explosion energy of $10. \times 10^{51}$ erg = 10 B and a metal yield of $20.6\,M_\odot$.

3. Results

The $60\,M_\odot$ Pop III star is expected to form inside a minihalo, with a total mass of about $10^6\,M_\odot$ and a virial radius of ~ 100 pc, located in the region with the highest mass resolution at $z \sim 27$. This allows us to resolve key small-scale features of the ensuing stellar feedback. Once the gas density inside the star-forming cloud exceeds the threshold for sink creation, a single $60\,M_\odot$ Pop III star promptly forms and immediately represents fully developed Pop III stars, acting as sources of UV radiation. The gas inside the halo is rapidly photo-heated up to temperatures of $T \approx 2 \times 10^4$ K. This drives the sound speed up to $30\,\mathrm{km\,s^{-1}}$, whereas the escape velocity of the host halo is only about $3\,\mathrm{km\,s^{-1}}$. The photo-heated gas is thus blown out of the shallow potential well of the minihalo. UV photons not only heat up the gas but also ionize the neutral hydrogen and helium. The ionization-front (I-front) propagation begins with a short supersonic phase (R-type), then switches to a subsonic phase (D-type), because the I-front is trapped behind a hydrodynamical shock [36, 19]. The I-front eventually breaks out, jumping ahead of the shock, and supersonically propagates into the low-density IGM. A giant bubble of hot, ionized, gas is created around the central star, with an inner region that has reached temperatures in excess of 10^4 K. The shape of this bubble is very irregular due to the inhomogeneous and anisotropic distribution of the gas in the surrounding IGM. The bubble sizes reflect the strength of the UV emission rates, which highly depend on stellar mass. This $60\,M_\odot$ star creates a bubble with a size of about 5 kpc.

After about 3.7 Myr evolution, the star eventually dies as a hypernova. The explosion creates a strong shock wave, traveling with a velocity of $v_{\rm sn} \approx 10^4\,\mathrm{km\,s^{-1}}$. The energy carried by the shock is able to reheat the relic H II region for an additional $t_{\rm sn} \approx r_h/V_{\rm sn} \approx 0.4$ Myr, assuming H II region radii of $r_h \approx 4$ kpc. The shock heating in the simulation is roughly about 0.6 Myr because the shock velocity is slowed down due to radiative cooling. After the shock dissipates, a hot and metal-rich ejecta is left behind in the IGM and continues to expand for about another 5 million years, with an increasingly ill-defined boundary. Eventually, the thermal energy of the initial ejecta is radiated away, and the expansion stalls. The mixing of the metals with primordial gas predominantly occurs before stalling. This metal-enriched bubble has reached radial sizes of ~ 2 kpc with corresponding metallicities of $10^{-4} - 10^{-2}\,Z_\odot$. The strong SN blast waves substantially suppress the gas density in the host halo, resulting in $n < 0.01\,\mathrm{g\,cm^{-3}}$, similar to that of the background IGM at this redshift of $z \sim 26$. We show the pre and post feedback from a single $60\,M_\odot$ star in Figure 1. The resulting feedback significantly enhances the IGM temperature, smoothees out density structures in nearby halos, and enriches the primordial gas over regions of ~ 2 kpc.

4. Conclusions

We have presented the results from the cosmological simulations of the impact of a pop III hypernova. We improve on earlier simulations by using updated Pop III stellar models. A

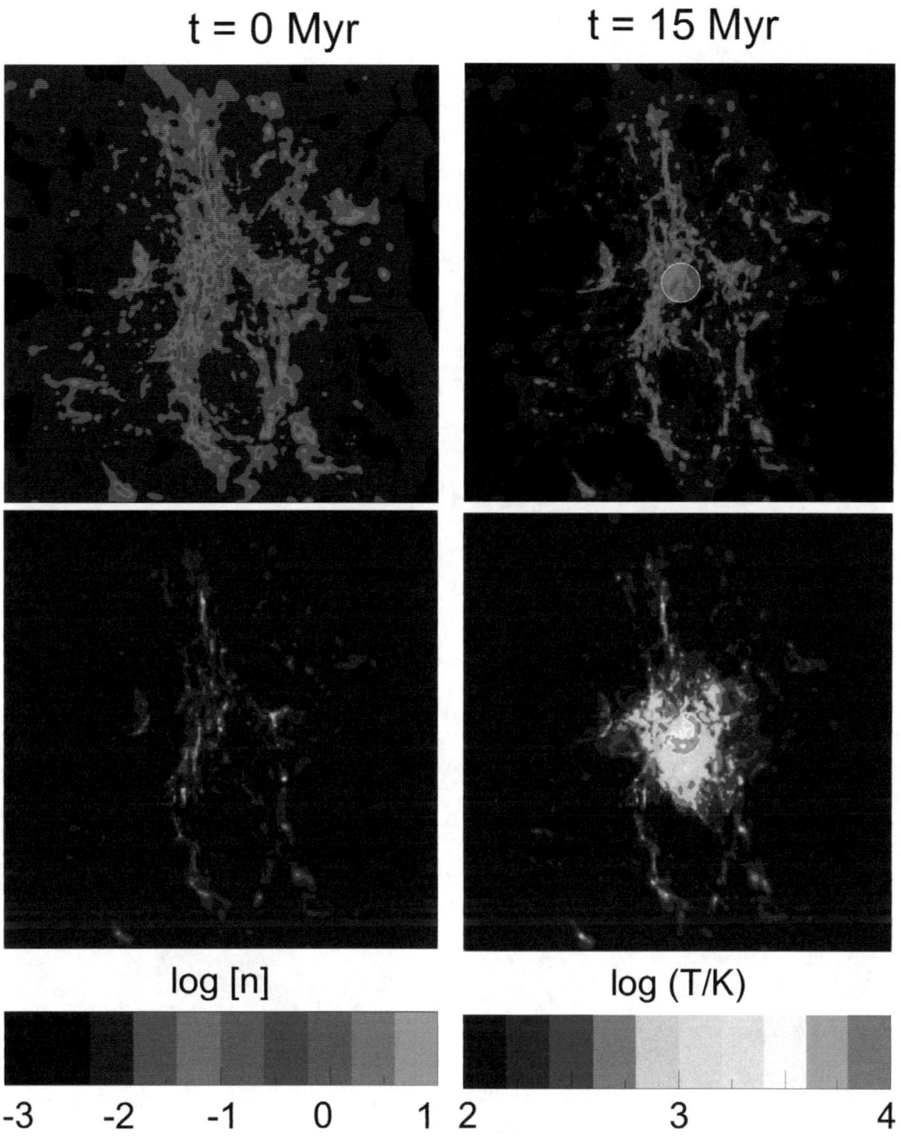

Figure 1. Cosmological impact of a 60 $M_\odot$ pop III star. Panels show temperatures and densities before/after the stellar feedback from a 60 $M_\odot$ star. The size of each panel is about $10 \times 10\,\mathrm{kpc}^2$. Some higher density clumps have been smoothed due to radiative and SN feedback which also chemically enriches the pristine gas within the orange circle of radius $\sim 1\,\mathrm{kpc}$. Both radiative and supernova feedback heat up the gas and change its chemistry on a scale of $3 - 4\,\mathrm{kpc}$.

single, energetic hypernova can impact the early universe on cosmological scales of several kpc. However, without a rapid rotation, a 60 $M_\odot$ Pop III star has a high probability of collapsing into a BH without any chemical enrichment. The imprint from the violent death of Pop III stars, in all their variety, might soon be amenable to empirical testing with the *James Webb Space Telescope (JWST)*, to be launched around 2018.

Acknowledgments

I thank Volker Bromm, Myoungwon Jeon, Stan Woosley, and Alexander Heger for many useful discussions. Work at UCSC has been supported by an IAU-Gruber Fellowship, the DOE HEP Program under contract DE-SC0010676 and the NASA Theory Program (NNX14AH34G). All numerical simulations were performed with allocations from the National Energy Research Scientific Computing Center (NERSC) and the Minnesota Supercomputing Institute (MSI) .

References

[1] Barkana R and Loeb A 2001 *Phys. Rep.* **349** 125–238
[2] Loeb A and Furlanetto S 2012 *The First Galaxies in the Universe* Princeton Series in Astrophysics (Princeton University Press) ISBN 9780691144924 URL `http://books.google.com/books?id=hj98RVlnRRgC`
[3] Karlsson T, Bromm V and Bland-Hawthorn J 2013 *Reviews of Modern Physics* **85** 809–848
[4] Mo H, van den Bosch F C and White S 2010 *Galaxy Formation and Evolution* (Cambridge University Press)
[5] Haiman Z, Thoul A A and Loeb A 1996 *ApJ* **464** 523
[6] Tegmark M, Silk J, Rees M J, Blanchard A, Abel T and Palla F 1997 *ApJ* **474** 1
[7] Bromm V, Yoshida N, Hernquist L and McKee C F 2009 *Nature* **459** 49–54
[8] Yoshida N, Oh S P, Kitayama T and Hernquist L 2007 *ApJ* **663** 687–707
[9] Greif T H, Glover S C O, Bromm V and Klessen R S 2009 *MNRAS* **392** 1381–1387
[10] Whalen D, Hueckstaedt R M and McConkie T O 2010 *ApJ* **712** 101–111
[11] Wise J H and Abel T 2008 *ApJ* **685** 40–56
[12] Greif T H, Glover S C O, Bromm V and Klessen R S 2010 *ApJ* **716** 510–520
[13] Alvarez M A, Wise J H and Abel T 2009 *ApJ* **701** L133–L137
[14] Jeon M, Pawlik A H, Greif T H, Glover S C O, Bromm V, Milosavljević M and Klessen R S 2012 *ApJ* **754** 34
[15] Bromm V, Coppi P S and Larson R B 2002 *ApJ* **564** 23–51
[16] Abel T, Bryan G L and Norman M L 2002 *Science* **295** 93–98
[17] Bromm V, Kudritzki R P and Loeb A 2001 *ApJ* **552** 464–472
[18] Schaerer D 2002 *A&A* **382** 28–42
[19] Whalen D, Abel T and Norman M L 2004 *ApJ* **610** 14–22
[20] Bromm V and Yoshida N 2011 *ARA&A* **49** 373–407
[21] Woosley S E, Heger A and Weaver T A 2002 *Reviews of Modern Physics* **74** 1015–1071
[22] Chen K J, Woosley S, Heger A, Almgren A and Whalen D J 2014 *ApJ* **792** 28
[23] Heger A and Woosley S E 2002 *ApJ* **567** 532–543
[24] Chatzopoulos E and Wheeler J C 2012 *ApJ* **748** 42
[25] Chen K J, Heger A, Woosley S, Almgren A and Whalen D J 2014 *ApJ* **792** 44
[26] Chen K J, Heger A, Woosley S, Almgren A, Whalen D J and Johnson J L 2014 *ApJ* **790** 162
[27] Wise J H, Turk M J, Norman M L and Abel T 2012 *ApJ* **745** 50
[28] Frebel A, Aoki W, Christlieb N, Ando H, Asplund M, Barklem P S and et al 2005 *Nature* **434** 871–873
[29] Heger A and Woosley S E 2010 *ApJ* **724** 341–373
[30] Springel V 2005 *MNRAS* **364** 1105–1134
[31] Komatsu E, Dunkley J, Nolta M R, Bennett C L, Gold B, Hinshaw G, Jarosik N, Larson D, Limon M, Page L, Spergel D N, Halpern M, Hill R S, Kogut A, Meyer S S, Tucker G S, Weiland J L, Wollack E and Wright E L 2009 *ApJS* **180** 330–376
[32] Johnson J L and Bromm V 2007 *MNRAS* **374** 1557–1568
[33] Greif T H, Johnson J L, Klessen R S and Bromm V 2009 *MNRAS* **399** 639–649
[34] Glover S C O and Jappsen A K 2007 *ApJ* **666** 1–19
[35] Sutherland R S and Dopita M A 1993 *ApJS* **88** 253–327
[36] Glover S C O and Brand P W J L 2001 *MNRAS* **321** 385–397

Numerical cosmology on the GPU with Enzo and Ramses

C. Gheller, P. Wang, F. Vazza, R. Teyssier

ETHZ-CSCS Via Trevano 131, Lugano, Switzerland
NVIDIA, Santa Clara, 95050, US
Hamburger Sternwarte, Gojenbergsweg 112, 21029 Hamburg, Germany
University of Zurich, Raemistrasse 71, CH-8006 Zurich

E-mail: cgheller@cscs.ch penwang@nvidia.com franco.vazza@hs.uni-hamburg.de romain.teyssier@gmail.com

Abstract.
A number of scientific numerical codes can currently exploit GPUs with remarkable performance. In astrophysics, Enzo and Ramses are prime examples of such applications. The two codes have been ported to GPUs adopting different strategies and programming models, Enzo adopting CUDA and Ramses using OpenACC. We describe here the different solutions used for the GPU implementation of both cases. Performance benchmarks will be presented for Ramses. The results of the usage of the more mature GPU version of Enzo, adopted for a scientific project within the CHRONOS programme, will be summarised.

1. Introduction

The availability of general purpose graphic accelerators (GPUs) has made possible their effective usage for scientific computing. Ultimate High Performance Computing (HPC) systems are increasingly equipped with GPUs employed not just as graphic accelerators but mainly as computational co-processors providing, on suitable classes of algorithms, outstanding performance with power efficiency significantly higher than standard CPUs. Supercomputers are thus increasingly populated with hundreds or thousands of accelerators that can overlap their computing capability with that of CPUs, minimising considerably overall times-to-solution of high-end scientific problems. An efficient exploitation of such hybrid architectures is not a trivial task, often requiring a full re-design and refactoring of the code in order to expose its massively parallel algorithmic components, that can be effectively accelerated by the GPU, and to minimize or hide any latency related to data transfer between the host and the device.

A number of codes for scientific numerical applications have been recently enabled to harness the power of GPUs. In astrophysics the list comprises (but is not limited to) the Bonsai, [1], Cholla [2] and Gamer [3] codes. In this paper, we will focus on the Enzo [4] and the Ramses [5] codes, which are extensively adopted by the computational astrophysics community. Enzo and Ramses are both Adaptive Mesh Refinement multi-species codes, designed to describe the evolution of cosmological structures, galaxies, gas clouds etc. The two codes can run on massive parallel HPC systems thanks to an effective MPI based implementation. An overview of the main features of Enzo and Ramses will be given in section 2.

Enzo and Ramses have been ported to GPUs adopting different strategies and programming models, Enzo exploiting CUDA and Ramses using OpenACC. The different solutions used for the GPU refactoring will be described in the following sections 3 and 4. Note that a CUDA based implementations of the radiative transfer module of Ramses has been developed by [6].

In section 3 we will focus on the re-design and refactoring work performed on Ramses, for which a number of benchmarks comparing the GPU and the CPU versions of the codes will be presented, highlighting the enhancements obtained with the refactoring and pointing out the main factors currently limiting the performance improvements.

The more mature Enzo's GPU implementation will be introduced in section 4. It is currently used in a project within the CHRONOS[1] programme, supporting high-end computational physics applications. The project aims at studying the origin and evolution of the magnetic field within large-scale structures and exploits the MHD CUDA enabled solver, allowing to overcome the bounds related to memory size, while maintaining the computational time reasonable. An overview of the achieved performance will be presented together with some notable scientific outcomes.

Both Ramses benchmarks and Enzo runs have been performed on the Piz Daint HPC system at ETHZ-CSCS, a Cray XC30 supercomputer accounting for more than 5000 computing nodes, each equipped with an 8-core Intel SandyBridge CPU (Intel Xeon E5-2670) and an NVIDIA Tesla K20X GPU.

Section 5 will summarize the work and draw the main conclusions.

2. Enzo and Ramses Overview

Enzo and Ramses are both Adaptive Mesh Refinement (AMR) codes designed to solve a broad variety of astrophysical problems. The AMR approach is the key feature for the two codes, also in terms of GPU enabling strategy. AMR provides high spatial resolution only where this is actually required, thus ensuring minimal memory usage and computational needs. The codes adopt different AMR solutions, namely Structured AMR (SAMR) by Enzo and Fully Threaded Tree (FTT [7]) by Ramses. With the SAMR approach, the computational domain is represented by a rectangular grid on which a solution of the underlying problem is computed. Regions requiring additional resolution are identified by some suitable criteria and covered by a disjoint union of rectangles (patches), which are then refined by integer factors (usually a factor of two is adopted). The solution is then computed on the superposition of patches. This process is used recursively and refinement are applied in space as well as in time. Similarly, FTT supports local refinements of the spatial domain, but with an unstructured approach, dynamically creating refinements on a cell-by-cell basis, the mesh being able to conform to complex boundaries. Spatial and time refinement criteria are the same as for SAMR.

The AMR hierarchy management is organized as a tree. In the case of Enzo, each node of the tree represents a patch which is pointed by the coarser father patch, and points to the higher resolution nested son patches. In the case of Ramses, each cell stores information about the parent (at lower resolution), neighbouring (at the same resolution) and children (at higher resolution) cells. In this way, the tree can be quickly traversed during various search operations. Comparing the two approaches, FTT optimizes memory usage and can provide efficient parallel implementations. However, memory access is not as efficient and simple as for the regular patches supported in SAMR.

On the AMR mesh, Enzo and Ramses solve the equations describing the dynamics of the two main matter component in the universe, dark and baryonic matter (DM and BM), driven by the gravitational field, generated by the combined mass distributions. The two codes use a particle-mesh N-body method to follow the dynamics of the collisionless DM component. The

[1] http://www.cscs.ch/user_lab/allocation_schemes/chronos_projects/index.html

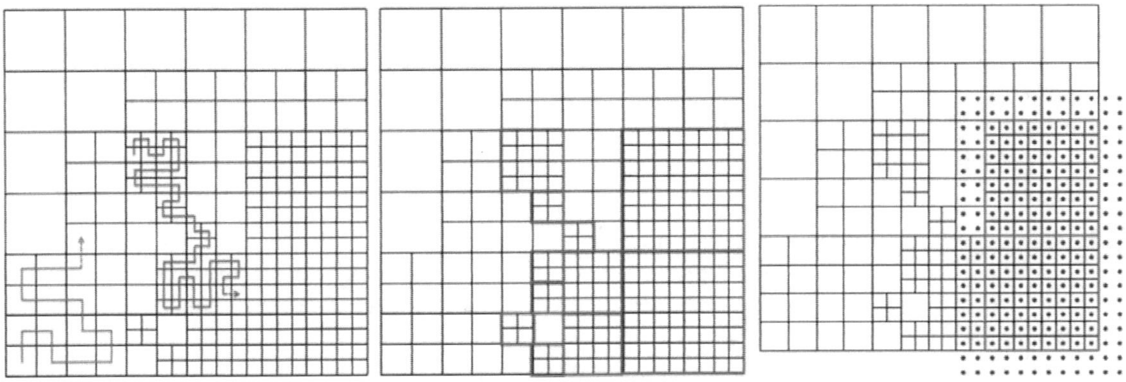

Figure 1. Ramses' hydro solver implementation on the GPU: step 1 (left), cells are ordered according to a space filling curve at each level; step 2 (centre), regular rectangular patches are created; step 3 (right), patches are gathered composing large chunks and hydro equations are solved.

BM component is represented as an ideal fluid, discretised on the AMR mesh. Its behaviour is described solving the Euler equations through different numerical approaches (e.g. the Piecewise Linear Method, PLM [8], the Piecewise Parabolic Method, PPM [9] or the TVD [10] schema). The gravitational potential is calculated solving the Poisson equation through a multigrid or a combined multigrid plus fast fourier transform approach. A variety of further physical processes is included in the two codes. Standard hydrodynamics is extended to magneto-hydrodynamics (MHD). Sub-grid phenomena, like star formation or supernovae explosions, are modelled. Additional basic physics, like cooling processes or radiative transfer, are supported.

The GPU enabling of the two codes follows distinct strategies, driven by the AMR approach, and programming models, as described in the following two sections.

3. Ramses OpenACC Implementation and Benchmarks

For Ramses, an incremental GPU porting strategy has been defined, extending and optimizing progressively the fraction of the code ported on the accelerator. The hydrodynamics solver has been identified as a primary target for Ramses' GPU implementation, being among the most computational demanding code's components and, at the same time, solving a local problem, ideally fitting the accelerators architecture. This first step is being followed by the progressive porting of all the other algorithmic components. This approach ideally fits the OpenACC programming model, which, furthermore, fully supports Fortran 90 and reduces the impact on the source code, minimizing the effort for its management and maintenance. OpenACC has been adopted as the solution for Ramses GPU refactoring.

The main challenge for an effective design of the hydro GPU algorithm is represented by the FTT based data management. The FTT approach allows a prompt identification of sibling and parent cells. However data results scattered in memory and its access, even in the case of neighbouring cells, is often on non contiguous memory locations. This represents a major issue in terms of performance for any processor relying on a cache based memory hierarchy. In order to circumvent this problem, Ramses gathers in a buffer all the data needed for the update of a few cells at a time. The resulting buffer's size is $N_v \times 6 \times 6 \times 6$ cells, N_v being the number of cells to update set to $N_v \sim 10$, in order to fit the cache memory hierarchy. This solution is not suitable to the GPU, since larger contiguous data chunks are necessary to maximise the GPU occupancy, minimizing at the same time the amount of memory accesses. A simple increase of

N_v would not be effective, preserving the number of accesses to main memory. Furthermore, data would be replicated unnecessarily, due to the 6^3 cubes building process. Hence, for the GPU a specific procedure has been designed and can be summarised as follows (see also Figure 1):

- at a given level of refinement, AMR cells are ordered according to a space filling curve based layout;
- cells are grouped in compact (i.e. with no holes) cubic patches of maximum possible size;
- neighbouring patches are grouped in large chunks collecting all the data necessary for the hydro equations integration, (i.e. inner cells plus boundary cells, properly calculated through interpolation if the patch adjoins cells at a different refinement level;
- hydro equations are solved on the chunk of data
- next chunk of data at the same level starts to be created or, if the level has been solved, calculation on the next coarser level begins.

Data transfers to and from the GPU have been minimised by transferring only necessary data i.e. data belonging the refinement level being updated) are moved to and from the GPU.

Table 1. Comparison of CPU and GPU performance on a representative cosmological run with a 256^3 base mesh and 8 levels of refinement. All the timings are in seconds.

Code version	Hydro (tot)	Hydro (calc)	Hydro (copy)	Total
CPU (1 core)	56218	56218	NA	155662
CPU (16 cores)	2918	2918	NA	8775
1 GPU	3009	2270	739	104811
16 GPUs	179	115	64	5718

In Table 1 we present some benchmarks performed to compare the CPU and OpenACC versions of the code in a representative cosmological simulation, with a 256^3 base grid and 8 levels of refinement. The tests were run on a single CPU core, on a full CPU socket (two CPUs sharing a common memory, for a total number of 16 cores), on a single GPU and on 16 GPUs (one GPU per computing node). Column 2 of the Table, compares the timings to run the hydro kernel. On overall, the GPU performs as 16 cores and about 18 times faster than a single core. A fraction of about 75% of the time on the GPU is spent on computation (column 3), while the remaining time is spent on data transfer. Due to the characteristics of the algorithm, this overhead cannot be hidden exploiting asynchronous processes, overlapping computation and data transfer. It will be reduced only porting a larger fraction of the code on the GPU, thus reducing the need for data transfers. The impact of the GPU on the total run time (column 5) is, of course, limited by the fact that only the hydro kernel is running on the accelerator, stressing, once more, the importance of having the majority of the code ported on the GPU. Finally, by comparing the results for 1 and 16 GPUs, we see how, for the case under investigation, the good scalability of the hydro kernel is preserved, despite the work being distributed among 16 distinct nodes, communicating through the network. Again, the copy time is a major source of overhead, the data to be transferred to the device increasing with the number of GPUs relatively to the amount of computation per device.

4. Enzo MHD GPU Simulations

Enzo's MHD solver, as well as the PLM/PPM hydro solver, have been ported to NVIDIA's CUDA framework (Wang et al. 2010; The Enzo Collaboration et al. 2013). Data is off-loaded to the GPU asynchronously on a patch-by-batch basis, effectively overlapping computation and data transfer. All fluxes necessary to update the patch's cells are computed on the GPU and only fluxes required for flux correction necessary for multilevel, adaptive time-step, calculations are moved back to the CPU, representing the only additional data transfer overhead. The porting onto GPUs replaced many shared temporary arrays of the CPU version into larger temporary arrays that are not shared among loop iterations, and exposed the massive parallelism of the algorithm. The larger memory usage is not a major issue, since updated patches are immediately transferred back to the CPU, freeing the corresponding GPU memory. For further details on the porting onto CUDA, we refer the reader to [11] and [4].

Enzo's GPU implementation is currently used in a CHRONOS project, aiming at studying the origin of cosmic magnetism in general and the observed distribution of magnetic fields in large scale structures. This is in fact an astrophysical puzzle. On one hand observations provide evidence for magnetic field strengths of up to a few $\sim \mu$G in galaxy clusters and galaxies (e.g. [12]), on the other the origin of such strong fields is unclear, given the low upper limits on the primordial magnetic field at the epoch of the Cosmic Microwave Background, and the uncertainty on fields generated by primordial mechanisms is still huge (e.g. [13]).
Cosmological simulations already proved that significant field amplification in galaxy clusters can be achieved by starting from weak cosmological fields and following turbulent small-scale dynamo (e.g. [14] and references therein). However, field amplification in the outer part of clusters and in filaments of the cosmic web is still poorly investigated. Recently, in [15], we tackled this problem. We uses the GPU-ported version of Enzo-MHD to achieve a high-resolution view of magnetic field amplification in the cosmic web. As we were mostly interested in the outer region of clusters and filaments, we primarily used large unigrid simulations, which are less affected by numerical dissipation compared to AMR runs and allow complete surveys of large cosmic volumes.
Running on Piz-Daint on 512 to 2048 computing nodes using, due to memory constraints, one to four tasks per node, we investigated various details of the numerical modelling of cosmic magnetism: from dependence on cluster/filament mass to the convergence with spatial resolution. Furthermore, we explored the role of additional seeding by magnetized jets from active galactic nuclei within clusters. In this project we produced the current largest cosmological MHD simulation to date (2400^3 cells and dark matter particles), as well as the most complete resolution study to date on the amplification in a cosmic filaments. In particular, the use of the GPU allowed us to complete such large unigrid run in $\approx 4.5 \cdot 10^6$ core hours with 512 nodes, cutting by a factor ~ 4 the total computational time compared to the CPU version of the code. Figure 2 (left) shows the projected magnetic field strength across a 1200^3 box simulating (25 Mpc)^3, at the final timestep (redshift $z = 0$). Magnetic fields are advected and amplified where gas matter concentrates, however their intensity never becomes dynamically relevant compared to the gas energy. For the small simulated galaxy clusters/groups, the maximum measured magnetic field is $\sim 0.1\mu G$, while the typical strength in filaments is instead $\sim 0.001\mu G$. The right panel in Figure 2 shows the potential role played by the additional release of magnetic fields from galaxies inside galaxy clusters. While this can boosts the maximum field inside clusters up to $\sim 1\mu G$ (i.e. close to observations), this additional seeding mechanisms does not increase significantly the magnetisation of filaments.

These results suggest that the memory of the seeding event(s) must be kept within filaments, as their evolution is dominated by compression and not by strong turbulence (which would erase any information on the initial seed field, as in galaxy clusters). We are presently working together with radio astronomers involved in the design and testing of the Square Kilometer Array, to assess which crucial information of cosmic magnetic fields will be accessible with the future configurations planned for the SKA, which might be able to give us crucial information of the initial seeding mechanisms of large-scale magnetic fields in the Universe [16].

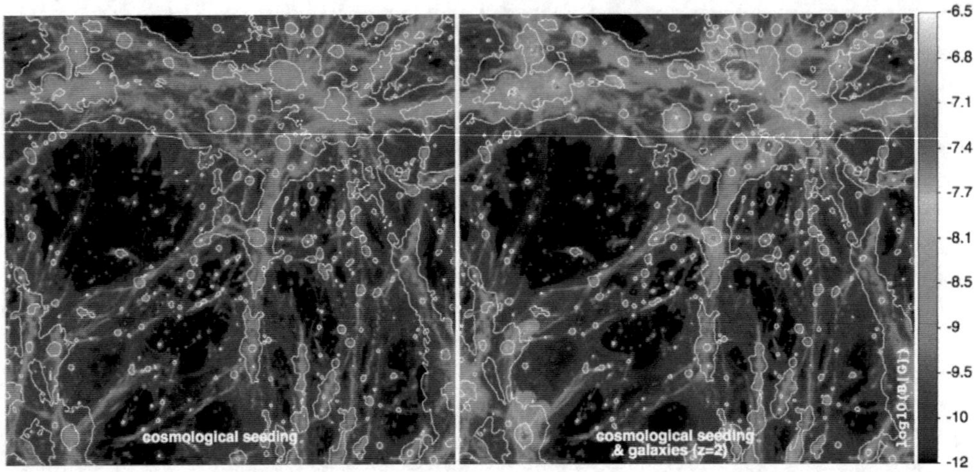

Figure 2. Projected maps of mass-weighted magnetic field intensity (with overlaid logarithmic contours of projected temperature) for a 1200^3 simulations of a $(25\ \mathrm{Mpc})^3$ volume at $z = 0$, with a cosmological weak magnetic field initialised at $z = 30$ (left) or with the additional release of magnetic loops from "galaxies" in the volume at $z = 2$ (right).

5. Conclusions

This paper has summarised the GPU implementation of the Enzo and Ramses codes. Although only part of both codes is currently implemented to exploit GPU empowered supercomputers, they have already proved the effectiveness of solutions capable of running on hybrid accelerated systems. Enzo has been used to run a number of large MHD simulations overcoming the limitations posed by memory constraints of current CPUs keeping, thanks to the GPU, the computational time acceptable. Ramses, despite the good results obtained for the hydro kernel, is pointing out how the porting of the whole application is often necessary for the full exploitation of the accelerator, minimizing the host-device data transfer overheads. On-going work is finalized to enable the cooling, MHD and gravity modules to the GPU.

References

[1] Bédorf J, Gaburov E and Zwart S P 2012
[2] Schneider E E and Robertson B E 2014 *ArXiv e-prints* (*Preprint* 1410.4194)
[3] Schive H Y, Tsai Y C and Chiueh T 2010 *ApJS* **186** 457–484 (*Preprint* 0907.3390)
[4] Enzo Collaboration T 2014 *ApJS* **211** 19 (*Preprint* 1307.2265)
[5] Teyssier R 2002 *A & A* **385** 337–364 (*Preprint* astro-ph/0111367)
[6] Aubert D and Teyssier R 2010 *ApJ* **724** 244–266 (*Preprint* 1004.2503)
[7] Khokhlov A 1998 *Journal of Computational Physics* **143** 519 – 543 ISSN 0021-9991
[8] Colella P and Glaz H M 1985 *Journal of Computational Physics* **59** 264–289
[9] Colella P and Woodward P R 1984 *Journal of Computational Physics* **54** 174–201
[10] Yee H, Warming R and Harten A 1985 *Journal of Computational Physics* **57** 327 – 360 ISSN 0021-9991 URL http://www.sciencedirect.com/science/article/pii/0021999185901834
[11] Wang P, Abel T and Kaehler R 2010 *New Astronomy* **15** 581–589 (*Preprint* 0910.5547)
[12] Ferrari C, Govoni F, Schindler S, Bykov A M and Rephaeli Y 2008 *Science & Space Review* **134** 93–118 (*Preprint* 0801.0985)
[13] Widrow L M, Ryu D, Schleicher D, Subramanian K, Tsagas C G and Treumann R A 2011 *ArXiv e-prints* (*Preprint* 1109.4052)
[14] Dolag K, Bykov A M and Diaferio A 2008 *Science & Space Review* **134** 311–335 (*Preprint* 0801.1048)
[15] Vazza F, Brüggen M, Gheller C and Wang P 2014 *MNRAS* **445** 3706–3722 (*Preprint* 1409.2640)
[16] Brown S, Emerick A, Rudnick L and Brunetti G 2011 *ApJL* **740** L28+ (*Preprint* 1109.0316)

Instability of counter-rotating stellar disks

R G Hohlfeld[1,2,4] and R V E Lovelace[3]
[1]Center for Computational Science, Boston University, Boston, MA 02215 USA
[2]Wavelet Technologies, Inc., Attleboro, MA 02703 USA
[3]Astronomy Department, Cornell University, Ithaca NY USA

E-mail: hohlfeld@bu.edu

Abstract. We use an N-body simulation, constructed using GADGET-2, to investigate an accretion flow onto an astrophysical disk that is in the opposite sense to the disk's rotation. In order to separate dynamics intrinsic to the counter-rotating flow from the impact of the flow onto the disk, we consider an initial condition in which the counter-rotating flow is in an annular region immediately exterior the main portion of the astrophysical disk. Such counter-rotating flows are seen in systems such as NGC 4826 (known as the "Evil Eye Galaxy"). Interaction between the rotating and counter-rotating components is due to two-stream instability in the boundary region. A multi-armed spiral density wave is excited in the astrophysical disk and a density distribution with high azimuthal mode number is excited in the counter-rotating flow. Density fluctuations in the counter-rotating flow aggregate into larger clumps and some of the material in the counter-rotating flow is scattered to large radii. Accretion flow processes such as this are increasingly seen to be of importance in the evolution of multi-component galactic disks.

1. Introduction

Most disk galaxies, including our own Milky Way, have vast majority most of their stars and gas on orbits (hereafter rotating) with the same sense of rotation about the center of the galaxy. However, there is a significant population of disk galaxies in which there are substantial components that orbit in the opposite sense (or at high inclination) to the general orbital motion of their host galaxy. Detailed studies of the inner structures of galaxies are required to clearly identify counter-rotating flows. Several of the well-observed cases of this phenomenon include NGC 4826 (also known as M64, and colloquially as the "Evil Eye Galaxy" or the "Black Eye Galaxy" due to prominent dust lanes associated with the counter-rotating flow, see figure 1), NGC 7217, NGC 4550, and NGC 5150 [1, 2, 3]. These examples in neighboring galaxies, where detailed studies of the inner structure required to clearly identify counter-rotating structures, make plausible a statistical argument that galaxies with such prominent counter-rotating structures are not uncommon. This motivates research to understand how these systems originate and the important dynamical instabilities controlling their evolution.

[4] Author to whom any correspondence should be addressed.

Figure 1. A view of NGC 4826. The boundary between counter-rotating and direct rotation is indicated by the dark dust lanes, approximately 3000 light years from the galaxy's center. Relative velocity of the two flows is 300 km/s.

Image Credit: NASA/ESA and the Hubble Heritage Team (AURA/ STScI)

Much work has been done in accretion flows onto black holes, neutron stars, and other condensed objects in which the accreting material is essentially or entirely gaseous. It is reasonable to consider the accretion as resulting from the dynamics of ordinary viscosity in the gas. We are considering a more complex scenario in this proposal in which the entire galaxies under study have stellar as well as gaseous material, and so it is necessary to consider instability mechanisms deriving from both particle (i.e., stellar and cluster dynamics and fluid flow mechanisms. With a galactic accretion flow composed of counter-rotating stellar flows interpenetrating, or adjacent with a bounding shear layer, it is most attractive to model the interaction as a two-stream instability [4]. These theoretical studies of galactic accretion disk flows concentrated on relatively low azimuthal mode numbers. The simulation study reported here indicates that, at least for disks with relatively flat rotation curves, the azimuthal mode number for well-developed instabilities is comparatively high, i.e. $m \geq 6$. This point will be discussed in more detail below, but the major thrust of the proposed research will be to extend theory and simulations to a common range of parameters and investigate correspondence with observational results.

2. Tools and Methodology

We describe here the computational tools used in carrying out and analysing the simulation reported here.

2.1. *GADGET*

This study, and most of our other simulation studies, are conducted using GADGET, developed by Springel and coworkers [5, 6]. GADGET has been used extensively for simulations of galaxy collisions and interactions, and thus is well-suited for our studies of galaxy accretion and associated evolution.

GADGET is a Langrangian code, which enables the treatment of processes spanning a large range of spatial and time scales. Numerous particle types are treated in a natural way, including hydrodynamic particles treated by smoothed particle hydrodynamic (SPH) techniques. This is important for our purposes because we wish to address cases in which instabilities involving stellar particles and instabilities involving gas occur.

GADGET has been implemented in parallel versions using a Message Passing Interface (MPI) suitable for use on cluster computers. Such implementations allow ambitious simulations with large numbers of particles, that we have found useful for our work.

2.2. Snapshot file analysis tools

In GADGET input and output is typically handled by "snapshot" files, defined in the GADGET user guide [7]. Many workers have written programs for creating and processing snapshot files and we have similarly adapted software for processing snapshot files for our needs.

Our snapshot file analysis programs are written in *Mathematica 10*, which we have found to be powerful for its integrated visualization capabilities as well as smoothly tying in with our theoretical analysis tools.[5] Statistical functions in Mathematica are useful for the estimation of particle distribution functions, which is important for identifying the operation of instabilities driven by velocity space effects. Furthermore, we have implemented modal analysis tools suitable for determining the azimuthal mode spectrum of instabilities.

Our analysis tools make it a straightforward task to trace both the stellar and interstellar gas populations of particles from any sources, as well as different populations of particles, which is necessary for distinguishing particles originating from an accreting body relative to the particles in the accreting galaxy.

3. Initial Conditions for the Simulation

The simulation described here is designed to demonstrate the operation of two-stream instability in a accretion in a galactic disk made of stars. Accordingly, we use GADGET particles and do not make use of GADGET's capabilities for simulations using smoothed particle hydrodynamics. Future work will address gaseous systems and systems with mixtures of gas and stars. A total of 500,000 particles was used to obtain adequate resolution of features in the accretion flow.

These simulations used relatively flat (constant velocity) rotation curves because of our interest in Mestel disks and in large disk galaxies [8, 9]. We used a modification of the standard Mestel disk to avoid some of the singularities and related numerical problems associated with this equilibrium solution. Specifically, we pick a mass profile and rotation curve of the form

$$\sigma(r) = \frac{\sigma_0}{\sqrt{1+r^2/\lambda^2}} \tag{1}$$

$$v_c^2(r) = 2\pi G\lambda\sigma_0\left[1 - \frac{1}{\sqrt{1+r^2/\lambda^2}}\right] \tag{2}$$

where σ is the mass surface density, σ_0 is the mass surface density at radius $r = 0$, λ is a nuclear scale length, v_c is the circular velocity, and G is the gravitational constant. Our initial condition is an infinitesimal disk, but initial velocities are three dimensions, so that the disk rapidly acquires a finite thickness. The initial velocity dispersion is chosen to be 20% of the circular velocity (and locally isotropic), with is sufficient to suppress the usual Toomre instability [10] except in the innermost regions not of interest for these simulations.

The disk extends out to $r = 100$ units (so that edge effects are not important at the shear layer), while the boundary between the rotating and counter-rotating flows is a 2 unit wide annulus centered at $r = 25$ units. In the boundary region there is a linear transition between the rotating and counter-rotating flows. The accreting flow, exterior to the bounding annulus rotates in the CW sense, while the galaxy interior to the bounding annulus rotates in the CCW sense. The orbital period at $r = 25$ is 1.18 GADGET time units. A halo of mass equal to the disk has been included to suppress the bar-forming instability.

[5] *Mathematica* is a registered trademark of Wolfram Research, Inc.

4. Simulation showing development of instability in time

In figure 2 we show the progression of the simulation at selected times. (Times are indicated in each pane in GADGET units.) In the online version of this paper, the particles in the accreting flow are shown in red, which will show as grey in the print version of the paper. The particles in the accreting galaxy are shown in black.

By a time of 7.5 a well-developed spiral disturbance appears in the accreting galaxy that has a trailing wave pattern, while a spiral disturbance is beginning to develop near the shear layer in the accreting flow. These patterns initially have a very high azimuthal mode number that decreases as simulation time advances. The spiral opening angles in both the interior and exterior flows open up in time. At time 40, a ring composed mostly of accreting material has developed with spiral disturbances penetrating the ring to make material clumps. Accreting material is transported interior to the shear layer as the instability develops, but some of the accreting material is also scattered to large radii, presumably due to gravitational scattering by clumps of material in the ring and spiral structures.

The two-stream instability for galactic accretion flows [4] was originally directed towards understanding flows with relatively small azimuthal mode number. Motivated by these initial simulation results, work is underway to extend that theory to higher azimuthal mode numbers so as to make detailed comparisons between the theory and simulations, as well as with observational results.

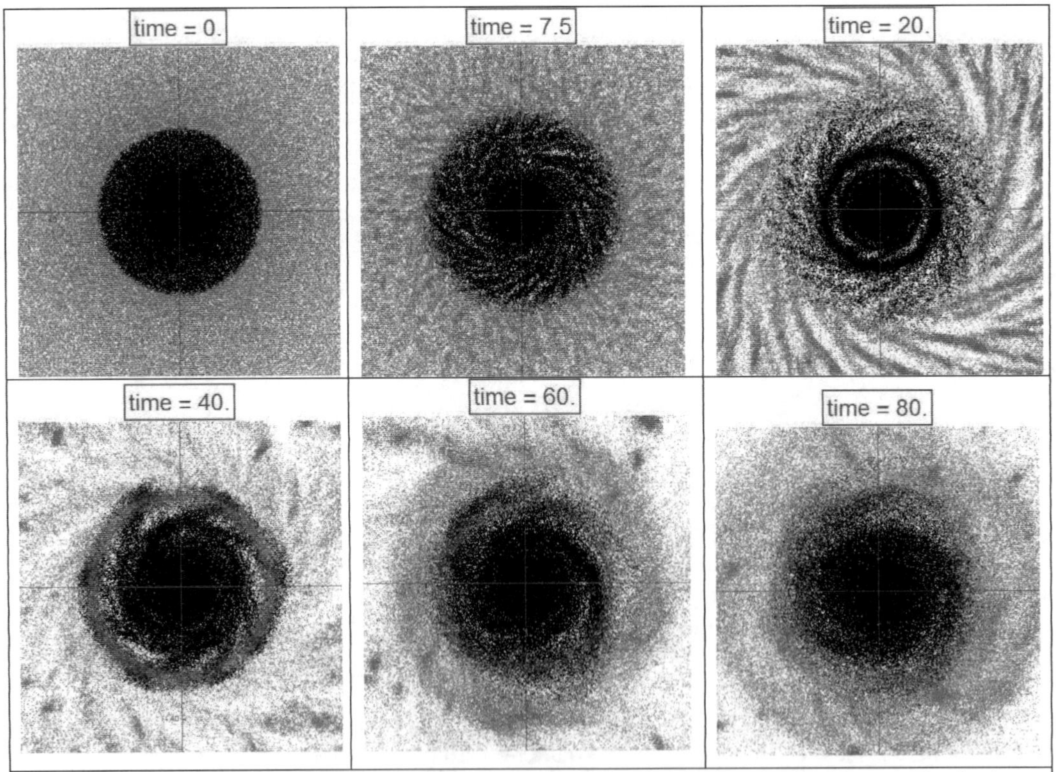

Figure 2. Selected images in the simulation showing the development of the instability in time. Times are in GADGET units. Accreting material is moving in the CW sense, while the accreting galaxy rotates in the CCW sense. A ring composed mostly of accreting material develops in the fourth pane, which shows clumping at a high azimuthal mode number. Accreting material is transported both to larger and smaller radii relative to the shear layer radius. The orbital period at $r = 25$ is 1.18 GADGET time units.

5. Azimuthal mode analysis

It is apparent by naked eye examination of the frames in figure 2 that the instability has a comparatively high mode number, which is changing in time. However, it is also quite likely that different modes are present at any one time. Quantifying the azimuthal mode spectrum will be important in making connections between these simulations and the extensions of the theory under development.

We chose a simulation frame at a time of 32, when the ring structure is well-developed (as shown in the left-hand side of figure 3). We collected stars within an annulus in the simulation and collected them in 128 bins based on their angular coordinates. We performed a Fast Fourier Transform of this binned angular data and then found the complex amplitude squared of each bin of the transform to obtain the angular power spectrum of the instability in the annulus. In the top rightmost bin of figure 3, we show the angular mode spectrum for the annulus between $r = 24$ to $r = 26$. Note the prominent peak in the spectrum for a mode number of 8 to 10. The pane below this was generated from the annulus from $r = 22$ to $r = 24$, i.e. interior to the prominent ring. The peak at mode numbers 8 to 10 is much reduced and the mode power can be seen to be significantly less. Further studies will determine the development of the azimuthal mode spectrum in time and radius.

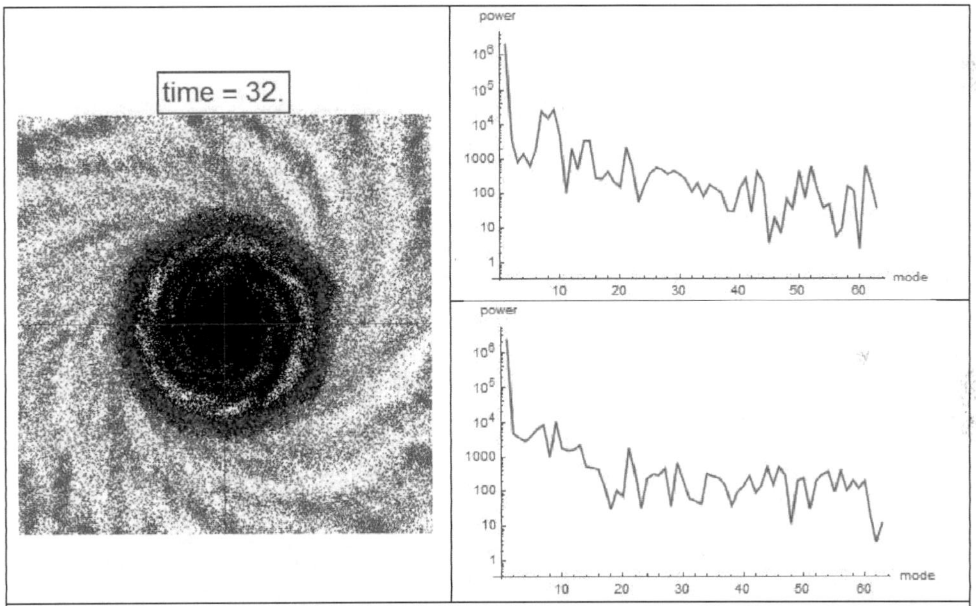

Figure 3. Azimuthal mode spectra for the simulation at time 32. The upper spectrum was collected in the annulus from radius 24 to 26, while the lower spectrum was collected from radius 22 to 24, interior to the prominent ring. Azimuthal mode numbers of 8 to 10 are important in the ring structure.

6. Conclusions

We list here the primary conclusions of this paper:

- Adjacent counter-rotating regions are unstable and interact by the two-stream instability.
- The instability has a high azimuthal mode number.
- A ring of material develops which contains particles with both senses of rotation.
- Clumps develop that are reminiscent of the star forming regions of NGC4862.

Journal of Physics: Conference Series **640** (2015) 012059 doi:10.1088/1742-6596/640/1/012059

- Significant radial transport of accreting material occurs, both inward and outward.

Acknowledgments
We thank Professor Neil Comins for useful discussions. We also wish to acknowledge a generous allocation of computer resources on the Massachusetts Green High Performance Computing Center from Boston University's Research Computing Services and the Center for Computational Science.

References

[1] Rubin, V 1994 Kinematics of NGC 4826: a sleeping beauty galaxy, not an evil eye, *Astron. J.* **107** 173-183.
[2] Rix H-W R, Franx M, Fisher D and Illingworth G 1992 NGC 4550: a laboratory for testing galaxy formation *Ap. J.* **400** L5-L8.
[3] Rix H-W R, Kennicutt R C, Braun R and Walterbos R A M 1995 Placid stars and excited gas in NGC 4826 *Ap. J.* **438** 155-169.
[4] Lovelace R V E, Jore K P and Haynes M P 1997 Two-stream instability of counterrotating galaxies *Ap. J.* **475** 83-96.
[5] Springel V, Yoshida N and White S D M 2001, GADGET: a code for collisionless and gasdynamical cosmological simulations *New Astronomy* **6** 79-117.
[6] Springel V 2005 The cosmological simulation code GADGET-2 *Mon. Not Roy. Astron. Soc.* **364** 1105-1134.
[7] Springel V 2005 User guide for GADGET-2, Max-Planck-Institute for Astrophysics (Garching: Germany).
[8] Mestel L 1963 On the galactic law of rotation *Mon. Not. Roy. Astron. Soc.* **126** 553-575.
[9] Binney J and Tremaine S 1987 *Galactic Dynamics* (Princeton University Press: Princeton NJ).
[10] Toomre A 1964 On the gravitational stability of a disk of stars *Ap. J.* **139** 1217-1238.

Teaching Density Functional Theory Through Experiential Learning

Shobhana Narasimhan

Theoretical Sciences Unit and Sheikh Saqr Laboratory, International Centre for Materials Science, Jawaharlal Nehru Centre for Advanced Scientific Research, Bangalore, 560064, India

E-mail: shobhana@jncasr.ac.in

Abstract. Today, quantum mechanical density functional theory is often the method of choice for performing accurate calculations on atomic, molecular and condensed matter systems. Here, I share some of my experiences in teaching the necessary basics of solid state physics, as well as the theory and practice of density functional theory, in a number of workshops held in developing countries over the past two decades. I discuss the advantages of supplementing the usual mathematically formal teaching methods, characteristic of graduate courses, with the use of visual imagery and analogies. I also describe a successful experiment we carried out, which resulted in a joint publication co-authored by 67 lecturers and students participating in a summer school.

1. Introduction

Today, quantum mechanical density functional theory is often the method of choice for performing accurate calculations on atomic, molecular and condensed matter systems. Such calculations provide data that can be used to interpret experiments, gain insight into structure-property relationships, and design novel materials. With recent rapid advances in computational power, it is now possible to perform such calculations on a desktop PC or laptop (or even, in some cases, on a mobile phone! [1]) With little need for infrastructure, this is an area of advanced research that can be easily embraced by students and scientists in the developing world. However, despite the widely prevalent view among many scientists that the teaching and practice of science is free of cultural influences, I (and, indeed, many others before me) have found that this is not true, and one has to take cultural factors into account even when teaching something that is as technical as physics.

2. Science, Culture, Teaching and Learning

The ways in which even subjects as seemingly "universal" as science or physics are taught, vary across cultures and countries: these differences are already manifested at the kindergarten and grade school level, where one observes either an instructionist or a constructionist method of teaching, with an emphasis placed either on information gathering or critical thinking.

In much of the developing world today, even science subjects are taught in a way that emphasizes rote memorization; this is certainly true of India, where I have spent a large part of my scientific career. There are two main possible reasons for this: one is that for millenia, the traditional way of teaching in India has involved learning texts "by heart", the other is that this is a colonial legacy – it did not suit colonial rulers to educate a native populace to think independently and critically; rather, their interests would be better served by having an "educated" cadre composed of individuals who would meekly carry out a prescribed set of instructions without stopping to analyze or critique them.

As a physics student in India in the 1970s and 1980s, I was often made to memorize physical laws and formulae (students who got a word wrong in the statement of a law were made to stand outside the class!); less emphasis was placed on problem solving or on understanding underlying physical principles.

It is clear that such training does not develop in students the critical skills necessary for a good scientist, and thus, efforts have to be made at the undergraduate and graduate level to help students make the transition from "knowledge" to "synthesis" [2, 3]. Among institutions in India that have won worldwide renown for their excellence in undergraduate science education are the Indian Institutes of Technology (IITs) and, more recently, the Indian Institutes of Science Education and Research (IISERs). However, these elite institutions can cater to only a small percentage of the student body across the country. Many students at smaller universities or colleges are desperate to learn, but lack mentors and/or access to resources.

3. Computational Physics and Density Functional Theory

I wish to argue that despite the well-known "digital divide" between the developed and developing countries, computational physics presents an ideal area where students of physics in developing countries can hope to do cutting-edge research.

This is because first, computational physics is relatively cheap. It is true that many computational physics groups worldwide use high-performance computing platforms, but with the increase in the speed and memory capacities of computers, and the accompanying decrease in the price of computers, today it is possible to tackle many interesting problems on a desktop PC or a laptop. Accordingly, this means that there are relatively low infrastructure requirements: electricity, water and air-conditioning can all be problems in developing countries, making it rather difficult to do high-quality experimental physics, for example.

As already mentioned in the Introduction, density functional theory (DFT) is, today, one of the most popular methods for treating condensed matter systems as well as other many-electron systems. The field has exploded in recent years, both due to the advances in computer technology, and due to methodological advances made in various aspects of how to map the many-electron Schrödinger equation "exactly" onto an equivalent system of one-particle equations. The 16 most-cited papers of all time in the Physical Review journals all deal with various aspects of DFT, and Walter Kohn was awarded the Nobel Prize in 1998 for his role in developing the field.

The number of lines in a typical DFT code is of the order of 100,000, and therefore few users of DFT write their own codes, and instead make use of standard packages. Today, there is a proliferation of DFT codes, some of which have to be purchased, and others which are freely downloadable. To mention the names of just a few DFT packages: QUANTUM ESPRESSO, VASP, Wien2K, Abinit, Siesta, FLEUR, CP2K, CPMD, etc. Many of these codes have user-friendly graphical user interfaces (GUIs). The apparent (misleading) ease of use of these interfaces and codes leads to a large number of students and scientists, all over the world, using them to perform calculations with little knowledge of what they are doing, using the code as a black box. Unfortunately, in such a situation, it is often possible, or even likely, that one will obtain completely meaningless results. There is therefore a great need for workshops where students can be trained to use such codes sensibly and appropriately.

4. QUANTUM ESPRESSO – The Distribution and Workshops

My experience of teaching DFT through experiential learning, especially in the developing world, has been gained through two main (and overlapping) platforms: the QUANTUM ESPRESSO schools, and ASESMA.

QUANTUM ESPRESSO [4, 5] is a distribution of codes that can perform standard DFT, density functional perturbation theory (DFPT) and time-dependent DFT calculations; it also comes equipped with many advanced features that have been developed in recent years. It is free and open-source, and comes equipped with the GNU public licence. An important activity of the QUANTUM ESPRESSO initiative is conducting training workshops, at both elementary and advanced levels. In part due to the close ties between the QUANTUM ESPRESSO developers and the Abdus Salam International Centre for Theoretical Physics in Trieste, there has been a special emphasis on conducting schools in developing countries. More than thirty such workshops have been conducted by the code developers; in addition, several workshops have also been conducted locally by various users of the codes.

The African Schools for Electronic Structure Methods and Applications (ASESMA) [6, 7] have been held biennially at various sites in Africa since 2010. They aim to train African students in electronic structure methods, following roughly the same pattern as that followed in the QUANTUM ESPRESSO schools.

While each training school or workshop may differ slightly in length and details, the typical duration is one to two weeks, with concepts and theory being covered in lectures in the mornings, and hands-on computer sessions in the afternoons, in which students apply the concepts learnt in the morning sessions. There is also often a project, in which students work in groups.

5. Teaching Through Visuals and Analogies

Since, in my experience, most students (and, often, especially those who have been trained in the developing world) are quite familiar with equations and mathematical formulations of problems, but have trouble translating this to conceptual knowledge, I find it helpful to make a lot of use of visuals and analogies while teaching the basics of solid state physics and DFT.

As an example of encouraging visual thinking, instead of just using the mathematical definition of a Bravais lattice:

$$\mathbf{R} = n_1\mathbf{a}_1 + n_2\mathbf{a}_2 + n_3\mathbf{a}_3, \tag{1}$$

I have students imagine that they are walking around on a sketched lattice, and picture the view they see – a Bravais lattice is one in which the view from each lattice point is the same (without turning around). Similarly, to understand the importance of the "plane-wave cutoff" used when performing DFT calculations in a plane-wave basis, I have them draw unit cells of various sizes, see how the Brillouin zones change accordingly in size, and thus the number of plane waves within the cutoff sphere, which in turn determines the size of the matrix that has to be diagonalized, and thus the computational time.

When it comes to analogies: my experience has been, the sillier the analogy, the more effective it is! One example that has proved popular is comparing the Kohn-Sham approach [8] of mapping the many-body problem to an effective one-body problem, to the (admittedly not very gender-neutral) problem of finding the "optimally happy" configuration of an ensemble of males and females with complicated interpersonal interactions (see Fig. 1); this is clearly much more easily tackled if one can replace it by an effective problem where only one female has to be moved while five males remain stationary. Another silly analogy that makes an effective point is comparing the use of pseudopotentials in electronic structure calculations to the effects of placing a cardboard cutout of a traffic policeman by the highway, in an effort to fool motorists into slowing down: it pretty much works, if done well enough, but is much cheaper!

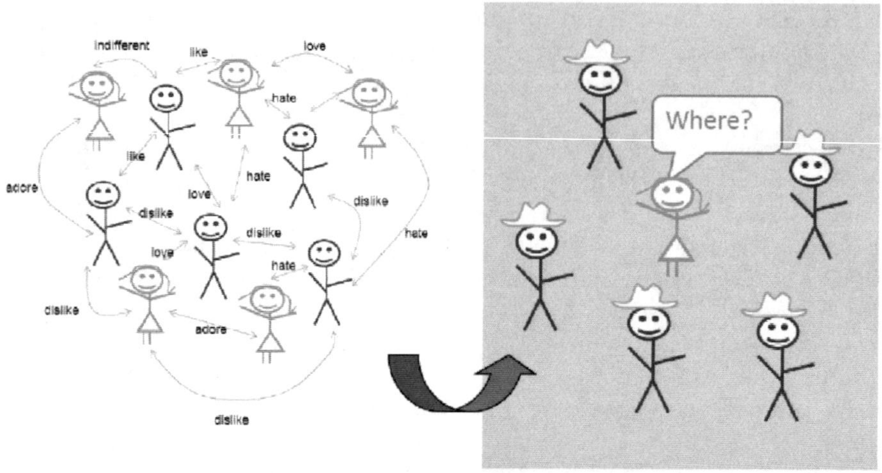

Figure 1. Analogy for the Kohn-Sham equations: mapping a complicated many-body problem onto an effective one-body problem.

6. Asking Questions

It goes without saying that, for students to learn effectively, it is important that they be encouraged to ask questions, and get into the habit where they examine critically whatever they have been taught. For both of these purposes, I have found it useful to make use of on-line sites such as "socrative" [9]. Despite the already mentioned digital divide, mobile phones are now extremely widespread throughout the developing world, and many developing countries now have a density of mobile phones (per population) that is similar to that seen in developed countries. With the advent of smart phones, both the instructor and the student can log on to such a website, which can then be used for rapid polling of the class as the lecture or hands-on session progresses, using either multiple choice questions or true/false questions, with immediate feedback in the terms of bar charts (similar to the operation of "clickers" which are used in some schools and colleges in the West).

For the teacher, the advantage of making use of such a tool is that one can get continuous feedback about whether or not concepts have been conveyed across adequately, and grasped accurately. One can then use this feedback to tailor the rest of the lecture accordingly. The responses from students to making use of such polling tools have been uniformly positive. Students say that they like the fact that it is interactive, and also the fact that it offers the opportunity to answer anonymously, since many students say they are otherwise reluctant to speak up in class, being put off by the fear of public "shame" or ridicule.

7. Hands-on Sessions and a "Massively Parallel" Calculation

The hands-on sessions require a room where students can work on computers either singly or in pairs, while guided through exercises by a tutor. Since students tend to have widely different paces of progress, it is important to have self-contained and clear instructions online, that students can follow at their own speed, if they wish to work faster, or are unable to keep up with the tutor.

The generally followed practice is to have a room full of students working through routine exercises in order to compute standard properties such as the lattice constant of silicon, or the vibrational frequency of carbon monoxide. This can seem like a huge waste of intellectual and computational power! For this reason, one can instead experiment with having students work on similarly simple but novel calculations – the typical such "massively parallel" research problem would be one that would require performing a large number of quick calculations, each of which can be done by either one student or a small group of students.

At a summer school held in Bangalore in 2006, we followed this approach, asking students to address the problem of whether carbonia (the extended form of CO_2) can be stabilized by doping with silicon. Within a few days, the students were exposed to all the different steps one has to go through in order to tackle such a problem, from reading the literature, to formulating the problem and planning out calculations, performing various convergence tests with respect to basis set size and Brillouin zone sampling, etc. One question that arises is what would happen if students (who, after all, are beginners) should make a mistake? We decided to handle this by assigning every calculation to two to four pairs of students – if all the results agreed, it would be assumed that the result was correct, whereas if there was any discrepancy, then the calculation would be repeated and checked by a tutor or lecturer. We found that the students were incredibly enthusiastic about this project, and everyone involved was extremely delighted when this work resulted in a research publication [10], co-authored by 67 students, lecturers and tutors from all over the world!

8. Follow Up

In today's digital society, it obviously makes sense to make full use of social media in order to keep students and lecturers connected during and after workshops. The participants of several workshops have stayed in touch through sites such as Facebook, and collaborations have been established through the contacts made. Many of the lectures from these workshops are also available online, and hundreds of students worldwide are making use of these resources.

9. Conclusions

In conclusion, I wish to reiterate that computational physics is an ideal field where students and researchers from developing countries can quickly and easily acquire the expertise to perform work that is of very high quality. However, there are traps and pitfalls for the uninitiated, and thus there is a great need for workshops and online resources that can help interested scientists to use the widely available computer packages appropriately and effectively. In this article, I have given some thoughts about how to do this, based on my experience in teaching in such workshops over the past two decades.

10. Acknowledgments

I wish to acknowledge the many people I have worked with in the QUANTUM ESPRESSO and ASESMA schools, especially Stefano Baroni, Sandro Scandolo, Stefano de Gironcoli, Nicola Marzari, Ralph Gebauer, Nithaya Chetty, Daniel Joubert, Richard Martin, George Amolo, Nicholas Makau and Alison Hatt. I also wish to thank all my students, past and present, who have acted as tutors and guinea pigs for many of these schools and workshops.

11. References

[1] Marzari N http://mml.materials.ox.ac.uk/Main/QEN900
[2] Bloom B S (ed.) 1956 *Taxonomy of Educational Objectives: The Classification of Educational Goals: Handbook I, The Cognitive Domain* Longmans, Green (New York, Toronto)
[3] Narasimhan S 2004 *Journal of Materials Education* **26** 75
[4] www.quantum-espresso.org
[5] Gianozzi P *et al.* 2009 *J. Phys.: Condens. Matter* **21** 39550
[6] Chetty N, Martin R M and Scandolo S 2010 *Nature Physics* **6** 830
[7] Feder T 2011 *Phys Today* **64** 28
[8] Kohn W and Sham L J 1966 *Phys Rev* **140** A1133
[9] www.socrative.com
[10] Aravindh A *et al.* 2007 *Solid State Comm.* **144** 273

The role of computational physics in the liberal arts curriculum

Rachele Dominguez

Department of Physics, Randolph-Macon College, P.O. Box 5005, Ashland, VA 23005, US

E-mail: racheledominguez@rmc.edu

Benjamin Huff

Department of Philosophy, Randolph-Macon College, P.O. Box 5005, Ashland, VA 23005, US

Abstract. The role of computational physics education varies dramatically from department to department. We will discuss a new computational physics course at Randolph-Macon College and our attempt to identify where it fits (or should fit) into the larger liberal arts curriculum and why. In doing so, we will describe the goals of the course, and how the liberal arts curriculum conditions the exploration of computational physics.

1. Introduction

Institutions of higher education in the United States have been under scrutiny in recent years. Popular books and articles have questioned the stability and effectiveness of colleges as well as the value of getting a college degree, where the increasingly high cost of tuition may not be worth the uncertain returns [1–3]. Among this general crisis of higher education, there is a sub-crisis among schools that characterize themselves as liberal arts colleges, prompting defenses of liberal education [4].

The humanities and fine arts sometimes take center stage in discussions of the value of the liberal arts, but the natural sciences, particularly physics, are clearly integral to the liberal arts, properly understood. The humanities stand out partly because to understand human beings and the human condition is one of the most important kinds of knowledge, but also because the humanities often seem the most in need of justification. In the eyes of a society concerned with practical outcomes, the sciences are much easier to justify. Yet, ironically, within an institution devoted to the liberal arts, scientists do not always have the vocabulary to fully connect their work with the institution's mission.

Liberal arts colleges have the weighty task of responding to changing student populations, developing job markets, and perceptions of the value of higher education, while holding true to their core values. As courses and curricula evolve in response to student and college needs, tensions inevitably arise as to how limited resources can best serve students. For faculty, these factors suggest a need to reflect on the goals of a liberal arts curriculum and examine how individual courses support this curriculum.

At Randolph-Macon College (where the authors teach), college relevance and the value of a liberal arts education are frequent topics of discussion. Our identity as a 'small liberal arts college' drives this discussion. Our introductory physics classes are capped at 24 students and upper division physics courses are capped at 12 (though a typical upper division course has half this number.) Randolph-Macon has around 1400 students total, all of whom are undergraduates. In contrast, Boston University (a research university, the host institution to this conference, and the *alma-mater* of one of the authors) has about 15,000 undergraduates, not including graduate students [5]. Randolph-Macon has three full time physics faculty members, while Boston University has about 40 [6]. These statistics support the image that Randolph-Macon fosters close-knit and personal learning communities in contrast to larger institutions.

More generally, liberal arts institutions tend to market themselves as providing a formative, student-centered experience, whereas research institutions advertise an opportunity to experience the creation of new knowledge. Liberal arts institutions tend to devote more required courses to general education requirements, emphasizing breadth across curricula over depth within the major. Depending on how they are fulfilled, the general education requirements at Randolph-Macon may occupy as much as 70% of a student's four year course of study. This number is roughly 40% in the College of Arts and Sciences at Boston University [7]. (These percentages can be mitigated by factors such as placement credit from high school, and taking courses that count for multiple requirements.)

In this paper, we explore how a computational physics class can uniquely contribute to the goals of a liberal arts curriculum.

2. Liberal education

The proper form and content of liberal education has been a subject of debate for centuries, and today the answer is often ambiguous even within a single institution. The term 'liberal education' describes an education that is related to liberty, or freedom, in some way; but there are several ways to understand both the kind of freedom involved and the way the process of education is related to it.

First, an education may be liberal in the sense that it is pursued freely. This could mean simply that the student is free to pursue his or her interests, and many faculty today will emphasize this; but traditionally liberal education refers to an education pursued for its own sake, rather than to meet some external need, such as the need to find a job. For Aristotle the defining feature of a liberal education is that it prepares the student for activities that are pursued for their own sake because they are good in themselves [8]. The traditional liberal arts disciplines fit this description straightforwardly: the humanities and natural and social sciences are traditionally understood to pursue knowledge for its own sake, while the fine arts particularly pursue beauty. Liberal activities, in this sense, are distinguished from those that are merely useful or necessary for some other purpose, and in particular from paid labor.

Second, an education may be liberal in the sense that it liberates, or makes the student free. As J.S. Mill discusses, one kind of intellectual freedom comes from acquiring the skills and foundational principles that allow the student to draw freely and intelligently on the full range of human knowledge, as she may need to do in many ways throughout her life [9, 10]. This requires a fluency in all the major kinds of knowledge, a sense of how various disciplines contribute to our understanding, and also the ability to integrate knowledge from across disciplines. Ultimately the world is an integrated whole that transcends disciplinary boundaries, as are most of the challenges with which it confronts us. Liberal education, in this sense, is comprehensive and holistic.

Another aspect of intellectual freedom is the ability to think for oneself, and form sound judgments of one's own–'critical thinking' in today's vocabulary. Enlightenment thinkers such as Kant emphasize relying on one's own reason and being freed from reliance on the authority of others [11]. As human knowledge expands, increasing reliance on others for their expertise becomes unavoidable, but the ability to choose sources wisely, discern legitimate expertise, and

Journal of Physics: Conference Series **640** (2015) 012061 doi:10.1088/1742-6596/640/1/012061

assess its limits is all the more essential. Such qualities of intellectual independence have a special value for citizens of a self-governing society, which led Thomas Jefferson to found a university.

These traditional approaches to liberal education continue to be valuable and relevant in the 21st century; but they are also under pressure. The breadth of a comprehensive approach to knowledge comes into competition with the depth of specialization needed to master any one area. Meanwhile, with the explosive growth of knowledge and information, a comprehensive education may seem more of an aspiration than a real possibility.

Today, college and university professors may think of their jobs mainly in terms of the discovery of new knowledge. This goal can compete with the goal of passing on knowledge to the next generation, as teaching comes to seem a mere distraction from research. Both learning what has already been discovered, and discovering new knowledge, are paradigmatically liberal activities, but we sometimes struggle to reconcile the two at a practical level.

Perhaps most urgently, with higher education no longer merely a privilege for the elite, the idea that education should be pursued primarily for its own sake becomes difficult to sustain. With college becoming a standard expectation for the middle class, most students can only afford an education if it will bring economic rewards. There is great pressure for a vocational approach to higher education.

Traditional conceptions of liberal education need a certain amount of rethinking in this context. Fortunately, this is a healthy part of any good tradition. One point to reconsider is the traditional opposition between liberal activities and paid work. The idea that paid employment is necessarily without intrinsic value is an oversimplification that seems out of place in our democratic and industrious culture. Some labor is merely useful, but many activities and achievements have great intrinsic value, whether they are paid or not.

This adjustment leads to a third conception of liberal education: an education may be liberal by preparing students for a self-directed life of action and accomplishment. Today's college-bound students have an enormous range of possible choices; their education should prepare them to choose their path well and succeed in it. For this aim, intellectual skills and the ability to learn new things are at least as important as any knowledge acquired in school.

The traditional liberal arts disciplines, approached in the right way, can be an excellent preparation in this sense as well, however. Skills such as written and oral communication, problem-solving, research, and creative application of prior knowledge to new situations are essential to a wide range of activities both in and outside the workplace in an advanced economy. One may learn other knowledge and technical skills, distinctive to one's chosen path, either in school or afterward. Liberal education should enable one to meet one's material needs through a career whose value goes beyond money.

3. Computational physics course

Turning to a more specific consideration of the place of computational physics in liberal arts education, we discuss here a course taught at Randolph-Macon. The course uses computational math techniques to solve sometimes advanced physics problems, where content is complementary to the core physics content of the major. A primary goal of any stand-alone computational physics course ought to be to teach content unique to the interdisciplinary field of computational physics. This may include methods such as numerical calculus, methods of solving differential equations, and Monte Carlo methods. However, computational physics generally can be used within the physics curriculum in a variety of ways.

On one end of the spectrum is a black box approach, where computers use software (such as Mathematica [12]) as tools to do calculations or solve differential equations. With this approach, there is often minimal effort to teach the students how the tool is doing the computation. Many physics departments use some kind of black box approach in traditional theoretical classes as a problem solving tool. On the other end of the spectrum is a comprehensive approach to teaching computational physics, essentially integrating computational physics (including math, physics, and computer science) throughout the physics curriculum [13]. This approach may result in a separate computational physics degree, or concentration, distinct from the traditional physics degree. For example Oregon State and Illinois State [14, 15] (both large, public universities) have separate majors in computational physics including several computational physics courses integrated throughout the four year program. Requiring a considerable amount of resources, this 'all in' approach is not currently within the scope of the capabilities of Randolph-Macon (and probably most small, liberal arts institutions.) The course described below is somewhere in the middle of this spectrum.

3.1. Course logistics

The course is three credit hours (though the next time the course is taught it will be a three credit lecture plus a three hour laboratory). The time is used for a variety of activities, including lecture, tutorial, and student-led seminars. Despite the low number of students enrolled in this class (four to five in a semester, which is a typical number for advanced physics courses at Randolph-Macon), a large number of majors have been represented including physics, engineering physics, mathematics, computer science, chemistry, and environmental science. The prerequisites include only introductory physics and calculus, allowing sophomores through seniors to take the course. Therefore, the students come to the course with varied skill sets and backgrounds. Opening advanced courses to a diverse group of students is encouraged at Randolph-Macon to boost enrollments and allow students to explore new topics. The course fulfills a general education requirement for a natural science as well as a computing requirement.

3.2. Course structure and content

The coursework begins with tutorial-type exercises, where students learn basic programming skills. The homework consists of physics content-driven project-style problems. The capstone of the course is a final independent project. Most of this work comes from the text, "Computer Simulation Methods" by Harvey Gould, Jan Tobachnik, and Wolfgang Christian [16]. The current version of the book uses Java and the Open Source Physics library, (though there is now an Easy Java Simulations (EJS) version of the text [17].) This library contains code that allows students to create graphical user interfaces (GUIs), plots and animations. Though the students do not have to write this code, they must learn to implement existing code. This approach allows the course to focus more on physics content than on computer science content.

The independent project is essential to the goals of the course. The student must extend a practice problem from the text with an original line of inquiry and develop a simulation to investigate a physical problem. The student must then design an investigation to study the problem by collecting, analyzing, and visualizing data, as well as presenting preliminary and final results in an informal classroom and a formal presentation setting. The projects completed in this class spanned a wide range of topics, including the Magnus force of a soccer ball, irregular quantum wells, environmental physics, critical systems, general relativity, complex networks, and thermal systems.

3.3. Liberal arts skills

A computational physics course can be quite valuable for advancing a number of skills connected to the goals of a liberal arts curriculum. Some of these skills are associated more directly to the content of computational physics than others. For example, clearly there are a number of technical skills required to perform the tasks of the course, such as basic computer skills, programming skills, and the collection and analysis of data using a variety of software tools. All of these skills are certainly assets to attaining a technical or scientific vocation. Such a course also introduces students to computation as a unique and legitimate problem solving technique.

However, problem solving in a computational course, such as the one described here, may provide a unique opportunity to focus on the development of higher level thinking skills, another important goal of a liberal education. A robust approach to problem solving encourages students to think critically about scientific investigation and the relationship of knowledge to inquiry. Investigatory activities in any science context can promote this goal [18], and reforms are underway to remove constraints such as cookbook laboratory protocols that hinder this goal [19].

However, there are always necessary constraints to creative inquiry in physics courses. Theoretical problems are often based on limited information and student mathematical abilities, while measurement tools and physical limitations restrict laboratory activities. Indeed, such constraints are an important part of the creative process in science, and specific constraints promote certain skills and limit the development of others.

Problem solving using computational physics techniques has inherent restrictions as well, including computational limitations, but many other physical and mathematical constraints are lifted. The relative ease with which students may adjust physical parameters (one cannot change the gravitational constant in a lab!), and analyze simulated measurements makes the computational investigatory process a unique environment for students to creatively explore new questions. In particular, students have control over, and think critically about, the investigatory process.

It was evident in the student led seminars of the Randolph-Macon course that students were thinking creatively and critically about their investigations. The evidence came through verbal explanations, questions, reasoning, and thoughtful interactions. This unique skill of oral communication as part of the investigation process within a learning community was practiced in a semi-formal way (something beyond struggling to get homework done in the middle of the night with friends). This practice of science is especially valuable in small college settings, where it is unlikely that students will be members of sizable research groups. The Randolph-Macon course also stressed more formal modes of scientific communication, including written laboratory reports and formal research talks as part of a college-wide research day.

3.4. Connection to the liberal arts curriculum

Intellectual freedom requires a broad command of human knowledge and an understanding of the relationships between disciplines. Computational physics is itself interdisciplinary, integrating knowledge and techniques from mathematics, computer science, and physics. The study of computational physics underscores that science itself is comprehensive, emphasizing that theory, modeling, mathematical algorithms, and empirical data have a place in building scientific knowledge.

The open-ended investigations in the course described above provide opportunities for students to freely explore content of their own interests within the relatively broad bounds accessible to computational investigation. They develop their investigations with supervision from the instructor and collaboration with their classmates. This free exploration of content and method increases the relevance of the course to liberal education and provides motivation to students. It encourages students to take responsibility for their own learning by investing time and effort in their educational pursuits.

Such investment is necessary for students to achieve intellectual freedom through scientific independence. This course provides an opportunity for students to develop investigative autonomy with manageably-sized projects at the undergraduate level. Such independence can empower students to achieve their goals, especially in scientific fields, preparing them for a life of self directed action and accomplishments. More directly, the nature of computational physics realistically models the complex tools used by scientists, providing valuable preparation for more substantial research projects in the future. Ruben Panoff expresses the need for this type of background: "The substantial role of computational approaches to physics research is not currently reflected proportionately in how we prepare future physicists" [20].

In essence, a computational physics course can teach students to think like scientists. Scientific thinking as discussed here not only aids students in achieving their goals, but helps to develop free individuals more generally, which is a good in itself. Scientifically literate individuals with a robust appreciation for the development of scientific knowledge thrive in a democratic society and enable such a society to thrive.

4. Conclusions

A computational physics course can contribute to the goals of a liberal arts curriculum in a substantial and unique way. Indeed, a liberal arts environment conditions the exploration of computational physics, setting parameters for prioritizing class and work time. Such a course is quite ambitious, and pursuing the goals set forth in this paper in a meaningful way necessarily requires sacrifices in content. However, the course described contributes to the full development of a free individual, and their ability to pursue knowledge both for its own sake and with a sense of its place and purpose in community.

4.1. Acknowledgments

We would like to thank the Rashkind Family Endowment as well as the Chenery Research Grant for making this work possible. We would also like to thank Scott Girdner for useful discussions.

References

[1] Smith D B 2011 The university has no clothes URL `http://nymag.com/news/features/college-education-2011-5/`
[2] Selingo J J 2013 *College Unbound: The Future of Higher Education and What It Means for Students* (Boston: New Harvest)
[3] Arum R and Roksa J 2010 *Academically Adrift: Limited Learning on College Campuses* 1st ed (Chicago: University Of Chicago Press)
[4] Ferrall V E 2011 *Liberal Arts at the Brink* (Cambridge, Mass.; London: Harvard University Press)
[5] Boston university admissions `http://www.bu.edu/admissions/about-bu/`
[6] Boston University Physics URL `http://physics.bu.edu/people/by_type/faculty-lecturers`
[7] Bachelor of arts degree overview URL `http://www.bu.edu/academics/cas/programs/degree-overview/`
[8] Aristotle 1998 *Politics* (Indianapolis, Ind: Hackett Publishing Company)
[9] Mill J S 1996 Inaugural address *Collected Works of John Stuart Mill: XXI. Essays on Equality, Law and Education* ed Robson J M (London: Routledge)
[10] Berkowitz P 2006 *Policy Review* **140** 47–67
[11] Kant I 1983 *Perpetual Peace and Other Essays* (Indianapolis: Hackett Publishing Company, Inc.)
[12] Wolfram Research I Mathematica
[13] Landau R 2006 *Computing in Science Engineering* **8** 22–30 ISSN 1521-9615
[14] Computational Physics for Undergraduates URL `http://www.physics.orst.edu/~rubin/CPUG/index.html`
[15] Physics- Computer Physics, Illinois State URL `http://illinoisstate.edu/majors/details.php?f=physics_computer.xml&p=%2Fmajors%2Fcas%2F`
[16] Gould H, Tobochnik J and Christian W 2006 *An Introduction to Computer Simulation Methods: Applications to Physical Systems* 3rd ed (San Francisco: Addison-Wesley)
[17] Gould H, Tobochnik J and Christian W 2011 *An Introduction to Computer Simulation Methods – EJS Edition* URL `http://www.compadre.org/OSP/filingcabinet/share.cfmUID=4856&FID=17708&code=873A9F5BC1`
[18] Etkina E and Heuvelen A V 2007 Investigative science learning environment - a science process approach to learning physics *Research-Based Reform of University Physics* vol 1
[19] Redish E F 2000 New models of physics instruction based on physics education research URL `http://eric.ed.gov/?id=ED438184`
[20] Panoff R 2011 Computational (physics education) vs. (computational physics) education: Many body for anybody APS March Meeting URL `http://meetings.aps.org/Meeting/MAR11/Session/V21.1`

Scalable Software for Multivariate Integration on Hybrid Platforms

E de Doncker[1], **F Yuasa**[2], **J Kapenga**[1] **and O Olagbemi**

[1] Department of Computer Science, Western Michigan University, Kalamazoo MI 49008, U S A
[2] High Energy Accelerator Research Organization (KEK), Oho 1-1, Tsukuba, Ibaraki, 305-0801, Japan

E-mail:
elise.dedoncker@wmich.edu, fukuko.yuasa@kek.jp, john.kapenga@wmich.edu,

Abstract. The paper describes the software infrastructure of the PARINT package for multivariate numerical integration, layered over a hybrid parallel environment with distributed memory computations (on MPI). The parallel problem distribution is typically performed on the region level in the adaptive partitioning procedure. Our objective has been to provide the end-user with state of the art problem solving power packaged as portable software. We will give test results of the multivariate PARINT engine, with significant speedups for a set of 3-loop Feynman integrals. An extrapolation with respect to the dimensional regularization parameter (ε) is applied to sequences of multivariate PARINT results $Q(\varepsilon)$ to obtain the leading asymptotic expansion coefficients as $\varepsilon \to 0$. This paper further introduces a novel method for a parallel computation of the $Q(\varepsilon)$ sequence as the components of the integral of a vector function.

1. Introduction

The PARINT parallel integration package supports the computation of multivariate integrals over hyper-rectangular and simplex regions. The algorithms and tools in the parallel integration engine include:
(i) an adaptive region subdivision algorithm, equipped with a load balancing strategy to handle localized integrand difficulties such as peaks and singularities;
(ii) a non-adaptive Quasi-Monte Carlo (QMC) method based on Korobov lattice rules;
(iii) Monte Carlo (MC) methods for high dimensions and/or erratic integrands;
(iv) a user interface based on the PARINT plugin compiler which pre-processes the user problem specification and integrand function to be linked with the ParInt executable;
(v) 1D rules corresponding to those in QUADPACK [1], which can be used for repeated integration in successive coordinate directions.
Repeated integration with QUADPACK programs has been proven extremely effective for applications to Feynman loop integrals with severe singularities in high-energy physics [2]. However, repeated or iterated integration has an impeding drawback with respect to the expense of the method for higher-dimensional problems (say, dimensions ≥ 6). In a sequential setting, multivariate adaptive integration is considered suitable for moderate dimensionality (say, up to 10) while possibly dealing with mildly irregular integrand behavior, and for higher dimensions with well-behaved functions. These limits can be moved up considerably in a parallel environment by allowing finer subdivisions with higher numbers of integrand evaluations.

Written in C and layered over MPI [3], the PARINT methods are implemented as tools for *automatic* integration, where the user defines the integrand function and the domain, and specifies a relative and absolute error tolerance for the computation (t_r and t_a, respectively). The integrand is defined as a vector function over the domain $\mathcal{D}$. Denoting the (exact) integral by

$$I\vec{f} = \int_{\mathcal{D}} \vec{f}(\vec{x})\, d\vec{x}, \tag{1}$$

it is then the objective to return an approximation $Q\vec{f}$ and absolute error estimate $E_a\vec{f}$ such that $||\ Q\vec{f} - I\vec{f}\ || \leq ||\ E_a\vec{f}\ || \leq \max\{\ t_a, t_r||\ I\vec{f}\ ||\ \}$, or indicate with an error flag that the requested accuracy cannot be achieved. When a relative accuracy (only) needs to be satisfied we set $t_a = 0$ (and vice-versa). If both $t_a \neq 0$ and $t_r \neq 0$, the weaker of the two error tolerances is imposed; if both $t_a = t_r = 0$ then the program will reach an abnormal termination, typically when the maximum number of function evaluations is reached. Optionally the PARINT installation can be configured to use long doubles instead of doubles.

This paper focuses on the distributed adaptive algorithm in the PARINT multivariate integration package combined with extrapolation techniques as addressed in Section 2, and presents test results for Feynman loop integrals in Section 3.

2. Adaptive integration and extrapolation

2.1. PARINT *parallel adaptive integration*

Layered over MPI [3] for distributed computing, the PARINT adaptive code assigns the roles of controller and worker processes for adaptive region partitioning (see also [4] and [5] for an explanation of the algorithm). The integration domain is divided initially among the workers. Each on its own part of the domain, the workers engage in an adaptive partitioning strategy similar to that of DQAGE in QUADPACK [1], DCUHRE [6] and HALF [7] by successive bisections, thus generating a local priority queue of subregions as a task pool. The priority queue is implemented as a max-heap keyed with the estimated integration error over the subregions, so that the subregion with the largest estimated error is stored in the root of the heap, or as a *deap* or *double-ended heap*, to allow for efficient deletions of the minimum as well as the maximum element in case the user specifies a maximum size to be retained for the heap structure.

An important mechanism of the distributed integration algorithm is the load balancing strategy, which is receiver-initiated and geared to keeping the loads on the worker task pools balanced, particularly for problems with irregular integrand behavior. The message passing is performed in a non-blocking and asynchronous manner, and permits overlapping of computation and communication. As a result of the asynchronous processing and message passing on MPI, PARINT executes on a hybrid platform (multi-core and distributed) by assigning multiple processes to each node. The parallelization may, however, lead to *breaking loss* or extra work performed at the end of the computation, due to the time elapsed while workers continue to generate subregions after the termination message has been issued but before receiving it. This may increase the total amount of work (compared to the sequential work), particularly when the number of processes is large.

2.2. *Use of* PARINT

PARINT can be invoked from the command line, or by calling the `pi_integrate()` function in a program for computing an integral of the form (1). A user guide is provided in [8]. The call sequence passes a pointer to the integrand function, typed as a pointer to a function that returns an integer, and where the parameters of the integrand function correspond to the integral dimension, argument vector $\vec{x}$, number of component functions `nfuncs` and the resulting values of the function $\vec{f}(x)$. Further parameters of `pi_integrate()` are (on input): `nfuncs`, an integer identifying the cubature/quarature rule to use, the maximum number of function evaluations allowed, the region type (hyper-rectangle or simplex) and specification; and (on output): the integral and error approximations `result[]` and `error[]`, and a user-declared pointer to a status structure.

When PARINT is used as a stand-alone executable, it uses the PARINT Plug-in Library (PPL) mechanism to specify integrand functions. The functions are written by the user, added to the library (along with related attributes), and then compiled using a PARINT-supplied compiler into *plug-in modules* (.ppl files). A single PPL file is loaded at runtime by the PARINT executable. Using a function library enables quick access to a predefined set of functions and lets PARINT users add and remove integrand functions dynamically without re-compiling the PARINT binary. Once these functions are stored in the library, they can be selected by name for integration.

For an execution on MPI, the MPI host file (`myhostfile`) contains lines of the form: `node_name slots=ppn` where ppn is the number of processes to be used on each participating node. A typical MPI run from the command line may be of the form

```
mpirun -np 64 --hostfile myhostfile ./parint -f fcn -lf 10000000 -ea 0.0 -er 5.0e-10
```

For example, with 4 nodes listed in `myhostfile` and `ppn = 16`, a total of 64 processes is requested. The integrand function of this run is named `fcn` in the user's library; the maximum number of function evaluations is 10000000, and the absolute and relative error tolerances are 0 and 5.0e-10, respectively.

2.3. *Use of extrapolation*

We apply numerical extrapolation to integrals with an asymptotic expansion in the dimensional regularization parameter ε, of the form

$$S(\varepsilon) \sim \sum_{k \geq K} C_k \varphi_k(\varepsilon) \qquad \text{as } \varepsilon \to 0. \tag{2}$$

For example, the $\varphi_k(\varepsilon)$ functions may be integer powers of ε, $\varphi_k(\varepsilon) = \varepsilon^k$, $K \geq k$, with $K = 0$ for finite integrals. If the $\varphi_k(\varepsilon)$ functions are known, we can apply a linear extrapolation by approximating $S(\varepsilon)$ for decreasing values of $\varepsilon = \varepsilon_\ell$, and truncating Eq (2) after $2, 3, \ldots, \nu$ terms to form linear systems of increasing size to be solved for the C_k variables.

When the $\varphi_k(\varepsilon)$ functions are unknown we perform a non-linear numerical extrapolation with the ϵ-algorithm [9, 10, 11, 12]. In that case we generate a sequence of integral approximations for $S(\varepsilon_\ell)$, using a geometric progression of ε_ℓ that tends to zero with increasing ℓ (see also [13, 2]), and employ a version of the ϵ-algorithm code (DQEXT) from QUADPACK. In between calls, the DQEXT implementation retains the last two lower diagonals of the triangular extrapolation table. When a new element of the input sequence is provided, the algorithm calculates a new lower diagonal, together with an estimate or measure of the *distance* of each newly computed element from preceding neighboring elements. With the location of the "*new*" element in the table relative to e_0, e_1, e_2, e_3 pictured as:

$$\begin{matrix} & e_0 & \\ e_3 & e_1 & \textit{new} \\ & e_2 & \end{matrix}$$

we assign $\textit{new} = e_1 + 1/(1/(e_1 - e_3) + 1/(e_2 - e_1) - 1/(e_1 - e_0))$, and set the distance measure for the *new* element as $|e_2 - e_1| + |e_1 - e_0| + |e_2 - \textit{new}|$. The new lower diagonal element with the smallest value of the distance measure is returned as the result for this call to the extrapolation code.

3. Results for Feynman loop integrals

One prominent application of multivariate integration is to the computation of Feynman loop integrals, which contribute higher order corrections for accurate theoretical predictions of the cross-section for particle interactions. The integral corresponding to an n-dimensional scalar Feynman diagram with L loops and N internal lines can be represented in Feynman parameter space as

$$I = \frac{\Gamma\left(N - \frac{nL}{2}\right)}{(4\pi)^{nL/2}} (-1)^N \int_0^1 \prod_{j=1}^{N} dx_j\, \delta(1 - \sum x_i) \frac{C^{N-n(L+1)/2}}{(D - i\varrho C)^{N-nL/2}}. \tag{3}$$

Here the functions C and D are polynomials determined by the topology of the corresponding Feynman diagram [14]. After removing the δ-function in (3) and eliminating one of the

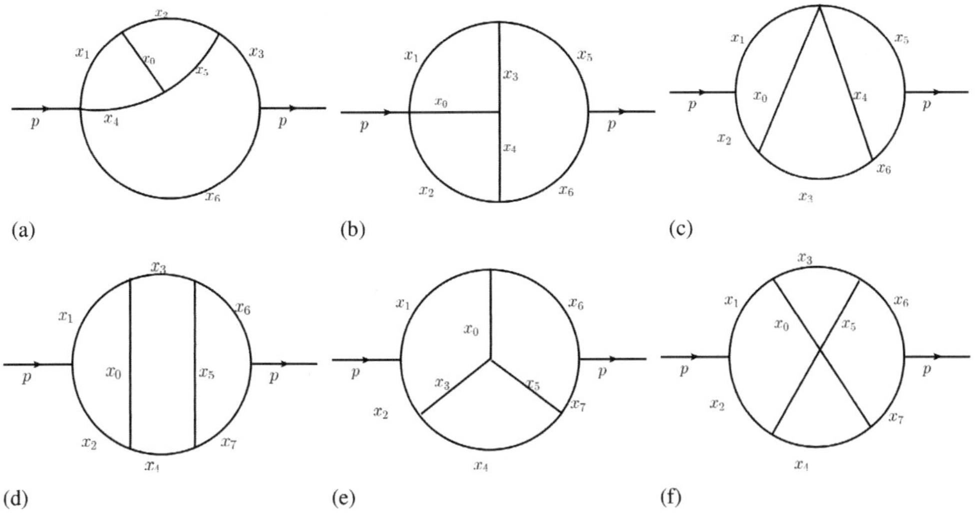

Figure 1. 3-loop diagrams (a) $N = 7$ (Laporta[15] Fig 2(q)), (b) $N = 7$ (Laporta[15] Fig 2(r)), (c) $N = 7$ (Laporta[15] Fig 2(s)), (d) $N = 8$ (Laporta[15] Fig 2(t)), (e) $N = 8$ (Laporta[15] Fig 2(u), (f) $N = 8$ (Laporta[15] Fig 2(v))

variables in view of $\sum_{j=1}^{N} x_j = 1$, the integration domain is a d-dimensional simplex $\mathcal{S}_d = \{(x_1, x_2, \ldots, x_d) \in \mathbb{R}^d \mid \sum_{j=1}^{d} x_j \leq 1 \text{ and } x_j \geq 0\}$, with $d = N - 1$. The term $i\varrho C$ prevents the integral in (3) from diverging if D vanishes in the interior of the domain. As this does not happen for the problems under consideration in this paper we can set $\varrho = 0$. We set $n = 4 - 2\varepsilon$ to compute the leading order coefficients in an asymptotic expansion of the integral with respect to the dimensional regularization parameter ε.

Test results are presented below for a set of 3-loop self-energy diagrams from Laporta [15] (given in Fig 1). In order to compare our integral approximations with Laporta's we set all masses $m_r = 1$ and $s = 1$, and furthermore divide the integral by $\Gamma^3(1 + \varepsilon)$. For the numerical approximations and timings we use the HPCC (High Performance Computation Center) cluster at WMU. The tests are run on 16-core cluster nodes with Intel Xeon E5-2670, 2.6GHz processors and 128GB of memory, and using the Infiniband interconnect for message passing via MPI. Tables 1-4 show various types of results obtained by calling the `pi_integrate()` function and running the executable with 16 processes on the HPCC cluster. The integrals are transformed from the (unit) simplex to the (unit) cube and the integration is taken over the cube, using the basic integration rule of polynomial degree 9 [6]. Tables 1

Table 1. Integral and leading order expansion coefficients using PARINT with 16 procs., and linear extrapolation for 3-loop integral of Laporta Fig 2(r), err. tol. $t_a = 10^{-12}$; max. # evals = 20B, $T(s)$ = elapsed time (s); E_a = integration estim. abs. error

	INTEGRAL FIG 2(r)			EXTRAPOLATION		
ε_ℓ	INTEGRAL	E_a	T(S)	RESULT C_0	RESULT C_1	RESULT C_2
2^{-3}	0.89462319318517	6.33e-10	356.9			
2^{-4}	1.07605987265074	8.56e-10	426.3	1.257496552116	-2.9029868714	
2^{-5}	1.19524813881849	1.15e-09	426.2	1.333416355943	-4.7250621633	9.7177349
2^{-6}	1.26445377191768	1.34e-09	426.2	1.341017173944	-5.1507079713	16.5280678
2^{-7}	1.30188593759114	1.46e-09	426.2	1.341390110905	-5.1954604067	18.1988254
2^{-8}	1.32137252564963	1.52e-09	426.2	1.341399132800	-5.1976978366	18.3778198
2^{-9}	1.33131707386056	1.55e-09	426.2	1.341399240859	-5.1977522983	18.3868241
2^{-10}	1.33634079003748	1.57e-09	426.2	1.341399241503	-5.1977529523	18.3870438
2^{-11}	1.33886565298449	1.58e-09	426.2	1.341399241505	-5.1977529584	18.3870480
		Eq (2), Laporta [15]:		1.341399241447	-5.1977529559	18.3870466

Table 2. Integral and leading order expansion coefficients using PARINT with 16 procs., and extrapolation with ϵ-algorithm for 3-loop integral of Laporta Fig 2(r), err. tol. $t_a = 10^{-12}$; max. # evals = 20B, $T(s)$ = elapsed time (s); E_a = integration estim. abs. error

	INTEGRAL FIG 2(r)			EXTRAPOLATION		
ε_ℓ	INTEGRAL	E_a	T(S)	RESULT C_0	RESULT C_1	RESULT C_2
2^{-3}	0.89462319318517	6.33e-10	356.9			
2^{-4}	1.07605987265074	8.56e-10	426.3			
2^{-5}	1.19524813881849	1.15e-09	426.2	1.423460265674	-2.0676022044	-7.9521322
2^{-6}	1.26445377191768	1.34e-09	426.2	1.360275447540	-3.2163693548	-23.710490
2^{-7}	1.30188593759114	1.46e-09	426.2	1.339480501116	-6.2931480198	5.5420653
2^{-8}	1.32137252564963	1.52e-09	426.2	1.341163816983	-5.4086365404	9.8856275
2^{-9}	1.33131707386056	1.55e-09	426.2	1.341410985041	-5.1746466051	25.0634683
2^{-10}	1.33634079003748	1.57e-09	426.2	1.341399965444	-5.1950222482	19.5001958
2^{-11}	1.33886565298449	1.58e-09	426.2	1.341399223875	-5.5197895049	18.6327973
2^{-12}	1.34013135392416	1.58e-09	426.2	1.341399240952	-5.1977615908	18.3726055
2^{-13}	1.34076502405465	1.59e-09	426.2	1.341399241506	-5.1977527356	18.3878520
	Eq (2), Laporta [15]:			1.341399241447	-5.1977529559	18.3870466

Table 3. Integral and leading order expansion coefficients using PARINT with 16 procs., and linear extrapolation for 3-loop integral of Laporta Fig 2(u), err. tol. $t_a = 10^{-12}$; max. # evals = 20B, $T(s)$ = elapsed time (s); E_a = integration estim. abs. error

	INTEGRAL FIG 2(u)			EXTRAPOLATION		
ε_ℓ	INTEGRAL	E_a	T(S)	RESULT C_0	RESULT C_1	RESULT C_2
2^{-3}	0.176698722960541	1.01e-09	363.8			
2^{-4}	0.179083545661235	9.17e-10	437.2	0.181468368362	-0.0381571632	
2^{-5}	0.180693790881676	9.24e-10	437.3	0.182582592016	-0.0648985309	0.14262063
2^{-6}	0.181617679292608	9.34E-10	437.2	0.182626195317	-0.0673403158	0.18168919
2^{-7}	0.182111420125171	9.79e-10	437.7	0.182627225039	-0.0674638824	0.18630234
2^{-8}	0.18236652627441	9.84e-10	437.6	0.182627237225	-0.0674669046	0.18654412
2^{-9}	0.182496175680214	9.86e-10	437.7	0.182627237219	-0.0674669014	0.18654358
2^{-10}	0.182561529243301	9.86e-10	437.8	0.182627237221	-0.0674669038	0.18654439
	Eq (2), Laporta [15]:			0.182627237539	-0.0674669097	0.18654624

Table 4. Integral and leading order expansion coefficients using one call to PARINT with vector function, 16 procs. and linear extrapolation for 3-loop integral of Laporta Fig 2(u), err. tol. $t_a = 10^{-12}$; max. # evals = 20B, $T(s)$ = elapsed time (s); E_a = integration estim. abs. error

	INTEGRAL FIG 2(u)		EXTRAPOLATION		
ε_ℓ	INTEGRAL	E_a	RESULT C_0	RESULT C_1	RESULT C_2
2^{-3}	0.176698722966533	1.57e-09			
2^{-4}	0.179083545593469	1.63e-09	0.181468368220	-0.0381571620	
2^{-5}	0.1806937908054	1.71e-09	0.182582591950	-0.0648985315	0.1426206455
2^{-6}	0.181617679222644	1.76e-09	0.182626195261	-0.0673403170	0.1816892047
2^{-7}	0.182111420130547	1.80e-09	0.182627225208	-0.0674639106	0.1863033669
2^{-8}	0.182366526276538	1.81e-09	0.182627237153	-0.0674668729	0.1865403500
2^{-9}	0.182496175679009	1.82e-09	0.182627237223	-0.0674669081	0.1865461701
2^{-10}	0.182561529242407	1.83e-09	0.182627237223	-0.0674669083	0.1865462415
	Eq (2), Laporta [15]:		0.182627237539	-0.0674669097	0.1865462421

Table 5. Parallel performance of PARINT (on MPI) for 3-loop integrals (C_0) of Fig 1, abs. tolerance $t_a = 5 \times 10^{-10}$, and max. number of evaluations = 10B

3-loop diag.	N	Result Laporta [15]	Result $p = 1$	Result $p = 64$	$T_1[s]$	$T_{64}[s]$	S_{64}
Fig 1 (a)	7	1.32644820827	1.326448206	1.32644819	902.7	15.8	57.1
Fig 1 (b)	7	1.34139924145	1.34139924147	1.3413992416	1026.2	14.4	71.3
Fig 1 (c)	7	2.00250004111	2.00250004113	2.0025000412	879.3	13.4	65.6
Fig 1 (d)	8	0.27960892328	0.2796089227	0.279608920	1019.7	15.9	64.1
Fig 1 (e)	8	0.18262723754	0.1826272372	0.1826272368	1018.3	15.8	64.4
Fig 1 (f)	8	0.14801330396	0.1480133036	0.1480133026	976.6	16.4	59.5

and 3 are computed with consecutive calls to `pi_integrate()` in a loop and linear extrapolation, for the functions of (Laporta) Fig 2(r) with $N = 7$, and Fig 2(u) with $N = 8$, respectively (depicted in Fig 1). The values of C_0, C_1 and C_2 are listed ($K = 0$ in Eq (2)). The abbreviation "B" refers to *billion* with regard to the number of integrand evaluations, and $T(s)$ represents the integration time in seconds. Table 2 is obtained similarly for Fig 2(r), but with non-linear extrapolation by the ϵ-algorithm.

The results in Table 4 for Fig 2(u) are computed with one call to `pi_integrate()` from the main program, integrating a vector function that contains the ε-dependence in the definition of its components. The integration time for the vector function (at 2180 seconds) is considerably less than the total integration time through the loop in Table 3, with comparable or slightly better accuracy for the same problem. If less accuracy is needed overall, the computation will take less time with both methods. The extrapolations in Tables 1-4 show good agreement with the values of the Eq (2) coefficients from Laporta [15]. Table 5 presents timing results on the HPCC cluster, corresponding to the six diagrams in Fig 1. Since the integrals with kinematics specified by Laporta [15] are finite and only the integral needs to be computed, we set $\varepsilon = 0$ in the integrand codes. The speedup $S_p = T_1/T_p$ (where T_p is the integration time incurred with p processes) indicates good scalability of the parallel implementation (see also [16]). For a total of $p = 64$ MPI processes on 4 cluster nodes, 16 procs are spawned per node.

4. Concluding remarks

In this paper we explored *distributed, automatic* integration with the PARINT multivariate integration package, in particular its adaptive integration engine. We demonstrated its accuracy and parallel scalability for the computation of the leading order coefficients in an asymptotic expansion with respect to the dimensional regularization parameter ε, as $\varepsilon \to 0$, for a class of 3-loop self-energy integrals from [15]. This procedure relies on obtaining approximations of a sequence of integrals as $\varepsilon \to 0$, and an extrapolation for convergence acceleration of the sequence. In view of the compute-intensive nature for moderate integral dimensions, the parallelization has been proven beneficial for obtaining considerable accuracy.

Acknowledgments

We acknowledge the support from the National Science Foundation under Award Number 1126438. This work is further supported by Grant-in-Aid for Scientific Research (24540292) of JSPS, and the Large Scale Simulation Program No.13/14-13 of KEK.

References

[1] Piessens R, de Doncker E, Überhuber C W and Kahaner D K 1983 *QUADPACK, A Subroutine Package for Automatic Integration* (*Springer Series in Computational Mathematics* vol 1) (Springer-Verlag)
[2] de Doncker E, Fujimoto J, Hamaguchi N, Ishikawa T, Kurihara Y, Shimizu Y and Yuasa F 2011 *Journal of Computational Science (JoCS)* **3** 102–112 doi:10.1016/j.jocs.2011.06.003
[3] Open-MPI http://www.open-mpi.org
[4] de Doncker E, Kaugars K, Cucos L and Zanny R 2001 *Proc. of Comp. Particle Physics Symp. (CPP 2001)* pp 110–119
[5] de Doncker E and Yuasa F 2014 *XV Adv. Comp. and Anal. Tech. in Phys. Res., Journal of Physics: Conference Series (ACAT 2013)* **523** doi:10.1088/1742-6596/523/1/012052
[6] Berntsen J, Espelid T O and Genz A 1991 *ACM Trans. Math. Softw.* **17** 437–451
[7] De Ridder L and Van Dooren P 1976 *Journal of Computational and Applied Mathematics* **2** 207–210
[8] ParInt http://www.cs.wmich.edu/parint
[9] Shanks D 1955 *J. Math. and Phys.* **34** 1–42
[10] Wynn P 1956 *Mathematical Tables and Aids to Computing* **10** 91–96
[11] Sidi A 1996 *Journal of Approximation Theory* **86** 21–40
[12] Sidi A 2011 *Numer. Math.* **119** 725–764
[13] de Doncker E, Shimizu Y, Fujimoto J and Yuasa F 2004 *Computer Physics Communications* **159** 145–156
[14] Nakanishi N 1971 *Graph Theory and Feynman Integrals* (Gordon and Breach, New York)
[15] Laporta S 2000 *Int. J. Mod. Phys. A* **15** 5087–5159 arXiv:hep-ph/0102033v1
[16] de Doncker E, Yuasa F, Kato K, Ishikawa T and Olagbemi O *XVI Adv. Comp. and Anal. Tech. in Phys. Res., Journal of Physics: Conference Series (ACAT 2014)* Accepted

Toward a Proof of Concept Cloud Framework for Physics Applications on Blue Gene Supercomputers

Patrick Dreher[1,2], William Scullin[3], Mladen Vouk[1]

(1)-NC State University (2) Massachusetts Institute of Technology, (3) Argonne Leadership Computing Facility

E-mail: padreher@ncsu.edu, wscullin@alcf.anl.gov, vouk@csc.ncsu.edu

Abstract. Traditional high performance supercomputers are capable of delivering large sustained state-of-the-art computational resources to physics applications over extended periods of time using batch processing mode operating environments. However, today there is an increasing demand for more complex workflows that involve large fluctuations in the levels of HPC physics computational requirements during the simulations. Some of the workflow components may also require a richer set of operating system features and schedulers than normally found in a batch oriented HPC environment. This paper reports on progress toward a proof of concept design that implements a cloud framework onto BG/P and BG/Q platforms at the Argonne Leadership Computing Facility. The BG/P implementation utilizes the Kittyhawk utility and the BG/Q platform uses an experimental heterogeneous FusedOS operating system environment. Both platforms use the Virtual Computing Laboratory as the cloud computing system embedded within the supercomputer. This proof of concept design allows a cloud to be configured so that it can capitalize on the specialized infrastructure capabilities of a supercomputer and the flexible cloud configurations without resorting to virtualization. Initial testing of the proof of concept system is done using the lattice QCD MILC code. These types of user reconfigurable environments have the potential to deliver experimental schedulers and operating systems within a working HPC environment for physics computations that may be different from the native OS and schedulers on production HPC supercomputers.

1. Introduction

Physics researchers running applications needing state-of-the-art computational hardware and software usually select supercomputer systems to run those applications. Although these systems have the capability to deliver the highest level of CPU power and network connectivity, what gets sacrificed in such a selection is the flexibility of choices for operating systems and user customized software. Over the past several years cloud computing [1] has emerged as an alternate hardware architecture for computation. From a technical and operational perspective, users now have a spectrum of choices in the design and configuration of customized system and application software stacks for various types of physics computations and analysis.

Today cloud computing options have been applied to situations where users have constraints on facility access to computational resources, small prototype computations needing many short calculations with different parameters, and large computations with minimal communications requirements between processors. Some of these types of computations have been successfully implemented in both private or commercial cloud computing systems ([2],[3]). The overall consensus is that today's cloud technologies have now advanced to the point where they can provide end-users with considerable flexibility and reliability to self-provision resources, either explicitly or implicitly, and provide on-demand computational capabilities and services. As a result, companies, academic institutions, organizations and individuals are seriously considering and experimenting with cloud computing as a platform for computation and data analysis.

Despite all of these advances, cloud computing has only had mixed success when attempting to implement physics supercomputing applications onto these types of platforms. Users explicitly requiring high performance computing favor systems that allow them to operate "close to the metal" with the ability to tune both the hardware and storage in order to optimize computational performance and overall throughput. Early efforts to re-create these HPC capabilities in cloud systems selected the most straightforward option of deploying these physics supercomputing applications onto existing cloud platforms. Although this method did show some promise for codes with minimal inter-processor communications requirements, the more tightly coupled HPC physics applications suffered degraded performance. Alternative approaches were then explored that involved constructing small groups of HPC clouds with more robust uniform hardware architectures and network connections. This design provided "spill-over provisioning" from the HPC supercomputer to a cloud system when the HPC system became saturated [4]. Although these implementations did provide some overall acceleration, the underlying shortcomings of delivering HPC supercomputer level computational throughput with commodity cloud cluster hardware still remained problematic.

The basic difficulty is that general cloud computing systems lack the specialized HPC architectural infrastructure needed to deliver the required high throughput. There are many examples of tightly coupled HPC codes that require state-of-the-art network interconnects to provide maximum computational throughput and minimum latency. For example physics applications that need lattice based infrastructures and communications requirements between nearest and next nearest neighbor sites usually have some of the highest memory bandwidth and network interconnect requirements. Running such applications on standard cloud computing systems generally results in degraded performance. Additional performance degradation is also attributed to a lack of uniformity in the computational hardware.

The motivation for this proof of concept project was to work toward building and provisioning a secure cloud framework tuned to a computational physics code based on a customized system software and application stack not found or able to be installed on current supercomputer systems. That stack would be embedded in a host's supercomputer hardware architecture in a way that would address many of the shortcomings noted with the current cloud computing implementations. Beyond the novelty of constructing such a system, this type of construct would capitalize on the infrastructure advantages and enhanced capabilities of a supercomputer, allowing the cloud to take advantage of the installed high speed network. This proof of concept cloud framework on a supercomputer would provide the low latency interconnects between processors and allowing for the cloud computing cluster to capitalize on the localized homogeneous and uniform HPC supercomputer architecture.

This type of design may also potentially assist with the issue of data migration and transfer. In many instances data from physics simulations and/or data input from sensors/devices are stored on disk farms at the supercomputer center. The post-processing and analysis of that data usually requires customized software stacks sometimes located at sites remote to the production supercomputer centers. Having the flexibility to change/customize the software stacks and operating systems within an HPC architecture using a cloud environment may alleviate the need to move data from one physical location to another to access the appropriate computational system.

In addition, these ideas offer the potential for designing frameworks that may lead to better hybrid workflows. This work may also allow for experimentation with new schedulers and operating systems within a working HPC environment that may be different from the native

OS and schedulers on the HPC supercomputer itself. Finally, from an economic perspective, the idea of capitalizing on the overall operational cost effectiveness of a supercomputer, such as the IBM Blue Gene architecture, and merging it with the elasticity of a cloud computing architecture may be a very cost effective design for the future.

Section 2 summarizes the key HPC and cloud computing design characteristics and the proof of concept implementation on the BG/P and Blue Gene/Q architectures. Section 3 provides a brief introduction into the properties of the lattice QCD physics code selected to test the cloud framework on the Blue Gene. Section 4 discusses some of the initial testing of physics applications on these systems. Finally section 5 summarizes possible next steps and suggested future trends for applying physics applications to cloud technologies and supercomputers.

2. Proof of Concept Cloud Computing Framework on the BG/P and BG/Q

The idea for the original project was to select a suitable supercomputer in which to embed a cloud computing framework. The IBM Blue Gene/P was selected as the HPC architecture because of its network topologies and excellent hardware infrastructure. In addition, an IBM research group ([5],[6]) developed an open source software utility called Kittyhawk. This software utility was implemented on the IBM Blue Gene/P architecture and showed promise for this project because it incorporated some of the key building blocks needed for embedding a cloud system into the BG/P with the needed access to internal supercomputer networks and communications. The cloud computing system selected for this project was a proven open source production level cloud architecture, called the Virtual Computing Laboratory (VCL) ([10], [11], [12], [13]). The software for VCL is available through the Apache Software Foundation [14]. This proof of concept design was demonstrated at the Department of Energy's Argonne Leadership Computing Facility (ALCF). Dreher, Vouk and Mathew recently published detailed reports ([8], [9]) documenting a successful demonstration of this proof of concept implementation of a secure cloud framework embedded in a BG/P in such a way that it allowed the VCL cloud system to capitalize on the host's supercomputer infrastructure. Based on the results from that work, these proof of concept ideas and framework were extended to the new Blue Gene/Q architecture. Dreher, Vouk and Scullin [7] recently reported on the initial the proof of concept cloud framework installation on the BG/Q.

3. Lattice QCD Physics Application

The physics application selected to test this new proof of concept cloud framework on the Blue Gene is a physics code used for lattice Quantum Chromodynamics (LQCD). This application is especially computationally demanding with large memory bandwidth and low latency requirements. LQCD is a non-perturbative implementation of a quantum field theory using the Feynman path integral approach. The method for lattice QCD calculations proceeds analogously as if the quantum field theory was being solved analytically.

The starting point for the LQCD approach is the partition function in Euclidean space-time

$$Z = \int \mathcal{D}A_\mu \, \mathcal{D}\psi \, \mathcal{D}\bar{\psi} \, e^{-S} \tag{1}$$

where S is the QCD action

$$S = \int d^4x \mathcal{L} \tag{2}$$

and the Lagrange density $\mathcal{L}$ represents the continuum Euclidean QCD. The expanded equations for the Lagrange density are

$$\mathcal{L} = \sum_r \sum_i^{n(r)} \bar{\psi}_{ri} i \not{D} \psi_{ri} - \frac{1}{4} F^a_{\mu\nu} F^{\mu\nu}_a \tag{3}$$

For fermions coupled to the strong nuclear force through gluon interactions in a particular gauge group the Dirac operator is

$$\not{D} = \not{\partial} - ig\not{A}_a T_r^a \tag{4}$$

and the field strength tensor $F_{\mu\nu}^a$ is

$$F_{\mu\nu}^a = \partial_\mu A_\nu^a - \partial_\nu A_\mu^a + gf^{abc}(A_\mu)_b(A_\nu)_c \tag{5}$$

The matrices T_r^a represent the generators in the representation r of the gauge group to which the fermions are assigned. The index a runs over the generators of the gauge group, i runs over all $n(r)$ fermion flavors belonging to the representation r. The f^{abc} are the structure constants of the gauge group. The fermions are represented by Grassmann variables ψ and $\bar{\psi}$. These can be integrated exactly with the result

$$Z = \int \mathcal{D}A_\mu \ \det M \ e^{\int d^4x \ (-\frac{1}{4}F_{\mu\nu}F^{\mu\nu})}. \tag{6}$$

After integration over the fermions, the fermionic contribution is now contained in the highly non-local term $\det M$ where M is the Dirac operator. The partition function at this point is an integral over only background gauge configurations that can be written as $S = S_{gauge} + S_{quarks} = \int d^4x \ (\frac{1}{4}F_{\mu\nu}F^{\mu\nu}) - \sum_i \log(\mathrm{Det}M_i)$ where the sum is over the quark flavors, distinguished by the value of the bare quark mass.

One of the objective of the computations in lattice QCD is to obtain values for experimentally observed quantities by calculating expectation values of operators

$$< \mathcal{O} > = \frac{1}{Z}\int \mathcal{D}A_\mu \ \mathcal{O} \ e^{-S} \ . \tag{7}$$

where $\mathcal{O}$ is any given combination of operators expressed in terms of time-ordered products of gauge and quark fields.

At present, the only means of carrying out such non-perturbative QCD calculations is through large scale physics first principles numerical simulations within the framework of lattice gauge theory (lattice QCD). These HPC simulations calculate expectation values of operators representing quarks and gluons that are believed to approximate the physical observables in the laboratory. Examples of such calculations include efforts to calculate the mass spectrum of the elementary particles (such as the proton) from these theoretical physics first principles.

The method summarized above is the only known technique for solving this type of relativistic quantum field theory. For readers interested in a complete in-depth discussion of the physics and computational methods of Lattice Quantum Chromodynamics the book by DeGrand and DeTar [15] is an excellent reference.

4. Testing the Proof of Concept Cloud Framework on The Blue Gene Using Lattice QCD

The first tests performed on this design were basic "hello world" codes. The basic "hello world" tests utilizing MPI functions were successful with the BG/P design using Kittyhawk and the BG/Q design using the FusedOS operating system described in [7]. The more interesting and complex tests involve installing production physics application codes using these proof of concept design based on the FusedOS and VCL cloud computing system installed on the BG/Q.

The Kittyhawk software worked well on the BG/P and the utility was able to be successfully integrated with the VCL system. The BG/Q architecture differed sufficiently from the BG/P and so a different approach of using an experimental heterogeneous operating system (FusedOS) was selected for testing. Some tests of the FusedOS environment have already been conducted by other groups. The authors in [19] tested the FusedOS using the LAMMPS code. LAMMPS is an HPC benchmark that simulates nearest neighbor interactions by constructing "particle-like" models of systems such as molecular dynamics. The specific tests within LAMMPS employed a three dimensional Lennard-Jones melt benchmark to test the code under the FusedOS heterogeneous environment. The detailed results are contained within their paper. The authors extend their comments regarding FusedOS in [20] suggesting that the FusedOS approach can accommodate existing codes already running in a compute node kernel (CNK) environment on a BG/Q without modifications.

Although the testing of the simple "hello world" codes utilizing MPI functions has been successful under FusedOS on blocks built of up to 128 nodes and I/O nodes, the more realistic lattice QCD physics application code has encountered difficulties. There were several solid production open source lattice QCD codes available from which to choose because theoretical physicists in the high energy and nuclear physics communities have banded together in various collaborations to write the necessary computer codes to do these numerical simulations on the fastest supercomputers. One of these lattice QCD groups that has developed QCD codes is the MIMD Lattice Computation (MILC) Collaboration [16]. This collaborations has written a fully open source implementation of lattice QCD that has been optimized in production for both the IBM BG/P and BG/Q supercomputers. Our project utilized a small test code within MILC (ks-spectrum) on a 4 node BG/Q to analyze the particle physics spectrum based on some Lattice QCD generated configurations.

When the MILC code from the MIMD Lattice Computation collaboration was implemented under FusedOS, several issues were encountered. The two primary issues noted involve handling of alignment issues and network geometry. When using the PowerPC A2 CPU's QPX vector unit, all addresses should be aligned to the nearest 32-byte boundary. If the alignment of an address is off for any reason, an exception is generated. The handling of this exception is software dependent. Under the CNK operating system, these alignment exceptions are trapped and handled, though a performance penalty is incurred each time. For this reason, CNK uses a user-tunable environment variable, $BG_MAXALIGNEXP$, to allow the user to set a maximum number of alignment exceptions to handle before forcing the application to terminate and core dump. Under FusedOS, the handling of these alignment errors is not handled by the operating system and application failure is assured.

It was determined that the design of the Blue Gene /Q hardware and FusedOS software have presented obstacles to the successful execution of production codes such as MILC, though it is believed that these observed difficulties are not insurmountable. It is suspected that most of those issues likely trace back to alignment exceptions. A second issue that arose was an incompatibility between the geometry of the available block and the MILC binary. While the MILC binary had no restrictions on geometry, it had been suggested that the smallest sub-block under FusedOS should be at least eight contiguous nodes in five dimensions. When attempted runs failed, crashing both application and operating system, we found it difficult to debug, relying on logging and hardware bugging interfaces in an attempt to de-construct the origin.

5. Observations and Next Steps

The continued development of these heterogeneous cloud frameworks within a supercomputer host environment still offer potential interesting alternatives for embedding cloud framework with supercomputer host architectures to develop hybrid workflows, schedulers and potential HPC cloud computing environments. In the short term, we continue to work with developers and will continue to try to target MILC as a sample application as it has been successfully run at all node counts in a Linux environment and under CNK. Over the longer term, continued research into advanced heterogeneous OS environments such as FusedOS may play a role toward the development of exascale level capabilities for physics computations and may provide cost effective options for combining the HPC capabilities of supercomputers with the provisioning flexibility and elasticity of cloud computing.

6. Acknowledgements

The authors would like to thank Dr. Yoonho Park from IBM, Dr. Eric Van Hensbergen from ARM and Marius Hillenbrand from Karlsruhe Institute of Technology for helpful conversations on FusedOS. This work is supported in part through NSF grant 0910767, 1318564, 1330553, the U.S. Army Research Office (ARO) grant W911NF-08-1-0105 managed by the NCSU Science of Security Initiative and the Science of Security Lablet, by the IBM Share University Research and Fellowships program funding, and the Argonne Leadership Computing Facility at Argonne National Laboratory, which is supported by the Office of Science of the U.S. Department of Energy under con- tract DE-AC02-06CH11357. One of us (Patrick Dreher) gratefully acknowledges support with an IBM Faculty award.

References

[1] Mell, P., and Grance, T. (2011) *The NIST Definition of Cloud Computing* **800-145** (Gaithersburg: National Institute of Standards and Technology)
[2] UberCloud Experimet, https://www.theubercloud.com/hpcexperiment/
[3] Amazon Web Services, http://aws.amazon.com
[4] – *http : //www.anl.gov/Mediacenter/News/*2009/*news*090402.*html*
[5] Appavoo, J., U, V., Rosenburg, B., DaSilva, D., and Moreira, J. (2009). Kittyhawk: Enabling Cooperation and Competition in a Global Shared Computational System. *IBM Journal of Research and Development*, **vol 53 no 4** 4 1-15.
[6] http://kittyhawk.bu.edu/kittyhawk/Kittyhawk.html
[7] Dreher P., Scullin W., Vouk M., (2014) Toward a Proof of Concept Implementation of a Cloud Infrastructure on the Blue Gene/Q, *International Journal of Grid and High Performance Computing* (submitted)
[8] Dreher, P., and Mathew, G. (2014) Toward Implementation of a Software Defined Cloud on a Supercomputer *IEEE International Conference on Cloud Engineering (IC2E)* IEEE Computer Society Conference Publishing Services, **10.1109/IC2E.2014.57**, 580-585
[9] G. Mathew, P. Dreher, M. Vouk, "Integrating VCL into a Re-Configurable Cloud Computing Layer on an IBM BG/P Supercomputer", (2014) 2nd International IBM Cloud Academy Conference, Atlanta, GA
[10] Vouk, M. (2008) Cloud Computing - Issues, Research and Implementations *Journal of Computing and Information Technology*, **vol 16 no 4** 235-246
[11] Vouk, M., Rindos, A., Averitt, S., Bass, J., Bugaev, M., Peeler, A., and Valenzisi, V. (2009). Using VCL Technology to Implement Distributed Reconfigurable Data Centers and Computational Services for Educational Institutions. *IBM Journal of Research and Development*, **vol 53 no 4** 1-18
[12] Dreher, P., Vouk, M., Sills, E., and Averitt, S. (2009). Evidence for a Cost Effective Cloud Computing Implementation Based Upon the NC State Virtual Computing Laboratory Model *Advances in Parallel Computing; High Speed and Large Scale Scientific Computing* **18** 236-250. Amsterdam: IOS Press
[13] Schaffer, H., Averitt, S., Hoit, M., Peeler, A., Sills, E., and Vouk, M. (2009). NCSUs Virtual Computing Laboratory: A Cloud Computing Solution. *IEEE Computer Society* **vol 42 no 7** 94-97
[14] Apache VCL, *http : //vcl.apache.org, alsohttp : //vcl.ncsu.edu*
[15] DeGrand, T. and DeTar, C. (2006) *Lattice methods for Quantum Chromodynamics* (London: World Scientific)
[16] *http : //physics.indiana.edu/ sg/milc.html*
[17] Blue Gene/Q: By Co-design, (2013), *Computer Science - Research and development*, **vol 28, no 2-3**, 127-135
[18] IBM Research, 2014, FusedOS, GitHub repository, https://github.com/ibm-research/fusedos
[19] Park, Y., Van Hensbergen, E., Hillenbrand, M., Inglett, T., Rosenburg, B., Ryu, K., Wisniewski, R., (2012), FusedOS: Fusing LWK Performance with FWK Functionality in a Heterogeneous Environment, *Proceedings of the 2012 IEEE 24th International Symposium on Computer Archixdtecture and High Performance Computing*, **8**, 211-218
[20] Park, Y., Hensbergen, E.V., Hillenbrand, M., Inglett, T., Rosenburg, B.S., Ryu, K.D., Wisniewski, R.W.: (2012) Poster: FusedOS: A Hybrid Approach to Exascale Operating Systems *High Performance Computing, Networking, Storage and Analysis (SCC), 2012 SC Companion:* 1417

Journal of Physics: Conference Series **640** (2015) 012064 doi:10.1088/1742-6596/640/1/012064

First experience with particle-in-cell plasma physics code on ARM-based HPC systems

Xavier Sáez[1], Alejandro Soba[2], Edilberto Sánchez[3], Mervi Mantsinen[1,4], Sergi Mateo[1], José M. Cela[1] and Francisco Castejón[3]

[1] Barcelona Supercomputing Center (BSC-CNS), Barcelona, Spain
[2] Centro de Simulación Computacional para Aplicaciones Tecnológicas (CONICET), Argentina
[3] Laboratorio Nacional de Fusión (CIEMAT), Madrid, Spain
[4] Institució Catalana de Recerca i Estudis Avançats (ICREA), Barcelona, Spain

E-mail: xavier.saez@bsc.es

Abstract. In this work, we will explore the feasibility of porting a Particle-in-cell code (EUTERPE) to an ARM multi-core platform from the Mont-Blanc project. The used prototype is based on a system-on-chip Samsung Exynos 5 with an integrated GPU. It is the first prototype that could be used for *High-Performance Computing* (HPC), since it supports double precision and parallel programming languages.

1. Introduction

During the last two decades, supercomputers have grown rapidly in performance to provide scientists the required computing power, at the cost of a similar growth in power consumption. The most used metric for evaluating supercomputers performance has been the speed of running benchmark programs (as Linpack [1]).

However, nowadays the computer's performance is limited by power consumption and power density. Energy has become one of the most expensive resources and, as a consequence, the energy expenses in a HPC center can easily exceed the cost of infrastructure after a few years in operation. Therefore, energy efficiency is already a primary concern for the design of any computer system and will define the maximum achievable performance. This change of the point of view is reflected in the increasing popularity of the Green500 list [2], in which supercomputers are ranked in terms of their power efficiency.

Looking to future, new developed platforms to build a sustainable exaflop supercomputer will have to be based on the power efficiency. The *Mont-Blanc project* [3] has the aim to design computer architectures capable of delivering an Exascale performance using 15 to 30 time less energy than present architectures. The reduction of energy consumption will be achieved by developing a HPC prototype based on the energy-efficient technology originally designed for mobile and embedded systems.

Particle-in-cell (PIC) is one of the most used methods in plasma physics simulations [4]. The quality of results achieved by this method relies on tracking a very large number of particles. Therefore, PIC codes require intensive computation and need to be adapted to new computing platforms constantly. *EUTERPE* is a parallel gyrokinetic PIC code for global linear and non-linear simulations of fusion plasma instabilities in three-dimensional geometries [5], specifically in tokamaks and stellarators [6, 7]. It has been written to target traditional HPC clusters using Message Passing Interface (MPI).

In this paper we study the portability of EUTERPE to the ARM-based prototype within the Mont-Blanc project. The rest of the paper is organized as follows. In section 2, we describe the PIC method basics and the EUTERPE code. In section 3, we describe the Mont-Blanc project and the architecture of the ARM-based prototype [8]. Next we explain the first experiences on porting EUTERPE to an ARM platform. Finally some conclusions are given.

2. Particle-in-cell methods

PIC methods are used to model physical systems whose behavior varies over different ranges of spatial scales. Macroscopically the dynamics is described by a system of partial differential equations (continuous model), while microscopically is modeled by a set of discrete particles. Explicitly, a PIC method follows the individual particles (or fluid elements) in a continuous phase space, whereas moments of the distribution (such as densities and currents) are computed concurrently on stationary mesh points.

PIC methods are one of the most popular approaches in plasma physics to simulate the interaction of independent charged particles with each other and with electromagnetic fields [9].

A *full kinetic description* using the PIC method is implemented by replacing the distribution function f_s by a number of *macroparticles*, which represents a cloud of particles. The charges and densities of macroparticles are accumulated by interpolation on the spatial mesh and then the field equations are solved on the mesh. Finally, the forces acting on macroparticles are obtained by the interpolation of the fields at the macroparticles positions [10].

After an initialization phase, it is possible to summarize a *PIC algorithm* with three steps repeated at each time step [11]:

- **pull**: particle properties are interpolated to neighboring points in the computational mesh.
- **solve**: moment equations are solved on the mesh.
- **push**: momentum of each particle is calculated by interpolation on the mesh. The particles are repositioned under the influence of the momentum and the particle properties are updated.

In general, many PIC codes designed to simulate various aspects of the plasma behavior [12, 13] were not originally developed for new supercomputers based on multi-core architectures and basically only exploit the *task level parallelism*. The inclusion of new parallel programming techniques (as hybridization) allows to make the best of new multiprocessor supercomputers under development. There are several codes written in this way [14, 15], but in general they are simplified versions of production codes used by scientific groups in the simulation of real plasma physics environments. Nevertheless, this trend is reinforcing to provide more power computing to these applications.

2.1. EUTERPE code

In the EUTERPE code, the distribution function f_s of each kinetic species (s) is discretized using particles and a control variables scheme (δf) is used to reduce noise. The electrons are assumed to respond adiabatically and only electrostatic perturbations are taken into account. The evolution of the distribution function of each kinetic species is given by the gyrokinetic equation:

$$\frac{\partial f_s}{\partial t} + \frac{\mathrm{dv}_{\|}}{\mathrm{dt}}\frac{\partial f_s}{\partial v_{\|}} + \frac{\mathrm{d}\mathbf{R}}{\mathrm{dt}}\frac{\partial f_s}{\partial \mathbf{R}} = C_s, \tag{1}$$

where C_s represents collisions with all species. The evolution in time of $v_{\|}$ and $\mathbf{R}$ is integrated from equations depending of the fields through the Poisson-Ampère equations [16].

Initially, EUTERPE was parallelized applying domain decomposition and domain cloning [17]. These strategies only allow to parallelize at task level using MPI. For that reason, we developed a hybrid version of the code to take advantage of all the levels of parallelism that a multi-core architecture can offer [18], using OpenMP in the most time-consuming routines and developing a hybrid solver (mixing MPI and OpenMP) for the quasi-neutrality equation.

The OpenMP was introduced to parallelize the movement of the particles inside a domain (*push phase*). It is a suitable approach because the computation of the movement of any particle is independent from the rest of particles, so several threads can read simultaneously the electric field on the nearest grid points to a given particle without conflicts (figure 1a). OpenMP was also introduced in the computation of the charge density on the grid points (*pull phase*). This time there is a conflict because several threads can update the same grid point at the same time (figure 1b). Depending on the number of cores and the available amount of memory this issue is solved by atomic arithmetic operations or a mesh copy per thread.

The hybrid solver developed was an implementation of the Block Jacobi Preconditioning of the Conjugate Gradient method [19], where the preconditioner is a set of diagonal blocks.

3. The Mont-Blanc project

Transitions in the HPC world are not casual facts. The highest-volume commodity market (desktop computers) tends to drive lower-volume HPC market, because the design of an HPC chip has to be amortized selling the maximum number of units to minimize its final price.

For example, figure 2 shows how systems based on special-purpose HPC systems were replaced (from 1990 to 2000) by RISC microprocessors in the Top500 list [20]. This transition took place due to the cheapness of the RISC microprocessors although they were slower.

Currently, we observe a similar trend: low-power microprocessors (used in mobile devices) are improving their performance and are including features needed to run HPC applications (as an on-chip floating-point unit). It is reasonable to consider that the same market forces that replaced vector processors by RISC processors will be able to replace present options with mobile processors [3].

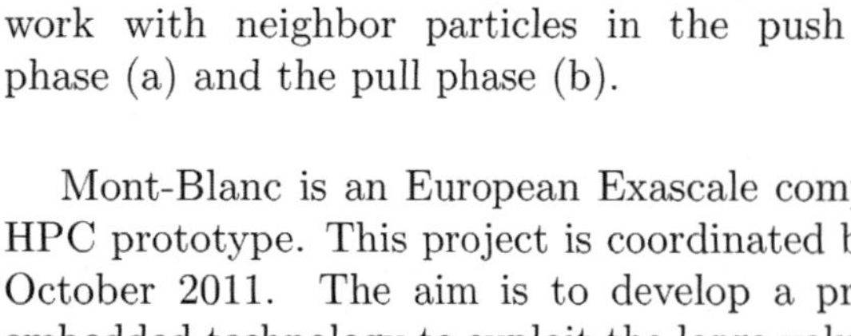

Figure 1. Two threads (*th0* and *th1*) that work with neighbor particles in the push phase (a) and the pull phase (b).

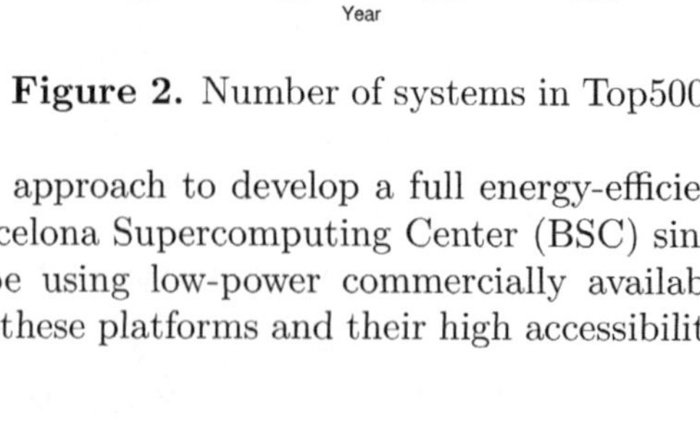

Figure 2. Number of systems in Top500.

Mont-Blanc is an European Exascale computing approach to develop a full energy-efficient HPC prototype. This project is coordinated by Barcelona Supercomputing Center (BSC) since October 2011. The aim is to develop a prototype using low-power commercially available embedded technology to exploit the large volume of these platforms and their high accessibility.

3.1. The ARM-based prototype

This work has been developed using a prototype named *Arndale*. It is based on a system-on-chip (SoC) Samsung Exynos 5 which contains an ARM Cortex-A15 dual core (at 1.7GHz) and an

ARM Mali T604 GPU.

ARM (Acorn RISC Machines) is a company which designs a RISC architecture. This kind of architectures reduces costs, heat and power use, which are desirable traits for embedded systems (smartphones, tablets, ...). Moreover, its simple design provides efficient multi-core CPUs with large number of cores at low cost and improved energy efficiency for HPC. Although ARM processors do not provide a sufficient level of performance for HPC yet, it is worthwhile to explore their potential considering ARM has a promising roadmap ahead.

The Samsung Exynos 5 with integrated GPU accelerator is the first embedded SoC that has the potential for HPC, since it supports 64-bit floating point arithmetic and provides support for parallel programming languages (such as OpenCL 1.1).

4. Work done on ARM-based platform

In order to obtain the best possible results on the prototype, all the available resources have to be used by the application: the dual core CPU (using OpenMP version developed previously) and the GPU (using a new OpenCL version).

OpenCL (Open Computing Language) [21] defines an application programming interface (API) that provides a homogeneous view of all computational resources through the following abstractions: the platform model, the memory model, the execution model and the programming model. It also support data-based and task-based parallel programming models. The execution model divides the computing system into a *host* (CPU) and a set of *compute devices* (GPU in our case). The host sends portions of the application (*kernels*) to the compute devices which execute many instances of them (*work-items*). The work-items are grouped by *work-group* and the work-items inside a group share resources such as memory.

We note here that the Mali GPU (in the prototype) has two special features which affect the common OpenCL programming. Firstly, the local and private memory are physically global memory. So, moving data from global to local or private memory typically does not improve performance. Secondly, all GPU threads have individual program counters. This means that all the threads are independent and can diverge without any performance impact.

In order to exploit the accelerator (Mali GPU) of the prototype, the selected routines have to be written in OpenCL. This mission involves some new tasks to make: the translation of the code from FORTRAN to C, the distribution of work between work-items and the creation and initialization of the OpenCL components (context, kernels, command queues and memory objects).

As a result of the previous work [18], we know which sections in the code are the most compute intensive: pushing the particles (*push*) and depositing their charge (*pull*). In this work our first aim is to minimize the necessary changes to these routines in OpenCL as compared with the previous version in order to increase the productivity.

In the *push* part, the work distribution is simple: one work-item is assigned to each particle. Since each work-item works on a different particle, all work-items write in different memory locations (as figure 1a shows but replacing threads with work-items). Therefore, the adaptation of the routine was reduced to a straightforward translation.

The *pull* part was more challenging to implement, since different particles can contribute to the charge density on the same mesh point. As we maintain the same work distribution than in push part, several threads can update the same memory location (as figure 1b shows but replacing threads with work-items). To avoid these memory conflicts, the two previously mentioned solutions used for the OpenMP implementation do not work properly due to the high number of threads on a GPU: atomic operations become inefficient because of lock contention that serializes the execution, and copies of the mesh require too much memory.

The strategy to solve this issue is inspired by the way in which OpenCL distributes data processing. A mesh copy is created per work-group, so only the work-items in the same work-group accumulate contributions on the same mesh copy. As a consequence, the lock contention is far minor and the reduction of memory conflicts makes the use of atomic operations feasible.

Nevertheless, a new step is required to reduce all these copies to the final mesh. The final global reduction is minor and a single kernel is enough to perform it since the number of mesh copies is limited by the number of work-groups. This time each work-item will collect the result of the same point in all the copies of the mesh.

The next step was to take advantage of all resources, so a convenient decision was to distribute the particles between CPU and GPU. Since the two devices shares the memory, all the particles are distributed between them without any additional data transfer.

Figure 3 shows the results of the ARM hybrid version (OpenMP + OpenCL). The test contains 1 million particles. We can see that as more particles are sent to GPU, the execution time of kernels decreases. Being the optimal configuration when the particles sent to the GPU are 40% in the *push* kernel and 60% in the *pull* kernel. At these minimum points, the combination of the CPU and GPU devices yields a modest but significant improvement over the CPU time only: about 30% in the *push* kernel and near 50% in the *pull* kernel.

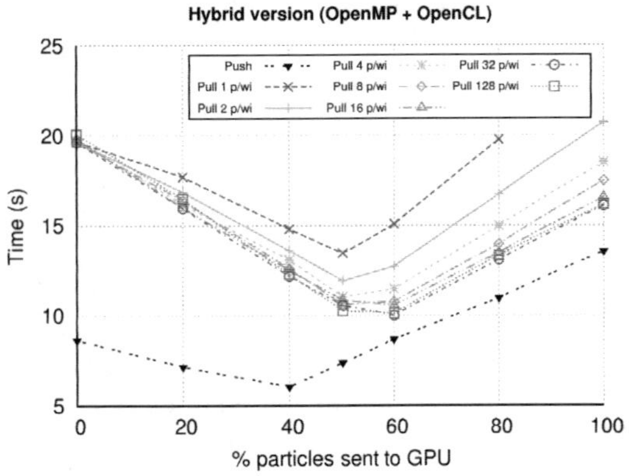

Figure 3. Execution time of the *push* and *pull* kernels for different combinations of particles sent to GPU and particles per work-item (*p/wi*).

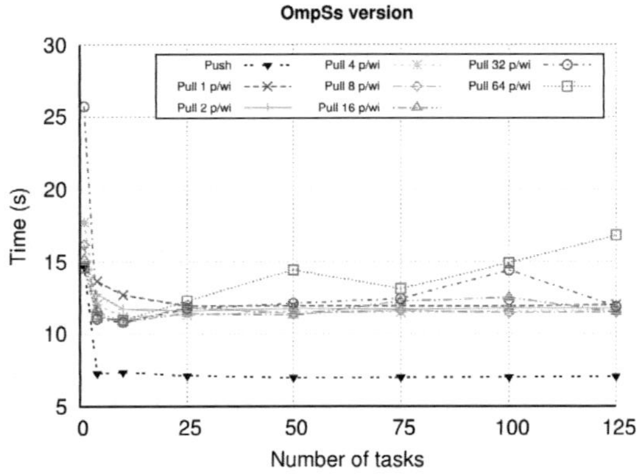

Figure 4. Execution time of the *push* and *pull* kernels for different combinations of number of tasks and particles per work-item (*p/wi*).

In order to port an application to a certain platform, besides the possibility to reach the maximum performance of it, the easiness of programming it is also important. The main drawback of OpenCL is the low programmability because is a low-level programming language. Therefore, it can be very time consuming to develop a code (meaning a low productivity).

To address this shortcoming, we ported the code to *OmpSs* [22] that is a task-based programming model developed at Barcelona Supercomputing Center (BSC). It provides an abstraction to the user which reduces programmer effort and unifies the SMP, heterogeneous and cluster programming in one model. The programmer annotates sections of the code to parallelize (*tasks*) with special directives. Over these tasks, the user can define *data dependencies* among them and specify the device where these tasks can be executed. The runtime system will analyze this information and automatically will schedule the execution of the tasks and perform the data transfers.

Figure 4 shows the results of the ARM OmpSs version. We can see that the performance is very stable and few tasks are enough to get the best performance for both kernels. Moreover, OmpSs offers a reduction in programming complexity, since the number of OmpSs directives included is far lower than the OpenCL calls included in the hybrid version (table 1).

Table 1. Comparison between the hybrid version (OpenMP+OpenCL) and the OmpSs version.

Kernel	Performance - best time (s)		Programmability - Productivity	
	Hybrid	OmpsSs	OpenCL API calls	OmpSs Directives
Push	6.02	6.92	161	12
Pull	10.31	10.83	167	18

5. Conclusions

This work confirmed the feasibility of porting a PIC code to an ARM-based platform. We have developed a hand-tuned hybrid version (OpenMP + OpenCL) and an OmpSs version of the most time-consuming kernels. Although OmpSs version is a bit slower than the hybrid version, it is a simpler version and its productivity has improved considerably. We can say that OmpSs simplifies the porting of codes to this new platform.

As future work we plan to code several state-of-art techniques to improve access to the data (in this case, particle data) in the GPU. In particular, we will explore if the benefits of sorting makes up for the extra sorting cost. Further, we will extend the implementation to several nodes using the parallelization with MPI on a new prototype.

Acknowledgments

The research leading to these results has received funding from the European Community's Seventh Framework Programme (FP7/2007-2013) under the Mont-Blanc Project (http://www.montblanc-project.eu), grant agreement nº 288777. This work was supported in part by grant ENE2012-30832, Ministerio de Economía y Competitividad and carried out within the framework of the EUROfusion Consortium, receiving funding from the European Union's Horizon 2020 research and innovation programme under grant agreement number 633053. The views and opinions expressed herein do not necessarily reflect those of the European Commission.

References

[1] Dongarra J, Luszczek P *et al.* 2003 *Concurrency and Computation: Practice and Experience* **15** 803–820
[2] The green 500: Ranking the world's most energy-efficient supercomputers `http://www.green500.org`
[3] Rajovic N *et al.* 2013 *Proc. of the Int. Conf. on HPC, Networking, Storage and Analysis* (ACM)
[4] Birdsall C K *et al.* 1991 *Plasma Phys. via Computer Simulation* 1st ed Plasma Phys. Series (Stanford Press)
[5] Jost G, TM T *et al.* 1999 *26th EPS Conference on Controlled Fusion and Plasma Physics*
[6] Kornilov V, Kleiber R, Hatzky R, Villard L and Jost G 2004 *Physics of Plasmas (1994-present)* **11** 3196–3202
[7] Sánchez E, Kleiber R, Hatzky R *et al.* 2013 *Plasma Physics and Controlled Fusion* **55** 014015
[8] ARM Ltd Cortex-a15 processor `http://www.arm.com/products/processors/cortex-a/cortex-a15.php`
[9] Dawson J M 1983 *Rev. Mod. Phys.* **55**(2) 403–447
[10] Pritchett P 2003 *Space Plasma Simu.* (*Lecture Notes in Phys.* vol 615) (Springer Berlin Heidelberg) pp 1–24
[11] Akarsu E *et al. Proceedings of the 1996 ACM/IEEE Conf. on Supercomputing* (IEEE Computer Society)
[12] Jolliet S, Bottino A, Angelino P, Hatzky R *et al.* 2007 *Computer Physics Communications* **177** 409 – 425
[13] Heikkinen J, Janhunen S, Kiviniemi T and Ogando F 2008 *Journal of Computational Physics* **227** 5582
[14] Burau H, Widera R *et al.* 2010 *Plasma Science, IEEE Transactions on* **38** 2831–2839
[15] Ragan-Kelley M 2010 Cs267 report particle simulation on a gpu with pycuda Tech. rep.
[16] Hahm T S 1988 *Physics of Fluids (1958-1988)* **31** 2670–2673
[17] Hatzky R 2006 *Parallel Computing* **32** 325 – 330
[18] Sáez X, Soba A, Sánchez E *et al.* 2011 *Proc. of the 19th Euromicro Conf. on Parallel, Distributed and Network-based Processing (PDP)* (IEEE) pp 385–389
[19] Saad Y 2003 *Iterative Methods for Sparse Linear Systems* (Society for Industrial and Applied Mathematics)
[20] TOP500org Top500 the lists. top500 supercomputer sites `http://www.top500.org/`
[21] OpenCL: The open standard for parallel programming of heterogeneous systems `https://www.khronos.org`
[22] Duran A, Ayguadé E, Badia R M, Labarta J, *et al.* 2011 *Parallel Processing Letters* **21** 173–193

Relation of Parallel Discrete Event Simulation algorithms with physical models

L.N. Shchur[1,2] **and L.V. Shchur**[1]

[1]Science Center in Chernogolovka, 142432 Chernogolovka, Russia
[2]National Research University Higher School of Economics, 101000 Moscow, Russia

E-mail: shchur@chg.ru

Abstract. We extend concept of local simulation times in parallel discrete event simulation (PDES) in order to take into account architecture of the current hardware and software in high-performance computing. We shortly review previous research on the mapping of PDES on physical problems, and emphasise how physical results may help to predict parallel algorithms behaviour.

1. Introduction

During the last decades high-performance computing evolved from big mainframes with a complicated central processing unit (CPU) into parallel processing of large numbers of CPUs and GPGPUs (general-purpose graphics processing units). Orchestrating thousands of them in a single task is one of the challenging problems of high-performance computing (HPC). We will focus our discussion on parallel discrete event simulations (PDES). PDES is the execution of a single discrete event simulation program on a parallel computer or on a cluster of computers [1]. It is widely used and is more and more applied in physics and computer science, as well as in economics and engineering. The system that should be simulated is divided into disjoint subsystems, which, nevertheless, are not isolated during the simulations, and dependencies between them should be resolved properly. There are three main essentials of PDES. Firstly, it is assumed that changes of possible, or potential, dependencies occur not at arbitrary, but at some particular moments of time. It is supposed that these changes' moments of time are spread on a scale which is large enough if compared to an elementary unit of time. Therefore, changes are considered as discrete (although random) in time, and are called *discrete events*.Subsystems evolve independently in time, as soon as there are no dependencies generated. We need to have some protocol in order to process dependencies correctly, in other words we would like to keep causality.This is done with the concept of local virtual time [2] associated with each subsystem, and which is the second essential feature of PDES. As soon as a new time event occurs in subsystem, the message containing necessary information is generated and stamped with the local simulation time. This message is sent to all other subsystems. This is the third essential feature of PDES - there is no information exchange through common variables and no access to shared memory. Instead all the information is spread via messages. The ensemble of messages generates a profile of local simulation times. The evolution of the profile in time depends on the particular scheme of managing causality, and the analysis of profile properties is the main subject of the present paper.

2. Extension of PDES concept and classification of PDES schemes

Originally [1], subsystems were associated with particular processing elements (PEs), which were assumed to be just one CPU. In paper [3] the concept was generalised taking into account that CPU may run a number of threads, and subsystems may be run in several parallel tasks. We generalise this problem further and analyse the current state of the typical HPC hardware and software.

The number of hybrid HPC systems built up with CPU and GPGPU, grows in Top500 list [1] during last years. Typically, each CPU can have several processing elements (cores) which can run concurrently two threads. We keep in mind the following mapping of a physical problem on a computing algorithm. Firstly, we associate each subsystem with a subtask. Secondly, we associate a processing element (PE) with each subtask. Thirdly, we assign local simulation time (LST) to each PE, $\tau_i(t)$, $i = 1, 2, ...N$, where t is the simulation time, and N is the total number of PEs. In the modern HPC system, we can think of PE as a core running by CPU or a kernel running by GPGPU. We do not discuss here optimisation and efficiency of simulations which depend strongly on a given and particular hardware architecture. We take into account only common and universal properties of parallel simulations on advanced hybrid HPC systems.

PEs within the PDES task can communicate conservatively or optimistically [1]. Under optimistic scheme all PEs run while assuming that all the causalities are fulfilled. Simulation is going on, and PE_i sends a message with the time stamp $T_i = \tau_i(t)$, i.e. at which event occurs. Suppose that at some moment of simulation it turns out that T_i is smaller than the current local simulation time τ_j of the processing element PE_j, that is $\tau_j > T_i$, and that PE_j depends on the corresponding event with the time stamp T_i. Therefore, causality gets broken, and the optimistic scheme provides a rollback mechanism to overcome this problem by sending anti-messages to recover causality [2]. It is clear that the recovery process typically is not local and can produce multiple anti-message processes, which can lead to an avalanche in LSTs. The time horizon will evolve back in time until no more anti-messages are generated. After that, simulations will resume and the system will proceed forward in time.

In conservative scheme, alternatively, PEs do not run until they receive all the information from dependant PEs, and causality is never broken.

Let us start our analysis with a simplified case in which our system is one-dimensional and each subsystem is dependent on local (left and right) subsystems. Let us map our system on the simple HPC architecture with N nodes, each of these nodes has M CPUs, and each CPU has c cores. This gives us hierarchy of memory access times. The fastest communication is between cores within CPU which have access to the on-chip memory. The next level of communication speed is within a node, its CPUs share the same node memory, onboard memory. The slowest level of the communication speed is the communication between nodes which interact usually over Infiniband hardware. Let us divide the task on $c * M * N$ subsystems, from which $c * M$ subsystems will run on one node.

Figure 1 presents the simplified model of hardware. Processing elements 1,2,3, and 4 are mapped on cores 1,2,3, and 4 of CPU1. Processing elements 5,6,7, and 8 are mapped on cores 1,2,3, and 4 of CPU2, etc. Solid arrows correspond to the communication within CPU, and they can be efficiently realised with the conservative scheme.

PEs within nodes (shown in Figure 1 with dashed arrows) can communicate conservatively, and we suppose that this communication is much faster than the communication between nodes (shown in Figure 1 with dotted arrows), which is usually the case[2]. Sometimes it can happen

[1] The list which ranks HPCs is updated twice a year [4] and is announced at International Conferences on Supercomputing which are held in May and November.

[2] In this case we have a mechanism which allows to run a particular group of threads within one CPU. It is reasonable for them to communicate without sending messages, and rather use a mechanism for thread synchronisation, as shown in Figure 1 with solid arrows.

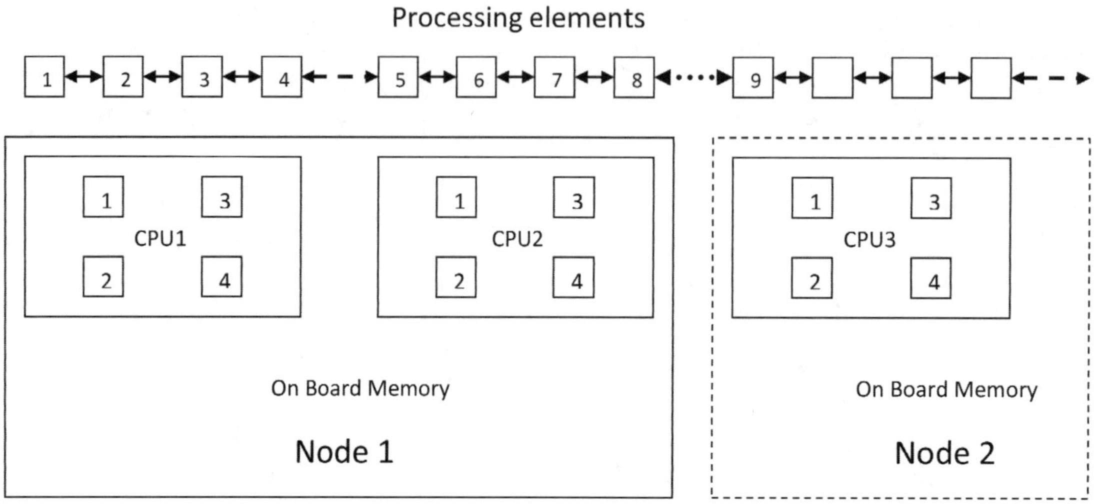

Figure 1. Mapping of processing elements on the cores, CPUs, and nodes.

that the average simulation time between events is much longer than sending messages via communication channels and analysing them. In this case the communication between nodes can be realised by any scheme, conservative or optimistic.

In fact, there are third possibility introduced in paper [3], where the mapping of PDES conservative scheme on the physical problem of surface growth in molecular beam epitaxy [5] is used for the following classification. The surface growth may be described with a nonlinear differential equation, known as KPZ equation [6] (see for details in the next section). Boundary conditions of KPZ equation can be associated with the PDES schemes such that free boundary conditions can be associated with the optimistic algorithm, and periodic boundary conditions with the conservative algorithm, and fixed boundary conditions leads to the FaS algorithm proposal [3]. Schematically, it can be explained by Figure 1. In FaS algorithm there is no communication between PEs 8 and 9 (i.e., between those PEs which are on the boundary between two nodes) for some time interval T_F, and PEs communicate conservatively within each node, leading to the propagation of freezing from the boundary PEs to the intermediate PEs (which is PE_4 and PE_5 in Figure 1). After that, Shift process initiated in a way that PEs are cyclically shifted to the right between CPUs and nodes such that PEs 1, 2, 3 and 4 are mapped on the cores 1, 2, 3 and 4 of CPU2, PEs 5, 6, 7 and 8 will be mapped on the cores 1, 2, 3 and 4 of CPU3, and so on. Then Freeze step of FaS algorithm starts again. It is argued in [3] that such algorithm may be efficient for hardware architectures where memory exchange and computations can be done in parallel.

Generalisation of our extension for the hybrid architecture of supercomputers, built up with either GPGPU with CUDA or OpenCL, or Intel ϕ architecture, is straightforward.

3. Conservative algorithm are deadlock-free

Let us describe in more details the mapping of conservative algorithm on the problem of molecular epitaxy surface growth [5].

Let us associate the value of a local simulated time $\tau_i(t)$ with PE number i, which is the function of global time t. For $t \geq 1$ the LST evolves iteratively as

$$\begin{aligned} \tau_i(t+1) &= \tau_i(t) + \eta_i(t) \quad \text{if} \quad \tau_i(t) \leq \min\{\tau_{i-1}(t), \tau_{i+1}(t)\}\,; \\ \tau_i(t+1) &= \tau_i(t) \quad \text{else} \quad , \end{aligned} \tag{1}$$

where η_i are random exponential variables. Each time the PE number i advances in time it sends messages to the right $(i+1)$ and left $(i-1)$ PEs with the time stamp of its LST $\tau_i(t)$.

The iterative process (1) can be rewritten as

$$\begin{aligned} \tau_i(t+1) = \quad & \tau_i(t) + \Theta\left(\tau_{i-1}(t) - \tau_i(t)\right) \\ & \times\Theta\left(\tau_{i+1}(t) - \tau_i(t)\right)\eta_i(t) \end{aligned} \tag{2}$$

using the Heaviside step function Θ.

It was argued by Korniss *et al* [5] that the coarse-grained time-horizon $\hat{\tau}$, $\left(\hat{\phi} = \partial\hat{t}/\partial x\right)$ obeys the Kardar-Parisi-Zhang (KPZ) equation

$$\frac{\partial\hat{\tau}}{\partial\hat{t}} = \frac{\partial^2\hat{\tau}}{\partial x^2} - \lambda\left(\frac{\partial\hat{\tau}}{\partial x}\right)^2 , \tag{3}$$

which should be extended with the noise to capture the fluctuations.

Introducing the local slopes $\phi_i = \tau_i - \tau_{i-1}$, the density of local minima can be written as

$$u(t) = \frac{1}{L}\sum_{i=1}^{L}\Theta\left(-\phi_i(t)\right)\Theta\left(-\phi_{i+1}(t)\right) \tag{4}$$

and its average

$$\langle u(t)\rangle = \langle\Theta\left(-\phi_i(t)\right)\Theta\left(\phi_{i+1}(t)\right)\rangle \tag{5}$$

is the mean velocity of the time horizon, equal to 0.246410(7). Hence, the efficiency of the algorithm (in this worst-case scenario) is about 25 per cent. The algorithm is free of deadlock, since at least the PE with the absolute minimum LST can proceed.

It is argued in [7] that the width of the LST profile growth as $\sqrt{N}$ with the number N of processing elements, thus the synchronisation in conservative algorithm becomes a problem.

4. Optimistic algorithm and directed percolation problem

The model of time evolution of LST profile in optimistic algorithm can be described as simulation process consisting of two steps [8]. The first step is the optimistic unrestricted growth at which a system evolves forward in time and the second is the backward algorithm of sending anti-messages. For this purpose one may introduce two parameters, J and K associated with two steps of optimistic scheme. Namely, at the first step one evaluates the time horizon by updating all the randomly chosen LST $\tau_i(t+1)$ with J following Poisson distribution, and every PE can be chosen with the non-zero probability. Then we have to relax K times the LST of all PE_i to LST of the nearest left (right) PE_{i-1} (PE_{i+1}), with K following Poisson distribution. It was found in [8] that the average of time profile $u(t)$ evolves proportionally to $(q-q_c)^\nu$, with $q = J/(J+K)$, with a value of $\nu = 1.74$ close to the critical exponent of directed percolation [9]. The value of $q_c \approx 0.23$ corresponds to the relation between the parameters of the model $K \approx 2.39J$. It should be mentioned that the width of the LST profile acts better for values of q faraway from the critical value q_c. When close to q_c the system practically does not evolve in time. The growth of width provides an analogy with the roughening transition, which is in the same universality class as a directed percolation [9].

Journal of Physics: Conference Series **640** (2015) 012065 doi:10.1088/1742-6596/640/1/012065

5. Nonlocal message exchange and stabilisation of the width growth

What happens to the evolution of LSTs profile in the case of long-range interactions? This question was answered in relation to the conservative algorithm in paper [10] through the introduction of random interactions of PEs throughout the whole system. This construction corresponds to the famous small-world network problem [11]. It was shown in [10] that an addition of only 10 per cents of random and non-local interactions between processing elements i and j, chosen randomly from $\{1, 2, \ldots, N-1, N\}$ processing elements, does stabilise the width of the LSTs profile. This scheme keeps property of finite evolution speed, typical for conservative scheme, and therefore, it is dead-lock free.

6. Discussion

In this paper we extend the concept of processing elements and local simulation times in order to take into account the current state of high performance hardware and software computing systems. We review the results of the association of PDES evolution with the physical processes, and emphasise that the results received from physical problems may be useful for the analysis of some classes of parallel algorithms. It seems that PDES algorithms may be extended to the simulation of systems with non-local interactions, and this investigation is in progress.

7. Acknowledgments

This work was supported by the grant of the Russian Science Foundation 14-21-00158.

References

[1] Fujimoto R M 1990 *Commun. ACM.* **33** 31
[2] Jefferson D R 1985 *Assoc. Comput. Mach. Trans. Programming Languages and Systems* **7** 404
[3] Shchur LN and Novotny M A 2004 *Phys. Rev. E* **70** 026703
[4] Top500 List `http:\\top500.org`
[5] Korniss G, Toroczkai Z , Novotny M A, and P.A. Rikvold P A 2000 *Phys. Rev. Lett.* **84** 1351
[6] Kardar M, Parisi G, and Zhang Y C 1986 *Phys. Rev. Lett* **56** 889
[7] Guclu H , Korniss G, Novotny M A, Toroczkai Z, and Rácz Z 2006 *Phys. Rev. E* **73** 066115
[8] Shchur L N and M. A. Novotny M A *unpublished.*
[9] Alon U, Evans M R, Hinrinchsen H, and Mukamel D 1996 *Phys. Rev. Lett.* **76** 2746
[10] Korniss G, Novotny M A, Guclu H, Toroczkai Z, Rikvold P A 2003 *Science* **299** 677
[11] Watts D J and Strogatz S H 1998 *Nature* **393** 440

Real-time and real-space program tuned in K-computer

Y. Zempo[1], N. Akino[2], M. Ishida[2], E. Tomiyama[3] and H. Yamamoto[3]

[1] Computer and Information Sciences, Hosei University, 3-7-2 Kajino, Koganei 184-8584 Japan
[2] Sumitomo Chemical, Co., Ltd., 6 Kitahara, Tsukuba 300-3264 Japan
[3] Research Organization for Information Science and Technology, 1-5-2 Minatojima-Minami, Chuo, Kobe 650-0047 Japan

E-mail: zempo@hosei.ac.jp

Abstract. We have transported the code from Earth Simulator (ES) to K-Computer (KC) keeping a scalability. The base code achieved remarkable linear scalability and significant parallelization efficiency in EC before. The reason of success is that, taking into account the architectural differences between ES and KC, we improve the program step by step. Actually, the program showed a 37% improvement in the total performance with the implementations such as hybrid OpenMP and MPI parallelization, and the optimal rank mapping, and the MPI communication concealment, including removing the unnecessary processes such as zero clear and data copy.

1. Introduction

Time-dependent density functional theory (TDDFT) is one of the most prominent and widely used methods for calculating excited states of medium-to-large molecules, and it is recognized as a powerful tool for studying electronic transitions of molecules [1]. In our calculations, the real-time and real-space technique is adopted in solving equation by the finite difference approach [2]. In particular, the real-space approach is suitable for large-scale parallel computing, and also allows the capture of a clear physical image. Within the framework of this approach, we can solve for the wavefunctions on the grid with a fixed domain, which encompasses the physical system of interest [3].

We have transported the code from the first generation Earth Simulator (ES) to K-Computer (KC). There are architectural differences between ES and KC: In ES each node has eight vector processors, and there is a high speed inter and intra-node communication; In KC, each nodes has eight cores per processor and each node connected by a special interconnect. Taking account of these differences, we have applied the following implementations and obtained the 37% better total performance.

First, we have stated with OpenMP and MPI hybrid parallelization, and optimal rank mapping, where the zyx-mapping is adopted in the 3D parallelization. It is fairly effective to get rid of the unnecessary processes such as zero clear and data copy. In calculation loops, the improvement of thread efficiency provides an efficient OpenMP process. Introducing the MPI_test call, communication concealment in MPI processes works effectively in the code. It takes a role of a trigger for MPI_isend and MPI_irecv processes. The detail will be introduced in the presentation.

The paper is organized as follows. The base code and the numerical detail is presented in the next section. The parallelization techniques and their achievements are shown in section 3. Finally, we summarize and discuss our conclusions in the last section.

2. Base code and numerical details

2.1. Time-dependent density-functional theory

In this section, we briefly present our TDDFT calculation procedure. One of the most successful ways of obtaining the electronic structure from the first-principles is the use of computational approaches based on the density functional theory (DFT) [4] with the local density approximation (LDA). The total energy of the ground state can be derived from the Kohn-Sham equation (KS) [5]. Although, for the excited states, it is much less successful in describing the optical responses and the excitation spectra, this difficulty is, in principle, solved by the extension of DFT to the time-dependent theory. Its foundation was established by Runge and Gross [1]. In analogy to the time-independent case, the TDDFT equation of motion coupled with pseudopotentials is given by

$$\Big\{ -\frac{1}{2}\nabla^2 + V_{ion}^{ps}(\mathbf{r}) + V_H(\mathbf{r},t) + V_{XC}[\rho(\mathbf{r},t)] + V_{ext}(\mathbf{r},t)\Big\}\psi_i(\mathbf{r},t) = i\frac{\partial}{\partial t}\psi_i(\mathbf{r},t)\ , \tag{1}$$

where V_{ion}^{ps} is an ionic pseudopotential, V_H is the Hartree potential, and V_{XC} is the exchange correlation potential. Since the exact time-dependent xc kernel is not known, the originally non-local time-dependent xc kernel is replaced with a time-independent local one. This is considered as reasonable when the density varies slowly with time. This approximation allows the use of a standard local ground-state xc functional in the TDDFT frame work. The Hartree and exchange-correlation potentials can be determined from the electronic charge density, $\rho(\mathbf{r},t) = \sum_i |\psi_i(\mathbf{r},t)|^2$. The summation is over all occupied states i. The Hartree potential is determined by $\nabla^2 V_H = -4\pi\rho$, and as the xc potential V_{XC}, the usual local density approximation (LDA) is used in our study. For the ionic potential, we employ the pseudopotential V_{ion}^{ps} in the separable form so that the only valence electrons are considered [6]. Prior to the calculation of optical responses, we first solve the usual, time-independent formulation of the pseudopotential-DFT method [7] to obtain the optimized electronic structure [2]. Then, we apply an external field V_{ext} to the system as a perturbation and follow the linear responses of the system in real time.

This method is effectively used for the cases where the potential is time-dependent, e.g. the time-dependent behaviors of electrons in oscillating electric and magnetic fields, and excited-state reactions.Empirically, we know that these can be described fairly well by TDDFT. Although the xc functional we adopted has been developed with respect to the electronic ground states, we have also employed in our practical calculations.

In our calculations, the real-time and real-space technique is employed in solving eq.(1) by the finite difference approach [8] without using explicit bases such as plane waves.and gaussians. Within the frame work of this approach, we can solve for the wave functions on the grid with a fixed domain, which encompasses the physical system of interests. The uniform grid is used in our study for the simplicity.

The wave functions are evolved by the time evolution operator, $\psi(t) = \exp[-iHt]\psi(0)$, with the initial wave function at $t = 0$,

$$\psi_i|_{t=0} = e^{ik_z z}\psi_i(0), \tag{2}$$

where H is the Hamiltonian of the system, and k_z is a small wave number corresponding to the external perturbation in the z direction. In the linear response, the time-dependent polarizability is proportional to the dipole matrix element $\mu(t) = \langle\psi(t)|\sum_i z_i|\psi(t)\rangle$, where z_i with $i = 1$, 2 and

3 represents x, y and z, respectively. The frequency-dependent polarizability in the z direction $\alpha_z(\omega)$ is then obtained as the time-frequency Fourier transformation of $\mu_z(t)$,

$$\alpha_z(\omega) = \frac{1}{k}\int dte^{-i\omega t}\mu_z(t). \tag{3}$$

The polarizability $\alpha(\omega)$ is given by the orientation average, $\alpha = (\alpha_x + \alpha_y + \alpha_z)/3$. The optical strength function, $S(\omega)$, is related to the imaginary part of the polarizability,

$$S(\omega) = \frac{2\omega}{\pi}\text{Im}\alpha(\omega). \tag{4}$$

2.2. Achievement of the base code running on ES
Base code was already tuned in the first generation Earth Simulator using MPI, and showed a good linearity in Fig. 1, which was applied to the calculation of fluoren six mulitimer. It performed at about 8 TFlops using 243 nodes in ES.

The ES is consisted of 640 processor nodes (PNs) connected by 640×640 single-stage crossbar switches, which control with inter-node data communications at 12.3 GB/s bidirectional transfer rate [9]. Thus the total bandwidth of inter-node network is about 8 TB/s. In each node, There were eight vector processors, which has the memory bandwidth at 256 GB/s. All CPUs in 640 nodes in ES seemed flatly connected. When we tuned our code in ES, we only paid an attention to usual vectorization for vector processors, and simple MPI libraries in the parallelization.

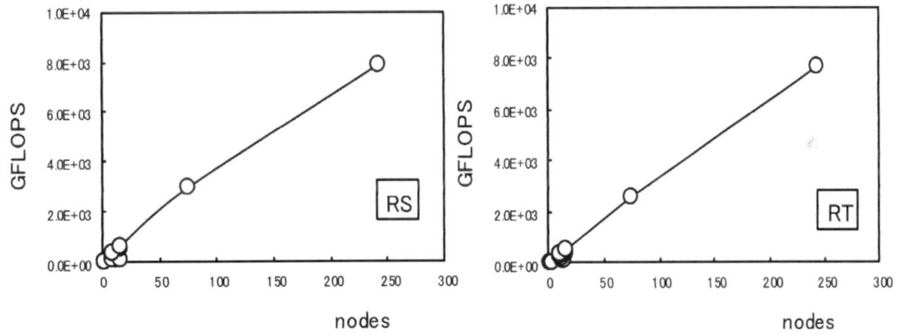

Figure 1. Scalability of the base code in ES. RS means the ground stat calculation, before the time evolution calculation RS. Almost linear scaling was observed up to 243 nodes.

3. Tuning detail
The system uses scalar CPUs using SPARC64 VIIIfx, which has 8 cores and shows the performance 128 Gflops. The memory bandwidth is 64 GBs, which is fairly small to the CPU performance, compared to that of ES. To achieve high performance and high scalability to this ultra-large-scale system, the CPU additionally supports SIMD processing [10].

In this section, we would like to focused on those which significantly work for improve our calculation code, although there are, of course, other various interesting features involved in the system. One of them is its network topology called Tofu.

Journal of Physics: Conference Series **640** (2015) 012066 doi:10.1088/1742-6596/640/1/012066

3.1. MPI and OpenMP hybrid
As the basis of our parallelization, we have stated with OpenMP and MPI hybrid parallelization. In each node, we used OpenMP We have set up a sample run, which takes 510 s witout taking into account I/O. Our tuning results are summarized in Fig.5. Each technique we adopted to improve our code is described in the following subsections.

3.2. MPI rank mapping
This thesis focuses on techniques of task mapping for solving problems on parallel computers with many processors on multi-dimensional networks like K-computer. The application problems can be decomposed to subtasks with the computational load. Our real space calculation is usually devided by the xyz directions into a rectangle shape of a group of meshes. According to the direction, we may use several groups which can be mapped to MPI ranks. This MPI rank mapping should be effectively adjust to the calculation loop in practical program, which is different to the natural order of the direction. In our optimization, the zyx rank mapping was significantly worked. It reduces 13% the calculation time, which is 75 s from 510 s mentioned in the previous section.

3.3. Removing unnecessary zero-clear
Traditionally, arrays are zero-cleared prior to MPI communications because of the safety in the data. In Fortran90, it is too easy to write this memory swipe with one line. Sometimes, one is not aware of this type of simple wrong implementation. However, the large array may be handled on the fairly small band width ration. It takes a large time to realize it. Removing this type of zero clear process from the algorithm, we can obtain the program, which runs 8 % faster (40 s shorter) than above.

3.4. Software operational optimization
In this section, we would like to mention about the software optimization such as redundant arithmetic calculation, getting rid of valuables from several loops, which can be considered as constant valuables inside, and replace the directive of OpenMP to work OpenMP thread to be optimized. In particular, we have introduce the specific CPU feature, i.e. called the SECTOR cache. It is a selectable mechanism in the user's program for reusable data. The SECTOR cache can be divided into two sectors. Reusable data by allocated on the SECTOR1 is prevented to be replaced on the cache memory. With these improvements, we reduced 46 s (9 %) faster.

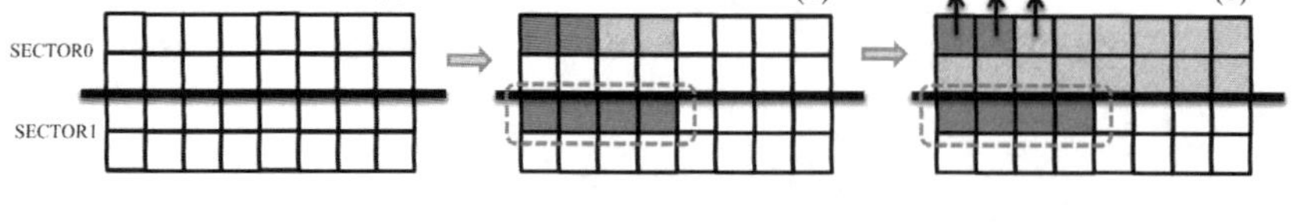

Figure 2. Image of using SECTOR cache. (a) Initial sage of both SECTOR caches, (b) The cache SECTORs are getting filled with data, but SECTOR1 is keeping the data. (c) When SECTOR0 is completely filled with data, the data in SECTOR0 will be swiped out. On the other hand, the data allocated in SECTOR1 is not affected by SECTOR0.

3.5. MPI communication concealment
MPI concealment is expected to make the parallel calculation efficient. During the operation, MPI_isend and MPI_ireceive, the other MPI process is expected to work. However, this kind

of operation, we need some trigger. Otherwise, the communication is still waiting for the next MPI_isend and MPI_ireceive call. To realize the concealment, we have place a MPI_test call each different type of MPI_isend or MPI_ireceive call. With this MPI communication concealment, the code runs 2.3 % faster before. It actually corresponds to the reduction of 12 s.

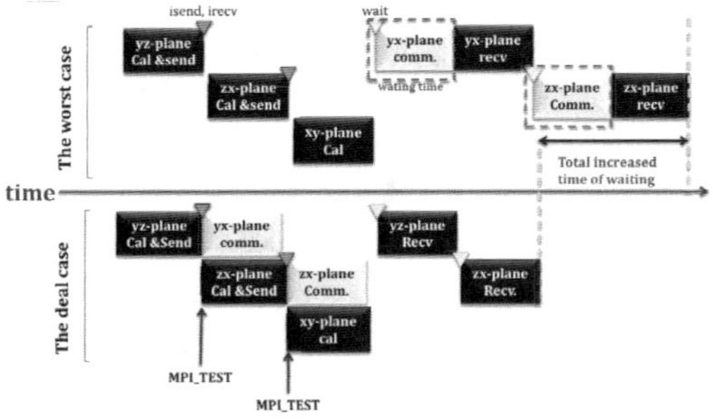

Figure 3. Image of MPI communication concealment. MPI_test will be a trigger to start the background communication.

3.6. Removing unnecessary copies

The memory copies is practically not a big matter in ES. However, it quite affects the calculation in K-computer, which has relatively small bandwidth. In this type of computer system, the leapfrog calculation in the time evolution provides significant effects. It is also related the memory access in Sec.3.3. It will be a trend in the recent parallel computation using huge number of CPUs, in which the interconnect network is getting multiple-stage. In the improvement shown in Fig.4, we removed unnecessary copies in time evolutions. Thus, we made up for the lost time 25 s (4.9%). Including the previous Sec.3, we have totally 37 % improvement. It corresponds to

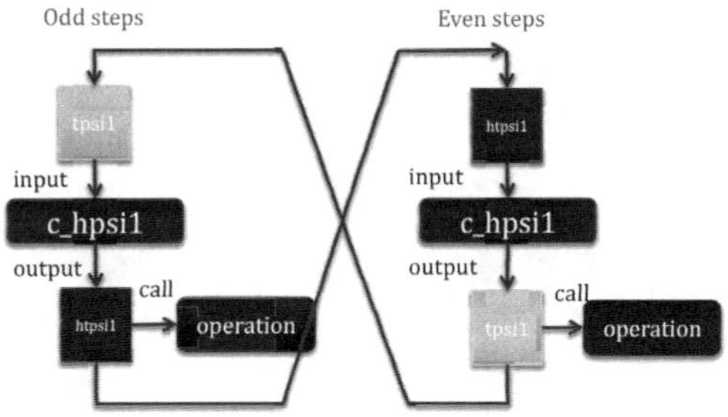

Figure 4. Time evolution scheme with leapfrogging. Calculated data is not necessary to be copied for the next step.

the reduction of the calculation time from 510 s to 322 s, as summarized in Fig. 5.

4. Concluding remarks

We transported our code from ES to K-computer with fairly good efficiency. There are several architectural differences between ES and K-computer system. Some of them are due to the memory bandwidth and the operational optimization. In the trend of the large scale computer system, in particular, many CPUs are required to realize HPC (High Performance Computing) in various scientific fields. Then, the more larger number of CPUs, the system might be the smaller memory bandwidth. In the parallel programing, we have to take care of this tendency.

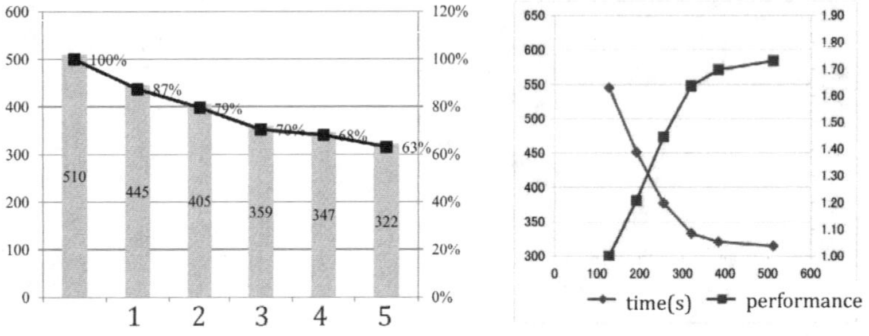

Figure 5. Tuning techniques and the performance of the code. (Left) The tuning technique and its effect. (Right) Scaling with the increase number of the nodes. The performance is measured by the results using 128 nodes.

Acknowledgments
The authors would like to express their gratitude to Professor K. Yabana for providing his code as our base code and also for useful discussions and encouragement. This work was partially supported by JSPS Grants-in-Aid for Scientific Research (C) Grant number 25390158, Sumitomo Chemical Co., Ltd. and Simulatio Corporation.

References
[1] Runge E and Gross E K U 1984 *Phys. Rev. Lett.* **52**, 997
[2] Yabana K and Bertisch G F 1996 *Phys. Rev.* B**54**, 4484
[3] Zempo Y, Akino N, Ishida M, Ishitobi M and Kurita Y 2008 *J. Phys. Condens. Matter* **20**, 064231
[4] Hohenberg P and Kohn W 1964 *Phys. Rev.* **136**, B864
[5] Kohn W and Sham L. J 1965 *Phys. Rev.* **140**, A1133
[6] Troullier N and Martins J L 1991 *Phys. Rev.* B**43**, 1993
[7] e.g. Pickett W E 1989 *Comput. Phys. Reports* **9** , 115
[8] Chelikowsky J, Troullier N, Wu K and Saad Y 1994 *Phys. Rev.* *B***50**, 11355
[9] http://www.jamstec.go.jp/es/jp/es1/system/index.html
[10] http://www.aics.riken.jp/en/k-computer/system

Comparison of meshfree and mesh-based methods for boundary value problems in physics

P. Niraula, Y. Han and J. Wang
Department of Physics, University of Massachusetts Dartmouth, Dartmouth, MA 02747, USA
E-mail: jwang@umassd.edu

Abstract. Meshfree methods using basis expansion are common in computational physics, but their efficiency and accuracy depend on the quality of basis sets. We describe the application of radial basis functions (RBF), a class of basis functions attracting considerable interest and showing promising results. In the RBF approach, the solutions are interpolated over a chosen set of RBFs scattered over the domain. The value of a given RBF depends on the distance to the center, in contrast to the usual approach where the basis functions depend on the position. The advantage of the RBF method is that it can be efficient, simple to implement, and easily adaptable to irregular domains or higher dimensions. We study the application of this meshfree method to boundary value problems, including electrostatics and quantum systems. We show that the method is robust and flexible, and can be a compelling alternative to mesh-based methods.

1. Introduction

Computational modeling of physical systems often requires us to solve problems expressed in partial differential equations (PDE) such as the Laplace or Schrödinger equations. One family of numerical methods involves the discretization of space into regular grids or meshes. Common examples include the finite difference and finite element methods (FEM). Such mesh-based methods are direct and usually easy to use, except for irregular domains or high accuracy control. Nonetheless, they form a baseline for further improvement.

Another family of methods takes a meshfree approach. Rather than discretizing space, the solutions are approximated by a set of basis functions. Eigenfunction expansion is a familiar example. Recently, radial basis functions (RBF) have been shown to be good candidates for meshfree solutions.

We describe a meshfree RBF method in which the solutions are interpolated with a set of RBFs at scattered centers. We first discuss the interpolation of known functions in terms of RBFs to analyze general properties such as the size and shape of the RBF basis. We then apply the RBF method to solve PDEs such as the Poisson equation by collocation method which evaluates the differential operators at the RBF centers of the interpolant. The results show that the method is robust and flexible, and can be advantageous in comparison to mesh-based methods such as FEM in some cases, including scalability to higher dimensions.

2. Radial basis function interpolation

The basic idea of meshfree methods is scattered data interpolation [1, 2]. We assume that, like the FEM method [3, 4], the solution u can be approximated by expansion in a distance-based

RBF set $\phi_i(\vec{r})$ as

$$u(\vec{r}) = \sum_{i=1}^{N} a_i \phi_i(\vec{r}). \tag{1}$$

where a_i are the expansion coefficients.

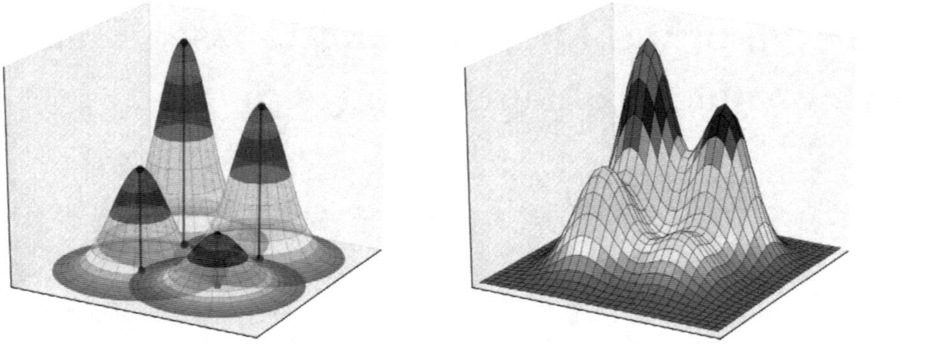

Figure 1. Scattered RBFs (left) and the resultant interpolation (right).

There are many possible functional forms for ϕ_i. We list two commonly used RBFs as example, the Gaussian (GA) and the multiquadric (MQ),

$$\text{GA:} \quad \phi_i(\vec{r}) = e^{-\epsilon^2|\vec{r}-\vec{r}_i|^2}, \qquad \text{MQ:} \quad \phi_i(\vec{r}) = \sqrt{1+\epsilon^2|\vec{r}-\vec{r}_i|^2}, \tag{2}$$

where ϵ is the shape (control) parameter, and $\vec{r}_i$ are the nodes (centers) scattered over the domain of interest.

We can use Eq. (1) for scattered data interpolation (Fig. 1). For example, suppose we have four ($N = 4$) known data points, u_i at $\vec{r}_i$, $i = 1$ to N. The scattered data points are not restricted to any particular geometry. The coefficients a_i can be determined by requiring

$$u(\vec{r}_i) = u_i, \tag{3}$$

where we have taken the nodes to coincide with the scattered data points. The result can be expressed as a matrix equation

$$\begin{bmatrix} \phi_1(\vec{r}_1) & \phi_2(\vec{r}_1) & \phi_3(\vec{r}_1) & \phi_4(\vec{r}_1) \\ \phi_1(\vec{r}_2) & \phi_2(\vec{r}_2) & \phi_3(\vec{r}_2) & \phi_4(\vec{r}_2) \\ \phi_1(\vec{r}_3) & \phi_2(\vec{r}_3) & \phi_3(\vec{r}_3) & \phi_4(\vec{r}_3) \\ \phi_1(\vec{r}_4) & \phi_2(\vec{r}_4) & \phi_3(\vec{r}_4) & \phi_4(\vec{r}_4) \end{bmatrix} \begin{bmatrix} a_1 \\ a_2 \\ a_3 \\ a_4 \end{bmatrix} = \begin{bmatrix} u_1 \\ u_2 \\ u_3 \\ u_4 \end{bmatrix}. \tag{4}$$

Eq. (4) can be solved for RBF expansion coefficients a_{1-4}. The matrix entries are defined at the RBF collocation points given by

$$\phi_i(\vec{r}_j) = \phi(|\vec{r}_i - \vec{r}_j|) = \phi_j(\vec{r}_i), \tag{5}$$

where ϕ is either a GA or MQ RBF in Eq. (2). Because the RBFs are based on the radial distance $r_{ij} = |\vec{r}_i - \vec{r}_j|$ from the nodes, the system matrix is positive and definite. It can also be shown that the determinant is invariant under exchange of any pair of nodes because the scalar distances r_{ij} stay the same. Therefore, irrespective of node locations, the system matrix is well-conditioned, and the solution is assured to be nonsingular. In fact, the tent basis functions in 1D FEM are linear RBFs, and share some of above traits, except they are not infinitely differentiable like the GA or MQ. In general, however, if the basis functions were position (rather than distance) based, the above mentioned properties would no longer hold. In this respect, scattered data interpolation by RBFs has unique advantages.

Fig. 1 illustrates an example consisting of four Gaussian RBFs centered at the scattered data points. After obtaining the expansion coefficients a_{1-4} (vertical lines at the nodes in Fig. 1, left), the resultant interpolant Eq. (1) can be used to approximate the solution over the whole domain.

Specifically, we analyze RBF interpolation in terms of the number of RBFs, N, and the shape parameter ϵ in a 1D test case. We use the test function $u(x) = \exp(\sin \pi x)$, with $|x| \leq 1$. For N nodes, we solve the $N \times N$ linear system Eq. (4) with the GA RBFs.

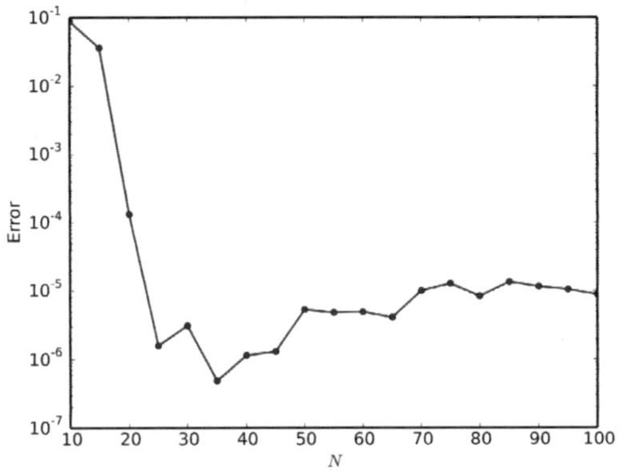

Figure 2. Interpolation error as a function of N, the number (size) of the basis set, at a given shape parameter.

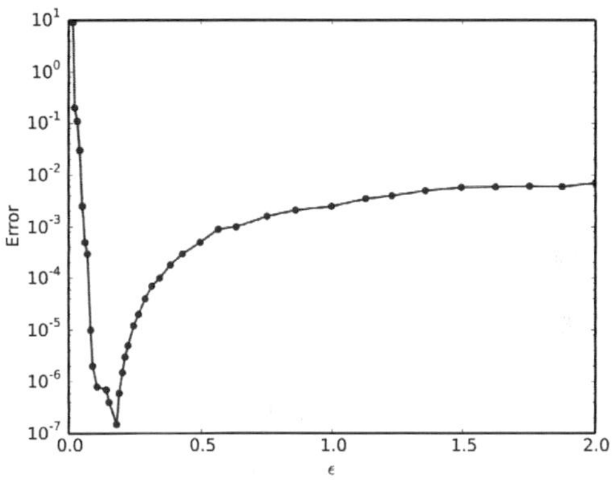

Figure 3. Interpolation error as a function of the shape parameter ϵ, at a given basis size N.

In Fig. 2, we show the maximum absolute error (upper bound) in terms of the number of nodes, N, for a fixed shape parameter $\epsilon = 1$. For simplicity, the nodes are uniformly distributed in the interval $x \in [-1, 1]$. Initially, the error decreases rapidly with increasing N. Denser nodes result in more accurate representation as expected. However, after certain $N \sim 40$, the error remains flat, and accuracy does not improve. We see that there is a limit to the size of the basis set beyond which no benefit can be obtained.

We can also keep the size N constant and vary the shape parameter ϵ. Fig. 3 displays the interpolation error as a function of ϵ for a fixed $N = 40$. The error is large for both large and small values of ϵ. There is an optimal shape parameter which minimizes the error, $\epsilon \sim 0.2$ in this case. In general, the optimal ϵ value depends on N. For best results, we need to find the best combination of N and ϵ in practice.

3. RBF collocation method

The RBF collocation method is based on differentiating a RBF interpolant, from which the differentiation operator matrices can be built [5]. As an example, we use the collocation method to solve the 2D Poisson equation with the Gaussian RBF (equally applicable to 3D).

We consider a region surrounded by an arbitrary boundary (the domain). Let N be the total number scattered nodes. Of these, we assume the number of internal nodes is n, and the number of boundary nodes is $N - n$.

Substituting the interpolant Eq. (1) into the Poisson equation $\nabla^2 u = f$, we obtain

$$\nabla^2 u = \sum_i a_i \nabla^2 \phi_i(\vec{r}) = f(\vec{r}). \tag{6}$$

The collocation method, also known as Kansa's method [6], stipulates that the Poisson equation Eq. (6) be satisfied at the internal nodes $\vec{r}_j$, $j \in [1, n]$, and simultaneously, the boundary conditions be also satisfied at the boundary nodes $\vec{r}_j$, $j \in [n+1, N]$.

To be concrete, we assume Dirichlet boundary conditions. We have for the collocation method

$$\sum_i a_i \nabla^2 \phi_i(\vec{r})|_{\vec{r}=\vec{r}_j} = f(\vec{r}_j), \quad j = 1, 2, ..., n, \tag{7a}$$

$$\sum_i a_i \phi_i(\vec{r}_j) = b_j, \quad j = n+1, n+2, ..., N, \tag{7b}$$

where b_j are the potential values on the boundary. Casting Eq. (7a)–(7b) into matrix form we have

$$A\mathbf{a} = \mathbf{b}, \tag{8}$$

where $\mathbf{a} = [a_1, a_2, ..., a_N]^T$ is the column vector to be solved for, and

$$A = \begin{bmatrix} D_{11} & D_{12} & \cdots & D_{1N} \\ \vdots & \vdots & \vdots & \vdots \\ D_{n1} & D_{n2} & \cdots & D_{nN} \\ \phi_{n+1,1} & \phi_{n+1,2} & \cdots & \phi_{n+1,N} \\ \vdots & \vdots & \vdots & \vdots \\ \phi_{N1} & \phi_{N2} & \cdots & \phi_{NN} \end{bmatrix}, \quad \mathbf{b} = \begin{bmatrix} f_1 \\ \vdots \\ f_n \\ b_{n+1} \\ \vdots \\ b_N \end{bmatrix}, \tag{9}$$

where $f_j = f(\vec{r}_j)$ is the charge density at node j, and the differentiation operators and RBF values are defined as

$$D_{ij} = \nabla^2 \phi_i(\vec{r})|_{\vec{r}=\vec{r}_j} \equiv \nabla_j^2 \phi_i(\vec{r}_j), \quad \phi_{ij} \equiv \phi_i(\vec{r}_j). \tag{10}$$

The system matrix A is square but asymmetric. It can be evaluated given the specific RBF (GA or MQ) and the nodes. If the operator is different than the Laplacian, it can be modified in a straightforward way. The vector $\mathbf{b}$ consists of the charge densities f_j and the boundary values b_j. Once the necessary parameters are known, we can obtain the solutions $\mathbf{a}$ by solving Eq. (8) with a standard matrix solver.

4. Application and results

Let us apply the RBF collocation method to solve the special case of $f = 0$ in Eq. (6) (Laplace equation) with Gaussian RBFs. The system geometry is a disk (roughly an octagon at the chosen spacing) in the center of a square box, Fig. 4. We assume the box is kept at zero potential and the disk at a constant potential (say 1). We simply distribute the nodes uniformly over the domain (23×23), though other choices are also possible, e.g., the Halton sequence. The number of internal nodes is $n = 344$, and the number of boundary nodes is $N - n = 185$. The $\mathbf{b}$ vector Eq. (9) consists of zeros in the top n rows and ones in the bottom $N - n$ rows.

The basic differentiation operator is

$$\nabla^2 \phi_i(\vec{r}) = e^{-\epsilon^2 |\vec{r}-\vec{r}_i|^2} = 4\epsilon^2 e^{-\epsilon^2 |\vec{r}-\vec{r}_i|^2} \left(\epsilon^2 |\vec{r} - \vec{r}_i|^2 - 1\right). \tag{11}$$

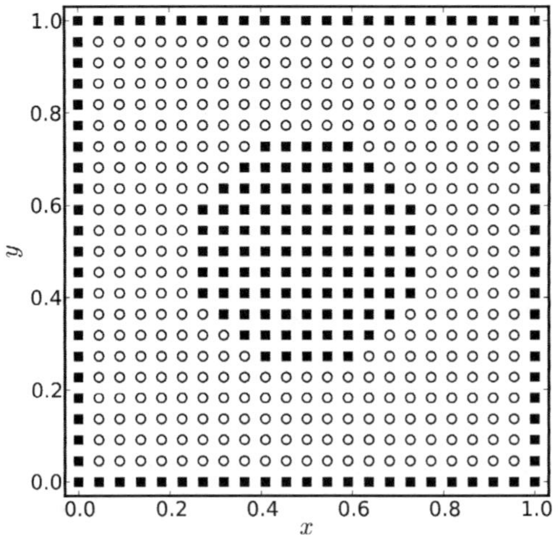

Figure 4. The disk-in-a-box configuration. There are $n = 344$ internal nodes ($\circ$) and 185 boundary nodes ($\blacksquare$) for a total of $N = 529$ nodes. The potential is constant on the octagon and zero along the walls.

It can be readily evaluated to form the system matrix A given the nodes and the shape parameter ϵ. As seen from Fig. 3, the value of ϵ is critical to the accuracy and success of the RBF method. The accuracy depends on ϵ, sometimes very sensitively, given the choice of RBF type and the placement of the nodes [7]. Typically, the extreme values, either very small or very large ϵ, lead to very bad approximations. It is unknown, however, where the sweet spot is in general. Sometimes we must find it by trial and error (such as in Fig. 3 for the test function). From the applied physicists' perspective, we resort to physics to guide us. Suppose h is the typical distance between the nodes. We expect that the resolution should be such that $\epsilon h \sim 1$, or $\epsilon \sim 1/h$.

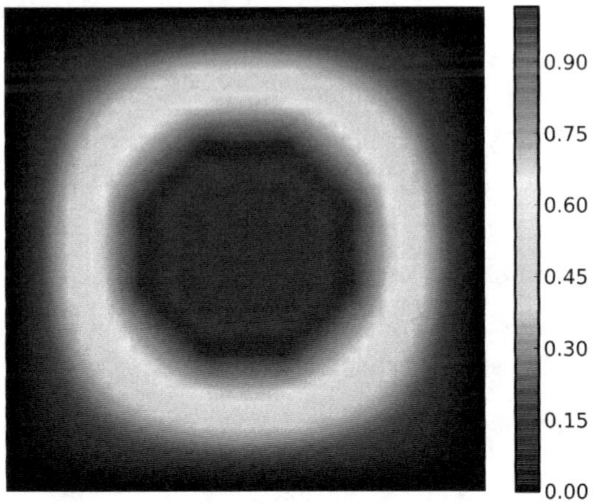

Figure 5. The potential for the disk-in-a-box configuration.

Fig. 5 shows the results of the RBF collocation method. The value $\epsilon = 20$ is used (the actual spacing is $1/22$). However, test shows that values from $10-25$ works equally well. The potential is highest, and approximately constant, within the disk (or the octagon), and decreases to zero toward the walls as expected. It also shows correct symmetry from the configuration.

We also apply the collocation method to a different geometry in Fig. 6 consisting of two opposing circular arcs held constant potential inside a grounded box. It is much easier to

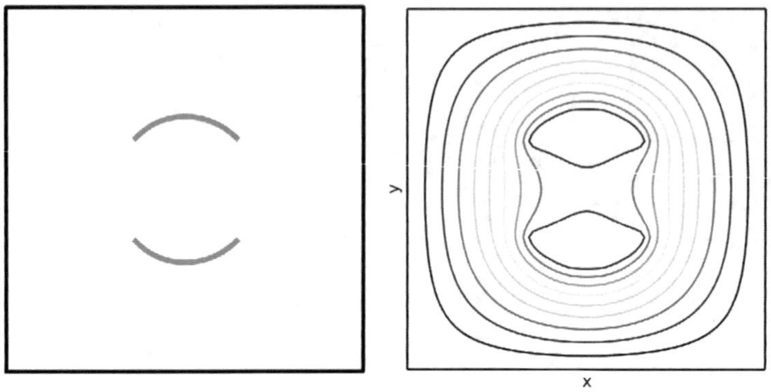

Figure 6. Circular arcs held at constant potential (left) and the potential contours (right).

distribute nodes along the arcs and in the domain with the RBF method than with a mesh-based method. The advantage to other irregular boundaries is evident.

Compared to FEM, the accuracy of the RBF method is good. Considering that sharp corners usually require higher density of nodes for the scattered data interpolation to be accurate, the RBF results with a modest number of uniform nodes in the present case are satisfactory. The RBF method is fast, and easier to use than FEM. The disadvantage is it is more difficult to control accuracy systematically.

With FEM, we can progressively reduce the error with smaller meshes (at least down to the limit before machine accuracy becomes a factor). In the RBF method, we expect that increasing the number of nodes will lead to more accurate results. However, if the number of nodes changes, we need to change the shape parameter accordingly, whose optimal value is not known a priori. And the situation may be different for different RBFs.

In summary, the meshfree RBF method can be an efficient method for problems involving PDEs. It is especially suitable for irregular boundaries and higher dimensions. For the case study with the Laplace equation, the RBF method performs equally well relative to mesh-based methods such as the FEM, but is easier to use. We are currently investigating the RBF method for other boundary value problems including quantum systems.

Acknowledgments

We would like to thank W. Li for coding help and useful discussions. The work was partially supported by TCR Engineering.

References

[1] M. D. Buhmann, *Radial basis functions: theory and implementations.* (Cambridge University Press, Cambridge, 2003).
[2] H. Wendland, *Scattered data approximation.* (Cambridge University Press, Cambridge), 2005.
[3] J. N. Reddy, *An introduction to the finite element method.* (McGraw-Hill, New York), 1993.
[4] J. Golde, J. Shertzer, and P. Oxley, Finite element solution of Laplaces equation for ion-atom chambers. *Am. J. Phys.*, **77**:81–86, (2009).
[5] G. E. Fasshauer, *Meshfree approximation methods with MATLAB.* (World Scientific, Hackensack, NJ), 2007.
[6] E. J. Kansa, Multiquadrics – a scattered data approximation scheme with applications to computational fluid dynamics II: Solutions to parabolic, hyperbolic, and elliptic partial differential equations. *Comp. Math. Appl.*, **19**:147–161, (1990).
[7] E. Larsson and B. Fornberg, A numerical study of some radial basis function based solution methods for elliptic PDEs. *Comp. Math. Appl.*, **46**:891–902, (2003).

Scar State on Time-evolving Wavepacket

Mitsuyoshi Tomiya[1], Hiroyoshi Tsuyuki[1], Kentaro Kawamura[1], Shoichi Sakamoto[1], Eric J. Heller[2]

[1] Faculty of Science and Technology, Seikei University, Kichijoji-Kitamachi 3-3-1, Musashino-shi, Tokyo, 180-8633, JAPAN

[2] Department of Physics, Harvard University, Cambridge, Massachussetts 02138

E-mail: tomiya@st.seikei.ac.jp

Abstract. The scar-like enhancement is found in the accumulation of the time-evolving wavepacket in stadium billiard. It appears close to unstable periodic orbits, when the wavepackets are launched along the orbits. The enhancement is essentially due to the same mechanism of the well-known scar states in stationary eigenstates. The weighted spectral function reveals that the enhancement is the pileup of contributions from scar states on the same periodic orbit. The availavility of the weighted spectrum to the semiclassical approximation is also disscussed.

1. Introduction

The time-average of absolute squares of a time-evolving wavefunction in the stadium billiard is investigated, launching a Gaussian wavepacket as the initial state. It's been an unique tool to detect dynamical properties of quantum systems. Nowadays nano-electronic devices, e.g. single electron transisters, quantum ratchets etc., are getting more and more realistic. Then dynamical behavior of an electron inside the (sub)nano-structure has been actively studied to clarify the physical properties of the structure[1, 2].

The novel feature of the nodal pattern of the stationary eigenstates in closed quantum systems was found about three decades ago[3]. It is drastically different in chaotic and integrable systems. In particular, the eigenstates of the chaotic systems often have the scars, which are the enhancement of quantum wavefunctions along classical unstable isolated periodic orbits.

Soon after the discovery, the semiclassical approximation has paved the way to understand the scars. Theories in phase space [4, 5] and in coordinate space[6] elucidated the contribution of the unstable periodic orbits in the scar states. Furthermore, in the theory of Heller[4], the propagation of a wavepacket near the unstable periodic orbit also uncovers the nature of the scar.

2. Gaussian Wavepacket

To investigate the properties of the time-evolution, the Gaussian wavepacket has been often used[7, 8, 9, 10]. Its initial form in 2D region is

$$\Psi_0(\mathbf{r}) = \frac{1}{\sigma_0\sqrt{\pi}} exp\left[\frac{i}{\hbar}\mathbf{p}_0(\mathbf{r}-\mathbf{r}_0) - \frac{(\mathbf{r}-\mathbf{r}_0)^2}{2{\sigma_0}^2}\right], \tag{1}$$

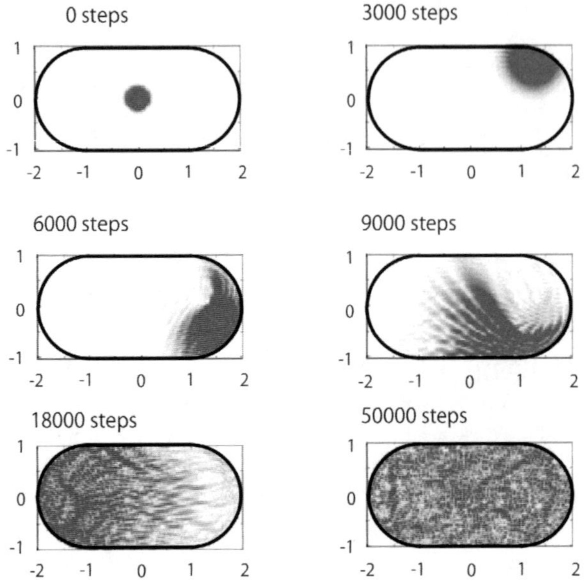

Figure 1. An example of the time-evolution of the Gaussian wavepacket is presented. The initial Gaussian wavepacket (Eq.1) is located at the center of the stadium. The launching angle is 30°. The angle is measured in the counterclockwise direction from the long axis of the stadium. After about 50000 steps the wavefuction almost diffuses all over the stadium.

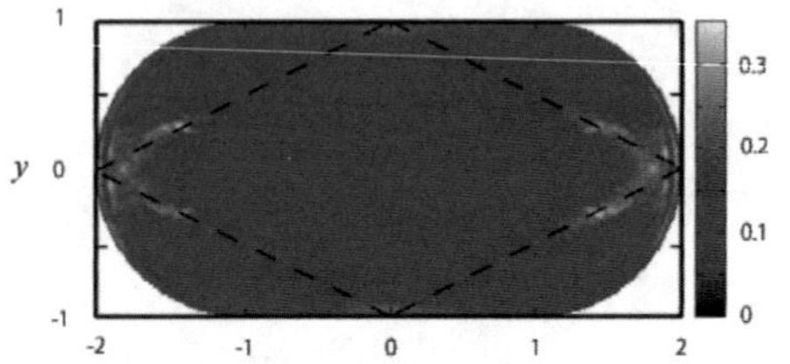

Figure 2. The accumulation $A(\mathbf{r})$ of the time-evolving wavepacket in stadium billiard under the condition $|\mathbf{p}_0| = 50$, $\sigma_0 = 0.1$. It is launched from (1, 0.5) and the launching angle θ is set by $tan\theta = -1/2$, which is defined in the counterclockwise direction from the long axis of the stadium. The broken lines represents the corresponding unstable periodic orbit. It shows a rhombus shape.

where $\mathbf{r} = (x, y)$ is a point inside the nano-structure and $\mathbf{r_0} = (x_0, y_0)$ is the initial location of the center of the wavepacket, and $\mathbf{p}_0 = (p_{0x}, p_{0y})$ is the initial momentum of the center. The standard deviation of the Gaussian: σ_0 determines the size of the wavepacket. The time-average of the absolute square of the evolving wavepacket serves its coordinate-dependence. Therefore the scar-like phenomena, i.e. the weak localizations, are much easier to be detected.

The wavepacket has to follow the time-dependent Schrödinger equation

$$i\hbar\frac{\partial\Psi}{\partial t} = (-\frac{\hbar^2}{2m}\nabla^2 + V)\Psi = H\Psi. \tag{2}$$

Adopting the 2×4 stadium billiard[11] as the 2D structure, the potential is set $V = 0$ inside the billiard and $V = \infty$ outside. The length of its long horizontal axis is 4 and the diameters of the circular parts are 2. For numerical calculation, the Schrödinger equation has to be discretized in both space and time, then the Hamiltonian H is realized by a large sized and yet very sparse matrix. We prepare the rectangle that circuscribes the stadium and install a lattice in it. The numbers of lattice points in the horizontal direction and the longitudinal direction are $N_x = 400$ and $N_y = 200$. Thus the lattice constant is $d = 0.01$. The time step is set $\Delta t = 10^{-4}$. Then the very short time-evolution can be well appoximated as

$$\Psi(\mathbf{r}, t + \Delta t) = exp(-\frac{i}{\hbar}H\Delta t)\Psi(\mathbf{r}, t) \approx \{1 - \frac{i}{\hbar}H\Delta t\}\Psi(\mathbf{r}, t). \tag{3}$$

We also checked that absence of the higher order does not affect the result of this work. The potential in the rectagle and outside the stadium billiard is set $V = 10^{300}$, instead of infinity which is unrealizable in numerical computation. This is almost the largest number for a double-precision floating-point number.

If the Gaussian wavepacket(1) is launched in 2D infinite flat space, at first it travels as a bunch with the initial velocity of the center of the wavepacket $\mathbf{v}_0 = \mathbf{p}/m$. Then, observing the absolute value of the wavepacket, its shape is always the Gaussian, however, its size is getting broader as $|\sigma(t)| = \sigma_0\sqrt{1+\left(\frac{\hbar t}{m\sigma_0{}^2}\right)^2}$. If time is large enough, $\sigma(t) \approx \frac{\hbar}{m\sigma_0}t$. In this work the wavepacket travels in the finite region, and repeated reflections on the boundary make the wavepacket eventually diffuse all around the billiard(Fig.1)(see also [12, 13]).

3. Dynamical Scar

Consequently, the time-evolving wavepacket seems to show no particular signature after diffusing all around the billiard(e.g., a snapshot of 5000th step in Fig.1). The wavefunction just billows all over and shows irregular and granular nodal patterns. On the contrary, the auto-correlation function already has revealed the long time recurrence[8]. Then we should turn to the accumulation, or the time-average of absolute square of the wavefunction: $A(\mathbf{r}) = \frac{1}{T}\int_0^T |\Psi(\mathbf{r},t)|^2 dt$. Numerically it is discretized as

$$A(\mathbf{r}_i) = \frac{1}{N_t}\sum_{j=1}^{N_t} |\Psi(\mathbf{r}_i, t_j)|^2, \tag{4}$$

at each mesh point $\mathbf{r}_i = (x_i, y_i)$, where the discretized time $t_j = j\Delta t$ and Δt is a time step. The integer N_t represents the number of the time steps. For our simulations, the natural system of units $\hbar = m = 1$ are adopted. Fig.2 shows the numerical results of the accumulation during $N_t = 10^6$ steps.

In spite of no significant characteristics in the snapshots of the evolving wavepacket, the accumulation (4) shows clear enhancement along an unstable periodic orbit. It is apparently similar to the scars of stationary wavefunction[3]. In various cases with different launching conditions the same phenomena are also found(Fig.3). The enhancement appears around the periodic orbit, if the initial location of the center of the wavepacket and its velocity are on and along the orbit. We shall call them the "dynamical scars" to distinguish from the scar states in stationary eigenstates. It is the enhancement in the accumulation of time-dependent wavefunction.

4. Weighted Spectral Function

Any state in quantum systems can be expanded by its eigenfunctions as

$$\Psi(\mathbf{r},t) = \sum_n c_n\psi_n(\mathbf{r},t) = \sum_n c_n\phi_n(\mathbf{r})\exp(-\frac{i}{\hbar}E_n t), \tag{5}$$

where $\psi_n(\mathbf{r},t) = \phi_n(\mathbf{r})\exp(-\frac{i}{\hbar}E_n t)$ is the n-th eigenstate of the system with the energy E_n. Then the time-average of $|\Psi(\mathbf{r},t)|^2$ becomes

$$\begin{aligned} A(\mathbf{r}) &= \lim_{T\to\infty}\frac{1}{T}\int_0^T |\Psi(\mathbf{r},t)|^2 dt \\ &= \lim_{T\to\infty}\frac{1}{T}\int_0^T \Big[\sum_n |c_n|^2|\phi_n(\mathbf{r})|^2 + \sum_{n\neq m} c_m^* c_n \phi_m^* \phi_n exp\Big\{\frac{i}{\hbar}(E_m - E_n)t\Big\}\Big]dt \\ &= \sum_n |c_n|^2|\phi_n(\mathbf{r})|^2, \end{aligned} \tag{6}$$

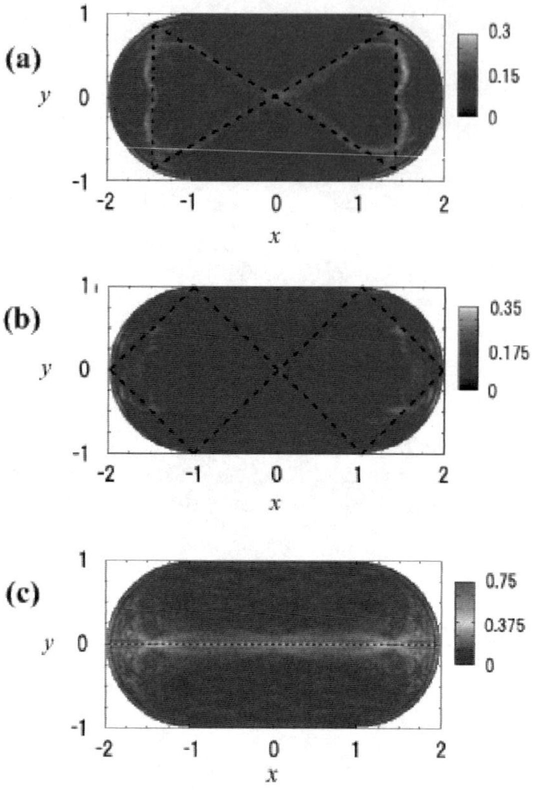

Figure 3. The accumulations $A(\mathbf{r})$ of time-evolving wavepacket in stadium billiard with different launching angles are presented. In all cases the initial Gaussian wavepackets are located at the center of the stadium, with $|\mathbf{p}_0| = 50$ and $\sigma_0 = 0.1$. The launching angles are (a)30° (the same as Fig.2), (b)45°, and (c)0°. The angles are measured in the counterclockwise direction from the long axis of the stadium. The broken lines represent the corresponding unstable periodic orbits.

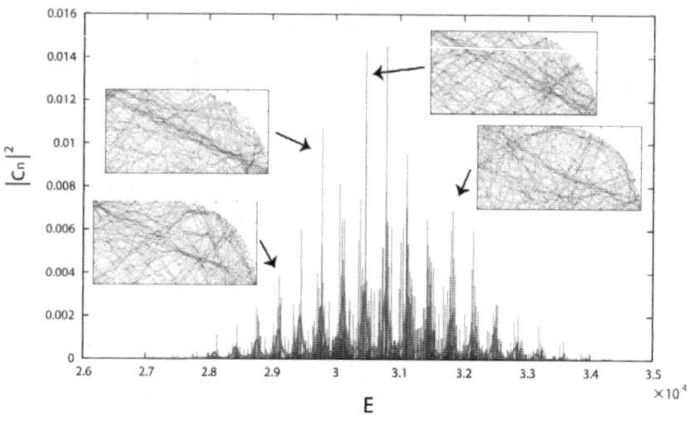

Figure 4. The weighted spectrum of the Gaussian wavepacket ($\sigma_0 = 0.15, |\mathbf{p}_0| = 50$) in desymmetrized stadium billiard is presented. The wavepacket is launched from the center of the circumscribed rectangle of the billiard and its launching angle is the same as Fig.2. Insets show the eigenstates corresponding to the high peaks. They often have "dynamical scars", a quarter of the rhombus. The large coefficient $|c_n|^2$ implies that it must have larger contribution on the rhombus, even if its eigenstates at the high peaks have no clear corresponding scars. Its smoothed estimated behavior is drawn in red solid curve defined by eq.(6) with $\lambda = 0.4614|\mathbf{p}_0|$. The constant 0.4614 is just the geometric Lyapunov exponent of the periodic orbit.

assuming $E_n \neq E_m$, if $n \neq m$. In other words, by eq.(6), if the coefficients c_n have larger values for the scar eigenstates on the same periodic orbit, the "dynamical scars" of the periodic orbit in the accumulation $A(\mathbf{r})$ should be observed[12, 13]. Obviously, also following eq.(6), it is expected to be persistent in longer time accumulation.

The accumulation has close relation to the weighted spectral function $C(E)$. The typical form of the function in chaotic systems has already been studied cautiously [7, 8, 9, 10]. It may be expected as

$$C(E) \approx \sqrt{2\pi}\frac{\sigma_0}{v} exp\left\{-\frac{\sigma_0^2}{2v^2\hbar^2}(E - E_0)^2\right\} \times \frac{\hbar}{\pi}\sum_{n=-\infty}^{+\infty}\frac{\lambda/2}{(E - E_n)^2 + (\lambda/2)^2}. \quad (7)$$

The Gaussian part forms the envelop function, whose shape is determined by the parameters of the inital wavepacket (1): σ_0, v and $E_0 = \frac{\mathbf{p}_0^2}{2m}$. The summation of Lorentzians corresponds to the shaper peaks of dominant eigenstates. The energy peaks have almost the same energy spacings and $E_n = E_p + n\Delta$ would represent the energies at the peaks. Here E_p is the highest maximum of the serial local peaks with width λ, and Δ is the energy spacing between local peaks. The Bohr-Sommerfeld quantization condition for the semiclassical action explains the nearly equal enegy spacings[14]. The width parameter λ is just the Lyapnov exponent of the periodic orbit that we concern. This curve $C(E)$ (7) is imposed on the histogram of $|c_n|^2$s in Fig.4. The setting is the same as Fig.2, in which the rombus shape of the "dynamical scar" appears clearly, and also the rombus-shaped scars can be observed in most of corresponding eigenstates at the local peaks of $C(E)$.

The existence of the scar states in chaotic billiard systems makes relatively smaller mount of eigenstates, which represent scars along the corresponding periodic orbit, have dominant coeffcients(Fig.4). Thus it turns out to be a "totalitarian" case of [9]. And simultaneously it also lets the "dynamical scars" emerge. Then the histogram must be spiky, though, it still has a clear remnant of the typical structure of $C(E)$(see Fig.4). Furthermore, by averaging over the energy range which is sufficiently larger than the energy spacing of levels and still much smaller than the energy that we concern, the smoothed shape of the histogram of $|c_n|^2$ becomes strikingly close to the weighted spectrum $C(E)$(Fig.5).

In Fig.4(and also in Fig.5), the local peaks of the histogram are found to be located at almost equal energy intervals $\Delta = 0.0337$. It agrees well with the semiclassical estimation $\Delta_{th} = \frac{\hbar^2}{m}(\frac{2\pi}{L})|p_0| = 0.0351$ for the rombus shaped periodic orbit, where $L = 4.472$ is its length.

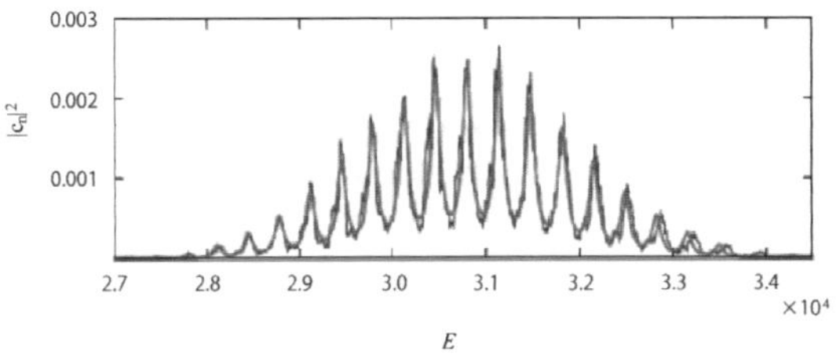

Figure 5. The weighted spectrum of the Gaussian wavepacket $\tilde{w}(E)$ (a red curve) and the averaged behavior of the expansion coefficients $|c_n|^2$(a blue curve) are presented. Here the averaging is performed in the energy range of 20ϵ. These lines matches considerably well.

5. Discussion

The Green's function provides the expression for the accumulation

$$
\begin{aligned}
A(\mathbf{r}) &= \int \sum_n |c_n|^2 |\phi_n(\mathbf{r})|^2 \delta(E - E_n) dE \approx \int w(E) \sum_n |\phi_n(\mathbf{r})|^2 \delta(E - E_n) dE \\
&= -\frac{1}{\pi} \int_{-\infty}^{\infty} w(E) \mathrm{ImG}(\mathbf{r}, \mathbf{r}; \mathrm{E}) \mathrm{dE}, \qquad (8)
\end{aligned}
$$

where the window function $w(E)$ is introduced[15]. Mathematically, if the condition $w(E) = |c_n|^2$ at $^\forall E = E_n$ is exactly satisfied, the approximation (8) becomes an exact equation. Practically it is concluded that the window function can be an approximated by the weighted spectrum: $w(E) \approx C(E)$, assuming that $A(\mathbf{r})$ is estimated in wide energy range which covers at least several serial scar states.

If we choose the window size small enough to suppose that the only one eigenstate would be in the window simultaneously. It is essentially the result of ref.[6] for the scar states. In this work the window size is much larger, because the initial wavepacket has to take the contribution of eigenstates in broader energy range. Thus we may not see the scar in the snapshot of time-dependent wavefunctions. The "dynamical scar" is the superposition of many states in the window.

Then the semiclassical expansion [6] is still available as

$$A(\mathbf{r}) = \langle \rho_0(\mathbf{r}, E) \rangle + Im \langle G_{osc}(\mathbf{r}, \mathbf{r}; E) \rangle, \tag{9}$$

where

$$\langle G_{osc}(\mathbf{r}, \mathbf{r}; E) \rangle = \frac{1}{i(2\pi)^{1/2}\hbar^{3/2}} \sum_n \int C(E) \frac{D_n(\xi)^{1/2}}{|\dot{\mathbf{r}}|} \left\{ exp \left[\frac{i}{\hbar} (S_n(\xi, \xi; E) + \frac{W_n(\xi)}{2} \eta^2) \right] - i \frac{\pi \mu_n}{2} \right\} dE, \tag{10}$$

and the notation $\langle \cdots \rangle$ stands for the average over the energy range that the window function covers, $\rho_0(\mathbf{r}, E)$ is just the averaged density of states derived by the Thomas-Fermi approximation. The ξ axis is set along the orbit, and the η axis perpendicular to it at the point ξ. The classical action $S(\xi, \xi; E) = \oint_C \mathbf{p} d\mathbf{r}$ of the n-fold repeated orbit can be derived $S_n = nS$ from the action $S(\xi, \xi; E)$ of the primitive orbit C. Then its period is $T_n(\mathbf{r}, E) = nT(\mathbf{r}, E)$, and T is the period of the primitive orbit. Its maximal number of conjugate points $\mu_n = n\mu$ on the orbit can be derived from the primitive one μ. And also $D = -(\frac{\partial^2 S_n}{\partial \eta' \partial \eta''})_{\eta'=\eta''=0}$ and $W(\xi) = (\frac{\partial^2 S}{\partial \eta'^2} + \frac{\partial^2 S}{\partial \eta' \partial \eta''} + \frac{\partial^2 S}{\partial \eta''^2})_{\eta'=\eta''=0}$ of the n-fold periodic orbit $W_n(\xi)$, $D_n(\xi)$ can be derived from D of the primitive :$D_n(\xi) = D \frac{\lambda_1 - \lambda_2}{\lambda_1^n - \lambda_2^n}$, $W_n = D_n(\lambda_1^2 + \lambda_2^2 - 2)$. Note that λ_1, $\lambda_2 = \lambda_1^{-1}$ are the eigenvalues of the monodromy matrix of the orbit.

Therefore, following the semi-classical approximation[6, 15], it can be expected that the enhancement that is similar to the scar also in the accumulation $A(\mathbf{r})$. Its argument is also close to the discussion for the scar states. The time average emphasizes the periodic orbit contribution and washes away just oscillatory part of the wavefunctions. The "dynamical scars" emerge from just the billowing texture.

6. Acknowledgement

This work was in part supported by the sabbatical program of Seikei University. Without the support, this work would never be started.

References

[1] J. Bird, Rep. Prog. Phys., 66, 583, 2003.
[2] T. Gorin, T. Prosen, T. H. Seligman and M. Žnidarič, Phys. Rep., 435, 33, 2006.
[3] E. J. Heller, Phys. Rev. Lett., 53, 1515, 1984.
[4] E. J. Heller, *Chaos and Quantum Physics*, edited by M. J. Giannoni, A. Voros, and J. Zinn-Justin, Les Houches Session LII, 1989 (Elsevier, Amsterdam, 1991).
[5] M. V. Berry, Proc. R. Soc. Lond. A, 423, 219, 1989
[6] E. B. Bogomolny, Physica D, 3, 169, 1988
[7] S. Tomsovic and E. J. Heller, Phys. Rev. Lett., 70, 1405, 1993.
[8] S. Tomsovic and E. J. Heller, Phys. Rev. E, 47, 282, 1993.
[9] L. Kaplan and E. J. Heller, Annals of Physics, 264, 171, 1998
[10] L. Kaplan and E. J. Heller, Phys. Rev. E, 62, 409, 2000
[11] L. A. Bunimovich, Funct. Anal. Appl., 8, 254, 1974
[12] H. Tsuyuki, M. Tomiya, S. Sakamoto and M. Nishikawa, e-J. Surf. Sci. Nanotech., 7, 721, 2009.
[13] M. Tomiya, H. Tsuyuki and S. Sakamoto, Comp. Phys. Comm., 182, 245, 2010.
[14] M. Tomiya, H. Tsuyuki, K. Kawamura, S. Sakamoto and E. J. Heller, in preparation.
[15] H-J. Stöckmann, *Quantum Chaos an Introduction*, (Cambridge University Press, Cambridge, 1999)

Journal of Physics: Conference Series **640** (2015) 012069 doi:10.1088/1742-6596/640/1/012069

The Maximum Entropy Method for Optical Spectrum Analysis of Real-Time TDDFT

M. Toogoshi, S. S. Kano and Y. Zempo

Computer and Information Sciences, Hosei University, 3-7-2 Kajino, Koganei 184-8584 Japan

E-mail: 14t0007@cis.k.hosei.ac.jp

Abstract. The maximum entropy method (MEM) is one of the key techniques for spectral analysis. The major feature is that spectra in the low frequency part can be described by the short time-series data. Thus, we applied MEM to analyse the spectrum from the time dependent dipole moment obtained from the time-dependent density functional theory (TDDFT) calculation in real time. It is intensively studied for computing optical properties. In the MEM analysis, however, the maximum lag of the autocorrelation is restricted by the total number of time-series data. We proposed that, as an improved MEM analysis, we use the concatenated data set made from the several-times repeated raw data. We have applied this technique to the spectral analysis of the TDDFT dipole moment of ethylene and oligo-fluorene with $n = 8$. As a result, the higher resolution can be obtained, which is closer to that of FT with practically time-evoluted data as the same total number of time steps. The efficiency and the characteristic feature of this technique are presented in this paper.

1. Introduction

Time-dependent density-functional theory (TDDFT) is widely used to calculate optical properties of medium-to-large molecules, and it is recognized as a powerful tool for analyzing electronic behaviors of excited states to a certain level [1]. In our calculations, we employed the real-time and real-space technique to solve the time-dependent equation of motion [2, 3]. Within the framework of this approach, the wave functions are calculated by the finite difference method on real spatial grids [4] without using explicit bases such as plane waves or Gaussian. In the time domain, we evolve the perturbed initial wave function until we obtain enough dynamic motions. In the traditional method, the optical strength function is calculated from the Fourier transform (FT) of the dynamic dipole moment.

In the frequency domain, however, the maximum entropy method (MEM) is frequently applied to the spectrum analysis with a relatively small number of time-series data to obtain a fairly high resolution and accuracy [5]. The MEM is based on the relationship that the known autocorrelation are connected with the power spectrum by the Fourier transform. We apply this method to the spectrum analysis of the dynamic dipole moment, which is calculated by real-time TDDFT. In the comparison of MEM and FT spectrum for several molecules, we realize that MEM needs less time steps than FT to obtain the optical spectrum with same resolution [6].

We propose that, for MEM, we use the several-times repeated time-series data in stead of single time-series data, in order to reduce the total number of time steps significantly. Since the spectral resolution of MEM depends on the maximum lag M of the autocorrelation where M is limited by the total number of time steps N, we need a large enough N for a reasonable resolution. With the repeated data, we can select the maximum lag M large enough without an additional computational cost. We have found that the MEM of the repeated data shows higher resolution than the FT spectra with the double size data. In this paper, we present the effectiveness and characteristic features of our new MEM, comparing with the FT and the conventional MEM.

The paper is organized as follows. Our new technique is presented in the next section, and the observed effectiveness and characteristics of our MEM technique follow along with our conclusions in the last section.

2. Method

In this section, we briefly describe our new technique. We propose that, as an improved MEM analysis, we use the concatenated data set made from the several-times repeated raw data. The resolution of the MEM spectra depends on not only the total number of time steps N but also the maximum lag M of autocorrelation. The larger value of M provide the well-developed peaks. Generally, the high frequency components of time-series data are taken into account, even for smaller M. But for the analysis of lower frequency components, M must be large. Since the maximum lag M limits the low frequency resolution where M is limited by N, we need a large enough N for a reasonable spectral resolution, and the total machine time could not be shortened than we hope. To solve this difficulty, we duplicate the data and repeat it many times as shown in Fig. 1 (a).

A framework of the present study is based on Burg's method [5], where the samples are given by the real-time TDDFT. Given that the time-series data obeys Gaussian distribution, the entropy h of a Gaussian process is given by

$$h \propto \int_{-\pi/\Delta t}^{\pi/\Delta t} \log P(\omega) d\omega, \tag{1}$$

where Δt is a sampling interval, and $P(\omega)$ is the power spectrum. We find $\hat{P}(\omega)$ which Eq.(1) is maximized under the constraints of the Wiener-Khintchin's theorem. The solution is

$$\hat{P}(\omega)/\Delta t = \left| \frac{\beta}{1 + \sum_{k=1}^{M} a_k \exp(-ik\omega\Delta t)} \right|^2, \tag{2}$$

where M is the maximum length of autocorrelation C_M. The parameters a_k and β are the Lagrange multipliers and they are the solution of Yule-Walker equation [7]

$$\begin{pmatrix} C_0 & C_1 & \cdots & C_M \\ C_1 & C_0 & \cdots & C_{M-1} \\ \vdots & \vdots & & \vdots \\ C_M & C_{M-1} & \cdots & C_0 \end{pmatrix} \begin{pmatrix} 1 \\ a_1 \\ \vdots \\ a_M \end{pmatrix} = \begin{pmatrix} |\beta|^2 \\ 0 \\ \vdots \\ 0 \end{pmatrix}. \tag{3}$$

We adopt the Levinson-Durbin algorithm to solve Eq.(3) efficiently.

3. Comparison of our new MEM and conventional methods

3.1. Ethylene

As a simple example to show the advantage of our proposed method, we applied our MEM to a small molecule ethylene. The molecular structure is based on the ground state. The time evolution is carried out according to the real-time TDDFT [2, 3]. The spectrum $S(E)$ is calculated from the dynamic dipole moment $\mu(t)$ in both FT and our MEM technique. The calculation procedure as shown Figure 1 (a) is that, after the real calculation up to the total

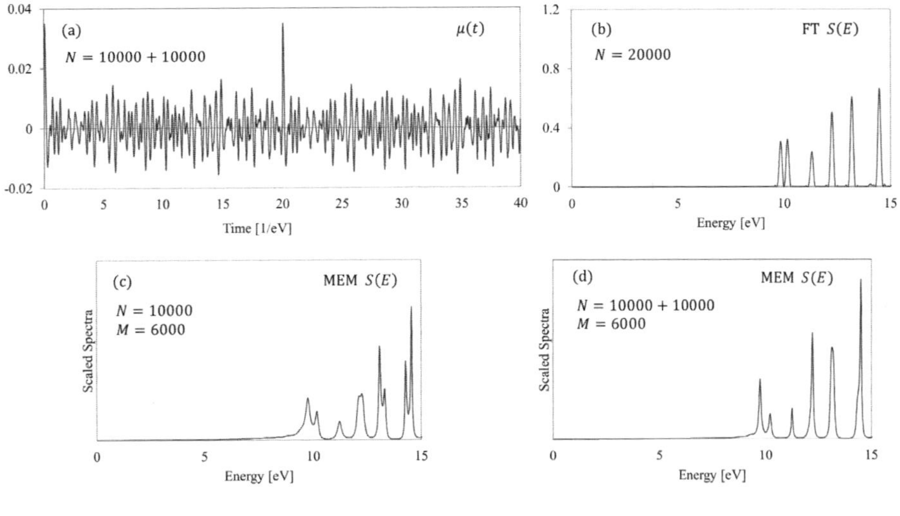

Figure 1. Ethylene absorption spectra. (a) the dynamic dipole moment of the twice repeated data, (b) the spectra of FT with the real time steps $N = 20000$, and the spectra calculated by our MEM with (c) $N = 10000$ and (d) the repeated data $N = 10000 + 10000$. For both MEM calculations, the same maximum lag $M = 6000$ are adopted. The MEM data are scaled to allow comparison with those of FT.

time steps 10000, we used the twice repeated data of the dipole moment, i.e. $N = 10000+10000$, for our MEM calculation. The time step is $\Delta t = 0.002$ eV^{-1}. The absorption spectra by FT of $\mu(t)$ is shown in Fig. 1 (b) with the number of real time steps $N = 20000$. Figure 1 (c) shows the result of MEM with the real time steps $N = 10000$ and the maximum lag $M = 6000$. Although the traditional MEM provides the higher resolution compared to that of FT [6], there are some peak splits around 15 eV. In the result of the repeated data as shown in Fig. 1 (d), on the other hand, there is no this kind of peak splits observed. In addition to it, our MEM makes the spectrum similar resolution to that of the FT with the real data $N = 20000$.

3.2. Fluorene

Poly(9,9-dialkyl-fluorene) and their substituted derivatives are expected to be basic materials for a blue emitting polymer diodes. Their electronic structures have been extensively studied [8, 9, 10, 11, 12]. We employed the oligomer of fluorene (FL) with $n = 8$ and carried out the MEM calculation. The calculation for this kind size molecule is quite expensive to realize enough time evolution. If we can save the calculation cost to figure out the lower energy spectrum, it is quite effective. As similar as the calculation of ethylene, we used the concatenated repeated dipole moment data. Thus, we compared the efficiency of MEM. The calculation was done by MEM with $N = 5000$ and $M = 2500$, and the four times repeated data $N = 5000 \times 4$ with $M = 2500$. The results were compared to that by FT with for the real times steps $N = 20000$.

The results are shown in Fig. 2. We can clearly see that MEM has a significant advantage to figure out the lower energy part of the spectrum. In particular, the concatenated four times repeated data of the dipole moment can be used for MEM, which is much more effective to the conventional MEM.

The reason why the repeated data is efficient for the lower energy part of the spectrum analysis can be explained as follows. The lower energy resolution is attributed to the maximum lag M of autocorrelation. The lager number of M provides the information of the lower energy

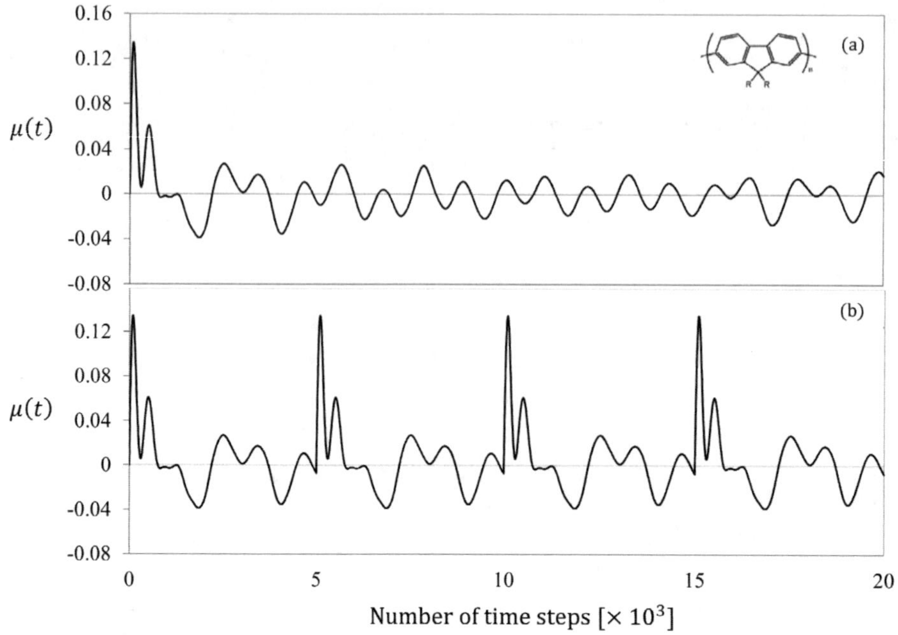

Figure 2. Concatenated four times repeated data of the dipole moment. (a) Real time evolution up to $N = 20000$ for FT. (b) Four times repeated data of the the dipole moment with $N = 5000 \times 4$ and $M = 2500$ for MEM.

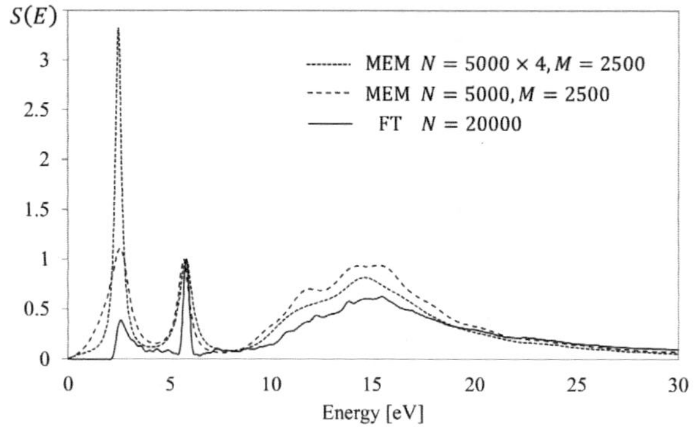

Figure 3. Comparison of absorption spectrum of oligo-FL_8 obtained by FT and MEM. The solid line is the result of FT with $N = 20000$, the dashed line is that of MEM with $N = 5000$ and $M = 2500$, and the dotted line is that of MEM with $N = 5000 \times 4$ and $M = 2500$. Spectrum are normalized at the second low energy peak.

part of the spectrum. However, the unphysical results such as peak splits and false peaks are appeared by selecting too large value of M from our experience. Also, M is restricted by the total number of time-series data N. These are caused by the few contributions of the higher-order autocorrelations compared to the small-order ones.

Figure 4 graphically shows the elements of the Toepliz matrices of ethylene in the Yule-Walker equation as stated in Eq.(3) with the total time steps (a) $N = 10000$ and (b) $N = 10000+10000$, respectively. The maximum lag M is used as the same value $M = 9999$ for both cases. The color is meant for the magnitude of C_M. All the values of the elements are normalized so as to make the diagonal value unity, which is corresponding to the zero-lag autocorrelation, colored red. In the case of the real data use, the values of high-order autocorrelations are almost close to zero for $M > 5000$ as shown in Fig. 4 (a). On the other hands, the result of repeated data provides the effective values of M from about $M = 5000$ to $M = 7500$ as shown in Fig. 4 (b). With the repeated data, the effective autocorrelations expressed by the maximum lag M can be increased. In other words, the method enhances the information of the low energy part for time-series data without additional calculations, which save our calculation cost.

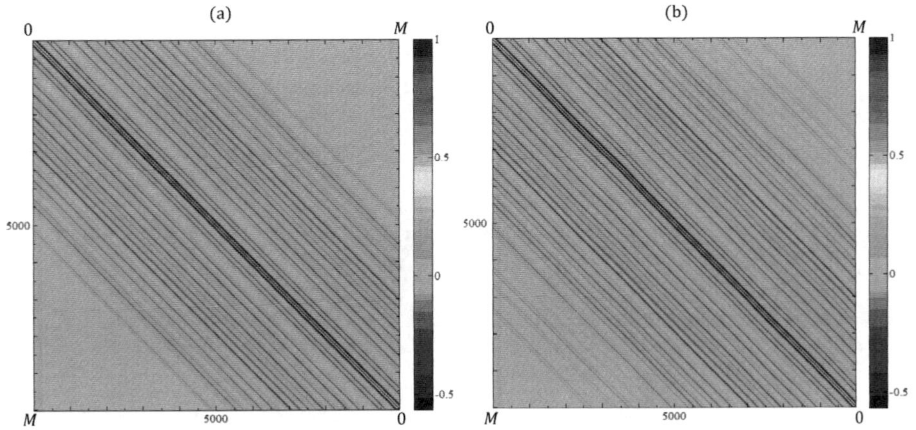

Figure 4. Magnitude of the matrix elements of Eq.(3) in the cases of (a) $N = 10000$ and $M = 9999$, and (b) $N = 10000 + 10000$ and $M = 9999$.

From these results above, MEM has several advantages. First, there is no the periodicity appeared to the signal in MEM, which is different from FT. When we applied FT to the repeated data in Fig. 1 (a), the periodicity was appeared to the spectrum as artificial frequency components, which disturbed the signal detection. Second, there is a phase jump for each frequency component at the connection point. However, MEM is originally used of the method for analysing the natural phenomena, which involve a lot of noises. This phase jump can be perceived as one of the noises. MEM is quite robust to detect signals from the system with noises. Thus, we observed the signal clearly in Fig. 1 (d)

4. Concluding remarks

By the use of MEM, the spectrum is generally calculated with the small number of time-series data, in comparison with the conventional FT [6]. We proposed the new MEM technique to obtain the spectrum with much smaller number of time-series data especially in the lower frequency part. In the new technique, we make a quasi-long size data, which is made by several times repeating the small size of time-series data. In this calculation, there are many higher-order autocorrelations involved, which are relevant to the lower frequency part of the spectrum. The resolution of the spectrum is dependent on the total number of time steps N and the autocorrelation lag M. In our calculation, we use the quasi-longer N. Thus, we can apparently increase the higher-order autocorrelation lag M. As a result, the spectrum is efficiently enhanced in the lower frequency part.

The new technique is applied to the spectrum analysis of the dynamic dipole moment of several molecules such as ethylene and oligo-fluorene with $n = 8$. We confirm that it is quite efficient to obtain spectra from TDDFT. With a fairly small number of time-series data, we obtained much better resolutions of the low frequency part of the spectrum. This means that we can save our calculation cost efficiently.

Acknowledgments
Authors would like to express their acknowledgements to Professor H. Hanaizumi for useful discussions and encouragements. This work was partially supported by JSPS Grants-in-Aid for Scientific Research (C) Grant number 25390158, Sumitomo Chemical, Co. Ltd. and Simulatio Corporation.

References
[1] Runge E and Gross E K U 1984 *Phys. Rev. Lett.* **52**, 997
[2] Yabana K and Bertisch G F 1996 *Phys. Rev.* B**54**, 4484
[3] Zempo Y, Akino, Ishida M, Ishitobi M and Kurita Y 2008 *J. Phys. Condens. Matter* **20**, 064231
[4] Chelikowsky J, Troullier N, Wu K and Saad Y 1994 *Phys. Rev. B***50**, 11355
[5] Burg J P 1968 *Advanced Study Institute on Signal Processing*, 15-0, NATO, Erschede, Netherlands
[6] Toogoshi M, Kano S S, Zempo Y 2014 *J. Phys. Conf. Ser.* **510**, 012027
[7] Haykin S and Kesler S 1979 *"Non linear Methods of Spectral Analysis"*, ed S. Haykin, Springer(Berlin)
[8] e.g. Hummer K, Puschnig P, Sagmeister S and Ambrosch-Draxl C 2006 *Mod. Phys. Lett.* B**20**, 261 , and the references therein.
[9] Yu W L, Cao Y, Pei J, Huang W and Heeger A J 1990 *Appl. Phys. Lett.* **75** 3270
[10] Wang J F, Feng J K, Ren A M, Liu X D, Ma Y G, Lu P and Zhang H X 2004 *Macromolecules* **37**, 3451
[11] Brière J F and Côté M 2004 *J. Chem. Phys.* B**108**, 3123
[12] Grice A W, Bradley D D C, Bernius, M T, Inbasekaran M, Woo EP and Wu W W 1998 *Appl. Phys. Lett.* **73**, 629

Algorithm of Shaping Multiple-beam Braggs Acousto-optic Diffraction Laser Field Into 1D and 2D Patterns

S. Zakharchenko, A. Baturin

E-mail: zakharchenko.sv@mipt.ru
Moscow Institute of Physics and Technology; Russia, Moscow Region, Dolgoprudniy, Institutskiy pereulok, 9

Abstract. Algorithm of solving a direct problem of acousto-optic interaction between laser emission and acoustic signal consisting of a set of equidistant frequency components is proposed. An infinite system of coupled wave differential equations is reduced to eigenvalue problem. The contribution of the higher rediffraction orders is analyzed separately. Inverse problem of finding an optimal set of equidistant frequency components of a driving acoustic signal to form the objective diffraction pattern is also considered and a few optimization approaches are analyzed. A naïve heuristic method of splitting 2D pattern into subframes, each suitable for simultaneous projection by two acousto-optical deflectors driven by multifrequency composite signal, is developed.

1. Introduction

Precise shaping of a laser beam has become of a great technological importance due to its possible applications in various fields of science and technology. For instance, the shaping a laser beam into confining potentials found its application for creation and subsequent manipulation of Bose-Einstein condensate [1]. Fiber-optic switch-multiplexer, based on high-effective multi-beam Bragg acousto-optic diffraction, was shown [2] to have low switching time (less than $10\mu s$), enhanced number of channels and the unique ability to transmit input signal to any group of channels simultaneously. All these applications require new rigorous algorithms for near real-time data processing with the delay between incoming pattern and calculated acoustic drive signal of up to a few seconds.

A traveling acoustic wave creates a perturbation of refractive index in a form of periodic structure similar to phase diffraction grating. Propagating light scatters on this structure and forms a diffraction pattern in the far field [3]. Since a period of the grating and a modulation depth of the refractive index depends on a frequency and an amplitude of the acoustic wave, the propagating light could be shaped into a required pattern by means of applying the proper set of acoustic wave modes.

The Bragg diffraction regime occurs when only one (principal) diffraction order is being produced, while others destructively interfere since they are not phase matched. If an acoustic field consists of N acoustic modes with a set of frequencies $f_1 \dots f_N$, the diffraction pattern becomes more complex: the output will have N principal diffraction orders corresponding to each acoustic mode. However, light from each principal beam could be rediffracted further by another acoustic mode. This results in cross modulation and generates parasite intermodulation modes corresponding to various sum and difference frequencies [4].

An acousto-optic interaction is described by the following infinite system of differential equations [5]:

$$\begin{gathered}
\frac{dG_0}{dx} = -j\sum_{n_1} c_{n_1} G_{n_1} \exp(-j\eta_{0\leftrightarrow n_1} x), \\
\frac{dG_{n_1}}{dx} = j[\sum_{n_2\neq n_1} c^*_{n_2} G_{n_1 n_2} \exp(j\eta_{n_1 n_2\leftrightarrow n_1} x) + c^*_{n_1} G_0 \exp(j\eta_{0\leftrightarrow n_1} x)], \\
\frac{dG_{n_1 n_2}}{dx} = -j[\sum_{n_3\neq n_2} c_{n_3} G_{n_1 n_2 n_3} \exp(-j\eta_{n_1 n_2\leftrightarrow n_1 n_2 n_3} x) + c_{n_2} G_{n_1} \exp(-j\eta_{n_1 n_2\leftrightarrow n_1} x)], \\
\dots \\
\frac{dG_{n_1 n_2 n_3 \dots n_\mu}}{dx} = -j[\sum_{n_{\mu+1}\neq n_\mu} c_{n_{\mu+1}} G_{n_1 n_2 n_3 \dots n_\mu n_{\mu+1}} \exp(-j\eta_{n_1 n_2 n_3 \dots n_\mu \leftrightarrow n_1 n_2 n_3 \dots n_\mu n_{\mu+1}} x) + \\
+ c_{n_\mu} G_{n_1 n_2 n_3 \dots n_{\mu-1}} \exp(-j\eta_{n_1 n_2 n_3 \dots n_\mu \leftrightarrow n_1 n_2 n_3 \dots n_{\mu-1}} x)],\ \mu = 2L,\ L\in\mathbb{Z} \\
\frac{dG_{n_1 n_2 n_3 \dots n_\mu n_{\mu+1}}}{dx} = j[\sum_{n_{\mu+2}\neq n_{\mu+1}} c^*_{n_{\mu+2}} G_{n_1 n_2 n_3 \dots n_{\mu+2}} \exp(j\eta_{n_1 n_2 n_3 \dots n_{\mu+2} \leftrightarrow n_1 n_2 n_3 \dots n_{\mu+1}} x) + \\
c^*_{n_{\mu+1}} G_{n_1 n_2 n_3 \dots n_\mu} \exp(j\eta_{n_1 n_2 n_3 \dots n_\mu \leftrightarrow n_1 n_2 n_3 \dots n_{\mu+1}} x)],\ \mu = 2L+1,\ L\in\mathbb{Z}, \\
\dots
\end{gathered} \tag{1}$$

where $G_{n_1 n_2 n_3 \dots n_\mu}(x)$ is the complex amplitude of the laser spatial mode $n_1 n_2 n_3 \dots n_\mu$, which corresponds to all photons that have successively interacted with phonons from acoustic waves with the following frequencies $f_1, f_2, f_3 \dots, f_\mu$; c_n — the complex amplitude of the acoustic wave with the frequency f_n; $\eta_{n_1 n_2 n_3 \dots n_\mu \leftrightarrow n_1 n_2 n_3 \dots n_\mu n_{\mu+1}}$ is the phase mismatch between the corresponding modes (it depends on the acousto-optic interaction geometry, thus considered to be constant and known); j – imaginary unit; c^*_n – complex conjugate of the amplitude.

Without loss of generality, we assume the amplitude of input light is equal to 1, then the boundary conditions are

$$\begin{cases} G_0\,|_{x=0} = 1; \\ G_{n_1 \dots n_\mu}\,|_{x=0} = 0; \quad \forall\mu\in\mathbb{Z}, \mu>0;\ n_1,\ \dots, n_\mu \in [0, N]. \end{cases} \tag{2}$$

Suppose the acoustic modes have equal frequency spacing between one another and $f_0 \dots f_{max}$ is the frequency range, where diffraction losses are relatively small. Then, the frequencies of N acoustic modes are

$$f_n = f_0 + n\Delta f = f_0 + n\frac{f_0 - f_{max}}{N}. \tag{3}$$

The angle through which the light beam corresponding to the spatial mode $n_1 n_2 n_3 \dots n_\mu$ is deviated depends on frequencies of acoustic waves it has interacted with. Suppose a linear approximation is valid, then

$$\theta_{n_1 n_2 n_3 \dots n_\mu} = \begin{cases} \theta_{Br} + \kappa \sum_{i=0}^{L} n_{2i} - \kappa \sum_{i=0}^{L} n_{2i+1} = \theta_{Br}^{k};\ k = \sum_{i=0}^{L} n_{2i} - \sum_{i=0}^{L} n_{2i+1}, \\ \qquad \mu = 2L+1,\ L \in \mathbb{Z}; \\ \theta_0 + \kappa \sum_{i=0}^{L} n_{2i} - \kappa \sum_{i=0}^{L-1} n_{2i+1} = \theta_0^{l};\ l = \sum_{i=0}^{L} n_{2i} - \sum_{i=0}^{L} n_{2i+1}, \\ \qquad \mu = 2L,\ L \in \mathbb{Z}; \end{cases} \tag{4}$$

where κ is the constant defined by the interaction geometry, $\theta_{Br} = \theta_{Br}^{0}$ – diffraction angle of the principal beam produced by the acoustic wave $f = f_0$, $\theta_0 = \theta_0^0$ – incidence angle of the undiffracted light beam and upper indices l and k are normalized angle. Henceforth, upper indices l and k are to be referred as normalized angles in zeroth and Bragg's field respectively.

Different spatial modes traveling at the same angle are inseparable; thus, there is no possibility to measure their individual amplitudes experimentally. Instead, we focus on overall beam intensities from Bragg's and zeroth-order fields $I(\theta_{Br}^{k})$ and $I(\theta_0^{l})$. They have more practical value since they define the diffraction pattern intensity distribution, which could be captured by an array of photo detectors. Summing amplitudes of all spectral modes from single beam at the specific angle θ_{Br}^{k} or θ_0^{l}, we get

$$G_{Br}^{k} = \sum_{\mu}^{\mu \le [\mu_{max}/2]} \left(\sum_{\forall \{n_1, n_2 \dots n_{2\mu+1}\}:\ \sum_{i=0}^{\mu} n_{2i} - \sum_{i=0}^{\mu} n_{2i+1} = k} G_{n_1 n_2 n_3 \dots n_{2\mu+1}} \right), \tag{5}$$

$$G_0^{l} = \sum_{\mu}^{\mu \le [\mu_{max}/2]} \left(\sum_{\forall \{n_1, n_2 \dots n_{2\mu}\}:\ \sum_{i=0}^{\mu} n_{2i} - \sum_{i=0}^{\mu-1} n_{2i+1} = k} G_{n_1 n_2 n_3 \dots n_{2\mu}} \right), \tag{6}$$

where μ_{max} is the maximum number of successive photon-phonon interaction that will be taken into the model and $0 \le n_i \le N$, $\forall i$.

The phase mismatch $\eta_{n_1 n_2 n_3 \dots n_\mu \leftrightarrow n_1 n_2 n_3 \dots n_\mu n_{\mu+1}}$ is defined by the interaction geometry, so if the spatial mode $n_1 n_2 n_3 \dots n_\mu$ has the same angle as $n_1^* n_2^* n_3^* \dots n_\mu^*$ and the spatial mode $n_1 n_2 n_3 \dots n_\mu, n_{\mu+1}$ has the same angle as $n_1^* n_2^* n_3^* \dots n_\mu^*, n_{\mu+1}^*$, then

$$\eta_{n_1 n_2 n_3 \dots n_\mu \leftrightarrow n_1 n_2 n_3 \dots n_\mu n_{\mu+1}} = \eta_{n_1^* n_2^* n_3^* \dots n_\mu^* \leftrightarrow n_1^* n_2^* n_3^* \dots n_\mu^* n_{\mu+1}^*} = \\ = \eta_{\theta_0^l \leftrightarrow \theta_{Br}^k},\ \mu = 2 * L,\ L \in \mathbb{Z}; \tag{7}$$

Substituting (7), (5), (6) in (1), we obtain:

$$\frac{dG_0^l}{dx} = -j[\sum_{n=0}^{N} c_{n3} G_{Br}^{l+n} \exp(-j\eta_{\theta_0^l \leftrightarrow \theta_{Br}^{l+n}} x)], \tag{8}$$

$$\frac{dG_{Br}^k}{dx} = j[\sum_{n=0}^{N} c_{n3} G_{Br}^{k-n} \exp(j\eta_{\theta_{Br}^k \leftrightarrow \theta_0^{k-n}} x)], \tag{9}$$

and the boundary conditions are

$$\begin{cases} G_0^0 |_{x=0} = 1; \\ G_0^l |_{x=0} = 0,\ \forall l \in \mathbb{Z}; \\ G_{Br}^k |_{x=0} = 0,\ \forall k \in \mathbb{Z}. \end{cases} \tag{10}$$

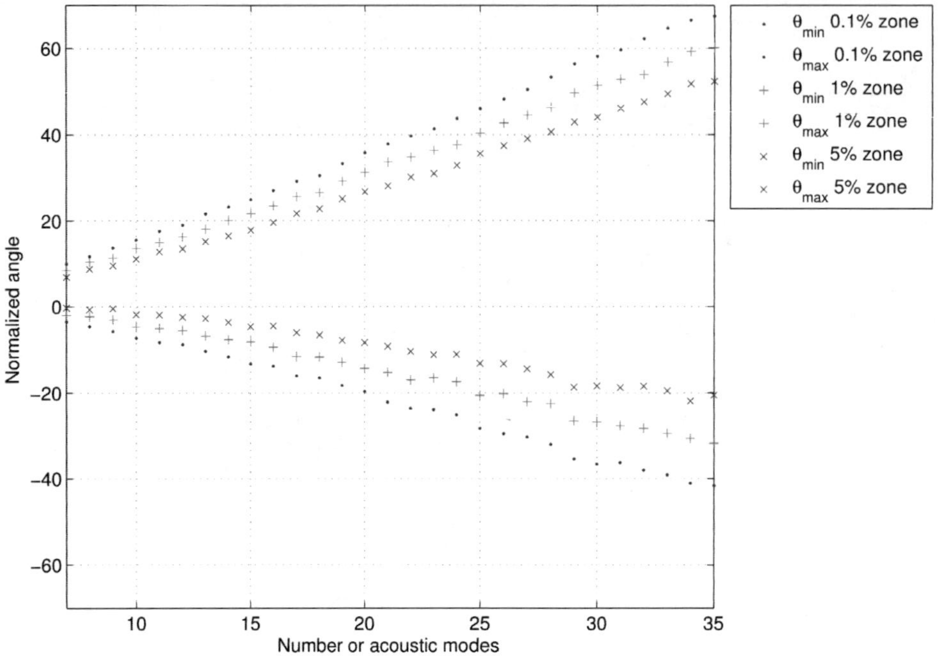

Figure 1. Diffraction region, where all intensities are greater then the specific threshold.

2. Spectral algorithm

Let us truncate the system (8) by limiting the normalized angle coefficients $-L \le l \le L$ and $-L \le k \le L$. The stiffness ratio of the truncated system is relatively high (≥ 1000). Because of that, let us build a spectral method by writing the solution as the following linear combination of base functions:

$$G^k_{Br}(x) = \sum_s V^s g^s_{Br,k} \exp(jx\xi^s_{Br,k}) \tag{11}$$

$$G^l_0(x) = \sum_s V^s g^s_{0,l} \exp(jx\xi^s_{0,l}), \tag{12}$$

where $\xi^s_{Br,k}$ is a spatial frequency; s, V^s, $g^s_{0,l}$ and $g^s_{Br,k}$ are unknown coefficients; $-L \le l \le L$ and $-L \le k \le L$; $k, l, s \in \mathbb{Z}$.

Substituting (11), (12) in (8), we obtain

$$\xi^s_{0,l} = \xi^s_{0,l+n} - \eta_{\theta^l_0 \leftrightarrow \theta^{l+n}_{Br}} \tag{13}$$

$$\xi^s_{Br,k} = \xi^s_{Br,k-n} + \eta_{\theta^k_{Br} \leftrightarrow \theta^{k-n}_0}. \tag{14}$$

Substituting (13), (14), (11), (12) in (8), we get eigenvalue equation

$$\left[\mathbf{M} - (-\xi^s_{0,0}\mathbf{E})\right] \begin{pmatrix} g^s_{0,-L} \\ \cdots \\ g^s_{0,L} \\ g^s_{Br,-L} \\ \cdots \\ g^s_{Br,L} \end{pmatrix} = 0, \tag{15}$$

Journal of Physics: Conference Series **640** (2015) 012070 doi:10.1088/1742-6596/640/1/012070

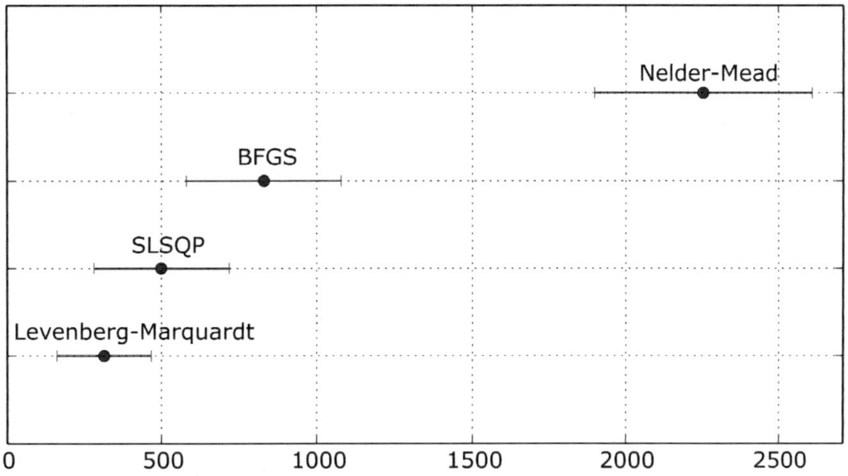

Figure 2. The number of spectral algorithm executions done by various optimization methods during minimization of the residual (18), averaged from 50 trials with random objective pattern I^*.

where **M** is some hermitian matrix[1] and **E** is identity matrix. Solving (15), we obtain spatial frequencies $\xi^s_{0,0}$ from eigenvalues and $\{g^s_{0,l}\}$, $\{g^s_{Br,k}\}$ from eigenvalues. Substituting (11), (12) in 10, we get the system of equations for calculating V^s:

$$\begin{cases} \sum_s V^s g^0_0 \,|_{x=0} = 1; \\ \sum_s V^s g^s_{0,l}|_{x=0} = 0, \ \forall l \in \mathbb{Z}; \\ \sum_s V^s g^s_{Br,l}|_{x=0} = 0, \ \forall k \in \mathbb{Z}. \end{cases} \tag{16}$$

This completes solving direct problem by means of spectral algorithm. However, by truncating the original system (8) we have artificially constrained which optical modes could exist and which could not. To make this assumption reasonable an amount of energy that could be transfered into the "forbidden" modes must be estimated. Let us select a neglectable intensity level I_n and start increasing the normalized angle cut-off coefficient L until $\theta_{min} = \min\left(\{k \mid G^k_{Br}(G^k_{Br})^* < I_n\}\right)$ and $\theta_{max} = \max\left(\{k \mid G^k_{Br}(G^k_{Br})^* < I_n\}\right)$ stop changing. According to fig. 1, the diffraction region $[\theta_{min} \dots \theta_{max}$ within which all the intensities are greater then the specific threshold, depends linearly on the number of acoustic modes. Thus, the following rough estimation is valid:

$$L \geq 2.5N. \tag{17}$$

3. Reverse problem

A reverse problem is the optimization problem of finding complex amplitudes $c = \{c_i\}$ which produce the diffraction pattern $I = \{I(\theta^k_{Br})\}$ with the least residual $f(\{c_i\}) = \|I - I^*\|$ to the objective pattern $I^* = \{I^*(\theta^k_{Br})\}$.

We have compared a performance of different optimization algorithms applied to this problem (Fig. 2). Benchmarking was conducted with the following parameters of the model: the number of acoustic modes $N = 5$, the normalized angle limitation $L = 50$, the objective function is

$$f(\{c_i\}) = \sqrt{\sum_{k=-L}^{L} \left(I(\theta^k_{Br}) - I^*(\theta^k_{Br})\right)^2}. \tag{18}$$

[1] For brevity sake, we omit exact form of this matrix since it is trivial to calculate it

4. 2D patterns

All the discussion above was related to the shaping of a light into 1D pattern. Let us consider attaching second successive acousto-optical deflector to the output of the first one. Suppose it is splitting each beam from the incoming 1D pattern in the orthogonal direction, thus forming a 2D pattern. It is easily shown that such setup is capable of projecting only specific class of 2D patterns. We say that 2d pattern defined by $N \times M$ binary matrix $\mathbf{G}$ is *projectable* iif it has rank-one factorization into outer product of some binary vectors $\mathbf{C}$ and $\mathbf{R}$:

$$\mathbf{S} = \mathbf{CR} = \begin{pmatrix} c_1 \\ c_2 \\ \dots \\ c_N \end{pmatrix} (r_1, r_2, \dots, r_M). \quad (19)$$

Therefore, we have a problem of decomposing an arbitrary 2d pattern defined by $N \times M$ binary matrix $\mathbf{G}$ into the sum of the least number of projectable 2d patterns $\mathbf{S_i}$. Detailed discussion of this problem is the object of another paper, so we shall limit ourselves to the consideration of iterative naïve greedy heuristics. Let us consider k-th step of the algorithm; let $\mathbf{G_k}$ be the input for this step and $\mathbf{G_1}$ for the first step be equal to $\mathbf{G}$.

Suppose $\mathbf{r^i}$, $1 \le i \le N$ is a row vector of the matrix $\mathbf{G_k}$. Let us introduce matrix $\mathbf{H_k}$, whose elements are the normalized inner product of the corresponding rows: $h_{ij} = \frac{\mathbf{r^i} \cdot \mathbf{r^j}}{\mathbf{r^i} \cdot \mathbf{r^i}}$, if $\mathbf{r^i} \cdot \mathbf{r^i} \ge 0$; 0 otherwise. Consider the row $\mathbf{r_w}$ that has the biggest value of $\mathbf{r^w} \cdot \mathbf{r^w} \ge 0$. Then, find a set $\mathcal{J}_k$, such as $\mathcal{J}_k = \{j \mid h_{wj} \ge \eta\}$, where $0 < \eta < 1$ is a parameter of heuristics. Finally, the elements of $\mathbf{R}$ and $\mathbf{C}$ from (19) are

$$r_i = \prod_{j \in \mathcal{J}_k} \mathbf{r_i^j}; \quad c_i = \mathbf{1}_{\mathcal{J}_k}(i), \quad (20)$$

where $\mathbf{1}_{\mathcal{J}_k}(x)$ is the indicator function. Substituting (20) in (19), we get projectable 2D pattern $\mathbf{S_k}$. If the residual $\mathbf{G_{k+1}} = \mathbf{G_k} - \mathbf{S_k}$ is a zero matrix, the algorithm finished the decomposition; otherwise repeat the procedure above for $\mathbf{G_{k+1}}$.

5. Conclusion

- The spectral algorithm for solving direct problem of acousto-optical interaction is proposed. This approach is more favorable than explicit and implicit methods because it doesn't suffer from high stiffness of the governing equations, doesn't accumulate errors and provides a convenient way to calculate partial derivatives of the intensity distribution via matrix perturbation.
- The reverse problem of finding the acoustic field that produce diffraction pattern close to the objective pattern is discussed. A few optimization approaches was compared.

Acknowledgments

This work was performed using the equipment provided by MIPT Center of Collective Usage (CCU MIPT) with the financial support from the Ministry of Education and Science of Russian Federation. (Grant ID RFMEFI59414X0009)

6. References

[1] Trypogeorgos D, Harte T, Bonnin A and Foot C 2013 *Opt Express.* **21(21):24837-46**
[2] Vainer A, Antonov S, Proklov V and Rezvov Y 2009 *Applied Optics* **48(7)**
[3] Korpel A 1988 *Acousto-optics / Adrian Korpel* (M. Dekker New York) ISBN 082477891
[4] Hecht D, Petrie G and Wofford S 1979 Multifrequency acousto-optic diffraction in optically birefringent media *1979 Ultrasonics Symposium* pp 46–50
[5] Gazalet M, Kastelik J, Bruneel C, Bazzi O and Bridoux E 1993 *Appl. Opt.* **32** 2455–2460

The statistical investigation of amplitude Scintillations at Indian high latitude Station Maitri, Antarctica

Prakash Khatarkar[1], Parvaiz A Khan[1], Shivangi Bhardwaj[2], P K Purohit[2], Roshni Atulkar[2] and A K Gwal[1]
[1]*Space Science Laboratory, Department of Physics & Electronics, Barkatullah University, Bhopal-462026, India*
[2]*National Institute of Technical Teachers' Training & Research, Bhopal-462002, India*

E-mail: purohit_pk2004@yahoo.com

Abstract. We have investigated the occurrence characteristics of ionospheric scintillations, using dual frequency GPS receiver, installed and operated at Indian scientific base station Maitri (71.45S and 11.45E) Antarctica, during December 2009 to December 2010. The scintillation morphology is described in terms of S4 Index. The scintillations are classified into four main categories as Weak (0.2<S4<0.4), Moderate (0.4<S4<0.6), Strong (0.6<S4<1.0) and Saturated (S4>1.0). From the analysis we found that the percentage of weak, moderate, strong and saturated scintillations were 96%, 80%, 58% and 7% respectively. The maximum percentage of all types of scintillation was observed in the summer season, followed by equinox and the least in winter season. As the year 2010 was a low solar activity period, consequently the maximum occurrences of scintillations were those of weak and moderate and only four cases of saturated scintillation were observed.

1. Introduction

The structure of the auroal and high latitude ionosphere is very complicated and varied. Dramatic changes take place very frequently in the auroral and polar regions. In this region, irregularities at different scale are quite common, which cause phase and amplitude fluctuations in the trans-ionospheric signals as these signals encounter the irregularities. The amplitude and phase fluctuations of radio signals are commonly referred to as scintillations. The ionospheric scintillation is a very important parameter for describing the morphological features of ionosphere. The ionospheric disturbances or irregularities produce significant effects on satellite signals used for communication and navigation; hence disturb communication and navigation applications. Therefore scintillation studies involving trans-ionospheric propagation is of considerable interest in understanding the physical processes controlling high latitude ionospheric dynamics viz. generation and transport mechanisms of ionospheric irregularities, soft particle precipitation with energies approximately 100eV, penetration of magnetospheric electric fields and auroral current systems [1].

A number of studies, concerning the GPS scintillations at high latitudes, have been conducted in the recent past [2, 3]. Using GPS observations from 11 high-latitude stations, Aarons [4] noted that phase fluctuation activity has a daily pattern mainly controlled by the motion of the receiver location into the auroral oval. De Franceschi et al. [5] using the observations from a chain of GPS ionospheric scintillation and TEC receivers in Northern Europe during the storm events of 30

October and 20 November 2003 found a strong influence of IMF on the formation and movement of plasma patches.

Many researchers have studied the morphology of L band GPS scintillation [6, 7, 8] in terms of S4 and σφ indices. Their morphological studies have shown that amplitude scintillation activity produced due to ionospheric irregularities varies with 11 year solar cycle, geomagnetic activity, geographical location, seasons as well as local time of the day.

The present study is carried out to statistically examine the diurnal and seasonal variations of scintillation at high latitude station Maitri, Antarctica during the low solar activity period 2010.

2. Data and method of analysis

To study the occurrence characteristics of high latitude L-band GPS scintillations, we have used the data recorded by NovAtel's dual frequency GPS receiver GSV4004A installed and operated at Indian Antarctic station Maitri, during December 2009 to December 2010. The Ionospheric Total Electron Content data were recorded with 30 second sampling rate in order to reduce processing time. The GPS receiver was set to track GPS signals at 1 second sampling rate and cutoff of elevation angle was set to 40 degree. The amplitude scintillation was monitored by computing the S4 index, which is defined as the standard deviation of the received signal power normalized to the average signal power. It is calculated for each 60 second period based on a 50 Hz sampling rate. The S4 index is computed over 60 second intervals and stored in the Ionospheric Scintillation Monitor Receiver (ISMR) data log along with the phase measurements.

The corrected S4, with the effects of ambient noise removed, is computed as

$$S_4 = \sqrt{\left(S_{4T}^2 - S_{4No}^2\right)}$$

Where S_{4T} is total S4 and S_{4No} is correction due to ambient noise over 60s intervals.

3. Results and discussion

The S4 scintillation index is the normalized variance of the intensity and is a commonly accepted measure of signal fluctuation caused by ionosphere irregularities. Figure 1 shows the diurnal variation of scintillation occurrence in the auroral region during December 2009 to December 2010. The twelve different panels show the occurrence of scintillations diurnally during all the twelve month of the year 2010. In the figure S4 index is plotted against the Universal Time. The different colors represent the different days of month. The different thresholds of S4 index were chosen to categorize the intensity level of scintillations. From the figure we find that all the four types of scintillation viz. weak, moderate, strong and saturated are observable. However, during the months of June and July 2010, very low level of scintillations were observed with only few cases of strong and moderate scintillations. The strong scintillation activity was observed during the months of November, December, February and March 2010. Only four cases of saturated condition were observed which occurred in the month of November, December, January and February. Figure 2 describes the percentage of scintillation observed during three different seasons of the year 2010. From the Figure 2 we find that maximum percentage of all types of scintillations is observed during summer season followed by equinox season and the least during winter season. The percentage of week, moderate, strong and saturated scintillations was observed to be 96%, 80%, 58% and 7% respectively. The only four cases of saturated scintillation were observed during summer season.

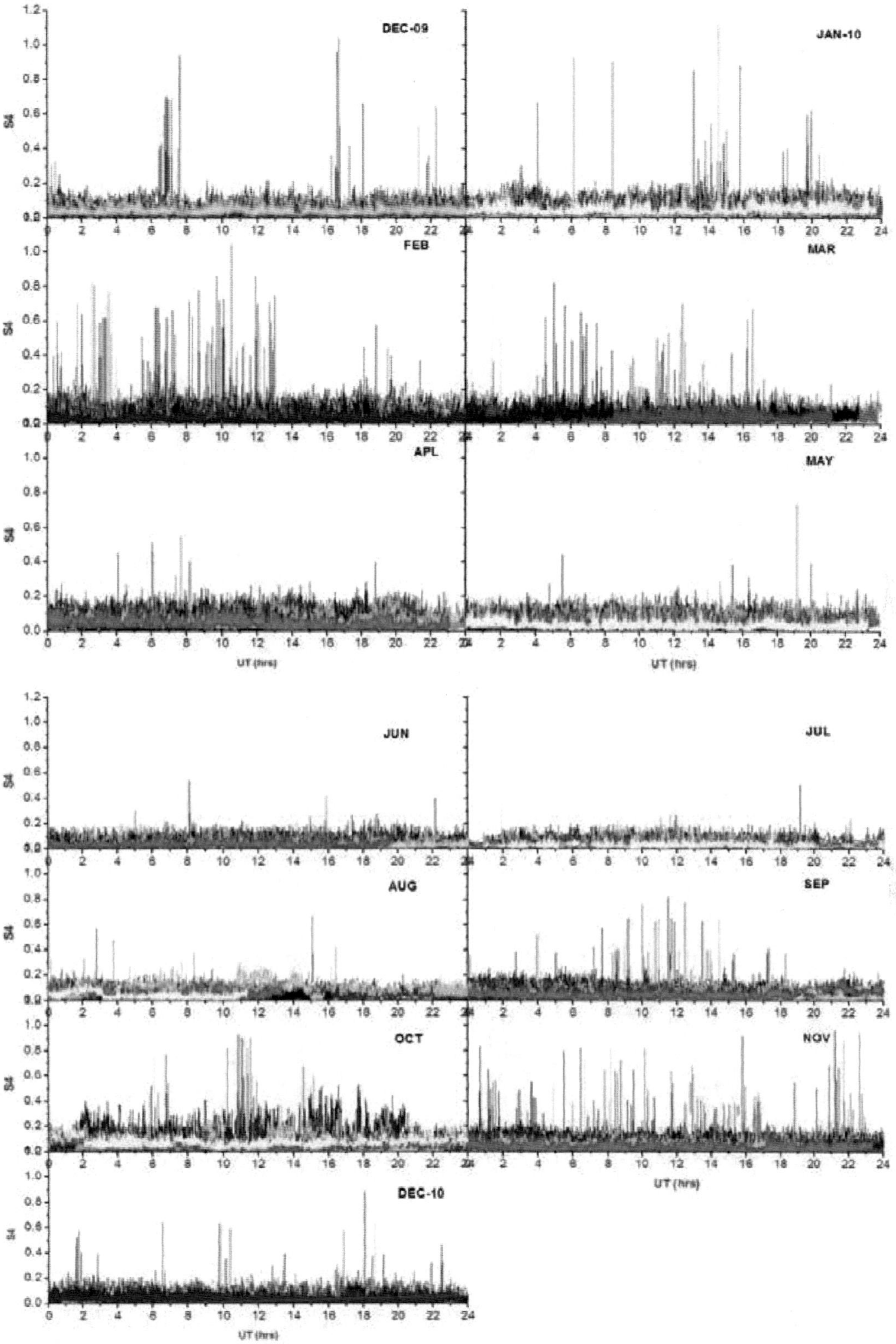

Figure 1. Mass plot of diurnal variation of scintillation during Dec.2009 to Dec.2010.

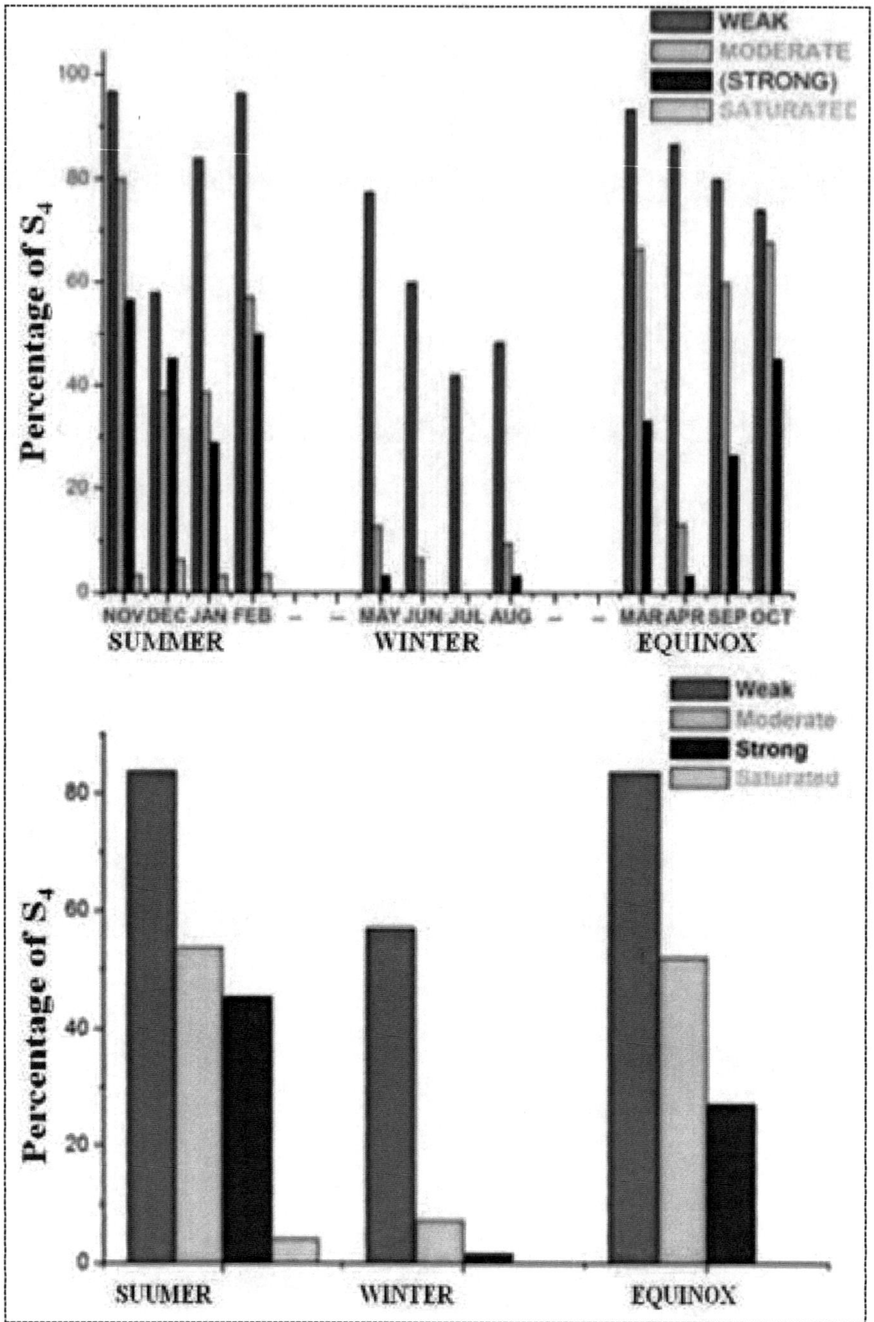

Figure 2. Seasonal percentage of amplitude scintillations with four distinct thresholds.

4. Conclusions

From our analysis to statistically examine the occurrence characteristics of GPS scintillation at high latitude station, Miatri during the low solar activity period 2010, we conclude our findings as follows: The percentage of weak, moderate, strong and saturated scintillations were 96%, 80%, 58% and 7% respectively. Thus, the maximum percentage was those of weak and moderate scintillations. Only four cases of saturated scintillation were observed during the observational period. The maximum occurrences of all types of scintillations were observed in summer season followed by equinox season and the least in winter season. In terms of monthly occurrences the maximum occurrence of scintillations was observed during the months of October, November, February and March.

5. Acknowledgements

The author (Prakash Khatarkar) is highly thankful to National Physical Laboratory (NPL), New Delhi and National Centre for Antarctic and Oceanic Research (NCAOR), Goa for providing opportunity to be a member of 29Th Indian Scientific Expedition to Maitri, Antarctica. The financial assistance received from University Grants commission, New Delhi, India, through Rajiv Gandhi National Fellowship (SRF) programme is also acknowledged.

6. References

[1] Kelley M C and Vickrey J F 1982 On the origin and spatial extent of high-latitude F-region irregularities *J. Geophys. Res.* **87** 4469-4475.

[2] De Franceschi G, Alfonsi L and Romano V 2006 Isacco: an Italianproject to monitor the high latitude ionosphere by means of GPS receivers *GPS Solut.* **10** 263– 267. doi:10.1007/s10291-006-0036-6.

[3] Meggs R W, Mitchell C N and Honary F 2008 GPS scintillation over the European Arctic during the November 2004 storms *GPS Solut.* **12** 281–287 doi:10.1007/s10291-008-0090-3.

[4] Aarons J 1997 Global Positioning System phase fluctuation at auroral latitudes *J. Geophys. Res.* **102** (A8) 17219 – 17231.

[5] De Franceschi G, Alfonsi L, Romano V, Aquino M, Dodson A,Mitchell C N, Spencer P, Wernik AW 2008 Dynamics of high latitude patches and associated small-scale irregularities during the October and November 2003 storms *J Atmos Solar-Terr Phys* **70** 879–888.

[6] Gwal A K, Dubey S, wahi R, and Feliziani A 2006 Amplitude and Phase scintillation study at Chiang Rai, Thailand *Adv. Space Res.* **38** (11) 2361-2365.

[7] Li G Z, Ning B Q, Zhao B Q, Liu L B, Liu J Y and Yumoto K 2008 Effects of geomagnetic storm on GPS ionospheric scintillation at Sanya *J. Atmos. Terr. Phy.* **70**(7) 1034-1045.

[8] Purohit P K, Bhattacharya S, Tiwari R and Gwal A K 2010 Study of space weather on GPS performance at low latitude station Bhopal and high latitude station Maitri, Antarctica, *International Journal of Eng. Science and technology* **2** 7856- 7865.

Evaluation of geomagnetic storm effects on the GPS derived Total Electron Content (TEC)

P. K. Purohit[1], Azad A. Mansoori[2], Parvaiz A. Khan[2], Roshni Atulkar[1], Purushottam Bhawre[2], Sharad C. Tripathi[2], Prakash Khatarkar[2], Shivangi Bhardwaj[2], Aslam A. M. [2], Malik A. Waheed[2] and A. K. Gwal[2]

[1]*National Institute of Technical Teachers' Training and Research, Bhopal-462002, India*
[2]*Space Science Laboratory, Department of Physics, Barkatullah University, Bhopal-462026, India*

E-mail: purohit_pk2004@yahoo.com

Abstract. The geomagnetic storm represents the most outstanding example of solar wind-magnetospheric interaction, which causes global disturbances in the geomagnetic field as well as triggers ionospheric disturbances. We study the behaviour of ionospheric Total Electron Content (TEC) during the geomagnetic storms. For this investigation we have selected 47 intense geomagnetic storms (Dst ≤ -100nT) that were observed during the solar cycle 23 i.e. during 1998- 2006. We then categorized these storms into four categories depending upon their solar sources like Magnetic Cloud (MC), Co-rotating Interaction Region (CIR), SH+ICME and SH+MC. We then studied the behaviour of ionospheric TEC at a mid latitude station Usuda (36.13N, 138.36E), Japan during these storm events produced by four different solar sources. During our study we found that the smooth variations in TEC are replaced by rapid fluctuations and the value of TEC is strongly enhanced during the time of these storms belonging to all the four categories. However, the greatest enhancements in TEC are produced during those geomagnetic storms which are either caused by Sheath driven Magnetic cloud (SH+MC) or Sheath driven ICME (SH+ICME). We also derived the correlation between the TEC enhancements produced during storms of each category with the minimum Dst. We found the strongest correlation exists for the SH+ICME category followed by SH+MC, MC and finally CIR. Since the most intense storms were either caused by SH+ICME or SH+MC while the least intense storms were caused by CIR, consequently the correlation was strongest with SH+ICME and SH+MC and least with CIR.

1. Introduction

Energy emitted from sun drives the earth's magnetosphere, thermosphere and ionosphere. The most powerful solar events like Coronal Mass Ejections (CME) are a result of plasma outbursts from active region of the sun [1]. CMEs interact with solar wind and Interplanetary Magnetic Field (IMF) during their propagation and disrupt the solar wind flow. Geomagnetic storms are largely associated with CMEs from the sun. CMEs faster than ~500 Km/s eventually drive shock waves which normally strike the earth's magnetosphere in 24 to 36 hours after the event onset on sun.

The solar wind energy deposited into the magnetosphere during geomagnetic storms will eventually be dissipated into the ionosphere and thermosphere. Therefore various physical and energy transport processes takes place in the ionosphere [2]. The high latitude electric field can penetrate into mid, low and equatorial ionosphere during a geomagnetic storm causing significant disturbances in ionospheric conditions [3, 4, 5]. The disturbed ionosphere is manifested as a large increase or depletion of electron density from their normal level. This kind of response of ionosphere to the geomagnetic storms is known as ionospheric storm. Description and prediction of the features of the ionospheric storms and understanding of the related processes has been a topic of intensive research from decades [6,7, 8, 9, 10].

Ionospheric storms are of two types: positive and negative. During the geomagnetic storms, ionospheric, peak electron density (Nmax), electron density (Ne) and Total Electron Content (TEC) increase or decrease dramatically from their normal level, known as positive or negative ionospheric storms. The positive ionospheric storms can cause many serious problems such as time delay, range error and scintillation in satellite communication and navigation. The positive ionospheric storms have been studied by many researchers [11, 12, 13]. Negative ionospheric storms can cause radio blackouts in ground based HF radio communication. Physical mechanisms of the negative storms are more or less well understood [14, 15].

Since every geomagnetic storm has its unique character and therefore prediction, forecasting and now casting of the response of ionosphere during geomagnetic storms is an interesting topic of study for researchers from decades.

2. Data and Methodology

We have selected the 47 geomagnetic storm with minimum Dst $\leq$ -100nT for the study, observed during solar cycle 23 i.e. 1998 – 2006. To study the effect of these geomagnetic storms on the ionosphere we have selected a mid latitude station Usuda (36.13N, 138.36E), Japan. The storm intensity is characterized by the Dst index. The data of Dst is available at various websites for downloading. However, for our study we have downloaded the data of Dst index from Space Physics Data Facility OMNI website (http://omniweb.gsfc.nasa.gov/).We have used the hourly values of Dst for the investigation.

We have also taken the IMF Bz from the measurements of ACE satellite at www.srl.caltech.edu/ACE with 1h resolution to characterize the Interplanetary Magnetic Field conditions.

We identified the solar and interplanetary sources that were responsible for causing each storm. For doing so we have taken help form various online catalogs and properties of various magnetic cloud structures. On the basis of solar sources we then categorized these storms into four categories.

The state of ionosphere is described by a number of parameters derived from the measurements of different instruments. The most widely used instruments for studying the ionospheric behaviour are the Ionosonde and GPS. From both ionosonde and GPS measurements we can define and derive a number of parameters that describe the state of ionosphere. However for the present investigation we have used the Total Electron Content (TEC) derived from the GPS measurement

A network of GPS receivers is spread over the globe and data is recorded regularly. The data recorded at all the stations which form the part of International GPS Service (IGS) is freely available to users and can be downloaded from the URL http://sopac.ucsd.edu/dataArchive/.

The data downloaded from the web is in RINEX format, which is then processed to get the required Total Electron Content (TEC). For our investigation we have used the TEC data of the Usuda (36.13N, 138.36E) station of Japan. The temporal resolution of the data is usually 30s. However, for our study we have constructed the hourly averages of TEC. We calculated the deviation of TEC from the quietest day of the month.

3. Results and Discussion

We studied the ionospheric response to geomagnetic storm produced by four different solar and interplanetary sources. For doing so we have selected 47 geomagnetic storm associated with four types of solar sources namely Magnetic Cloud (MC), Sheath and Magnetic Cloud (SH+MC), Sheath and ICME (SH+ICME) and Corotating Interaction Regions (CIR). The occurrences of geomagnetic storms vary considerably from year to year during a particular solar cycle as well as from cycle to cycle also. Figure 1 shows the yearly distribution of geomagnetic storms during solar cycle 23. From the figure we find that maximum number of intense geomagnetic storms (11) occurred during year 2001 followed by 2002 (09). The least number of intense storms were observed during 2006 (01). The occurrence of geomagnetic storms depends on the state of solar activity. Since maximum number of active regions appear on the sun during solar maximum, consequently the frequency of storms increase during these years. Similarly during solar minimum the solar activity conditions remain quite low and accordingly during these less number of storms are observed.

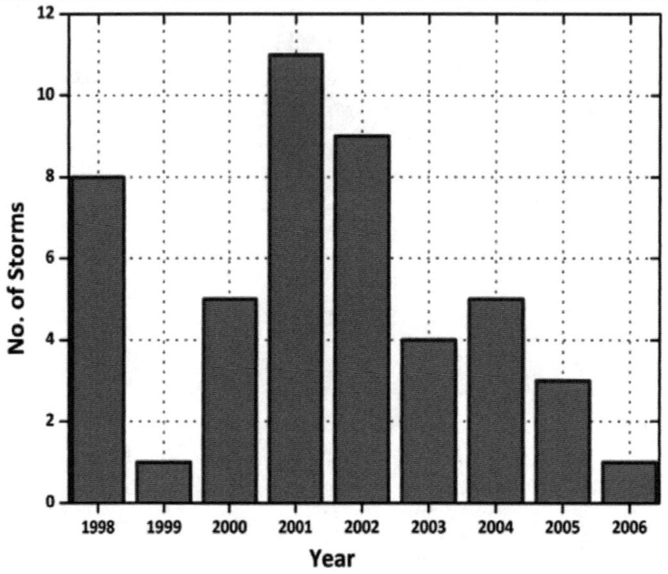

Figure 1. Yearly distribution of geomagnetic storms during solar cycle 23 i.e. 1998-2006.

We then identified the solar source of each of the 47 geomagnetic storms that were responsible for causing them. To identify the solar sources of selected geomagnetic storms we have taken the help from various online catalogs and properties of interplanetary structures. The distribution of geomagnetic storms with various solar sources is shown in figure 2. The major solar sources of selected geomagnetic storms were identified as Magnetic Cloud (MC), Sheath and Magnetic Cloud (SH+ MC), Sheath and ICME (SH+ICME) and Corotating Interaction Region (CIR). The maximum number of storms (17) were found to be produced by SH+ICME followed by SH+ MC, CIR and MC. It therefore follows from the figure that the effective drivers of geomagnetic storms are Sheath driven ICME or Magnetic Cloud.

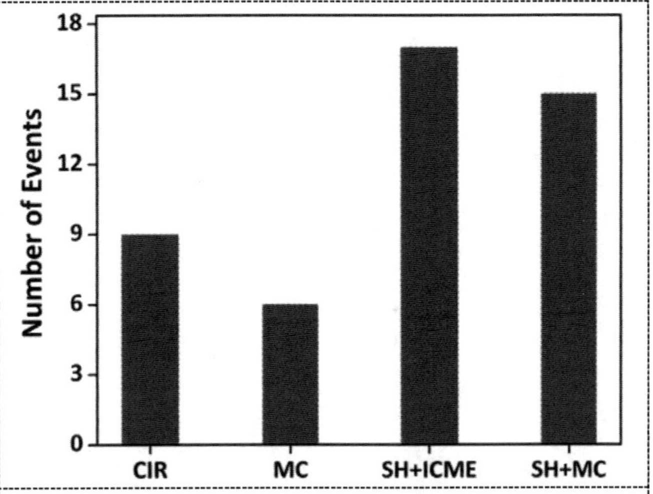

Figure 2. Distribution of selected geomagnetic storms with their solar and interplanetary sources.

We then performed the single regression analysis to quantify the magnitude of effect of storms produced by different sources on the ionospheric TEC. To draw the scatter plots and derive the correlation between TEC and storm index for four different categories of storms, we took the peak values of TEC and peak values of Dst. The scatter plot of peak values of TEC and Dst for all the storms of each category is shown in figure 3. From the figure we found that the scatter of points between Dst and TEC is quite large both for storms produced either by CIR or MC, while opposite pattern is observed for storms produced by SH+ICME and SH+MC. We also derived the correlation coefficients between the Dst and TEC for all the four categories of storms and are shown in figure 3. The correlation coefficient between Dst and TEC is 0.28, 0.39, 0.52 and 0.68 for storms produced by CIR, MC, SH+MC and SH+ICME respectively.

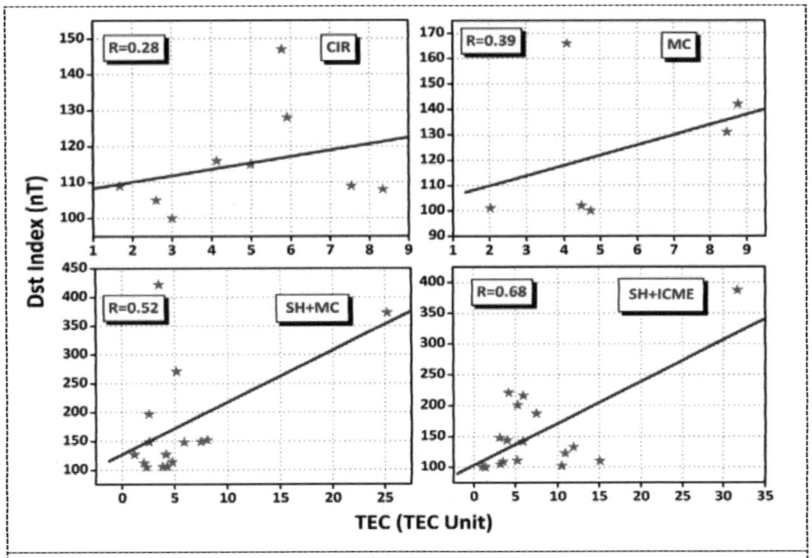

Figure 3. Correlation between TEC and Dst during storms produced by four different solar sources.

4. Conclusions

We studied the effect of geomagnetic storms caused by different solar sources, on the ionosophric TEC. The main conclusions of the study are enumerated below:

- The geomagnetic storms significantly affect the ionosphere having any of the solar origin. A positive deviation of TEC was observed during all the storms having different solar sources.
- The strongest effect is produced by those geomagnetic storms which are produced by SH+ICME and SH+MC. The TEC undergoes strong and rapid fluctuations as well as exhibits largest deviations during the storms produced by these two solar drivers.
- The effect of geomagnetic storms which have CIR and MC as their solar sources have a less impact on ionosphere.

The TEC exhibits a strong correlation with Dst during geomagnetic storms produced either by SH+ICME and SH+MC.

5. Acknowledgements

The authors are thankful to University Grants Commission New Delhi, India for providing financial assistance. The authors also acknowledge different online sources from where the data sets used in the study were obtained.

6. References

[1] D'ujanga F. M., et al. 2013 Total electron content of the ionosphere at two stations in East Africa during 24-25 October 2011 geomagnetic storm *Adv. Space Res.* **51** 712-721.

[2] Buonsanto M J and Fuller-Rowell T J 1997 Strides made in understanding space weather at Earth *EOS Transactions American Geophysical Union* **78** (1) 1-7.

[3] Rastogi, R G 1977 Geomagnetic storms and electric fields in the equatorial ionosphere *Nature* **268** 422–424.

[4] Kikuchi T et al. 2003 Equatorial counter electrojets during a substorm *J. Geophys. Res.* **108** (A11) 1406. doi:10.1029/2003JA009915.

[5] Fejer B G et al. 2007 Equatorial ionospheric electric fields during the November 2004 magnetic storm *J. Geophys. Res.* **112** (A10304) doi:10.1029/2007JA012376.

[6] Kane R P 1973 Global evolution of F2-region storm *J. Atmos. & Solar Terres. Phys.* **35** 1953–1966.

[7] Kutiev I and Muhtarov P 2001 Modeling of mid-latitude F region response to geomagnetic activity *J. Geophys. Res.* **106** (A8) 15501–15509.

[8] Fejer B G 2002 Low latitude ionospheric electrodynamics *J. Atm. Terr. Phys.* **64** 1401– 1408.

[9] Liu L et al. 2004 The low latitude ionospheric effects of the April 2000 magnetic storm near 1200E *Earth Planet Space* **56** 607– 612.

[10] Lin C H et al. 2005 Large-scale variations of the low-latitude ionosphere during the October–November 2003 superstorm: Observational results *J. Geophys. Res.* **110** (A09S28) doi:10.1029/2004JA010900.

[11] Balan N and Rao P B 1990 Dependence of ionospheric response on the local time of sudden commencement and intensity of storms *J. Atm. Terr. Phys.* **52** 269–275.

[12] Basu S et al. 2001 Response of the equatorial ionosphere in the South Atlantic region to the great magnetic storm of July 15, 2000 *Geophys. Res. Lett.* **28** 3577–3580.

[13] Pincheira X T et al. 2002 An investigation of ionospheric responses, and disturbance thermospheric winds, during magnetic storms over South America sector *J. Geophys. Res.* **107** (A11) 1379, doi:10.1029/2001JA000263.

[14] Rishbeth H 1991 F-region storms and thermospheric dynamics J*ournal of Geomagnetism and Geoelectricity* **43** 513–524.

[15] Mendillo M and Narvaez C 2010 Ionospheric storms at geophysicallyequivalent sites-Part 2: Local time patterns for sub-auroral ionospheres *Ann. Geophys.* **28** 1449–1462.